Selected Papers on COHERENCE AND FLUCTUATIONS OF LIGHT

WITH BIBLIOGRAPHY

EDITED BY
L. Mandel and E. Wolf
Department of Physics and Astronomy
University of Rochester

IN TWO VOLUMES

Vol. I

1850–1960

Dover Publications, Inc., New York

Published in Canada by General Publishing Company, Ltd., 30 Lesmill Road, Don Mills, Toronto, Ontario.
Published in the United Kingdom by Constable and Company, Ltd., 10 Orange Street, London WC 2.

First published in 1970, this selection of papers appears here in collected form for the first time. Authors' corrections have been made in the following papers: Nos. 1, 29, 32, 38, 48, 50, 60, 63, 65, 70, 71, 72, 74, 76, 80, 81, 85, 87, 88 and 92. With these exceptions, the papers are reproduced without change from the original journals.

The editors and publisher are indebted to the authors and original publishers for permission to reproduce these papers.

International Standard Book Number: 0-486-62570-2
Library of Congress Catalog Card Number: 74-111608

Manufactured in the United States of America
Dover Publications, Inc.
180 Varick Street
New York, N.Y. 10014

PREFACE

During the last few years there has been a great deal of interest in coherence and fluctuations of light. Although certain aspects of coherence were already discussed during the last century, and much of the formal development of the classical theory of coherence occurred prior to 1960, the invention of the laser in the first half of this decade, and its increasing importance in science and technology, provided a strong impetus for further development. In view of the current interest, we thought that it would be desirable to reprint a selection of original papers on the subject of optical coherence, together with an extensive bibliography on the whole field.

We have done our best to choose those papers that have been of particular significance in contributing either to the development or to the understanding of the subject, but we are only too aware that any selection from the wealth of available material is bound to be somewhat subjective.

We interpret the term coherence in its broad sense, as descriptive of a measure of correlation in optical fields. The extreme case of complete coherence, with which traditional optics has mainly been concerned, is an idealization that is never realized in practice. For this reason papers dealing with complete coherence have been included only if they are concerned with some fundamental aspect of the problem; otherwise much of the literature covering conventional interferometry and light scattering might have been included in these volumes, either among the reprints or in the bibliography. Similar remarks apply also to the subject of lasers. Since the laser is the most important source of coherent light, any one of the many papers dealing with the subject might legitimately have been a candidate for inclusion. We have therefore limited ourselves to a few papers on lasers in which the coherence or fluctuation properties of light were emphasized. The subjects falling under the general heading of optical pumping also presented a special problem. Here coherence is again an essential part of the phenomenon, although the emphasis in the published literature is usually on other aspects. We have compromised by including in the bibliography some of the early classic papers on the subject, but not the more recent work.

Within the foregoing limitations we believe that the bibliography is reasonably complete. A few papers are also listed which fall outside the domain of optics. These deal with the mathematical apparatus of coherence theory and some closely related topics; we felt they would be of value to anyone wishing to take a broader view of the subject. In the bibliography, papers that appeared in the period 1850–1950 are grouped together; papers that appeared between 1951 and 1966 are grouped year by year. In each group the publications are listed in alphabetical order of authors. Those papers which are reprinted in these volumes are indicated by asterisks in the bibliography.

Several early papers, though not explicitly employing the concept of coherence, were nevertheless of basic importance in the formulation of the early theories of partial coherence. Lack of space prevented us from reprinting them, but they may still be read with profit today. This applies particularly to some papers of G. G. Stokes and A. A. Michelson.

Several review articles dealing with the general field of optical coherence have appeared in recent years. We have chosen not to reprint them in these volumes, but they are listed below in chronological order, for the benefit of readers who might wish to obtain some general orientation in the subject.

E. Wolf, The Coherence Properties of Optical Fields, *Proceedings of a Symposium on Astronomical Optics and Related Subjects*, edited by Z. Kopal (North-Holland Publishing Company, Amsterdam 1956) p. 177.

L. Mandel, Fluctuations of Light Beams, *Progress in Optics*, Vol. 2, edited by E. Wolf (North-Holland Publishing Company, Amsterdam 1963; John Wiley & Sons, Inc., New York 1963) p. 181.

E. Wolf, Basic Concepts of Optical Coherence Theory, *Optical Masers* (Proceedings of the Symposium on Optical Masers, New York 1963) edited by J. Fox (J. Wiley & Sons, New York 1963) p. 29.

E. Wolf, Recent Researches on Coherence Properties of Light, *Quantum Electronics III* (Proceedings of the Third International Conference on Quantum Electronics, Paris 1963), edited by P. Grivet and N. Bloembergen (Dunod, Paris 1964; Columbia University Press, New York 1964) p. 13.

R. J. Glauber, Optical Coherence and Photon Statistics, *Quantum Optics and Electronics* (Les Houches Lectures 1964), edited by C. deWitt, A. Blandin and C. Cohen-Tannoudji (Gordon and Breach, New York 1965) p. 63.

Ernst Ledinegg, Klassiche und Quantentheorie kohärenter und nichtkohärenter elektromagnetischer Felder, *Acta Physica Austriaca* **20** (1965) 253.

L. Mandel and E. Wolf, Coherence Properties of Optical Fields, *Rev. Mod. Phys.* **37** (1965) 231.

C. L. Mehta, Coherence and Statistics of Radiation, *Lectures on Theoretical Physics*, edited by W. E. Brittin (University of Colorado Press, Boulder, Colo. 1965) Vol. 7c, p. 345.

F. T. Arecchi, Proprieta Statistiche della Radiazione Coerente, *Nuovo Cimento* Suppl. IV (1966) 756.

E. Wolf, Some Recent Research on Coherence and Fluctuations of Light, *Opt. Acta* **13** (1966) 281.

H. Paul, Ein Beitrag zur Quantentheorie der optischen Kohärenz, *Fortschritte der Physik* **14** (1966) 141.

We would like to thank all of the authors whose papers are reproduced in these volumes, and the editors of the journals where the papers first appeared, for granting us permission for reprinting. Some authors have kindly supplied us with lists of misprints and errors. These have been corrected in the present edition.

Finally we wish to acknowledge our indebtedness to Dr. Gabriel Bédard for much help in the preparation of the bibliography.

Leonard Mandel
Emil Wolf

Department of Physics and Astronomy
University of Rochester
March 7, 1969

CONTENTS

Early Papers

1951–1954

PAPER NO. 1

Reprinted from *Physica*, Vol. 1, pp. 201–210, 1934.

DIE WAHRSCHEINLICHE SCHWINGUNGSVERTEILUNG IN EINER VON EINER LICHTQUELLE DIREKT ODER MITTELS EINER LINSE BELEUCHTETEN EBENE

von P. H. VAN CITTERT

Mitteilung aus dem Physikalischen Institut der Universität, Utrecht

Zusammenfassung

Es wird die wahrscheinliche Schwingungsverteilung in einer von einer Lichtquelle entweder direkt oder mittels einer Linse beleuchteten Ebene untersucht. Es zeigt sich, dass diese Verteilung in beiden Fällen völlig identisch ist und nur abhängt von dem Öffnungswinkel des Lichtbündels, welches die Ebene beleuchtet. Die Korrelation zwischen den Schwingungsvektoren in verschiedenen Punkten der Ebene wird bestimmt durch eine Funktion, welche mit der Beugungsfunktion der Linse identisch ist.

Die Lichterregung in einer von einer Lichtquelle beleuchteten Fläche setzt sich aus den Schwingungsvektoren, welche von den verschiedenen Elementen der Lichtquelle herrühren, zusammen. Ist die Lichtquelle genügend weit von der Fläche entfernt, so gibt jedes Element dieser Lichtquelle in einem Punkt der Fläche einen Schwingungsvektor, dessen Grösse von der Lage des Punktes unabhängig, dessen Phase jedoch von dieser Lage abhängig ist. Die Lichterregung in den verschiedenen Punkten setzt sich also zusammen aus einer grossen Anzahl Schwingungsvektoren gleicher Grösse, jedoch mit Phasenunterschieden, welche nicht nur von der zufälligen Phasenverteilung in der Lichtquelle, sondern auch von der Lage der betrachteten Punkte abhängig sind.

Eine scheinbar ganz andere Beleuchtungsweise erzielt man, wenn man eine ausgedehnte Lichtquelle mittels einer Linse scharf in der Fläche abbildet. Dann verursacht jedes Element der Lichtquelle ein Beugungsbild: die verschiedenen Punkte der Fläche erhalten also von dem betrachteten Element Schwingungsvektoren von verschiedener Grösse, deren Phase jedoch in allen Punkten dieselbe ist [1]). Die Lichterregung in einem Punkte der Fläche setzt sich in diesem Falle zusammen aus sehr vielen Schwingungsvektoren verschiedener Grösse, deren Phasenverteilung jedoch nur von der zufälligen Phasenverteilung in der Lichtquelle abhängig ist.

In beiden Fällen muss jedoch eine Korrelation zwischen den Schwingungszuständen benachbarter Punkte bestehen. Es wird sich herausstellen, dass diese Korrelation für beide Fälle identisch ist, und nur von dem Öffnungswinkel abhängt, unter welchem die Lichtquelle bzw. die Linie von der beleuchteten Fläche aus gesehen wird.

Betrachten wir erst die wahrscheinliche Lichterregung durch eine ausgedehnte Lichtquelle in einem Punkte hervorgerufen. Setzen wir voraus, dass das Element $d\alpha$ der sich von o bis α_0 ausdehnenden Lichtquelle in dem Punkte O die Energie $id\alpha$, also den Schwingungsvektor $\sqrt{id\alpha}$ mit Fase δ hervorruft. Der resultierende Schwingungsvektor in O sei ρ, ϑ mit Komponenten $x = \rho \cos \vartheta$, $y = \rho \sin \vartheta$ [2]).

Die Wahrscheinlichkeit

$$w\,(x, \alpha_0)\,dx$$

dass der resultierende Schwingungsvektor die Komponente x hat, berechnet sich aus:

$$w\,(x, \alpha_0 + d\alpha_0) = \int_0^{2\pi} \frac{d\vartheta}{2\pi}\, w\,(x - \sqrt{id\alpha_0}\cos\vartheta, \alpha_0)$$

Entwicklung von w nach α_0 und x gibt:

$$\frac{\partial w}{\partial \alpha_0} = \frac{i}{4}\frac{\partial^2 w}{\partial x^2}$$

Also:

$$w = \frac{1}{\sqrt{\pi i \alpha_0}}\, e^{-\frac{x^2}{i\alpha_0}}$$

1) Der Irrtum, dass die Phase in den verschiedenen Punkten des Beugungsbildes verschieden ist, rührt daher, dass man bei der Berechnung des Beugungsbildes, Wellenflächen benutzt, welche sich in dem Rande der Linse, statt diejenigen, welche sich im Zentrum der Linse schneiden.

2) Es wird eine polarisierte Lichtquelle vorausgesetzt; ist dies nicht der Fall, so ändern sich die Resultate nicht wesentlich. Ebenso werden wir von vornherein die Komponenten x und y des Schwingungsvektors als völlig unabhängig von einander betrachten. Dies geschieht um unnötige verwickeltere Berechnungen vorzubeugen. Wenn man diese Voraussetzung nicht macht, bekommt man dieselben Resultate.

Setzen wir $i\alpha_0 = I$ gleich der mittleren Intensität, so wird:

$$w(x)\, dx = \frac{1}{\sqrt{\pi I}} e^{-\frac{x^2}{I}} dx$$

Ebenso:

$$w(y)\, dy = \frac{1}{\sqrt{\pi I}} e^{-\frac{y^2}{I}} dy$$

also:

$$w(x, y)\, dx\, dy = \frac{1}{\pi I} e^{-\frac{x^2+y^2}{I}} dx\, dy\ ^{1)}$$

oder:

$$w(\rho,\vartheta)\, \rho\, d\rho\, d\vartheta = \frac{1}{\pi I} e^{-\frac{\rho^2}{I}} \rho\, d\rho\, d\vartheta \tag{1}$$

Ist die Intensität der Lichtquelle noch von α abhängig, so muss $I = i\alpha_0$ durch

$$I = \int_0^{\alpha_0} i(\alpha)\, d\alpha \tag{1a}$$

ersetzt werden.

Setzt man $\rho^2 = A$ gleich der resultierenden Intensität in O, so gibt Integration nach ϑ:

$$w(A)\, dA = \frac{1}{I} e^{-\frac{A}{I}} dA.$$

Betrachten wir jetzt den ersten Fall: Eine ziemlich weit von der Ebene $\xi\eta$ entfernte, in allen Punkten gleich intensive Lichtquelle mit Winkelabmessungen $2\alpha_0$, $2\beta_0$. Ein Element mit Koordinaten α, β dieser Lichtquelle gibt in O als ($\xi = 0, \eta = 0$) den Schwingungsvektor $\sqrt{i\, d\alpha\, d\beta}, \delta$, in $O_1(\xi = \alpha', \eta = 0)$ jedoch $\sqrt{i\, d\alpha\, d\beta}, \delta + (2\pi a'/\lambda)\, \sin\alpha \cos\beta$, oder wenn wir voraussetzen, dass α_0 und β_0 so klein sind, dass $\sin\alpha = \alpha$ und $\cos\beta = 1$ ist:

$$\sqrt{i\, d\alpha\, d\beta},\ \delta \text{ bzw. } \sqrt{i\, d\alpha\, d\beta},\ \delta + a\alpha$$

worin $a = 2\pi a'/\lambda$.

Die Rechnungen werden bedeutend vereinfacht, wenn wir erst

1) Vlg. v. Laue, Enz. d. Math. Wiss. *V*3, 393, 1915.

nach der Wahrscheinlichkeit fragen, dass die Differenz, bzw. die Summe der Schwingungsvektoren in O und O_1 gleich v, ϑ (v_x, v_y) bzw. s, ϑ' (s_x, s_y) sind. Bezeichnen wir die Wahrscheinlichkeit, dass die Komponente v_x auftritt, mit

$$W\,(v_x, \alpha_0, \beta_0)\,dv_x.$$

Ändert sich nun α_0 um einen Betrag $d\alpha_0$, so dehnt sich die Lichtquelle an beiden Seiten um einen Streifen $2\beta_0\,d\alpha_0$ aus. Wenn diese Streifen zu der X-Komponente des Lichtvektors in O den Betrag $\rho \cos \vartheta$ bzw. $\rho' \cos \vartheta'$ beitragen, so tragen sie zu der X-Komponente der Differenz den Betrag

$$\rho\,\{\cos \vartheta - \cos(\vartheta + a\alpha_0)\}\ \text{bzw.}\ \rho'\,\{\cos \vartheta' - \cos(\vartheta' - a\alpha_0)\}$$

bei. Die Wahrscheinlichkeit, dass in O der Betrag $\rho \cos \vartheta$ beigetragen wird ist nach (1):

$$w\,(\rho\vartheta)\,\rho\,d\rho\,d\vartheta = \frac{1}{2\pi\,\beta_0\,i\,d\alpha_0}\,e^{-\frac{p^2}{2\beta_0\,i\,d\alpha_0}}\,\rho\,d\rho\,d\vartheta$$

und also ist:

$$W(v_x, \alpha_0 + d\alpha_0, \beta_0) = \int_0^\infty \int_0^{2\pi} w\,(\rho\,\vartheta)\,\rho\,d\rho\,d\vartheta \,.\int_0^\infty \int_0^{2\pi} w\,(\rho'\,\vartheta')\,\rho'\,d\rho'\,d\vartheta'$$

$$W\,[v_x - \rho\,\{\cos\vartheta - \cos(\vartheta + a\alpha_0)\} - \rho'\,\{\cos \vartheta' - \cos(\vartheta' - a\alpha_0)\}, \alpha_0, \beta_0].$$

Entwickelung gibt:

$$\frac{\partial W}{\partial \alpha_0} = 4\beta_0\,i\,\sin^2 \frac{a\alpha_0}{2}\,\frac{\partial^2 W}{\partial v_x^2}.$$

Oder

$$W\,(v_x) = \frac{1}{\Phi\sqrt{\pi}}\,e^{-\frac{v_x^2}{\phi^2}}$$

worin

$$\Phi^2 = 8\alpha_0\,\beta_0\,i\left(1 - \frac{\sin a\alpha_0}{a\alpha_0}\right) = 2\,I\,(1 - k_a),$$

wo

$$4\alpha_0\,\beta_0\,i = I\ \text{und}\ \frac{\sin a\alpha_0}{a\alpha_0} = k_a.$$

Also ist:

$$W(v_x)\, dv_x = \frac{1}{\sqrt{2\pi I\,(1 - k_a)}}\, e^{-\frac{v_x^2}{2I\,(1-k_a)}}\, dv_x \tag{2}$$

und ebenso:

$$W(v_y)\, dv_y = \frac{1}{\sqrt{2\pi I\,(1 - k_a)}}\, e^{-\frac{v_y^2}{2I\,(1-k_a)}}\, dv_y \tag{2'}$$

Analog findet man für die Wahrscheinlichkeit, dass die Summe der Schwingungsvektoren in O und O_1 die Komponenten s_x und s_y hat:

$$W(s_x)\, ds_x = \frac{1}{\sqrt{2\pi I\,(1 + k_a)}}\, e^{-\frac{s_x^2}{2I\,(1+k_a)}}\, ds_x \tag{3}$$

$$W(s_y)\, ds_y = \frac{1}{\sqrt{2\pi I\,(1 + k_a)}}\, e^{-\frac{s_y^2}{2I\,(1+k_a)}}\, ds_y \tag{3'}$$

Die Wahrscheinlichkeit $W(x, x_1)\, dx\, dx_1$, dass zu gleicher Zeit die Komponenten x und x_1 auftreten, findet sich aus der Wahrscheinlichkeit, dass zu gleicher Zeit die Summe $x + x_1 = s_x$ und die Differenz $x - x_1 = v_x$ ist:

$$W(v_x, s_x)\, dv_x\, ds_x = \frac{1}{2\pi I\,\sqrt{1 - k_a^2}}\, e^{-\left\{\frac{v_x^2}{2I(1-k_a)} + \frac{s_x^2}{2I(1+k_a)}\right\}}\, dv_x\, ds_x$$

oder weil $dv_x\, ds_x = 2dx\, dx_1$:

$$W(x, x_1)\, dx\, dx_1 = \frac{1}{\pi I\,\sqrt{1 - k_a^2}}\, e^{-\frac{x^2 - 2k_a\, x x_1 + x_1^2}{I\,(1-k_a^2)}}\, dx\, dx_1 \tag{4}$$

Ebenso:

$$W(y, y_1)\, dy\, dy_1 = \frac{1}{\pi I\,\sqrt{1 - k_a^2}}\, e^{-\frac{y^2 - 2k_a\, y y_1 + y_1^2}{I\,(1-k_a^2)}}\, dy\, dy_1. \tag{4'}$$

Die Wahrscheinlichkeit, dass die Vektoren ρ und ρ_1 mit der Phasendifferenz ψ auftreten, wird also:

$$W(\rho, \rho_1, \psi)\, \rho\rho_1 d\rho\, d\rho_1 d\psi = \frac{1}{\pi^2 I^2 (1 - k_a^2)}\, e^{-\frac{\rho^2 - 2\rho\rho_1\, k_a \cos\psi + \rho_1^2}{I\,(1-k_a^2)}}\, \rho\rho_1 d\rho\, d\rho_1 d\psi \tag{5}$$

Betrachten wir jetzt den zweiten Fall: Die Beleuchtung der Ebene durch Abbildung einer ausgedehnten homogenen Lichtquelle mittels einer Linse. Wir denken uns die zu betrachtenden Punkte O und O_1 so benachbart, dass das Bild der Lichtquelle bedeutend grösser als der Abstand dieser Punkte ist, und wir also die Lichtquelle als unendlich gross ansehen dürfen. Das Element der Lichtquelle, dessen Bild in der Bildebene durch die Koordinate p' und q' bestimmt wird, gibt in dem Punkt $\xi = a'$, $\eta = b'$ den Schwingungsvektor

$$\frac{2}{\pi}\sqrt{i}\,\frac{\sin(p-a)\alpha_0}{p-a}\cdot\frac{\sin(q-b)\alpha_0}{q-b}\,dp\,dq,$$

wo $2\alpha_0$ bzw. $2\beta_0$ die Winkelabmessungen der rechteckig gedachten Linse vorstellen, und $a = 2\pi a'/\lambda$, u.s.w. ist [1]). Die Phase dieses Vektors ist unabhängig von der Lage des Punktes ab.

Denken wir uns anfänglich die Abmessungen des Bildes der Lichtquelle $2p_0'$, $2q_0'$ und betrachten wir wieder die Wahrscheinlichkeit dass die Differenz bzw. die Summe der Lichtvektoren in O ($\xi = 0$, $\eta = 0$) und O' ($\xi = a'$, $\eta = 0$) v bzw. s beträgt. Wenn wir p_0 um einen kleinen Betrag dp_0 vermehren, dehnt sich die Lichtquelle wieder um zwei Streifen $2q_0\,dp_0$ aus. Wenn ein Element $dq\,dp_0$ irgendeines dieser Streifen in O den Schwingungsvektor ρ, ϑ beiträgt, so ist derselbe in O_1

$$\rho\,\frac{\dfrac{\sin(p_0-a)\alpha_0}{p_0-a}}{\dfrac{\sin p_0\alpha_0}{p_0}},\ \vartheta \quad \text{bzw.} \quad \rho\,\frac{\dfrac{\sin(p_0+a)\alpha_0}{p_0+a}}{\dfrac{\sin p_0\alpha_0}{p_0}},\ \vartheta$$

Die Wahrscheinlichkeit, dass ein *ganzer* Streifen in O den Vektor ρ, ϑ gibt, ist nach (1):

$$w(\rho\vartheta)\,\rho\,d\rho\,d\vartheta = \frac{1}{\pi f}\,e^{-\frac{\rho^2}{f}}\,\rho\,d\rho\,d\vartheta$$

wo

$$f = \int_{-q_0}^{+q_0} dq\cdot\frac{4i}{\pi^2}\,\frac{\sin^2 p_0\alpha_0}{p_0^2}\cdot\frac{\sin^2 q_0\beta_0}{q_0^2}\,dp_0$$

oder für Lim $q = \infty$:

$$f = \frac{4\beta_0 i}{\pi}\,\frac{\sin^2 p_0\alpha_0}{p_0^2}\,dp_0.$$

1) Vorsetzung des Faktors $2/\pi$ bewirkt, dass die mittlere Intensität wieder gleich $I = 4\alpha_0\beta_0 i$ ist.

Bezeichnen wir wieder die Wahrscheinlichkeit für die X-Komponente der Differenz mit $W(v_x)$, dann ist wieder:

$$W(v_x, p_0 + dp_0) = \int_0^\infty \int_0^{2\pi} w(\rho\vartheta)\, \rho\, d\rho\, d\vartheta \int_0^\infty \int_0^{2\pi} w(\rho'\vartheta')\, \rho'\, d\rho'\, d\vartheta' :$$

$$. W\left[v_x - \rho \left\{1 - \frac{\frac{\sin (p_0 - a)\alpha_0}{p_0 - a}}{\frac{\sin p_0 \alpha_0}{p_0}}\right\} \cos\vartheta - \rho' \left\{1 - \frac{\frac{\sin (p_0 + a)\alpha_0}{p_0 + a}}{\frac{\sin p_0 \alpha_0}{p_0}}\right\} \cos\vartheta', p_0\right]$$

oder:

$$\frac{\partial W}{\partial p_0} = f(p_0)\, \frac{\partial W}{\partial v_x^2}$$

worin

$$f(p_0) = \frac{\beta_0 i}{\pi} \left\{\frac{\sin^2 p_0 \alpha_0}{p_0^2} - 2\frac{\sin p_0 \alpha_0}{p_0} \cdot \frac{\sin (p_0 - a)\alpha_0}{p_0 - a} + \frac{\sin^2 (p_0 - a)}{(p_0 - a)^2} + \right.$$

$$\left. + \frac{\sin^2 p_0 \alpha_0}{p_0^2} - 2\frac{\sin p_0 \alpha_0}{p_0} \cdot \frac{\sin (p_0 + a)\alpha_0}{p_0 + a} + \frac{\sin^2 (p_0 + a)}{(p_0 + a)^2}\right\}$$

Die Lösung ist wieder:

$$W = \frac{1}{\Phi\sqrt{\pi}}\, e^{-\frac{v_x^2}{\phi^2}}$$

wo

$$\Phi^2 = 4 \int_0^{p_0} f(p)\, dp.$$

Integration gibt für Lim $p_0 = \infty$:

$$\varphi^2 = 8\alpha_0\, \beta_0\, i \left(1 - \frac{\sin a\alpha_0}{a\alpha_0}\right) = 2I\,(1 - k_a)$$

wo wieder

$$I = 4\alpha_0\, \beta_0\, i \text{ und } k_a = \frac{\sin a\alpha_0}{a\alpha_0}.$$

Die hier gefundene Wahrscheinlichkeit ist identisch mit der in (2) angegebenen Wahrscheinlichkeit für den ersten Fall. Ebenso sind die Wahrscheinlichkeiten für die Komponenten v_y, s_x und s_y identisch mit den korrespondierenden Wahrscheinlichkeiten im ersten Fall. Die Beleuchtung der Ebene ist also in beiden Fällen völlig identisch.

Es macht also gar keinen Unterschied, ob man die Ebene mit einer Lichtquelle direkt beleuchtet oder diese Lichtquelle mittels einer Linse in der Ebene abbildet: die Korrelation zwischen den resultierenden Lichtschwingungen in den verschiedenen Punkten der Ebene ist nur abhängig von dem Öffnungswinkel des auf die Ebene treffenden Lichtbündels und unabhängig von der Art wie dieses Bündel erhalten ist.

In beiden Fällen ist also die Wahrscheinlichkeit für das Zusammentreffen zweier bestimmten Lichterregungen in O ($\xi = o$, $\eta = o$) bzw. in O_1 ($\xi = a'$, $\eta = o$) bestimmt durch:

$$W(xx_1) = \frac{1}{\pi I \sqrt{(1-k_a^2)}}\, e^{-\frac{x^2 - 2k_a xx_1 + x_1^2}{I(1-k_a^2)}}$$

$$W(yy_1) = \frac{1}{\pi I \sqrt{(1-k_a^2)}}\, e^{-\frac{y^2 - 2k_a yy_1 + y_1^2}{I(1-k_a^2)}} \quad \text{wo } k_a = \frac{\sin a\alpha_0}{a\alpha_0}.$$

Ebenso gilt für den Fall, dass für den Punkt O_1 der Punkt $\xi = o$, $\eta = b'$ gewählt wird:

$$W(xx_1) = \frac{1}{\pi I \sqrt{(1-k_b^2)}}\, e^{-\frac{x^2 - 2k_b xx_1 + x_1^2}{I(1-k_b^2)}}$$

$$W(yy_1) = \ldots\ldots\ldots\ldots\ldots\ldots\ldots\ldots$$

$$\text{wo } k_b = \frac{\sin b\beta_0}{b\beta_0}$$

und schliesslich für den Fall, dass wir für O_1 den Punkt $\xi = a'$, $\eta = b'$ wählen:

$$W(xx_1) = \frac{1}{\pi I \sqrt{(1-k^2)}}\, e^{-\frac{x^2 - 2k xx_1 + x_1^2}{I(1-k^2)}} \tag{6}$$

$$W(yy_1) = \ldots\ldots\ldots\ldots\ldots\ldots\ldots\ldots$$

worin

$$k = k_a k_b = \frac{\sin a\alpha_0}{a\alpha_0} \cdot \frac{\sin b\beta_0}{b\beta_0}. \tag{7}$$

Die Verteilung der Korrelation, welche durch (7) bestimmt wird, ist also mit dem Beugungsbilde der Linse identisch. Dies ist um so merkwürdiger, als im ersten Fall bei der betrachteten inkohärenten Lichtquelle von Beugung oder Interferenz nicht die Rede sein kann.

Im Allgemeinen kann man sagen, dass die Korrelation zwischen den Schwingungsvektoren in zwei Punkten O_1 und O_2 durch Formel (6) bestimmt wird, worin

$$k = \frac{\sin (a_1 - a_2)\alpha_0}{a_1 - a_2} \cdot \frac{\sin (b_1 - b_2)\beta_0}{b_1 - b_2}.$$

Die Schwingungsvektoren in O_1 und O_2 sind identisch für $k = 1$, also nur für unmittelbar benachbarte Punkte; sie sind völlig von einander unabhängig für $k = 0$, also für die Nullstellen der Beugungsfigur. Die Lichterregung rund um ein Punkt O ist also mit der Erregung in O korreliert, diese Korrelation nimmt jedoch mit wachsender Entfernung von O nach Null ab, und zwar um so schneller je grösser der Öffnungswinkel des auffallenden Bündels ist. Bei noch grösserer Entfernung jedoch tritt eine kleine negative Korrelation auf, welche darauf hindeutet, dass in diesem Gebiete die entgegengesetzte Phase wie diejenige in O etwas grössere Wahrscheinlichkeit hat, u.s.w.

Hierauf beruht z.B. die bekannte M i c h e l s o n sche Methode zur Bestimmung von Sterndurchmessern. Denken wir uns zwei Spalte in der beleuchteten Ebene in der Entfernung a' in der ξ-Richtung. Es lässt sich leicht zeigen, dass, wenn die Differenz bzw. die Summe der Lichtvektoren in den Spalten v, δ bzw. s, δ' sind, das Interferenzbild bestimmt wird durch

$$v^2 \sin^2 \varphi + s^2 \cos^2 \varphi + 2\, v\, s \sin \varphi \cos \varphi \sin (\delta - \delta').$$

Multiplizieren wir diesen Ausdruck mit den Wahrscheinlichkeiten dass v und s beide auftreten, und integrieren wir nach v, δ, s und δ', so bekommen wir:

$$2\, I\, (1 + k_a \cos 2\, \varphi).$$

Die Sichtbarkeit der Interferenzen ist also gleich k_a. (Negative Sichtbarkeit bedeutet eine Umwechslung der hellen und der dunklen Streifen). Mit wachsender Abstand zwischen den Spalten verschwinden also die Interferenzen für

$$a' = \frac{\lambda}{2\alpha_0}, \quad \frac{2\lambda}{2\alpha_0}, \quad \frac{3\lambda}{2\alpha_0}, \quad \ldots\ldots\ldots \text{ u.s.w.}$$

Ein zweiter Beispiel liefert die maximal zulässige Spaltbreite in der Spektroskopie. Wie schon früher vom Verfasser[1]) gezeigt wurde, ist

1) ZS. f. Phys. **65**, 547, 1930; **69**, 298, 1931.

die Intensitätsverteilung in den Spektrallinien tatsächlich unabhängig davon ob der Spalt direkt beleuchtet wird oder ob die Lichtquelle mittels einer Linse in der Spaltebene abgebildet wird. In derselben Arbeit ist gezeigt worden, dass die maximal zulässige Spaltbreite a' in Wellenlängen gemessen etwa $1^1/_2$ mal so gross wie der umgekehrte Öffnungswinkel des Kollimators ist, und weiter, dass man den Kollimator nicht mehr als etwa $^2/_3$ ausfüllen darf. Bezeichnen wir den Öffnungswinkel des Kollimators mit Ω und denjenigen des eintretenden Bündels mit $2\alpha_0$, dann soll also:

$$\frac{a'}{\lambda} = \frac{1^1/_2}{\Omega} \text{ und } 2\alpha_0 = {}^2/_3\,\Omega$$

also

$$a' = \frac{\lambda}{2\alpha_0}$$

sein. Man soll also den Spalt höchstens so weit machen, dass die Lichtvektoren an beiden Spaltbacken für das erste Mal nicht korreliert sind.

Ist die Lichtquelle im ersten Fall nicht rechteckig und ist sie nicht überall gleich stark leuchtend, dann kann man doch in analoger Weise die Korrelation berechnen. Mann muss dabei i als eine Funktion von α und β betrachten und sich die Lichtquelle in einem Rechteck eingeschlossen denken. Zwischen den wahren Begrenzungen und den angenommenen rechteckigen setzen wir dann i gleich o. Mann findet dann in ganz analoger Weise dieselben Ausdrücke für die Wahrscheinlichkeiten, jedoch muss man dann für k statt:

$$\frac{\sin a\alpha_0}{a\alpha_0} \cdot \frac{\sin b\beta_0}{b\beta_0}$$

den Wert:

$$k = \frac{\int_{-\alpha_0}^{+\alpha_0} d\alpha \int_{-\beta_0}^{+\beta_0} d\beta\, i\,(\alpha\beta) \cos\,(a\alpha - b\beta)}{\int_{-\alpha_0}^{+\alpha_0} d\alpha \int_{-\beta_0}^{-\beta_0} d\beta\, i\,(\alpha\beta)}$$

einsetzen.

Dieselbe Korrelation erzielt man wieder mit einer Linse von derselben Form wie die Lichtquelle, jedoch mit einer Durchlässigkeit, welche mit $i(\alpha\beta)$ proportional ist.

PAPER NO. 2

Reprinted from *Physica*, Vol. 6, pp. 1129–1138, 1939.

KOHAERENZ-PROBLEME

von P. H. VAN CITTERT

Mitteilung aus dem Physikalischen Institut der Reichs-Universität Utrecht

Zusammenfassung

Es wird gezeigt dass der kürzlich von Zernike[1]) eingeführte Begriff „Kohärenz" mit dem schon vor fünf Jahren vom Verfasser[2]) eingeführten Begriff „Korrelation" identisch ist. Der Begriff wird auf nicht monochromatische Lichtquellen erweitert, und gezeigt wird dass auch in diesem Fall der schon früher vom Verfasser bewiesene Satz gültig ist, nämlich dass die Korrelation identisch ist mit der Sichtbarkeit der Interferenzen. Allgemeine Formeln werden abgeleitet um den Einfluss der Grösse der Lichtquelle und der Zusammensetzung des Lichtes bei optischer Abbildung in Rechnung zu ziehen.

Die Lichtschwingungen in den verschiedenen Punkten einer Ebene die von einer (monochromatischen) Lichtquelle beleuchtet wird, sind nicht von einander unabhängig, sondern korreliert. Die Grösse dieser Korrelation hängt von den Abmessungen der Lichtquelle ab: eine punktförmige Lichtquelle z.B. gibt eine völlig kohärente Beleuchtung; bei sehr grossen Lichtquellen jedoch sind die Schwingungen in benachbarten Punkten nahezu unabhängig von einander.

Bekanntlich kann die Lichtschwingung

$$s = \rho e^{i\phi} e^{i\nu t}$$

in einem XH-Diagram vorgestellt werden durch einen Vektor ρ, welcher einen Winkel φ mit der X-Achse einschliesst, und dessen Komponente $\xi = \rho \cos \varphi$, bzw. $\eta = \rho \sin \varphi$ sind. Betrachten wir einen festen Punkt P_1 (x_1, y_1) und einen veränderlichen Punkt $P(x, y)$ in der beleuchteten Ebene, so kommt die Wahrscheinlichkeit im Frage, dass gleichzeitig in P_1 die Komponente ξ_1 und η_1 und in P die Komponente ξ und η auftreten. Vor einigen Jahren ist vom Verfasser (l.c.) für den Fall dass eine Ebene durch eine ausgedehnte symmetrische Lichtquelle beleuchtet wird, gezeigt worden, dass diese Wahrscheinlichkeit abhängig ist von einer Funktion $k(x—x_1, y—y_1)$, welche er die *Korrelation* benannt hat: $k = 1$ bedeutete völlig kohärente Beleuchtung, $k = 0$ völlig inkohärente Beleuchtung.

Für diese Korrelation wurden vom Verfasser folgende Sätze bewiesen:

1°. *Die Funktion k ist mit der Beugungsfunktion einer Blende, welche in Amplitüdo-Durchlässigkeit, Form und Stelle mit der Lichtstärke, Form und Stelle der Lichtquelle übereinstimmt, identisch.*

2°. *Wenn man die Lichtschwingungen in den betrachteten Punkten zur Interferenz bringt, bekommt man ein Interferenzsystem mit einer Sichtbarkeit gleich dem mit der gegenseitigen Lage der Punkte übereinstimmenden Wert der Korrelation k.*

Dieser Satz wurde vom Verfasser angewandt um neues Licht auf die bekannte M i c h e l s o n sche Methode zur Bestimmung von Sterndurchmessern zu werfen.

3°. *Die Korrelationsverteilung in einer Ebene, die beleuchtet wird dadurch dass man mittels einer Linse eine ausgedehnte Lichtquelle in der Ebene abbildet, ist identisch mit der Korrelationsverteilung, welche man bekommt, wenn die Linse durch eine Lichtquelle gleicher Abmessungen ersetzt wird.*

M.a.W.: *Es macht gar keinen Unterschied ob man eine Ebene direkt oder mittels einer Linse beleuchtet* *).

Nun hat Z e r n i k e (l.c.) neulich den Begriff „Kohärenz" eingeführt. Diesen Begriff definierte Z e r n i k e als die Sichtbarkeit der Interferenzen, ergeben durch die Zusammenwirkung der Lichtschwingungen in den Punkten P_1 und P. *Es leuchtet unmittelbar ein, dass dieser von Z e r n i k e eingeführte Begriff „Kohärenz" mit dem vom Verfasser eingeführten Begriff „Korrelation" identisch ist.*

Es kann folglich nicht wundern, dass es Z e r n i k e gelingt mit seinem Begriff „Kohärenz" die Sätze 1 und 3 aufs Neue zu beweisen.

Im Folgenden wollen wir einige Anwendungen des Begriffs Korrelation (oder Kohärenz) besprechen welche sich beziehen auf Abbildungs- bzw. Interferenzprobleme. Um unnötig verwickelten Berechnungen vorzubeugen, wollen wir annehmen, dass alle Abmessungen der Lichtquelle u.s.w. in der Y-Richtung sehr gross sind im Gegensatz zu den Abmessungen in der X-Richtung. Ist dies nicht der Fall, so müssen die früher vom Verfasser beschriebenen Methoden angewendet werden.

3) Schon in 1931, Z. Phys. **69**, 298, 1931, wurde dieser Satz von Verfasser publiziert.

Fangen wir an mit der Berechnung der Korrelation in einer Ebene, die von einer von der Ebene weit entfernten Lichtquelle L_1L_2 (Fig. 1) beleuchtet wird. Wir denken uns die Lichtquelle aufgebaut durch n von einander unabhängige Elemente, jedes in der Ebene eine Amplitude ρ und also eine Intensität proportional mit ρ^2 verursachend. Die totale mittlere Intensität $n\rho^2$ stellen wir gleich I. Wir setzen weiter voraus, dass diese Elemente so über die Lichtquelle verteilt sind,

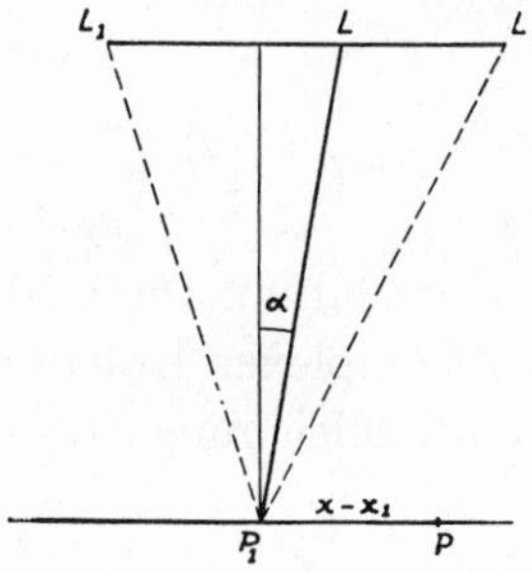

Fig. 1. L_1L_2 = Lichtquelle, P_1P = beleuchtete Ebene.

dass der Teil der Lichtquelle der in der Richtung α liegt, in P_1 die mittlere Intensität $f(\alpha)d(\alpha)$ ergibt. Es ist also auch:

$$\int_{\alpha_1}^{\alpha_2} f(\alpha)d(\alpha) = I$$

Wir fragen nun nach der Wahrscheinlichkeit $W(\xi_1, \eta_1, \xi, \eta, n)$ $d\xi_1 d\eta_1 d\xi d\eta$ dass diese Lichtquelle zu gleicher Zeit in P_1 den Vektor $\xi_1\eta_1$ und in P den Vektor $\xi\eta$ verursacht. Die F o k k e r-Pl a n c k-sche Methode ergibt unmittelbar die Gleichung:

$$W(\xi_1, \eta_1, \xi, \eta, n+1) = \int_{\alpha_1}^{\alpha_2} \frac{f(\alpha)}{I} d\alpha \int_0^{2\pi} \frac{d\varphi}{2\pi} W\Big[\xi_1 - \rho\cos\varphi,\ \eta_1 - \rho\sin\varphi,$$
$$\xi - \rho\cos\Big\{\varphi + \frac{2\pi}{\lambda}(x - x_1)\alpha\Big\},\ \eta - \rho\sin\Big\{\varphi + \frac{2\pi}{\lambda}(x - x_1)\alpha\Big\},\ n\Big] \quad (1)$$

Entwicklung ergibt:

$$\frac{\partial W}{\partial n} = \frac{\rho^2}{4}\Big\{\frac{\partial^2 W}{\partial \xi_1^2} + \frac{d^2 W}{\partial \eta_1^2} + \frac{\partial^2 W}{\partial \xi^2} + \frac{\partial^2 W}{\partial \eta^2} + 2c\frac{\partial^2 W}{d\xi \partial \xi_1} + 2c\frac{\partial^2 W}{\partial \eta \partial \eta_1} -$$
$$- 2s\frac{\partial^2 W}{\partial \xi \partial \eta_1} + 2s\frac{\partial^2 W}{\partial \eta \partial \xi_1}\Big\} \quad (2)$$

wo

$$c(x - x_1) = \frac{1}{I} \int_{a_1}^{a_2} f(\alpha) \cos \frac{2\pi}{\lambda} (x - x)\alpha \, d\alpha$$

und

$$s(x - x_1) = \frac{1}{I} \int_{a_1}^{a_2} f(\alpha) \sin \frac{2\pi}{\lambda} (x - x_1)\alpha \, d\alpha.$$

$$k = c + is = \frac{1}{I} \int_{a_1}^{a_2} f(\alpha) \, e^{(2\pi i/\lambda)(x - x_1)\alpha} \, d\alpha \tag{3}$$

ist die (hier komplexe) Korrelation. Diese ist identisch mit der von Z e r n i k e eingeführten komplexen Kohärenz.

Die Lösung der Differentialgleichung ist, weil $n\rho^2 = I$:

$$W(\xi_1, \eta_1, \xi, \eta) =$$

$$= \frac{1}{\pi^2(1 - c^2 - s^2)I^2} \, e^{-\{\xi^2_1 + \eta^2_1 + \xi^2 + \eta^2 - 2c(\xi\xi_1 + \eta\eta_1) + 2s(\xi\eta_1 - \eta\xi_1)\}/(1 - c^2 - s^2)I} \tag{4}$$

Wenn jetzt der Schwingungsvektor $\xi_1\eta_1$ im Punkte P_1 mit dem Vektor $\xi\eta$ im Punkte P zusammenwirkt, bekommt man das Interferenzsystem

$$s(\psi) = (\xi_1 + i\eta_1) \, e^{i\nu t} + (\xi + i\eta) \, e^{i\{\nu t - (2\pi/\lambda)(x - x_1)\psi\}}$$

und die Intensitätsverteilung:

$$I(\psi) = s(\psi)s^*(\psi)$$

Multiplikation mit der Wahrscheinlichkeit

$$W(\xi_1, \eta_1, \xi, \eta) \, d\xi_1 d\eta_1 d\xi d\eta$$

dass diese Vektoren zugleich auftreten, und Integration nach ξ_1, η_1, ξ und η gibt für die mittlere Intensitätsverteilung:

$$\overline{I(\psi)} = 2I \left\{ 1 + c \cos \frac{2\pi}{\lambda} (x - x_1)\psi + s \sin \frac{2\pi}{\lambda} (x - x_1) \, \psi \right\}$$

Die Sichtbarkeit der Interferenzen z ist also:

$$z = \sqrt{c^2 + s^2} = | \, k \, |$$

Betrachten wir jetzt einen durch die Lichtquelle L_1L_2 beleuchteten Spalt R_1R_2, welcher mittels irgendeines optischen Systems O in der

Ebene Q abgebildet wird (Fig. 2). Setzen wir voraus dass die Schwingung

$$s_P = (\xi + i\eta)\, e^{i\nu t}$$

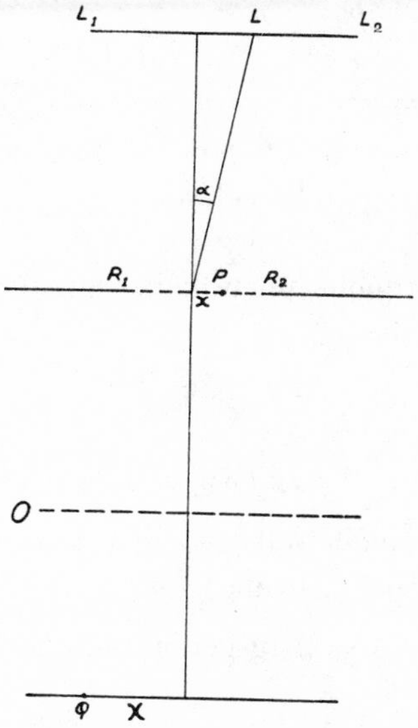

Fig. 2. L_1L_2 = Lichtquelle, R_1R_2 = Spalt, Q = beliebiges optischer System, Q = Bildebene.

in dem Punkte P der Spaltebene, im Punkte Q der Bildebene die Schwingung

$$s_Q = (\xi + i\eta)\, A(xX) e^{i\nu t}$$

ergibt, so ist die totale Amplitüde in Q

$$S(X) = \int_{Spalt} (\xi + i\eta)\, A(xX)\, dx$$

und die Intensität

$$I(X) = \int_{Sp} (\xi + i\eta)\, A(xX)\, dx \,.\, \int_{Sp} (\xi_1 - i\eta_1)\, A^*(x_1X)\, dx_1$$

oder

$$I(X) = \int_{Sp}\int_{Sp} B\, dx\, dx_1.$$

Die Wahrscheinlichkeit, dass der Integrant

$$B = (\xi\xi_1 + \eta\eta_1 - i\xi\eta_1 + i\eta\xi_1)\, A(xX)\, A^*(x_1X)$$

auftritt, ist wiederum:

$$W(\xi_1, \eta_1, \xi, \eta)\, d\xi_1 d\eta_1 d\xi d\eta$$

Die mittlere Intensität ist also:

$$\overline{I(X)} = \int dx \int dx_1 \iiiint W \,.\, B \,.\, d\xi_1 d\eta_1 d\xi d\eta$$

Integration ergibt:

$$\overline{I(X)} = \mathrm{I} \int_{Sp} dx \int_{Sp} dx_1 k(x - x_1)\, A(xX)\, A^*(x_1 X) \;^{*)} \tag{5}$$

Diese Beziehung ermöglicht den Einfluss der Grösse der Lichtquelle auf die Abbildung zu berechnen.

Ins Besondere ist:

1°. wenn die Lichtquelle unendlich schmal ist (völlig kohärente Spaltbeleuchtung),

$$k(x - x_1) = 1$$

und also

$$\overline{I(X)} = I \int dx \int dx_1\, A(xX)\, A^*(x_1 X)$$

2°. wenn die Lichtquelle sehr gross ist, und $f(\alpha) =$ konstant (völlig inkohärente Spaltbeleuchtung),

$$k(x - x_1) = 0 \text{ ausgenommen } k(0) = 1,$$

und also

$$\overline{I(X)} = I \int dx\, A(xX)\, A^*(xX)$$

---

*) Diese Beziehung kann auch so abgeleitet werden:

Der Punkt L der Lichtquelle ergibt in P die Schwingung:

$$s'_P = \sqrt{f(\alpha)}\, e^{i(\nu t + (2\pi/\lambda) x \alpha)} dx\, d\alpha$$

und in Q:

$$s'_Q = \sqrt{f(\alpha)}\, A(xX)\, e^{i(\nu t + (2\pi/\lambda) x \alpha} dx\, d\alpha$$

Die totale Schwingung durch L in Q verursacht, ist also:

$$S'(X) = \int_{Sp} \sqrt{f(\alpha)}\, A(xX)\, e^{i(\nu t + (2\pi/\lambda) x \alpha)} dx\, d\alpha$$

und die Intensität

$$I'(X) = \int_{Sp} dx \int_{Sp} dx_1\, f(\alpha)\, A(xX)\, A^*(x_1 X)\, e^{(2\pi i/\lambda)(x - x_1)\alpha} d\alpha$$

Die totale Intensität in X ist also:

$$\overline{I(X)} = \int_{a_1}^{a_2} I'(X)\, d\alpha$$

oder, weil

$$\frac{1}{I} \int_{a_1}^{a_2} d\alpha f(\alpha)\, e^{(2\pi\, /\lambda)(x - x_1)\alpha} = k(x - x_1)$$

$$\overline{I(X)} = I \int_{Sp} dx \int_{Sp} dx_1\, k(x - x_1)\, A(xX)\, A^*(x_1 X).$$

Im Vorigen ist für monochromatische Beleuchtung die Korrelation berechnet worden, welche besteht zwischen den Lichtschwingunge in *zwei verschiedenen Punkten*. Wenn, dagegen, die Lichtquelle nicht monochromatisch ist, ändert sich die Schwingung in *einem bestimmten Punkt* jeden Augenblick, und man kann in diesem Falle fragen nach der Korrelation zwischen der Schwingung in einem bestimmten Moment t_1 an einem Punkt und die Schwingung am gleichen Punkt τ Sekunden später ($\tau = t - t_1$). Auch diese Korrelation (oder Kohärenz) ist für einige Fälle zu berechnen.

1. Denken wir uns eine Lichtquelle, welche ein schmales Frequenzgebiet um die mittlere Frequenz ν_0 aussendet (eine zusammengesetzte oder verbreiterte Spektrallinie). In diesem Falle kann man sich die Lichtquelle aufgebaut denken von n unabhängigen Elementen, welche jedes eine Amplitüde ρ und also eine mittlere Intensität (:) ρ^2 haben, und welche derart über die verschiedenen Frequenzen verteilt sind, dass es $f(\mu)d\mu$ Elemente im Frequenzgebiet zwischen $\nu_0 + \mu$ und $\nu_0 + \mu + d\mu$ ergibt.

Wir nehmen wieder:

$$n\rho^2 = \int_{-\mu}^{+\mu} f(\mu)d\mu = I.$$

Die Frage nach der Wahrscheinlichkeit, dass der Schwingungsvektor $\xi_1\eta_1$ zur Zeit t_1 auftritt und der Vektor $\xi\eta$ zur Zeit $t = t_1 + \tau$, ist völlig analog mit der oben gestellten Frage nach der Wahrscheinlichkeitsverteilung in einer Ebene, welche beleuchtet wird durch eine ausgedehnte Lichtquelle mit Intensitätsverteilung $f(\alpha)d\alpha$.

Die F o k k e r-Pl a n c k sche Methode ergibt sofort wieder Gleichung (1), mit dem blossen Unterschied das jetzt statt $(2\pi/\lambda)(x - x_1)\alpha$ stehen muss $\mu(t - t_1)$ oder $\mu\tau$.

Die Wahrscheinlichkeit wird also auch jetzt bestimmt durch Gleichung (3), wo ebenso statt $(2\pi/\lambda)(x - x_1)\alpha$, $\mu\tau$ geschrieben wird.

In diesem Falle wird die Korrelation also (in analoger Weise wie (3)):

$$k_1 = \frac{1}{I}\int_{-\infty}^{+\infty} f(\mu)\, e^{i\mu(t-t_1)}\, d\mu = \frac{1}{I}\int_{-\infty}^{+\infty} f(\mu)\, e^{i\mu\tau}\, d\mu \tag{6}$$

Hieraus findet man gleich dass die Sichtbarkeit der Interferenzen, welche man bekommt mit dem Michelson Interferometer, beträgt:

$$z_1(\tau) = | k_1(\tau) | = \sqrt{c_1^2(\tau) + s_1^2(\tau)}.$$

Dieses Resultat ist völlig in Übereinstimmung mit den bekannten Formeln von Michelson für die Sichtbarkeitskurven.

Zu gleicher Zeit führt dieses Resultat zu der Stellung dass Beleuchtung eines Spaltes mittels einer ausgedehnten monochromatischen Lichtquelle mit einer Intensitätsverteilung $f(\alpha)d\alpha$, ein Interferenzsystem verursacht, welches bei Änderung des Spaltabstandes den gleichen Sichtbarkeitsverlauf ergibt wie das bekannte Michelsonexperiment mit einer Lichtquelle mit Intensitätsverteilung $f(\mu)d\mu$ *).

2. An zweiter Stelle kann das Licht bestehen aus unabhänglichen Pulsen mit Frequenz ν_0, deren Amplitüdo und Phase sich langsam ändern (natürliche Linienform) und welche man also schreiben kann

$$s = f(t)\, e^{i\nu_0 t}.$$

Mit Hilfe der Methode Fokker-Planck findet man jetzt:

$$k_1 = \frac{1}{I} \int_{-\infty}^{+\infty} f(p)\, f^*(p - \tau)\, dp \tag{7}$$

wo

$$I = \int_{-\infty}^{+\infty} f(p)\, f^*(p)\, dp.$$

Man kann dieses Resultat auch in folgender Weise finden:

Die Fourier-entwicklung ergibt:

$$s = f(t)\, e^{i\nu_0 t} = \frac{1}{2\pi} \int_{-\infty}^{+\infty} d\nu \int_{-\infty}^{+\infty} f(p)\, e^{i(\nu_0 - \nu)p}\, dp \,.\, e^{i\nu t}.$$

Folglich kann man sich die Pulse aufgebaut denken aus Schwingungen $e^{i\nu t}$, mit der Amplitüde:

$$a = \frac{1}{2\pi} \int_{-\infty}^{+\infty} f(p)\, e^{i\mu p}\, dp$$

und mit der Energie:

$$F(\mu) = \frac{1}{4\pi^2} \int\int dp\, dp_1\, f(p)\, f^*(p_1)\, e^{i\mu(p - p_1)}.$$

*) Wenn im Algemeinen für die Intensitäte die Werte α_0 und μ_0 korrespondieren, so korrespondieren für die Sichtbarkeiten die Werte $(2\pi/\lambda)(x - x_1)\alpha_0$ und $\mu_0\tau$.

Nach Gleichung (6) ist jetzt:

$$k_1 = \int_{-\infty}^{+\infty} F(\mu)\, e^{i\mu\tau}\, d\mu : \int_{-\infty}^{+\infty} F(\mu)\, d\mu.$$

Integration nach μ gibt:

$$k_1 = \frac{1}{I} \int_{-\infty}^{+\infty} f(p)\, f^*(p - \tau)\, dp \tag{7}$$

oder, weil $\tau = t - t_1$:

$$k_1 = \frac{1}{I} \int_{-\infty}^{+\infty} dp\, f(p + t)\, f^*(p + t_1).$$

Jetzt nehmen wir als Beispiel irgendeinen Interferenzapparat, bei dem die Interferenzen sich bilden durch die Zusammenwirkung der Schwingungen:

$$e^{i\nu_0 t},\ r\, e^{i\nu_0(t+\tau)},\ r^2\, e^{i\nu_0(t+2\tau)},\ \ldots .\ r^{l-1}\, e^{i\nu_0\{t+(l-1)\tau\}}.$$

Aus dieser Zusammenwirkung resultiert die Schwingung:

$$s = \sum_{q=0}^{q=l-1} r^q\, e^{i\nu_0(t+q\tau)} = \frac{1 - r^l\, e^{il\nu_0\tau}}{1 - r\, e^{i\nu_0\tau}}\, e^{i_0\nu t}$$

also die Intensität:

$$I = \sum_{q=0}^{q=l-1} \sum_{q_1=0}^{q_1=l-1} r^q\, r^{q_1}\, e^{i\nu_0(q-q_1)\tau} = \frac{1 + r^{2l} - 2r^l \cos l\nu_0\tau}{1 + r^2 - 2r \cos \nu_0\tau}.$$

Wenn aber das Licht nicht monochromatisch ist, sondern die Korrelation k_1 ist, dann wird, wie (3):

$$\bar{I} = I \sum_{q=0}^{q=l-1} \sum_{q_1=0}^{q_1=l-1} k_1\, \{(q - q_1)\tau\}\, r^q\, r^{q_1}\, e^{i\nu_0(q-q_1)\tau}$$

Wenn, im speziellen Falle:

$$k_1(q - q_1)\tau = \frac{1}{I} \int_{-\infty}^{+\infty} f(\mu)\, e^{i(q-q_1)\mu\tau}\, d\mu$$

dann findet man, wie zu erwarten:

$$\bar{I} = \int_{-\infty}^{+\infty} f(\mu)\, d\mu\, \Sigma\, \Sigma\, r^q\, r^{q_1}\, e^{i(\nu_0+\mu)(q-q_1)\tau}$$

$$= \int_{-\infty}^{+\infty} f(\mu)\, d\mu\, \frac{1 + r^{2l} - 2r^l \cos l(\nu_0 + \mu)\tau}{1 + r^2 - 2r \cos (\nu_0 + \mu)\tau}.$$

Wenn aber

$$k_1\{(q-q_1)\tau\} = \frac{1}{I}\int_{-\infty}^{+\infty} f(p+q\tau)\, f^*(p+q_1\tau)\, dp,$$

wird

$$\bar{I} = \int dp\, \Sigma\, f(p+q\tau)\, r^q\, e^{i\nu_0 q\tau} \,.\, \Sigma\, f^*(p+q\tau)\, r^{q_1}\, e^{-i\nu_0 q_1\tau}.$$

Auch dieses Resultat ist einleuchtend.

Denken wir uns schliesslich wieder eine Ebene, beleuchtet durch eine ausgedehnte Lichtquelle, welche eine Intensitätsverteilung $f(\alpha)d\alpha$ hat und eine Totalintensität $I = \int_{a_1}^{a_2} f(\alpha)d\alpha$. Das Licht ist jetzt nicht mehr monochromatisch, sondern hat eine Frequenzverteilung $g(\mu)d\mu$.

Die Korrelation nach der Zeit ist dann:

$$k_1 = c_1 + is_1 = \frac{1}{I_1}\int_{-\infty}^{+\infty} g(\mu)\, e^{i\mu\tau}\, d\mu,$$

wo

$$I_1 = \int_{-\infty}^{+\infty} g(\mu)d\mu$$

Wenden wir jetzt wieder die Methode F o k k e r-P l a n c k an, so führt dieses aufs neue zu der Differentialgleichung (2), in dem Sinne dass jetzt:

$$k = c + is = \frac{1}{II_1}\int_{-\infty}^{+\infty} d\mu \int_{a_1}^{a_2} d\alpha\, f(\alpha)\, g(\mu)\, e^{i(\nu_0+\mu)(x-x_1)\,\alpha/c}$$

ist, und deswegen

$$k = \frac{1}{I}\int_{a_1}^{a_2} f(\alpha)\, k_1\left\{\frac{(x-x_1)\alpha}{c}\right\} e^{i\nu_0(x-x_1)\,\alpha/c}$$

oder

$$k = \frac{1}{I}\int_{a_1}^{a_2} f(\alpha)\, k_1\left\{\frac{(x-x_1)\alpha}{c}\right\} e^{(2\pi i/\lambda_0)\,(x-x_1)\alpha}\, d\alpha.$$

Eingegangen am 3. Oktober 1939.

LITERATURVERZEICHNIS

1) F. Z e r n i k e, Physica, **5**, 785, 1938.
2) P. H. v a n C i t t e r t, Physica, **1**, 201, 1934.

PAPER NO. 3

Reprinted from *Physikalische Zeitschrift*, Vol. 10, pp. 185–193, 1909.

Zum gegenwärtigen Stand des Strahlungsproblems.

Von A. Einstein.

In der letzten Zeit sind in dieser Zeitschrift von den Herren H. A. Lorentz[1]), Jeans[2]) und Ritz[3]) Meinungsäußerungen erschienen, die geeignet sind, den heutigen Stand dieses ungemein wichtigen Problems erkennen zu lassen. In der Meinung, daß es von Vorteil sei, wenn alle, die über diese Sache ernsthaft nachgedacht haben, ihre Ansichten mitteilen, auch wenn sie zu endgültigem Resultat nicht haben vordringen können, teile ich das Folgende mit.

1. Die einfachste Form, in der wir die bisher erkannten Gesetzmäßigkeiten der Elektrodynamik ausdrücken können, ist durch die Maxwell-Lorentzschen partiellen Differentialgleichungen gegeben. Diejenigen Formen, in denen retardierte Funktionen vorkommen, sehe ich, im Gegensatz zu Herrn Ritz[3]), nur als mathematische Hilfsformen an. Ich sehe mich dazu in erster Linie dadurch gezwungen, daß jene Formen das Energieprinzip nicht in sich schließen, indem ich glaube, daß wir an der strengen Gültigkeit des Energieprinzips so lange festhalten sollen, bis wir gewichtige Gründe gefunden haben, auf diesen Leitstern zu verzichten. Es ist ja gewiß richtig, daß die Maxwellschen Gleichungen für den leeren Raum, für sich allein genommen, gar nichts aussagen, daß sie nur eine Zwischenkonstruktion darstellen; genau das gleiche läßt sich ja bekanntlich auch von den Newtonschen Bewegungsgleichungen sagen, sowie von jeder Theorie, die noch der Ergänzung durch andere Theorien bedarf, um ein Bild für einen Komplex von Erscheinungen liefern zu können. Was die Maxwell-Lorentzschen Differentialgleichungen gegenüber Formen, welche retardierte Funktionen enthalten, auszeichnet, das ist der Umstand, daß sie für jeden Augenblick, und zwar relativ zu jedem unbeschleunigten Koordinatensystem, einen Ausdruck für die Energie und für die Bewegungsgröße des betrachteten Systems liefern. Bei einer Theorie, die mit retardierten Kräften operiert, kann man den Momentanzustand eines Systems überhaupt nicht beschreiben, ohne für diese Beschreibung frühere Zustände des Systems zu benützen. Hat z. B. eine Lichtquelle A einen Lichtkomplex gegen den Schirm B hin abgesandt, dieser den Schirm B aber noch nicht erreicht, so ist nach den mit retardierten Kräften operierenden Theorien der Lichtkomplex durch nichts repräsentiert als durch die Vorgänge, welche bei der vorhergegangenen Emission im aussendenden Körper stattgefunden haben. Energie und die Bewegungsgröße müssen dann — wenn man auf diese Größen nicht überhaupt verzichten will — als Zeitintegrale dargestellt werden.

Herr Ritz behauptet nun zwar, daß wir durch die Erfahrung dazu gezwungen seien, die Differentialgleichungen zu verlassen und die retardierten Potentiale einzuführen. Indessen scheint mir seine Begründung nicht stichhaltig zu sein.

Setzt man mit Ritz:

$$f_1 = \frac{1}{4\pi} \int \frac{\varphi\left(x', y', z', t - \frac{r}{c}\right)}{r} dx', dy', dz'$$

und

1) H. A. Lorentz, diese Zeitschr. **9**, 562—563, 1908.
2) J. H. Jeans, diese Zeitschr. **9**, 853—855, 1908.
3) W. Ritz, diese Zeitschr. **9**, 903—907, 1908.

$$f_2 = \frac{1}{4\pi} \int \frac{\varphi\left(x', y', z', t + \frac{r}{c}\right)}{r} dx', dy', dz,$$

so ist sowohl f_1 wie f_2 eine Lösung der Gleichung

$$\frac{1}{c^2} \frac{\partial^2 f}{\partial t^2} - \Delta f = \varphi(x y z t),$$

es ist also auch

$$f_3 = a_1 f_1 + a_2 f_2$$

eine Lösung, wenn $a_1 + a_2 = 1$. Es ist aber nicht richtig, daß die Lösung f_3 eine allgemeinere Lösung ist als f_1, und daß man die Theorie spezialisiert, indem man $a_1 = 1$, $a_2 = 0$ setzt. Setzt man

$$f(x, y, z, t) = f_1,$$

so kommt dies darauf hinaus, daß man die elektromagnetische Wirkung im Punkte x, y, z berechnet aus denjenigen Bewegungen und Konfigurationen der elektrischen Mengen, welche vor dem Augenblick t stattgefunden haben. Setzt man

$$f(x, y, z, t) = f_2,$$

so benützt man zur Bestimmung jener elektromagnetischen Wirkung diejenigen Bewegungen und Konfigurationen, welche nach dem Augenblick t stattfinden.

Im ersteren Fall berechnet man das elektromagnetische Feld aus der Gesamtheit der es erzeugenden, im zweiten Fall aus der Gesamtheit der es absorbierenden Vorgänge. Wenn der ganze Vorgang in einem allseitig begrenzten (endlichen) Raume vor sich geht, kann man ihn ebensowohl in der Form

$$f = f_1$$

wie in der Form

$$f = f_2$$

darstellen. Wenn nun ein vom Endlichen ins Unendliche emittiertes Feld betrachtet wird, kann man naturgemäß nur die Form

$$f = f_1$$

anwenden, weil eben die Gesamtheit der absorbierenden Vorgänge nicht in Betracht gezogen wird. Aber es handelt sich hier um ein irreführendes Paradoxon des Unendlichen. Es lassen sich stets beide Darstellungsweisen anwenden, wie entfernt man sich auch die absorbierenden Körper denken mag. Man kann also nicht schließen, daß die Lösung $f = f_1$ spezieller sei, als die Lösung $a_1 f_1 + a_2 f_2$, wobei $a_1 + a_2 = 1$.

Daß ein Körper nicht „Energie aus dem Unendlichen empfängt, ohne daß irgend ein anderer Körper ein entsprechendes Quantum Energie verliert", kann nach meiner Meinung ebenfalls nicht als Argument angeführt werden. Zunächst können wir, wenn wir bei der Erfahrung bleiben wollen, nicht vom Unendlichen reden, sondern nur von Räumen, die außerhalb des betrachteten Raumes liegen. Ferner aber kann aus der Nichtbeobachtbarkeit eines derartigen Vorgangs eine Nichtumkehrbarkeit der elektromagnetischen Elementarvorgänge ebensowenig geschlossen werden, als eine Nichtumkehrbarkeit der elementaren Bewegungsvorgänge der Atome aus dem zweiten Hauptsatz der Thermodynamik gefolgert werden darf.

2. Man kann der Jeansschen Auffassung entgegenhalten, daß es vielleicht unzulässig sei, die allgemeinen Ergebnisse der statistischen Mechanik auf mit Strahlung gefüllte Hohlräume anzuwenden. Indessen kann man auch auf folgendem Wege zu dem von Jeans gefolgerten Gesetz gelangen[1]).

Ein Ion, welches um eine Gleichgewichtslage in Richtung der X-Achse zu oszillieren vermag, emittiert und absorbiert nach der Maxwellschen Theorie nur dann im Mittel gleiche Mengen Strahlung pro Zeiteinheit, wenn zwischen der mittleren Schwingungsenergie $\overline{E_\nu}$ und der Energiedichte der Strahlung ϱ_ν bei der Eigenfrequenz ν des Oszillators die Beziehung

$$E_\nu = \frac{c^3}{8\pi\nu^2} \varrho_\nu, \tag{I}$$

besteht, wobei c die Lichtgeschwindigkeit bedeutet. Wenn das oszillierende Ion auch mit Gasmolekülen (oder überhaupt mit einem mittels der Molekulartheorie darstellbaren System) in Wechselwirkung zu treten vermag, so muß nach der statistischen Theorie der Wärme notwendig

$$\overline{E_\nu} = \frac{RT}{N} \tag{II}$$

sein (R = Konstante der Gasgleichung, N = Zahl der Atome in einem Grammatome, T = absolute Temperatur), wenn im Mittel keine Energie vom Gas durch den Oszillator auf den Strahlungsraum übertragen wird[2]).

Aus diesen beiden Gleichungen folgt

$$\varrho_\nu = \frac{R}{N} \frac{8\pi}{c^3} \nu^2 T, \tag{III}$$

also genau das auch von den Herren Jeans und H. A. Lorentz gefundene Gesetz[3]).

3. Daran, daß unsere heutigen theoretischen Ansichten zu dem von Herrn Jeans vertretenen Gesetz mit Notwendigkeit führen, ist nach meiner Meinung nicht zu zweifeln. Aber als

---

1) Vgl. A. Einstein, Ann. d. Phys. (4) **17**, 133—136, 1905.

2) M. Planck, Ann. d. Phys. 1, 99, 1900. M. Planck, Vorlesungen über die Theorie der Wärmestrahlung. III. Kapitel.

3) Es sei ausdrücklich bemerkt, daß diese Gleichung eine unabweisbare Konsequenz der statistischen Theorie der Wärme ist. Der im soeben zitierten Planckschen Buche auf S. 178 enthaltene Versuch, die Allgemeingültigkeit der Gleichung II in Frage zu stellen, beruht — wie mir scheint — nur auf einer Lücke in Boltzmanns Betrachtungen, welche unterdessen durch die Gibbsschen Untersuchungen ausgefüllt wurde.

nicht viel weniger sicher erwiesen können wir es ansehen, daß die Formel (III) nicht mit den Tatsachen vereinbar ist. Warum senden denn die festen Körper nur von einer gewissen, ziemlich scharf ausgesprochenen Temperatur an sichtbares Licht aus? Warum wimmelt es nicht überall von ultravioletten Strahlen, wenn doch beständig solche bei gewöhnlicher Temperatur erzeugt werden? Wie ist es möglich, höchst empfindliche photographische Platten lange Zeit in Kassetten aufzubewahren, wenn diese beständig kurzwellige Strahlen erzeugen? Bezüglich weiterer Argumente verweise ich auf § 166 des mehrfach zitierten Planckschen Werkes. Wir werden also wohl sagen müssen, daß uns die Erfahrung dazu zwingt, entweder die von der elektromagnetischen Theorie geforderte Gleichung (I) oder die von der statistischen Mechanik geforderte Gleichung (II) oder endlich beide Gleichungen zu verwerfen.

4. Wir müssen uns fragen, in welcher Beziehung steht die Plancksche Strahlungstheorie zu der unter 2. angedeuteten, auf unseren gegenwärtig anerkannten theoretischen Grundlagen ruhenden Theorie? Die Antwort auf diese Frage wird nach meiner Meinung dadurch erschwert, daß der Planckschen Darstellung seiner eigenen Theorie eine gewisse logische Unvollkommenheit anhaftet. Ich will im folgenden dies kurz auseinander zu setzen versuchen.

a) Wenn man sich auf den Standpunkt stellt, daß die Nichtumkehrbarkeit der Naturvorgänge nur eine scheinbare ist, und daß der nichtumkehrbare Vorgang in einem Übergang zu einem wahrscheinlicheren Zustand bestehe, so muß man zunächst eine Definition der Wahrscheinlichkeit W eines Zustandes geben. Die einzige solche Definition, die nach meiner Meinung in Betracht kommen kann, wäre die folgende:

Es seien $A_1, A_2 \cdots\cdots A_l$ alle Zustände, welche ein nach außen abgeschlossenes System bei bestimmtem Energieinhalt anzunehmen vermag, bzw. genauer gesagt, alle Zustände, welche wir an einem solchen System mit gewissen Hilfsmitteln zu unterscheiden vermögen. Nach der klassischen Theorie nimmt das System nach einer bestimmten Zeit einen bestimmten dieser Zustände (z. B. A_l) an, und verharrt darauf in diesem Zustand (thermodynamisches Gleichgewicht). Nach der statistischen Theorie nimmt aber das System in unregelmäßiger Folge alle Zustände $A_1 \ldots\ldots A_l$ immer wieder an[1]). Beobachtet man das System eine sehr lange Zeit Θ hindurch, so wird es einen gewissen Teil τ_ν dieser Zeit geben, so daß das System während τ_ν und zwar nur während τ_ν den Zustand A_ν inne hat. Es wird $\frac{\tau_\nu}{\Theta}$ einen bestimmten Grenzwert besitzen, den wir die Wahrscheinlichkeit W des betreffenden Zustandes A_ν nennen.

Ausgehend von dieser Definition kann man zeigen, daß für die Entropie S die Gleichung bestehen muß

$$S = \frac{R}{N} \lg W + \text{konst},$$

wobei die Konstante für alle Zustände gleicher Energie dieselbe ist.

b) Weder Herr Boltzmann noch Herr Planck haben eine Definition von W gegeben. Sie setzen rein formal $W =$ Anzahl der Komplexionen des betrachteten Zustandes.

Verlangt man nun, daß diese Komplexionen gleich wahrscheinlich sein sollen, wobei man die Wahrscheinlichkeit der Komplexion analog definiert, wie wir unter a) die Wahrscheinlichkeit des Zustandes definiert haben, so gelangt man genau zu der unter a) gegebenen Definition der Zustandswahrscheinlichkeit; man hat nur das logisch unnötige Element Komplexion in der Definition mit verwendet.

Obwohl nun die angegebene Beziehung zwischen S und W nur gilt, wenn die Komplexionswahrscheinlichkeit in der angegebenen oder in gleichbedeutender Weise definiert wird, hat weder Herr Boltzmann noch Herr Planck die Komplexionswahrscheinlichkeit definiert. Aber Herr Boltzmann hatte doch klar erkannt, daß das von ihm gewählte molekulartheoretische Bild ihm die von ihm getroffene Wahl der Komplexionen in ganz bestimmter Weise vorschrieb; er hat dies in seiner in den Wiener Sitzungsberichten des Jahres 1877 erschienenen Arbeit „Über die Beziehung . . . . ." auf S. 404 und 405 dargelegt[1]). Auch bei der Resonatorentheorie der Strahlung wäre Herr Planck in der Wahl der Komplexionen nicht frei gewesen. Er hätte das Gleichungspaar

$$S = \frac{R}{N} \lg W$$

und

$$W = \text{Zahl der Komplexionen}$$

nur ansetzen dürfen, wenn er die Bedingung hinzugefügt hätte, daß die Komplexionen so gewählt werden müssen, daß sie in dem von ihm gewählten theoretischen Bilde auf Grund statistischer Betrachtungen als gleich wahrscheinlich befunden werden. Er wäre auf diesem Wege zu der von Jeans verteidigten Formel gelangt. So sehr sich jeder Physiker darüber freuen muß, daß sich Herr Planck in so glücklicher Weise über diese Forderung hinweg-

1) Daß diese letztere Auffassung die allein haltbare ist, geht unmittelbar aus den Eigenschaften der Brownschen Bewegung hervor.

1) Vgl. auch L. Boltzmann, Vorlesungen über Gastheorie, 1. Bd., S. 40, Zeile 9—23.

setze, so wenig wäre es angebracht, zu vergessen, daß die Plancksche Strahlungsformel mit der theoretischen Grundlage, von welcher Herr Planck ausgegangen ist, unvereinbar ist.

5. Es ist einfach zu sehen, in welcher Weise die Grundlagen der Planckschen Theorie abgeändert werden können, damit die Plancksche Strahlungsformel wirklich als Konsequenz der theoretischen Grundlagen resultiert. Ich gebe hier nicht die betreffenden Ableitungen, sondern verweise nur auf meine diesbezüglichen Abhandlungen[1]). Das Resultat ist folgendes: Man gelangt zur Planckschen Strahlungsformel, wenn man

1. an der von Planck aus der Maxwellschen Theorie hergeleiteten Gleichung (I) zwischen Resonatorenergie und Strahlungsdichte festhält[2]),
2. die statistische Theorie der Wärme durch folgende Annahme modifiziert: Ein Gebilde, welches mit der Frequenz ν Schwingungen auszuführen vermag, und welches dadurch, daß es eine elektrische Ladung besitzt, Strahlungsenergie in Energie der Materie und umgekehrt zu verwandeln vermag, vermag nicht Schwingungszustände jeder beliebigen Energie anzunehmen, sondern nur solche Schwingungszustände, deren Energie ein Vielfaches von $h \cdot \nu$ ist. h ist dabei die von Planck so benannte, in seiner Strahlungsgleichung auftretende Konstante.

6. Da die soeben mitgeteilte Mod fikation der Grundlagen der Planckschen Theorie zu sehr tiefgreifenden Änderungen unserer physikalischen Theorien mit Notwendigkeit hinführt, ist es sehr wichtig, möglichst einfache, voneinander unabhängige Interpretationen der Planckschen Strahlungsformel sowie überhaupt des Strahlungsgesetzes, soweit dasselbe als bekannt vorausgesetzt werden darf, aufzusuchen. Zwei diesbezügliche Betrachtungen, die sich durch ihre Einfachheit auszeichnen, seien im folgenden kurz mitgeteilt.

Die Gleichung $S = \frac{R}{N} \lg W$ wurde bisher hauptsächlich derart angewendet, daß man auf Grund einer mehr oder weniger vollständigen Theorie die Größe W und aus dieser die Entropie berechnete. Man kann diese Gleichung aber auch umgekehrt dazu benutzen, um aus den mit Hilfe der Erfahrung ermittelten Entropiewerten S_ν die statistische Wahrscheinlichkeit der einzelnen Zustände A_ν eines nach außen abgeschlossenen Systems zu ermitteln. Eine Theorie, welche andere als die so ermittelten Werte für die Zustandswahrscheinlichkeit liefert, ist offenbar zu verwerfen.

Eine Betrachtung der angedeuteten Art zur Ermittlung gewisser statistischer Eigenschaften von in einen Hohlraum eingeschlossener Wärmestrahlung habe ich bereits in einer früheren Arbeit[1]) durchgeführt, in der ich die Theorie der Lichtquanten zuerst darlegte. Da ich aber damals von der nur in der Grenze (für kleine Werte von $\frac{\nu}{T}$) gültigen Wienschen Strahlungsformel ausging, will ich hier eine ähnliche Betrachtung angeben, welche eine einfache Deutung des Inhalts der Planckschen Strahlungsformel liefert.

Es seien V und v zwei miteinander kommunizierende Räume, die durch diffus vollkommen reflektierende Wände begrenzt seien. In diese Räume sei Wärmestrahlung vom Frequenzbereich $d\nu$ eingeschlossen. H sei die momentan in V, η die momentan in v befindliche Strahlungsenergie. Nach einiger Zeit gilt dann mit gewisser Annäherung dauernd die Proportion $H_0 : \eta_0 = V : v$. In einem beliebig herausgegriffenen Zeitpunkt wird η von η_0 abweichen nach einem statistischen Gesetz, das sich aus der Beziehung zwischen S und W unmittelbar ergibt, indem man zu den Differentialen übergeht

$$dW = \text{konst}\, e^{\frac{N}{R} \cdot S}\, d\eta .$$

Bezeichnet man mit Σ bzw. σ die Entropie der in den beiden Räumen befindlichen Strahlung und setzt man $\eta = \eta_0 + \varepsilon$, so hat man

$$d\eta = d\varepsilon$$

und

$$S = \Sigma + \sigma = \Sigma_0 + \sigma_0 + \left\{\frac{d(\Sigma + \sigma)}{d\varepsilon}\right\}_0 \varepsilon + \frac{1}{2}\left\{\frac{d^2(\Sigma + \sigma)}{d\varepsilon^2}\right\}_0 \varepsilon^2 \ldots .$$

Die letztere Gleichung geht wegen

$$\left\{\frac{d(\Sigma + \sigma)}{d\varepsilon}\right\}_0 = 0,$$

wenn man annimmt, daß V sehr groß ist gegen v, über in

$$S = \text{konst} + \frac{1}{2}\left\{\frac{d^2\sigma}{d\varepsilon^2}\right\}_0 \varepsilon^2 + \cdots\cdots$$

Begnügt man sich mit dem ersten nicht verschwindenden Glied der Entwicklung, was einen um so kleineren Fehler bedingt, je größer v, gegenüber dem Kubus der Strahlungswellenlänge ist, so erhält man

$$dW = \text{konst} \cdot e^{-\frac{1}{2}\frac{N}{R}\left(\frac{d^2\sigma}{d\varepsilon^2}\right)_0 \varepsilon^2} \cdot d\varepsilon .$$

1) A. Einstein, Ann. d. Phys. (4) **20**, 1906 und Ann. d. Phys (4) **22**, 1907, § 1.

2) Es kommt dies darauf hinaus, daß man annimmt, daß die elektromagnetische Theorie der Strahlung wenigstens richtige zeitliche Mittelwerte liefert. Daran läßt sich aber angesichts der Brauchbarkeit der Theorie in der Optik kaum zweifeln.

1) Ann. d. Phys. (4) **17**, 132—148, 1905.

Für den Mittelwert $\overline{\varepsilon^2}$ des Quadrates der Energieschwankung der in v befindlichen Strahlung erhält man daraus

$$\overline{\varepsilon^2} = \frac{1}{\frac{N}{R}\left\{\frac{d^2\sigma}{d\varepsilon^2}\right\}_0}.$$

Ist die Strahlungsformel bekannt, so kann man σ aus derselben berechnen[1]). Betrachtet man als Ausdruck der Erfahrung die Plancksche Strahlungsformel, so erhält man nach einfacher Rechnung

$$\overline{\varepsilon^2} = \frac{R}{Nk}\left\{\nu h \eta_0 + \frac{c^3}{8\pi\nu^2 d\nu}\cdot\frac{\eta_0^2}{v}\right\}.$$

Wir haben so einen leicht zu interpretierenden Ausdruck für die mittlere Größe der Schwankungen der in v befindlichen Strahlungsenergie erlangt. Wir wollen nun zeigen, daß die jetzige Theorie der Strahlung mit diesem Resultat unvereinbar ist.

Nach der jetzigen Theorie rühren die Schwankungen lediglich daher, daß die unendlich vielen, den Raum durchsetzenden Strahlen, welche die Strahlung von v konstituieren, miteinander interferieren und so einen Wert der Momentanenergie liefern, der bald größer, bald kleiner ist, als die Summe der Energie, welche die einzelnen Strahlen liefern würden, wenn sie gar nicht miteinander interferierten. Man könnte so die Größe $\overline{\varepsilon^2}$ durch eine mathematisch etwas komplizierte Betrachtung exakt ermitteln. Wir begnügen uns hier mit einer einfachen Dimensionalbetrachtung. Es müssen folgende Bedingungen erfüllt sein:

1. Die Größe der mittleren Schwankung hängt nur von λ (Wellenlänge), $d\lambda$, σ und v ab, wobei σ die auf Wellenlängen bezogene Strahlungsdichte bedeutet ($\sigma d\lambda = \varrho d\nu$).

2. Da sich die Strahlenenergien benachbarter Wellenlängenbereiche und Volumina[2]) einfach addieren, und die betreffenden Schwankungen voneinander unabhängig sind, muß $\overline{\varepsilon^2}$ bei bestimmtem λ und ϱ den Größen $d\lambda$ und v proportional sein.

3. $\overline{\varepsilon^2}$ hat die Dimension des Quadrates einer Energie.

Dadurch ist der Ausdruck für $\overline{\varepsilon^2}$ bis auf einen Zahlenfaktor (von der Größenordnung 1) vollkommen bestimmt. Man gelangt auf diese Weise zum Ausdruck $\sigma^2\lambda^4 v\, d\lambda$, der bei Einführung der oben benützten Variabeln in den zweiten Term der vorhin für $\overline{\varepsilon^2}$ entwickelten Formel übergeht. Diesen zweiten Term aber hätten wir allein für $\overline{\varepsilon^2}$ erhalten, wenn wir von der Jeansschen Formel ausgegangen wären.

1) Vgl. z. B. das mehrfach zitierte Plancksche Werk Gleichung (230).

2) Natürlich nur, wenn diese genügend groß sind.

Man hätte dann noch $\frac{R}{Nk}$ gleich einer Konstanten von der Größenordnung 1 zu setzen, was der Planckschen Bestimmung des Elementarquantums entspricht[1]). Das erste Glied des obigen Ausdrucks für $\overline{\varepsilon^2}$, das bei der sichtbaren Strahlung, die uns allenthalben umgibt, einen weitaus größeren Beitrag liefert als das zweite, ist also mit der jetzigen Theorie nicht vereinbar.

Setzt man mit Planck $\frac{R}{Nk} = 1$, so würde das erste Glied, wenn es allein vorhanden wäre, eine solche Schwankung der Strahlungsenergie liefern, wie wenn die Strahlung aus voneinander unabhängig beweglichen, punktförmigen Quanten von der Energie $h\nu$ bestünde. Es läßt sich dies durch eine einfache Rechnung zeigen. Es sei ausdrücklich daran erinnert, daß das erste Glied einen um so größeren Beitrag zur mittleren prozentischen Energieschwankung

$$\left(\sqrt{\frac{\overline{\varepsilon^2}}{\eta_0^2}}\right)$$

liefert, je kleiner die Energie η_0 ist, und daß die Größe dieser vom ersten Glied gelieferten prozentischen Schwankung davon unabhängig ist, über einen wie großen Raum v die Strahlung verteilt ist; ich erwähne dies, um zu zeigen, wie grundverschieden die tatsächlichen statistischen Eigenschaften der Strahlung sind von denjenigen, welche wir nach unserer jetzigen Theorie, die sich auf lineare, homogene Differentialgleichungen stützt, erwarten sollten.

7. Im vorigen haben wir die Schwankungen der Energieverteilung berechnet, um Aufschlüsse über die Natur der Wärmestrahlung zu erhalten. Im folgenden soll kurz gezeigt werden, wie man durch Berechnen der Schwankungen des Strahlungsdruckes (also von Schwankungen der Bewegungsgröße) zu ganz entsprechenden Resultaten gelangen kann.

Es befinde sich in einem allseitig von Materie von der absoluten Temperatur T umgebenen Hohlraum ein in Richtung senkrecht zu seiner Normalen frei beweglicher Spiegel[2]). Denken wir uns diesen von Anfang an mit einer gewissen Geschwindigkeit bewegt, so wird infolge dieser Bewegung an seiner Vorderseite mehr Strahlung reflektiert, als an seiner Rückseite; es ist daher der auf seiner Vorderseite wirkende Strahlungsdruck größer als der auf die Rückseite wirkende. Es wird also auf den Spiegel infolge seiner Bewegung relativ zur

1) Bei Durchführung der oben angedeuteten Interferenzbetrachtung würde man wohl $\frac{R}{Nk} = 1$ erhalten.

2) Die Bewegungen des Spiegels, von denen hier die Rede ist, sind der sogenannten Brownschen Bewegung suspendierter Teilchen durchaus analog.

Hohlraumstrahlung eine der Reibung vergleichbare Kraft wirken, welche nach und nach die Bewegungsgröße des Spiegels aufzehren müßte, wenn nicht andererseits eine bewegende Ursache bestünde, welche die durch jene Reibungskraft verlorene Bewegungsgröße im Mittel gerade ersetzte. Den im vorigen studierten unregelmäßigen Schwankungen der Energie eines Strahlungsraumes entsprechen nämlich auch unregelmäßige Schwankungen der Bewegungsgröße bzw. unregelmäßige Schwankungen der von der Strahlung auf den Spiegel ausgeübten Druckkräfte, die den Spiegel in Bewegung versetzen müßten, auch wenn er anfänglich ruhte. Die mittlere Bewegungsgeschwindigkeit des Spiegels ist nun aus der Entropie-Wahrscheinlichkeitsbeziehung, das Gesetz der obengenannten Reibungskräfte aus dem als bekannt angenommenen Strahlungsgesetz zu ermitteln. Aus diesen beiden Resultaten berechnet man dann die Wirkung der Druckschwankungen und ist in der Lage, aus diesen wieder Schlüsse in betreff der Konstitution der Strahlung oder — genauer gesprochen — in betreff der Elementarvorgänge der Reflexion der Strahlung am Spiegel zu ziehen.

Es sei mit v die Geschwindigkeit des Spiegels zur Zeit t bezeichnet. Infolge der obenerwähnten Reibungskraft nimmt im darauf folgenden Zeitteilchen τ diese Geschwindigkeit um $\frac{Pv\tau}{m}$ ab, wenn mit m die Masse des Spiegels, mit P die verzögernde Kraft bezeichnet wird, welche der Einheit der Geschwindigkeit des Spiegels entspricht. Wir bezeichnen ferner mit Δ diejenige Geschwindigkeitsänderung des Spiegels während τ, welche den unregelmäßigen Schwankungen des Strahlungsdruckes entsprechen. Die Geschwindigkeit des Spiegels zur Zeit $t+\tau$ ist

$$v - \frac{P\tau}{m} v + \Delta .$$

Als Bedingung dafür, daß v im Mittel, während τ ungeändert bleibt, erhalten wir

$$\overline{\left(v - \frac{P\tau}{m} v + \Delta\right)^2} = \overline{v^2}$$

oder, indem man relativ unendlich kleines wegläßt und berücksichtigt, daß der Mittelwert von $v\Delta$ offenbar verschwindet:

$$\overline{\Delta^2} = \frac{2P\tau}{m} \overline{v^2} .$$

In dieser Gleichung läßt sich zunächst $\overline{v^2}$ mittels der aus der Entropie-Wahrscheinlichkeitsgleichung ableitbaren Gleichung

$$\frac{m\overline{v^2}}{2} = \frac{1}{2} \frac{RT}{N}$$

ersetzen. Bevor wir ferner den Wert der Reibungskonstante P angeben, spezialisieren wir das behandelte Problem durch die Annahme, daß der Spiegel Strahlung von dem bestimmten Frequenzbereich (zwischen ν und $\nu + d\nu$) vollkommen reflektiere, für Strahlung anderer Frequenz aber vollkommen durchlässig sei. Durch eine Rechnung, welche ich hier der Kürze halber nicht angebe, erhält man durch eine rein elektrodynamische Untersuchung, die für jede beliebige Strahlungsverteilung gültige Gleichung

$$P = \frac{3}{2c}\left[\varrho - \frac{1}{3}\nu \frac{d\varrho}{d\nu}\right] d\nu f ,$$

falls man mit ϱ wieder die Strahlungsdichte bei der Frequenz ν, mit f die Fläche des Spiegels bezeichnet. Durch Einsetzen der für $\overline{v^2}$ und P ermittelten Werte erhält man

$$\frac{\overline{\Delta^2}}{\tau} = \frac{RT}{N} \cdot \frac{3}{c}\left[\varrho - \frac{1}{3}\nu \frac{d\varrho}{d\nu}\right] d\nu f .$$

Indem wir unter Benutzung der Planckschen Strahlungsformel diesen Ausdruck umformen, erhalten wir

$$\frac{\overline{\Delta^2}}{\tau} = \frac{1}{c}\left[h\varrho\nu + \frac{c^3}{8\pi} \frac{\varrho^2}{\nu^2}\right] d\nu f .$$

Die nahe Verwandtschaft dieser Beziehung mit der im vorigen Abschnitt für die Energieschwankung $(\overline{\varepsilon^2})$ abgeleiteten ist unmittelbar zu sehen[1]), und man kann an sie genau entsprechende Betrachtungen anknüpfen wie an jene. Wieder müßte sich nach der jetzigen Theorie der Ausdruck auf das zweite Glied reduzieren (Schwankung durch Interferenz). Wäre das erste Glied allein vorhanden, so ließen sich die Schwankungen des Strahlungsdruckes vollständig erklären durch die Annahme, daß die Strahlung aus voneinander unabhängig beweglichen, wenig ausgedehnten Komplexen von der Energie $h\nu$ bestehe. Auch hier besagt die Formel, daß nach der Planckschen Formel die Wirkungen der beiden genannten Schwankungsursachen sich verhalten wie Schwankungen (Fehler), welche voneinander unabhängigen Ursachen entspringen (additive Verknüpfung der Terme, aus denen sich das Schwankungsquadrat zusammensetzt).

8. Aus den letzten beiden Betrachtungen geht nach meiner Meinung unwiderlegbar hervor, daß die Konstitution der Strahlung eine andere sein muß, als wir gegenwärtig meinen. Unsere gegenwärtige Theorie liefert zwar, wie die treffliche Übereinstimmung von Theorie und Experiment in der Optik beweist, die

1) Man kann jene in der Form schreiben (wobei $\frac{R}{Nk} = 1$ gesetzt ist:

$$\overline{\varepsilon^2} = \left\{h\varrho\nu + \frac{c^3\varrho^2}{8\pi\nu^2}\right\} v\, d\nu .$$

allein direkt wahrnehmbaren zeitlichen Mittelwerte in richtiger Weise, führt aber mit Notwendigkeit zu mit der Erfahrung unvereinbaren Gesetzen über die thermischen Eigenschaften der Strahlung, sobald man nur an der Entropie-Wahrscheinlichkeit-Beziehung festhält. Die Abweichung der Erscheinungen von der Theorie tritt desto stärker hervor, je größer ν und je kleiner ϱ ist. Es sind bei kleinem ϱ die zeitlichen Schwankungen der Strahlungsenergie eines bestimmten Raumes bzw. der Druckkraft der Strahlung auf eine bestimmte Fläche viel größer als unsere jetzige Theorie erwarten läßt.

Wir haben gesehen, daß das Plancksche Strahlungsgesetz sich begreifen läßt unter Heranziehung der Annahme, daß Oszillationsenergie von der Frequenz ν nur auftreten kann in Quanten von der Größe $h\nu$. Es genügt nach dem Vorigen nicht die Annahme, daß Strahlung nur in Quanten von dieser Größe emittiert und absorbiert werden könne, daß es sich also lediglich um eine Eigenschaft der emittierenden bzw. absorbierenden Materie handle; die Betrachtungen 6 und 7 zeigen, daß auch die Schwankungen in der räumlichen Verteilung der Strahlung und diejenigen des Strahlungsdruckes derart erfolgen, wie wenn die Strahlung aus Quanten von der angegebenen Größe bestünden. Es kann nun zwar nicht behauptet werden, daß die Quantentheorie aus dem Planckschen Strahlungsgesetze als Konsequenz folge, und daß andere Interpretationen ausgeschlossen seien. Man kann aber wohl behaupten, daß die Quantentheorie die einfachste Interpretation der Planckschen Formel liefert.

Es ist hervorzuheben, daß die angegebenen Überlegungen im wesentlichen keineswegs ihren Wert verlieren würden, falls sich die Plancksche Formel noch als ungültig erweisen sollte; gerade der von der Erfahrung genügend bestätigte Teil der Planckschen Formel (das für große $\frac{\nu}{T}$ in der Grenze gültige Wiensche Strahlungsgesetz) ist es, welcher zur Lichtquantentheorie führt.

9. Die experimentelle Erforschung der Konsequenzen der Lichtquantentheorie ist nach meiner Meinung eine der wichtigsten Aufgaben, welche die Experimentalphysik der Gegenwart zu lösen hat. Die bis jetzt gezogenen Konsequenzen kann man in drei Gruppen ordnen.

a) Es ergeben sich Anhaltspunkte für die Energie derjenigen Elementarvorgänge, welche mit Absorption bzw. Emission von Strahlung bestimmter Frequenz verknüpft sind (Stokessche Regel; Geschwindigkeit der durch Licht oder Röntgenstrahlen erzeugten Kathodenstrahlen; Kathodolumineszenz usw.). Hierher gehört auch die interessante Anwendung, die Herr Stark von der Lichtquantentheorie gemacht hat, um die eigentümliche Energieverteilung im Spektrum einer von Kanalstrahlen emittierten Spektrallinie zu erklären [1]).

Die Schlußweise ist hier immer folgende: Erzeugt ein Elementarvorgang einen andern, so ist die Energie des letzteren nicht größer als die des ersteren. Die Energie eines der beiden Elementarvorgänge ist aber bekannt (von der Größe $h\nu$), wenn letzterer in der Absorption oder Emission von Strahlung bestimmter Frequenz besteht.

Besonders interessant wäre das Studium der Ausnahmen vom Stokesschen Gesetz. Zur Erklärung dieser Ausnahmen muß angenommen werden, daß ein Lichtquant erst dann emittiert wird, wenn das betreffende Emissionszentrum zwei Lichtquanten absorbiert hat. Die Häufigkeit eines derartigen Ereignisses, also auch die Intensität des emittierten Lichtes von kleinerer Wellenlänge als das erzeugende, wird in diesem Falle bei schwacher Bestrahlung (nach dem Massenwirkungsgesetz) dem Quadrat der erregenden Lichtstärke proportional sein müssen, während bei Gültigkeit der Stokesschen Regel bei schwacher Bestrahlung Proportionalität mit der ersten Potenz der erregenden Lichtintensität zu erwarten ist.

b) Wird bei Absorption [2]) jedes Lichtquants ein Elementarvorgang gewisser Art bewirkt, so ist $\frac{E}{h\nu}$ die Anzahl dieser Elementarvorgänge, falls die Energiemenge E von Strahlung der Frequenz ν absorbiert wird.

Wird also z. B. die Menge E einer Strahlung von der Frequenz ν unter Ionisierung eines Gases von diesem absorbiert, so ist zu erwarten, daß $\frac{E}{Nh\nu}$ Grammoleküle des Gases dabei ionisiert werden. Diese Beziehung setzt nur scheinbar die Kenntnis von N voraus; schreibt man nämlich die Plancksche Strahlungsformel in der Form

$$\varrho = \alpha\nu^3 \frac{1}{e^{\frac{\beta\nu}{T}} - 1},$$

so ist $\frac{E}{R\beta\nu}$ die Anzahl der ionisierten Grammmoleküle.

Diese Beziehung, welche ich bereits in meiner ersten Arbeit [3]) über diesen Gegenstand angab, ist leider bisher unbeachtet geblieben.

c) Das unter 5 Mitgeteilte führt zu einer

1) J. Stark, diese Zeitschr. **9**, 767, 1908.

2) Die analoge Betrachtung gilt natürlich auch umgekehrt für die Lichterzeugung durch Elementarvorgänge (z. B. durch Ionenstöße).

3) Ann. d. Phys. (4) **17**, 132—148, 1905, § 9.

Modifikation der kinetischen Theorie der spezifischen Wärme[1]) und zu gewissen Beziehungen zwischen optischem und thermischem Verhalten der Körper.

10. Es erscheint schwierig, ein theoretisches System aufzustellen, welches die Lichtquanten in vollständiger Weise deutet, wie unsere heutige Molekularmechanik in Verbindung mit der Maxwell-Lorentzschen Theorie die von Herrn Jeans vertretene Strahlungsformel zu deuten vermag. Daß es sich nur um eine Modifikation unserer heutigen Theorien, nicht um ein vollständiges Verlassen derselben handeln wird, scheint schon daraus hervorzugehen, daß das Jeanssche Gesetz in der Grenze (für kleine $\frac{\nu}{T}$) gültig zu sein scheint. Einen Hinweis darauf, wie jene Modifikation durchzuführen sein dürfte, liefert eine von Herrn Jeans vor einigen Jahren durchgeführte, nach meiner Meinung höchst wichtige Dimensionalbetrachtung, die ich im folgenden — in einigen Punkten modifiziert — kurz wiedergebe.

Wir denken uns, daß in einem abgeschlossenen Raume ein ideales Gas, Strahlung sowie Ionen vorhanden seien, welch letztere vermöge ihrer Ladung einen Energieaustausch zwischen Gas und Strahlung zu vermitteln vermögen. Es ist zu erwarten, daß in einer an die Betrachtung dieses Systems geknüpften Strahlungstheorie folgende Größen eine Rolle spielen, also in dem zu ermittelnden Ausdruck für die Strahlungsdichte ϱ auftreten werden:

a) die mittlere Energie η eines molekularen Gebildes (bis auf einen unbenannten Zahlenfaktor gleich $\frac{RT}{N}$),
b) die Lichtgeschwindigkeit c,
c) das Elementarquantum ε der Elektrizität,
d) die Frequenz ν.

Aus der Dimension von ϱ kann man nun unter ausschließlicher Berücksichtigung der Dimensionen der vier ebengenannten Größen in einfacher Weise ermitteln, welche Gestalt der Ausdruck für ϱ haben muß. Man erhält, indem man für η den Wert $\frac{RT}{N}$ setzt:

$$\varrho = \frac{\varepsilon^2}{c^4} \nu^3 \psi(\alpha),$$

wobei

$$\alpha = \frac{R\varepsilon^2}{Nc} \frac{\nu}{T},$$

wobei ψ eine unbestimmt bleibende Funktion bezeichnet. Diese Gleichung enthält das Wiensche Verschiebungsgesetz, dessen Gültigkeit kaum mehr bezweifelt werden kann. Man hat dies als eine Bestätigung dafür aufzufassen, daß außer den oben eingeführten vier Größen in dem Strahlungsgesetz keine weiteren Größen eine Rolle spielen, die eine Dimension haben.

Daraus schließen wir, daß die in der Gleichung für ϱ auftretenden Koeffizienten $\frac{\varepsilon^2}{c^4}$ und $\frac{R\varepsilon^2}{Nc}$ bis auf bei theoretischen Entwicklungen auftretende dimensionslose Zahlenfaktoren, welche sich natürlich aus einer Dimensionalbetrachtung nicht ergeben können, numerisch gleich sein sollen den in der Planckschen (oder Wienschen) Strahlungsformel auftretenden Koeffizienten. Da jene sich nicht ergebenden dimensionslosen Zahlenfaktoren die Größenordnung kaum wesentlich ändern dürften, so kann man der Größenordnung nach setzen[1]):

$$\frac{h}{c^3} = \frac{\varepsilon^2}{c^4} \quad \text{und} \quad \frac{h}{k} = \frac{R}{N}\frac{\varepsilon^2}{c},$$

also

$$h = \frac{\varepsilon^2}{c} \quad \text{und} \quad k = \frac{N}{R}.$$

Die zweite dieser Gleichungen ist die, mittels welcher Herr Planck die Elementarquanta der Materie oder Elektrizität bestimmt hat. Zum Ausdruck für h ist zu bemerken, daß

$$h = 6 \cdot 10^{-27}$$

und

$$\frac{\varepsilon^2}{c} = 7 \cdot 10^{-30}.$$

Es fehlt ja hier um 3 Dezimalen. Aber dies dürfte wohl darauf zurückzuführen sein, daß die dimensionslosen Faktoren unbekannt sind.

Das Wichtigste dieser Ableitung liegt darin, daß durch sie die Lichtquantenkonstante h auf das Elementarquantum ε der Elektrizität zurückgeführt wird. Es ist nun daran zu erinnern, daß das Elementarquantum ε ein Fremdling ist in der Maxwell-Lorentzschen Elektrodynamik[2]). Man muß fremde Kräfte heranziehen, um in der Theorie das Elektron zu konstruieren; man pflegt ein starres Gerüst einzuführen, das verhindern soll, daß die elektrischen Massen des Elektrons unter dem Einfluß ihrer elektrischen Wechselwirkung auseinanderfahren. Es scheint mir nun aus der Beziehung $h = \frac{\varepsilon^2}{c}$ hervorzugehen, daß die gleiche Modifikation der Theorie, welche das Elementarquantum ε als Konsequenz enthält, auch die Quantenstruktur der Strahlung als Konsequenz

1) A. Einstein, Ann. d. Phys. (4) **22**, 1907, S. 180—190 und S. 800.

1) Die Plancksche Formel lautet:

$$\varrho = \frac{8\pi h \nu^3}{c^3} \frac{1}{e^{\frac{h\nu}{kT}} - 1}$$

2) Vgl. Levi-Civita, Comptes Rendus 1907: „Sur le mouvement etc.".

enthalten wird. Es wird die Fundamentalgleichung der Optik

$$D(\varphi)=\frac{1}{c^2}\frac{\partial^2\varphi}{\partial^2 t}-\left(\frac{\partial^2\varphi}{\partial x^2}+\frac{\partial^2\varphi}{\partial y^2}+\frac{\partial^2\varphi}{\partial z^2}\right)=0$$

zu ersetzen sein durch eine Gleichung, in der auch die universelle Konstante ε (wahrscheinlich das Quadrat derselben) in einem Koeffizienten auftritt. Die gesuchte Gleichung (bzw. das gesuchte Gleichungssystem) muß in den Dimensionen homogen sein. Es muß bei Anwendung der Lorentz-Transformation in sich selbst übergehen. Sie kann nicht linear und homogen sein. Sie muß — wenigstens falls das Jeanssche Gesetz wirklich in der Grenze für kleine $\frac{v}{T}$ gültig ist — für große Amplituden in der Grenze auf die Form $D(\varphi)=0$ führen.

Es ist mir noch nicht gelungen, ein diesen Bedingungen entsprechendes Gleichungssystem zu finden, von dem ich hätte einsehen können, daß es zur Konstruktion des elektrischen Elementarquantums und der Lichtquanten geeignet sei. Die Mannigfaltigkeit der Möglichkeiten scheint aber nicht so groß zu sein, daß man vor der Aufgabe zurückschrecken müßte.

Nachtrag.

Aus dem unter 4. in der vorstehenden Abhandlung Gesagten könnte der Leser leicht einen unzutreffenden Eindruck gewinnen über den Standpunkt, welchen Herr Planck seiner eigenen Theorie der Temperaturstrahlung gegenüber einnimmt. Deshalb halte ich es für angezeigt, das Folgende zu bemerken.

Herr Planck hat in seinem Buche an mehreren Stellen hervorgehoben, daß seine Theorie noch nicht als etwas Fertiges, Abgeschlossenes aufzufassen sei. Er sagt z. B. am Schluß der Vorrede wörtlich: „Es liegt mir aber daran, auch an dieser Stelle noch besonders hervorzuheben, was sich im letzten Paragraphen des Buches näher ausgeführt findet, daß die hier entwickelte Theorie keineswegs den Anspruch erhebt, als vollkommen abgeschlossen zu gelten, wenn sie auch, wie ich glaube, einen gangbaren Weg eröffnet, um die Vorgänge der Energiestrahlung von dem nämlichen Gesichtspunkt aus zu überblicken wie die der Molekularbewegung."

Die betreffenden Auseinandersetzungen in meiner Abhandlung sind nicht als ein Einwand (im eigentlichen Sinne des Wortes) gegen die Plancksche Theorie aufzufassen, sondern lediglich als ein Versuch, das Entropie-Wahrscheinlichkeitsprinzip etwa schärfer zu fassen und anzuwenden, als man es bisher getan hat. Eine schärfere Fassung dieses Prinzips war notwendig, weil ohne eine solche die folgenden Entwicklungen in der Abhandlung, in welchen auf die molekulare Struktur der Strahlung geschlossen wird, nicht genügend begründet gewesen wären. Damit meine Fassung des Prinzips nicht als etwas ad hoc Gewähltes, Willkürliches erscheine, mußte ich zeigen, warum mich die bisherige Formulierung des Prinzips noch nicht vollkommen befriedigte.

Bern, Januar 1909.

(Eingegangen 23. Januar 1909.)

PAPER NO. 4

Reprinted from *Journal of the Optical Society of America*, Vol. 35, pp. 525–531, 1945.

The Statistical Properties of Unpolarized Light

Henry Hurwitz, Jr.
Cornell University, Ithaca, New York

(Received November 1, 1944)

In a beam of monochromatic unpolarized light the electric field vector at a point traces out an ellipse whose size, eccentricity, and orientation are slowly varying functions of time. The statistical properties of the parameters of this ellipse are investigated. It is shown that the quantity S which is defined as twice the product of the principle axes of the ellipse divided by the sum of the squares is uniformly distributed between zero and one. It therefore has median value $\frac{1}{2}$ which corresponds to a ratio of minor to major axis equal to .268. Hence fairly thin ellipses predominate. The square root of the sum of the squares of the semi-major and semi-minor axes, R, is statistically independent of S and has the distribution function $(r^3/2p^4)\exp(-r^2/2p^2)$ where $2p^2$ is the average value of R^2.

INTRODUCTION

Many textbooks and teachers of optics are rather vague in their descriptions of unpolarized light. They pass over the subject with a cryptic statement of the sort "unpolarized light is a mixture of plane-polarized light in all directions" or "unpolarized light is a random mixture of light with all kinds of polarizations." Students find difficulty in obtaining from these brief remarks a clear picture of how the electric field vector at a point varies as a function of time. In this note we shall perform some simple statistical calculations whose results might serve as a pedagogical aid in making the situation clear to students.

By natural, or unpolarized light we shall mean radiation which is emitted by molecules which have been randomly excited (e.g. thermally, or by an electrical discharge), and which are not in a region of space where large external fields provide a preferred direction. We shall assume that the beam of light under consideration is monochromatic in the sense that its line breadth is small compared to the central frequency ν_0. It is immaterial whether the monochromatic character arises because the light has been passed through a narrow filter, or because it consists of a single spectral line. In order to avoid the necessity of considering quantum effects we shall limit the discussion to fairly intense beams. This means effectively that the number of photons reaching the detector in a time interval equal to the reciprocal line breadth must be large compared to one.

As will be seen from the formulas given below, a beam of unpolarized monochromatic light would, if observed over a time interval short compared to the reciprocal of the line breadth, appear to be elliptically polarized. But the type of polarization changes continually so that in any experiment which requires a longer period of observation one measures only the average effect of a large variety of different polarizations. In other words, in a beam of unpolarized monochromatic light the electric vector at a point traces out an ellipse whose size, eccentricity, and orientation gradually change. After a time somewhat larger than the reciprocal of the line breadth the new ellipse is completely unrelated to the original one. The purpose of our calculation is to find the statistical properties of the various parameters of the "instantaneous ellipses" traced out by the electric vector.

STATISTICAL FOUNDATIONS

Our statistical calculations will be based on the following mathematical representation for a monochromatic beam of unpolarized light:[1]

$$\begin{aligned} E_x &= X(t)\cos\omega_0 t + U(t)\sin\omega_0 t = A(t)\cos(\omega_0 t - \delta_x(t)),\\ E_y &= Y(t)\cos\omega_0 t + V(t)\sin\omega_0 t = B(t)\cos(\omega_0 t - \delta_y(t)),\\ A(t) &= [X^2(t)+U^2(t)]^{\frac{1}{2}};\quad B(t) = [Y^2(t)+V^2(t)]^{\frac{1}{2}};\\ \delta_x(t) &= \tan^{-1}[U(t)/X(t)];\quad \delta_y(t) = \tan^{-1}[V(t)/Y(t)];\quad \omega_0 = 2\pi\nu_0. \end{aligned} \tag{1}$$

[1] We assume that the electric field is constant over the part of the cross section of the beam under consideration.

Here E_x and E_y are the x and y components of the electric field at a point P in the beam. The quantities $X(t)$, $Y(t)$, $U(t)$, and $V(t)$ are statistically independent Gaussianly distributed random functions. The probability that a particular one of them has a value between x and $x+dx$ is

$$P_X(x)dx = \frac{1}{p} P_G\left(\frac{x}{p}\right)dx$$

$$= \frac{1}{(2\pi)^{\frac{1}{2}}} \exp\left(-\frac{x^2}{2p^2}\right)\frac{dx}{p}, \quad (2)$$

where p^2 is the average of the square of each one of these quantities. In accordance with Eq. (2) and the independence of $X(t)$, $Y(t)$, $U(t)$, and $V(t)$, the quantities $A(t)$, $B(t)$, $\delta_x(t)$, and $\delta_y(t)$ are all independent. Furthermore $\delta_x(t)$ and $\delta_y(t)$ are uniformly distributed between 0 and 2π, and $A(t)$ and $B(t)$ are distributed according to the law

$$P_A(x)dx = P_B(x)dx = \frac{x}{p^2} \exp\left(-\frac{x^2}{2p^2}\right)dx. \quad (3)$$

(In this notation $P_A(x)dx$ is the probability that $A(t)$ lies between x and $x+dx$, etc.) The eight functions $X(t)$, $Y(t)$, $U(t)$, $V(t)$, $A(t)$, $B(t)$, $\delta_x(t)$, and $\delta_y(t)$ vary slowly compared to $\cos \omega_0 t$ and $\sin \omega_0 t$. Thus the values of one of these functions at two different times are not statistically independent unless the time difference is large compared to the reciprocal line breadth. For smaller time differences the correlation depends on the shape of the line.

We shall indicate two different methods of obtaining the representation given in Eqs. (1) to (3). The first is based on a simple classical model of light emission while the second makes use of a natural independence assumption.

In the classical model which we shall employ in the first derivation the molecules are assumed to emit radiation in the form of wave trains which contain only Fourier components with frequencies close to ν_0 and which last for a time interval of order of magnitude of the reciprocal line breadth. (In accordance with quantum theory the pulses may be imagined to contain an amount of energy equal to $h\nu_0$.) The nature of the polarization of the individual pulses is unimportant so long as it is sufficiently random. For example, if the pulses are plane polarized, then all planes of polarization must be equally probable, while if the pulses are circularly polarized, right- and left-handed polarizations must be equally likely. A typical field component in a pulse will be equal to $f(t-t_i)$ where t_i is the time when the pulse begins. Expanding $f(t)$ in a Fourier series in the long time interval 0 to T we may write

$$f(t) = \sum_{k=0}^{\infty} (\alpha_k \cos \omega_k t + \beta_k \sin \omega_k t),$$
$$\omega_k = 2\pi k/T. \quad (4)$$

The quantity $\alpha_k^2+\beta_k^2$ is small if ω_k differs from ω_0 by an amount large compared to the line breadth. The problem of finding the resultant of a large number of pulses of the sort described above has been studied in great detail in other connections.[2] Therefore without going into details about the mathematical calculation we shall simply quote the result. From the fact that the times t_i when the individual pulses are emitted are randomly distributed it can be shown that the x and y components of the electric field at a point in the beam may be written in the following form:

$$E_x = \sum_{k=0}^{\infty} p_k (X_k \cos \omega_k t + U_k \sin \omega_k t),$$
$$E_y = \sum_{k=0}^{\infty} p_k (Y_k \cos \omega_k t + V_k \sin \omega_k t). \quad (5)$$

[2] H. Hurwitz and M. Kac, Ann. Math. Stat. **15**, 173 (1944); S. O. Rice, Bell Sys. Tech. J. **23**, 282 (1944), **24**, 46 (1945). Some insight as to why the quantities in Eq. (5) are Gaussianly distributed can be had from the following considerations. Let

$$F(t) = \sum_1^N f(t-t_i) = \sum_{k=0}^N (a_k \cos \omega_k t + b_k \sin \omega_k t).$$

From Eq. (4)

$$a_k = \sum_1^N (\alpha_k^2+\beta_k^2)^{\frac{1}{2}} \cos(\omega_k t_i - \varphi_k),$$

$$b_k = \sum_1^N (\alpha_k^2+\beta_k^2)^{\frac{1}{2}} \sin(\omega_k t_i - \varphi_k), \quad \varphi_k = \tan^{-1}\frac{\beta_k}{\alpha_k}.$$

If the t_i's are randomly distributed the quantities a_k and b_k can be regarded as the two components of a vector which is the sum of N two-dimensional vectors of constant length but random direction. Hence from the familiar results of the problem of "random walk" we can immediately conclude that a_k and b_k are Gaussianly distributed in the limit that N becomes infinite.

The quantities p_k^2 are proportional to $\alpha_k^2+\beta_k^2$ and are equal to $2\pi/c$ times the radiant-energy current in the beam per unit frequency range at frequency ω_k. The quantities X_k, Y_k, U_k, and V_k are independently distributed Gaussian random variables such that

$$\begin{aligned}\langle X_k\rangle_{Av}=\langle Y_k\rangle_{Av}=\langle U_k\rangle_{Av}=\langle V_k\rangle_{Av}=0,\\ \langle X_k^2\rangle_{Av}=\langle Y_k^2\rangle_{Av}=\langle U_k^2\rangle_{Av}=\langle V_k^2\rangle_{Av}=1.\end{aligned} \tag{6}$$

Thus the probability that a particular one of these variables has a value between x and $x+dx$ is

$$P_G(x)dx=[1/(2\pi)^{\frac{1}{2}}]\exp(-x^2/2)dx. \tag{7}$$

The representation given in Eqs. (5) and (7) is strictly correct only in the limit that the number of pulses per unit time becomes infinite. However it is a sufficiently good approximation if the number of pulses emitted in a time equal to the reciprocal line breadth is large compared to one.

Equation (5) can be immediately written in the form of Eq. (1) by using the definitions

$$\begin{aligned}X(t)&=\sum p_k[X_k\cos(\omega_k-\omega_0)t\\&\qquad+U_k\sin(\omega_k-\omega_0)t],\\ U(t)&=\sum p_k[-X_k\sin(\omega_k-\omega_0)t\\&\qquad+U_k\cos(\omega_k-\omega_0)t],\\ Y(t)&=\sum p_k[Y_k\cos(\omega_k-\omega_0)t\\&\qquad+V_k\sin(\omega_k-\omega_0)t],\\ V(t)&=\sum p_k[-Y_k\sin(\omega_k-\omega_0)t\\&\qquad+V_k\cos(\omega_k-\omega_0)t].\end{aligned} \tag{8}$$

The quantities $X(t)$, $Y(t)$, $U(t)$, and $V(t)$, being linear combinations of Gaussianly distributed variables, are themselves Gaussianly distributed. Furthermore they are independently distributed because of the orthogonality property of the coefficients in their expansions in terms of X_k, Y_k, U_k, and V_k. The quantity p appearing in Eq. (2) is given by

$$p^2=\sum_{k=0}^{\infty}p_k^2. \tag{9}$$

Thus p^2 is $2\pi/c$ times the average power flow per unit area in the beam. The only remaining property of the functions $X(t)$, $Y(t)$, $U(t)$, and $V(t)$ to be proved is that they are slowly varying compared to $\cos\omega_0 t$ and $\sin\omega_0 t$. But this follows immediately from Eq. (8) and the fact that p_k is appreciable only if ω_k is close to ω_0.

Our second method of obtaining the fundamental representation given in Eqs. (1) to (3) is similar to the derivation of the Maxwellian distribution of velocities in a gas from the assumption that the Cartesian components of the velocities are independently distributed. The analogous assumption which we shall make is that E_x and E_y are independently distributed. More specifically we shall assume that the functions $X(t)$ and $U(t)$ in Eq. (1) are distributed independently of the functions $Y(t)$ and $V(t)$.[3] This independence property followed directly from the model assumed in the above derivation. However it is more general than the model there assumed and so may logically be taken as the starting point in our second derivation. In this derivation we begin by arbitrarily writing the field components in the form given in Eq. (1). Since the electric field is assumed to have appreciably large Fourier components only for angular frequencies close to ω_0 it follows immediately that the functions $X(t)$, $Y(t)$, $U(t)$, and $V(t)$ are slowly varying compared to $\cos\omega_0 t$ and $\sin\omega_0 t$. It remains to be proved that these four functions are Gaussianly distributed and that $X(t)$ is independent of $U(t)$ and $Y(t)$ is independent of $V(t)$. In addition to the fundamental assumption that $X(t)$ and $U(t)$ are independent of $Y(t)$ and $V(t)$ we shall assume that the statistical properties of the field components are independent of the orientation of the x and y axes. This means, for example, that for any value of θ the quantity $X(t)\cos\theta+Y(t)\sin\theta$ must have the same distribution function as the quantity $X(t)$. The fact that $X(t)$, $Y(t)$, $U(t)$, and $V(t)$ are Gaussianly distributed is then an immediate consequence of the following theorem:

[3] The assumption that E_x and E_y are statistically independent can be justified in terms of the somewhat more physical assumption that the statistical properties of an intense beam of unpolarized light cannot be changed by passing the light through a wave plate. By the use of a wave plate of suitable thickness, one field component can be retarded with respect to the other by any desired amount. It is evident that no possible type of correlation between E_x and E_y could remain invariant under this general class of transformations.

If A and B are independently distributed random variables and for every θ the variable $A\cos\theta+B\sin\theta$ has the same distribution as A (or B), then A and B are Gaussianly distributed. This theorem was proved in utmost generality by M. Kac and H. Steinhaus.[4] Their proof proceeds by obtaining a simple functional equation for the characteristic function of the distribution. Since A and $A\cos\theta+B\sin\theta$ have the same distribution it follows that for every real ξ the average value of $e^{i\xi A}$ satisfies the relation

$$f(\xi)=\langle e^{i\xi A}\rangle_{\text{Av}}=\langle e^{i\xi(A\cos\theta+B\sin\theta)}\rangle_{\text{Av}}. \quad (10)$$

But since A and B are independently distributed Eq. (10) can be rewritten

$$f(\xi)=\langle e^{i\xi A\cos\theta}\rangle_{\text{Av}}\langle e^{i\xi B\sin\theta}\rangle_{\text{Av}}$$

$$=f(\xi\cos\theta)f(\xi\sin\theta); \quad (11)$$

or

$$f((x^2+y^2)^{\frac{1}{2}})=f(x)f(y), \quad (12)$$

where

$$x=\xi\cos\theta;\quad y=\xi\sin\theta. \quad (13)$$

It is easy to show that the only continuous solution of this well known equation is of the form

$$f(\xi)=\exp(\lambda\xi^2). \quad (14)$$

Furthermore λ is negative since

$$|f(\xi)|=|\langle e^{i\xi A}\rangle_{\text{Av}}|\leqslant 1. \quad (15)$$

Thus, since $f(\xi)$ is the characteristic function of a Gaussian distribution the theorem is proved. In our application of this theorem the random variable A corresponds to $X(t)$ (or $U(t)$) and the random variable B corresponds to $Y(t)$ (or $V(t)$).

The proof that $X(t)$ is independent of $U(t)$ and $Y(t)$ is independent of $V(t)$ is based on the fact that $X(t)$, $Y(t)$, $U(t)$, and $V(t)$ are Gaussianly distributed, and the additional assumption that $A(t)$ is independent of $\delta_x(t)$ and $B(t)$ is independent of $\delta_y(t)$. This latter assumption is clearly justified since $\delta_x(t)$ and $\delta_y(t)$ depend on the origin of time which is arbitrary. Let $P_A(a)da$ be the probability that $A(t)$ lies between a and $a+da$. Then, since $A(t)$ and $\delta_x(t)$ are independent, and in view of the relations given in Eq. (1), the probability that $X(t)$ lies between x and $x+dx$, and simultaneously $U(t)$ lies between u and $u+du$ is

$$P_{X,U}(x,u)dxdu=\frac{P_A((x^2+u^2)^{\frac{1}{2}})}{2\pi(x^2+u^2)^{\frac{1}{2}}}dxdu$$

$$=g(x^2+u^2)dxdu. \quad (16)$$

Hence

$$P_X(x)=\frac{1}{p(2\pi)^{\frac{1}{2}}}\exp\left(-\frac{x^2}{2p^2}\right)$$

$$=\int_{-\infty}^{+\infty}P_{X,U}(x,u)du$$

$$=\int_{-\infty}^{+\infty}g(x^2+u^2)du. \quad (17)$$

Replacing x^2 by x^2+z^2 we have

$$\frac{1}{p(2\pi)^{\frac{1}{2}}}\exp\left(-\frac{(x^2+z^2)}{2p^2}\right)$$

$$=\int_{-\infty}^{+\infty}g(x^2+z^2+u^2)du; \quad (18)$$

so that

$$\frac{1}{p(2\pi)^{\frac{1}{2}}}\exp\left(-\frac{x^2}{2p^2}\right)\int_{-\infty}^{+\infty}\exp\left(-\frac{z^2}{2p^2}\right)dz$$

$$=\int_{-\infty}^{+\infty}\int_{-\infty}^{+\infty}g(x^2+z^2+u^2)dudz, \quad (19)$$

or

$$\exp\left(-\frac{x^2}{2p^2}\right)=\pi\int_{x^2}^{\infty}g(y)dy. \quad (20)$$

By differentiating both sides of the above equation with respect to x^2 we find

$$\frac{1}{2p^2}\exp\left(-\frac{x^2}{2p^2}\right)=\pi g(x^2), \quad (21)$$

so that

$$P_{X,U}(x,u)dxdu$$

$$=\frac{1}{2\pi p^2}\exp\left(-\frac{(x^2+u^2)}{2p^2}\right)dxdu. \quad (22)$$

Since the joint distribution function of $X(t)$ and $U(t)$ is the product of the individual distribution functions the variables $X(t)$ and $U(t)$ are independent. This completes our second derivation of Eqs. (1) to (3).

[4] M. Kac and H. Steinhaus, Studia Mathematica 6, 89 (1936).

The difference between polarized and unpolarized light can be easily understood in terms of our fundamental representation as given in Eqs. (1) to (3). For light which is elliptically polarized in the usual sense the quantities $A(t)$, $B(t)$, $\delta_x(t)$, and $\delta_y(t)$ still are functions of time, but they vary in such a way that $A(t)/B(t)$ and $\delta_x(t)-\delta_y(t)$ remain constant. Hence though the amplitude and phase of the ellipse traced by the electric vector vary with time, the eccentricity and orientation remain constant. This can be accomplished in the laboratory by first removing one of the components of the electric field by a Nicol prism or Polaroid disk and then passing the resultant plane polarized beam through a wave plate. Clearly variations in the phase and amplitude of the plane polarized beam cannot cause variations in phase differences and amplitude ratios of components of the field after the wave plate.

An interesting paradox is encountered if one considers the fluctuations in the energy flowing past a point in the beam in a time interval τ. Let

$$W_\tau(t)=\int_t^{t+\tau}(A^2(t')+B^2(t'))dt'. \qquad (23)$$

The statistical behavior of the quantity $W_\tau(t)$ has been carefully investigated in another connection by M. Kac. His calculations are based on an equation corresponding to Eq. (5). It is easy to show that the quantity

$$\Delta=\frac{\langle W_\tau^2(t)\rangle_{\text{Av}}-\langle W_\tau(t)\rangle_{\text{Av}}^2}{\langle W_\tau(t)\rangle_{\text{Av}}^2}$$

depends only on τ and not on $\langle W_\tau(t)\rangle_{\text{Av}}$. Its limiting value for small τ is $\frac{1}{2}$ while for large τ it approaches zero. For values of τ of order of magnitude of the reciprocal line breadth the value of Δ depends on the line shape. According to quantum theory the quantity $W_\tau(t)$ is proportional to the number of photons incident in the time interval τ. One would expect this number n to have a Poisson distribution which would mean that Δ should be inversely proportional to n. As was pointed out by V. Weisskopf, the explanation of this paradox lies in the fact that the photons obey Einstein-Bose statistics. The photons therefore have a greater tendency for bunching than they would have if they were independent.[5] Hence the fluctuations in $W_\tau(t)$ are increased so that if the intensity is sufficiently large Δ is independent of $\langle W_\tau(t)\rangle_{\text{Av}}$ instead of being inversely proportional to it.

STATISTICAL PROPERTIES OF THE INSTANTANEOUS ELLIPSE

The distribution function of the parameters of the instantaneous ellipse can be calculated quite easily from Eqs. (1) to (3). Let $M(t)$ and $N(t)$ be respectively the semi-major and semi-minor axes of the instantaneous ellipse. Then[6]

$$M^2(t)+N^2(t)=A^2(t)+B^2(t)=R^2(t),$$
$$S(t)=\frac{2M(t)N(t)}{R^2(t)}=\frac{2A(t)B(t)}{R^2(t)}|\sin\delta(t)|, \qquad (24)$$
$$\delta(t)=\delta_x(t)-\delta_y(t).$$

Let

$$\Phi(t)=\tan^{-1}\frac{B(t)}{A(t)} \quad 0\leqslant\Phi(t)\leqslant\frac{\pi}{2}. \qquad (25)$$

Then the probability that $R(t)$ lies between r and $r+dr$ and simultaneously $\Phi(t)$ lies between φ and $\varphi+d\varphi$ is

$$P_{R,\Phi}(r,\varphi)drd\varphi=P_A(r\cos\varphi)P_B(r\sin\varphi)rdrd\varphi$$
$$=\frac{r^3}{2p^4}\sin 2\varphi\exp\left(-\frac{r^2}{2p^2}\right)drd\varphi. \qquad (26)$$

The quantity $rdrd\varphi$ represents the element of area in the A-B plane, and we have used Eq. (3) for $P_A(x)$ and $P_B(x)$. From Eq. (24)

$$S(t)=\sin 2\Phi(t)|\sin\delta(t)|. \qquad (27)$$

Since $\delta(t)$ is uniformly distributed between zero and 2π the probability that $|\sin\delta(t)|$ lies between σ and $\sigma+d\sigma$ is

$$P_{|\sin\delta|}(\sigma)d\sigma=\frac{2}{\pi}d\sin^{-1}\sigma=\frac{2}{\pi}\frac{d\sigma}{(1-\sigma^2)^{\frac{1}{2}}}. \qquad (28)$$

If the values of R and Φ are fixed, the probability that S lies between s and $s+ds$, which is obtained by substituting $ds/\sin 2\Phi$ for $d\sigma$ and $s/\sin 2\Phi$ for

[5] L. Brillouin, *Die Quantenstatistik* (Julius Springer, Berlin, 1931), p. 111, 177.
[6] M. Born, *Optik* (Julius Springer, Berlin, 1933), p. 23.

σ in Eq. (28), is

$$P_{|\sin\delta|}\left(\frac{s}{\sin 2\Phi}\right)\frac{ds}{\sin 2\Phi}=\frac{2}{\pi}\frac{ds}{\sin 2\Phi\left(1-\frac{s^2}{\sin^2 2\Phi}\right)^{\frac{1}{2}}}. \quad (29)$$

Hence the probability that R lies between r and $r+dr$ while S lies between s and $s+ds$ is

$$P_{S,R}(s,r)dsdr=dsdr\int\frac{2}{\pi}\frac{P_{R,\Phi}(r,\varphi)d\varphi}{\sin^2\varphi\left(1-\frac{s^2}{\sin^2 2\varphi}\right)^{\frac{1}{2}}}. \quad (30)$$

The integral is to be extended over that part of the first quadrant for which $\sin 2\varphi$ is greater than s. From Eq. (26)

$$P_{S,R}(s,r)dsdr=dsdr\frac{r^3}{\pi p^2}\exp\left(-\frac{r^2}{2p^2}\right)\times\int\frac{d\varphi}{\left(1-\frac{s^2}{\sin^2 2\varphi}\right)^{\frac{1}{2}}}. \quad (31)$$

But

$$\int\frac{d\varphi}{\left(1-\frac{s^2}{\sin^2 2\varphi}\right)^{\frac{1}{2}}}=-\frac{1}{2}\int\frac{dy}{(1-s^2-y^2)^{\frac{1}{2}}}=\frac{\pi}{2}. \quad (32)$$

$$y=\cos 2\varphi$$

since the range of integration is clearly from $y=(1-s^2)^{\frac{1}{2}}$ to $y=-(1-s^2)^{\frac{1}{2}}$. Thus

$$P_{S,R}(s,r)dsdr=\frac{r^3}{2p^4}\exp\left(-\frac{r^2}{2p^2}\right)dsdr. \quad (33)$$

Equation (33) shows that S is uniformly distributed between zero and one, and that R is distributed independently of S according to the law

$$P_R(r)dr=\frac{r^3}{2p^4}\exp(-r^2/2p^2)dr. \quad (34)$$

The distribution of the ratio

$$L(t)=N(t)/M(t) \quad (35)$$

can be found immediately from Eq. (33). We have

$$S=2L/(1+L^2), \quad (36)$$

so that

$$dS=2(1-L^2)dL/(1+L^2)^2. \quad (37)$$

Hence the joint distribution function of $L(t)$ and $R(t)$ is

$$P_{L,R}(l,r)dldr=\frac{(1-l^2)}{(1+l^2)^2}\frac{r^3}{p^4}\times\exp\left(-\frac{r^2}{2p^2}\right)dldr, \quad (38)$$

and the distribution function of $L(t)$ is

$$P_L(l)dl=[2(1-l^2)/(1+l^2)^2]dl. \quad (39)$$

The function $P_L(l)$ has its maximum value for l equal to zero, and decreases monotonically to zero when l equals one. The median value of $L(t)$ is related to the median value of $S(t)$ by Eq. (36). Since the median value of $S(t)$ is $\frac{1}{2}$ we find for the median value of $L(t)$

$$L_m=.268. \quad (40)$$

We therefore have the rather surprising result that more than half the time, the ellipse being traced out by the electric vector has a major axis more than three and one half times as large as the minor axis. The quantities $L(t)$ and $M(t)$ are not independently distributed but have the joint distribution function

$$P_{L,M}(l,m)dldm=(1-l^2)\frac{m^3}{p^4}\exp\left[-\frac{m^2(1+l^2)}{2p^2}\right]dldm. \quad (41)$$

The quantities $M(t)$ and $N(t)$ have the joint distribution function

$$P_{M,N}(m,n)dmdn=\frac{(m^2-n^2)}{p^4}\exp\left[-\frac{(m^2+n^2)}{2p^2}\right]dmdn. \quad (42)$$

It is interesting to note that the distribution function $P_S(s)$ of $S(t)$ can be derived directly from the requirement that it be unchanged if the light is passed through a quarter wave plate. This condition leads to an integral equation for $P_S(s)$ which has a constant as its only solution. An alternate and elegant derivation of the same result using Poincaré's representation for polarized light has been suggested by H. Mueller and L. Tisza.[7]

CONCLUDING REMARKS

The statistical properties of unpolarized light which are discussed above cannot be measured with ordinary optical equipment such as wave plates and Nicol prisms. The standard experiments cannot be performed in sufficiently short times and so give only the average values of the field components. However the difficulty is only one of experimental technique for the parameters of the instantaneous ellipse are measurable in principle if the beam is sufficiently intense. In fact an actual experiment might be possible if radiation in the radio or microwave region were used. Although a spectral line in the microwave region could conceivably be employed, it would probably be more convenient to use a narrow spectral range of high temperature black body radiation. The experimental procedure would be to employ two narrow band receivers fed by antennas responsive to electric fields in perpendicular directions. The output of the receivers (possibly after beating with a standard frequency) would be connected to the two pairs of deflection plates of an oscilloscope. The pattern on the oscilloscope screen would then correspond to the pattern traced out by that part of the incident electric field in the narrow frequency range to which the receivers were responsive. (The fact that the narrow frequency range is singled out by filters in the receivers rather than by a filter acting on the radiation before it reaches the antennas does not alter the final results.)

An interesting feature of the experiment just described is that if the receiver noise were large enough to completely overshadow the signal due to the incoming radiation, the statistical behavior of the pattern on the oscilloscope screen would be unchanged. The reason is that the most important forms of receiver noise are fundamentally due to shot effect and therefore have the statistical properties corresponding to Eqs. (5) and (7). This suggests that a simple way to obtain a model of the behavior of the electric vector in natural monochromatic light is to feed two narrow band amplifiers with shot noise from two diodes and put their outputs on the deflection plates of an oscilloscope. The electron beam will then trace out an ellipse whose parameters fluctuate at a rate inversely proportional to the band width and have the statistical properties described in the preceding section.

[7] Private communication.

PAPER NO. 5

Reprinted from *Annalen der Physik*, Vol. 23, pp. 1–43, 1907.

Die Entropie von partiell kohärenten Strahlenbündeln;

von M. Laue.

Einleitung.

In einer im Vorjahr erschienenen Abhandlung „Zur Thermodynamik der Interferenzerscheinungen"[1]) wurde gezeigt, daß sich bei kohärenten Strahlenbündeln das Additionstheorem der Entropie: „Die Entropie eines Systems ist die Summe der Entropien seiner (räumlich getrennten) Teile" nicht mit dem Prinzip der Zunahme der Entropie verträgt. Aus dem Zusammenhang zwischen Entropie und Wahrscheinlichkeit wurde gefolgert, daß es das Additionstheorem ist, welches man aufgeben muß. Es wurde auch die Formel für die Entropie eines Systems monochromatischer Strahlenbündel gegeben, welche alle durch reguläre Reflexion und Brechung an selbst nicht strahlenden Körpern aus einem einzigen entstanden sind. Dort war demnach nur von vollständig kohärenten Strahlenbündeln die Rede. Entsendet aber einer dieser Körper selbst Strahlung von derselben Schwingungszahl wie das einfallende Licht, so vermischen sich mit dem reflektierten und dem gebrochenen Strahlenbündel zwei zu ihnen inkohärente, so daß die auf diese Weise entstehenden weder untereinander noch mit etwaigen dem einfallenden Licht kohärenten Strahlenbündeln vollkommen kohärent bleiben. Es wäre unmöglich, bei Interferenzerscheinungen zwischen Strahlenbündeln dieser Art absolut dunkele Minimalstellen zu erhalten, wie sie bei der Superposition gleich starker, vollkommen kohärenter Strahlen auftreten; andererseits addieren sich dabei die Energiegrößen nicht, wie bei vollständig in-

---

1) M. Laue, Ann. d. Phys. **20**. p. 365. 1906.

kohärenten Strahlen. Vielmehr bilden solche partiell kohärente Strahlenbündel den stetigen Übergang zwischen diesen beiden Extremen. Die Formel für ihre Entropie muß dementsprechend zwischen der sich aus dem Additionstheorem ergebenden, für absolut inkohärente Strahlenbündel und der für vollständig kohärente gültigen den stetigen Übergang vermitteln.

Um sie abzuleiten, bedürfen wir zunächst eines quantitativen Maßes für die Interferenzfähigkeit zweier Strahlenbündel. Dies, die „Kohärenz", wie wir es nennen wollen, muß eine von den Intensitäten unabhängige, zudem meßbare, d. h. nur durch die zeitlichen Mittelwerte gewisser Energiequanten bestimmte Größe sein. Wir werden sie so wählen, daß sie stets ein positiver echter Bruch ist, dessen Grenzwerte 0 und 1 absolute Inkohärenz und vollständige Kohärenz bedeuten. Es ist jedoch bemerkenswert, daß nicht die Kohärenz selbst, sondern die sie zu 1 ergänzende Größe, die „Inkohärenz", in den Formeln für die Entropie sowie in den die Kohärenzverhältnisse von drei partiell kohärenten Strahlenbündeln beherrschenden Relationen auftritt.

Von seiten der Thermodynamik brauchen wir nur den Satz zu Hilfe zu ziehen, daß die Reflexion und Brechung an der Grenze nicht absorbierender Körper ein umkehrbarer Vorgang ist; da sich zwei inkohärente Strahlenbündel, deren Entropie bekannt ist, durch gemeinsame Reflexion und Brechung in zwei partiell kohärente verwandeln lassen, läßt sich auch die Entropie der letzteren berechnen (vgl. § 5). Nun wurde dieser Satz in der früheren Abhandlung zwar sehr wahrscheinlich gemacht, indem das ihm entgegenstehende Additionstheorem der Entropie beseitigt und die Umkehrbarkeit der Reflexion und Brechung an einer planparallelen Platte für monochromatisches Licht und der Wert $^1/_2$ des Reflexionsvermögens unmittelbar gezeigt wurde. Doch blieb die Beweiskraft des letzteren Grundes für die Thermodynamik deswegen zweifelhaft, weil es fraglich schien, ob man aus einem kontinuierlichen Spektrum — nur von solchen wissen wir bestimmt, daß sie Entropie besitzen — hinreichend homogene Strahlung spektralanalytisch aussondern und dabei noch merkliche Energiemengen in der Hand behalten könnte — ganz abgesehen von der Beschränkung auf einen bestimmten Wert des Reflexionsvermögens. Deshalb wollen wir zuerst die letztere Frage entscheiden (§ 1) und dann die Umkehrbarkeit der Spiegelung und Brechung sowohl für die planparallele Platte, als die Grenze zweier Medien beweisen.

Erster Teil.

Die Umkehrbarkeit der Reflexion und Brechung.

§ 1. Die relative Breite der schmalsten aus einem kontinuierlichen Spektrum auszusondernden Bereiche.

Daß sich aus einem kontinuierlichen Spektrum von hoher Temperatur Bereiche isolieren lassen, die nicht wesentlich breiter als manche schmale Spektrallinien sind, zeigt der Rowlandsche Atlas des Sonnenspektrums, in welchem[1]) noch Fraunhofersche Linien getrennt erscheinen, deren Differenz der Wellenlängen $\Delta\lambda$ zwischen $^1/_{10}$ und $^1/_{100}$ Å.-Einh., d. h. zwischen 10^{-9} und 10^{-10} cm, liegt. Die relative Breite des zwischen zwei solchen Linien befindlichen Spektralbereiches ist $\Delta\lambda/\lambda$, also etwa 10^{-5}. Andererseits geht aus den neuerdings von Hrn. Schönrock[2]) übersichtlich zusammengestellten Resultaten Michelsons über die Sichtbarkeit der Interferenzen als Funktion des Gangunterschiedes hervor, daß die relative Breite bei manchen Spektrallinien von derselben Größenordnung ist. Denn mag man an der dort behaupteten Intensitätsverteilung, welche aus der (übrigens nur auf 10 Proz. genau bestimmten) Sichtbarkeitskurve keineswegs eindeutig hervorgeht[3]), zweifeln, so ist doch diese Größenordnung sichergestellt; eine Änderung an ihr würde den Betrag des größten Gangunterschiedes, bei welchem die Interferenzen gerade noch sichtbar sind, wesentlich verändern. Nun liegt nach den Tabellen der Schönrockschen Arbeit die beobachtete relative Halbweite δ/λ_0 stets zwischen 10^{-5} und 10^{-6}, mit Ausnahme der dort genannten Linien des Quecksilbers ($\delta/\lambda_0 = 7{,}8 \,.\, 10^{-7}$), des Thalliums ($\delta/\lambda_0 = 5{,}5 \,.\, 10^{-7}$) und des Wismuts ($\delta/\lambda_0 = 7{,}5 \,.\, 10^{-7}$). Um zur relativen Breite zu gelangen, hat man aber δ/λ_0 noch mit einem zwischen 4

1) Vgl. H. Kayser, Handbuch der Spektroskopie **1.** p. 123.
2) O. Schönrock, Ann. d. Phys. **20.** p. 995, 1906.
3) Lord Rayleigh, Phil. Mag. (5.) **31.** p. 407. 1892.

und 5 liegenden Faktor zu multiplizieren[1]), so daß man dafür stets die Größenordnung 10^{-5} findet.

Natürlich soll hiermit nicht gesagt sein, daß auch das Licht der Spektrallinien Entropie besitzen müßte. Nur daß man in der Strahlungsthermodynamik von fast ebenso homogenem Licht reden darf, wie in den anderen Teilen der Optik, geht aus unserer Betrachtung hervor.

Häufig wird es aber gar nicht nötig sein, die vorausgesetzte Homogenität als so extrem hoch aufzufassen. In den Betrachtungen des dritten Paragraphen wird z. B. ein Prisma die Hauptrolle spielen, dessen Dispersion ohne Einfluß sein soll. Bei einem Flintglasprisma, das die beiden D-Linien eben noch trennt, muß[2]) die ausgenutzte Dicke aber mindestens 1 cm betragen. Es wäre nun ein Leichtes, sie wesentlich unter diese Grenze herunterzusetzen. Dann können wir das ganze Intervall zwischen den D-Linien als homogen betrachten, obwohl seine relative Breite ziemlich genau gleich 10^{-3} ist.

§ 2. Die Umkehrbarkeit der Spiegelung und Brechung an einer planparallelen Platte.

Nach § 2 der früheren Arbeit läßt sich die Veränderlichkeit des Reflexionsvermögens r einer planparallelen Platte durch Verminderung ihrer Dicke so herabsetzen, daß es selbst für so breite Spektralbereiche, wie der soeben erwähnte, einen konstanten Wert hat. Wird nun ein in oder senkrecht zur Einfallsebene polarisiertes, hinreichend homogenes Strahlenbündel von der spezifischen Intensität $\mathfrak{K}$ (vgl. Fig. 1) durch Spiegelung und Brechung an der Platte P in zwei neue zerlegt, deren Intensitäten $r\mathfrak{K}$ und $(1-r)\mathfrak{K}$ sind, und werden diese darauf von den vollkommen reflektierenden, ebenen, zur Platte P symmetrisch

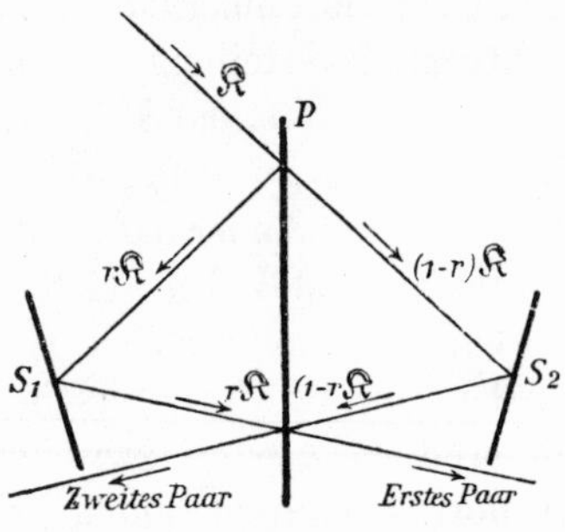

Fig. 1.

1) Vgl. § 6 der Schönrockschen Arbeit.

2) Vgl. P. Drude, Lehrbuch der Optik, II. Aufl., Leipzig, Hirzel, 1906. p. 220.

liegenden Spiegeln S_1 und S_2 so auf die Platte zurückgeworfen, daß sie das zweite Mal unter anderem Einfallswinkel zu ihr gelangen, so entstehen vier Strahlenbündel von den spezifischen Intensitäten $r(1-r')\mathfrak{K}$, $r'(1-r)\mathfrak{K}$ und $rr'\mathfrak{K}$, $(1-r)(1-r')\mathfrak{K}$. Von diesen überlagern sich die beiden ersteren sowie die beiden letzteren und interferieren miteinander; der Gangunterschied beträgt beim ersten Paar 0, beim zweiten π, ganz unabhängig von der Größe des Öffnungswinkels (vgl. § 2 der ersten Abhandlung). Wählen wir nun, was stets möglich ist,

$$r' = 1 - r,$$

so werden die Strahlenbündel des letzteren Paares an Intensität einander gleich, heben sich also auf. Die ganze Energie des einfallenden Strahlenbündels muß sich dann in dem aus der Interferenz des ersten Paares hervorgehenden Strahlenbündel wiederfinden; dies hat also, wie man auch auf rechnerischem Wege leicht bestätigt, wieder die volle Intensität $\mathfrak{K}$. Damit ist die Umkehrung der ersten Reflexion und Brechung vollzogen.

Durch diese Überlegung ist aber die Umkehrbarkeit *jeder* regulären, absorptionsfreien Spiegelung und Brechung bewiesen. Im allgemeinen befinden sich zwar der gespiegelte und gebrochene Strahl in verschiedenen Mitteln; doch ist das unwesentlich, da man ein Strahlenbündel unter Benutzung des Polarisationswinkels ohne Verlust an Energie und Kohärenz aus jedem beliebigen Mittel in jedes andere austreten lassen kann. Auch die Spiegelung und Brechung an der Grenze kristallinischer Körper ist hier einbegriffen; denn die dabei auftretende Zerspaltung eines Strahlenbündels in mehr als zwei läßt sich immer durch eine Reihe von gewöhnlichen Spiegelungen und Brechungen ersetzen. Eine etwaige Krümmung der spiegelnden Flächen ändert hieran nichts.

Trotzdem wollen wir im folgenden Paragraphen die Umkehrbarkeit dieses Vorganges für die Grenze zweier isotropen Medien noch besonders beweisen. Bei der Neuheit des Gegenstandes kann eine doppelte Sicherung der Grundlagen nützlich scheinen.

§ 3. Die Umkehrbarkeit der Spiegelung und Brechung an der Grenze von zwei verschiedenen nicht absorbierenden Körpern.

Nehmen wir an, daß auf das bei A (Fig. 2) gelegene Flächenstück f der ebenen Grenze eines für Strahlung der betrachteten Wellenlänge durchsichtigen Körpers gegen das Vakuum ein hinreichend homogenes, in oder senkrecht zur Einfallsebene polarisiertes Strahlenbündel aus dem letzteren kommend auffällt. Dann gehen von f zwei Strahlenbündel aus; das reflektierte schreitet in das Vakuum hinein fort, das gebrochene dringt in den genannten Körper ein. Da nun aber selbst der durchsichtigste ponderabele Stoff die Strahlung auf längeren Strecken durch Absorption und Zerstreuung erheblich schwächt, wollen wir das letztere sogleich wieder in das Vakuum austreten lassen. Dies gelingt ohne nochmalige Zerlegung, wenn wir die Lage der zweiten Grenzfläche so wählen, daß das Strahlenbündel sie unter dem Polarisationswinkel und senkrecht zur Einfallsebene polarisiert erreicht; dies soll in Fig. 2 bei B der Fall sein. Die Substanz bildet dann ein Prisma, doch liegt der Strahlengang im allgemeinen nicht, wie in der Figur angenommen ist, in einer Ebene.

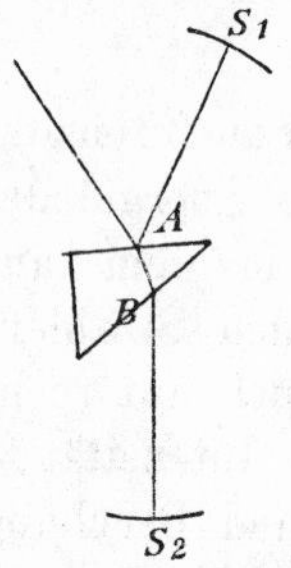

Fig. 2.

Das reflektierte Strahlenbündel lassen wir nun auf einen vollkommen spiegelnden Kugelspiegel S_1 auftreffen, welcher einen Punkt A der Fläche f zum Mittelpunkt hat; sein Radius sei R, seine Brennweite φ; bekanntlich ist

$$\varphi = \frac{R}{2}.$$

Es entwirft ein Bild von der Fläche f, dessen Lage und Form wir untersuchen wollen.

Beziehen wir die Koordinaten x, y, z und x', y', z' eines Punktes und seines Bildes auf ein Achsenkreuz, dessen x-Achse mit der Richtung des Strahlenbündels nach der Spiegelung an S_1 zusammenfällt, und dessen Anfang im Brennpunkt des Spiegels liegt, so lauten die Gesetze der geometrischen Optik

$$xx' = \varphi^2, \quad \frac{y'}{y} = \frac{z'}{z} = -\frac{\varphi}{x}.$$

Für den Punkt A ist $x = x'$, da er offenbar in sich selbst abgebildet wird; also $x = x' = \varphi$. Für die seitliche und die Tiefenvergrößerung, die das ihn umgebende Flächenstück f bei der Abbildung erfährt, gilt demnach

$$\frac{y'}{y} = \frac{z'}{z} = -1,$$

$$\frac{dx'}{dx} = -\frac{\varphi^2}{x^2} = -1.$$

Ein Punkt C von f wird in denjenigen Punkt C' der Ebene von f abgebildet, welcher zu ihm in bezug auf A symmetrisch liegt (Fig. 3). Das Bild von f liegt in derselben Ebene wie f und unterscheidet sich von ihm nur durch eine Drehung vom Betrage π um den Punkt A. Wir fragen nun nach der optischen Weglänge W von C über einen Punkt des Spiegels S_1 nach C'.

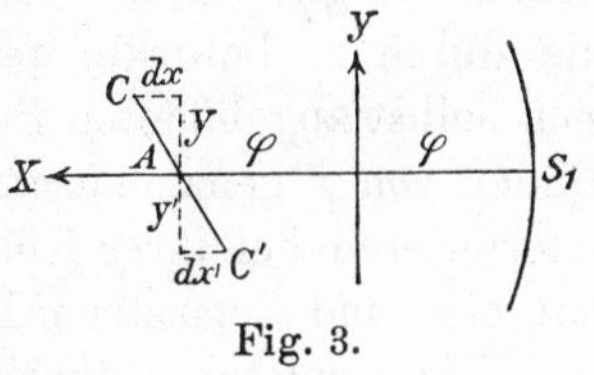

Fig. 3.

Zunächst ist unmittelbar einzusehen, daß W für alle Punkte von S_1 denselben Betrag hat. Führen wir sodann in der Ebene von f ein Koordinatensystem ξ, η ein, dessen Anfang in A liegt, so folgt aus der Reziprozität zwischen Objekt- und Bildpunkt, daß W sich bei Vertauschung von C und C', d. h. wenn man die Vorzeichen von ξ und η umkehrt, nicht ändert. Es gilt demnach in erster Näherung die Reihenentwickelung:

$$W = W_0 + \frac{1}{R}(a\,\xi^2 + b\,\eta^2 + c\,\xi\,\eta).$$

Nun ist W eine Länge; die einzigen ihm an Dimension gleichen unter seinen Bestimmungsstücken sind ξ, η und R. Die Koeffizienten a, b, c sind aber von ξ und η, daher auch von R unabhängig. Das Verhältnis der Differenz $W - W_0$ zur Wellenlänge λ,

$$\frac{W - W_0}{\lambda} = \frac{1}{R\lambda}(a\,\xi^2 + b\,\eta^2 + c\,\xi\,\eta),$$

läßt sich also durch Vergrößerung von R so klein machen, als man nur will, so groß die Dimensionen von f auch gegen λ sein mögen. Eine gegen die Wellenlänge kleine Strecke darf in der Optik stets gleich Null gesetzt werden. Wir werden daher W als längs der ganzen Fläche f konstant betrachten.

Auch das bei A gebrochene, bei B das Prisma verlassende

Strahlenbündel soll nun auf einen absolut reflektierenden Hohlspiegel S_2 auftreffen. Im Gegensatz zu S_1 darf dieser aber nicht Kugelform haben, da durch die Brechung bei B das Strahlenbündel astigmatisch geworden ist; er muß vielmehr so beschaffen sein, daß er jede Brennlinie des letzteren in sich selbst abbildet. Dann kehrt das von ihm zurückgeworfene Strahlenbündel nach B zurück, tritt dort wiederum ohne Spiegelung in das Prisma ein, wird gleichzeitig homozentrisch und entwirft bei A ein Bild von f. Da die Gesetze der Abbildung unabhängig von der Art sind, wie wir sie verwirklichen, so gilt für die Abbildung durch S_2 alles, was wir für die durch S_1 bewirkte gesagt haben. Wählen wir als den in sich selbst abgebildeten Punkt wiederum A, so fallen die beiden Bilder von f genau zusammen und die beiden Strahlenbündel interferieren bei ihrer Rückkehr in jedem Punkt der Bildfläche mit ein- und demselben Phasenunterschied.

Bei A entstehen durch abermalige Spiegelung und Brechung im allgemeinen zwei neue Strahlenbündel, von denen das eine in die Prismensubstanz eindringt. In ihm superponieren sich zwei Strahlenbündel von gleicher Intensität; denn jedes von ihnen ist durch eine Reflexion und eine Brechung bei A geschwächt. Wählen wir aber die optischen Weglängen von A nach S_1 und S_2 einander gleich, so ist ihr Phasenunterschied gleich π; denn einen Phasensprung von diesem Betrage hat das eine bei der Reflexion in A erfahren. Sie heben sich also auf. Die ganze Energie des einfallenden Strahlenbündels muß sich demnach in dem zweiten, von A aus in das Vakuum hinein fortschreitenden Strahlenbündel finden. Dies besitzt daher außer derselben Brennfläche f, der gleichen Öffnung und dem gleichen Einfallswinkel wie das einfallende auch die gleiche spezifische Intensität. Die erste Reflexion und Brechung bei A ist damit vollkommen rückgängig gemacht.

Zweiter Teil.

Die Entropie von zwei partiell kohärenten Strahlenbündeln.

§ 4. Das Maß der Interferenzfähigkeit.

Es mögen zwei monochromatische, in den geometrischen Bestimmungsstücken, d. h. in der Größe der Brennfläche und des Öffnungswinkels, sowie in der Neigung gegen die Normale der Brennfläche, übereinstimmende Strahlenbündel unbekannten Ursprungs gegeben sein. Wir sollen die Frage nach ihrer Interferenzfähigkeit beantworten. Nun ist diese im allgemeinen gerade so, wie die Intensitäten, langsam veränderlich; um die Vorstellung ein wenig zu vereinfachen, wollen wir den beiden Strahlenbündeln die geringste Länge zuschreiben, welche sie haben können, wenn sie in der Thermodynamik überhaupt noch eine selbständige Rolle spielen sollen. Sie müssen dann immer noch so lang sein, daß sie ihre Brennfläche während der kürzesten möglichen Dauer einer optischen Energiemessung beleuchten. Dann hat ihre Kohärenz ebenso wie die Intensitäten einen einzigen, bestimmten Wert.

Um sie zu ermitteln, müssen wir sie, da ihr Ursprung uns nach Voraussetzung unbekannt ist, durch Spiegelungen an vollkommenen Spiegeln so leiten, daß sie sich einmal kreuzen. Aber mangels jeden Urteils über ihren Gangunterschied dürfen wir keineswegs sogleich beim ersten Mal erwarten, die etwa vorhandene Interferenzfähigkeit zu entdecken. Vielmehr wird der Gangunterschied im allgemeinen oberhalb der Grenze liegen, bei welcher infolge der Inhomogenität der Strahlung die Interferenzfähigkeit selbst vollkommen kohärenter Strahlen aufhört. Wir müssen deshalb den Versuch oft wiederholen, indem wir den Gangunterschied jedesmal um einen unterhalb dieser Grenze liegenden Betrag verändern. Dabei muß dann einmal die Kohärenz zutage treten, wenn sie überhaupt vorhanden ist. Freilich wäre die Zahl der notwendigen Wiederholungen meist außerordentlich groß; denn selbst bei den feinsten Spektrallinien liegt die genannte Grenze des Gangunterschiedes unterhalb von 10^2 cm, während die Länge unserer Strahlenbündel sehr viel größer ist. Immerhin ist die Zahl der möglichen Wiederholungen eine ganz bestimmte, endliche, so daß mir gegen die prinzipielle Möglichkeit des Verfahrens nichts einzuwenden zu sein scheint.

Dem sich zunächst vielleicht erhebenden Einwurf, daß man die fraglichen Interferenzen wie die Strahlung überhaupt nur wahrnehmen könnte, wenn man sie absorbieren läßt, ist entgegenzuhalten, daß wir im Druck auf vollkommene Spiegel eine von Absorption unabhängige Wirkung der Strahlung auf die Materie kennen.

Haben wir dann das Vorhandensein von Kohärenz nachgewiesen, so müssen wir zur Feststellung ihres Maßes den Gangunterschied innerhalb engerer Grenzen variieren, bis wir die größte mögliche Deutlichkeit der Interferenzen erreichen; dann, wissen wir, ist der Gangunterschied so klein, daß die Inhomogenität der Strahlung keine Rolle mehr spielt. Solche Gangunterschiede wollen wir im folgenden allein in Betracht ziehen.

Der praktisch einzig in Betracht kommende Fall ist natürlich der, daß man Ursprung und Geschichte beider Strahlenbündel kennt; dann läßt sich, wie wir an mehreren Beispielen sehen werden, ihre Kohärenz leicht berechnen. Es mußte nur erst einmal festgestellt werden, daß sie, wie die Intensitäten, eine der Messung wenigstens prinzipiell stets zugängliche Größe ist. Anderenfalls dürfte sie in den Formeln der Thermodynamik nicht auftreten.

Statt von Strahlenbündeln, die aus vielen Wellen bestehen, sprechen wir in diesem Paragraphen nur von einzelnen Wellen; der Übergang zu Strahlenbündeln erfolgt dann so, daß wir zwischen je zwei einander entsprechenden Wellen beider Strahlenbündel dieselben Kohärenzverhältnisse voraussetzen. Dies ist gerechtfertigt; denn erlitten bei den Änderungen, denen das Strahlenbündel unterworfen wird, nicht alle ihm angehörenden Wellen das gleiche Schicksal, so müßten wir das Strahlenbündel in mehrere Teile zerlegt denken, für welche dies zutrifft.

Diese Wellen setzen wir nun als physikalisch homogen, die Funktionen $f(t)$ und $g(t)$, welche die in ihnen stattfindenden Schwingungen darstellen, also als nahezu periodisch voraus. Nun ist es in der Optik zwar üblich geworden, in solchen Fällen mit einzelnen Sinusfunktionen zu rechnen; doch läßt sich dann der Unterschied zwischen kohärenten und inkohärenten Wellen nur nachträglich und etwas gewaltsam einführen. Wir ziehen deswegen die exaktere Darstellung durch Fouriersche Integrale vor:

$$(1) \quad \begin{cases} f(t) = \int d\nu\, F_\nu \cos(2\pi\nu t - \varphi_\nu), \\ g(t) = \int d\nu\, G_\nu \cos(2\pi\nu t - \gamma_\nu). \end{cases}$$

Die Integrationsbereiche, in welchen F_ν und G_ν merkliche

Werte besitzen, sind nach Maßgabe der spektralen Reinheit der Schwingungen schmal; doch brauchen wir diese Voraussetzung erst später einzuführen.

Bei der Superposition beider Wellen werden alle Energiegrößen proportional zu dem Mittelwerte

$$\overline{(f+g)^2} = \overline{f^2} + \overline{g^2} + 2\overline{fg},$$

der für eine mit der Dauer der kürzesten optischen Messungen vergleichbare Zeit τ zu bilden ist. Wir berechnen $\overline{fg}$, indem wir die von Hrn. Planck[1]) gegebene Ableitung des Wertes von $\overline{f^2}$ ein wenig verallgemeinern.

Aus (1) folgt unmittelbar:

$$\overline{fg} = \frac{1}{\tau}\int\limits_t^{t+\tau} dt \iint d\nu\, d\nu'\, F_\nu\, G_{\nu'} \cos(2\pi\nu t - \varphi_\nu)\cos(2\pi\nu' t - \gamma_{\nu'})$$

$$= \frac{1}{2\tau}\int\limits_t^{t+\tau} dt \iint d\nu\, d\nu'\, F_\nu\, G_{\nu'} \Big\{\cos(2\pi(\nu'-\nu)t - (\gamma_{\nu'} - \varphi_\nu))$$
$$+ \cos(2\pi(\nu'+\nu)t - (\gamma_{\nu'} + \varphi_\nu))\Big\}.$$

Führen wir die Integration nach der Zeit aus, so folgt hieraus:

$$\overline{fg} = \tfrac{1}{2}\iint d\nu\, d\nu'\, F_\nu\, G_{\nu'} \left\{\frac{\sin\pi(\nu'-\nu)\tau}{\pi(\nu'-\nu)\tau}\cos\big(\pi(\nu'-\nu)(2t+\tau) - (\gamma_{\nu'}-\varphi_\nu)\big)\right.$$
$$\left. + \frac{\sin\pi(\nu'+\nu)\tau}{\pi(\nu'+\nu)\tau}\cos\big(\pi(\nu'+\nu)(2t+\tau) - (\gamma_{\nu'}+\varphi_\nu)\big)\right\}.$$

Soll ein von der Zeit τ unabhängiger Mittelwert $\overline{fg}$ existieren, was erfahrungsgemäß der Fall ist, so muß die zweite, $\nu' + \nu$ enthaltende Hälfte dieses Integrals verschwindend klein sein; denn da τ gegen die Lichtperiode $1/\nu$ sehr groß ist, ist $(\nu + \nu')\tau$ eine große Zahl, der in Rede stehende Teil also sicher Funktion von τ. Das gleiche gilt für die erste Hälfte in den Bereichen des Integrationsintervalls, in welchen nicht $\nu' - \nu$ klein gegen ν und ν' ist; auch diese Bereiche dürfen nicht in Betracht kommen. Führen wir

$$\lambda = \tfrac{1}{2}(\nu' + \nu),$$
$$\mu = \tfrac{1}{2}(\nu' - \nu)$$

---

1) M. Planck, Ann. d. Phys. **1.** p. 69. 1900 oder M. Planck, Theorie der Wärmestrahlung p. 190. J. A. Barth, Leipzig 1906.

als neue Integrationsvariable ein, so können wir dies dahin ausdrücken, daß nur ein schmaler Streifen in der Umgebung der Gerade $\mu = 0$ für die Integration in Betracht kommen darf. Da in ihm

$$\frac{\sin 2\pi\mu\tau}{2\pi\mu\tau} = 1,$$

so finden wir die Gleichung:

(2) $$\overline{fg} = \tfrac{1}{4}\iint d\lambda\, d\mu\, F_{\lambda-\mu}\, G_{\lambda+\mu} \cos\left(4\pi\mu t - (\gamma_{\lambda+\mu} - \varphi_{\lambda-\mu})\right).$$

Alle diese Annahmen sind in der Hypothese der natürlichen Strahlung enthalten.

Schließen sich die Spektralbereiche der Schwingungen f und g, d. h. die Integrationsintervalle in den Gleichungen (1) gegenseitig aus, so folgte hieraus

$$\overline{fg} = 0$$

als Ersatz für die sonst häufig angewandte Formel:

$$\overline{\cos(2\pi\nu t - \varphi_\nu)\cos(2\pi\nu' t + \varphi_{\nu'})} = 0,$$

wenn ν und ν' voneinander verschieden sind. Ebenso ist $\overline{fg} = 0$ bei inkohärenten Schwingungen, auch wenn sie dem gleichen Spektralbereich angehören. Die Differenz der Phasen,

$$\gamma_{\lambda+\mu} - \varphi_{\lambda-\mu},$$

schwankt dann ebenso schnell und unregelmäßig hin und her, wie jede dieser Phasenfunktionen selbst, so daß sich die in unregelmäßigster Weise bald positiven, bald negativen Werte der in (2) zu integrierenden Funktion bei der Integration gegenseitig aufheben.

Anders aber, wenn es sich um vollständig kohärente, homogene Wellen von gleicher Schwingungszahl handelt. Da die betrachteten Gangunterschiede nach Voraussetzung so klein sind, daß die Inhomogenität ohne Einfluß ist, besteht zwischen ihnen eine Beziehung

(3) $$\gamma_\nu - \varphi_\nu = a,$$

wo a als Konstante betrachtet werden darf. Ebenso ist, wenn ϱ eine andere Konstante ist, G_ν mit F_ν durch eine Relation

(3a) $$G_\nu = \varrho F_\nu$$

verbunden; denn wäre ϱ noch innerhalb so schmaler Integrationsbereiche, wie sie in (1) homogener Strahlung entsprechen, veränderlich, so könnte man im Widerspruch zur Erfahrung aus der Geschichte der Strahlenbündel nicht auf ihre relative Stärke schließen; man müßte vielmehr noch die Funktionen F_ν und G_ν selbst kennen. Definiert man nun die Funktion $f^*(t)$ dahin, daß sie aus $f(t)$ hervorgeht, wenn man in (1) statt der Kosinusfunktion den Sinus setzt, also:

$$f^*(t) = \int d\nu\, F_\nu \sin(2\pi\nu t - \varphi_\nu), \tag{4}$$

so lassen sich die Gleichungen (3) und (3a) zusammenfassen in die Beziehung

$$g = \varrho\,(f\cos a + f^* \sin a), \tag{5}$$

so daß

$$\overline{fg} = \varrho\,(\overline{f^2}\cos a + \overline{ff^*}\sin a) \tag{6}$$

wird.

Setzen wir jetzt in (2) $g = f^*$, d. h.

$$\varrho = 1, \quad a = \gamma_\nu - \varphi_\nu = \frac{\pi}{2},$$

$$\gamma_{\lambda+\mu} - \varphi_{\lambda-\mu} = \frac{\pi}{2} + (\varphi_{\lambda+\mu} - \varphi_{\lambda-\mu}),$$

so finden wir:

$$\overline{ff^*} = \tfrac{1}{4}\iint d\lambda\, d\mu\, F_{\lambda+\mu} F_{\lambda-\mu} \sin\left(4\pi\mu t - (\varphi_{\lambda+\mu} - \varphi_{\lambda-\mu})\right).$$

Hier wechselt die zu integrierende Funktion mit μ ihr Zeichen. Der Integrationsbereich ist aber zur Geraden $\mu = 0$ symmetrisch. Also ist

$$\overline{ff^*} = 0; \tag{7}$$

diese Gleichung ist das Analogon zu

$$\overline{\cos 2\pi\nu t \sin 2\pi\nu t} = 0.$$

Bevor wir (7) anwenden, leiten wir noch einige Rechnungsregeln ab. Der Übergang von f zu f^* setzt $\varphi_\nu + \pi/2$ an die Stelle von φ_ν; die Wiederholung dieser Operation an f^* macht daraus $\varphi_\nu + \pi$, so daß in (1) alle Kosinusfunktionen ihr Zeichen wechseln; daraus folgt

$$(f^*)^* = -f. \tag{7a}$$

Ist ferner

$$f(t) + g(t) = k(t) = \int d\nu\, K_\nu \cos(2\pi\nu t - \varkappa_\nu),$$

so bestimmen sich nach (1) K_ν und $\varkappa_\nu$ aus der komplexen Gleichung

$$K_\nu e^{-i\varkappa_\nu} = F_\nu e^{-i\varphi_\nu} + G_\nu e^{-i\gamma_\nu}.$$

Aus ihr schließen wir, daß

$$\int d\nu\, K_\nu \sin(2\pi\nu t - \varkappa_\nu) = \int d\nu\, F_\nu \sin(2\pi\nu t - \varphi_\nu) + \int d\nu\, G_\nu \sin(2\pi\nu t - \gamma_\nu),$$

$$(f + g)^* = f^* + g^* \tag{7b}$$

ist.

Aus (6) und (7) folgt unmittelbar

$$\overline{fg} = \varrho \cos a\, \overline{f^2}, \tag{8}$$

und aus (5), (7a) und (7b) geht hervor

$$g^* = \varrho(-f \sin a + f^* \cos a),$$

$$\overline{fg^*} = -\varrho \sin a\, \overline{f^2}. \tag{9}$$

Der gleichzeitige Übergang von f zu f^* und von g zu g^*, welcher sowohl γ_ν als φ_ν um $\pi/2$ vergrößert, läßt aber die Phasendifferenz

$$\gamma_{\lambda+\mu} - \varphi_{\lambda-\mu}$$

in (2) ungeändert; es ist deshalb

$$\overline{f^* g^*} = \overline{fg}, \tag{10}$$

speziell

$$\overline{f^{*2}} = \overline{f^2}. \tag{11}$$

Aus (5) und (9) können wir daher schließen:

$$\overline{f^* g} = -\overline{fg^*}. \tag{12}$$

In den Gleichungen (8), (9) und (12) waren f und g als vollkommen kohärent gedacht. Jetzt soll sich aber dem zu f vollständig kohärenten Anteil

$$g_f = \varrho(f \cos a + f^* \sin a) \tag{13}$$

an g die dazu vollkommen inkohärente Schwingung g_f' gemäß der Gleichung

$$g = g_f + g_f' \tag{14}$$

überlagern, so daß

$$\overline{g^2} = \overline{g_f^2} + \overline{g_f'^2} \tag{15}$$

wird. Dann sind f und g nur noch partiell kohärent. Trotzdem bleiben die Gleichungen (8), (9) und (12) in Kraft, weil man nach (13) in ihnen überall g_f an die Stelle von g setzen kann. Wir machen jetzt von den folgenden Gleichungen Gebrauch:

$$\overline{g_f^2} = \varrho^2 \overline{f^2} \quad \text{(vgl. (13), (7) und (11))},$$

$$\text{(16)} \qquad \overline{fg}^2 = \varrho^2 \cos^2 a \, \overline{f^2}^2 = \cos^2 a \, \overline{f^2} . \overline{g_f^2} \quad \text{(vgl. (8))},$$

$$\text{(17)} \quad \overline{f^* g}^2 = \overline{f g^*}^2 = \varrho^2 \sin^2 a \, \overline{f^2}^2 = \sin^2 a \, \overline{f^2} . \overline{g_f^2} \quad \text{(vgl. (9) u. (12))}.$$

Durch Addition der beiden letzteren folgt:

$$\text{(18)} \qquad \frac{\overline{fg}^2 + \overline{fg^*}^2}{\overline{f^2} . \overline{g^2}} = \frac{\overline{g_f^2}}{\overline{g^2}} .$$

Dies Verhältnis der Intensität des zu f kohärenten Anteiles an g zur Gesamtintensität nennen wir die Kohärenz i_{fg}. Die Reihenfolge der Indizes ist gleichgültig, denn eine Vertauschung von f und g läßt nach (17) die linke Seite von (18) unberührt, so daß der Reziprozitätssatz

$$\text{(19)} \qquad \frac{\overline{g_f^2}}{\overline{g^2}} = \frac{\overline{f_g^2}}{\overline{f^2}}$$

besteht. Auch macht es nichts aus, wenn man f durch eine lineare Kombination $\alpha f + \beta f^*$ ersetzt; nach (7), (10), (11) und (12) ist nämlich

$$\begin{aligned}
\overline{(\alpha f + \beta f^*) g}^2 &= \alpha^2 \overline{fg}^2 + 2 \alpha \beta \overline{fg} . \overline{f^* g} + \beta^2 \overline{f^* g}^2 , \\
\overline{(\alpha f + \beta f^*) g^*}^2 &= \beta^2 \overline{fg}^2 - 2 \alpha \beta \overline{fg} . \overline{f^* g} + \alpha^2 \overline{f^* g}^2 , \\
\overline{(\alpha f + \beta f^*)^2} &= (\alpha^2 + \beta^2) \overline{f^2} ,
\end{aligned}$$

so daß man dabei den alten Wert wiederfindet. Hierin drückt sich die Unabhängigkeit der Kohärenz von der Intensität und der Phasendifferenz aus. Experimentell kann man sie ermitteln, indem man bei in dem angegebenen Sinn kleinen Gangunterschieden die Helligkeit an zwei Stellen eines Interferenzstreifens mißt, die um $^1/_4$ Streifenbreite voneinander abstehen. Ist ihr Betrag an der einen Stelle

$$I_1 = \overline{(f + g)^2} = \overline{f^2} + \overline{g^2} + 2 \overline{fg} ,$$

so ist er an der anderen (vgl. (11))

$$I_2 = \overline{(f \pm g^*)^2} = \overline{f^2} + \overline{g^2} \pm 2 \overline{fg^*} ,$$

so daß sich die Kohärenz nach der Formel

$$i_{fg} = \frac{\overline{fg}^2 + \overline{fg^*}^2}{\overline{f^2} \cdot \overline{g^2}} = \frac{1}{4\,\overline{f^2} \cdot \overline{g^2}} \left\{ \left(J_1 - (\overline{f^2} + \overline{g^2})\right)^2 + \left(J_2 - (\overline{f^2} + \overline{g^2})\right)^2 \right\}$$

berechnet. Wegen (15) ist

$$(20) \qquad i_{fg} = \frac{\overline{g_f^2}}{\overline{g^2}}$$

stets ein positiver echter Bruch. Die Grenzfälle $i = 0$ und $i = 1$ bedeuten absolute Inkohärenz und vollständige Kohärenz.

In den späteren Formeln wird übrigens statt der Kohärenz i_{fg} meist die „*Inkohärenz*"

$$(21) \qquad j_{fg} = 1 - i_{fg} = \frac{\overline{g_f'^2}}{\overline{g^2}} = \frac{\overline{f_g'^2}}{\overline{f^2}} = 1 - \frac{\overline{fg}^2 + \overline{fg^*}^2}{\overline{f^2} \cdot \overline{g^2}}$$

auftreten.

Um ein Beispiel für die Verwendung des quantitativen Kohärenzbegriffes zu geben, denken wir uns von zwei vollständig kohärenten Strahlenbündeln das eine zweimal an selbst emittierenden Körpern gespiegelt. Wir fragen nach der Kohärenz des dann entstehenden Strahlenbündels zu dem nicht gespiegelten.

Bei der ersten Reflexion geht die Schwingung f in die ihr partiell kohärente

$$g = g_f + g_f'$$

über, diese wiederum bei der zweiten Reflexion in

$$h = h_g + h_g'.$$

f und h_g' sind in dem betrachteten Fall absolut inkohärent, darum auch h_g' und f_g', wenn man

$$f = f_g + f_g'$$

setzt. Aus den beiden letzten Gleichungen und (7b) folgt aber:

$$\overline{fh} = \overline{f_g h_g},$$

$$\overline{fh^*} = \overline{f_g h_g^*},$$

$$i_{fh} = \frac{\overline{fh}^2 + \overline{fh^*}^2}{\overline{f^2} \cdot \overline{h^2}} = \frac{\overline{f_g h_g}^2 + \overline{f_g h_g^*}^2}{\overline{f_g^2} \cdot \overline{h_g^2}} \cdot \frac{\overline{f_g^2}}{\overline{f^2}} \cdot \frac{\overline{h_g^2}}{\overline{h^2}}.$$

Der erste dieser drei Brüche ist die Kohärenz der vollständig kohärenten Wellen f_g und h_g, also 1, so daß (vgl. (20))

$$i_{fh} = i_{fg} \cdot i_{gh}$$

folgt.

Haben die spiegelnden Körper und die auffallenden Strahlenbündel gleiche Temperatur (isotherme Reflexion), so wird die Intensität bei der Reflexion nicht geändert; es ist vielmehr

$$\overline{f^2} = \overline{g^2} = \overline{h^2}.$$

Sind andererseits r_1 und r_2 die Reflexionsvermögen, so ist

$$\overline{g_f^2} = r_1 \overline{f^2},$$
$$\overline{h_g^2} = r_2 \overline{g^2},$$

so daß

$$i_{fg} = r_1, \quad i_{gh} = r_2, \quad i_{fh} = r_1 r_2$$

folgt.

Man sieht ohne weiteres, nach welchem Gesetz die Kohärenz zu dem anderen Strahlenbündel bei weiteren isothermen Spiegelungen abnimmt. Erleiden zwei partiell kohärente Strahlenbündel beide isotherme Spiegelungen mit den Reflexionsvermögen $r_1, r_2, \ldots r_n$ und $\varrho_1, \varrho_2, \ldots \varrho_m$, so wird ihre Kohärenz dadurch auf den Bruchteil des Anfangswertes herabgesetzt, welcher durch das Produkt $r_1 r_2 \ldots r_n \, \varrho_1 \varrho_2 \ldots \varrho_m$ gegeben ist.

§ 5. Die Entropie von zwei partiell kohärenten Strahlenbündeln.

Wir sprechen im folgenden von linear polarisierten Strahlenbündeln, welche sich im allgemeinen zwar in verschiedenen Medien befinden, aber so beschaffen sind, daß man sie durch reguläre Spiegelungen und Brechungen in dasselbe Mittel und in ihm zur Deckung miteinander bringen kann. Nach dem Sinussatz der geometrischen Optik müssen dann ihre Brennflächen f, ihre Öffnungswinkel ω und ihre Neigungswinkel ϑ zur Normalen der Brennfläche mit dem Brechungsindex n des Mittels in der Beziehung stehen, daß der Ausdruck

$$n^2 f \omega \cos \vartheta$$

für alle einen und denselben Wert hat. Ihre Längen l müssen sich außerdem wie die Lichtgeschwindigkeiten q verhalten. Ferner sollen sie dieselbe Schwingungszahl ν und dieselbe spektrale Breite $d\nu$ besitzen. Wir setzen

$$n^2 f \omega \cos \vartheta \frac{l}{q} d\nu = \sigma, \tag{22}$$

und haben dann σ als für alle Strahlenbündel gleich zu betrachten.

Die Energie eines Strahlenbündels von der spezifischen Intensität $\mathfrak{K}$ beträgt

$$f\,\omega \cos\vartheta \,\frac{l}{q}\,\mathfrak{K}\,d\,\nu = \sigma\,\frac{\mathfrak{K}}{n^2}\,.$$

Bequemer aber als mit der spezifischen Intensität rechnet es sich mit der „*reduzierten spezifischen Intensität*"

$$\varkappa = \frac{\mathfrak{K}}{n^2}\,;$$

denn es wird dabei nicht nur der Ausdruck für die Energie einfacher, sondern es enthalten auch die Temperatur- und die Entropieformel $\mathfrak{K}$ stets in der Verbindung $\mathfrak{K}/n^2$. Die letztere Formel lautet, waren h und k die Konstanten des Planckschen Energieverteilungsgesetzes, c die Lichtgeschwindigkeit und ν die Schwingungszahl bedeutet:

$$(23)\quad \left\{\begin{aligned} E &= \sigma\,L(\varkappa)\,,\\ L(\varkappa) &= \frac{k\,\nu^2}{c^2}\left[\left(1+\frac{c^2\,\varkappa}{h\,\nu^3}\right)\log\left(1+\frac{c^2\,\varkappa}{h\,\nu^3}\right)-\frac{c^2\,\varkappa}{h\,\nu^3}\log\left(\frac{c^2\,\varkappa}{h\,\nu^3}\right)\right]. \end{aligned}\right.$$ [1]

Jetzt sollen zwei derartige, in oder senkrecht zur Einfallsebene polarisierte, partiell kohärente Strahlenbündel gleichzeitig an der ebenen Grenze zweier diathermanen Medien so gespiegelt und gebrochen werden, daß die vier dabei entstehenden Strahlenbündel sich paarweise genau überdecken. Sie müssen dann von verschiedenen Seiten auf die Grenze auftreffen und die Sinus ihrer Einfallswinkel müssen sich umgekehrt wie die Brechungsindizes verhalten (vgl. Fig. 4). Das Energieprinzip verlangt die Konstanz der Gesamtenergie $\sigma(\varkappa_1 + \varkappa_2)$, in Verbindung mit dem Sinussatz also, daß

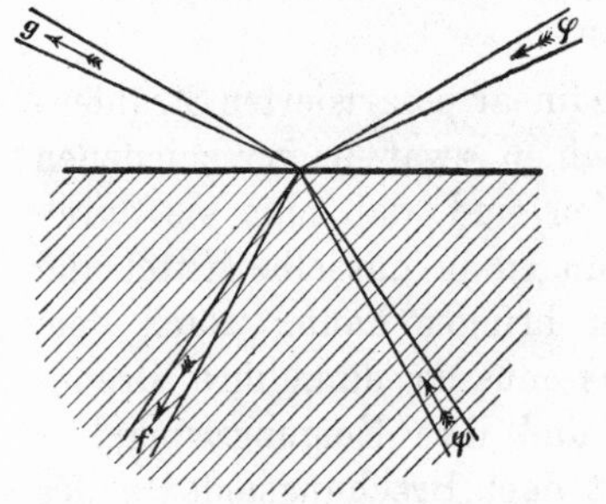

Fig. 4.

$$(24)\qquad \varkappa_1 + \varkappa_2 = \Gamma\,.$$

bei dem Vorgang invariant bleibt.

---

1) M. Planck, Ann. d. Phys. **4.** p. 553; **6.** p. 818. 1901, sowie M. Planck, Theorie der Wärmestrahlung p. 156, Gleichung (229). Leipzig, J. A. Barth 1906.

Nun ist der betrachtete Vorgang umkehrbar; das Entropieprinzip fordert daher, daß auch die Entropie eine solche Invariante ist. Sie hängt im allgemeinen nicht von der Gesamtenergie allein ab, wie wir am Beispiel zweier inkohärenten Strahlenbündel sehen. Daher muß der Vorgang noch eine zweite, von I' unabhängige Invariante I'' besitzen, aber auch nicht mehr. Dann existierte noch eine dritte, so müßten alle drei Bestimmungsstücke des betrachteten Systems, nämlich die beiden reduzierten spezifischen Intensitäten $\varkappa_1$ und $\varkappa_2$, sowie die Inkohärenz j, unverändert bleiben, während wir doch offenbar in der Wahl des Reflexionsvermögens einen Freiheitsgrad besitzen. Die Entropie hängt also sicherlich außer von σ nur von zwei Invarianten I' und I'' ab. Wir behaupten, daß

$$j\varkappa_1\varkappa_2 = I'' \tag{25}$$

diese zweite Invariante ist.

Um dies zu beweisen, gehen wir auf die Schwingungsvorgänge in entsprechenden Wellen der vier Strahlenbündel ein. Als Lichtvektor betrachten wir aber keine der beiden Feldstärken, sondern eine ihnen proportionale Größe, deren Quadrat im zeitlichen Mittelwert unmittelbar der Intensität $\varkappa$ gleich ist. In den beiden einfallenden Strahlenbündeln sollen die Funktionen $\varphi(t)$ und $\psi(t)$ die Schwingung unmittelbar an der Grenze darstellen, in den beiden neu entstehenden die Funktionen $f(t)$ und $g(t)$. Wäre $\psi = 0$, so wäre (vgl. Fig. 4)

$$f = \delta\varphi, \quad g = -\varrho\varphi;$$

da

$$\overline{f^2} + \overline{g^2} = \overline{\varphi^2}$$

sein muß, ist

$$\delta^2 + \varrho^2 = 1. \tag{26}$$

Wäre $\varphi = 0$, so gälte aber

$$f = \varrho\psi, \quad g = \delta\psi.$$

Die Größe ϱ, deren Quadrat das Reflexionsvermögen mißt, muß hier das entgegengesetzte Vorzeichen erhalten, wie das erste Mal, weil mit der einen dieser Spiegelungen der Phasensprung π verknüpft ist. Im allgemeinen ist daher

$$\begin{cases} f = \delta\varphi + \varrho\psi, \\ g = -\varrho\varphi + \delta\psi. \end{cases} \tag{27}$$

Da nun

$$j\varkappa_1\varkappa_2 = (1-i)\varkappa_1\varkappa_2 = \overline{f^2}.\overline{g^2} - (\overline{fg}^2 + \overline{fg^*}^2)$$

ist, so ist zu zeigen, daß

$$\overline{f^2}.\overline{g^2} - (\overline{fg}^2 + \overline{fg^*}^2) = \overline{\varphi^2}.\overline{\psi^2} - (\overline{\varphi\psi}^2 + \overline{\varphi\psi^*}^2)$$

ist.

Nach (27), (26), (7) und (7 b) ist

$$\overline{f^2} = \delta^2\overline{\varphi^2} + \varrho^2\overline{\psi^2} + 2\delta\varrho\overline{\varphi\psi},$$

$$\overline{g^2} = \varrho^2\overline{\varphi^2} + \delta^2\overline{\psi^2} - 2\delta\varrho\overline{\varphi\psi},$$

$$\overline{fg} = (\delta^2-\varrho^2)\overline{\varphi\psi} + \delta\varrho(\overline{\psi^2}-\overline{\varphi^2}),$$

$$\overline{fg^*} = (\delta^2+\varrho^2)\overline{\varphi\psi^*} = \overline{\varphi\psi^*}.$$

Daraus schließen wir unter wiederholter Berücksichtigung von (26):

$$\begin{aligned}\overline{f^2}\cdot\overline{g^2} - (\overline{fg}^2 + \overline{fg^*}^2) &= (\delta^2\overline{\varphi^2} + \varrho^2\overline{\psi^2})(\varrho^2\overline{\varphi^2} + \delta^2\overline{\psi^2})\\ &+ 2\delta\varrho(\delta^2-\varrho^2)\overline{\varphi\psi}(\overline{\psi^2}-\overline{\varphi^2}) - 4\delta^2\varrho^2\overline{\varphi\psi}^2\\ &- (\delta^2-\varrho^2)^2\overline{\varphi\psi}^2 - \delta^2\varrho^2(\overline{\psi^2}-\overline{\varphi^2})^2\\ &- 2\delta\varrho(\delta^2-\varrho^2)\overline{\varphi\psi}(\overline{\psi^2}-\overline{\varphi^2}) - \overline{\varphi\psi^*}^2\\ &= \overline{\varphi^2}.\overline{\psi^2} - (\overline{\varphi\psi}^2 + \overline{\varphi\psi^*}^2),\end{aligned}$$

was zu beweisen war.

Die Entropie ist also Funktion von σ, von

$$I' = \varkappa_1 + \varkappa_2$$

und

$$I'' = j\varkappa_1\varkappa_2.$$

Um sie zu bestimmen, denken wir uns die Strahlenbündel inkohärent; dann ist

$$E = \sigma\left(L(\varkappa_1) + L(\varkappa_2)\right),$$

und wegen $j = 1$

$$\varkappa_1 + \varkappa_2 = I'$$

$$\varkappa_1\varkappa_2 = I'',$$

also

$$\varkappa_1 = \tfrac{1}{2}(I' + \sqrt{I'^2 - 4I''}),$$

$$\varkappa_2 = \tfrac{1}{2}(I' - \sqrt{I'^2 - 4I''}),$$

oder umgekehrt. Daher ist die Entropie

$$E = \sigma\left[L\left(\tfrac{1}{2}(I' + \sqrt{I'^2 - 4I''})\right) + L\left(\tfrac{1}{2}(I' - \sqrt{I'^2 - 4I''})\right)\right].$$

Bei beliebigem Werte der Inkohärenz j geht diese Gleichung über in

$$(28)\quad \begin{cases} E = \sigma \Big\{ L\left(\tfrac{1}{2}\left[(\varkappa_1 + \varkappa_2) + \sqrt{(\varkappa_1 + \varkappa_2)^2 - 4 j \varkappa_1 \varkappa_2}\,\right]\right) \\ \qquad + L\left(\tfrac{1}{2}\left[(\varkappa_1 + \varkappa_2) - \sqrt{(\varkappa_1 + \varkappa_2)^2 - 4 j \varkappa_1 \varkappa_2}\,\right]\right)\Big\}. \end{cases}$$

Diese Formel stellt die Entropie als Funktion der beiden reduzierten spezifischen Intensitäten und der Inkohärenz j dar.

Da nach (21)

$$(\varkappa_1 + \varkappa_2)^2 - 4 j \varkappa_1 \varkappa_2 = (\varkappa_1 - \varkappa_2)^2 + 4 i \varkappa_1 \varkappa_2 ,$$

sind die Argumente der beiden L-Funktionen stets reell und positiv. Für vollständig kohärente Wellen ($j = 0$) geht (28), da nach (23) $L(0) = 0$ ist, in

$$E = \sigma\, L(\varkappa_1 + \varkappa_2)$$

über, was mit dem in § 4 der ersten Abhandlung[1]) Gesagten übereinstimmt.

Wir wollen dies Resultat graphisch diskutieren. Zu diesem Zweck setzen wir

$$\frac{c^2 \varkappa_1}{h \nu^3} = a + x , \qquad \frac{c^2 \varkappa_2}{h \nu^3} = a - x .$$

E wird dann nach (28) und (23) proportional zu der Funktion

$$(29)\quad \begin{cases} \Phi(x, i) = f\left(a + \sqrt{a^2 - (1 - i)(a^2 - x^2)}\right) \\ \qquad - f\left(a + \sqrt{a^2 - (1 - i)(a^2 - x^2)}\right), \end{cases}$$

wenn

$$f(x) = (1 + x) \log (1 + x) - x \log x$$

ist. Um eine anschaulichere Figur zu erhalten, haben wir statt j die Kohärenz i eingeführt; die Größe

$$a = \frac{c^2 (\varkappa_1 + \varkappa_2)}{2 h \nu^3}$$

bleibt bei der gemeinsamen Spiegelung und Brechung unverändert, wir betrachten sie als Konstante und fragen nach der Fläche, welche durch (29) bestimmt ist, wenn man

$$x = \frac{c^2 (\varkappa_1 - \varkappa_2)}{2 h \nu^3}, \quad i \text{ und } \Phi(x, i)$$

als rechtwinklige Koordinaten im Raum betrachtet. Da in (29) x nur im Quadrat auftritt, ist sie zur Ebene $x = 0$ symmetrisch.

---

1) M. **Laue**, Ann. d. Phys. **20**. p. 365. 1906.

Eine physikalische Bedeutung kommt ihn aber nur in dem durch die Ungleichungen

$$0 \leqq i \leqq 1,$$

$$-a \leqq x \leqq a$$

abgegrenzten Bereich zu. Für $i = 0$ wird

$$\Phi = f(a + x) + f(a - x).$$

Den Verlauf dieser Kurve haben wir schon in § 1 der früheren Arbeit diskutiert; Φ nimmt für $x = \pm a$ den Wert $f(2a)$ an und steigt mit abnehmendem Absolutwert von x bis zu dem für $x = 0$, d. h. für zwei gleich starke Strahlenbündel erreichten Betrage $2f(a)$. Sie ist in der oberen Hälfte von Fig. 5 zu

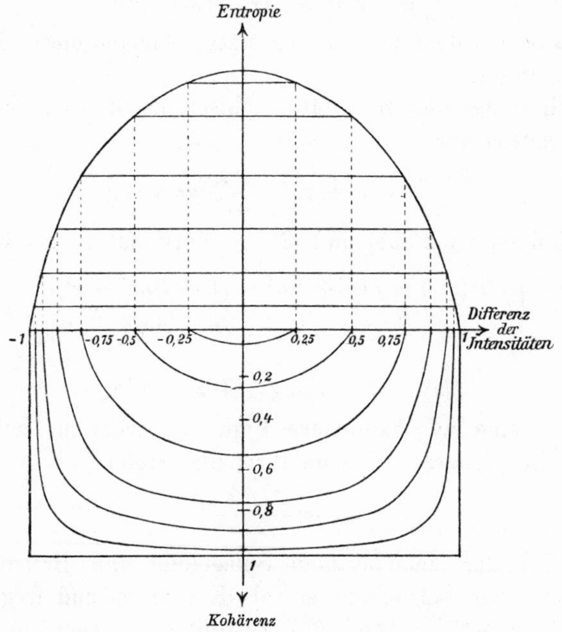

Fig. 5.

sehen, wo nach oben die Größe $\Phi - f(2a)$ in einem willkürlichen Maßstabe, und als Abszisse nach rechts die Größe x aufgetragen ist (a ist gleich 1 gesetzt). Für $i = 1$, sowie für

$x = \pm a$ wird $\Phi = f(2a)$. In den Rahmen dieser drei Geraden und der genannten Kurven ist die Entropiefläche eingespannt. Da

$$\frac{\partial \Phi}{\partial i} = \left\{ f'\left(a + \sqrt{a^2 - (1-i)(a^2 - x^2)}\right) - f'\left(a - \sqrt{a^2 - (1-i)(a^2 - x^2)}\right) \right\} \frac{a^2 - x^2}{\sqrt{a^2 - (1-i)(a^2 - x^2)}}$$

ist, und

$$f'(x) = \log\left(1 + \frac{1}{x}\right)$$

mit wachsendem x abnimmt, ist in dem ganzen physikalisch bedeutsamen Bereich der Klammerausdruck negativ, dagegen sein Multiplikator positiv, demnach:

$$\frac{\partial \Phi}{\partial i} < 0;$$

dies gilt auch im Punkt $x = 0$, $i = 0$, wo

$$\frac{\partial \Phi}{\partial i} = \lim_{i=0} \frac{f'\left(a(1+\sqrt{i})\right) - f'\left(a(1-\sqrt{i})\right)}{\sqrt{i}} = 2af''(a) = -\frac{2}{1+a}$$

wird. *Mit wachsender Kohärenz nimmt die Entropie bei konstanten Intensitäten ab.* Nur längs der Grenzlinien $x = \pm a$ wird $\partial \Phi / \partial i = 0$, weil hier, wo $\varkappa_1$ oder $\varkappa_2$ Null ist, der Kohärenzbegriff keine Bedeutung mehr hat. Für $i = 1$ wird $\partial \Phi / \partial i$ zugleich mit $-f'\left(a - \sqrt{a^2 - (1-i)(a^2 - x^2)}\right)$ negativ unendlich; desgleichen

$$\frac{\partial \Phi}{\partial x} = \left\{ f'\left(a + \sqrt{a^2 - (1-i)(a^2 - x^2)}\right) - f'\left(a - \sqrt{a^2 - (1-i)(a^2 - x^2)}\right) \right\} \frac{x(1-i)}{\sqrt{a^2 - (1-i)(a^2 - x^2)}}$$

für $x = \pm a$. Die Entropiefläche fällt also zu den drei sie begrenzenden Geraden senkrecht ab.

Die beste Übersicht über die Form der Fläche gewährt aber die Betrachtung der Kurven gleicher Entropie, deren Gleichung nach (29)

$$(1-i)(a^2 - x^2) = (a^2 - x_0^2)$$

autet; x_0^2 ist der Parameter der Schar. Fig. 5 zeigt in ihrer unteren Hälfte einige dieser Kurven; als Ganzes gibt sie eine Darstellung der Entropiefläche nach Art der darstellenden Geometrie. Denkt man sie längs der x-Achse um 90° geknickt und jede der Niveaulinien bis zur Höhe der ihr entsprechenden Geraden gehoben, so gelangt sie dabei auf die Entropiefläche.

Da die Differentialgleichung dieser Schar

$$\frac{di}{dx} = -\frac{2x(1-i)}{a^2 - x^2}$$

lautet, sind die Punkte $x = \pm a$, $i = 1$, in welchen di/dx unbestimmt wird, singuläre Punkte. In der Tat setzt sich die durch sie führende Kurve, für welche $x_0^2 = a^2$ wird, aus den Geraden $x = \pm a$ und $i = 1$ zusammen. Diese drei Begrenzungsgeraden der Entropiefläche bilden die tiefste Niveaulinie. Die höchste hier auftretende Niveaulinie ist dagegen durch $x_0^2 = 0$ definiert, sie liegt aber nur mit dem Punkt $x = 0$, $i = 0$ in dem physikalisch bedeutsamen Bereich. Von allen Strahlenpaaren gleicher Gesamtenergie haben also zwei inkohärente, gleich starke Strahlenbündel die größte Entropie. Auf den anderen Niveaulinien erreicht i seinen Maximalwert x_0^2/a^2 für $a = 0$, während x seinen größten im genannten Bereich liegenden Absolutwert $|x_0|$ für $i = 0$ annimmt.

Diese Kurven gleicher Entropie geben nun unmittelbar an, wie sich die Kohärenz eines Strahlenpaares bei gemeinsamer Spiegelung und Brechung ändert. Alle Zustände, welche durch Punkte derselben Kurve dargestellt sind, lassen sich dabei durch passende Wahl des Reflexionsvermögens ineinander überführen. Zwei inkohärente Strahlen lassen sich z. B. in zwei partiell kohärente verwandeln, wenn man ihre Intensitäten ganz oder teilweise ausgleicht. Die maximale Kohärenz wird bei vollständigem Ausgleich erreicht; waren die Intensitäten der inkohärenten Strahlen $\varkappa_1'$ und $\varkappa_2'$, so beträgt sie

$$i_{\text{max.}} = \frac{x_0^2}{a^2} = \left(\frac{\varkappa_1' - \varkappa_2'}{\varkappa_1' + \varkappa_2'}\right)^2.$$

Umgekehrt läßt sich jedes Paar partiell kohärenter Strahlenbündel durch gemeinsame Reflexion und Brechung in ein vollkommen inkohärentes Paar verwandeln, wenn man die Differenz der Intensitäten möglichst vergrößert. Ausgenommen sind nur vollständig kohärente Strahlenbündel, die wegen

$$I'' = j\varkappa_1\varkappa_2 = 0$$

stets wieder in zwei vollkommen kohärente übergehen, und

zwei vollständig inkohärente Strahlenbündel von gleicher Intensität, die stets unverändert bleiben.

§ 6. Partiell polarisierte Strahlung.

Eine partiell polarisierte Welle läßt sich in den verschiedensten Arten als Superposition zweier senkrecht zueinander polarisierten Wellen auffassen. Herrschen in einem Paar von aufeinander senkrechten Richtungen die Schwingungen $\varphi(t)$ und $\psi(t)$, so herrschen in einem anderen Paar die Schwingungen

$$f = \cos\omega \,.\, \varphi + \sin\omega \,.\, \psi\,,$$
$$g = -\sin\omega \,.\, \varphi + \cos\omega \,.\, \psi\,,$$

wo ω der Winkel zwischen entsprechenden Richtungen beider Paare ist. Diese Gleichungen unterscheiden sich aber von (27) wegen (26) nur in der Bezeichnungsweise; wir können ohne weiteres

$$\delta = \cos\omega\,, \quad \varrho = \sin\omega$$

setzen. Daher sind die Ausdrücke

$$I' = \varkappa_1 + \varkappa_2\,,$$
$$I'' = j\,\varkappa_1\,\varkappa_2$$

von der Wahl der Schwingungsrichtungen unabhängig. Wir können der Berechnung der Entropie eines polarisierten Strahlenbündels

$$E = \sigma\left\{L\left[\tfrac{1}{2}\left((\varkappa_1 + \varkappa_2) + \sqrt{(\varkappa_1 + \varkappa_2)^2 - 4j\,\varkappa_1\,\varkappa_2}\right)\right] + L\left[\tfrac{1}{2}\left((\varkappa_1 + \varkappa_2) - \sqrt{(\varkappa_1 + \varkappa_2)^2 - 4j\,\varkappa_1\,\varkappa_2}\right)\right]\right\},$$

jedes beliebige Paar zueinander senkrechter Richtungen zugrunde legen. Ein partiell polarisiertes Strahlenbündel läßt sich auch thermodynamisch als Superposition zweier senkrecht zueinander polarisierten, partiell kohärenten Strahlenbündeln auffassen; daß diese in der Fortpflanzungsrichtung zusammenfallen, macht nichts aus.

Hieraus fließt eine Bestätigung unserer Theorie, welche sich wohl auch zu einem zweiten Beweise ausgestalten ließe. Diese logische Zerlegung in zwei partiell kohärente, senkrecht zueinander polarisierte Schwingungen läßt sich nämlich ohne Verlust an Kohärenz oder Intensität verwirklichen, wenn man das

Strahlenbündel durch eine doppelbrechende Substanz hindurchgehen läßt. Die Reflexion beim Ein- und Austritt kann man dadurch beliebig herabgesetzt denken, daß man die Brechungsindizes des Kristalles als nur wenig von dem der Umgebung verschieden annimmt. Freilich ist dann auch die Doppelbrechung gering, doch läßt sich trotzdem auf hinreichend langen Strecken die vollständige Trennung der beiden Strahlenbündel erzielen. Da der Vorgang umkehrkar ist, lassen sich zwei inkohärente, linear polarisierte Strahlenbündel auch auf dem Wege ohne Entropiezunahme in zwei partiell kohärente verwandeln, daß man sie zunächst durch Doppelbrechung zu einem partiell polarisierten Strahlenbündel zusammensetzt und dann dies nach zwei anderen Richtungen durch Doppelbrechung zerlegt.

Die Niveaukurven in Fig. 5 zeigen die Kohärenzverhältnisse in einem partiell polarisierten Strahl. Die Kohärenz ist Null für die beiden Hauptrichtungen, da die entsprechenden Intensitäten, die Hauptintensitäten $\varkappa_1'$ und $\varkappa_2'$, den größtmöglichen Unterschied besitzen. Da die Intensitäten für zwei andere, aufeinander senkrechte Schwingungen sich nach Gleichungen

$$\varkappa_1 = \varkappa_1' \cos^2\omega + \varkappa_2' \sin^2\omega\,,$$
$$\varkappa_2 = \varkappa_1' \sin^2\omega + \varkappa_2' \cos^2\omega$$

aus dem Azimut ω gegen die Hauptrichtungen berechnen, und

$$j\,\varkappa_1\,\varkappa_2 = \varkappa_1'\,\varkappa_2'$$

sein muß, ist ihre Kohärenz

$$i = 1 - j = 1 - \frac{\varkappa_1'\,\varkappa_2'}{\varkappa_1\,\varkappa_2} = \frac{(\varkappa_1' - \varkappa_2')^2 \cos^2\omega \sin^2\omega}{(\varkappa_1' \cos^2\omega + \varkappa_2' \sin^2\omega)(\varkappa_1' \sin^2\omega + \varkappa_2' \cos^2\omega)}\,.$$

Ihren Maximalwert

$$i_{\text{max.}} = \left(\frac{\varkappa_1' - \varkappa_2'}{\varkappa_1' + \varkappa_2'}\right)^2$$

erreicht sie nach § 4, wenn $\varkappa_1 = \varkappa_2$, also $\omega = \pm \pi/4$ wird.

Es liegt nahe, $i_{\text{max.}}$ zur Messung des Polarisationsgrades zu verwenden und die „*Polarisation*"

$$p = \sqrt{i_{\text{max.}}} = \left|\frac{\varkappa_1' - \varkappa_2'}{\varkappa_1' + \varkappa_2'}\right|$$

zu setzen.[1]) Durch Angabe der Gesamtintensität

$$\varkappa = \varkappa_1' + \varkappa_2'$$

1) Dies Maß hat schon Chr. Jensen (Beiträge zur Photometrie des Himmels, Diss. Kiel 1898; Metereol. Zeitschr. **36.** p. 545. 1901) benutzt.

und der Polarisation ist dann ein Strahlenbündel vollkommen definiert. Seine Entropie beträgt

$$E = \sigma\left\{L\left(\varkappa\frac{1+p}{2}\right) + L\left(\varkappa\frac{1-p}{2}\right)\right\}, \tag{31}$$

da

$$\varkappa_1' = \varkappa\frac{1+p}{2}, \quad \varkappa_2' = \varkappa\frac{1-p}{2}$$

oder umgekehrt ist.

Eine Ausnahmestellung nimmt elliptisch polarisiertes Licht ein, welches stets vollkommen polarisiert ist, so daß es sich in keine zwei inkohärente Hauptschwingungen zerlegen läßt; ihm ist der Wert $p = 1$ zuzuschreiben, welcher sich aus (30) für vollkommene lineare Polarisation ergibt. Seine Entropie berechnet sich dann richtig nach (31) zu

$$E = \sigma L(\varkappa).$$

Dritter Teil.

Die Entropie von drei partiell kohärenten Strahlenbündeln.

§ 7. Die Kohärenzbeziehungen zwischen drei partiell kohärenten Strahlenbündeln.

Am Ende von § 4 lernten wir ein Beispiel kennen, in welchem zwischen den Kohärenzen i_{fg}, i_{gh} und i_{hf} eine Gleichung bestand; dies war aber nur durch die Annahme vollkommener Inkohärenz zwischen den Schwingungen f_g' und h_g' bedingt. Jetzt wollen wir die Frage nach dem Zusammenhang dieser drei Kohärenzen in voller Allgemeinheit lösen.

Wir gehen zu diesem Zweck wiederum von den Gleichungen

$$\begin{cases} f = f_g + f_g' \\ h = h_g + h_g' \end{cases} \tag{32}$$

aus, in denen die Schwingungen f_g' und h_g' zu g vollständig inkohärent, dagegen f_g und h_g zu g vollkommen kohärent sein sollen. Die letzteren sind daher auch zueinander vollkommen kohärent, d. h. es besteht eine Gleichung

$$h_g = \lambda(f_g \cos a + f_g^* \sin a), \tag{33}$$

während die Schwingungen f_g' und h_g' untereinander noch alle Grade der Kohärenz, von 0 bis 1, haben können. Wir werden im allgemeinen

$$h_g' = (h_g')_{f_g'} + (h_g')'_{f_g'} \tag{34}$$

zu setzen haben, wo

$$(35) \qquad (h_g')_{f_g'} = \mu (f_g' \cos b + f_g'^* \sin b)$$

zu f_g' vollkommen kohärent, während $(h_g)'_{f_g'}$ zu f_g' absolut inkohärent ist.

Aus den Gleichungen (32) folgt aber

$$(36) \qquad \begin{cases} \overline{fh} = \overline{f_g h_g} + \overline{f_g' h_g'} \\ \overline{f^* h} = \overline{f_g^* h_g} + \overline{f_g'^* h_g'}, \end{cases}$$

und aus (33), (34), (35):

$$\overline{f_g h_g} = \lambda \cos a \overline{f_g^2} = \cos a \sqrt{\overline{f_g^2} . \overline{h_g^2}},$$

$$\overline{f_g^* h_g} = \lambda \sin a \overline{f_g^2} = \sin a \sqrt{\overline{f_g^2} . \overline{h_g^2}},$$

$$\overline{f_g' h_g'} = \mu \cos b \overline{f_g'^2} = \cos b \sqrt{\overline{f_g'^2} . \overline{(h_g')^2_{f_g'}}},$$

$$\overline{f_g'^* h_g'} = \mu \sin b \overline{f_g'^2} = \sin b \sqrt{\overline{f_g'^2} . \overline{(h_g')^2_{f_g'}}}.$$

Nach Einsetzung dieser Werte in (36) ergibt sich im Hinblick auf (20) und (21)

$$i_{fh} = \frac{\overline{fh}^2 + \overline{f^* h}^2}{\overline{f^2} . \overline{h^2}} = \frac{\overline{f_g^2}}{\overline{f^2}} \cdot \frac{\overline{h_g^2}}{\overline{h^2}} + \frac{\overline{(h_g)^2_{f'g}}}{\overline{h_g'^2}} \cdot \frac{\overline{f_g'^2}}{\overline{f^2}} \cdot \frac{\overline{h_g'^2}}{\overline{h^2}}$$

$$+ 2 \cos (a - b) \sqrt{\frac{\overline{(h_g')^2_{f'g}}}{\overline{h_g'^2}} \cdot \frac{\overline{f_g^2}}{\overline{f^2}} \cdot \frac{\overline{h_g^2}}{\overline{h^2}} \cdot \frac{\overline{f_g'^2}}{\overline{f^2}} \cdot \frac{\overline{h_g'^2}}{\overline{h^2}}}.$$

$$(37) \quad i_{fh} = i_{fg} i_{gh} + i_{f_g' h_g'} j_{fg} j_{gh} + 2 \cos (a - b) \sqrt{i_{f_g' h_g'} i_{fg} i_{gh} j_{fg} j_{gh}}.$$

Zwischen den drei Kohärenzen i_{fh}, i_{fg} und i_{gh} besteht also im allgemeinen keine Gleichung, sondern nur eine Ungleichung; da nämlich sowohl $i_{f_g' h_g'}$ als $\cos(a - b)$ höchstens gleich 1 sind, so ist

$$(38) \qquad i_{fh} \leqq \left(\sqrt{i_{fg} i_{gh}} + \sqrt{j_{fg} j_{gh}}\right)^2.$$

Man überzeugt sich leicht, daß der rechtsstehende Ausdruck höchstens den Wert 1 erreichen kann, indem man $i_{fg} = \cos^2 \alpha$, $i_{gh} = \cos^2 \beta$ setzt; diese Ungleichung beschränkt i_{fh} also stets auf einen Teil des ihm an sich zur Verfügung stehenden Bereiches von 0 bis 1.

Nach (37) kann man offenbar, abgesehen von der Einschränkung durch (38), die vier Kohärenzen i_{fh}, i_{fg}, i_{gh} und $i_{f_g' h_g'}$ willkürlich bestimmen. Es könnte zunächst scheinen, als wären die Kohärenzbeziehungen zwischen drei Schwingungen sogar erst durch sechs Größen, außer den genannten nämlich durch $i_{f'_h g'_h}$ und $i_{g'_f h'_f}$, vollsländig festgelegt. Doch bestehen zwischen diesen Größen zwei stets indentisch erfüllte Gleichungen, so daß nun vier von ihnen unabhängig sind.

Betrachten wir nämlich die schon in (32) und (34) angewandte Zerlegung:

$$h = h_g + h_g',$$
$$h_g' = (h_g')_{f_g'} + (h_g)'_{f_g'}.$$

Die Summe

$$h_g + (h_g')_{f_g'}$$

läßt sich hier nach (34) und (35) als homogene lineare Funktion von g, g^*, f_g' und $f_g'^*$ darstellen; oder wenn man vermöge der Gleichungen

$$f = f_g + f_g',$$
$$f^* = f_g^* + f_g'^*,$$

f_g' und $f_g'^*$ eliminiert, als Funktion von f, f^*, g und g^*, oder, was wiederum dasselbe sagt, als homogene lineare Funktion von f, f^*, g_f' und $g_f'^*$. Im Gegensatz hierzu ist $(h_g)'_{f_g'}$ nach Definition inkohärent zu f_g', aber als Differenz den zu g inkohärenten Schwingungen h_g' und $(h_g')_{f_g'}$ auch inkohärent zu g; daher auch inkohärent zu f_g und zu f, und ebenso zu g_f und g_f', der Differenz von g und g_f.

Wir haben demnach eine Zerlegung

$$h = \alpha_1 g_f' + \beta_1 g_f'^* + \gamma_1 f + \delta_1 f^* + h_1$$

gefunden, wo h_1 zu g_f' und f inkohärent ist. Es kann nur *eine* solche Zerlegung geben. Setzen wir nämlich noch eine andere Zerlegung

$$h = \alpha_2 g_f' + \beta_2 g_f'^* + \gamma_2 f + \delta_2 f^* + h_2$$

an, wo h_2 wiederum zu g_f' und f inkohärent ist, so wäre durch die Relation

$$0 = (\alpha_1 - \alpha_2) g_f' + (\beta_1 - \beta_2) g_f'^* + (\gamma_1 - \gamma_2) f + (\delta_1 - \delta_2) f^* + (h_1 - h_2)$$

die Differenz $h_1 - h_2$ durch g_f' und f ausgedrückt, während sie doch zu beiden inkohärent sein muß. Dieser Widerspruch zwingt zu dem Schluß, daß die Koeffizienten der letzten Gleichung identisch Null sind, daß also $\alpha_1 = \alpha_2$ etc. und $h_1 = h_2$ ist.

Dennoch finden wir eine zweite solche Zerlegung, wenn wir von den Gleichungen

$$h = h_f + h_f'$$

$$h_f' = (h_f')_{g_f'} + (h_f')'_{g_f'}$$

ausgehen. Denn hier ist

$$h_f + (h_f')_{g_f'}$$

eine homogene lineare Funktion von f, f^*, g_f' und $g_f'^*$, dagegen $(h_f')'_{g_f'}$ nach Definition zu g_f' und als Differenz von h_f' und $(h_f')_{g_f'}$ zu f inkohärent. Daher muß

$$h_g + (h_g')_{f_g'} = h_f + (h_f')_{g_f'}$$

$$(h_g')'_{f_g'} = (h_f')'_{g_f'}$$

sein. Und da nach (21)

$$\text{(39a)} \qquad \overline{(h_g')'^2_{f_g'}} = j_{f_g' h_g'} \overline{h_g'^2} = j_{f_g' h_g'} j_{gh} \overline{h^2}$$

ist, folgt aus der letzteren Beziehung die Gleichung

$$j_{f_g' h_g'} j_{gh} = j_{g_f' h_f'} j_{fh},$$

welche sich wegen der Gleichberechtigung der Schwingungen f, g, h unmittelbar zu der Doppelgleichung

$$\text{(39)} \qquad \frac{j_{f_h' g_h'}}{j_{fg}} = \frac{j_{g_f' h_f'}}{j_{gh}} = \frac{j_{h_g' f_g'}}{j_{hf}}$$

erweitern läßt. Der Symmetrie halber führen wir als vierte unabhängige Variable zur Charakterisierung der Kohärenzverhältnisse neben den drei Inkohärenzen „zweiter Ordnung" j_{fg}, j_{gh}, j_{hf} die „*Inkohärenz dritter Ordnung*"

$$\text{(40)} \qquad J = j_{fg} j_{gh} j_{h_g' f_g'} = j_{fg} j_{g_f' h_f'} j_{hf} = j_{f_h' g_h'} j_{gh} j_{hf}$$

ein. Sie ist ein positiver echter Bruch, welcher Null wird, wenn auch nur zwei der Schwingungen vollkommen kohärent sind, und welcher für drei vollständig inkohärente Schwingungen den Wert 1 annimmt. Denn setzt man in (37)

$$i_{fg} = i_{gh} = i_{hf} = 0,$$

so folgt

$$i_{f_g' h_g'} = 0, \quad j_{f_g' h_g'} = 1$$

und aus (40)

$$J = 1.$$

Ohne der Frage nach der einfachsten Art, J zu bestimmen, näher zu treten, wollen wir wenigstens zeigen, daß es prinzipiell meßbar ist. Kennen wir nämlich die Inkohärenz j_{fg}, so wissen wir, in welchem Stärkeverhältnis wir f und g zu superponieren haben, um f_g' aus f zu isolieren; die dazu nötige Phasendifferenz läßt sich durch Probieren bestimmen. Ebenso können wir auch h_g' isolieren, wenn j_{gh} bekannt ist, und dann die Inkohärenz $j_{f_g' h_g'}$, aus welchen sich nach (40) J berechnen läßt, nach der in § 4 erwähnten Methode messen.

Die Zahl der unabhängigen Variablen, welche (bei gleichen Werten von σ) ein System von drei partiell kohärenten Strahlenbündeln bestimmen, beträgt 7; es sind dies nämlich die drei reduzierten spezifischen Intensitäten $\varkappa_1$, $\varkappa_2$, $\varkappa_3$, die drei Inkohärenzen zweiter Ordnung j_{12}, j_{23}, j_{31} und die Inkohärenz dritter Ordnung J. Um die Anzahl der Invarianten zu ermitteln, von welchen die Entropie eines solchen Systems abhängt, müssen wir nun zunächst fragen, wieviel Freiheitsgrade wir besitzen, wenn wir es auf umkehrbarem Wege in ein anderes verwandeln.

§ 8. Die Anzahl der Freiheitsgrade.

Ein optischer Vorgang, welcher drei Schwingungen ψ, φ, χ in drei andere, f, g, h, überführt, ist dem Superpositionsprinzip gemäß stets durch Gleichungen von der Form

$$(41)\quad \left\{\begin{aligned} f &= a_1(\varphi\cos\alpha_1 + \varphi^*\sin\alpha_1) + b_1(\psi\cos\beta_1 + \psi^*\sin\beta_1) \\ &\qquad + c_1(\chi\cos\gamma_1 + \chi^*\sin\gamma_1), \\ h &= a_2(\varphi\cos\alpha_2 + \varphi^*\sin\alpha_2) + b_2(\psi\cos\beta_2 + \psi^*\sin\beta_2) \\ &\qquad + c_2(\chi\cos\gamma_2 + \chi^*\sin\gamma_2), \\ g &= a_3(\varphi\cos\alpha_3 + \varphi^*\sin\alpha_3) + b_3(\psi\cos\beta_3 + \psi^*\sin\beta_3) \\ &\qquad + c_3(\chi\cos\gamma_3 + \chi^*\sin\gamma_3) \end{aligned}\right.$$

dargestellt. Verläuft er ohne Absorption und macht er aus drei Strahlenbündeln von gleichem Werte von σ drei von *demselben* Wert σ (d. h. gilt der Sinussatz der geometrischen Optik), so muß die gesamte Energie $\sigma\Sigma\varkappa$, also auch $\Sigma\varkappa$ ungeändert bleiben, d. h. es muß

$$(42)\qquad \overline{f^2} + \overline{g^2} + \overline{h^2} = \overline{\varphi^2} + \overline{\psi^2} + \overline{\chi^2}$$

sein. Für die Momentanwerte f^2 etc. läßt sich eine solche Gleichung nicht aufstellen. Bei gemeinsamer Spiegelung und

Brechung von zwei Strahlenbündeln an einer planparallelen Platte z. B. wäre

$$(f^2+g^2)-(\varphi^2+\psi^2)$$

proportional zur Abnahme der in der Platte befindlichen Energie, also nur im zeitlichen Mittel gleich Null. Nach (41), sowie nach (7), (10), (11) und (12) ist aber

$$\overline{f^2}+\overline{g^2}+\overline{h^2}=(a_1^2+b_1^2+c_1^2)\overline{\varphi^2}+\dots$$
$$+2\left(a_1 b_1 \cos(\alpha_1-\beta_1)+a_2 b_2 \cos(\alpha_2-\beta_2)+a_3 b_3 \cos(\alpha_3-\beta_3)\right)\overline{\varphi\psi}+\dots$$
$$+2\left(a_1 b_1 \sin(\alpha_1-\beta_1)+a_2 b_2 \sin(\alpha_2-\beta_2)+a_3 b_3 \sin(\alpha_3-\beta_3)\right)\overline{\varphi\psi^*}+\dots$$

Gleichung (42) fordert daher die Gültigkeit der folgenden neun Identitäten:

$$(43)\quad\left\{\begin{aligned}
&a_1^2+b_1^2+c_1^2=1,\\
&a_2^2+b_2^2+c_2^2=1,\\
&a_3^2+b_3^2+c_3^2=1,\\
&a_1 b_1 \cos(\alpha_1-\beta_1)+a_2 b_2 \cos(\alpha_2-\beta_2)+a_3 b_3 \cos(\alpha_3-\beta_3)=0,\\
&b_1 c_1 \cos(\beta_1-\gamma_1)+b_2 c_2 \cos(\beta_2-\gamma_2)+b_3 c_3 \cos(\beta_3-\gamma_3)=0,\\
&c_1 a_1 \cos(\gamma_1-\alpha_1)+c_2 a_2 \cos(\gamma_2-\alpha_2)+c_3 a_3 \cos(\gamma_3-\alpha_3)=0,\\
&a_1 b_1 \sin(\alpha_1-\beta_1)+a_2 b_2 \sin(\alpha_2-\beta_2)+a_3 b_3 \sin(\alpha_3-\beta_3)=0,\\
&b_1 c_1 \sin(\beta_1-\gamma_1)+b_2 c_2 \sin(\beta_2-\gamma_2)+b_3 c_3 \sin(\beta_3-\gamma_3)=0,\\
&c_1 a_1 \sin(\gamma_1-\alpha_1)+c_2 a_2 \sin(\gamma_2-\alpha_2)+c_3 a_3 \sin(\gamma_3-\alpha_3)=0.
\end{aligned}\right.$$

Von den 18 Parametern $a_1, a_2 \dots \alpha_1, \alpha_2 \dots$ in (41) sind demnach neun willkürlich zu wählen. Doch entspricht nicht jedem von diesen ein Freiheitsgrad zur Veränderung der Intensitäten und Kohärenzgrößen. Vielmehr kann man aus den Gleichungen (41) auch drei Beziehungen zwischen den Funktionen

$$\begin{aligned}
F&=f\cos\lambda+f^*\sin\lambda, & \Phi&=\varphi\cos\lambda'+\varphi^*\sin\lambda',\\
G&=g\cos\mu+g^*\sin\mu, & \Psi&=\psi\cos\mu'+\psi^*\sin\mu',\\
H&=h\cos\nu+h^*\sin\nu, & X&=\chi\cos\nu'+\chi^*\sin\nu'
\end{aligned}$$

ableiten, welche lauten:

$$\begin{aligned}
F=a_1&\left[\Phi\cos(\alpha_1+\lambda-\lambda')+\Phi^*\sin(\alpha_1+\lambda-\lambda')\right]\\
&+b_1\left[\Psi\cos(\beta_1+\lambda-\mu')+\Psi^*\sin(\beta_1+\lambda-\mu')\right]\\
&+c_1\left[X\cos(\gamma_1+\lambda-\nu')+X^*\sin(\gamma_1+\lambda-\nu')\right]\ \text{etc.}
\end{aligned}$$

Hier treten die Differenzen der sechs Phasenverschiebungen λ, μ, ν, λ', μ', ν' auf, von denen fünf voneinander unabhängig sind; um die Gleichungen (41) zu vereinfachen, können wir beispielsweise

$$\alpha_1 + \lambda - \lambda' = 0, \qquad \beta_1 + \lambda - \mu' = 0,$$
$$\beta_2 + \mu - \mu' = 0, \qquad \gamma_2 + \mu - \nu' = 0,$$
$$\gamma_3 + \nu - \nu' = 0$$

setzen. Die Anzahl der Gleichungen (43) bleibt davon unberührt, wie man leicht sieht. Nun haben aber die Schwingungen F, G, H, Φ, Ψ, X dieselben Intensitäten wie f, g, h, φ, ψ, χ; sie stehen auch in denselben Kohärenzbeziehungen zueinander, wie die letzteren. Zur Änderung der Intensitäten und Kohärenzbeziehungen stehen uns also nur vier Freiheitsgrade zur Verfügung.

Dieselbe Überlegung für n Strahlenbündel ergibt $2n^2$ Parameter in den n zu (41) analogen Gleichungen, dagegen n^2 Beziehungen, welche den Identitäten (43) entsprechen; für bloße Phasenveränderungen werden dabei $2n-1$ Freiheitsgrade verbraucht. Zur Veränderung der Intensitäten und Kohärenzbeziehungen bleiben demnach

$$2n^2 - n^2 - (2n-1) = (n-1)^2$$

Freiheitsgrade übrig; für $n = 2$ nur einer, in Übereinstimmung mit § 5, welcher uns für die drei Variabelen $\varkappa_1$, $\varkappa_2$, j_{12} zwei Invarianten kennen lehrte.

Dies Ergebnis hat aber nur dann für uns Bedeutung, wenn ein Vorgang von der betrachteten Art in der Natur möglich ist, und wenn wir sicher sind, daß er umkehrbar verläuft. Denken wir uns nun ein System von drei partiell kohärenten Strahlenbündeln nacheinander drei gemeinsamen Spiegelungen und Brechungen von je zwei Strahlenbündeln unterworfen, welche durch die Gleichungen (vgl. (26) und (27))

$$(44)\quad \begin{cases} \varphi_1 = \delta\varphi + \varrho\psi, & \\ \psi_1 = -\varrho\varphi + \delta\psi, & \chi_1 = \chi, \\ \psi_2 = \varepsilon\psi_1 + \sigma\chi_1, & \\ \chi_2 = -\sigma\psi_1 + \varepsilon\chi_1, & \varphi_2 = \varphi_1, \\ f = \zeta\chi_2 + \tau\varphi_2, & \\ g = -\tau\chi_2 + \zeta\varphi_2, & h = \psi_2, \\ \delta^2 + \varrho^2 = \varepsilon^2 + \sigma^2 = \zeta^2 + \tau^2 = 1 & \end{cases}$$

dargestellt sind. Deuten wir φ, ψ, χ; $\varphi_1, \psi_1, \chi_1$; $\varphi_2, \psi_2, \chi_2$; f, g, h als rechtwinklige Koordinaten im Raum, so stellt jeder dieser drei Vorgänge die Drehung des Koordinatenkreuzes um eine seiner Achsen dar. Bekanntlich läßt sich jede Drehung um den Anfangspunkt durch drei solche Drehungen ersetzen, während eine vierte zu keinem Ergebnis führt, das nicht auch schon mit dreien zu erhalten wäre. Der durch (44) dargestellte Vorgang gibt uns also drei Freiheitsgrade in die Hand.

Offenbar erhalten wir aber einen vierten, wenn z. B. nicht $\varphi_2 = \varphi_1$, sondern

$$\varphi_2 = \varphi_1 \cos \alpha + \varphi_1^* \sin \alpha$$

ist. Daher läßt sich jeder absorptionslose, mit dem Sinussatz verträgliche Vorgang, welcher ein System von drei partiell kohärenten Strahlenbündeln in ein anderes umwandelt, durch drei gemeinsame Spiegelungen und Brechungen von je zwei Strahlenbündeln ersetzen. Dieser Satz ist auch auf mehr als drei Strahlenbündel zu übertragen (nur ist dann die Zahl der Spiegelungen und Brechungen größer) und beweist, daß Absorptionslosigkeit und Gültigkeit des Sinussatzes hinreichende Bedingungen für die Umkehrbarkeit eines optischen Vorganges sind.

§ 9. Die Entropie.

Nach § 7 haben wir sieben unabhängige Variabele, nach § 8 vier Freiheitsgrade. Es muß also drei Invarianten geben, welche die einzigen Kombinationen darstellen, in denen $\varkappa_1, \varkappa_2, \varkappa_3$, j_{12}, j_{23}, j_{31} und J in der Formel für die Entropie auftreten können. Da wir jede mögliche Umwandlung unseres Systems durch eine Reihe gemeinsamer Spiegelungen und Brechungen von zwei Strahlenbündeln ersetzen können, und die drei Invarianten naturgemäß bezüglich der Indizes 1, 2, 3 symmetrisch sind, genügt für jede Invariante der Nachweis der Unveränderlichkeit bei gemeinsamer Spiegelung und Brechung der Strahlenbündel 1 und 2.

Bei diesem Vorgang sind aber nach § 5 unveränderlich die Größen

$$\varkappa_1 + \varkappa_2 \tag{45}$$

und

$$j_{12} \varkappa_1 \varkappa_2 ; \tag{46}$$

ferner die Intensität $\varkappa_3$ des unverändert bleibenden Strahlenbündels 3, sowie die Größe

(47) $$j_{13}'\, j_{23}\, \varkappa_3.$$ [1])

Setzen wir nämlich wieder

$$\varkappa_1 = \overline{f^2}, \quad \varkappa_2 = \overline{g^2}, \quad \varkappa_3 = \overline{h^2},$$

so ist nach (39a)

$$j_{13}'\, j_{23}\, \varkappa_3 = j_{f_g' h_g'}\, j_{g\,h}\, \overline{h^2} = \overline{(h_g')'^2_{f_g'}}\,;$$

und $(h_g')'_{f_g'}$ ist nach § 7 der zu f und g inkohärente Anteil an h, während

$$h - (h_g')'_{f_g'} = h_g + (h_g')_{f_g'}$$

eine lineare Kombination von f und g darstellt; also auch eine lineare Kombination der Schwingungen, in welche f und g bei der Spiegelung und Brechung übergehen, während $(h_g')'_{f_g'}$ auch zu den letzteren inkohärent ist. Eine derartige Zerlegung ist aber (ebenfalls nach § 7) eindeutig; daher bleibt $(h_g')'_{f_g'}$ bei Spiegelung und Brechung unverändert, und ebenso der Ausdruck

$$j_{13}'\, j_{12}\, \varkappa_3.$$

Schließlich behält dabei noch der Ausdruck

(48) $$\begin{cases} j_{23}\, \varkappa_2\, \varkappa_3 + j_{31}\, \varkappa_3\, \varkappa_1 = \overline{g^2} \,.\, \overline{h^2} - (\overline{g\,h}^2 + \overline{g\,h^*}^2) \\ \qquad\qquad + \overline{f^2} \,.\, \overline{h^2} - (\overline{f\,h}^2 + \overline{f\,h^*}^2) \end{cases}$$

seinen Wert. Denn setzt man nach (26) und (27)

$$f = \delta\varphi + \varrho\psi, \qquad h = \chi,$$
$$g = -\varrho\varphi + \delta\psi,$$
$$\delta^2 + \varrho^2 = 1,$$

so erhält man:

$$\overline{f^2} = \delta^2\overline{\varphi^2} + \varrho^2\overline{\psi^2} + 2\delta\varrho\,\overline{\varphi\psi}, \qquad \overline{g^2} = \varrho^2\overline{\varphi^2} + \delta^2\overline{\psi^2} - 2\delta\varrho\,\overline{\varphi\psi},$$

$$\overline{f\,h} = \delta\,\overline{\varphi\chi} + \varrho\,\overline{\psi\chi}, \qquad \overline{g\,h} = -\varrho\,\overline{\varphi\chi} + \delta\,\overline{\psi\chi},$$

$$\overline{f\,h^*} = \delta\,\overline{\psi\chi^*} + \varrho\,\overline{\psi\chi^*}, \qquad \overline{g\,h^*} = -\varrho\,\overline{\varphi\chi^*} + \delta\,\overline{\psi\chi^*},$$

$$\begin{aligned} \overline{f^2}\,.\,\overline{h^2} - (\overline{f\,h}^2 + \overline{f\,h^*}^2) &= \delta^2\,[\overline{\varphi^2}\,.\,\overline{\chi^2} - (\overline{\varphi\chi}^2 + \overline{\varphi\chi^*}^2)] \\ &+ \varrho^2\,[\overline{\psi^2}\,.\,\overline{\chi^2} - (\overline{\psi\chi}^2 + \overline{\psi\chi^*}^2)] \\ &+ 2\,\delta\,\varrho\,[\overline{\varphi\psi}\,.\,\overline{\chi^2} - (\overline{\varphi\chi}\,.\,\overline{\psi\chi} + \overline{\varphi\chi^*}\,.\,\overline{\psi\chi^*})], \end{aligned}$$

---

1) j_{13}' soll die früher als $j_{f_g' h_g'}$ bezeichnete Größe sein.

$$\overline{g^2} \cdot \overline{h^2} - (\overline{g h}^2 + \overline{g h^*}^2) = \varrho^2 [\overline{\varphi^2} \cdot \overline{\chi^2} - (\overline{\varphi \chi}^2 + \overline{\varphi \chi^*}^2)]$$
$$+ \delta^2 [\overline{\psi^2} \cdot \overline{\chi^2} - (\overline{\psi \chi}^2 + \overline{\psi \chi^*}^2)]$$
$$- 2 \delta \varrho [\overline{\varphi \psi} \cdot \overline{\chi^2} - (\overline{\varphi \chi} \cdot \overline{\psi \chi} + \overline{\varphi \chi^*} \cdot \overline{\psi \chi^*})] .$$

Daraus folgt durch Addition die zu beweisende Gleichung:

$$\overline{g^2} \cdot \overline{h^2} - (\overline{g h}^2 + \overline{g h^*}^2) + \overline{f^2} \cdot \overline{h^2} - (\overline{f h}^2 + \overline{f h^*}^2)$$
$$= \overline{\psi^2} \cdot \overline{\chi^2} - (\overline{\psi \chi}^2 + \overline{\psi \chi^*}^2) + \overline{\varphi^2} \cdot \overline{\chi^2} - (\overline{\varphi \chi}^2 + \overline{\varphi \chi^*}^2).$$

Gelingt es uns jetzt, aus den Ausdrücken (45), (46), (47), (48) und $\varkappa_3$ drei Kombinationen zu bilden, die sich durch Vertauschung der Indizes 1, 2, 3 nicht ändern, so sind dies die gesuchten Invarianten. Addieren wir nun zu (45) $\varkappa_3$, so erhalten wir die erste Größe

$$I' = \varkappa_1 + \varkappa_2 + \varkappa_3 ,$$

auf welche dies Kriterium zutrifft; addieren wir ferner (46) und (48), so finden wir die zweite dieser Art, nämlich

$$I'' = j_{12} \varkappa_1 \varkappa_2 + j_{23} \varkappa_2 \varkappa_3 + j_{31} \varkappa_3 \varkappa_1 ;$$

und multiplizieren wir schließlich (46) und (47), so finden wir in Rücksicht auf Gleichung (40), nach welcher

$$J = j_{12} j_{23} j'_{13}$$

in den Indizes 1, 2, 3 symmetrisch ist, als dritte Invariante

$$I''' = J \varkappa_1 \varkappa_2 \varkappa_3 .$$

Für drei inkohärente Strahlenbündel ist nun nach §§ 4 und 7

$$j_{12} = j_{23} = j_{31} = J = 1.$$

Läßt sich unser System auf umkehrbarem Wege auf solche zurückführen, so bestimmen sich ihre Intensitäten K_1, K_2, K_3 aus den Gleichungen:

$$K_1 + K_2 + K_3 = I',$$
$$K_1 K_2 + K_2 K_3 + K_3 K_1 = I'',$$
$$K_1 K_2 K_3 = I'''.$$

Daraus folgt:

Die Entropie eines Systems von drei partiell kohärenten Strahlenbündeln beträgt:

(49) $$E = \sigma [L(K_1) + L(K_2) + L(K_3)] ,$$

wobei K_1, K_2, K_3 *die Wurzeln der kubischen Gleichung*

$$K^3 - I' K^2 + I'' K - I''' = 0 \tag{50}$$

sind, und die Koeffizienten I', I'', I''' *die Werte*

$$\begin{cases} I' = \varkappa_1 + \varkappa_2 + \varkappa_3 , \\ I'' = j_{12} \varkappa_1 \varkappa_2 + j_{23} \varkappa_2 \varkappa_3 + j_{31} \varkappa_3 \varkappa_1 , \\ I''' = J \varkappa_1 \varkappa_2 \varkappa_3 \end{cases} \tag{51}$$

haben. Dabei sind, um es zu wiederholen, die $\varkappa$ die reduzierten spezifischen Intensitäten der Strahlenbündel, die j die in § 4 definierten Inkohärenzen zweiter und J die Inkohärenz dritter Ordnung.

Ist $\varkappa_3 = 0$, so verschwindet I''', also auch eine von den drei Wurzeln von (50); da sich zugleich I'' auf $j_{12} \varkappa_1 \varkappa_2$ reduziert, findet man unmittelbar die Formel (28) für die Entropie zweier partiell kohärenter Strahlen wieder. Ist $J = 0$, so verschwindet ebenfalls I''' und eine der Wurzeln. Das System läßt sich dann auf *zwei* inkohärente Strahlenbündel zurückführen. Hierher gehört auch der Fall, daß von den drei Strahlenbündeln zwei vollkommen kohärent sind. Dann ist eine der Größen j, etwa j_{12}, gleich Null, nach (40) also auch J, während aus (37), indem man die Indizes f, g, h mit 1, 2, 3 vertauscht, $j_{13} = j_{23}$ folgt, so daß

$$I'' = j_{13} (\varkappa_1 + \varkappa_2) \varkappa_3$$

wird. Man sieht hieraus, daß zwei vollkommen kohärente Strahlenbündel einem einzigen thermodynamisch äquivalent sind, dessen spezifische Intensität die Summe ihrer spezifischen Intensitäten ist; denn $\varkappa_1$ und $\varkappa_2$ treten dann nur noch in der Verbindung $\varkappa_1 + \varkappa_2$ auf. Sind endlich alle drei Strahlenbündel vollkommen kohärent, so verschwinden alle Inkohärenzen, mit ihnen die Inverianten I'' und I''' und zwei der Wurzeln von Gleichung (50), während die dritte gleich I' wird. Die Entropie eines solchen Systems ist also nach (49) und (51)

$$E = \sigma L (\varkappa_1 + \varkappa_2 + \varkappa_3) .$$

Schon an diesen Beispielen und ebenso an dem trivialen Fall dreier inkohärenter Strahlenbündel, in welchem $\varkappa_1$, $\varkappa_2$, $\varkappa_3$ selbst die Wurzeln der Gleichung (50) sind, sieht man, daß diese reelle positive Wurzeln haben kann. Negative reelle

Wurzeln sind auch schon allein dadurch ausgeschlossen, daß ihre Koeffizienten abwechselnd positiv und negativ sind. Dagegen ist mir der Beweis nicht gelungen, daß komplexe Wurzeln unmöglich sind. Träten solche einmal auf, so ließe sich das gegebene System auf keinem umkehrbaren Wege in drei vollkommen inkohärente Strahlenbündel verwandeln, und der Beweis von Formel (49) würde insofern hinfällig, als die Art der Abhängigkeit der Entropie von den Invarianten I', I'', I''' unbestimmt bliebe, obwohl diese Formel auch dann noch einen reellen Wert lieferte.

Ebenso lassen wir es zweifelhaft, wie sich unser Ergebnis auf mehr als drei Strahlenbündel überträgt. Die erste Invariante hat dann natürlich den Wert

$$I' = \sum^{l} \varkappa_l ,$$

und die zweite

$$I'' = \frac{1}{2!} \sum^{l} \sum^{m} I_{lm} \varkappa_l \varkappa_m . \quad l \pm m .$$

Die Doppelsumme hier setzt sich nämlich zusammen aus dem Glied

$$j_{12} \varkappa_1 \varkappa_2 ,$$

ferner aus Termen, die wie

$$j_{34} \varkappa_3 \varkappa_4$$

die Indizes 1 und 2 nicht enthalten, und aus Ausdrücken von der Form

$$j_{1l} \varkappa_1 \varkappa_l + j_{2l} \varkappa_2 \varkappa_l ;$$

alle diese Summanden bleiben aber bei gemeinsamer Spiegelung und Brechung der Strahlenbündel 1 und 2 unverändert Der Faktor $\frac{1}{2!}$ ist hinzugesetzt, weil jeder Summand $j_{lm} \varkappa_l \varkappa_m$ auch noch in der Form $j_{ml} \varkappa_m \varkappa_l$ auftritt.

Man kann vermuten, und die Betrachtung am Schluß der nächsten Paragraphen stützt diese Annahme, daß

$$I''' = \frac{1}{3!} \sum^{l} \sum^{m} \sum^{n} J_{lmn} \varkappa_l \varkappa_m \varkappa_n ,$$

$$I'''' = \frac{1}{4!} \sum^{l} \sum^{m} \sum^{n} \sum^{o} J_{lmno} \varkappa_l \varkappa_m \varkappa_n \varkappa_o \text{ etc.}$$

wird, wo J_{lmno} eine „Inkohärenz vierter Ordnung“ ist; bei allen diesen Summenbildungen sind nur voneinander verschiedene Werte der Indizes l, m, n, o zuzulassen. Die Entropie wäre dann

$$E = \sigma \sum^{l} L(K_l),$$

wo die K_l die n Wurzeln der Gleichung

$$K^n - I' K^{n-1} + I'' K^{n-2} - \ldots \pm I^{(n)} = 0$$

bedeuteten. Doch scheinen mir auf diesem Gebiete noch manche Schwierigkeiten zu liegen.

Allgemeine Folgerungen.

In § 4 wurde gezeigt, daß die isotherme Spiegelung, d. h. die Reflexion eines Strahlenbündels an einem Körper von gleicher Temperatur, die Kohärenz des ersteren zu einem anderen Strahlenbündel herabsetzt; und in § 5 wurde abgeleitet, daß mit abnehmender Kohärenz die Entropie zunimmt. Demnach ist die isotherme Spiegelung, welche als umkehrbar zu bezeichnen ist, wenn es sich nur um den reflektierenden Körper und das gespiegelte Strahlenbündel handelt, irreversibel, sobald man noch ein ihm partiell kohärentes Strahlenbündel mit in Betracht zieht. Der eigentliche Grund dafür liegt aber nicht in dem Vorgang an der Grenzfläche, sondern allein in der Absorption und Zerstreuung des in den Körper eindringenden Strahlenbündels; denn dies büßt dabei zwar nicht seine Intensität, wohl aber die Kohärenz zu dem genannten ein.

Ferner ist in einem von vollkommen spiegelnden Wänden eingeschlossenen, absorbierende und diathermane Körper enthaltenden Hohlraum Gleichheit der Temperatur aller Körper und Strahlenbündel noch nicht hinreichende Bedingung für den Zustand maximaler Entropie. Vielmehr ist es denkbar, daß trotz des vollzogenen Temperaturausgleiches noch partielle Kohärenz zwischen einigen Strahlenbündeln besteht. Bei jeder Absorption und Zerstreuung wird diese aber herabgesetzt, während neue Kohärenzen nicht geschaffen werden; denn gleichtemperierte, inkohärente Strahlen liefern nach § 5 auch bei gemeinsamer Spiegelung und Brechung stets wieder inkohärente Strahlen. Daher nähert sich der Zustand mit der Zeit sicher dem durch absolute Inkohärenz gekennzeichneten Entropie-maximum. Die Theorie ist also vollkommen im Recht, wenn sie auf Gleichgewichtszustände das Additionstheorem anwendet. Nur falls der Hohlraum gar keine absorbierenden Substanzen enthält, können sich in ihm — wie auch andere instabile Strahlungszustände[1]) — Kohärenzen dauernd halten.

Eine andere Folgerung betrifft die Fortpflanzung der Strahlung in dispergierenden Körpern. Bei fehlender Absorption lassen sich dabei drei Stadien unterscheiden[2]): Im ersten schreitet die Welle mit Gruppengeschwindigkeit fort und erleidet periodische Formänderungen, indem an die Stelle der Schwingung f

$$f \cos \alpha + f^* \sin \alpha$$

tritt, wobei α zur zurückgelegten Strecke proportional ist; Kohärenzverhältnisse werden dadurch nicht berührt. Im zweiten bewirkt die Dispersion der Gruppengeschwindigkeit eine stärker und stärker werdende Formänderung, welche die Kohärenz natürlich immer mehr herabsetzt. Im dritten Stadium endlich ist die Welle in einen langen Zug angenäherter Sinusschwingungen aufgelöst, deren Periode sich nur langsam ändert. Sinusschwingungen gleicher Periode sind aber stets interferenzfähig, ganz unabhängig von ihrem Ursprung. Man könnte dann also die Strahlung verschiedener Körper zur Interferenz bringen, was sich unmittelbar zur Konstruktion eines Perpetuum mobile zweiter Art verwenden ließe.[3]) Der zweite Hauptsatz fordert daher, daß die Strahlung praktisch vollkommen absorbiert und zerstreut ist, bevor sie das letzte Stadium erreicht.

Jetzt können wir aber zeigen, daß dem Entropieprinzip zufolge nicht einmal das zweite, viel früher eintretende Stadium ohne wesentliche Schwächung erreicht werden darf. Leiten wir nämlich das eine von zwei kohärenten, anfangs im Vakuum befindlichen Strahlenbündeln durch ein dispergierendes Mittel, so würde im zweiten Stadium mit abnehmender Kohärenz die Entropie wachsen. Lassen wir es darauf wieder ins Vakuum austreten, und das andere in demselben Mittel dieselbe Strecke

1) M. Planck, Vorl. über Theorie der Wärmestrahlung, Leipzig 1906. § 51 und § 52.

2) M. Laue, Ann. d. Phys. **18.** p. 523. 1905.

3) M. Laue, Ann. d. Phys. **20.** p. 365. 1906, vgl. § 3.

zurücklegen, so erlitte dies dieselbe Änderung der Schwingungsform, die Kohärenz und mit ihr die Entropie nähme wieder den Anfangswert an. Auch diesem Widerspruch mit dem Entropieprinzip entgeht man nur durch die Annahme starker Absorption. Tatsächlich ist meines Wissens auch noch nie eine Abnahme der Interferenzfähigkeit bei der Fortpflanzung durch dispergierende Mittel beobachtet worden. Selbst bei einem Woodschen Versuch[1]), bei welchem das Licht der gelben Heliumlinie sehr stark anomal dispergierenden Natriumdampf durchsetzte, behielt es seine Kohärenz kollkommen.

Es ist bemerkenswert, daß diese zweite Einschränkung im Gegensatz zu der ersteren sich nicht auf das Prinzip von der Unmöglichkeit des Perpetuum mobile zweiter Art zurückführen läßt, sondern auf dem Boltzmannschen Gedanken des Zusammenhanges zwischen Entropie und Wahrscheinlichkeit und der daraus entspringenden Planckschen Idee der Nichtgültigkeit des Additionstheorems beruht; denn nur auf dieser Grundlage läßt sich überhaupt die Kohärenz thermodynamisch werten.

Zum Schluß wollen wir noch einen Schritt zur Deutung unseres Ergebnisses im Sinne dieser Theorie unternehmen. Sind die Intensitäten $\varkappa_1$ und $\varkappa_2$ zweier Strahlenbündel groß gegen $h\nu^3/c^2$ und ist ihre Inkohärenz j_{12} nicht gerade klein gegen 1, so sind in Gleichung (28) die Argumente der beiden L-Funktionen von derselben Größenordnung wie $\varkappa_1$ und $\varkappa_2$, und man kann Formel (23) durch die einfachere Beziehung

$$(52) \qquad L(\varkappa) = \frac{k\,\nu^2}{c^2}\log\left(\frac{c^2\,\varkappa}{h\,\nu^3}\right), \qquad E = \sigma\,\frac{k\,\nu^2}{c^2}\log\left(\frac{c^2\,\varkappa}{h\,\nu^3}\right)$$

ersetzen. Wir befinden uns dann, da für die Temperatur T hieraus

$$\frac{1}{T} = \frac{\partial L}{\partial \varkappa} = \frac{k^2\,\nu^2}{c^2\,\varkappa}$$

folgt, im Gültigkeitsbereich des Rayleighschen Strahlungsgesetzes[2]); d. h. wir vernachlässigen die Größe des elementaren Wirkungsquantums (Lichtquants). Da die Wahrscheinlich-

---

1) R. W. Wood, Phil. Mag. (6) 8. p. 324. 1904.

2) M. Planck, Theorie der Wärmestrahlung p. 159. Leipzig 1906.

keit $W(\varkappa)$ eines Strahlenbündels mit seiner Entropie durch die universelle Beziehung

$$E = k \log W$$

verknüpft ist, wird dabei

$$W(\varkappa) = \left(\frac{c^2 \varkappa}{h \nu^3}\right)^{\sigma \frac{\nu^2}{c^2}}. \tag{53}$$

Setzen wir nun (52) in (28) ein, so finden wir für die Entropie und die Wahrscheinlichkeit eines Systems von zwei partiell kohärenten Strahlenbündeln nach (53)

$$E = \sigma \frac{k \nu^2}{c^2} \log \left[\left(\frac{c^2}{h \nu^3}\right)^2 j_{12}\, \varkappa_1\, \varkappa_2\right],$$

$$W = \left[\left(\frac{c^2}{h \nu^3}\right)^2 j_{12}\, \varkappa_1\, \varkappa_2\right]^{\sigma \frac{\nu^2}{c^2}} = W(\varkappa_1) \,.\, W(j_{12}\, \varkappa_2).$$

Bei absoluter Inkohärenz wären demnach die Wahrscheinlichkeiten beider Strahlenbündel einfach zu multiplizieren, wie das selbstverständlich ist. Für partiell kohärente Strahlenbündel hat man dagegen gemäß unserer Definition der Inkohärenz j (Gleichung (21)) dabei von der Intensität des einen den Anteil abzuziehen, der von dem zum anderen Strahlenbündel kohärenten Anteil der Schwingung herrührt, und nur ihren dazu inkohärenten Anteil zu berücksichtigen; denn dessen Intensität ist

$$\overline{g_f'^{\,2}} = j_{fg}\, \overline{g^2} = j_{12}\, \varkappa_2 .$$

Bei drei Strahlenbündeln, für welche j_{12}, j_{23}, j_{31} und J nicht gerade klein gegen 1 sein dürfen, geht Formel (49) durch Substitution von (52) gemäß (50) und (51) über in

$$E = \sigma \frac{k \nu^2}{c^2} \log \left[\left(\frac{c^2}{h \nu^3}\right)^3 K_1\, K_2\, K_3\right] = \sigma \frac{k \nu^2}{c^2} \log \left[\left(\frac{c^2}{h \nu^3}\right)^3 J\, \varkappa_1\, \varkappa_2\, \varkappa_3\right].$$

Da nach (40)

$$J = j_{12}\, j_{23}\, j'_{31},$$

so wird nach (53)

$$W = \left[\left(\frac{c^2}{h \nu^3}\right)^3 \varkappa_1 \cdot j_{12}\, \varkappa_2 \cdot j_{23}\, j'_{31}\, \varkappa_3\right]^{\sigma \frac{\nu^2}{c^2}}$$

$$= W(\varkappa_1) \,.\, W(j_{12}\, \varkappa_2) \,.\, W(j_{23}\, j'_{31}\, \varkappa_3).$$

Es ist aber nach (39a)

$$j_{23}\, j'_{31}\, \varkappa_3 = j_{gh}\, j_{h_g' f_g'}\, \overline{h^2} = \overline{(h_g')_{f_g'}^2}$$

derjenige Anteil an der Intensität $\varkappa_3$, welcher allein von dem zu den beiden anderen Strahlenbündeln inkohärenten Anteil der Schwingung h herrührt.

Man sieht ohne weiteres, wie sich die Wahrscheinlichkeit von vier, fünf und mehr Strahlenbündeln berechnen wird; unsere Vermutung über die Entropie, die wir am Ende von § 9 aussprachen, findet hierin eine Stütze. Auch ahnt man, daß unsere rein phänomenologisch abgeleiteten Formeln im Lichte der Wahrscheinlichkeitstheorie einen einfachen Sinn erhalten werden. Diesen zu entdecken, muß das nächste Ziel der Thermodynamik der Interferenzerscheinungen sein.

Berlin, März 1907.

(Eingegangen 27. März 1907.)

PAPER NO. 6

Reprinted from *Annalen der Physik*, Vol. 23, pp. 795–797, 1907.

Die Entropie von partiell kohärenten Strahlenbündeln; Nachtrag; von M. Laue.

In einer kürzlich erschienenen Untersuchung[1]) habe ich gezeigt, daß die Entropie eines Systems von drei partiell kohärenten Strahlenbündeln den Betrag

$$E = \sigma\left(L(K_1) + L(K_2) + L(K_3)\right) \tag{49}$$

hat, wenn man unter K_1, K_2, K_3 die Wurzeln der Gleichung

$$K^3 - I' K^2 + I'' K - I''' = 0 \tag{50}$$

versteht. Doch enthielt die Ableitung dieser Formel die nur an Beispielen bestätigte Annahme, daß diese Gleichung stets drei reelle, positive Wurzeln hat, daß sich also jedes System der genannten Art auf umkehrbarem Wege in drei inkohärente Strahlenbündel verwandeln läßt. Ihr Beweis soll hier nachgeholt werden.

Zu diesem Zwecke stellen wir die Kurve

$$y = K^3 - I' K^2 + I'' K - I'''$$

graphisch dar. Verändern wir den letzten ihrer Koeffizienten, I''', so hebt oder senkt sie sich ohne Änderung ihrer Form. Ist daher

$$A_1 < I''' < A_2 ,$$

so liegen die Kurven

$$z_1 = K^3 - I' K^2 + I'' K - A_1$$

und

$$z_2 = K^3 - I' K^2 + I'' K - A_2$$

ganz über bez. unter der Kurve y. Wir wollen A_1 und A_2 nun so wählen, daß die beiden letzteren Kurven reelle Schnittpunkte mit der K-Achse haben; gelingt dies, so ist der Nachweis erbracht, daß auch die zwischen ihnen liegende y-Kurve drei reelle Punkte mit ihr gemeinsam hat.

1) M. Laue, Ann. d. Phys. **23.** p. 1. 1907. Alle Bezeichnungen und die Numerierung der Gleichungen sind von dort übernommen.

Bekanntlich hat jede Gleichung

$$z = -\begin{vmatrix} \varkappa_1 - K & l & m \\ l & \varkappa_2 - K & n \\ m & n & \varkappa_3 - K \end{vmatrix} = 0$$

reelle Wurzeln[1]), wenn die Koeffizienten reell sind. Rechnet man die Determinante aus, so nimmt sie die Form an:

$$\begin{aligned} z = K^3 &- (\varkappa_1 + \varkappa_2 + \varkappa_3) K^2 \\ &+ (\varkappa_1 \varkappa_2 + \varkappa_2 \varkappa_3 + \varkappa_3 \varkappa_1 - l^2 - m^2 - n^2) K \\ &- \left(\varkappa_1 \varkappa_2 \varkappa_2 - (l^2 \varkappa_3 + m^2 \varkappa_2 + n^2 \varkappa_1) + 2\, l m n\right). \end{aligned}$$

Setzen wir nun einmal

$$l = -\sqrt{i_{12} \varkappa_1 \varkappa_2}, \quad m = -\sqrt{i_{13} \varkappa_1 \varkappa_3}, \quad n = -\sqrt{i_{23} \varkappa_2 \varkappa_3},$$

das andere Mal

$$l = +\sqrt{i_{12} \varkappa_1 \varkappa_2}, \quad m = +\sqrt{i_{13} \varkappa_1 \varkappa_3}, \quad n = +\sqrt{i_{23} \varkappa_2 \varkappa_3},$$

so gehen nach (21) und (51) die Koeffizienten K^2 und K in I' und I'' über, während das absolute Glied im ersten Fall

$$A_1 = \varkappa_1 \varkappa_2 \varkappa_3 \left(1 - (i_{12} + i_{23} + i_{31}) - 2\sqrt{i_{12}\, i_{23}\, i_{31}}\right),$$

im zweiten aber

$$A_2 = \varkappa_1 \varkappa_2 \varkappa_3 \left(1 - (i_{12} + i_{23} + i_{31}) + 2\sqrt{i_{13}\, i_{23}\, i_{31}}\right)$$

wird. Wir haben jetzt nur noch zu zeigen, daß

$$A_1 \leqq I''' \leqq A_2;$$

oder, da nach (51)

$$I''' = J \varkappa_1 \varkappa_2 \varkappa_3$$

ist, daß

$$(53) \qquad \begin{cases} 1 - (i_{12} + i_{23} + i_{31}) - 2\sqrt{i_{12}\, i_{23}\, i_{31}} \leqq J \\ 1 - (i_{12} + i_{23} + i_{31}) + 2\sqrt{i_{12}\, i_{23}\, i_{31}} \geqq J. \end{cases}$$

ist.

Zu diesem Zwecke dient Gleichung (37), in welcher wir zunächst nach (21) und (40)

$$i_{f_g' h_g'} j_{fg}\, j_{gh} = j_{fg}\, j_{gh} - J$$

setzen. Vertauschen wir dann noch die Indizes f, g, h mit 1, 2, 3, so lautet sie:

$$(54) \quad i_{13} = i_{12}\, i_{23} + j_{12} j_{23} - J + 2\cos(a - b)\sqrt{i_{12}\, i_{23} (j_{12} j_{23} - J)}.$$

---

1) Man wird z. B. dann auf sie geführt, wenn man nach den Achsen der Fläche zweiter Ordnung $\varkappa_1 x^2 + \ldots + 2\, l x y + \ldots = 1$ fragt.

Aus ihrer Auflösung nach J, welche aussagt, daß

$$J = 1 - (i_{12} + i_{23} + i_{31}) + 2\sin^2(a-b)\, i_{12}\, i_{23}$$
$$\mp 2\cos(a-b)\sqrt{i_{12}\, i_{23}\,(i_{13} - \sin^2(a-b)\, i_{12}\, i_{23})}$$

ist, erkennt man unmittelbar, daß

$$J \geqq 1 - (i_{12} + i_{23} + i_{31}) - 2\sqrt{i_{12}\, i_{23}\, i_{31}}\,.$$

Die andere Hälfte der Ungleichung (53) leitet man aber einfacher ab, indem man aus (54) unmittelbar den Schluß zieht, daß

$$\sqrt{i_{12}\, i_{23}} - \sqrt{j_{12} j_{23} - J} \leqq \sqrt{i_{13}} \leqq \sqrt{i_{12}\, i_{23}} + \sqrt{j_{12} j_{23} - J} \tag{55}$$

ist. Gilt eins der beiden Gleichheitszeichen, so ist die quadratische Funktion von $\sqrt{i_{13}}$

$$1 - (i_{12} + i_{23} + i_{31}) + 2\sqrt{i_{12}\, i_{23}\, i_{31}} - J$$

gleich Null. Für unendlich große Werte ihres Arguments ist sie aber negativ unendlich, in dem durch (55) abgegrenzten Bereich daher positiv; d. h. es ist

$$1 - (i_{12} + i_{23} + i_{31}) + 2\sqrt{i_{12}\, i_{23}\, i_{31}} \geqq J\,.$$

Damit ist die Ungleichung (53) und gleichzeitig die Realität der Wurzeln der Gleichung (50) bewiesen. Daß diese nie negativ sind, geht aus dem Alternieren des Vorzeichens der Koeffizienten hervor.

Berlin, Juni 1907.

(Eingegangen 7. Juni 1907.)

PAPER NO. 7

Reprinted from *Journal of Mathematics and Physics* (*M.I.T.*), Vol. 7, pp. 109–125, 1927–1928.

COHERENCY MATRICES AND QUANTUM THEORY

By NORBERT WIENER

Introduction

The statistical method of determining the linear relations between a number of coördinates recorded in a sequence of simultaneous observations is the theory of correlation coefficients,[1] which may be cast into a matrix form. This method, however, pays no attention to the order of the observations constituting the sequence. Attempts have been made to apply the method of correlation to time-series, using such notions as that of the correlation of a function with itself under a time-lag.[2] These methods, while they have achieved a certain measure of success, are largely of an *ad hoc* nature, and as yet have failed to crystallize out any simple, general theory.

It is the purpose of the present paper to develop a theory which seems to the author to possess more of an intrinsic logical necessity than those already existing. This theory is also applied to the study of polarized light and of quantum phenomena. It is the conviction of the author that it explains the linear character of the Schrödinger equation, and that it serves as an important link between the Heisenberg and Schrödinger theories, the classical mechanics, as modified by the introduction of Kaluza's fifth dimension, and the Bohr quantum rules. The author reserves a more detailed discussion of these relations for a later paper.

§1. **Definition of the Coherency Matrix.** We start with a set of (possibly complex valued) functions $f_1(t)$, $f_2(t)$, $\cdots$, $f_n(t)$, all defined over the infinite range $-\infty < t < \infty$. We make the definition

$$\phi_{jk}(\tau) = \lim_{T\to\infty} \frac{1}{2T} \int_{-T}^{T} f_j(t+\tau)\bar{f}_k(t)dt. \tag{1}$$

Here, as throughout the sequel, the barring of a quantity denotes the formation of the conjugate complex quantity. We suppose that $\phi_{jk}(\tau)$ exists for $0 < j,\ k \leq n,\ -\infty < \tau < \infty$.

[1] Cf. G. U. Yule, *An Introduction to the Theory of Statistics* (London, 1919), Chapter IX-XII.
[2] Cf. F. C. Mills, *Statistical Methods* (New York, 1924), Chapter XI.

As we show in the appendix,

$$\phi_{kj}(\tau) = -\overline{\phi}_{jk}(\tau).$$

We also show that all the functions $\phi_{jk}(\tau)$ are bounded.

We now introduce the new definition

$$R_{jk}(u) = \int_{-1}^{1} \phi_{jk}(\tau) \frac{e^{iu\tau}-1}{i\tau} d\tau$$

$$+ \underset{A\to\infty}{\text{l.m.}} \left[\int_{1}^{A} + \int_{-A}^{-1} \right] \phi_{jk}(\tau) \frac{e^{iu\tau}}{i\tau} d\tau. \qquad (2)$$

Formally this is equivalent to

$$R'_{jk}(u) = \int_{-\infty}^{\infty} \phi_{jk}(\tau) e^{iu\tau} d\tau,$$

but the theory of the integrated function is more general. Of course, this definition only determines $R_{jk}(u)$ except for a set of values of u of measure zero. To make the definition more precise, we shall take $R_{jk}(u)$ to be the derivative of its integral wherever the latter exists, and this, by a well-known theorem, will be everywhere over the region where $R_{jk}(u)$ exists except for a set of points of measure zero. At other points, $R_{jk}(u)$ will be undefined. $R_{jk}(u-0)$ and $R_{jk}(u+0)$, however, may still have an unambiguous definition.

The function $R_{jk}(u)$ will always exist and be summable and of summable square over any finite range, as $(e^{iu\tau}-1)i\tau$ is finite near $\tau=0$, while $\phi_{jk}(\tau)/i\tau$ is quadratically summable over any infinite interval not including the origin. $R_{jk}(u)$ will in general be complex even when the $f_k(u)$'s are real, since $e^{iu\tau}$ is complex. We shall however have

$$R_{kj}(u) = \int_{-1}^{1} \phi_{kj}(\tau) \frac{e^{iu\tau}-1}{i\tau} d\tau + \underset{A\to\infty}{\text{l.m.}} \left[\int_{1}^{A} + \int_{-A}^{-1} \right] \phi_{kj}(\tau) \frac{e^{iu\tau}}{i\tau} d\tau$$

$$= \int_{-1}^{1} \overline{\phi}_{jk}(-\tau) \frac{e^{iu\tau}-1}{i\tau} d\tau + \underset{A\to\infty}{\text{l.m.}} \left[\int_{1}^{A} + \int_{-A}^{-1} \right] \overline{\phi}_{jk}(-\tau) \frac{e^{iu\tau}}{i\tau} d\tau$$

$$= \int_{-1}^{1} \overline{\phi}_{jk}(\sigma) \frac{e^{-iu\sigma}-1}{-i\sigma} d\sigma + \underset{A\to\infty}{\text{l.m.}} \left[\int_{1}^{A} + \int_{-A}^{-1} \right] \overline{\phi}_{jk}(\sigma) \frac{e^{-iu\sigma}}{-i\sigma} d\sigma$$

$$= \overline{R}_{jk}(u). \qquad (3)$$

In particular, $R_{jj}(u)$ is real. The matrix $\|R_{jk}(u+0)-R_{jk}(u-0)\|$ is Hermitian. We call it a *coherency matrix*.

To see the meaning of $R_{jk}(u)$, let us consider the particular case where

$$f_j(t)=\sum_1^m A_{jh}e^{i\Lambda_h t}. \quad [\Lambda_1<\Lambda_2<\cdots<\Lambda_n]$$

Then

$$\phi_{jk}(\tau)=\lim_{T\to\infty}\frac{1}{2T}\int_{-T}^{T}\sum_{h,0=1}^{m} A_{jh}\overline{A}_{kl}e^{i\Lambda_h\tau}e^{i(\Lambda_h-\Lambda_e)t}dt$$

$$=\sum_{h=1}^{m} A_{jh}\overline{A}_{kh}e^{i\Lambda_h\tau};$$

$$R_{jk}(u)=\int_{-1}^{1}\sum_{h=1}^{m} A_{jh}\overline{A}_{kh}\frac{e^{i\tau(u+\Lambda_h)}-e^{i\tau\Lambda_h}}{i\tau}d\tau$$

$$+\left[\int_1^\infty+\int_{-\infty}^{-1}\right]\sum_{h=1}^{m} A_{jh}\overline{A}_{kh}\frac{e^{i\tau(u+\Lambda_h)}}{i\tau}d\tau$$

$$=\sum_{h=1}^{m} A_{jh}\overline{A}_{kh}\left[K+\int_{-\infty}^{\infty}\frac{\sin\tau(u+\Lambda_h)}{\tau}d\tau\right], \tag{4}$$

where K is some constant. Thus

$$R_{jk}(v)-R_{jk}(u)=\pi\sum_{h=1}^{m} A_{jh}\overline{A}_{kh}[\operatorname{sgn}(v+\Lambda_h)-\operatorname{sgn}(u+\Lambda_h)]$$

$$=2\pi\sum_{h=p}^{h=q}A_{jh}\overline{A}_{kh}, \tag{5}$$

provided that

$$\cdots<\Lambda_{p-1}<-v<\Lambda_p<\cdots<\Lambda_q<-u<\Lambda_{q+1}<\cdots$$

In particular,

$$R_{jk}(-\Lambda_h+0)-R_{jk}(-\Lambda_h-0)=2\pi A_{jh}\overline{A}_{kh}$$

so that the coherency matrix $\|R_{jk}(-\Lambda_h+0)-R_{jk}(-\Lambda_h-0)\|$, or

$$2\pi\left\|\begin{matrix} |A_{1h}|^2 & A_{2h}\bar{A}_{1h} & \cdots \\ A_{1h}\bar{A}_{2h} & |A_{2h}|^2 & \cdots \\ \cdot & \cdot & \cdot \\ \cdot & \cdot & \cdot \end{matrix}\right\|$$

is manifestly Hermitian.

In the case where the $f_j(t)$'s are trigonometrical polynomials, the diagonal terms of $\|R_{jk}(v)-R_{jk}(u)\|$ indicate the sum of the squares of the amplitudes belonging to frequencies of the f_j's lying between $-u$ and $-v$. They are, roughly speaking, of the nature of energies. The non-diagonal terms may be regarded in the light of mutual energies, such as we find recorded in an integrating wattmeter when we superimpose a current from one source on an electromotive force from another. We shall find these ideas suggestive even when the f_j's are not trigonometrical polynomials.

§2. **Transformations of Coherency Matrices.** We now introduce the idea of a linear transformation of the coördinates $f_j(t)$, of the nature of what Weyl[3] calls a *unitary transformation*. Let

$$g_j(t)=\sum_1^n \alpha_{jk} f_k(t), \tag{6}$$

where

$$\sum_{k=1}^{n} \alpha_{jk}\bar{\alpha}_{hk}=S_{jh}. \qquad [1\leq j,\ h\leq n] \tag{7}$$

Let us put

$$\psi_{jk}(\tau)=\lim_{T\to\infty}\frac{1}{2T}\int_{-T}^{T} g_j(t+\tau)\bar{g}_k(t)dt;$$

$$S_{jk}(u)=\int_{-1}^{1}\psi_{jk}(\tau)\frac{e^{iu\tau}-1}{i\tau}d\tau$$

$$+\operatorname*{l.m.}_{A\to\infty}\left[\int_1^A+\int_{-A}^{-1}\right]\psi_{jk}(\tau)\frac{e^{iu\tau}}{i\tau}d\tau.$$

[3] H. Weyl, *Quantenmechanik und Gruppentheorie*, Zs. für Physik 46 (1927), p. 4.

Then

$$\psi_{jk}(\tau) = \sum_{h,\,l=1}^{n} \alpha_{jh}\bar{\alpha}_{kl}\phi_{hl}(\tau);$$

$$S_{jk}(u) = \sum_{h,\,l} \alpha_{jh}\bar{\alpha}_{kl}R_{hl}(u).$$

From the Hermitian matrix

$$\| R_{jk}(v) - R_{jk}(u) \|$$

let us construct the Hermitian form

$$\sum_{jk} [R_{jk}(v) - R_{jk}(u)]x_j\bar{x}_k.$$

The corresponding form derived from $S_{jk}(u)$ will be

$$\sum_{jkhl} [R_{hl}(v) - R_{hl}(u)]\alpha_{jh}x_j\bar{\alpha}_{kl}\bar{x}_k$$

$$= \sum_{jk} [R_{jk}(v) - R_{jk}(u)]y_j\bar{y}_k,$$

where

$$y_j = \sum_h \alpha_{hj}x_h.$$

That is, the vector y_j is derived from the vector x_j by a unitary transformation, and the S-form is transformed into the R-form by the inverse of this transformation, which is likewise unitary.

By a well-known theorem,[4] we may choose our unitary transformation so that

$$S_{jk}(v) - S_{jk}(u) = 0. \qquad [j \neq k]$$

We shall suppose the α_{hj} so chosen as to effect this result. By the results of a previous paper of the author,[5] $S_{jj}(u)$ is monotone increasing, and the matrix $\|S_{jk}(v) - S_{jk}(u)\|$ assumes the form

$$\left\| \begin{matrix} A & 0 & 0 & \cdots \\ 0 & B & 0 & \cdots \\ 0 & 0 & C & \cdots \\ \cdot & \cdot & \cdot & \cdot \\ \cdot & \cdot & \cdot & \cdot \end{matrix} \right\|$$

[4] H. Weyl, *loc. cit.*, p. 44.

[5] Norbert Wiener, *The Spectrum of an Arbitrary Function*, Proc. London Math. Soc. 27 (1928), p. 489.

where $A, B, C, \cdots$ are non-negative if $v \geq u$. The functions $S_{jk}(u)$ are also bounded. Since the form $Ax_1\bar{x}_1 + Bx_2\bar{x}_2 + \cdots$ is non-negative, if we transform our diagonal matrix into

$$\begin{Vmatrix} A_{11} & A_{21} & \cdots \\ A_{12} & A_{22} & \cdots \\ \cdot & \cdot & \cdot \\ \cdot & \cdot & \cdot \end{Vmatrix}$$

by a unitary transformation, we shall always have

$$\sum_{jk} A_{jk} x_j \bar{x}_k \geq 0$$

and the converse will hold. We shall call such an Hermitian form *positive definite*. The sum of two such forms will clearly be positive definite.

In the case of such a form, if we make all x_h's vanish except x_j and x_k, and then remember that the determinant of a matrix is an invariant under a unitary transformation, we see that

$$|A_{jk}|^2 \leq A_{jj} A_{kk}.$$

A fortiori,

$$|A_{jk}| \leq \frac{A_{jj} + A_{kk}}{2}.$$

It hence results that the total variation of $R_{jk}(u)$ cannot exceed half the sum of the total variations of $R_{jj}(u)$ and $R_{kk}(u)$. As these latter functions are bounded and monotone, they are of limited total variation, and $R_{jk}(u)$ is likewise of limited total variation. It is therefore continuous except for a denumerable set of finite jumps, and possesses a derivative except for a set of points of zero measure. Thus the matrix

$$\begin{Vmatrix} R'_{11}(u) & R'_{21}(u) & \cdots \\ R'_{12}(u) & R'_{22}(u) & \cdots \\ \cdot & \cdot & \cdot \\ \cdot & \cdot & \cdot \end{Vmatrix}$$

exists for almost all values of u, and may also be regarded as a coherency matrix. It is easy to see that it is positive definite, as it represents a limit of positive definite matrices.

§3. **Examples of Coherency Matrices.** We give below some examples of second order coherency matrices. We choose matrices of the form $\| R_{jk}(\Lambda+0)-R_{jk}(\Lambda-0) \|$, as these are simpler to illustrate. We shall write for purposes of abbreviation

$$F(x)=\frac{\sqrt{2}\cos \Lambda x}{\sqrt{\pi}};$$

$$G(x)=\frac{\sqrt{2}\cos \Lambda\,(x+\sqrt{|x|})}{\sqrt{\pi}}.$$

We put the functions f_1 and f_2 generating each matrix above the matrix it generates:

$$f_1(t)=F(t);\qquad f_2(t)=aF(t+\xi).$$

$$\begin{Vmatrix} 1 & ae^{i\Lambda\xi} \\ ae^{-i\Lambda\xi} & a^2 \end{Vmatrix} \tag{A}$$

$$f_1(t)=aF(t);\qquad f_2(t)=bG(t).$$

$$\begin{Vmatrix} a^2 & 0 \\ 0 & b^2 \end{Vmatrix} \tag{B}$$

$$f_1(t)=aF(t);\quad f_2(t)=bF(t+\xi)+cG(t).$$

$$\begin{Vmatrix} a^2 & abe^{i\Lambda\xi} \\ abe^{-i\Lambda\xi} & b^2+c^2 \end{Vmatrix} \tag{C}$$

§4. **Incoherent Addition of Matrices.** As a particular case of the effect of a linear transformation on a matrix, let us form the coherency matrix of $f_1(t), \cdots, f_n(t), g_1(t), \cdots, g_n(t)$ for a particular n, and let it have the form

$$\left\|\begin{matrix} A_{11} & A_{21} & \cdots & A_{n1} & 0 & 0 & \cdots & 0 \\ A_{12} & A_{22} & \cdots & A_{n2} & 0 & 0 & \cdots & 0 \\ \cdot & \cdot & \cdot & \cdot & \cdot & \cdot & \cdot & \cdot \\ \cdot & \cdot & \cdot & \cdot & \cdot & \cdot & \cdot & \cdot \\ A_{1n} & A_{2n} & \cdots & A_{nn} & 0 & 0 & \cdots & 0 \\ 0 & 0 & \cdots & 0 & B_{11} & B_{21} & \cdots & B_{n1} \\ 0 & 0 & \cdots & 0 & B_{12} & B_{22} & \cdots & B_{n2} \\ \cdot & \cdot & \cdot & \cdot & \cdot & \cdot & \cdot & \cdot \\ \cdot & \cdot & \cdot & \cdot & \cdot & \cdot & \cdot & \cdot \\ 0 & 0 & \cdots & 0 & B_{1n} & B_{2n} & \cdots & B_{nn} \end{matrix}\right\|.$$

Then the coherency matrix of $f_1(t)+g_1(t)$, $f_2(t)+g_2(t)$, $\cdots$, $f_n(t)+g_n(t)$ will have the form

$$\left\|\begin{matrix} A_{11}+B_{11} & A_{21}+B_{21} & \cdots & A_{n1}+B_{n1} \\ A_{12}+B_{12} & A_{22}+B_{22} & \cdots & A_{n2}+B_{n2} \\ \cdot & \cdot & \cdot & \cdot \\ \cdot & \cdot & \cdot & \cdot \\ A_{1n}+B_{1n} & A_{2n}+B_{2n} & \cdots & A_{nn}+B_{nn} \end{matrix}\right\|.$$

That is, if two phenomena are completely incoherent, the coherency matrix of their sum will be the sum of their coherency matrices.

§5. **Coherency Matrices and Polarized Light.** The theory of polarized light forms the simplest application of coherency matrices. The electric displacement in polarized light is a plane vector, a function of the time, which may be represented by two scalar functions of the time, its rectangular cartesian components. For each frequency, these yield a coherency matrix. A real unitary transformation of this matrix amounts simply to a rotation of our axes, together with a possible reflection. By such a transformation, the matrix may be brought into the form

$$\left\|\begin{matrix} A & -Bi \\ Bi & C \end{matrix}\right\|,$$

where

$$AC > B^2.$$

A matrix of this form may always be obtained by the incoherent addition of the two matrices

$$\begin{Vmatrix} \alpha^2 & -\alpha\beta i \\ \alpha\beta i & \beta^2 \end{Vmatrix}$$

and

$$\begin{Vmatrix} \gamma^2 & 0 \\ 0 & \gamma^2 \end{Vmatrix}.$$

The former represents completely elliptically polarized light, with the ellipse of polarization $\frac{x^2}{\alpha^2}+\frac{y^2}{\beta^2}=1$. The latter represents completely unpolarized light. The sum represents partly elliptically polarized light, with ellipse of polarization as above, percentage of polarization

$$\frac{\alpha^2+\beta^2}{100(\alpha^2+\beta^2+2\gamma^2)},$$

and total intensity $\alpha^2+\beta^2+2\gamma^2$.

As examples of simple types of polarized light, we may take

$$\begin{Vmatrix} 2 & 0 \\ 0 & 0 \end{Vmatrix} \quad \text{and} \quad \begin{Vmatrix} 0 & 0 \\ 0 & 2 \end{Vmatrix}.$$

Plane polarized light at 0° and 90°.

$$\begin{Vmatrix} 1 & 1 \\ 1 & 1 \end{Vmatrix} \quad \text{and} \quad \begin{Vmatrix} 1 & -1 \\ -1 & 1 \end{Vmatrix}.$$

Plane polarized light at 45° and 135°.

$$\begin{Vmatrix} 1 & -i \\ i & 1 \end{Vmatrix} \quad \text{and} \quad \begin{Vmatrix} 1 & i \\ -i & 1 \end{Vmatrix}.$$

Circularly polarized light with opposite senses of rotation.

$$\begin{Vmatrix} 1 & 0 \\ 0 & 1 \end{Vmatrix}.$$

Unpolarized light.

As is easily seen, the sum of the diagonal terms of a coherency matrix is an invariant under a unitary transformation. It represents the intensity of the light. Another invariant is the determinant of the matrix. If the latter vanishes, the matrix, if it does not vanish identically, assumes the form

$$\begin{Vmatrix} A & 0 \\ 0 & 0 \end{Vmatrix}$$

when reduced to diagonal form. Such a matrix will be called *completely polarized.* Two coherency matrices reducing to

$$\begin{Vmatrix} A & 0 \\ 0 & 0 \end{Vmatrix} \quad \text{and} \quad \begin{Vmatrix} 0 & 0 \\ 0 & B \end{Vmatrix}$$

under the same unitary transformations will be said to be complementary. Thus a Nicol prism analyses light into two complementary components; circularly polarized light is complementary to polarized light with an opposite sense of rotation, etc. The analysis of light in terms of two complementary types is performed as follows: its matrix is transformed by the unitary transformation reducing the latter to diagonal form, and the terms on the principal diagonals represent the relative intensities of the two components.

§6. **Coherency Matrices and Quantum Theory.** We now come to the applications of coherency matrices to quantum theory. Our coherency matrices are here considered with reference, not to time, but with respect to the fifth dimension x_0 of Kaluza.[6] Moreover, instead of depending on an index j assuming a finite set of values, our functions of x_0 whose coherency we investigate contain the four continuous dimensions of space and time as parameters. That is, our original functions have the form $f(x, y, z, t; x_0)$. The functions $\phi_{jk}(\tau)$ are replaced by expressions of the form

$$\begin{aligned}&\phi(x, y, z, t; x_1, y_1, z_1, t_1; \tau)\\ &\quad = \lim_{T\to\infty} \frac{1}{2T}\int_{-T}^{T} f(x, y, z, t; x_0+\tau)\, \bar{f}(x_1, y_1, z_1, t_1; x_0)dx_0\end{aligned}$$

[6] Th. Kaluza, *Sitzungsber. der Berliner Akad.* 1921, p. 966.

while the $R_{jk}(u)$ are replaced by functions

$$R(x, y, z, t; x_1, y_1, z_1, t_1; u) = \int_{-1}^{1} \phi(x, y, z, t; x_1, y_1, z_1, t_1; \tau)\frac{e^{iu\tau}-1}{i\tau}d\tau$$

$$+ \underset{A\to\infty}{\text{l.m.}} \left[\int_{1}^{A} + \int_{-A}^{-1}\right] \phi(x, y, z, t; x_1, y_1, z_1, t_1; \tau)\frac{e^{iu\tau}}{i\tau}\, d\tau.$$

In other words, our Hermitian *matrices* are replaced by Hermitian *kernels* in the sense of Hilbert.[7]

Another matter of importance is that in the quantum theory only one single value $\frac{2\pi}{h}$ of the frequency u ever comes into play. In other words, our coherency kernel is of the form

$$\left[\frac{\partial R}{\partial u}(x, y, z, t; x_1, y_1, z_1, t_1, u)\right] = K(x. v\ z, t; x_1, y_1, z_1, t_1).$$

We shall suppose this quantity K to exist.

In what follows, we shall make no attempt at mathematical rigor. To do so would require the entire theory of integral equations, non-singular and singular. To go into such intricacies would not serve in the last to illuminate the larger outlines of the quantum theory.

The problem of reducing an Hermitian kernel to diagonal form is already classical in the literature.[8] Under very extended hypotheses, the problem admits a unique solution, and may be treated by the aid of the so-called Hellinger integral. In the very simplest cases, the kernel is the kernel of a Fredholm equation, and may be expanded by the so-called bilinear formula in terms of the characteristic functions of the equation. These functions constitute a discrete set of normal and orthogonal functions, and the Hermitian kernel in the diagonal form is reduced to a matrix consisting of a discrete set of rows and columns. In the more general case, the discrete rows and columns of the diagonal matrix are supplemented by a continuous portion.

In general, the characteristic functions of the integral equation

[7] Cf. D. Hilbert, *Grundzüge einer Allgemeinen Theorie der linearen Integralgleichungen* (Leipzig and Berlin, 1912), pp. 162, 177.
[8] D. Hilbert, *loc. cit.*, p. 162.

$$\psi(x, y, z, t) = \lambda \int_{-\infty}^{\infty} dx_1 \int_{-\infty}^{\infty} dy_1 \int_{\infty}^{\infty} dz_1$$

$$\lim_{T\to\infty} \frac{1}{2T} \int_{-T}^{T} K(x, y, z, t; x_1, y_1, z_1, t_1) \psi(x_1, y_1, z_1, t_1) dt_1 \qquad (8)$$

do not depend in any peculiarly simple way on t. It is quite otherwise in the case when the coherency kernel is built up in the following manner: the function $f(x, y, z, t; x_0)$ is the sum of a large number of functions

$$f_1(x, y, z, t; x_0) + f_2(x, y, z, t; x_0) + \cdots$$

all having the same coherency kernel when referred to a proper time origin, and arranged at haphazard in time. When we reduce the coherency kernel of the sum function to a scale on which it is comparable to the coherency kernels of the partial functions composing it, it will clearly have the form

$$\lim_{T\to\infty} \frac{1}{2T} \int_{-T}^{T} K(x, y, z, t+w; x_1, y_1, z_1, t_1+w) dw$$

$$= \lim_{T\to\infty} \frac{1}{2T} \int_{-T}^{T} K(x, y, z, t-t_1+\mu; x_1, y_1, z_1, \mu) d\mu$$

$$= Q(x, y, z; x_1, y_1, z_1; t-t_1).$$

If we expand this kernel trigonometrically in $t-t_1$, it will assume the form

$$\Sigma Q_k(x, y, z; x_1, y_1, z_1) e^{i\Lambda_k(t-t_1)}.$$

Here, of course, the sum may be wholly or partly replaced by an integral. If we now expand $Q_k(x, y, z; x_1, y_1, z_1)$ in terms of its characteristic functions, we have

$$Q_k(x, y, z; x_1, y_1, z_1) = \Sigma A_{kl} \psi_{kl}(x, y, z) \bar{\psi}_{kl}(x_1, y_1, z_1),$$

and

$$Q(x, y, z; x_1, y_1, z_1; t-t_1)$$
$$= \sum_k \sum_l A_{kl} \psi_{kl}(x, y, z) e^{i\Lambda_k t} \bar{\psi}_{kl}(x_1, y_1, z_1) e^{-i\Lambda_k t_1}.$$

That is, the characteristic functions of the kernel $Q(x, y, z, t; x_1, y_1, z_1, t_1)$ are of the form

$$\psi_{kl}(x, y, z) e^{i\Lambda_k t},$$

and are orthogonal in a generalized sense over space-time. This is the same form as that of the characteristic functions of a steady-state Schrödinger equation.

If $f(x, y, z, t; x_0)$ is a solution of the five-dimensional Schrödinger equation, then $\phi(x, y, z, t; x_1, y_1, z_1, t_1; x_0)$ is also, as it is a mean of such solutions, displaced with respect to x_0, while this latter quantity does not occur in the coefficients of the five-dimensional Schrödinger equation. It follows readily that $K(x, y, z, t; x_1, y_1, z_1, t_1)$ is a solution of the four-dimensional Schrödinger equation. Hence

$$\psi_{kl}(x, y, z)e^{i\Lambda_k t} = \lambda \int_{-\infty}^{\infty} dx_1 \int_{-\infty}^{\infty} dy_1 \int_{-\infty}^{\infty} dz_1$$

$$\lim_{T\to\infty} \frac{1}{2T}\int_{-T}^{T} K(x, y, z, t; x_1, y_1, z_1, t_1)\psi_{kl}(x_1, y_1, z_1)e^{i\Lambda_k t_1}dt_1$$

is a solution of the same equation, and the characteristic functions of our coherency matrix are also characteristic functions of the Schrödinger equation, in the case where the coefficients do not contain the time explicitly. The exponents Λ_k are the corresponding characteristic numbers.

We thus see that in the case of a function satisfying Schrödinger's equation, the coherency kernel leads us back to Schrödinger's equation. On the other hand, any coherency kernel, even if it is not derived from a linear problem like that of the Schrödinger equation, gives rise to a linear integral equation determining a set of characteristic functions. This is not surprising, as the coherency matrices constitute a natural extension of correlation coefficients, while these latter quantities are only of use as a measure of the degree of *linear* dependence of one quantity on another. This suggests the conjecture that the linear character of the Schrödinger equation may be artificial; that it merely represents the characteristic function theory belonging to the coherency kernel of a problem of an essentially non-linear character. It may be that this is the sense in which it is possible to dispose of the wave-particle antinomy running through the whole of modern physics.

As an example of this possibility, let us reflect on a consideration put forward by Schrödinger himself in an early paper. He takes

$\phi_1(x_1, x_2, x_3, x_4)$, $\phi_2(x_1, x_2, x_3, x_4)$, $\phi_3(x_1, x_2, x_3, x_4)$, $\phi_4(x_1, x_2, x_3, x_4)$ as the components of the electromagnetic potential, and points out that if we choose a proper scale for our potentials

$$\int [\phi_1 dx_1 + \phi_2 dx_2 + \phi_3 dx_3 + \phi_4 dx_4]$$

over a period or quasiperiod will be an integral multiple of the Planck h when and only when the Bohr quantum condition is satisfied.[9] Now, the Kaluza five-dimensional quadratic form is

$$\sum_{j,k=1}^{4} g_{jk} dx_j dx_k + (\phi_1 dx_1 + \phi_2 dx_2 + \phi_3 dx_3 + \phi_4 dx_4 + dx_0)^2$$

and this vanishes for the paths of electrons. If our electron moves with a velocity low with respect to that of light, we may (with only a second order error) identify $\Sigma g_{jk} dx_j dx_k$ and $-dx_4^2$, and we get

$$dx_0 = -\phi_1 dx_1 - \phi_2 dx_2 - \phi_3 dx_3 - (\phi_4 - 1) dx_4.$$

If we choose a different value for our potential at infinity as the one belonging to the Schrödinger equation, we see that the change in x_0 over a period or quasiperiod is an integral multiple of h. It follows that if we put

$$x_1 = x_1(x_0), \cdots, x_4 = x_4(x_0)$$

the only permissible values of x_1, x_2, and x_3 are those containing $e^{\frac{2\pi i x_0}{h}}$ in their harmonic analysis.

If now we take the electron density over x_1, x_2, x_3, x_0 as our function $f(x_1, x_2, x_3, x_4, x_0)$, and build up its coherency kernel, no change in the number of those orbits not satisfying the Bohr frequency condition will affect this kernel. The characteristic numbers of the kernel under the hypothesis of steady state will not, however, represent the energies of the Bohr orbits, but their mechanical frequencies. We thus arrive at a characteristic function problem similar to that of the Schrödinger equation, but not identical with it. The nearness of our starting-point to that of the original Bohr theory is worthy of comment.

A word or two concerning the relation of our coherency kernels

[9] E. Schrödinger, Über eine bemerkenswerte Eigenschaft der Quantenbahnen eines einzelnen Elektrons. *Zs. für Physik* 12 (1923), pp. 13–24.

to the Heisenberg matrices may be in order. These latter matrices represent the transforms of the operators x, y, or z in the case of the q-matrices, $\frac{h}{2\pi i}\frac{\partial}{\partial x}$, $\frac{h}{2\pi i}\frac{\partial}{\partial y}$, or $\frac{h}{2\pi i}\frac{\partial}{\partial z}$ in the case of the p-matrices, by the transformations reducing our coherency kernel to diagonal form. It is interesting to pursue the analogy in the case of coherency matrices of the second order. In this case, the closest analogue to the x operator is the matrix

$$\left\| \begin{matrix} 1 & 0 \\ 0 & -1 \end{matrix} \right\|.$$

Then it may be verified that the Heisenberg matrix corresponding to the matrix

$$\left\| \begin{matrix} A & B-Ci \\ B+Ci & D \end{matrix} \right\|$$

is

$$\left\| \begin{matrix} \frac{A-D}{\sqrt{(A-D)^2+4(B^2+C^2)}} & \frac{2(B-Ci)}{\sqrt{(A-D)^2+4(B^2+C^2)}} \\ \frac{2(B+Ci)}{\sqrt{(A-D)^2+4(B^2+C^2)}} & \frac{D-A}{\sqrt{(A-D)^2+4(B^2+C^2)}} \end{matrix} \right\|$$

In particular, the Heisenberg matrices corresponding to plane polarized light at 0°, plane polarized light at 45°, and circularly polarized light in one direction, are

$$\left\| \begin{matrix} 1 & 0 \\ 0 & -1 \end{matrix} \right\|;$$

$$\left\| \begin{matrix} 0 & 1 \\ 1 & 0 \end{matrix} \right\|;$$

$$\left\| \begin{matrix} 0 & -i \\ i & 0 \end{matrix} \right\|.$$

It is perhaps worthy of comment that these three matrices play an important role in Weyl's treatment of the spinning electron by means of finite groups. If circularly polarized light or light polarized at 45° is analyzed by a Nicol prism at 0°, it appears unpolarized. The same remark applies if circularly polarized light or light polarized at 0° is examined by a Nicol prism at 45°, or if plane polarized light at 0° or at 45° is examined by a device which can only detect circular polarization. It is also true that light is completely characterized as to its state of polarization if it is analyzed by the instruments just mentioned and if its intensity is determined. This symmetry between the three sorts of light in question is purely formal and mathematical, and probably explains their relation to the Weyl group theory.

APPENDIX

By the Schwarz inequality,

$$|\phi_{jk}(\tau)| \leq \sqrt{\overline{\lim_{T\to\infty}} \frac{1}{2T}\int_{-T}^{T} |f_j(t+\tau)|^2 dt \; \overline{\lim_{T\to\infty}} \frac{1}{2T}\int_{T}^{T} |f_k(t)|^2 dt}.$$

Now,

$$\begin{aligned} \overline{\lim_{T\to\infty}} \int_{-T}^{T} |f_j(t+\tau)|^2 dt &= \overline{\lim_{T\to\infty}} \frac{1}{2T}\int_{-T+\tau}^{T+\tau} |f_j(t)|^2 dt \\ &\leq \overline{\lim_{T\to\infty}} \frac{1}{2(T+\tau)}\int_{-T-\tau}^{T+\tau} |f_j(t)|^2 dt \\ &= \phi_{jj}(0). \end{aligned}$$

Therefore

$$|\phi_{jk}(\tau)| \leq \sqrt{\phi_{jj}(0)\phi_{kk}(0)}. \tag{1}$$

Thus the functions $\phi_{jk}(\tau)$ are all bounded.

Again,

$$\phi_{kj}(\tau)=\lim_{T\to\infty}\frac{1}{2T}\int_{-T}^{T}\bar{f}_j(t)f_k(t+\tau)dt$$

$$=\lim_{T\to\infty}\frac{1}{2T}\int_{-T+\tau}^{T+\tau}\bar{f}_j(t-\tau)f_{k(t)}dt$$

$$=\bar{\phi}_{jk}(-\tau)+\lim_{T\to\infty}\frac{1}{2T}\int_{T}^{T+\tau}\bar{f}_j(t-\tau)f_k(t)dt$$

$$+\lim_{T\to\infty}\frac{1}{2T}\int_{-T-\tau}^{-T}\bar{f}_j(t-\tau)f_k(t)dt. \tag{2}$$

By the Schwarz inequality,

$$\left|\lim_{T\to\infty}\frac{1}{2T}\int_{T}^{T+\tau}\bar{f}_j(t-\tau)f_k(t)dt\right|\leq\overline{\lim_{T\to\infty}}\frac{1}{2T}\sqrt{\int_{T-\tau}^{T}|f_j(t)|^2dt\int_{T}^{T+\tau}|f_k(t)|^2dt}$$

$$\leq\sqrt{\overline{\lim}\left[\frac{1}{2T}\int_{-T}^{T}|f_j(t)|^2dt-\frac{1}{2(T-\tau)}\int_{-T+\tau}^{T-\tau}|f_j(t)|^2dt\right]\frac{1}{2(T+\tau)}\int_{-T-\tau}^{T+\tau}|f_k(t)|^2dt}$$

$$=0. \tag{3}$$

Similarly,

$$\lim_{T\to\infty}\frac{1}{2T}\int_{-T+\tau}^{-T}\bar{f}_j(t-\tau)f_k(t)dt=0. \tag{4}$$

Combining (2), (3), and (4), we get

$$\phi_{kj}(\tau)=\bar{\phi}_{jk}(-\tau). \tag{5}$$

PAPER NO. 8

Reprinted from *Physica*, Vol. 5, pp. 785–795, 1938.

THE CONCEPT OF DEGREE OF COHERENCE AND ITS APPLICATION TO OPTICAL PROBLEMS

by F. ZERNIKE, Groningen

Summary

The maximum visibility of the interferences obtainable from two points in a wave field is defined as their *degree of coherence* γ. By a simple statistical method general formulae are found for deducing γ from illumination data. For any extended lightsource γ is found equal to the amplitude in a certain diffraction image. It does not change by the use of a condensing lens, but depends only on the aperture of the illuminating cone. These properties are applied to the microscopic observation of objects in transmitted light.

Introduction. In the usual treatment of interference and diffraction phenomena, there is nothing intermediate between coherence and incoherence. Indeed, the first term is understood to mean complete dependence of phases, the second complete independence. To be sure of the first or of the second property, one must know that the vibrations in question originate in the same, respectively in different, radiating centres.

This is the reason why problems with an extended lightsource can only be treated in the following way. The pattern caused by each point of the source separately is first determined. Afterwards the intensities of the different patterns are superposed to get the final result.

It would be an improvement in many respects if intermediate states, of partial coherence, could be treated as well. Von Laue[1]) introduced a measure of partial coherence, which he used in energetic considerations. It is not suitable for use in interference phenomena. Berek[2]) has tried this by means of his „degree of consonance". For various reasons he could not give exact proofs of the properties foreseen in this way. Van Cittert[3]), on the other hand, calculated exact partition functions which show the correlation between the amplitudes in different points. He thus anticipated the theorems of our §§ 3 and 5, without introducing a degree of coherence. For discussing general problems his method would seem too intricate.

In this paper the difficulty is solved by starting from a fundamentally different definition of incoherence, which can easily be extended to partial coherence.

1. *Definitions. Two vibrations of light shall be called incoherent, if their superposition gives no visible interferences.* The interference experiment of Thomas Young may serve as an example. A screen is illuminated by a small light source. At two adjacent points P_1 and P_2 of the screen two pinholes are made. Behind the screen the light from the two holes shows interference fringes. These disappear with a somewhat broader source. As mentioned above, this is ordinarily explained by the overlapping of a great many mutually incoherent interference patterns.

According to our definition, we shall in this case say that the vibrations from P_1 and P_2 are *incoherent* or that P_1 and P_2 are incoherently illuminated by the source. For intermediate sizes of the source the fringes do not disappear, but they are less visible than with a point source because of a growing intensity in the minima. To express this in a consistent way, we shall then call P_1 and P_2 *partly coherent.*

Now it is well known that Michelson [4]) defined the *visibility* of interference fringes by the fraction

$$\frac{I_{max} - I_{min}}{I_{max} + I_{min}} \tag{1}$$

The same value will serve us as a measure of partial coherence. By definition *the „degree of coherence" of two light-vibrations shall be equal to the visibility of the interference fringes that may be obtained from them under the best circumstances,* that is, when both intensities are made equal and only small path-differences introduced. Clearly the degree of coherence may have any positive value from 0 to 1, the limiting values respectively meaning incoherence and coherence.

The usefulness of the new concept lies primarily in the fact, to be demonstrated further on, that the degree of coherence can be calculated from the illumination data, that is from the positions and the intensities of the different parts of the light-source. Also it does away with the necessity to retrace all interfering rays to the true source, enabling on the contrary to divide any problem into separate parts, consecutive in the direction of propagation of the light.

2. *Definition in statistical terms.* Let us consider two arbitrary points P_1 and P_2 in the same wave field. Putting aside polarisation effects, we shall simplify our problem by taking the amplitude of vibration to be a scalar. Further we make the usual assumption that a stationary phenomenon with practically monochromatic light is studied. The vibration in P_1 can then be represented by

$$a_1\, e^{i\phi_1}\, e^{i\omega t}$$

in which the amplitude a_1 and the phase φ_1, though practically constant during a small number of vibrations, change in a random way in any longer interval of time. As is usual in alternating current theory, we shall represent the factor of $e^{i\omega t}$ as a vector in the complex plane and designate it as the (complex) amplitude A_1.

$$a_1\, e^{i\phi_1} = A_1$$

By observation at the point P_1 it is of course impossible to find anything about the phase or about the momentary amplitude, the only observable quantity being the intensity J_1, which is proportional to the mean value of a_1^2. Omitting the constant we have

$$J_1 = \overline{a_1^2} = \overline{A_1\, A_1^*}$$

in which the * denotes the complex conjugate and the bar the mean value.

Considering now two points we may either intercept the waves at these points in order to observe the intensities J_1 and J_2, or let them proceed further — through two pinholes in a screen — and observe the intensities of the resulting interference pattern. Apart from irrelevant details, the second case means that the vibrations are added after introducing a variable path difference. In a formula, we want the mean of

$$(A_1^*\, e^{-i\beta} + A_2^*)\,(A_1\, e^{i\beta} + A_2) = A_1^* A_1 + A_2^* A_2 + A_1^* A_2 e^{-i\beta} + A_1 A_2^*\, e^{i\beta}.$$

The fourth term being the conjugate of the third, the result will be

$$J(\beta) = J_1 + J_2 + 2\Re(\overline{A_1^* A_2}\, e^{-i\beta})$$

where $\Re$ denotes the real part. Clearly this will be a maximum when the quantity between ( ) is real positive, that is when

$$\beta = \beta_{max} = \arg(\overline{A_1^* A_2}) \tag{2}$$

and

$$J_{max} = J_1 + J_2 + 2\,|\overline{A_1^* A_2}|$$

$$J_{min} = J_1 + J_2 - 2\,|\overline{A_1^* A_2}|$$

According to our definition the visibility calculated from these values by (1) will be equal to the degree of coherence γ only in case $J_1 = J_2$. This gives

$$\gamma = \frac{|\overline{A_1^* A_2}|}{J_1}$$

In an analogous way the general formula is found to be

$$\gamma = \frac{|\overline{A_1^* A_2}|}{\sqrt{J_1 J_2}} \tag{3}$$

In this calculation we have advanced the first vibration by the variable phase angle β before combining it with the second one, while (2) gives the value of β at which they re-inforce each other. We are therefore justified to say that the amount β_{max} brings the first into phase with the second, or that β_{max} is the phase difference second *minus* first.

For the further mathematics it is advantageous to combine the degree of coherence and the phase difference into a „complex degree of coherence''

$$\gamma\, e^{i\beta_{max}} = \frac{\overline{A_1^* A_2}}{\sqrt{J_1 J_2}} \tag{4}$$

according to (2) and (3). As it is rather the numerator of (4) that will occur in our formulae, a special name and symbol for $\overline{A_1^* A_2}$ will be useful. Let it be called the *mutual intensity* J_{12}

$$J_{12} = \overline{A_1^* A_2} \tag{5}$$

3. *Calculation from illumination data.* In order to calculate the degree of coherence for two arbitrary points P_1 and P_2 from the

positions and intensities of the different parts of the light source, a statistical method will be used that has proved succesful in similar cases [5]). It consists in treating the random causes of the problem as if they were exactly known functions. The problem can then be solved, so that the quantities of which we want the mean values are expressed as integrals containing the initial variables. After that the statistical mean value of the result is determined, taking into account the known statistical properties of the initial variables.

In our case the solution is easily obtained by superposing the contributions from the separate points of the source. Let the radiating surface be divided into elements of dimensions small compared with the wavelength. At a distance r from any such element, it will cause a vibration represented by

$$\frac{a}{r} e^{i\omega t - ikr}$$

where $k = 2\pi/\lambda$, while the complex number a determines amplitude and phase. Therefore, if the mth element has a distance r_{m1} from the point P_1 and a complex amplitude a_m, we have for the amplitude A_1 in P_1

$$A_1 = \sum_m a_m \, e^{-ikr_{m1}}/r_{m1}$$

and in the same way

$$A_1^* A_2 = \sum_m \sum_n a_m^* \, a_n \, e^{-ik(r_{n2} - r_{m1})}/r_{m1} \, r_{n2}$$

Clearly the a's are here the initial variables, which in reality are dependent on chance. Taking now the mean value, we have

$$J_{12} = \overline{A_1^* A_2} = \sum_m \sum_n \overline{a_m^* \, a_n} \, e^{-ik(r_{n2} - r_{m1})}/r_{m1} \, r_{n2} \tag{6}$$

Now the different parts of the light source are statistically independent, therefore

$$\overline{a_m^* \, a_n} = 0, \quad m \neq n$$

This might also be concluded from the considerations of § 2: if it were otherwise, lightbeams from the mth and from the nth element would on interference give fringes of non-vanishing visibility.

The double summation in (6) is thus reduced to a single one

$$J_{12} = \sum_m \overline{a_m^* \, a_m} \, e^{-ik(r_{m2} - r_{m1})}/r_{m1} \, r_{m2} \tag{7}$$

If P_2 coincides with P_1, we get the intensity at this point

$$J_1 = \sum_m \overline{a_m^* \, a_m} / r_{m1}^2$$

It is evident from this that $\overline{a_m^* a_m}$ simply represents the radiating intensity of the mth element of the source. Introducing therefore the *intensity per unit area* $j(x)$, where x stands for the co-ordinates on the surface of the source, the intensity at P_1 becomes

$$J_1 = \int j(x) \, d\sigma / r^2$$

while (7) changes into the final form

$$J_{12} = \int \frac{j(x) \, d\sigma}{r_1^2} \, e^{-ik(r_2 - r_1)} \tag{8}$$

where $r_1 r_2$ could be replaced by r_1^2, as P_1 and P_2 will never be far apart.

In order to get an idea of the dependence of J_{12} on the mutual positions of P_1 and P_2, it should be remarked that the same integral occurs in a totally different case, namely in the calculation of the diffraction image of a point source. Indeed, let the extended source of our problem be replaced by a transparant layer of the same form and situation, which transmits a variable fraction $\tau(x)$ of the light incident on it. Let a point source and a lens be placed behind this layer in such a way that an image of the source is projected at P_1. The diffracted amplitude in a neighbouring point P_2 may then be calculated by applying Huygens' principle to the absorbing surface. This gives

$$A_2 = C \int \frac{d\sigma \sqrt{\tau(x)}}{r_2} \, e^{i\phi(x)} \, e^{-ikr_2}$$

where C depends on the intensity of the incident light, while the phase $\varphi(x)$ can at once be determined by moving P_2 to the position P_1. As this is the geometric image, all waves arrive there in the same phase, or

$$\varphi(x) - kr_1 = \text{const.}$$

and

$$A_2 = C' \int \frac{d\sigma \sqrt{\tau(x)}}{r_1} \, e^{-ik(r_2 - r_1)}$$

This will be proportional to (8), if

$$\frac{\sqrt{\tau(x)}}{r_1} = \frac{j(x)}{r_1^2}$$

that is the *amplitudes* caused by the different parts of the diffracting surface must be equal to the *intensities* caused by the corresponding parts of the light source. Eliminating the constants we have

$$A_2/A_1 = J_{12}/J_1$$

In words: The *degree of coherence* between a variable point P_2 and a fixed point P_1 is represented by the *amplitude of a diffraction pattern* round the centre P_1. In most applications the brightness $j(x)$ of the light source will be a constant. The corresponding diffraction problem is then simply that of an aperture of the same form as the source in the original problem.

In this connection Michelson's method [6]) of measuring small angular diameters by the disappearence of interference fringes appears in a new light. The large „interferometer'' attachment together with the 100 inch telescope of the Mt. Wilson Observatory may be said only to determine the degree of coherence in the light waves coming from a star. Thus in the case of α *Orionis* incoherence was found to occur for the first time for any pair of points 307 cm apart. This corresponds to a circular diffraction pattern, the first dark ring of which would have a radius of 307 cm. Such a pattern would be formed by a circular aperture subtending an angle of 0,047 seconds.

4. *The propagation of the degree of coherence.* Suppose the complex degree of coherence on a certain surface S to be known, either by calculation in the above way or by observation. We shall now show that this, together with the distribution of intensity on S, is sufficient to calculate the coherence and intensity on another surface S' lying farther away in the direction of the light beam.

Let the complex amplitude on S for a moment be designated by $a(x)$. The corresponding amplitude $A(\xi)$ on the surface S' is then determined by the diffraction integral

$$A(\xi) = \frac{ik}{2\pi} \int_S \frac{a(x)}{r} e^{-ikr}\, dx$$

In the same way as above we deduce from this

$$A^*(\xi_1)\,A(\xi_2) = \frac{k^2}{4\pi^2}\int\limits_S\int\limits_S \frac{a^*(x_1)\,a(x_2)}{r_{11}\,r_{22}}\,e^{-ik(r_{22}-r_{11})}\,dx_1\,dx_2$$

or, taking the statistical mean value

$$J_{12} = \frac{k^2}{4\pi^2}\int\int \frac{j_{12}}{r_{11}\,r_{22}}\,e^{-ik(r_{22}-r_{11})}\,dx_1\,dx_2 \tag{9}$$

which formula expresses the mutual intensity J_{12} on the second surface in terms of the same quantity j_{12} on the first surface. According to (4) and (5) these mutual intensities are equivalent to the intensities and complex degrees of coherence.

If a diaphragm is situated between the surfaces S and S' this will intercept certain radii r, that is it will cause narrower limits of integration. When the effect of the diaphragm is introduced in this way, however, the diffraction at its edge is neglected. If we want to take it into account, it is of course necessary to apply Huygens' principle to the plane of the diaphragm. This simply means that the transition from S to S' must be made in two steps, $S \rightarrow$ diaphragm and diaphragm $\rightarrow S'$.

Formula (9) may also be applied in case one or more lenses lie between S and S', provided the straight distances r are replaced by the optical paths measured along the light rays.

5. *Coherence in the image of a light-source.* As an example in which the diffraction at an interposed aperture cannot be neglected, we shall treat more fully the case that the light-source is imaged on the plane in which the degree of coherence is wanted. The neglect of diffraction at the lens would here lead — and has often led — to the wrong conclusion that adjacent points in the image are incoherent.

It is therefore necessary to make the calculation in two steps, as mentioned above, first making up the mutual intensity on a surface S' directly behind the lens, essentially by the use of (8), then going to the final surface S by (9). This will result in a threefold integral.

Let coordinate axes be chosen with their origin near the points P_1 and P_2 in the surface S and with XY plane tangent to S. The degree of coherence will be calculated in the XY plane.

An arbitrary point Q of the intermediate surface S' may then be determined by the direction cosines ξ and η of the radius vector OQ, which may have a length $f(\xi, \eta)$. The distance from Q to a point P in the XY plane will then be

$$r_{PQ} = f(\xi, \eta) - \xi x - \eta y \tag{10}$$

as higher powers of the small quantities x and y may be neglected. Let a point of the light source be indicated by the position of its image P. Applying now (8) to calculate the mutual intensity j_{12} in the intermediate surface, the constancy of path from object to image point gives

$$r_1 + r_{PQ_1} = r_2 + r_{PQ_2}$$

and using (10), we have

$$r_2 - r_1 = f(\xi_1 \eta_1) - f(\xi_2 \eta_2) + (\xi_2 - \xi_1)x + (\eta_2 - \eta_1)y$$

Thus we get

$$j_{12} = C \int \frac{j(x)dx}{r^2} e^{ik(f_2-f_1)-ik(\xi_2-\xi_1)x} \tag{11}$$

omitting for shortness the dependence on y and η.

We will further introduce a variable transmission by different parts of the lens. If the *amplitude* is thereby changed by a factor $\tau(\xi, \eta)$, the mutual intensity will evidently be changed into

$$j_{12}\, \tau^*(\xi_1 \eta_1)\, \tau(\xi_2 \eta_2)$$

With this value we can now make the second step and find the mutual intensity J_{12} in the XY plane by (9). Again using (10), the exponential under the integral sign will become, with the same abbreviation as above,

$$e^{-ik(f_2-f_1)+ik(\xi_2 x_2-\xi_1 x_1)}$$

Substituting (11), the terms with $(f_1 - f_2)$ cancel and

$$J_{12} = C' \frac{k^2}{4\pi^2} \iiint j(x)\, \tau^*(\xi_1)\, \tau(\xi_2) e^{-ik(\xi_2 x-\xi_1 x-\xi_2 x_2+\xi_1 x_1)}\, dx\, d\xi_1\, d\xi_2 \tag{12}$$

Strictly speaking, the denominators and the factors originating from transformations of surface-elements ought to appear in this formula. Their treatment forms a photometric problem which is less important here. The result has been taken up in the constant C'.

Now the exponential in (12) may be written in the form

$$e^{ik\xi_2(x_2-x_1)}\, e^{ik(\xi_1-\xi_2)(x-x_1)}$$

where the first factor is independent of x and ξ_1. The integration with respect to these variables therefore gives the following simple result by the use of Fourier's theorem, *provided $j(x)$ is constant*

$$\iint \frac{1}{k} \tau^*(\xi_1) e^{ik(x-x_1)(\xi_1-\xi_2)} \, dk(x-x_1) \, d\xi_1 = \frac{2\pi}{k} \tau^*(\xi_2)$$

where the limits for x have been taken as $-\infty$ and ∞. Practically it will suffice if at the limits the exponent is many times $2\pi i$. Expressed in physical terms, this means that the image of the source, of constant brightness, must be large compared with the extent of the diffraction by the lens.

As half of the variables have been suppressed, the factor $2\pi/k$ should be squared, so that it does not appear in the final result, which becomes

$$J_{12} = C'' \int \tau^*(\xi_2) \, \tau(\xi_2) \, e^{ik\xi_2(x_2-x_1)} \, d\xi_2$$

By (10) and (8), this is again proportional to the amplitude of the diffraction image of the lens. or to the mutual intensity caused by an extended source with brightness equal to $|\tau(\xi)|^2$. We have thus proved the curious theorem: *The degree of coherence in a plane illuminated through a lens is the same, whether a source of uniform brightness be imaged on the plane, or placed directly behind the lens.* It should further be noted that the amplitude-factor $\tau(\xi)$ has not been restricted to real values. Of course complex values will mean that the lens introduces variable additional phase-differences, as is the case with aberrations or lack of focussing. In our result, however, only the squared modulus of $\tau(\xi)$ appears, so that *the phase-changing properties have no influence on the coherence.*

The only practical use of an illuminating lens therefore lies in its condensing property, that is, a much smaller source will be sufficient with the lens. Especially in the case of microscope condensers, it has often be presumed that a high degree of correction must be of advantage for the resolving power. Our result proves on the contrary that this has no influence. In ordinary microscopic observations, where the necessary brightness is easily attained, the condenser may even as well be replaced by an illuminated white surface, subtending the necessary angle as seen from the object.

The microscopic observation of transparent objects is thus influenced by the illuminating aperture as well as by the objective aperture The diffraction image corresponding to the first aperture determines the range of the coherence in the object, while that of the second aperture determines the overlapping of the object-points in the image. For equal illuminating and observing apertures the result will be that two points just separable by the objective are incoherently illuminated. The transparent object will then appear practically in the same way as a self-luminous one.

Received June 28th, 1938.

REFERENCES

1) M. von Laue, Ann. Physik **23**, 1, 1907, also Handb. Experimentalphysik XVIII, 277 (1928).
2) M. Berek, Z. Physik **36**, 675, 824, **37**, 287, **40**, 420, 1926, also C. Lakeman and J. Th. Groosmuller, Physica **8**, 193, 1928.
3) P. H. van Cittert. Physica **1**, 201, 1934.
4) A. A. Michelson. Phil. Mag. **30**, 1, 1890.
5) F. Zernike. Z. Physik **79**, 516, 1932.
6) A. A. Michelson and F. G. Pease. Astrophys. J. **53**, 249, 1921, J. A. Anderson, Astrophys. J. **55**, 48, 1922.

PAPER NO. 9

Reprinted from *Physical Review*, Vol. 93, pp. 99–110, 1954.

Coherence in Spontaneous Radiation Processes

R. H. Dicke
Palmer Physical Laboratory, Princeton University, Princeton, New Jersey
(Received August 25, 1953)

By considering a radiating gas as a single quantum-mechanical system, energy levels corresponding to certain correlations between individual molecules are described. Spontaneous emission of radiation in a transition between two such levels leads to the emission of coherent radiation. The discussion is limited first to a gas of dimension small compared with a wavelength. Spontaneous radiation rates and natural line breadths are calculated. For a gas of large extent the effect of photon recoil momentum on coherence is calculated. The effect of a radiation pulse in exciting "super-radiant" states is discussed. The angular correlation between successive photons spontaneously emitted by a gas initially in thermal equilibrium is calculated.

In the usual treatment of spontaneous radiation by a gas, the radiation process is calculated as though the separate molecules radiate independently of each other. To justify this assumption it might be argued that, as a result of the large distance between molecules and subsequent weak interactions, the probability of a given molecule emitting a photon should be independent of the states of other molecules. It is clear that this model is incapable of describing a coherent spontaneous radiation process since the radiation rate is proportional to the molecular concentration rather than to the square of the concentration. This simplified picture overlooks the fact that all the molecules are interacting with a common radiation field and hence cannot be treated as independent. The model is wrong in principle and many of the results obtained from it are incorrect.

A simple example will be used to illustrate the inadequacy of this description. Assume that a neutron is placed in a uniform magnetic field in the higher energy of the two spin states. In due course the neutron will spontaneously radiate a photon via a magnetic dipole transition and drop to the lower energy state. The probability of finding the neutron in its upper energy state falls exponentially to zero.[1,2]

If, now, a neutron in its ground state is placed near the first excited neutron (a distance small compared with a radiation wavelength but large compared with a particle wavelength and such that the dipole-dipole interaction is negligible), the radiation process would, according to the above hypothesis of independence, be unaffected. Actually, the radiation process would be strongly affected. The initial transition probability would be the same as before but the probability of finding an excited neutron would fall exponentially to one-half rather than to zero.

The justification for these assertions is the following: The initial state of the neutron system finds neutron 1 excited and neutron 2 unexcited. (It is assumed that the particles have nonoverlapping space functions, so that particle symmetry plays no role.) This initial state may be considered to be a superposition of the triplet and singlet states of the particles. The triplet state is capable of radiating to the ground state (triplet) but the singlet state will not couple with the triplet system. Consequently, only the triplet part is modified by the coupling with the field. After a long time there is still a probability of one-half that a photon has not been emitted. If, after a long period of time, no photon has been emitted, the neutrons are in a singlet state and it is impossible to predict which neutron is the excited one.

On the other hand, if the initial state of the two neutrons were triplet with $s=1$, $m_s=0$ namely a state with one excited neutron, a photon would be certain to be emitted and the transition probability would be just double that for a lone excited neutron. Thus, the presence of the unexcited neutron in this case doubles the radiation rate.

In recent years the excitation of correlated states of atomic radiating systems with the subsequent emission of spontaneous coherent radiation has become an important technique for nuclear magnetic resonance research.[3] The description usually given of this process is a classical one based on a spin system in a magnetic field. The purpose of this note is to generalize these results to any system of radiators with a magnetic or electric dipole transition and to see what effects, if any, result from a quantum mechanical treatment of the radiation process. Most of the previous work[4] was quite early and not concerned with the problems being considered here. In a subsequent article to be published in the *Review of Scientific Instruments* some of these results will be applied to the problem of instrumentation for microwave spectroscopy.

In this treatment the gas as a whole will be considered as a single quantum-mechanical system. The problem will be one of finding those energy states representing correlated motions in the system. The spontaneous emission of coherent radiation will accompany transitions between such levels. In the first problem to be considered the gas volumes will be assumed to have

[1] W. Heitler, *The Quantum Theory of Radiation* (Clarendon Press, Oxford, 1936), first edition, p. 112.

[2] E. P. Wigner and V. Weisskopf, Z. Physik **63**, 54 (1930).

[3] E. L. Hahn, Phys. Rev. **77**, 297 (1950); **80**, 580 (1950).

[4] E.g., W. Pauli, *Handbuch der Physik* (Springer, Berlin, 1933), Vol. 24, Part I, p. 210; G. Wentzel, *Handbuch der Physik* (Springer, Berlin, 1933), Vol. 24, Part I, p. 758.

dimensions small compared with a radiation wavelength. This case, which is of particular importance for nuclear magnetic resonance experiments and some microwave spectroscopic applications, is treated first quantum mechanically and then semiclassically, the radiation process being treated classically. A classical model is also described. In the next case to be considered the gas is assumed to be of large extent. The effect of molecular motion on coherence and the effect on coherence of the recoil momentum accompanying the emission of a photon are discussed. Finally, the two principal methods of exciting coherent states by the absorption of photons from an intense radiation pulse or the emission of photons by the gas are discussed. Calculations of these two effects are made for the gas system initially in thermal equilibrium. The effect of photon emission on inducing coherence is discussed as a problem in the angular correlation of the emitted photons.

DIPOLE APPROXIMATION

The first problem to be considered is that of a gas confined to a container the dimensions of which are small compared with a wavelength. It is assumed that the walls of the container are transparent to the radiation field. In order to avoid difficulties arising from collision broadening it will be assumed that collisions do not affect the internal states of the molecules. It will be assumed that the transition under question takes place between two nondegenerate states of the molecule. The assumption of nondegeneracy is made in order to limit the scope of the problem to its bare essentials. It might be assumed that nondegenerate states are present as a result of a uniform static electric or magnetic field acting on the gas. Actually, for many of the questions being discussed it is not essential that the degeneracies be split. Also, it will be assumed that there is insufficient overlap in the wave functions of separate molecules to require that the wave functions be symmetrized.

Since it is assumed that internal coordinates of the individual molecules are unaffected by collisions and but two internal states are involved for each molecule, the wave function for the gas may be written conveniently in a representation diagonal in the center-of-mass coordinates and the internal energies of the molecules. The internal energy coordinate takes on only two values. Omitting for the moment the radiation field, the Hamiltonian for an n molecule gas can be written

$$H=H_0+E\sum_{j=1}^{n} R_{j3}, \tag{1}$$

where $E=\hbar\omega=$ molecular excitation energy. Here H_0 acts on the center-of-mass coordinates and represents the translational and intermolecular interaction energies of the gas. ER_{j3} is the internal energy of the jth molecule and has eigenvalues $\pm\frac{1}{2}E$. H_0 and all the R_{j3} commute with each other. Consequently, energy eigenfunctions may be chosen to be simultaneous eigenfunctions of H_0, R_{13}, R_{23}, $\cdots$, R_{n3}.

Let a typical energy state be written as

$$\psi_{gm}=U_g(\mathbf{r}_1\cdots\mathbf{r}_n)[++-+\cdots]. \tag{2}$$

Here $\mathbf{r}_1\cdots\mathbf{r}_n$ designates the center-of-mass coordinates of the n molecules, and $+$ and $-$ symbols represent the internal energies of the various molecules. If the number of $+$ and $-$ symbols are denoted by n_+ and n_-, respectively, then m is defined as

$$\begin{aligned} m&=\tfrac{1}{2}(n_+-n_-),\\ n&=n_++n_-=\text{number of gaseous molecules.}\end{aligned} \tag{3}$$

If the energy of motion and mutual interaction of the molecules is denoted by E_g, then the total energy of the system is

$$E_{gm}=E_g+mE. \tag{4}$$

It is evident that the index m is integral or half-integral depending upon whether n is even or odd. Because of the various orders in which the $+$ and $-$ symbols can be arranged, the energy E_{gm} has a degeneracy

$$\frac{n!}{(\frac{1}{2}n+m)!(\frac{1}{2}n-m)!}. \tag{5}$$

This degeneracy has its origin in the internal coordinates only.

In addition, the wave function may have additional degeneracy from the center-of-mass coordinates. It should be noted in this connection that the degeneracy of the total wave function will depend upon whether or not the molecules are regarded as distinguishable or not.

If the molecules are indistinguishable, the symmetry of U_g will depend upon the symmetries of the wave function under interchanges of internal coordinates. For example, the states with all molecules excited are symmetric under an interchange of the internal coordinates of any two molecules. Consequently, for these states U_g must be symmetric for Bose molecules and antisymmetric for Fermi molecules. The limitations of symmetry are normally without physical significance as it is assumed that the gas is of such low density that the various molecules have nonoverlapping wave functions.

Of the Hamiltonian equation (1), H_0 operates on the center-of-mass coordinates only and gives

$$H_0U_g=E_gU_g, \tag{6}$$

whereas R_{j3} operates on the plus or minus symbol in the jth place corresponding to the internal energy of the jth molecule. Except for the factor $\frac{1}{2}$, it is analogous to one of the Pauli spin operators. As operators similar to the other two Pauli operators are also needed in this development, the properties of all three are listed here.

$$\begin{aligned} &\qquad\quad\overset{j}{\downarrow}\\ R_{j1}[\cdots&\pm\cdots]=\tfrac{1}{2}[\cdots\mp\cdots],\\ R_{j2}[\cdots&\pm\cdots]=\pm\tfrac{1}{2}i[\cdots\mp\cdots],\\ R_{j3}[\cdots&\pm\cdots]=\pm\tfrac{1}{2}[\cdots\pm\cdots].\end{aligned} \tag{7}$$

It is also convenient to define the operators

$$R_k = \sum_{j=1}^{n} R_{jk}, \quad k = 1, 2, 3, \tag{8}$$

and the operator

$$R^2 = R_1^2 + R_2^2 + R_3^2. \tag{9}$$

In this notation the Hamiltonian becomes

$$H = H_0 + ER_3, \tag{10}$$

and

$$R_3\psi_{gm} = m\psi_{gm}. \tag{11}$$

To complete the description of the dynamical system, there must be added to the Hamiltonian that of the radiation field and the interaction term between field and the molecular system.

For the purpose of definiteness the ineraction of a molecule with the electromagnetic field will be assumed to be electric dipole. The main results are actually independent of the type of coupling. The interaction energy of the jth molecule with the electromagnetic field can be written as

$$-\mathbf{A}(\mathbf{r}_j) \cdot \sum_{k=1}^{N-1} \frac{e_k}{m_k c} \mathbf{P}_k. \tag{12}$$

Here the configuration coordinates of the molecule are taken to be the center-of-mass coordinates and the coordinates relative to the center of mass of any $N-1$ of the N particles which constitute the jth molecule. e_k and m_k are the charge and mass of the kth particle, and $\mathbf{P}_k$ is the momentum conjugate to the position of the kth particle relative to the center of mass. The molecule is assumed electrically neutral.

Since $\mathbf{P}_k$ is an odd operator, it has only off-diagonal elements in a representation with internal energy diagonal. Hence the general form of Eq. (12) is

$$-\mathbf{A}(\mathbf{r}_j) \cdot (\mathbf{e}_1 R_{j1} + \mathbf{e}_2 R_{j2}). \tag{13}$$

$\mathbf{e}_1$ and $\mathbf{e}_2$ are constant real vectors the same for all molecules. The total interaction energy then becomes

$$H_1 = -\sum_j \mathbf{A}(\mathbf{r}_j) \cdot (\mathbf{e}_1 R_{j1} + \mathbf{e}_2 R_{j2}). \tag{14}$$

Since the dimensions of the gas cell are small compared with a wavelength, the dependence of the vector potential on the center of mass of the molecules can be omitted and the interaction energy (12) becomes

$$H_1 = -\mathbf{A}(0) \cdot (\mathbf{e}_1 R_1 + \mathbf{e}_2 R_2). \tag{15}$$

Since the interaction term Eq. (15) does not contain the center-of-mass coordinates, the selection rule on the molecular motion quantum number g is $\Delta g = 0$. Consequently there is no Doppler broadening of the transition frequency. This results solely from the small size of the gas container.[5]

The operators R_1, R_2, and R_3, apart from a factor of $\hbar$, obey the same commutation relations as the three components of angular momentum. Consequently, the interaction operator Eq. (15) obeys the selection rule $\Delta m = \pm 1$. In general, it has nonvanishing matrix elements between a given state Eq. (2) and a large number of states with $\Delta m = \pm 1$. In order to simplify the calculation of spontaneous radiation transitions, it is desirable that a set of stationary states be selected in such a way that the interaction term has matrix elements joining a given state with, at most, one state of higher and lower energy, respectively. Because of the very close analogy between this formalism and that of a system of particles of spin $\frac{1}{2}$, known results can be taken over from the spin formalism.

In a manner similar to an angular momentum formalism,[6] the operations H and R^2 commute; consequently, stationary states can be chosen to be eigenstates of R^2. These new states are linear combinations of the states of Eq. (2). The operator R^2 has eigenvalues $r(r+1)$. r is integral or half-integral and positive, such that

$$|m| \leqslant r \leqslant \tfrac{1}{2}n. \tag{16}$$

The eigenvalue r will be called the "cooperation number" of the gas. Denote the new eigenstates by

$$\psi_{gmr}. \tag{17}$$

Here

$$H\psi_{gmr} = (E_g + mE)\psi_{gmr}, \tag{18}$$

$$R^2\psi_{gmr} = r(r+1)\psi_{gmr}. \tag{19}$$

The degeneracy of the stationary states is not completely removed by introducing R^2. The state (g, m, r) has a degeneracy

$$\frac{n!(2r+1)}{(\frac{1}{2}n+r+1)!(\frac{1}{2}n-r)!}. \tag{20}$$

The complete set of eigenstates ψ_{gmr} may be specified in the following way: the largest value of m and r is

$$r = m = \tfrac{1}{2}n.$$

This state is nondegenerate in the internal coordinates and may be written as

$$\psi_{g,\frac{1}{2}n,\frac{1}{2}n} = U_g \cdot [++\cdots+]. \tag{20a}$$

All the states with this same value of $r = \frac{1}{2}n$, but with different values of m, are nondegenerate also and may be generated as[7]

$$\psi_{gmr} = [(R^2 - R_3^2 - R_3)^{-\frac{1}{2}}(R_1 - iR_2)]^{r-m}\psi_{grr}. \tag{21}$$

The operator $R_1 - iR_2$ reduces the m index by unity every time it is applied and the fractional power operator is to preserve the normalization of the wave function.[8] The fractional power operator is defined as having positive eigenvalues only.

[5] R. H. Dicke, Phys. Rev. **89**, 472 (1953).

[6] E. U. Condon and G. H. Shortley, *The Theory of Atomic Spectra* (Cambridge University Press, Cambridge, 1935), pp. 45–49.

[7] See reference 6, p. 48, Eq. (3).

[8] See reference 6, p. 48.

The state $\psi_{g,\frac{1}{2}n-1,\frac{1}{2}n}$ is one of n states with this value of m. The remaining $n-1$ states should be chosen to be orthogonal to this state, orthogonal to each other, and normalized. Since these remaining $n-1$ states are not states of $r=\frac{1}{2}n$, they must be states of $r=\frac{1}{2}n-1$, the only other possibility. Again the complete set of states with this value of r can be generated using Eq. (21), where now $r=\frac{1}{2}n-1$, and the operator in Eq. (21) is applied to each of the $n-1$ orthogonal states of $r=m=\frac{1}{2}n-1$. This procedure can be repeated until all possible values of r are exhausted, in which case all the stationary states have been defined.

With this definition of the stationary states, the interaction energy operator has matrix elements joining a given state of the gas to but two other states. Aside from the factor involving the radiation field operator, the matrix elements of the interaction energy may be written[8]

$$(g, r, m|\mathbf{e}_1R_1+\mathbf{e}_2R_2|g, r, m\mp 1)\\ =\tfrac{1}{2}(\mathbf{e}_1\pm i\mathbf{e}_2)[(r\pm m)(r\mp m+1)]^{\frac{1}{2}}. \quad (23)$$

Transition probabilities will be proportional to the square of the matrix elements. In particular, the spontaneous radiation probabilities will be

$$I=I_0(r+m)(r-m+1). \quad (24)$$

Here, by setting $r=m=\frac{1}{2}$, it is evident that I_0 is the radiation rate of a gas composed of one molecule in its excited state. I_0 has the value[9]

$$I_0=\frac{4}{3}\frac{\omega^2}{c}\left|\left(\sum_k \frac{e_k\mathbf{P}_k}{m_kc}\right)_{+-}\right|^2=\frac{1}{3}\frac{\omega^2}{c}|\mathbf{e}_1-i\mathbf{e}_2|^2\\ =\frac{1}{3}\frac{\omega^2}{c}(e_1^2+e_2^2). \quad (25)$$

If $m=r=\frac{1}{2}n$ (i.e., all n molecules excited),

$$I=nI_0. \quad (26)$$

Coherent radiation is emitted when r is large but $|m|$ small. For example, for even n let

$$r=\tfrac{1}{2}n, \quad m=0; \quad I=\tfrac{1}{2}n(\tfrac{1}{2}n+1)I_0. \quad (27)$$

This is the largest rate at which a gas with an even number of molecules can radiate spontaneously. It should be noted that for large n it is proportional to the square of the number of molecules.

Because of the fact that with the choice of stationary states given by Eq. (21) a given state couples with but one state of lower energy, this radiation rate [Eq. (27)], is an absolute maximum. Any superposition state will radiate at the rate

$$I=I_0\sum_{r,m}P_{r,m}(r+m)(r-m+1)\\ =I_0\langle(R_1+iR_2)(R_1-iR_2)\rangle, \quad (28)$$

where $P_{r,m}$ is the probability of being in the state r, m.

[9] Reference 1, p. 106.

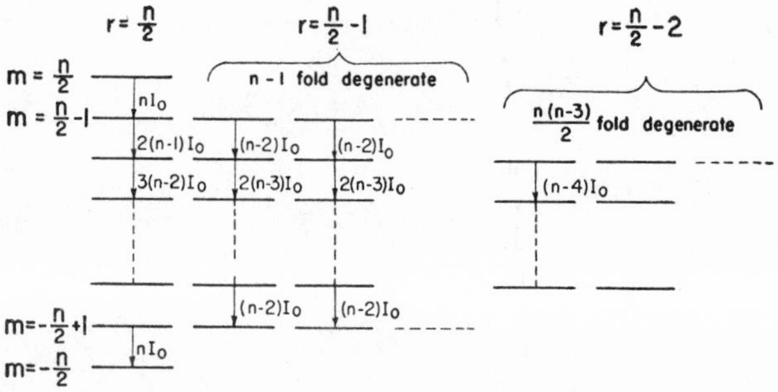

FIG. 1. Energy level diagram of an n-molecule gas, each molecule having 2 nondegenerate energy levels. Spontaneous radiation rates are indicated. $E_m=mE$.

There are no interference terms. Consequently, no superposition state can radiate more strongly than Eq. (27). An energy level diagram which shows the relative magnitudes of the various radiation probabilities is given in Fig. 1.

States with a low "cooperation number" are also highly correlated but in such a way as to have abnormally low radiation rates. For example, a gas in the state $r=m=0$ does not radiate at all. This state, which exists only for an even number of molecules, is analogous to a classical system of an even number of oscillators swinging in pairs oppositely phased.

The energy trapping which results from the internal scattering of photons by the gas appears naturally in the formalism. As an example, consider an initial state of the gas for which one definite molecule, and only this molecule, is excited. The gas at first radiates at the normal incoherent rate for a short time and thereafter fails to radiate. The probability of a photon's being emitted during the radiating period is $1/n$. These results follow from the fact that the assumed state is a linear superposition of the various states with $m=1-n/2$, and that $1/n$ is the probability of being in the state $r=\frac{1}{2}n$. The probability that the energy will be "trapped" is $(n-1)/n$. This is analogous to the radiation by a classical oscillator when $n-1$ similar unexcited oscillators are near. The solution of this classical problem shows that only $1/n$ of the excitation energy is radiated. The remainder appears in nonradiating normal modes of the system.

For want of a better term, a gas which is radiating strongly because of coherence will be called "super-radiant." There are two obvious ways in which a "super-radiant" state may be excited. First, if all the molecules be excited, the gas is in the state characterized by

$$r=m=\tfrac{1}{2}n. \quad (29)$$

As the system radiates it passes to states of lower m with r unchanged. This will take the system to the "super-radiant" region $m\sim 0$.

Another way in which such a state can be excited is to start with the gas in its ground state,

$$r=-m=\tfrac{1}{2}n, \quad (30)$$

and irradiate it with a pulse of radiation.[9a] If the pulse is sufficiently intense, the system is lifted to energy states with $m\sim 0$ but with r unchanged, and these states are "super-radiant."

Although the "super-radiant" states have abnormally large spontaneous radiation rates, the stimulated emission rate is normal. For example, with the system in the state m, r, the stimulated emission rate is proportional to

$$(r+m)(r-m+1)-(r+m+1)(r-m)=2m. \quad (30a)$$

With $m>0$ this is the normal incoherent stimulated emission rate. For $m<0$ this becomes the negative of the incoherent absorption rate.

As has been pointed out, the pulse technique for exciting "super-radiant" states is commonly used in nuclear magnetic resonance experiments. Here there is one important point that needs clarification, however. Instead of starting in the highly organized state given by Eq. (30) the pulse is applied to a system that is in thermal equilibrium at high temperatures. For example, if the system be a set of proton spins, the energy necessary to turn a spin over in the magnetic field may be about

$$E\sim 10^{-5}kT. \quad (31)$$

Under these conditions the two spin states of the proton are very nearly equally populated and it might be expected that thermal equilibrium would imply a badly disorganized system. The randomness in the initial state does not imply, however, complete randomness in m and r. For a gas with n, large states of low r have a high degeneracy. These states have a high statistical weight and are favored. However, Eq. (16) sets a lower bound on r for any m. The result is a relatively small range of values of m and r. For a system with n molecules in thermal equilibrium the mean square deviation from the mean of m is

$$n/4-\bar{m}^2/n. \quad (32)$$

Here $\bar{m}$ is the mean of m and is for high temperatures equal to

$$\bar{m}=-\tfrac{1}{4}nE/kT. \quad (33)$$

For a definite value of m the mean value of $r(r+1)$ is

$$m^2+\tfrac{1}{2}n, \quad (34)$$

and the mean square deviation is

$$\tfrac{1}{4}n^2-m^2. \quad (35)$$

The expression (32)–(35) may be easily derived using the density matrix formalism assuming the appropriate statistical ensemble.

It is hence clear that if

$$\bar{m}^2\gg n\gg 1, \quad (36)$$

the percentage deviation from the mean of m is small, that the percent deviation from the mean of $r(r+1)$ is small, and that the mean of $r(r+1)$ is approximately the smallest value compatible with the mean value of m. Thus, in the case of a gas system at high temperature, for sufficiently large n, values of m and r cluster to such an extent that the system may be considered as approximately in a state of definite $r=m=-nE/4kT$. If this gas is excited by a pulse of the proper intensity to excite states $m\sim 0$, the radiation rate after the pulse is approximately

$$I\cong I_0 r(r+1)\cong I_0 n^2(E/4kT)^2, \quad (37)$$

which is proportional to n^2 and hence coherent. A better calculation good for all temperatures gives the result [see Eq. (78) with $\theta=90°$]

$$I=\tfrac{1}{4}I_0 n(n-1)\tanh^2(E/2kT)+\tfrac{1}{2}nI_0. \quad (37a)$$

SEMICLASSICAL TREATMENT

For the spontaneous radiation from super-radiant states ($m\sim 0$) a semiclassical treatment is generally adequate. This method, which is a generalization of the well-known picture used in describing radiation from a nuclear spin system,[10] treats the molecular systems quantum mechanically but calculates the radiation process classically. In the following calculation the gas system will be assumed to be excited by a radiation pulse, which excites it from thermal equilibrium to a set of super-radiant states. To calculate the radiation rate, the expectation value of the electric dipole moment is treated as a classical dipole. When the gas contains a large number of molecules the dipole moment of the gas as a whole should be given by the sum of the expectation values of the individual dipole moments.

In thermal equilibrium the gas may be considered as having n_- molecules in the ground state and n_+ molecules in the excited state. A molecule which is initially in its ground state is assumed to be thrown into a superposition state of $+$ and $-$ by the radiation pulse. It is assumed that there is a unity probability ratio. The internal part of the wave function of the molecules after the pulse is given by

$$\psi_+=\frac{1}{\sqrt{2}}\left\{[+]\exp\left(-i\frac{\omega}{2}t\right)+[-]\exp i\left(\frac{\omega}{2}t+\delta\right)\right\}. \quad (38)$$

This is the most general form for ψ_+ apart from a possible multiplication phase factor. Here δ is a phase given by the phase of the exciting pulse. In a similar way a molecule in the excited state has its wave function converted to

$$\psi_-=\frac{1}{\sqrt{2}}\left\{[-]\exp i\frac{\omega}{2}t-[+]\exp\left(-i\frac{\omega}{2}t-i\delta\right)\right\}. \quad (39)$$

Instead of calculating the expectation value of the electric dipole moment it is more convenient to calculate the expectation value of the polarization current of the

[9a] See F. Bloch and I. I. Rabi, Revs. Modern Phys. 17, 237 (1945), for a discussion of the effect of a pulse on the analogous spin-$\frac{1}{2}$ system.

[10] F. Bloch, Phys. Rev. 70, 460 (1946).

jth molecule given by

$$\left(\sum_{k=1}^{N-1}\frac{e_k\mathbf{P}_k}{m_k}\right)=c\langle\mathbf{e}_1R_{j1}+\mathbf{e}_2R_{j2}\rangle$$

$$=\pm\tfrac{1}{2}c[\mathbf{e}_1\cos(\omega t+\delta)+\mathbf{e}_2\sin(\omega t+\delta)]. \quad (40)$$

The plus sign is obtained from the plus state, Eq. (38), and the negative sign from Eq. (39). Note the oscillating time dependence which results from the states being energy-superposition states. The polarization current for the gas as a whole is then

$$\mathbf{j}=(n_+-n_-)(c/2)[\mathbf{e}_1\cos(\omega t+\delta)+\mathbf{e}_2\sin(\omega t+\delta)]. \quad (41)$$

The radiation rate calculated classically is then[11]

$$I=\frac{2}{3}\frac{\omega^2}{c_3}|\mathbf{J}^2|=\frac{1}{12}\frac{\omega^2}{c}(n_+-n_-)^2(e_1^2+e_2^2). \quad (42)$$

In thermal equilibrium $n_+/n_-=\exp(-E/kT)$, from which

$$n_+-n_-=n\tanh(E/2kT). \quad (43)$$

Substituting into Eq. (42) gives the classical radiation rate

$$I=\frac{1}{12}\frac{\omega^2}{c}n^2(e_1^2+e_2^2)\tanh^2\left(\frac{E}{2kT}\right). \quad (44)$$

This may be compared with the quantum-mechanical result [Eq. (37a) and Eq. (25)]. For large n the two results are equal.

CLASSICAL MODEL

When the gas is in a state of definite "cooperation number" r which has a very large value, it is possible to represent it in its interaction with the electromagnetic field by a simple classical model. The energy-level spacing and the matrix elements joining adjacent levels are similar to those of a rotating top of large angular momentum and carrying an electric dipole moment. The details depend upon $\mathbf{e}_1$ and $\mathbf{e}_2$,which in turn depend on the nature of the original states. Let us consider a specific example. Assume that the radiators are atoms having a $^1P_1$ excited state and a $^1S_0$ ground state. Assume that the degeneracy of the excited state is split by a magnetic field in the z direction and that the $m_l=1$ excited level is being used. Under these conditions $\mathbf{e}_1$ and $\mathbf{e}_2$ are orthogonal to each other and the z axis, and the system has energy levels and interactions with the field identical with those of a spinning top having an electric dipole moment along its axis and precessing about the z axis as a result of an interaction with a static *electric* field in that direction. Consequently, since large quantum numbers are involved, to a good approximation the gas can be replaced by this classical model, which consists of a spinning top, in calculating both the interaction of the field on the gas and *vice versa*.

[11] Reference 1, p. 26.

RADIATION LINE BREADTH AND SHAPE

Under conditions for which the above "classical model" is valid, it is easy to calculate the natural line breadth and shape factor. This is of considerable importance in microwave spectroscopy. It has been customary to regard the natural line breadth as too small to be of any practical importance. However, as will be seen below, when coherence is properly taken into account the natural radiation breadth of the line may be far from negligible.

Using the above classical model, the angle between the spin axis and the z axis (the polar angle) will be designated as φ. In this approximation the quantum number m may be replaced by

$$m=r\cos\varphi, \quad (44a)$$

from which, using Eq. (24), the radiation rate becomes

$$I=I_0r^2\sin^2\varphi. \quad (44b)$$

Also, the internal energy of the gas is

$$mE=rE\cos\varphi. \quad (44c)$$

Balancing the radiation rate to the energy loss of the gas gives

$$\dot{\varphi}=(I_0r/E)\sin\varphi,$$

from which, assuming $\varphi=90°$ if $t=0$,

$$\sin\varphi=\operatorname{sech}(\alpha t),$$

where $\alpha=I_0r/E$. The radiated wave has the following form as a function of time:

$$A(t)=\begin{cases}e^{i\omega t}\sin\varphi, & t>0,\\ 0, & t<0,\end{cases}\qquad \hbar\omega=E.$$

The Fourier transform gives the line shape and has the value

$$\mathcal{A}(\beta)=\left(\frac{\pi}{2}\right)^{\frac{1}{2}}\frac{1}{\alpha}\operatorname{sech}\left(\frac{\pi}{2}\frac{\beta-\omega}{\alpha}\right). \quad (44d)$$

It should be noted that this is not of the usual Lorentz form. The line width at half-intensity points is

$$\Delta\omega=1.12I_0r/E=1.12\gamma r. \quad (44e)$$

Here γ is the line width at half-intensity points for the radiation from isolated single molecules. Putting in the maximum value of r gives a line breadth of $\Delta\omega=1.12\gamma n/2$, which is generally very substantially larger than γ.

RADIATION FROM A GAS OF LARGE EXTENT

A classical system of simple harmonic oscillators distributed over a large region of space can be so phased relative to each other that coherent radiation is obtained in a particular direction. It might be expected also that the radiating gas under consideration would have energy levels such that spontaneous radiation occurs coherently in one direction.

It will be assumed that the gas occupies a region having dimensions generally larger than radiation wavelength but small compared with the reciprocal of the natural line width,

$$\Delta k = \Delta\omega / c.$$

It is necessary to turn again to the general expression for the interaction term in the Hamiltonian equation (13). The vector potential operator can be expanded in plane waves:

$$\mathbf{A}(\mathbf{r}) = \sum_{\mathbf{k}'} [\mathbf{v}_{\mathbf{k}'} \exp(i\mathbf{k}' \cdot \mathbf{r}) + \mathbf{v}_{\mathbf{k}'}^* \exp(-i\mathbf{k}' \cdot \mathbf{r})]. \quad (45)$$

$\mathbf{v}_{\mathbf{k}'}$ and its Hermitian adjoint $\mathbf{v}_{\mathbf{k}'}^*$ are photon destruction and creation operators, respectively. After substituting Eq. (45) into (13), the interaction term becomes

$$H_1 = -\tfrac{1}{2} \sum_{\mathbf{k}'} \mathbf{v}_{\mathbf{k}'} \cdot (\mathbf{e}_1 - i\mathbf{e}_2) \sum_{j=1}^{n} R_{j+} \exp(i\mathbf{k}' \cdot \mathbf{r}_j)$$

$$-\tfrac{1}{2} \sum_{\mathbf{k}'} \mathbf{v}_{\mathbf{k}'}^* \cdot (\mathbf{e}_1 + i\mathbf{e}_2) \sum_{j=1}^{n} R_{j-} \exp(-i\mathbf{k}' \cdot \mathbf{r}_j), \quad (46)$$

where $R_{j\pm} = R_{j1} \pm iR_{j2}$. In this expression, terms involving the product of the photon creation operator and the "excitation operator" R_{j+}, etc., have been dropped as these terms do not lead to first-order transitions for which energy is conserved. The form of Eq. (46) suggests defining the operators:

$$\begin{aligned} R_{\mathbf{k}1} &= \textstyle\sum_j (R_{j1} \cos \mathbf{k} \cdot \mathbf{r}_j - R_{j2} \sin \mathbf{k} \cdot \mathbf{r}_j), \\ R_{\mathbf{k}2} &= \textstyle\sum_j (R_{j1} \sin \mathbf{k} \cdot \mathbf{r}_j + R_{j2} \cos \mathbf{k} \cdot \mathbf{r}_j). \end{aligned} \quad (47)$$

In terms of these operators the interaction energy becomes

$$H_1 = -\tfrac{1}{2} \textstyle\sum_{\mathbf{k}'} (\mathbf{v}_{\mathbf{k}'} \cdot \mathbf{e} R_{\mathbf{k}'+} + \mathbf{v}_{\mathbf{k}'}^* \cdot \mathbf{e}^* R_{\mathbf{k}-'}), \quad (48)$$

where

$$R_{\mathbf{k}'\pm} = R_{\mathbf{k}'1} \pm iR_{\mathbf{k}'2} = \sum_{j=1}^{n} R_{j\pm} \exp(\pm i\mathbf{k}' \cdot \mathbf{r}_j),$$

$$\mathbf{e} = \mathbf{e}_1 - i\mathbf{e}_2.$$

For every direction of propagation $\mathbf{k}$ there are two orthogonal polarizations $\mathbf{v}_{\mathbf{k}}$ of $\mathbf{A}$. By a proper choice of polarization basis, the dot product of one of the basic polarizations with $\mathbf{e}$ can be assumed zero. This radiation oscillator is never excited and can be ignored. The orthogonal polarization is the one which couples with the gas. The polarization of emitted or absorbed radiation is uniquely given by the direction of propagation and need not be explicitly indicated.

The operators of Eq. (47), together with R_3, obey the angular momentum commutation relations. The operator

$$R_{\mathbf{k}}^2 = R_{\mathbf{k}1}^2 + R_{\mathbf{k}2}^2 + R_3^2 \quad (49)$$

commutes with the operators of Eq. (47) and with R_3. In Eq. (49) $\mathbf{k}$ is regarded as a fixed index. This operator does not commute with another one of the same type having a different index. Omitting for a moment the translational part of the wave function, wave functions may be so chosen as to be simultaneous eigenfunctions of the internal energy ER_3 and $R_{\mathbf{k}}^2$. They may be written as ψ_{mr} and are generated by an expression analogous to Eq. (21):

$$R_{\mathbf{k}}^2 \psi_{mr} = r(r+1)\psi_{mr}, \quad ER_3\psi_{mr} = mE\psi_{mr}. \quad (50)$$

By analogy with the development leading to Eq. (24) it is clear that these states represent correlated states of the gas for which radiation emitted in the $\mathbf{k}$ direction is coherent. Thus, coherence is limited to a particular direction only, provided the initial state of the gas is given by a function of the same type as Eq. (50). The selection rules for the absorption or emission of a photon with momentum $\mathbf{k}$ are

$$\Delta r = 0, \quad \Delta m = \pm 1. \quad (51)$$

The spontaneous radiation rate in the direction $\mathbf{k}$ is given by Eq. (24), where I and I_0 are now to be interpreted as radiation rates per unit solid angle in the direction $\mathbf{k}$. This may be written as

$$I(\mathbf{k}) = I_0(\mathbf{k})[(r+m)(r-m+1)]. \quad (51a)$$

If a photon is emitted or absorbed having a momentum $\mathbf{k}' \neq \mathbf{k}$, the selection rules are

$$\Delta r = \pm 1, 0; \quad \Delta m = \pm 1. \quad (52)$$

To prove this, it may be noted that the commutation relations of the $2n$ operators

$$\begin{aligned} R_{j1}' &= R_{j1} \cos(\mathbf{k} \cdot \mathbf{r}_j) - R_{j2} \sin(\mathbf{k} \cdot \mathbf{r}_j), \\ R_{j2}' &= R_{j1} \sin(\mathbf{k} \cdot \mathbf{r}_j) + R_{j2} \cos(\mathbf{k} \cdot \mathbf{r}_j), \end{aligned} \quad (53)$$

with those of Eq. (47) are of the same type as denoted by Condon and Shortley[12] as $\mathbf{T}$. The selection rules satisfied by these operators are of the type given by Eq. (52).[13] The operators of Eq. (47), with $\mathbf{k} = \mathbf{k}'$, may be expressed as linear combinations of those of Eq. (53). Hence the operators of Eq. (47), with $\mathbf{k}$ replaced by $\mathbf{k}'$, satisfy the selection rules given by Eq. (52).

As was discussed previously in the dipole approximation, super-radiant states may be excited by irradiating the gas with radiation until states in the vicinity of $m=0$ are excited. In the present case the incident radiation is assumed to be plane with a propagation vector $\mathbf{k}$. After excitation the gas radiates coherently in the $\mathbf{k}$ direction. Because of the selection rules Eq. (52), radiation in directions other than $\mathbf{k}$ tends to destroy the coherence with respect to the direction $\mathbf{k}$ by causing transitions generally to states of lower r.

DOPPLER EFFECT

Because of the occurrence of the center-of-mass coordinates in the "cooperation" operator Eq. (49), it fails to commute with H_0 [Eq. (1)]; hence eigenstates of $R_{\mathbf{k}}^2$ are generally not stationary. This is equivalent to the fact that relative motion of classical oscillators will gradually destroy the coherence of the emitted radiation. If, on the other hand, a set of classical oscillators all move with the same velocity, the state of coherence

[12] Reference 6, p. 59.
[13] Reference 6, pp. 60–61.

is stationary. The corresponding question in the case of the quantum mechanical system is whether there exist simultaneous eigenstates of H and $R_{\mathbf{k}}^2$ such that coherent radiation is emitted in a transition from one state to another. By starting with the state defined by

$$\psi_{srr}=(\exp i\mathbf{s}\cdot\sum_j \mathbf{r}_j)\cdot[+++\cdots+], \quad r=n/2, \quad (54)$$

and using the method leading to Eq. (21), there is obtained the set of states

$$\psi_{smr}=[(R_{\mathbf{k}}^2-R_3^2-R_3)^{-\frac{1}{2}}(R_{\mathbf{k}1}-iR_{\mathbf{k}2})]^{r-m}\psi_{srr}. \quad (55)$$

If it is assumed that the gas is free, the functions Eq. (55) are simultaneous eigenfunctions of H and $R_{\mathbf{k}}^2$. Consequently, the coherence in the **k** direction is stationary.

These states are analogous to the classical oscillators all moving with the same speed. Note one important difference, however; from Eq. (55) the momentum of an excited molecule is always

$$\mathbf{p}_+=\hbar\mathbf{s}, \quad (56)$$

whereas if a molecule is in its ground state the momentum, as given by Eq. (55), is

$$\mathbf{p}_-=\hbar(\mathbf{s}-\mathbf{k}), \quad (57)$$

the difference being the recoil momentum of the photon. Thus, the coherent states Eq. (55) are always a superposition of states such that the excited molecules have one momentum and the unexcited have another. Hence it is clear that the recoil momentum given to a molecule when it radiates in the **k** direction does not produce a molecular motion which destroys the coherence but rather is required to preserve the coherence.

The gain or loss in photon energy which has its origin in the Doppler effect is equal to the loss or gain in the kinetic energy of a radiator which results from the photon-induced recoil. Expressed as a fractional shift in photon frequency, this is

$$\frac{\Delta\omega}{\omega}=\frac{\hbar(\mathbf{S}-\frac{1}{2}\mathbf{k})\cdot\mathbf{k}}{Mck}. \quad (58)$$

Here M is the molecular mass. For energy states such that $|m|\ll n/2$, Eq. (58) can be written as

$$\frac{\Delta\omega}{\omega}=\frac{\mathbf{v}\cdot\mathbf{k}}{ck}. \quad (59)$$

Where **v** is the total momentum of the gas divided by its total mass. Equation (59) is the usual classical expression for the Doppler shift for a radiator moving with a velocity **v**. Consequently, for the highly correlated states $|m|\sim 0$ the Doppler effect can be described in classical terms.

The stationary states Eq. (55) do not form a complete set. In particular, the final state, a photon being emitted or absorbed with a momentum not **k**, is not one of these states. The set of stationary states may be made complete by adding all the other possible orthogonal plane wave states, each being characterized by a definite momentum and internal energy for each molecule. With this set of orthogonal states, matrix elements can be easily calculated for transitions from the states given by Eq. (55) to states in which photons appear having momenta not equal to **k**. These matrix elements are found to have a magnitude characteristic of the incoherent radiation process. It should be noted that only for one magnitude of **k** as well as for direction are the matrix elements of a coherent transition obtained.

PULSE-INDUCED COHERENCE RADIATION

It will be assumed in this section that a gas initially in thermal equilibrium is illuminated for a short time by an intense radiation pulse. The intensity and angular dependence of the spontaneous radiation emitted after the pulse will be calculated. In order to avoid the difficulties associated with motional effects, the molecules will be assumed so massive that their center-of-mass coordinates can be represented by small stationary wave packets. The center-of-mass coordinates will be then treated as time-independent parameters in the equation. It is assumed that the intensity of the exciting radiation pulse is so great that the fields acting on the gas during the pulse can be considered as described classically. The spontaneous radiation rate after the exciting pulse will be calculated quantum mechanically.

Because the initial state of the gas is a mixed state describing thermodynamic equilibrium, it is convenient to use the density matrix formalism.[14] It will be assumed that one has an ensemble of gas systems statistically identical and that what one is calculating is certain ensemble averages.

For a pure state, Eq. (28) shows that the spontaneous radiation rate in the **k**′ direction can be written as the expectation value

$$I(\mathbf{k}')=I_0(\mathbf{k}')\langle R_{\mathbf{k}'+}R_{\mathbf{k}'-}\rangle. \quad (60)$$

For a state which may be mixed or pure using the density matrix formalism this becomes the trace

$$I(\mathbf{k}')=I_0(\mathbf{k}')\ \mathrm{tr}R_{\mathbf{k}'-}\rho R_{\mathbf{k}'+}. \quad (62)$$

Here the density matrix is defined as the ensemble mean

$$\rho=[\psi\psi^*]_{\mathrm{Av}}. \quad (63)$$

In Eq. (63) the wave function ψ is interpreted as a column vector and the * is the Hermitian adjoint. The symbol $[\ ]_{\mathrm{Av}}$ signifies an ensemble mean.

Assume that the exciting radiation pulse is in the form of a plane wave in the **k** direction. The fields which act on the various molecules differ only in their arrival time. The Hamiltonian of the system can be written

$$H=\hbar\omega R_3-\sum_j \mathbf{A}_j(t)\cdot(\mathbf{e}_1R_{j1}+\mathbf{e}_2R_{j2}). \quad (64)$$

Here $\mathbf{A}_j(t)$ is a classical field quantity and

$$\mathbf{A}_j(t)=0, \quad \begin{matrix} t<t_j, \\ t>t_j+\tau \end{matrix} \quad (65)$$

[14] R. C. Tolman, *The Principles of Statistical Mechanics* (Clarendon Press, Oxford, 1938), p. 325.

where t_j is the arrival time of the radiation pulse at the jth molecule. Neglecting for the moment the interaction term, the time dependence of the wave function can be given by the unitary transformation

$$\psi(t)=\exp(-i\omega tR_3)\cdot\psi(0). \quad (66)$$

In general, the wave function after the interaction with the electromagnetic field can be obtained through a unitary transformation on the wave function prior to the pulse. The wave function of the gas after the radiation pulse has passed completely over the gas can be related to that before by

$$\psi'(t)=\exp(-i\omega tR_3)T\psi(0). \quad (67)$$

Here T is a unitary matrix which represents the effect of the pulse on the gas. To find the most general form of T it is convenient to consider the effect of the pulse on a particular molecule. Since this molecule has only two internal states of interest, its wave function can be regarded as a spinor in a pseudo "spin space." Then, apart from a multiplicative phase factor which has no physical significance, any unitary transformation can be represented as a rotation in "spin space." Any arbitrary rotation can be represented as a rotation about the No. 3 axis followed by a rotation about an axis perpendicular to No. 3. Except for the arrival time the radiation pulse is identical in its effect on each molecule of the gas. The operator T can be written then as the product

$$T=\exp[i\omega \textstyle\sum_j t_jR_{j3}]$$
$$\cdot\prod_l \exp i\left[\frac{\theta}{2}(R_{l+}\alpha+R_{l-}\alpha^*)+\theta' R_{l3}\right]$$
$$\cdot\exp[-i\omega \textstyle\sum_j t_jR_{j3}]. \quad (67a)$$

The first and second rotations are through angles of θ' and θ, respectively, and the phase of α determines the direction of the 2nd rotation axis. It is assumed that $|\alpha|=1$ and that the arrival time at the jth molecule is

$$t_j=(1/\omega)\mathbf{k}\cdot\mathbf{r}_j. \quad (67b)$$

Equation (67a) becomes Eq. (68) after making use of (67b):

$$T=\exp i\frac{\theta}{2}(R_{\mathbf{k}+}\alpha+R_{\mathbf{k}-}\alpha^*)\cdot\exp i\theta' R_3. \quad (68)$$

It should be noted that the effect of the different times of arrival of the pulse at the various molecules is contained in $\mathbf{k}\cdot\mathbf{r}_j$ which appears in $R_{\mathbf{k}\pm}$ in Eq. (68).

The reason for choosing this transformation to be a rotation about No. 3 followed by a perpendicular rotation is that the rotation about No. 3 is the same as a time displacement and has no effect since the initial state is assumed to be one of thermal equilibrium.

Assume that the initial density matrix can be written as

$$\rho_0=\frac{\exp(-ER_3/kT)}{\text{tr}\exp(-ER_3/kT)}=2^{-n}\prod_i(1-\gamma R_{j3}), \quad (69)$$
$$\gamma=2\tanh(E/2kT).$$

The density matrix after the radiation pulse is

$$\rho(t)=\exp(-i\omega tR_3)\cdot T\rho_0T^{-1}\exp(i\omega tR_3). \quad (70)$$

The spontaneous radiation rate after the exciting pulse is given by Eq. (62) which becomes

$$I(\mathbf{k}')=I_0(\mathbf{k}')\,\text{tr}T\rho_0T^{-1}R_{\mathbf{k}'+}R_{\mathbf{k}'-}, \quad (71)$$

since R_3 commutes with $R_{\mathbf{k}'+}R_{\mathbf{k}'-}$. The radiation rate is thus independent of the time after the exciting pulse. This is because the effect of the radiated field on the gas has been neglected. Equation (71) is to be interpreted as the radiation rate immediately after the exciting pulse. Since ρ_0 and R_3 commute, Eq. (71) can be written as

$$I(\mathbf{k}')=I_0(\mathbf{k}')\,\text{tr}\exp[\tfrac{1}{2}i\theta(R_{\mathbf{k}+}\alpha+R_{\mathbf{k}-}\alpha^*)]\cdot\rho_0$$
$$\cdot\exp[-\tfrac{1}{2}i\theta(R_{\mathbf{k}+}\alpha+R_{\mathbf{k}-}\alpha^*)]\cdot R_{\mathbf{k}'+}R_{\mathbf{k}'-}. \quad (72)$$

It is desirable to transform ρ_0 before evaluating the trace

$$\rho'=\exp[\tfrac{1}{2}i\theta(R_{\mathbf{k}+}\alpha+R_{\mathbf{k}-}\alpha^*)]$$
$$\cdot\rho_0\exp[-\tfrac{1}{2}i\theta(R_{\mathbf{k}+}\alpha+R_{\mathbf{k}-}\alpha^*)]$$
$$=2^{-n}\textstyle\prod_j(1-\gamma R_{j3}^\dagger), \quad (73)$$

where

$$R_{j3}^\dagger=R_{j3}\cos\theta-\tfrac{1}{2}i(R_{j+}'\alpha-R_{j-}'\alpha^*)\sin\theta. \quad (74)$$

The primed operators are obtained from Eq. (53) as

$$R_{j\pm}'=R_{j1}'\pm iR_{j2}'=R_{j\pm}\exp(\pm i\mathbf{k}\cdot\mathbf{r}_j). \quad (75)$$

The trace in Eq. (72) can now be evaluated to give

$$I(\mathbf{k}')=I_0(\mathbf{k}')\sum_{jl}\text{tr}2^{-n}\prod_s(1-\gamma R_{s3}^\dagger)R_{j+}''R_{l-}''. \quad (76)$$

The double prime is Eq. (75) referred to the $\mathbf{k}'$ direction. To evaluate the trace the following relations are needed: For A_i and B_j functions of the R's of molecules i and j,

$$\text{tr}\,A_iB_j=2^{-n}\,\text{tr}A_i\,\text{tr}B_j,$$
$$\text{tr}R_{j3}=\text{tr}R_{j\pm}=0,\quad \text{tr}R_{j3}^2=2^{n-2}, \quad (77)$$
$$\text{tr}R_{j+}R_{j-}=\text{tr}R_{j-}R_{j+}=2^{n-1}.$$

The final result is

$$I(\mathbf{k}')=I_0(\mathbf{k}')\cdot\tfrac{1}{2}n[1-\cos\theta\cdot\tanh(E/2kT)$$
$$+\tfrac{1}{2}\sin^2\theta\cdot\tanh^2(E/2kT)$$
$$\cdot(n|[\exp i(\mathbf{k}-\mathbf{k}')\cdot\mathbf{r}]_{\text{Av}}|^2-1)]. \quad (78)$$

Here the symbol $[\ ]_{\text{Av}}$ signifies a mean over all the molecules of the gas. For the example considered in Eq. (37a) this mean is unity, and Eq. (37a) follows by integrating over all directions of the emitted radiation. Aside from the factor $I_0(\mathbf{k}')$, the directional dependence of the emitted radiation is given by this mean. This factor is identical with the distribution factor for radiation about a set of classical isotropic radiators which have been excited by a plane wave. Consequently, for a θ of 90° and $n\tanh^2(E/kT)$ large compared with unity, the angular distribution of radiation is just the classical one.

The physical significance of the angle θ is that $\sin^2\tfrac{1}{2}\theta$

is the probability of the pulse exciting a molecule in its ground state. Also, if the exciting pulse is a constant amplitude wave of frequency ω during the duration of the pulse, the angle θ is proportional to the product of pulse amplitude and duration.

If the radiating system consists of a set of particles of spin $\frac{1}{2}$ in a uniform magnetic field, the angle θ has a geometrical significance. The initial state of a particle will have spin parallel or antiparallel to the field. The radiofrequency pulse will change its state such that its spin axis will be tipped through an angle θ. Note that if $\theta = 180°$ the populations of the + and − populations have been just interchanged, corresponding to a transition from a positive temperature T to the negative temperature $-T$.[15] $\theta = 90°$ corresponds to the excitation of molecules to energy superposition states Eqs. (38) and (39) for which the gas is radiating coherently.

ANGULAR CORRELATION OF SUCCESSIVE PHOTONS

The system to be considered here is assumed to be initially in thermal equilibrium. It is allowed to radiate spontaneously. The angular correlation between successive photons is calculated. This correlation was implicit in some of the earlier development, for example in Eq. (51a). As an example, consider a gas composed of widely separated molecules, all excited. Assume that a photon is emitted in the **k** direction. The radiation rate for the second photon in this direction is by Eq. (51a).

$$I(\mathbf{k}) = I_0(\mathbf{k})2(n-1). \tag{79}$$

This is twice the incoherent rate. It is not hard to show that for an intermolecular spacing large compared with a radiation wavelength the radiation rate averaged over all directions is the incoherent rate. Hence from Eq. (79) the radiation probability in the direction **k** has twice the probability averaged over all directions.

In the problem to be considered, the system will consist initially of the gas in thermal equilibrium having a temperature T (possibly negative) and a photonless field. The molecules will be assumed fixed in position and with intermolecular distances large compared with a radiation wavelength. Photons are observed to be emitted in the directions $\mathbf{k}_1, \mathbf{k}_2, \cdots, \mathbf{k}_{s-1}$ and only these photons are emitted. The problem is one of finding the radiation rate in the $\mathbf{k}_s$ direction for the next photon.

Stated more exactly, it is assumed that there is an ensemble of gaseous systems, each with its own external radiation field. Every member of the ensemble which is capable of radiating will eventually radiate a photon. Those members which radiate their first photon into a small solid angle in the direction $\mathbf{k}_1$, are selected to form a new ensemble. For this second ensemble the time zero is taken to be the time that a photon was detected for each member of the ensemble.

It is convenient to calculate correlations for the gas systems forming a microcanonical distribution having an energy per gas system of m_0E. The results for a canonical distribution with a temperature T can subsequently be determined as an average over the microcanonical distributions.

Since the initial state of the system is assumed photonless, it is sufficient to give the explicit dependence of the initial density matrix on the molecular coordinates. Except for normalization this can be written as a projection operator for states of molecular energy m_0E. A particularly useful form for this density matrix is

$$\rho_0 = \frac{\sum_{q=1}^{n} \exp 2\pi i \frac{q}{n}(R_3 - m_0)}{\operatorname{tr} \sum_{q=1}^{n} \exp 2\pi i \frac{q}{n}(R_3 - m_0)}. \tag{80}$$

This is a convenient way to write the density matrix because of the relation

$$\exp\left(2\pi i \frac{q}{n} R_3\right) = \prod_j \exp\left(2\pi i \frac{q}{n} R_{j3}\right) = \prod_j \left[\cos\left(\pi \frac{q}{n}\right) + 2iR_{j3} \sin\left(\pi \frac{q}{n}\right)\right]. \tag{81}$$

Here the product is over $j = 1, \cdots, n$. To illustrate the importance of Eq. (81) the trace appearing in the denominator D of Eq. (80) will be calculated using the relations Eq. (77).

$$\begin{aligned} D &= \sum_{q=1}^{n} \exp\left(-2\pi i \frac{q}{n} m_0\right) \cdot \operatorname{tr} \prod_j \left[\cos\left(\pi \frac{q}{n}\right) + 2iR_{j3} \sin\left(\pi \frac{q}{n}\right)\right] \\ &= \sum_{q=1}^{n} 2^n \exp\left(-2\pi i \frac{q}{n} m_0\right) \cdot \cos^n\left(\pi \frac{q}{n}\right) \\ &= \frac{n!n}{(\frac{1}{2}n + m_0)!(\frac{1}{2}n - m_0)!}, \quad |m_0| < \frac{n}{2} \\ &= 2n \quad \text{for } |m_0| = n/2. \end{aligned} \tag{82}$$

After one photon has been emitted and absorbed in the photon detector, the system is again photonless and its density matrix is (see Appendix 1)

$$\rho_1 = (R_{\mathbf{k}_1-}\rho_0 R_{\mathbf{k}_1+}) / (\operatorname{tr} R_{\mathbf{k}_1-}\rho_0 R_{\mathbf{k}_1+}). \tag{83}$$

After $s-1$ photons it is

$$\rho_{s-1} = \frac{R_{\mathbf{k}_{s-1}-} \cdots R_{\mathbf{k}_1-}\rho_0 R_{\mathbf{k}_1+} \cdots R_{\mathbf{k}_{s-1}+}}{\operatorname{tr} R_{\mathbf{k}_{s-1}-} \cdots R_{\mathbf{k}_1-}\rho_0 R_{\mathbf{k}_1+} \cdots R_{\mathbf{k}_{s-1}+}}. \tag{84}$$

The R's are defined in Eqs. (48) and (47) or (46). The radiation rate in the $\mathbf{k}_s$ direction immediately after the $s-1$ photon is from Eq. (62)

$$I(\mathbf{k}_s) = I_0(\mathbf{k}_s) \operatorname{tr} R_{\mathbf{k}_s-}\rho_{s-1} R_{\mathbf{k}_s+}. \tag{85}$$

Note that $s \leqslant \frac{1}{2}n + m_0$. For any l, $R_{l\pm}^2 = 0$. Consequently,

[15] E. M. Purcell and R. V. Pound, Phys. Rev. 81, 279 (1951).

the numerator of Eq. (84) can be written

$$\frac{1}{(s-1)!}\sum_{u,v\cdots=1}^{s-1}\sum_{u',v'\cdots=1}^{s-1}\sum_{j,l\cdots=1}^{n} \times \exp i[(\mathbf{k}_u-\mathbf{k}_{u'})\cdot\mathbf{r}_j+(\mathbf{k}_v-\mathbf{k}_{v'})\cdot\mathbf{r}_l+\cdots]. \quad (86)$$
$$R_{j-}R_{l-}\cdots\rho_0\cdots R_{l+}R_{j+}.$$

Each of the above sums is over $s-1$ indices, including only terms for which all $s-1$ indices take on different values. The trace of the expression appears in the denominator of Eq. (84). In order to evaluate this trace it is necessary first to evaluate

$$\operatorname{tr}R_{j-}R_{l-}\cdots\rho_0\cdots R_{l+}R_{j+}=\operatorname{tr}\rho_0\cdots R_{l+}R_{l-}R_{j+}R_{j-} = \operatorname{tr}\rho_0\cdots(\tfrac{1}{2}+R_{l3})(\tfrac{1}{2}+R_{j3}). \quad (87)$$

If Eqs. (80), (81), and (82) are substituted into Eq. (87), and use is made of Eq. (77) and the equality

$$\operatorname{tr}[\cos(\pi q/n)+2iR_{j3}\sin(\pi q/n)](\tfrac{1}{2}+R_{j3}) = 2^{n-1}\exp(i\pi q/n), \quad (87a)$$

Eq. (87) becomes

$$=\frac{2^{n-s+1}}{D}\sum_{q=1}^{n}\exp\left[i\pi\frac{q}{n}(s-1-2m_0)\right]\cdot\cos^{n-s+1}\left(\pi\frac{q}{n}\right)$$
$$=\frac{(n-s+1)!(\frac{1}{2}n+m_0)!}{n!(\frac{1}{2}n+m_0-s+1)!},\quad |m_0|<\tfrac{1}{2}n \text{ or } |m_0|=\tfrac{1}{2}n,\ s=1$$
$$=\tfrac{1}{2},\quad |m_0|=\tfrac{1}{2}n,\ s>1. \quad (88)$$

Making use of Eq. (88) the denominator of Eq. (84) can be written as

$$=P_{s-1}\frac{(n-s+1)!(\frac{1}{2}n+m_0)!}{n!(\frac{1}{2}n+m_0-s+1)!},\quad |m_0|<\tfrac{1}{2}n \text{ or } |m_0|=\tfrac{1}{2}n,\quad s=1$$
$$=\tfrac{1}{2}P_{s-1},\quad m_0=\tfrac{1}{2}n,\ s>1, \quad (89)$$

where

$$P_{s-1}=\frac{1}{(s-1)!}\sum_{u,v\cdots=1}^{s-1}\sum_{u',v'\cdots=1}^{s-1}\sum_{j,l\cdots=1}^{n} \times\exp i[(\mathbf{k}_u-\mathbf{k}_{u'})\cdot\mathbf{r}_j+(\mathbf{k}_v-\mathbf{k}_{v'})\cdot\mathbf{r}_l+\cdots],\quad s>1$$
$$P_0=1. \quad (90)$$

Here, as before, each of the above sums is over $s-1$ indices, including only terms for which all $s-1$ indices take on different values. If Eq. (84) is substituted into Eq. (85), the numerator is Eq. (89) with s increased by one unit. Consequently, substituting Eq. (89) into Eq. (85),

$$I(\mathbf{k}_s)=I_0(\mathbf{k}_s)\frac{P_s(\frac{1}{2}n+m_0-s+1)}{P_{s-1}(n-s+1)}. \quad (91)$$

To restate the meaning of this equation, $I(\mathbf{k}_s)$ is the radiation probability per unit time per unit solid angle in the direction $\mathbf{k}_s$; $I_0(\mathbf{k}_s)$ is the corresponding radiation probability for a single isolated excited molecule. It has been assumed that the gas was initially in the energy state m_0E [see Eq. (3)] with a random distribution over the degeneracy of this state. The gas was observed to radiate photons $\mathbf{k}_1$, $\mathbf{k}_2$, $\cdots\mathbf{k}_{s-1}$ previously to $\mathbf{k}_s$. Equation (91) is the radiation rate immediately after the $\mathbf{k}_{s-1}$ photon was observed. As a check on the correctness of this expression, note that the incoherent rate is obtained if $s=1$. Also, for $m_0=\frac{1}{2}n$ and $\mathbf{k}_1=\mathbf{k}_2=\cdots=\mathbf{k}_s=\mathbf{k}$, the radiation rate Eq. (91) agrees with Eq. (51a).

It should be noted that Eq. (91) is independent of the ordering of the subscripts 1, $\cdots$, $s-1$. Consequently, the angular distribution of the s photon is dependent upon the direction of a previous photon but is independent of the previous photon's position in the sequence of prior photons.

For a gas which contains a large number of randomly positioned molecules and for which previous photons have either been emitted in the direction $\mathbf{k}_3$ or in quite different directions, the radiation rate [Eq. (91)] is approximately equal to the incoherent rate times the number of photons previously emitted in this direction plus one.

Perhaps the case of most physical interest is where $s=2$. In this case Eq. (91) becomes

$$I(\mathbf{k}_2)=I_0(\mathbf{k}_2)\frac{\frac{1}{2}n+m_0-1}{n-1}[n|[\exp i\Delta\mathbf{k}\cdot\mathbf{r}]_{\text{Av}}|^2+n-2],$$
$$\Delta\mathbf{k}=\mathbf{k}_2-\mathbf{k}_1. \quad (92)$$

The symbol $[\ ]_{\text{Av}}$ signifies an average over all the molecular positions.

In case of a gas system at a temperature T, Eq. (91) must be averaged over all possible values of m_0 to give

$$\bar{I}(\mathbf{k}_s)=I_0(\mathbf{k}_s)\frac{P_s\displaystyle\sum_{m_0=s-\frac{1}{2}n-1}^{\frac{1}{2}n}(\tfrac{1}{2}n+m_0+1-s)\frac{n!}{(\frac{1}{2}n+m_0)!(\frac{1}{2}n-m_0)!}\exp\left(-\frac{m_0E}{kT}\right)}{(n-s+1)P_{s-1}\displaystyle\sum_{m_0=s-\frac{1}{2}n-1}^{\frac{1}{2}n}\frac{n!\exp(-m_0E/kT)}{(\frac{1}{2}n+m_0)!(\frac{1}{2}n-m_0)!}}. \quad (93)$$

For $|E/kT|\ll1$ and $s\ll n$, Eq. (93) can be approximated by

$$\bar{I}(\mathbf{k}_s)=I_0(\mathbf{k}_s)\frac{(\frac{1}{2}n+\bar{m}_0+1-s)P_s}{(n-s+1)P_{s-1}}, \quad (94)$$

where

$$\bar{m}_0=-\tfrac{1}{4}nE/kT.$$

It is a pleasure to acknowledge the assistance of the author's colleague, Professor A. S. Wightman, who read the manuscript and made a number of helpful suggestions.

APPENDIX I

It is assumed that the system consists initially of a gas with an energy m_0E and a photonless radiation field. A photon and only one photon is observed to be emitted. The effect of the photon emission on the state of the system is required.

There are two separate effects to be considered. First there is the effect on the state of the system which has its origin in the interaction between the field and gas. Second there is the effect of the observation which determines that a photon and one photon only has been emitted, that this photon was emitted in the **k** direction, and that the photon was absorbed in the detector. The first part of the problem is solved using Schrödinger's equation. The Hamiltonian of the system is

$$H=\hbar\omega R_3+H_0+H', \quad H_0=\sum_{\mathbf{k}'} H_{\mathbf{k}'},$$
$$H'=-\tfrac{1}{2}\sum_{\mathbf{k}'}[\mathbf{v}_{k'}\cdot\mathbf{e}R_{\mathbf{k}'+}+\mathbf{v}_{\mathbf{k}'}{}^*\cdot\mathbf{e}^*R_{\mathbf{k}'-}]. \quad (95)$$

Here $H_{\mathbf{k}'}$ is the energy of the $\mathbf{k}'$ radiation oscillator. Assume a pure state represented by a wave function ψ_0 at a time $t=0$. Assume that ψ_0 is an eigenstate of R_3 and is photonless. At some later time it is

$$\psi(t)=\exp(-iHt/\hbar)\psi_0=\left(1-\frac{i}{\hbar}Ht-\frac{H^2}{2\hbar^2}t^2+\cdots\right)\psi_0. \quad (96)$$

For the quadratic and higher powers of t each term will be a sum of products of H' and $(H_0+\hbar\omega R_3)$. However, the interaction term H' consists of sums of terms of the type

$$U_{\mathbf{k}'}=\mathbf{v}_{\mathbf{k}'}\cdot\mathbf{e}R_{\mathbf{k}'+} \quad (97)$$

and its Hermitian adjoint. The operator $U_{\mathbf{k}'}$ consists of the product of a photon annihilation operator and a gas excitation operator. It converts an eigenstate of R_3 and H_0 into another such or it gives zero. The most general term operating on ψ_0 in Eq. (96) is therefore a product of powers of $H_0+\hbar\omega R_3$ and terms of the type $U_{\mathbf{k}'}$ and $U_{\mathbf{k}'}{}^*$ taken in various orders. In each of these terms $H_0+\hbar\omega R_3$ always operates on an eigenfunction and consequently can be moved to the end of the product as a number, the eigenvalue. Consequently $\psi(t)$ becomes

$$\psi(t)=[1+\sum_{\mathbf{k}'} g_{\mathbf{k}'}(t)U_{\mathbf{k}'}{}^*+\sum_{\mathbf{k}'} h_{\mathbf{k}'}(t)U_{\mathbf{k}'}U_{\mathbf{k}'}{}^*$$
$$+\sum_{\mathbf{k}'\mathbf{k}''} g_{\mathbf{k}'\mathbf{k}''}(t)U_{\mathbf{k}'}{}^*U_{\mathbf{k}''}{}^*+\cdots]\psi_0. \quad (98)$$

The g's and h's are numbers, functions of the time. It may be noted that since ψ_0 represents a photonless state, an annihilation operator for a given radiation oscillator $\mathbf{k}'$ appears only if preceded by the corresponding creation operator.

Assuming that at the time t a photon measurement is made which indicates the presence of photon **k** and no other photons, the wave function after the measurement is

$$\psi'=P_{\mathbf{k}}\psi, \quad (99)$$

where the operator $P_{\mathbf{k}}$ is a projection operator for the **k** photon state.

$$P_{\mathbf{k}}=\frac{H_{\mathbf{k}}}{\hbar\omega_k}\prod_{\mathbf{k}'}{}'\left(\frac{\hbar\omega_{k'}-H_{\mathbf{k}'}}{\hbar\omega_{k'}}\right). \quad (100)$$

The product is over all $\mathbf{k}'\neq\mathbf{k}$. Two-photon excitation of one radiation oscillator has been neglected.

$$\psi'=[g_{\mathbf{k}}(t)U_{\mathbf{k}}{}^*+\sum_{\mathbf{k}'} H_{\mathbf{k}\mathbf{k}'}(t)U_{\mathbf{k}}{}^*U_{\mathbf{k}'}U_{\mathbf{k}'}{}^*$$
$$+\sum_{\mathbf{k}} I_{\mathbf{k}\mathbf{k}'}(t)U_{\mathbf{k}'}U_{\mathbf{k}'}{}^*U_{\mathbf{k}}{}^*+\cdots]\psi_0. \quad (101)$$

In summing over the direction of $\mathbf{k}'$ in the second and third terms above, the expression

$$R_{\mathbf{k}'+}R_{\mathbf{k}'-}=\sum_{ab}\exp[i\mathbf{k}'\cdot(\mathbf{r}_a-\mathbf{r}_b)]\cdot R_{a+}R_{b-} \quad (102)$$

appears under the integral. By expanding the exponential in spherical harmonics it can be seen that for $a\neq b$ this integral vanishes, as it has been assumed that

$$\mathbf{k}'\cdot(\mathbf{r}_a-\mathbf{r}_b)\gg 1 \text{ for } a\neq b.$$

It should be indicated that the angular dependence is not wholly in the exponential in Eq. (102) but exists in part in the square of the dot product of **e** and $\mathbf{v}_{\mathbf{k}'}$. However, this contribution to the angular dependence includes only spherical harmonics of finite degree in fact with $l<3$. As the only terms which need to be included in Eq. (102) are $a=b$, Eq. (102) becomes

$$R_{\mathbf{k}'+}R_{\mathbf{k}'-}=\tfrac{1}{2}+R_3+(\text{terms from } a\neq b). \quad (103)$$

Independent of its position in a series of products of U's the expression on the right side of Eq. (103) will operate on an eigenfunction and becomes an eigenvalue which can be removed as a number. In the higher-order terms in Eq. (101) $U_{\mathbf{k}'}$ and $U_{\mathbf{k}'}{}^*$ may not appear adjacent to each other, but if they do not, some other pair such as $U_{\mathbf{k}''}U_{\mathbf{k}''}{}^*$ will appear, and after removing this as an eigenvalue another such pair will occur, and eventually the $\mathbf{k}'$ pair will be adjacent. Consequently, to all orders in the expansion

$$\psi'=f(t)U_{\mathbf{k}}{}^*\psi_0, \quad (104)$$

where f is a function of the time of observation. As the photon detector also absorbs the photon, the wave function must be multiplied by the annihilation operator $\mathbf{e}\cdot\mathbf{v}_{\mathbf{k}}$. This gives, except for the time factor,

$$\psi''\sim R_{\mathbf{k}-}\psi_0, \quad (105)$$

which is another photonless state but with one quantum less energy.

If the initial density matrix ρ_0 contains only photonless states of the same energy m_0E, then from Eqs. (63) and (105) it is transformed to

$$\rho_1=R_{\mathbf{k}-}\rho_0R_{\mathbf{k}+}/\mathrm{tr}(R_{\mathbf{k}-}\rho_0R_{\mathbf{k}+}), \quad (106)$$

representing the photonless state of the ensemble of systems after the emission, detection, and absorption of photon described by **k**.

PAPER NO. 10

Reprinted from *Physical Review*, Vol. 93, pp. 121–123, 1954.

A Stokes-Parameter Technique for the Treatment of Polarization in Quantum Mechanics

U. Fano
National Bureau of Standards, Washington, D. C.
(Received September 4, 1953)

A technique is presented which adapts the Stokes method to quantum mechanics and serves to calculate polarization effects by means of Pauli matrices. Illustrative examples indicate that this technique may be not only convenient mathematically but also physically more transparent than previous methods.

1. INTRODUCTION

THE treatment of the polarization of electromagnetic radiation in terms of Stokes parameters[1] has been drawing increased attention, probably because it relies on operational concepts and therefore is particularly suited to quantum physics.[2] Recent successes in experimenting with the polarization of high-energy radiations[3] have stimulated theoretical studies of polarization effects and thereby increased the desirability of more powerful techniques for these studies.

No general technique for the quantum-mechanical application of the Stokes method seems to be available in the literature, even though the ideas of this method are being utilized by an increasing number of workers. Since such a technique is in fact amenable to a rather simple formulation, it may be worthwhile to present it in this paper. The method constitutes no more than a transcription of established quantum mechanical theory. A few very simple examples will illustrate the operation of the technique.

2. THE STOKES PARAMETER METHOD

(a) Stokes Parameters

The intensity and the kind and degree of polarization of a beam of light (or other electromagnetic radiation) can be represented by four parameters as follows. Take two orthogonal unit vectors $\mathbf{A}_1$ and $\mathbf{A}_2$ perpendicular to the beam direction as a frame of reference. The parameters are:

(1) the total light intensity I_0;

(2) the difference I_1 between the light intensity transmitted by a filter (Nicol prism) which accepts the linear polarization $\mathbf{A}_1$ and the intensity transmitted by the "opposite" filter which accepts $\mathbf{A}_2$;

(3) the difference I_2 between the light intensities transmitted by a filter which accepts the linear polarization $(\mathbf{A}_1+\mathbf{A}_2)/\sqrt{2}$ and by the opposite filter;

(4) the difference I_3 between the intensities transmitted by a filter which accepts circular polarization rotating from $\mathbf{A}_1$ to $\mathbf{A}_2$ and by the opposite filter.

For convenience in the following applications we write

$$I_1=I_0P_\zeta,\quad I_2=I_0P_\xi,\quad I_3=I_0P_\eta \tag{1}$$

and regard the set of parameters as a four component vector $I_0(1, \mathbf{P})$, where $\mathbf{P}=(P_\zeta, P_\xi, P_\eta)$ is a vector of the three-dimensional Poincaré representation,[2] which describes the kind and degree of polarization $(P\leq 1)$.[4]

(b) Density Matrix

Since two independent states of light polarization constitute a complete set, the quantum-mechanical treatment of polarization is mathematically equivalent (isomorphic) to the treatment of the orientation of a spin $\frac{1}{2}$ particle. Therefore the density matrix which represents the polarization state of a light beam according to quantum mechanics is a 2×2 matrix, which can be resolved into the sum of a unit matrix $\mathcal{I}$ and of Pauli matrices $(\omega_\zeta, \omega_\xi, \omega_\eta)=\boldsymbol{\omega}$. (These matrices are taken in the usual representation, with ω_ζ diagonal.) The coefficients of this sum are the Stokes parameters and we write the density matrix[5] in the form

$$\mathcal{J}=\tfrac{1}{2}I_0(\mathcal{I}+\mathbf{P}\cdot\boldsymbol{\omega}). \tag{2}$$

(c) Response of a Light Detector

A detector that serves as a polarization analyzer responds to light of different polarizations with different efficiency. Maximum and minimum efficiencies ϵ_M and ϵ_m correspond to completely polarized beams which have opposite polarizations with Poincaré vectors $\mathbf{P}$ equal, respectively, to $\mathbf{Q}$ and $-\mathbf{Q}$ $(Q=1)$. Quantum-mechanically, the detector is represented by an operator with the eigenstates $\mathbf{Q}$ and $-\mathbf{Q}$ and with the eigen-

[1] G. G. Stokes, Trans. Cambridge Phil. Soc. **9**, 399 (1852).
[2] See, e.g., U. Fano, J. Opt. Soc. Am. **39**, 859 (1949).
[3] Hoover, Faust, and Dohne, Phys. Rev. **85**, 58 (1952) (double Compton scattering); E. Bleuler and H. L. Bradt, Phys. Rev. **73**, 1938 (1948); R. C. Hanna, Nature **162**, 332 (1948); C. S. Wu and I. Shaknov, Phys. Rev. **77**, 136 (1950); F. L. Hereford, Phys. Rev. **81**, 482, 627 (1951) (all on polarization of annihilation quanta); F. Metzger and M. Deutsch, Phys. Rev. **78**, 551 (1950) (polarization-direction correlation of gamma-quanta); A. P. French and J. O. Newton, Phys. Rev. **85**, 1041 (1952); S. B. Gunst and L. A. Page, Phys. Rev. **92**, 970 (1953) (polarization by transmission through magnetized iron); F. L. Hereford and J. P. Keuper, Phys. Rev. **90**, 1043 (1953) (polarization effects in photoelectric effect); K. Phillips, Phil. Mag. **44**, 169 (1953) (polarization of bremsstrahlung); D. H. Wilkinson, Phil. Mag. **43**, 659 (1952); L. W. Fagg and S. S. Hanna, Phys. Rev. **88**, 1205 (1952) (polarization analysis by deuteron disintegration).

[4] Unitary transformations of the frame of reference $(\mathbf{A}_1, \mathbf{A}_2)$ are accompanied by rotations of the Poincaré axes (ζ, ξ, η). The formalism described in this paper is independent of the choice of the frame of reference.

[5] The equivalence between the Stokes parameters and the density matrix has been pointed out by D. L. Falkoff and J. E. Macdonald, J. Opt. Soc. Am. **41**, 862 (1951).

values ϵ_M and ϵ_m, i.e., by the matrix[6]

$$\mathcal{O}=\tfrac{1}{2}[(\epsilon_M+\epsilon_m)\mathbf{1}+(\epsilon_M-\epsilon_m)\mathbf{Q}\cdot\boldsymbol{\omega}]. \quad (3)$$

The probability of response of this detector to a light beam with the density matrix (2) is given by the trace of the product of the matrices (2) and (3), namely,

$$\mathrm{Tr}(\mathcal{O}\mathcal{J})=\tfrac{1}{2}I_0[(\epsilon_M+\epsilon_m)+(\epsilon_M-\epsilon_m)\mathbf{Q}\cdot\mathbf{P}]. \quad (4)$$

(d) Probability of Interactions with Matter

The calculation of photon emission, absorption, and scattering will be adapted to the Stokes formalism in close analogy to a well known technique for calculating the collisions of free Dirac electrons.[7] The quantum-mechanical expression of interaction probabilities is usually proportional to the square of a perturbation matrix element V_{fi} which relates to the transition from an initial state i to a final state f. If the initial and/or final states are not "pure states" (e.g., not completely polarized), the transition probability is proportional to a sum $\sum_f{}^* \sum_i{}^* |V_{fi}|^2$; the asterisks indicate that the sums must be carried out according to rules implied by the statement of the problem (e.g., over both photon polarizations if the polarization is irrelevant to the problem).

The perturbation matrix element V_{fi} for an interaction of photons with matter is a linear function of the polarization vector $\mathbf{A}_i$ for any incident radiation and of the polarization vector $\mathbf{B}_f^\dagger$ for any outgoing radiation (the dagger denotes Hermitian conjugation). In turn, $\mathbf{A}_i$ may be expressed in terms of unit polarization vectors $\mathbf{A}_\alpha$ ($\alpha=1, 2$) of the incident radiation and of a two-component wave function $a_{i\alpha}$,[8] $\mathbf{A}_i=\sum_\alpha a_{i\alpha}\mathbf{A}_\alpha$; similarly $\mathbf{B}_f^\dagger=\sum_\beta b_{f\beta}^\dagger \mathbf{B}_\beta$.

When the $\sum_f{}^* \sum_i{}^* |V_{fi}|^2$ is formed, the $\sum_i{}^*$ may be factored out as $\sum_i{}^* a_{i\alpha}a_{i\alpha'}^\dagger$ and constitutes in fact the density matrix (2) of the incident radiation, $\mathcal{J}_{\alpha\alpha'}=\sum_i{}^* a_{i\alpha}a_{i\alpha'}^\dagger$. Similarly the $\sum_f{}^* b_{f\beta'}b_{f\beta}^\dagger$ may be factored out and regarded as the matrix (3), $\mathcal{O}_{\beta'\beta}$, of an ideal detector which accepts outgoing radiation with the kind and degree of polarization specified in the statement of the problem.

In a scattering problem, which involves both incident and outgoing radiation, the perturbation matrix element V_f, is a linear function of both $\mathbf{B}_f^\dagger$ and $\mathbf{A}_i$. It can be expressed as

$$V_{fi}(\mathbf{B}_f^\dagger, \mathbf{A}_i)=\sum_\beta \sum_\alpha b_{f\beta}^\dagger V(\mathbf{B}_\beta, \mathbf{A}_\alpha)a_{i\alpha}, \quad (5)$$

where $V(\mathbf{B}_\beta, \mathbf{A}_\alpha)$ may be regarded as a matrix $V_{\beta\alpha}$ and resolved, if desired, into a sum of standard matrices:

$$\begin{aligned}V(\mathbf{B}_\beta, \mathbf{A}_\alpha)=\tfrac{1}{2}\{&[V(\mathbf{B}_1, \mathbf{A}_1)+V(\mathbf{B}_2, \mathbf{A}_2)]\mathbf{1}\\&+[V(\mathbf{B}_1, \mathbf{A}_1)-V(\mathbf{B}_2, \mathbf{A}_2)]\omega_\zeta\\&+[V(\mathbf{B}_1, \mathbf{A}_2)+V(\mathbf{B}_2, \mathbf{A}_1)]\omega_\xi\\&+i[V(\mathbf{B}_1, \mathbf{A}_2)-V(\mathbf{B}_2, \mathbf{A}_1)]\omega_\eta\}. \quad (5')\end{aligned}$$

Accordingly, the $\sum_f{}^* \sum_i{}^* |V_{fi}|^2$ takes the form of the trace of a product of matrices, quite analogous to (27) of reference 7:

$$\begin{aligned}\sum_f{}^* \sum_i{}^* |V_{fi}|^2&=\sum_{\alpha\alpha'\beta\beta'} \mathcal{O}_{\beta'\beta}V_{\beta\alpha}\mathcal{J}_{\alpha\alpha'}V_{\alpha'\beta'}^\dagger\\&=\mathrm{Tr}(\mathcal{O}V\mathcal{J}V^\dagger)=\mathrm{Tr}(V^\dagger\mathcal{O}V\mathcal{J}). \quad (6)\end{aligned}$$

The core of the procedure suggested here is, then, to represent the perturbation matrix element as a polarization operator, according to (5) and (5′). In emission or absorption processes the procedure seems to fail because V_{fi} contains only one polarization vector, $B_f^\dagger$ or A_i, respectively, and cannot be reduced directly to a matrix $V_{\beta\alpha}$. However, in the event of emission, there is no density matrix $\mathcal{J}$ which depends on the polarization coordinates α, α'; the factors V and $V^\dagger$ in (6) constitute then a single polarization operator $V(\mathbf{B}_\beta)V^\dagger(\mathbf{B}_{\beta'})=(VV^\dagger)_{\beta\beta'}$ which can be resolved in the manner of (5′). In the event of absorption, the operator $\mathcal{O}$ disappears from (6) and one can construct the operator $(V^\dagger V)_{\alpha'\alpha}$.

A basic set of conventions for the treatment of polarization effects in terms of Stokes parameters and of matrices $\boldsymbol{\omega}$, according to (6), (2), (3), and (5′), has thus been completed. Products and traces of the matrices $\boldsymbol{\omega}$ are carried out according to the standard rules for Pauli matrices.

(e) Stokes Parameters of Emitted or Scattered Radiation

To display the intensity and the polarization of emitted or scattered radiation, it is not necessary to inquire about the probability of a specific event, such as the response of the detector represented by the matrix $\mathcal{O}$ in (6). Instead of constructing the full product $\mathcal{O}V\mathcal{J}V^\dagger$ in (6) and taking its trace, one may simply resolve the product $V\mathcal{J}V^\dagger$ into the sum of a unit matrix and of the polarization matrices $\boldsymbol{\omega}$. The coefficients of the matrices are proportional to the Stokes parameters of the emitted or scattered radiation. Indeed $V\mathcal{J}V^\dagger$ represents (here as well as in reference 7) the perturbed density matrix. Selection rules on the types of polarization resulting from particular processes become apparent upon inspection of the polarization operators which are contained in V and $V^\dagger$, as shown in the following examples.

3. EXAMPLES

(a) Dipole Emission of Light

In the theory of dipole emission, V is proportional to the displacement of electric charge, $\mathbf{r}$. In a transition

[6] If Q points along ζ, the matrix (3) is clearly diagonal with the stated eigenvalues; otherwise the diagonal form is achieved by a unitary transformation of ($\mathbf{A}_1, \mathbf{A}_2$), i.e., by a rotation of $\boldsymbol{\omega}$.

[7] See, e.g., W. Heitler, *Quantum Theory of Radiation* (Oxford University Press, London, 1944), second edition, p. 151 ff. The operator $(H_0+E_0)/4E_0$ of Eq. (27) of this reference represents, for unpolarized electrons of positive energy, the density matrix $\mathcal{J}$ which appears in (6); the operator $(H+E)/2E$ represents a detector $\mathcal{O}$ which accepts only positive energy electrons with any spin orientation.

[8] In the second quantization formalism, $a_{i\alpha}$ represents a destruction operator.

between fully specified states, $\mathbf{r}$ represents a specific matrix element; in general one may treat $\mathbf{rr}^\dagger$ as an operator whose expectation value can be determined later by a procedure such as (6). Proceeding as suggested in 2*e*, we resolve $VV^\dagger$, or $\mathbf{r}\cdot\mathbf{B}\,\mathbf{r}^\dagger\cdot\mathbf{B}^\dagger$, which amounts to the same, into a sum of Pauli matrices. Equation (5′) shows that the Stokes parameters of the emitted light are proportional, respectively, to

$$\mathbf{r}\cdot\mathbf{B}_1\,\mathbf{r}^\dagger\cdot\mathbf{B}_1+\mathbf{r}\cdot\mathbf{B}_2\,\mathbf{r}^\dagger\cdot\mathbf{B}_2=\mathbf{r}\cdot\mathbf{r}^\dagger-\mathbf{r}\cdot\mathbf{n}\,\mathbf{r}^\dagger\cdot\mathbf{n}, \tag{7}$$

$$\mathbf{r}\cdot\mathbf{B}_1\,\mathbf{r}^\dagger\cdot\mathbf{B}_1-\mathbf{r}\cdot\mathbf{B}_2\,\mathbf{r}^\dagger\cdot\mathbf{B}_2, \tag{7'}$$

$$\mathbf{r}\cdot\mathbf{B}_1\,\mathbf{r}^\dagger\cdot\mathbf{B}_2+\mathbf{r}\cdot\mathbf{B}_2\,\mathbf{r}^\dagger\cdot\mathbf{B}_1, \tag{7''}$$

$$i(\mathbf{r}\cdot\mathbf{B}_1\,\mathbf{r}^\dagger\cdot\mathbf{B}_2-\mathbf{r}\cdot\mathbf{B}_2\,\mathbf{r}^\dagger\cdot\mathbf{B}_1)=i\mathbf{r}\times\mathbf{r}^\dagger\cdot\mathbf{n}, \tag{7'''}$$

where $\mathbf{n}=\mathbf{B}_1\times\mathbf{B}_2$ indicates the direction of emission. The expression (7) for the total emission along $\mathbf{n}$ is trivial and familiar; the expression (7‴) for the degree of circular polarization may be somewhat more transparent than usual.

(b) Rayleigh Scattering

In the nonrelativistic approximation, the essential factor of V is the scalar product of the polarization vectors of the incident and scattered waves. If the unit vectors $\mathbf{A}_1$, $\mathbf{B}_1$ are laid in the plane of scattering, Eq. (5′) yields that V is proportional to

$$\tfrac{1}{2}[(1+\cos\vartheta)\mathcal{I}-(1-\cos\vartheta)\omega_\zeta], \tag{8}$$

where ϑ is the angle of scattering. The product $V\mathcal{J}V^\dagger$ is then proportional to

$$\begin{aligned}\tfrac{1}{8}[(1+\cos\vartheta)\mathcal{I}&-(1-\cos\vartheta)\omega_\zeta]I_0(\mathcal{I}+\mathbf{P}\cdot\omega)\\ &\times[(1+\cos\vartheta)\mathcal{I}-(1-\cos\vartheta)\omega_\zeta]\\ &=\tfrac{1}{4}[1+\cos^2\vartheta-\sin^2\vartheta P_\zeta]\mathcal{I}+\tfrac{1}{4}[(1+\cos^2\vartheta)P_\zeta-\sin^2\vartheta]\omega_\zeta\\ &\qquad+\tfrac{1}{2}\cos\vartheta P_\xi\omega_\xi+\tfrac{1}{2}\cos\vartheta P_\eta\omega_\eta.\end{aligned} \tag{9}$$

The results contained in this formula are rather well known. Notice how the anticommutation property of the ω's operates as a selection rule: the coefficient of each ω vanishes for the scattered wave whenever it vanishes for both the incident wave and the operator V; one might say loosely that the output contains any one kind of polarization only if that kind is present in the input or generated by the interaction.

(c) Bremsstrahlung

The bremsstrahlung cross section has been calculated relativistically in the Born approximation, taking into account polarization effects, by May[9] and by Gluckstern, Hull, and Breit.[10,11] Equation (15) of reference 11 shows this cross section as consisting of four terms. The last term, independent of polarization, arises from spin effects (these effects should also yield a polarization if the spin orientations were not averaged out in both the initial and final electron states). The first three terms depend on the components, in the direction of the polarization $\mathbf{B}$, of the electron momentum before and after the collision, p_{0l} and p_l; they are proportional, respectively, to p_{0l}^2, $2p_{0l}p_l$, and p_l^2, which are written in our notations as $\mathbf{p}_0\cdot\mathbf{B}\,\mathbf{p}_0\cdot\mathbf{B}^\dagger$, $\mathbf{p}_0\cdot\mathbf{B}\,\mathbf{p}\cdot\mathbf{B}^\dagger+\mathbf{p}\cdot\mathbf{B}\,\mathbf{p}_0\cdot\mathbf{B}^\dagger$, and $\mathbf{p}\cdot\mathbf{B}\,\mathbf{p}\cdot\mathbf{B}^\dagger$. These quantities can be expressed as a sum of Pauli matrices by means of (5′). Calling $\mathbf{n}$ the direction of the bremsstrahlung, ϑ_0 and ϑ the angles between $\mathbf{p}_0$ and $\mathbf{n}$ and between $\mathbf{p}$ and $\mathbf{n}$, and φ the angle between the planes $(\mathbf{p}_0\mathbf{n})$ and $(\mathbf{pn})$ and laying $\mathbf{B}_1$ in the plane $(\mathbf{p}_0\mathbf{n})$, with $\mathbf{p}_0\cdot\mathbf{B}_1>0$, we find

$$\begin{aligned}\mathbf{p}_0\cdot\mathbf{B}\,\mathbf{p}_0\cdot\mathbf{B}^\dagger&=\tfrac{1}{2}[\mathcal{I}+\omega_\zeta]\sin^2\vartheta_0,\\ \mathbf{p}_0\cdot\mathbf{B}\,\mathbf{p}\cdot\mathbf{B}^\dagger+\mathbf{p}\cdot\mathbf{B}\,\mathbf{p}_0\cdot\mathbf{B}^\dagger&\\ =\tfrac{1}{2}\sin\vartheta_0\sin\vartheta&[(\mathcal{I}+\omega_\zeta)\cos2\varphi+\omega_\xi\sin2\varphi],\\ \mathbf{p}\cdot\mathbf{B}\,\mathbf{p}\cdot\mathbf{B}^\dagger&=\tfrac{1}{2}\sin^2\vartheta[\mathcal{I}+\omega_\zeta\cos2\varphi+\omega_\xi\sin2\varphi].\end{aligned} \tag{10}$$

The Stokes parameters of the bremsstrahlung are obtained by replacing p_{0l}^2, $2p_{0l}p_l$, and p_l^2 in (15) of reference 11 with the expressions (10) and separating out the coefficients of the unit matrix and of the various Pauli matrices. The coefficient of the unit matrix is the Bethe-Heitler formula for the total intensity. The coefficients of ω_ζ and ω_ξ characterize the partial linear polarization (the coefficient of ω_ξ is bound to vanish when the direction of $\mathbf{p}$ is averaged out). The absence of ω_η in (10) indicates the total lack of circular polarization which is stressed in reference 11.

(d) Compton Effect

The relativistic treatment of Compton scattering yields a polarization which differs from that of Rayleigh scattering owing to the effect of electron spin. If the spin orientation before and after scattering is averaged out, the spin effects modify (9) only by the insertion of a new term in the coefficient of the unit matrix and in the coefficient of $P_\eta\omega_\eta$. The results for a nonrandom initial spin orientation are more complicated and are given in reference 2. The calculation for a nonrandom initial orientation combined with an analysis of the final orientation is still more complicated; it is being completed now by Tolhoek and Lipps[12] by means of a technique of the type presented in this paper.

[9] M. May, Phys. Rev. 84, 265 (1951).

[10] Gluckstern, Hull, and Breit, Science **114**, 480 (1951).

[11] Gluckstern, Hull, and Breit, Phys. Rev. **90**, 1026 (1953).

[12] I wish to thank Dr. Tolhoek and Mr. Lipps for information on their work and for participating, together with Dr. S. Berko and Dr. F. L. Hereford, in stimulating discussions.

PAPER NO. 11

Reprinted from *Proceedings of the Royal Society*, Ser. A, Vol. 208, pp. 263–277, 1951.

The concept of partial coherence in optics

BY H. H. HOPKINS, PH.D.

(*Communicated by Sir George Thomson, F.R.S.—Received* 16 *February* 1951)

It is shown that a 'phase-coherence factor' may be defined in a manner which leads, without recourse to explicit statistical analysis, to the theorems established by van Cittert (1934) and Zernike (1938) for analogous factors. An invalid approximation made in their calculations of the phase-coherence factor for a plane illuminated directly by a source is corrected. The new treatment is applied to the theory of Young's experiment, the stellar interferometer, and illumination in the microscope. The phase-coherence factor defined here enables a general theory of the formation of optical images to be formulated. Further, it is shown that the diameter of the area of coherence on a plane illuminated by a source of angular radius α is given by $d = \dfrac{0{\cdot}16\lambda}{N \sin \alpha}$, where N is the refractive index of the intervening medium.

1. INTRODUCTION

In classical wave theory it is usual to make a rigid distinction between coherent and incoherent disturbances. If two disturbances of complex amplitude $A_1\,e^{i\psi_1}, A_2\,e^{i\psi_2}$ are superposed at a given point, the resultant amplitude is written $u = A_1\,e^{i\psi_1} + A_2\,e^{i\psi_2}$, and the instantaneous intensity is

$$I = |\,u\,|^2 = A_1^2 + A_2^2 + 2A_1A_2\cos(\psi_1 - \psi_2). \tag{1}$$

Averaged over a finite interval of time, the mean intensity is given by the same expression (1), if the disturbances are coherent, that is, if they originate in the same element of the source. If they are incoherent, that is, if they originate in different elements of the source, the product term is absent, since, over a finite time, $\int \cos(\psi_1 - \psi_2)\,dt = 0$ when ψ_1, ψ_2 are random. This suggests that one might write

$$I = A_1^2 + A_2^2 + 2A_1A_2\gamma_{12} \tag{2}$$

for intermediate states, where the absolute amplitudes A_1, A_2 both receive contributions from a number of the same elements of the source and there is a partial correlation of phase. $\gamma_{12} = 0$ would then denote incoherence; and $\gamma_{12} = +1, -1$ would denote cophasal and antiphasal coherence respectively.

Earlier treatments of this question have employed explicit statistical calculations. For example, van Cittert (1934) defined a 'degree of consonance' and calculated functions expressing the correlation between the (complex) amplitudes at any two points. By means of rather elaborate calculations, he succeeded in establishing certain theorems which are given below using much simpler methods. The proofs given by van Cittert relate to the special case of a perfectly focused aberration-free optical system having a rectangular aperture. Zernike (1938) was able to generalize these proofs by employing a simpler definition of partial coherence.

Zernike based his definition on a consideration of Young's interference experiment. Suppose P_1, P_2 are two pinholes in a screen which is illuminated by a source Σ. If Σ subtends a small enough angle at the screen, fringes of high contrast will be seen in a plane some distance beyond the screen. If the angular size of Σ is increased, the visibility of the fringes decreases. This may be explained, in one way, by the overlapping of the (incoherent) interference patterns corresponding to the different points of the source, which patterns will be laterally displaced by amounts depending on the angular subtense of Σ at the screen. Alternatively, the disturbances at P_1, P_2 may be said to be only partially coherent, and a 'degree of coherence' may be defined by

$$V_{12} = \frac{I_{\text{max.}} - I_{\text{min.}}}{I_{\text{max.}} + I_{\text{min.}}}, \tag{3}$$

that is, by the visibility of the interference fringes as defined by Michelson (1890). For the disturbances at any two points P_1, P_2 Zernike defines the degree of coherence V_{12} to be the visibility of the interference fringes that may be obtained from them 'under the best circumstances'; in other words under optimum phase conditions, and with equal intensities. It is not clear from this definition that V_{12} has any immediate bearing on the superposition of two disturbances under any other conditions of phase and intensity. Again, if the coherence function is known for any domain, we may ask in what manner it can be employed in calculating (say) the superposed effects of disturbances from points of an extended area in the domain. In fact, as will be shown, both these questions may be treated by using what amounts to the complex form of Zernike's degree of coherence. It requires, however, to be redefined. When this is done, both the theory and its applications are immediately extended and become susceptible to simpler mathematical treatment. For example, no appeal need be made to statistical considerations beyond those implied in the elementary rules for the addition of coherent and incoherent disturbances.

2. The phase-coherence factor

Suppose the intensities I_1, I_2 at two points P_1, P_2 in a wave field are known (figure 1). If now disturbances from P_1 and P_2 reach a point P, either directly or by way of some system S (not necessarily the same for the two points), we may consider in what manner the two disturbances are to be combined at P. The time-averaged intensity at P will depend on the phase correlation between the disturbances at P_1 and P_2.

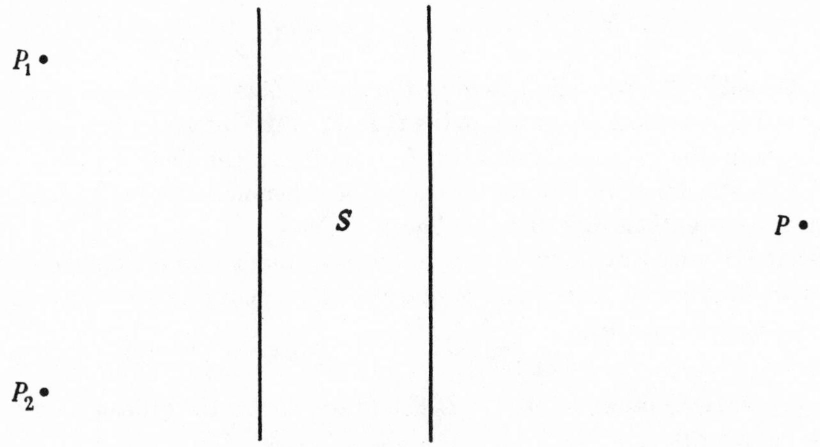

Figure 1.

To calculate the intensity at P an additional factor must therefore be known. Such a phase-coherence factor should be so defined that it enables one to combine the disturbances from P_1, P_2 knowing only the intensities I_1, I_2, and the imaging properties from P_1 to P, and from P_2 to P.

Suppose light from an extended source Σ to reach the points P_1, P_2 either directly or by way of some system. Let u_1, u_2 be the complex amplitudes produced at P_1, P_2 by an element $d\sigma$ of the source. u_1, u_2 are then coherent, but not necessarily in phase. Let f_1 be the complex amplitude produced at P by a disturbance of unit amplitude and zero phase originating at P_1 and reaching P. Let f_2 be similarly defined for a disturbance originating at P_2. Then f_1, f_2 express the imaging properties of the paths P_1P, P_2P. Since u_1, u_2 are coherent, the amplitude produced at P due to disturbances from $d\sigma$ is equal to $u_1 f_1 + u_2 f_2$. Hence the intensity at P due to the element $d\sigma$ is given by

$$dI_p = (u_1 f_1 + u_2 f_2)(\bar{u}_1 \bar{f}_1 + \bar{u}_2 \bar{f}_2)\, d\sigma,$$

that is,

$$dI_p = |u_1|^2 d\sigma\, |f_1|^2 + |u_2|^2 d\sigma\, |f_2|^2 + 2R\{u_1 \bar{u}_2\, d\sigma f_1 \bar{f}_2\}, \tag{4}$$

where a bar denotes a complex conjugate. In obtaining (1) the identity

$$z_1 \bar{z}_2 + \bar{z}_1 z_2 = 2R(z_1 \bar{z}_2)$$

has been used.

The different elements of the source are incoherent. To obtain the intensity at P by disturbances from the whole of the source Σ, the expression (4) has to be integrated over it. That is,

$$I_p = \left\{\int_\Sigma |u_1|^2 d\sigma\right\} |f_1|^2 + \left\{\int_\Sigma |u_2|^2 d\sigma\right\} |f_2|^2 + 2R\left\{\int_\Sigma u_1 \bar{u}_2\, d\sigma f_1 \bar{f}_2\right\}. \tag{5}$$

Now $|u_1|^2 d\sigma$, $|u_2|^2 d\sigma$ are simply the partial intensities at P_1, P_2 due to $d\sigma$. Hence

$$\int_\Sigma |u_1|^2 d\sigma = I_1, \quad \int_\Sigma |u_2|^2 d\sigma = I_2, \tag{6}$$

where I_1, I_2 are the total intensities at P_1, P_2 due to the whole of Σ. Equation (5) can thus be written

$$I_p = I_1 |f_1|^2 + I_2 |f_2|^2 + 2\sqrt{(I_1 I_2)}\, R\{\gamma_{12} f_1 \bar{f}_2\}, \tag{7}$$

where

$$\gamma_{12} = \frac{1}{\sqrt{(I_1 I_2)}} \int_\Sigma u_1 \bar{u}_2\, d\sigma. \tag{8}$$

γ_{12}, as defined by (8), we shall take to be the phase-coherence factor between P_1 and P_2. Then, if f_1, f_2 are known for the paths P_1P, P_2P, the intensity at P can be found from (7), knowing only the intensities I_1, I_2 and γ_{12}. The factor γ_{12} is in general complex. Its modulus we shall see measures the coherence of the vibrations, and its argument denotes their difference of phase.

It is often convenient to write f_1, f_2 in terms of moduli and exponential factors. If g_1, g_2 are the absolute amplitudes produced at P by disturbances of unit amplitude at P_1, P_2 respectively, then

$$f_1 = g_1 e^{ik\delta_1}, \quad f_2 = g_2 e^{ik\delta_2}, \tag{9}$$

where g_1, g_2 are the optical paths P_1P, P_2P, and $k = 2\pi/\lambda$ is the propagation constant. We may also write γ_{12} in the form

$$\gamma_{12} = V_{12} e^{i\beta_{12}}, \tag{10}$$

where $V_{12} = |\gamma_{12}|$ and $\beta_{12} = \arg\gamma_{12}$. Equation (7) now becomes

$$I_p = (g_1 A_1)^2 + (g_2 A_2)^2 + 2g_1 A_1 g_2 A_2 V_{12} R\{\exp[i(\beta + k\overline{\delta_1 - \delta_2})]\},$$

where $A_1 = \sqrt{I_1}$, $A_2 = \sqrt{I_2}$ are the absolute amplitudes at P_1, P_2. Clearly $g_1 A_1 = A_1'$, $g_2 A_2 = A_2'$ are the absolute amplitudes at P produced by the disturbances from P_1, P_2 respectively. Hence we may write

$$I_p = A_1'^2 + A_2'^2 + 2A_1' A_2' V_{12} R\{\exp[i(\beta + k\overline{\delta_1 - \delta_2})]\}. \tag{11}$$

The interpretation of (11) when applied to Young's experiment provides a simple basis for the discussion of the significance of different value of γ_{12}.

We shall show first that $|\gamma_{12}| \leqslant |$. If this is true, then $|\gamma_{12}|^2 \leqslant 1$. Now, from (8),

$$|\gamma_{12}|^2 = \frac{1}{I_1 I_2}\left|\int_\Sigma u_1 \bar{u}_2 d\sigma\right|^2 \leqslant \frac{1}{I_1 I_2}\left\{\int_\Sigma |u_1 \bar{u}_2| d\sigma\right\}^2,$$

and, using the values of I from (6),

$$I_1 I_2 = \left\{\int_\Sigma |u_1|^2 d\sigma\right\}\left\{\int_\Sigma |u_2|^2 d\sigma\right\} \geqslant \left\{\int_\Sigma |u_1 u_2| d\sigma\right\}^2.$$

the inequality being that of Schwarz. Since $|u_1 \bar{u}_2| = |u_1 u_2|$, it follows that $|\gamma_{12}|^2 \leqslant 1$. Hence γ_{12} lies either inside or on the unit circle in the complex plane.

Returning to equation (11), let A_1', A_2' be the absolute amplitudes at P of disturbances coming directly from two pinholes P_1, P_2. First let P be equidistant from P_1 and P_2. Then $\delta_1 = \delta_2$, and

$$I_p = A_1'^2 + A_2'^2 + 2A_1' A_2' V_{12} R(e^{i\beta_{12}}). \tag{12}$$

Suppose $|\gamma_{12}| = V_{12} = 1$. Then

$$I_p = A_1'^2 + A_2'^2 + 2A_1' A_2' \cos\beta_{12}.$$

The disturbances at P_1, P_2 are thus perfectly coherent when $|\gamma_{12}| = 1$, and have a difference of phase equal to β_{12}.

If P is now displaced parallel to the direction $P_1 P_2$, the path difference $(\delta_1 - \delta_2)$ is made to vary. The intensity, as given by (11), then goes through maximum and minimum values equal to

$$I_{\text{max.}} = A_1'^2 + A_2'^2 + 2A_1' A_2' V_{12},$$
$$I_{\text{min.}} = A_1'^2 + A_2'^2 - 2A_1' A_2' V_{12},$$

whence the visibility of the fringes is given by

$$V = \frac{2A_1' A_2' V_{12}}{A_1'^2 + A_2'^2}. \tag{13}$$

If the intensities are made equal, $A_1' = A_2'$ and V_{12} is then equal to the visibility of the fringes. Hence $|\gamma_{12}|$ amounts to precisely the same thing as Zernike's degree of coherence. The significance of $\arg\gamma_{12} = \beta_{12}$ will be seen to amount to a lateral displacement of the fringe system from the position which it occupies when $\beta_{12} = 0$ and the disturbances at P_1, P_2 are cophasal. Thus $|\gamma_{12}|$, $\arg\gamma_{12}$ denote respectively the coherence and phase difference of the disturbances at P_1, P_2. Arg γ_{12} gives the advance of the phase at P_2 relative to that at P_1. If $V_{12} = 0$, the product term is absent from (11), and I_p is then obtained as the sum of the separate intensities at P. In this case, therefore, the vibrations at P_1, P_2 are incoherent. Arg γ_{12} is indeterminate, as one would expect.

3. The calculation of γ_{12} for points in a plane illuminated by a source

Let (ξ, η) (figure 2) be a plane containing a source Σ, and let $I(\xi, \eta)$ be the intensity at a point A of Σ. P_1 and P_2 are two points in a plane (X, Y), distance R from the source. It will now be shown how to calculate γ_{12}, the phase-coherence factor between P_1 and P_2.

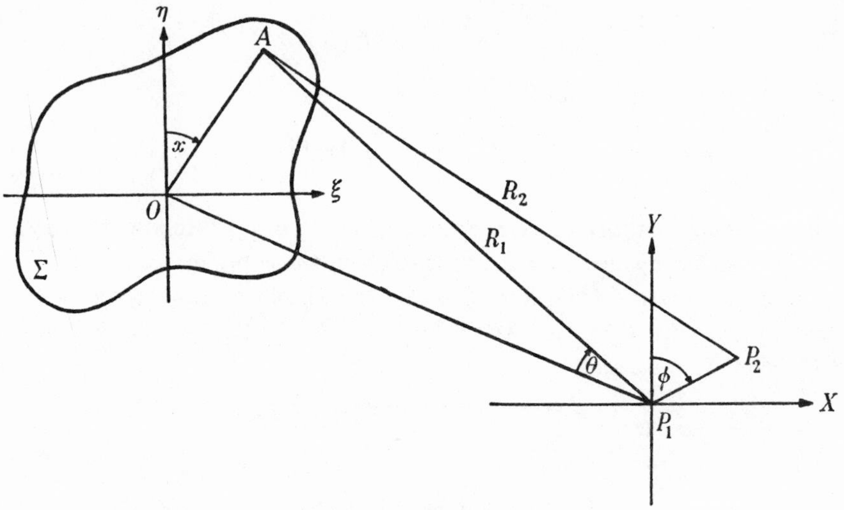

Figure 2.

The complex amplitudes at P_1, P_2 produced by the element of the source $d\sigma$, situated at A, are

$$u_1 = \frac{\sqrt{\{I(\xi,\eta)\}}}{R_1} e^{-ikR_1}, \quad u_2 = \frac{\sqrt{\{I(\xi,\eta)\}}}{R_2} e^{-ikR_2}, \tag{14}$$

where $R_1 = AP_1, R_2 = AP_2$ and $k = 2\pi/\lambda$. According to (4), the phase-coherence factor is now given by

$$\gamma_{12} = \frac{1}{\sqrt{(I_1 I_2)}} \int_\Sigma \frac{I(\xi,\eta)}{R_1 R_2} e^{ik(R_2-R_1)} d\sigma. \tag{15}$$

The form of the integral in (15) is well known. It represents the amplitude at P_2 in the diffraction pattern around P_1, associated with an aperture identical in form with Σ. The amplitudes in the diffraction integral are taken to be equal to the intensities in the source, $I(\xi, \eta)$. Equation (15) is equivalent to one obtained by Zernike (1938), using explicit statistical considerations. It shows that the phase-coherence factor between any point P_2 and a given point P_1 is the same as the complex amplitude at P_2 in the diffraction pattern associated with an aperture coincident with the source, the pattern being centred on P_1.

In obtaining (15) the radiation from an element of the source has been assumed to be angularly uniform. It will, in fact, vary with the inclination of AP_1, AP_2, to the normal to $d\sigma$. However, the contribution to the integral (10) will be mainly from the immediate neighbourhood of the point for which the phase (R_2-R_1) is stationary. For this region R_2, R_1 are very nearly normal to $d\sigma$, providing the distance P_1P_2 is small in comparison with the distances of these points from Σ. Hence no obliquity factor need appear in (15).

Suppose now that P_1 (figure 2) is at the origin in the (X, Y) plane, and let P_2 be the point (X, Y). Then, since A has co-ordinates (ξ, η),

$$R_2 - R_1 = \frac{\rho^2}{R_1+R_2} - \frac{2}{R_1+R_2}(X\xi + Y\eta), \tag{16}$$

where $\rho = P_1P_2 = \sqrt{(X^2+Y^2)}$. For small values of X, Y it is sufficiently accurate to write (16) as

$$R_2 - R_1 = -\frac{1}{R}(X\xi + Y\eta),$$

and (15) then becomes

$$\gamma_{12} = \frac{1}{R^2\sqrt{(I_1I_2)}}\int_\Sigma I(\xi,\eta)\exp\left[-i\frac{k}{R}(X\xi+Y\eta)\right]d\xi d\eta, \tag{17}$$

from which it will be seen that, apart from a constant factor, $\gamma_{12}(X, Y)$ is the Fourier transform of $I(\xi, \eta)$.

For our case, however, it is frequently necessary to retain the term in ρ^2, since, if Σ is very small, γ_{12} can be appreciably different from zero even for large values of (X, Y). For the case of a circular source of centre O, of angular radius α, it is convenient to write (ξ, η), (X, Y) in terms of polar co-ordinates (r, χ), (ρ, ϕ). The variable r is defined by $r = \sin\theta/\sin\alpha$, and ranges from 0 to 1. θ is the angle shown in figure 2.

In the integral of (15), one may write $\frac{1}{R_1R_2} = \frac{1}{R_1^2}$, since R_2 will differ but little from R_1. With this approximation

$$\frac{d\sigma}{R_1R_2} = \tan\theta\, d\theta\, d\chi = \frac{\sin^2\alpha}{\cos^2\theta} r dr\, d\chi. \tag{18}$$

From (16), the argument of the exponential term is seen to be

$$ik\left\{\frac{\rho^2\cos\theta}{2R} - \rho\sin\theta\cos(\chi-\phi)\right\} = ik\frac{\rho^2}{2R} - \frac{\lambda}{8\pi R}(zr)^2 - zr\cos(\chi-\phi), \tag{19}$$

where $z = k\sin\alpha\rho$, and the approximation $\cos\theta = 1-\frac{1}{2}\sin^2\theta$ has been used. This is valid, since $\rho^2/2R$ will be negligible except where α is small. Values of z greater than about 10 will never occur, and r is limited to the range 0·0 to 1·0. The term in $(zr)^2$ may thus be neglected, since $R \gg \lambda$.

The value of γ_{12} for a plane illuminated by a circular source is now, substituting (18) and (19) in (15),

$$\gamma_{12} = \frac{\sin^2\alpha}{\sqrt{(I_1I_2)}}\exp\left[ik\frac{\rho^2}{2R}\right]\int_0^1\int_0^{2\pi}\frac{I(\xi,\eta)}{\cos^2\theta}\exp\left[-izr\cos(\chi-\phi)\right] r dr d\chi. \tag{20}$$

From what has been said above in connexion with (15) it will be seen that the factor $1/\cos^2\theta$ may be omitted, since θ will be small for the neighbourhood of Σ for which the phase $(R_2 - R_1)$ is stationary. Moreover, we have assumed each element of the source to radiate uniformly in all directions. If the amplitude is taken to be reduced by a factor $\cos\theta$ in a direction making an angle θ with the normal to the plane of the source, then the intensity in (15) will appear as $I(\xi,\eta)\cos^2\theta$. Using this value in (20), and postulating a source of uniform intensity, this expression becomes

$$\gamma_{12} = \frac{I\sin^2\alpha}{\sqrt{(I_1 I_2)}}\exp\left[ik\frac{\rho^2}{2R}\right]\int_0^1\int_0^{2\pi}\exp\left[-izr\cos(\chi-\phi)\right]r\,dr\,d\phi,$$

or, using two well-known integrals,

$$\gamma_{12} = \frac{I\pi\sin^2\alpha}{\sqrt{(I_1 I_2)}}\exp\left[ik\frac{\rho^2}{2R}\right]\frac{2J_1(z)}{z}. \tag{21}$$

Since P_1, P_2 are close together, $I_1 = I_2$, and each of these intensities is equal to $I\pi\sin^2\alpha$. Hence, finally

$$\gamma_{12} = \frac{2J_1(z)}{z}\exp\left[ik\frac{\rho^2}{2R}\right], \tag{22}$$

where

$$z = \frac{2\pi}{\lambda}N\sin\alpha\rho, \tag{23}$$

and (23) now includes a refractive index N for the sake of generality.

Suppose the source is limitingly small, $\alpha \to 0$. Then $2J_1(z)/z = 1$, and

$$\gamma_{12} = \exp\left[ik\frac{\rho^2}{2R}\right]. \tag{24}$$

The points P_1, P_2 are thus coherently illuminated ($|\gamma_{12}| = 1$), and P_2 is advanced in phase by an amount $k(\rho^2/2R)$ relative to P_1. This must be so, since P_1, P_2 are now illuminated by a point source whose distance from P_2 is greater than that from P_1 by an amount equal to $\rho^2/2R$. The exponential factor in (22) does not appear in the corresponding results obtained by either van Cittert or Zernike. It will be equal to unity for all but very small values of α. Nevertheless, it must generally be retained when dealing with small sources.

The value of $2J_1(z)/z$ already drops to 0·880 when $z = 1$. Regarding this as a tolerable departure from perfect coherence, the diameter of the circle which is coherently illuminated by a source of angular radius α is, from (23),

$$d = \frac{0{\cdot}16\lambda}{N\sin\alpha}. \tag{25}$$

The larger the angular size of the source, therefore, the smaller is the coherently illuminated area. The smallest value that d can take is $0{\cdot}16\lambda/N$. Hence it is impossible to illuminate a finite area of any surface incoherently, although the area of coherence will be small for all but very small values of α. The formula (25) finds considerable application in deciding the size of source required in experiments using interference and diffraction.

4. The theory of the stellar interferometer

The theory of the stellar interferometer may be considered from the present standpoint. The stellar source, of angular radius α, illuminates the plane of the outer mirrors M_1, M_2 of the interferometer. The visibility and position of the fringes observed are thus determined by $|\gamma_{12}|$, $\arg \gamma_{12}$ respectively. Since R is so large, $\rho^2/2R$ is negligible. If M_1 is at the origin and M_2 has co-ordinates (X, Y), then, in the general case,

$$\gamma_{12}(X, Y) = \frac{1}{\sqrt{(I_1 I_2)}} \int_\Sigma I(u, v), \mathrm{e}^{-i2\pi(uX+vY)}\, du\, dv, \tag{26}$$

where, in (18), (ξ, η) have been replaced by the angular variables $u = \xi/\lambda R$, $v = \eta/\lambda R$. If $I(u, v)$ is defined to be zero outside Σ, the domain of integration may be extended to the whole (u, v) plane. By the Fourier inversion theorem, $I(u, v)$ can then be expressed as

$$I(u, v) = \int_{-\infty}^{+\infty} \int_{-\infty}^{+\infty} \gamma_{12}(X, Y)\, \mathrm{e}^{i2\pi(uX+vY)}\, dX\, dY, \tag{27}$$

from which the constant factor $\sqrt{(I_1, I_2)}$ has been omitted. Thus the distribution of intensity in the stellar source $I(u, v)$ may be obtained from a knowledge of $\gamma_{12}(X, Y)$. This would entail a measurement of both the visibility and position of the fringes seen in the telescope for different separations of the mirrors in different azimuths, since both the modulus and argument of γ_{12} must be known. Such measurements would meet with great instrumental difficulties, since exact equality of the two optical paths would have to be preserved. A self-compensating system might achieve this. With such an instrument the relative brightnesses and angular diameters and separation of double stars could be measured.

If the source has radial symmetry, we can use (θ, χ) instead of (u, v), and replace (X, Y) by the polar co-ordinates (ρ, ϕ). Equation (26) then becomes

$$\gamma_{12}(\rho) = \int_\Sigma I(\theta) \int_0^{2\pi} \exp\left[-i\frac{2\pi}{\lambda}\theta\rho\cos(\chi - \phi)\right] \theta\, d\theta\, d\chi,$$

or, integrating with respect to χ,

$$\gamma_{12}(\rho) = \int_\Sigma I(\theta)\, \theta J_0\left(\frac{2\pi}{\lambda}\theta\rho\right) d\theta, \tag{28}$$

from which a constant factor has been omitted. ρ is the distance between the two mirrors. γ_{12} is now wholly real. Apart from an ambiguity of sign, therefore, it is uniquely defined by its modulus; that is, by the fringe visibility. The equivalent of (27) may now be written

$$I(\theta) = \int_0^\infty \gamma_{12}(\rho)\, \rho J_0\left(\frac{2\pi}{\lambda}\theta\rho\right) d\rho. \tag{29}$$

Hence the radial distribution of intensity in the source may be determined solely from measurements of visibilities.

If the source is circular and of uniform intensity, (28) gives, with an appropriate factor to make $\gamma_{12}(0) = 1$,

$$\gamma_{12} = \frac{2J_1(z)}{z}, \tag{30}$$

where the variable (23) is now defined by $z = \frac{2\pi}{\lambda}\alpha\rho$. The angular radius of the source is α. The condition for the disappearance of the fringes ($|\gamma_{12}| = 0$) is now that z shall be a zero of $J_1(z)$ other than $z = 0$. If $\rho = \rho_n$ for the nth disappearance of the fringes the radius of the source is given by

$$\alpha = \frac{\lambda z_n}{2\pi\rho_n}, \tag{31}$$

where z_n is the nth root of $J_1(z)$, zero being excluded.

It will be noted that $\gamma_{12}(z)$ changes sign on passing through a zero, $z = z_n$. Thus a phase jump of π occurs, and the positions of the bright and dark fringes are interchanged.

5. The propagation of the phase-coherence factor

Zernike (1938) has shown how a knowledge of the coherence for pairs of points on any given surface S permits the calculation of this quantity for pairs of points on a second surface S', towards which the light is propagated after leaving S. Zernike obtained his result by again considering statistical mean values. It will now be shown that the same result may be obtained by means of the simpler method used in deriving (15) above.

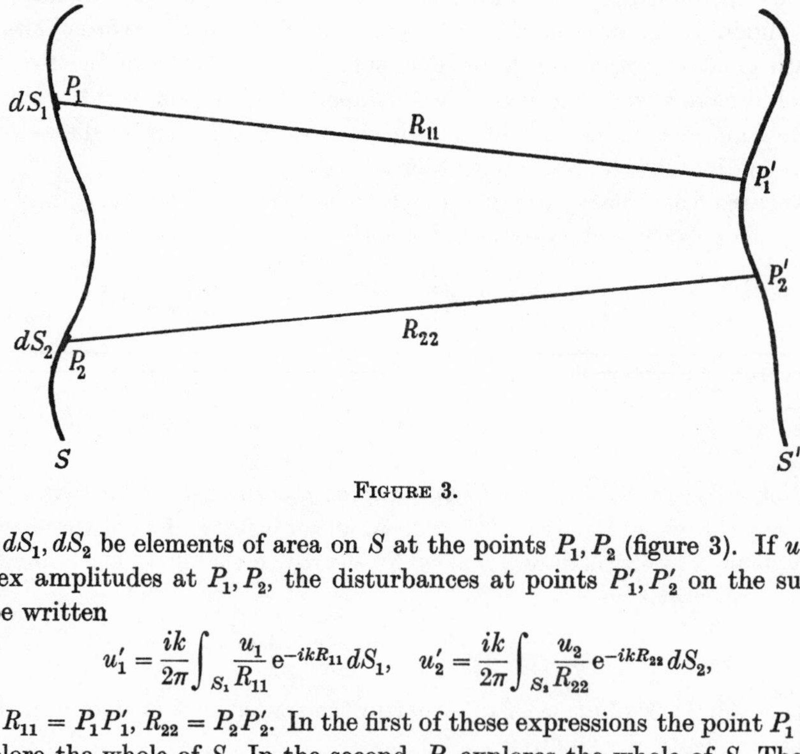

Figure 3.

Let dS_1, dS_2 be elements of area on S at the points P_1, P_2 (figure 3). If u_1, u_2 are complex amplitudes at P_1, P_2, the disturbances at points P_1', P_2' on the surface S' may be written

$$u_1' = \frac{ik}{2\pi}\int_{S_1}\frac{u_1}{R_{11}}e^{-ikR_{11}}\,dS_1, \quad u_2' = \frac{ik}{2\pi}\int_{S_2}\frac{u_2}{R_{22}}e^{-ikR_{22}}\,dS_2, \tag{32}$$

where $R_{11} = P_1P_1'$, $R_{22} = P_2P_2'$. In the first of these expressions the point P_1 is made to explore the whole of S. In the second, P_2 explores the whole of S. The phase-coherence factor between P_1', P_2' is, by definition,

$$\gamma_{12}' = \frac{1}{\sqrt{(I_1'I_1')}}\int_\Sigma u_1'\bar{u}_2'\,d\sigma.$$

Hence, using (32),

$$\gamma'_{12} = \frac{k^2}{4\pi^2 \sqrt{(I'_1 I'_2)}} \int_{S_1} \int_{S_2} \left\{ \int_{\Sigma} u_1 \bar{u}_2 d\sigma \right\} \frac{1}{R_{11} R_{22}} e^{ik(R_{22}-R_{11})} dS_1 dS_2,$$

the variables $d\sigma, dS_1, dS_2$ being independent. Thus

$$\gamma'_{12} = \frac{k^2}{4\pi^2} \frac{1}{\sqrt{(I'_1 I'_2)}} \int_{S_1} \int_{S_2} \frac{\sqrt{(I_1 I_2)}\,\gamma_{12}}{R_{11} R_{22}} e^{ik(R_{22}-R_{11})} dS_1 dS_2, \tag{33}$$

the integrations extending twice over the surface S. This is equivalent to the expression obtained by Zernike.

The presence of any diaphragm or screen between S and S' can be accounted for by limiting the integrations over S to those parts which can be seen from P'_1 and P'_2, unless the aperture in the diaphragm is very small. In this case it is necessary to obtain γ_{12} for the plane of the diaphragm and then to find γ'_{12} as a second step.

If S, S' are in the entrance and exit pupils of a lens system a different procedure is adopted. Let P_1, P_2 have co-ordinates $(\xi_1, \eta_1), (\xi_2, \eta_2)$, and suppose a disturbance at any point P, of unit amplitude and zero phase, produces a complex amplitude $\tau(\xi, \eta)$ at the corresponding point P'. If u_1, u_2 are the amplitudes at P_1, P_2 which are produced by an element of the source, then

$$u'_1 = \tau(\xi_1, \eta_1)\, u_1, \quad u'_2 = \tau(\xi_2, \eta_2)\, \eta_2$$

are the disturbances at P'_1, P'_2 associated with the same element. Hence

$$u'_1 \bar{u}'_2 = \tau(\xi_1, \eta_1)\, \tau^*(\xi_2, \eta_2)\, u_1 \bar{u}_2,$$

where * denotes a complex conjugate. Integration over the source then gives

$$\gamma'_{12} = \frac{\tau(\xi_1, \eta_1)\, \tau^*(\xi_2, \eta_2)}{|\tau(\xi_1, \eta_1)|\,|\tau(\xi_2, \eta_2)|} \gamma_{12}, \tag{34}$$

since $\sqrt{I'_1} = |\tau(\xi_1, \eta_1)| \sqrt{I_1}$ and a similar expression holds for $\sqrt{I'_2}$. The function $\tau(\xi, \eta)$, the complex transmission of the lens, expresses the aberration (including any defect of focus) and light transmission of the system. If the light transmission is uniform, then $|\tau(\xi_1, \eta_1)| = |\tau(\xi_2, \eta_2)|$. The arguments are likewise equal if the system is free from aberration. From (34) it will be seen that the coherence in the exit pupil is the same as that in the entrance pupil. The phase relationships may be changed. In the absence of aberration and defect of focus the phase relationships are also preserved. It will be clear that the above considerations do not apply to points in the neighbourhood of the edge of the pupil, nor in the case of very small apertures.

6. The theory of illumination in the microscope

The influence of the condenser on images formed in the microscope may now be considered from the point of view of the phase-coherence relationships between points in the illuminated object. In the case of so-called critical illumination an image of an extended source is formed in the plane of the object. It is then necessary to find the phase-coherence factor for the plane on which the light source is imaged. The argument which will now be given follows in part one given by Zernike (1938).

In figure 4, Σ is the plane of the light source, of which $d\sigma$ is an element having co-ordinates (x, y). The angular size of Σ is assumed to be large. Let $I(x, y)$ represent the distribution of intensity over Σ.

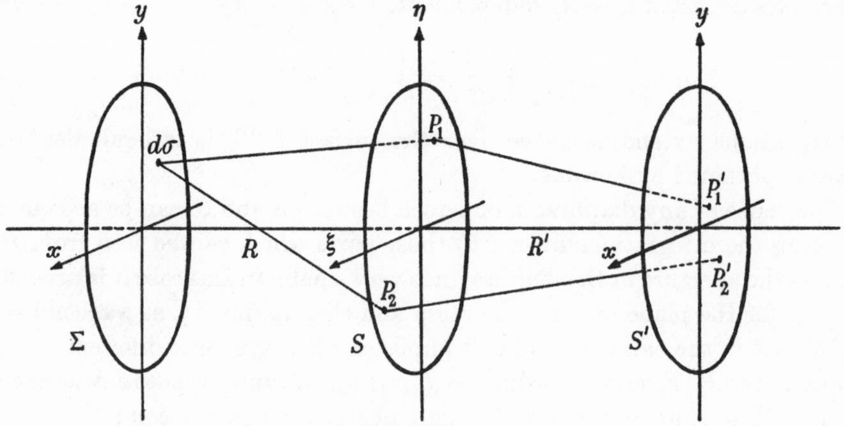

FIGURE 4

S is the plane of the condenser lens which images Σ on or near the plane S'. In the entrance pupil of the condenser, the phase-coherence factor between points $P_1(\xi_1, \eta_1)$, $P_2(\xi_2, \eta_2)$ is, according to (10), and omitting the factor $e^{ik\rho^2/2R}$,

$$\frac{1}{R^2\sqrt{(I_1 I_2)}}\int_\Sigma I(x, y)\, e^{ik\{x(\xi_1-\xi_2)+y(\eta_1-\eta_2)\}}\, d\sigma$$

where (x, y) are the angular co-ordinates of $d\sigma$ as seen from S: that is, the linear co-ordinates divided by R, the distance between Σ and S. Using (34), the phase-coherence factor in the exit pupil of the condenser is given by

$$\gamma_{12} = \frac{\tau(\xi_1, \eta_1)\,\tau^*(\xi_2, \eta_2)}{R^2\sqrt{(I_1 I_2)}}\int_\Sigma I(x, y)\, e^{ik\{x(\xi_1-\xi_2)+y(\eta_1-\eta_2)\}}\, d\sigma, \tag{35}$$

where I_1, I_2 now refer to points in the exit pupil. For this reason the factor $\dfrac{1}{|\tau(\xi_1, \eta_1)|\,|\tau(\xi_2, \eta_2)|}$ does not appear explicitly. Proceeding from S to S' by means of (33), gives for the phase-coherence factor on S' the expression

$$\gamma'_{12} = \frac{k^2}{4\pi^2}\frac{1}{\sqrt{(I'_1 I'_2)}}\int_{S_1}\int_{S_2}\frac{\sqrt{(I_1 I_2)}\,\gamma_{12}}{R'^2}\, e^{ik(R_{22}-R_{11})}\, dS_1\, dS_2, \tag{36}$$

in which one may write

$$R_{22} - R_{11} = (\xi_1 x_1 + \eta_1 y_1) - (\xi_2 x_2 + \eta_2 y_2),$$

where (x_1, y_1), (x_2, y_2) are the angular co-ordinates of P'_1, P'_2 as seen from S; that is, the linear co-ordinates divided by R', the distance between S and S'.

Substitution of (35) in (36) gives

$$\gamma'_{12} = \frac{k^2}{4\pi^2}\frac{1}{R^2 R'^2\sqrt{(I'_1 I'_2)}}\int_{S_1}\int_{S_2}\int_\Sigma \tau(\xi_1, \eta_1)\,\tau^*(\xi_2, \eta_2)\, I(x, y)$$
$$e^{ik\{x(\xi_1-\xi_2)+y(\eta_1-\eta_2)\}}\, e^{ik\{(\xi_1 x_1+\eta_1 y_1)-(\xi_2 x_2+\eta_2 y_2)\}}\, d\sigma\, dS_1\, dS_2, \tag{37}$$

where, it will be understood, each integral sign denotes a double integration. By virtue of the identities

$$x(\xi_1-\xi_2)+\xi_1 x_1-\xi_2 x_2 = (x+x_2)(\xi_1-\xi_2)+\xi_1(x_1-x_2),$$
$$y(\eta_1-\eta_2)+\eta_1 y_1-\eta_2 y_2 = (y+y_2)(\eta_1-\eta_2)+\eta_1(y_1-y_2),$$

two of the double integrals in (37) may be written

$$\int_{S_2}\int_{\Sigma} I(x,y)\,\tau^*(\xi_2,\eta_2)\,e^{ik\{(x+x_2)(\xi_1-\xi_2)+(y+y_2)(\eta_1-\eta_2)\}}\,d\sigma\,dS_2$$
$$= \frac{4\pi^2}{k^2}\iint_{S_2}\iint_{\Sigma} I(x,y)\,\tau^*(\xi_2,\eta_2)\,e^{i2\pi\{u(\xi_1-\xi_2)+v(\eta_1-\eta_2)\}}\,du\,dv\,d\xi_2\,d\eta_2, \qquad (38)$$

where $$2\pi u = k(x+x_2), \quad 2\pi v = k(y+y_2).$$

If $\tau(\xi,\eta)$ is defined to be zero outside the domain of the exit pupil, the limits of integration with respect to ξ_2, η_2 may be made infinite. Suppose now that $I(x,y) = I$ over the interior of Σ and is zero outside, the source being then assumed to be of uniform brightness. If Σ is large enough, the integrations with respect to u and v may be given infinite limits. The condition under which this is permissible is that the index of the exponential function shall be many times greater than $2\pi i$ at the limits†, (38) may now be written

$$\frac{4\pi^2 I}{k^2}\iiiint_{-\infty}^{+\infty} \tau^*(\xi_2,\eta_2)\,e^{i2\pi\{u(\xi_1-\xi_2)+v(\eta_1-\eta_2)\}}\,du\,dv\,d\xi_2\,d\eta_2 = \frac{4\pi^2 I}{k^2}\tau^*(\xi_1,\eta_1)$$

by the Fourier inversion theorem. It will be seen that the necessary condition will be satisfied providing the image of the light source is large compared with the extent of the Fraunhofer diffraction pattern of the lens aperture.

Using this result, (37) becomes

$$\gamma'_{12} = \frac{I}{R^2 R'^2\sqrt{(I'_1 I'_2)}}\int_{S_1}\tau(\xi_1,\eta_1)\,\tau^*(\xi_1,\eta_1)\,e^{ik\{\xi_1(x_1-x_2)+\eta_1(y_1-y_2)\}}\,dS_1.$$

The factor I/R^2 expresses the intensity in the entrance pupil, while

$$\tau(\xi_1,\eta_1)\,\tau^*(\xi_1,\eta_1) = |\tau(\xi_1,\eta_1)|^2$$

gives the transmission of the lens. Hence, writing $I(\xi,\eta)$ for the intensity at (ξ,η) in the exit pupil, there is obtained

$$\gamma'_{12} = \frac{1}{R'^2\sqrt{(I'_1 I'_2)}}\int_S I(\xi,\eta)\,e^{ik\{\xi(x_1-x_2)+\eta(y_1-y_2)\}}\,dS, \qquad (39)$$

which is identical in form with (15). It follows that the phase-coherence relationships in the object plane are the same as those which obtain if the condenser is replaced by a source of the same size and shape, the intensity at any point in the source being equal to the intensity at the corresponding point in the condenser aperture. Moreover, the result is independent of the aberrations and focusing of the condenser.

† If $I(x, y)$ = constant, (38) is a Fourier integral. In the case of a Fourier series this condition corresponds to a large number of terms being included.

A less rigorous, but very suggestive, proof of this result can be given. Thus, if the source subtends an angle α at the entrance pupil of the condenser such that α is greater than (say) 0·05, the coherence outside a diameter corresponding to $z = 3$ (23) will be very small. This diameter is equal to about 10λ. Hence the entrance pupil is approximately incoherently illuminated; and therefore, by (34), so is the exit pupil. The result then follows at once. A similar reasoning serves to establish the validity of the result for more complicated illuminating systems, for example, that due to Köhler.

Consider now a microscope objective of numerical aperture $N \sin \alpha$, and let $z = \frac{2\pi}{\lambda} N \sin \alpha \rho$ denote distances in the object plane. If this latter is illuminated by a condenser of numerical aperture $sN \sin \alpha$, the phase-coherence factor for the object plane will be

$$\gamma_{12}(z) = \frac{2J_1(sz)}{sz}. \tag{40}$$

If the illuminating aperture is equal to that of the objective $s = 1$, and γ_{12} is the zero for two points on the limit of resolution, since $\frac{J_1(z)}{z} = 0$ in this case. However, the images of (say) the periphery of an object will still be formed by the superposition of the diffraction images of points closer together than this. For these points γ_{12} will be of significant magnitude. Hence it is not true that a 'transparent object will then appear practically in the same way as a self-luminous one' (see Zernike 1938).

Using the result (40) in (7) we may find an expression for the light intensity in the microscopic image of two small pinholes when illuminated by a condenser having an aperture equal to s times that of the imaging objective. If the latter is perfectly corrected

$$f_1 = \frac{2J_1(z_1)}{z_1}, \quad f_2 = \frac{2J_1(z_2)}{z_2}, \tag{41}$$

where z_1, z_2 denote the distances of a current point from the geometric images of the two pinholes, whose distance apart is denoted by z. Putting $I_1 = I_2 = 1$, (7) gives

$$I(z_1, z_2) = \left\{\frac{2J_1(z_1)}{z_1}\right\}^2 + \left\{\frac{2J_1(z_2)}{z_2}\right\}^2 + 2\frac{2J_1(sz)}{sz}\left\{\frac{2J_1(z_1)}{z_1}\frac{2J_1(z_2)}{z_2}\right\}. \tag{42}$$

The result (42) was first obtained by Hopkins & Barham (1950) under more restricted conditions. A perfectly focused condenser, having no aberration, was treated by explicit integrations. The influence of the aperture of the condenser on resolution was studied using such an expression.

7. The theory of the formation of images

We shall conclude by showing how the phase-coherence factor may be applied to the study of the diffraction image of an extended transparent object when illuminated by transmitted light. Suppose an element of the source $d\sigma$ produces a complex amplitude $u(x, y)$ at the point (x, y) of the object plane, and let

$A(x,y) = |u(x,y)|$. If the complex transmission of the object is represented by $E(x,y)$, the amplitude $u(x,y)$ becomes $u(x,y)\,E(x,y)$ after passing through the object. The diffraction image of (x,y) will be centred on the point (x,y) in the image plane. Let $f(x'-x, y'-y)$ be the complex amplitude at the point (x',y') in the image plane, due to a disturbance of unit amplitude and zero phase at (x,y) in the object plane. Then the amplitude at (x',y') due to the disturbance from (x_1,y_1) associated with the element of source $d\sigma$ will be

$$u(x_1,y_1)\,E(x_1,y_1)f(x'-x_1,y'-y_1)\,dS_1,$$

where dS_1 is an element of area at (x_1,y). The total amplitude at (x',y') in the image plane due to all elements of the source will be

$$u'(x',y') = \int_{S_1} u(x_1,y_1)\,E(x_1,y_1)f(x'-x_1,y'-y_2)\,dS_1. \tag{43}$$

An equivalent expression can be obtained using an independent current point (x_2,y_2) in the object plane. The intensity at (x',y') due to the element $d\sigma$ is then given by

$$dI = \left\{\int_{S_1} u(x_1,y_1)\,E(x_1,y_1)f(x'-x_1,y'-y_1)\,dS_1 \int_{S_2} u^*(x_2,y_2) \times E^*(x_2,y_2)f^*(x'-x_2,y'-y_2)\,dS_2\right\}d\sigma,$$

where * denotes a complex conjugate. All the variables are independent. Hence integrating over the whole source, the total intensity at (x',y') is found to be

$$I(x',y') = \int_{S_1}\int_{S_2}\left\{\int_{\Sigma} u(x_1,y_1)\,u^*(x_2,y_2)\,d\sigma\right\}E(x_1,y_1)\,E^*(x_2,y_2) \times f(x'-x_1,y'-y_1)f^*(x'-x_2,y'-y_2)\,dS_1dS_2,$$

or, finally,

$$I(x',y') = \int_{S_1}\int_{S_2}\gamma_{12}(x_1-x_2,y_1-y_2)\,A(x_1,y_1)\,A(x_2,y_2)\,E(x_1,y_1) \times E^*(x_2,y_2)f(x'-x_1,y'-y_1)f^*(x'-x_2,y'-y_2)\,dS_1dS_2, \tag{44}$$

where γ_{12} is the phase-coherence factor.

If the object is coherently illuminated, $\gamma_{12} = 1$ and $I(x',y')$ is then found to be

$$I(x',y') = \left|\int_S A(x,y)\,E(x,y)f(x'-x,y'-y)\,dS\right|^2. \tag{45}$$

On the other hand, if the object is incoherent, $\gamma_{12} = 0$ except when $x_1 = x_2, y_1 = y_2$, for which $\gamma_{12} = 1$, (44) then gives

$$I(x',y') = \int_S I(x,y)\,|\,E(x,y)|^2\,|f(x'-x,y'-y)|^2\,dS, \tag{46}$$

where $I(x,y)$ is the intensity falling on (x,y). Only the modulus of $E(x,y)$ now appears, showing that the image is unaffected by the argument of $E(x,y)$. Since a transparent object can never be incoherently illuminated, (46) must apply only to a self-luminous object. In this case $I(x, y) \mid E(x, y)|^2$ represents simply the intensity at the point (x, y).

In the general theory (44) can be evaluated, using a form for γ_{12} such as that in (40). (45) and (46) then occur as the limits $s \to 0$, $s \to \infty$ respectively.

References

van Cittert, P. H. 1934 *Physica* **1**, 201.

Hopkins, H. H. & Barham, P. M. 1950 *Proc. Phys. Soc.* B, **63**, 737.

Michelson, A. A. 1890 *Phil. Mag.* **30**, 1.

Zernike, F. 1938 *Physica*, **5**, 785.

PAPER NO. 12

Reprinted from *Proceedings of the Royal Society*, Ser. A, Vol. 217, pp. 408–432, 1953.

On the diffraction theory of optical images

BY H. H. HOPKINS

Physics Department, Imperial College, London, S.W. 7

(*Communicated by N. F. Mott, F.R.S.—Received* 31 *December* 1952)

The theory of image formation is formulated in terms of the coherence function in the object plane, the diffraction distribution function of the image-forming system and a function describing the structure of the object. There results a four-fold integral involving these functions, and the complex conjugate functions of the latter two. This integral is evaluated in terms of the Fourier transforms of the coherence function, the diffraction distribution function and its complex conjugate. In fact, these transforms are respectively the distribution of intensity in an 'effective source', and the complex transmission of the optical system—they are the data initially known and are generally of simple form. A generalized 'transmission factor' is found which reduces to the known results in the simple cases of perfect coherence and complete incoherence. The procedure may be varied in a manner more suited to non-periodic objects.

The theory is applied to study *inter alia* the influence of the method of illumination on the images of simple periodic structures and of an isolated line.

1. INTRODUCTION

In a recent communication (Hopkins 1951) a 'phase-coherence' factor was introduced by means of which the correlation of phase between wave disturbances at any two points may be specified. For two points P_1, P_2 of an illuminated surface the phase-coherence factor is defined by means of the expression

$$\Gamma_{12} = \frac{1}{\sqrt{(I_1 I_2)}} \int_\Sigma U_1 U_2^* \, d\sigma, \tag{1}$$

where I_1, I_2 are the intensities at P_1, P_2 produced by the source Σ, and U_1, U_2 are the complex amplitudes at these points associated with an element $d\sigma$ of the source.

* denotes a complex conjugate. In the paper cited a number of propositions relating to the phase-coherence factor was established. It will be convenient to recall some of these results here.†

When an object is illuminated by transmitted or specularly reflected light there is always a finite degree of correlation between the phases at different points of the object. If a plane source has an intensity $\gamma(x, y)$ at the point (x, y), the phase-coherence factor between points P_1, P_2 of a plane illuminated by the source is given by

$$\Gamma(u_1 - u_2, v_1 - v_2) = \frac{1}{\sqrt{(I_1 I_2)}} \iint_{-\infty}^{+\infty} \gamma(x, y)\, e^{i[x(u_1-u_2)+y(v_1-v_2)]}\, dx\, dy, \tag{2}$$

in which $\gamma(x, y)$ is taken to be zero at points exterior to the source. In (2), (u_1, v_1), (u_2, v_2) are rectangular co-ordinates of P_1, P_2; and I_1, I_2 are the intensities at these points due to the source. The units of the co-ordinates (x, y), (u, v) will be defined below. The source is assumed to be parallel with and at some distance from the plane which it illuminates, and the line joining the origins of (x, y) and (u, v) is taken to be perpendicular to both the plane and the source. If we choose the unit of intensity to make $\sqrt{(I_1 I_2)} = 2\pi$, and assume that the plane is uniformly illuminated, (2) defines $\Gamma(u, v)$ to be the Fourier transform of $\gamma(x, y)$—the notation being that frequently employed (Titchmarsh 1937).

Generally the object plane will be illuminated by means of some arrangement of condenser lenses. By means of methods described earlier (Hopkins 1951) the phase-coherence factor in the object plane may be built up step by step, proceeding from the source to the entrance pupil of the first condenser, thence to the exit pupil and so on. When the phase-coherence function for the object plane $\Gamma(u, v)$, is thereby constructed, one may postulate an 'effective source', defined by the intensity distribution

$$\gamma(x, y) = \frac{1}{2\pi} \iint_{-\infty}^{+\infty} \Gamma(u, v)\, e^{-i(ux+vy)}\, du\, dv, \tag{3}$$

which source may replace the illuminating system. The effective source so defined may not be physically realizable, although $\gamma(x, y)$ must be wholly real.‡ However, it ensures that the correct phase-coherence conditions obtain in the object plane, and its use simplifies theoretical considerations.

To specify the conditions of illumination of any surface one needs to specify both the phase-coherence between all pairs of points of the surface and the intensity of illumination at all points of the surface. The former information is implicitly summarized in the postulation of an effective source; the latter constitutes a photometric problem. Usually the object plane will be uniformly illuminated, and even if this is not so it is convenient to assume uniform illumination and to suppose that the incident light passes through a screen of suitable non-uniform transparency which covers the object. The non-uniformity of illumination then appears as a real

† Certain changes of notation are introduced in what follows.

‡ This follows since $\Gamma_{21}^* = \Gamma_{12}$, which is an immediate consequence of the definition (1). I am indebted to Dr E. H. Linfoot for drawing my attention to this point, which has enabled me to correct a wrong statement in the original manuscript.

factor in the complex function which is used to specify the structure of the object. The simplifying assumption of uniform illumination will therefore be made in what follows.

The effective source will be considered to be placed at some distance from the object plane, as shown at E_o in figure 1. If this source is assumed to illuminate directly the object plane at O, the plane E_o constitutes the entrance pupil for the subsequent image-forming optical system, whose actual entrance and exit pupils are in fact at (say) E, E' respectively. Hence to simulate exactly the actual conditions of illumination and image-formation a field lens must be assumed to exist at O, which lens forms an image of the effective source E_o in the entrance pupil at E.

If the original source employed is of uniform intensity and of sufficient extent, it may be shown (Hopkins 1951) that the phase-coherence factor for any pair of points of the object plane is independent of the precise illuminating system employed. Moreover, the phase-coherence conditions are unchanged if the particular illuminating system is replaced by a source of the same size and shape as the exit pupil of the illuminator, and having an intensity $\gamma(x, y)$ at the point (x, y) equal to that which obtains at the same point of the exit pupil itself. Hence, with a broad original source, the effective source as defined above is known at once.

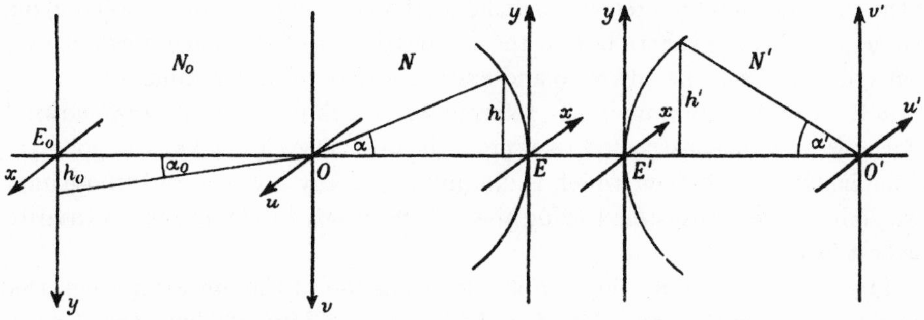

Figure 1. Arrangement of effective source and image-forming optical system.

The variables (x, y), (u, v) will now be defined with reference to figure 1. $E_oOEE'O'$ is the axis of an optical system. E_o is the (arbitrarily chosen) location of the effective source—explicit reference to which need, in fact, only be made if an approximate physical interpretation of the later theoretical results is required. O is the axial point of an object, whose image formed by an axially symmetrical optical system is at O'. The entrance and exit pupils of the optical system are at E, E' respectively.

Let a ray from O intersect a sphere of centre O and radius OE at a distance h from the axis. The refracted part of the ray, proceeding to the image point O', intersects a sphere of centre O' and radius $O'E'$ at a distance h' from the axis. These two spheres are the so-called reference spheres for the object and image spaces respectively. If the incident and refracted parts of the ray make angles α, α' with the axis, and (ξ, η) are geometrical distances in the object, the variables (u, v) are defined by the expressions

$$u = k(N \sin\alpha)\,\xi, \quad v = k(N \sin\alpha)\,\eta \tag{4}$$

where N is the refractive index of the object space, and $k = 2\pi/\lambda$ is the propagation constant. Similarly, for the image plane,

$$u' = k(N' \sin\alpha')\,\xi', \quad v' = k(N' \sin\alpha')\,\eta', \tag{5}$$

from which it will be seen that $u' = u$, $v' = v$ if (u', v') is the geometrical image of (u, v). Usually the optical system will have a circular diaphragm, and the particular ray of light will be that which passes through the rim of the diaphragm. This ray, continued backwards (after refraction if the refractive indices of the object and condenser spaces N, N_o are unequal), meets the effective source at a distance h_o from the axis, and the point (a_o, b_o) of the effective source will then be given rectangular co-ordinates (x_o, y_o) defined by

$$x_o = a_o/h_o, \quad y_o = b_o/h_o. \tag{6}$$

Similarly, points of the reference sphere having geometrical co-ordinates (a, b), (a', b') measured in planes perpendicular to the axis will be denoted by

$$x = a/h,\ y = b/h; \quad x = a'/h',\ y = b'/h'. \tag{7}$$

It then follows that the point (x, y) of the source, the object point O, and (x, y) of the entrance pupil reference sphere lie on the same incident ray, and (x, y) on the exit pupil reference sphere lies on the refracted part of the same incident ray. In terms of (x, y) both the entrance and exit pupils are of unit radius.

In the earlier communication referred to (Hopkins 1951) it was shown how a formula may be constructed to express the intensity $I'(u', v')$ at the point (u', v') of the image of an object, which is illuminated in any manner and whose image is formed by an optical system having aberrations of any kind and not necessarily set for best focus.

Thus, let $f(x, y)$ denote the complex transmission of the optical system, that is, between the reference spheres at E and E'. $f(x, y)$ will include both the aberrations of the system—assumed to have the same aberration for all points of the object—and a term corresponding to defect of focus. It is convenient to define $f(x, y)$ to be zero at points exterior to the pupils. The complex amplitude at (u', v') of the image plane associated with a wave originating at (u_1, v_1) of the object plane is then proportional to

$$F(u' - u_1, v' - v_1) = \frac{1}{2\pi}\iint_{-\infty}^{+\infty} f(x, y)\, e^{i\{(u'-u_1)x + (v'-v_1)y\}}\, dx\, dy. \tag{8}$$

It will be noted that $F(u, v)$, $f(x, y)$ are Fourier transforms of each other.

Let a complex function $E(u, v)$ specify the complex transmission of the object. If an element $d\sigma$ of the source produces a complex amplitude U_1 at (u_1, v_1), the amplitude in the image plane due to light from $d\sigma$ is given, apart from a constant factor, by

$$\iint_{-\infty}^{+\infty} U_1 E(u_1, v_1)\, F(u' - u_1, v' - v_1)\, du_1\, dv_1.$$

An equivalent expression can be obtained using an independent current point (u_2, v_2). The partial intensity in the image which is associated with light from $d\sigma$ is then given by

$$dI'(u', v') = d\sigma \iiiint_{-\infty}^{+\infty} U_1 E(u_1, v_1)\, F(u'-u_1, v'-v_1)\, U_2^* E^*(u_2, v_2) \times F^*(u'-u_2, v'-v_2)\, du_1 du_2 dv_1 dv_2.$$

Integrating over the whole source, and noting the definition (1), the total intensity in the image is found to be

$$I'(u', v') = \iiiint_{-\infty}^{+\infty} \Gamma(u_1-u_2, v_1-v_2)\, E(u_1, v_1)\, F(u'-u_1, v'-v_1)\, E^*(u_2, v_2) \times F^*(u'-u_2, v'-v_2)\, du_1\, du_2\, dv_1\, dv_2, \quad (9)$$

in obtaining which a uniform illumination of the object plane has been assumed, and a constant factor omitted. This is essentially the formula referred to above. The assumption of a uniform illumination of the object plane accounts for the absence of two additional factors which appear in the integrand of the corresponding earlier result.

It will be appreciated that, by means of (9), the light distributions in the images of objects in any plane of focus may be calculated, no matter what the aberrations or conditions of illumination. Moreover, by choosing $\Gamma(u_1-u_2, v_1-v_2)$ to be zero, except when $u_2 = u_1, v_2 = v_1$, in which case it is put equal to unity, the images of self-luminous (incoherent) objects may be studied. According to (3), this requires $\gamma(x, y) = \frac{1}{2\pi}$, for then $\Gamma(u_1-u_2, v_1-v_2) = \delta(u_1-u_2, v_1-v_2)$, where δ denotes the Dirac delta function. This would correspond to having a uniform effective source of infinite extent. Physically such a source would not illuminate a plane incoherently (Hopkins 1951), since the sine of its angular radius can never exceed unity. However, with the definitions (6), the variables (x, y) may formally take any values from $-\infty$ to $+\infty$ although physically they become meaningless when $\sqrt{(x^2+y^2)}\sin\alpha_0 > 1$.

In what follows it is shown that the multiple integral (9) may be evaluated in terms of a double integral involving the inverse Fourier transforms $\gamma(x, y), f(x, y)$ of the functions $\Gamma(u, v)$, $F(u, v)$. These latter functions themselves need not appear. Since it is $\gamma(x, y)$, $f(x, y)$ which are known initially—and they are usually of very simple form, whereas their transforms are often too difficult to obtain explicitly—a general solution of the formation of optical images is obtained. This general treatment is illustrated by application to some simple cases. The mathematical technique makes possible the solution of a wide range of problems, as an example of which the influence of the illuminating aperture on optical images may be mentioned. By giving suitable forms to $f(x, y)$, phase contrast and other 'artificial' images may be studied. An approximate physical interpretation of the results is found to be possible.

2. General formulae

We shall consider, first, the evaluation of the multiple integral

$$I'(u',v') = \iiiint_{-\infty}^{+\infty} \Gamma(u_1-u_2, v_1-v_2)\, E(u_1,v_1)\, F(u'-u_1, v'-v_1)\, G(u_2,v_2) \times H(u'-u_2, v'-v_2)\, \mathrm{d}u_1\, \mathrm{d}u_2\, \mathrm{d}v_1\, \mathrm{d}v_2 \quad (10)$$

in terms of the functions $\gamma(x,y), f(x,y), h(x,y)$, from which the transforms $\Gamma(u,v)$, $F(u,v)$, $H(u,v)$ derive. The specialization $G(u,v) = E^*(u,v)$, $H(u,v) = F^*(u,v)$ will be made later.

The functions $E(u,v)$, $G(u,v)$ may be replaced by their Fourier series

$$\left.\begin{aligned} E(u,v) &= \frac{1}{2\pi} \sum_m \sum_n a(m,n)\, \mathrm{e}^{\mathrm{i}(mu+nv)}, \\ G(u,v) &= \frac{1}{2\pi} \sum_p \sum_q b(p,q)\, \mathrm{e}^{\mathrm{i}(pu+qv)}, \end{aligned}\right\} \quad (11)$$

or, in cases where one or other of the functions is non-periodic, the corresponding Fourier integrals. If the expressions (11) are substituted for $E(u,v)$, $G(u,v)$, and the variables are then changed to $U_1 = u' - u_1$, $U_2 = u' - u_2$, with similar definitions for V_1, V_2, (10) becomes

$$I'(u',v') = \left(\frac{1}{2\pi}\right)^2 \sum_m \sum_n \sum_p \sum_q C(m,n;\,p,q)\, a(m,n)\, b(p,q)\, \mathrm{e}^{\mathrm{i}\{(m+p)u'+(n+q)v'\}}, \quad (12)$$

where

$$C(m,n;\,p,q) = \iiiint_{-\infty}^{+\infty} \Gamma(-U_1-U_2, -V_1-V_2)\, \mathrm{e}^{-\mathrm{i}(mU_1+nV_1)} F(U_1,V_1) \times \mathrm{e}^{\mathrm{i}(pU_2+qV_2)} H(-U_2,-V_2)\, \mathrm{d}U_1\, \mathrm{d}U_2\, \mathrm{d}V_1\, \mathrm{d}V_2, \quad (13)$$

(U_2, V_2) having been replaced by $(-U_2, -V_2)$. The general theorem (Titchmarsh 1937)

$$\int_{-\infty}^{+\infty} F(u)\, G(u)\, H(u)\, \mathrm{d}u = \frac{1}{\sqrt{(2\pi)}} \iint_{-\infty}^{+\infty} f(-x_1-x_2)\, g(x_1)\, h(x_2)\, \mathrm{d}x_1\, \mathrm{d}x_2$$

is easily modified and extended to two variables, leading to the result

$$\iiiint_{-\infty}^{+\infty} F(-U_1-U_2, -V_1-V_2)\, G(U_1,V_1)\, H(U_2,V_2)\, \mathrm{d}U_1\, \mathrm{d}U_2\, \mathrm{d}V_1\, \mathrm{d}V_2 = 2\pi \iint_{-\infty}^{+\infty} f)x,y)\, g(x,y)\, h(x,y)\, \mathrm{d}x\, \mathrm{d}y. \quad (14)$$

Using this theorem in (13) gives

$$C(m,n;\,p,q) = 2\pi \iint_{-\infty}^{+\infty} \gamma(x,y) f(x+m, y+n)\, h(p-x, q-y)\, \mathrm{d}x\, \mathrm{d}y, \quad (15)$$

since

$$\mathrm{e}^{-\mathrm{i}(mU_1+nV_1)} F(U_1,V_1)$$

is the transform of $f(x+m, y+n)$, and

$$e^{+i(pU_2+qV_2)}H(-U_2, -V_2)$$

is that of $h(p-x, q-y)$. The required result is now obtained if the value of $C(m,n;p,q)$ in (15) is used in (12).

To return to the formula (9), we write in (10), $G(u,v) = E^*(u,v), H(u,v) = F^*(u,v)$. In (11) we must then put $b(p,q) = a^*(p,q)$, and replace p and q in the exponential terms by $-p$ and $-q$. The complex transmission of the object is thus expressed by

$$E(u,v) = \frac{1}{2\pi}\sum_m \sum_n a(m,n)\,e^{i(mu+nv)} \tag{16}$$

or the corresponding integral. In place of (12) we have

$$I'(u',v') = \left(\frac{1}{2\pi}\right)^2 \sum_m \sum_n \sum_p \sum_q C(m,n;\,p,q)\,a(m,n)\,a^*(p,q)\,e^{i\{(m-p)u'+(n-q)v'\}}, \tag{17}$$

and (15) becomes

$$C(m,n;\,p,q) = 2\pi \iint_{-\infty}^{+\infty} \gamma(x,y) f(x+m, y+n) f^*(x+p, y+q)\,dx\,dy. \tag{18}$$

A general solution of the optical problem is given by the formulae (17) and (18).

The structure of the object is specified by the function $E(u,v)$ (16), and the distribution of intensity in the image is given by $I'(u',v')$. The intensity transmitted by the object is given by $|E(u,v)|^2$, and the 'perfect' image should reproduce this distribution of intensity. Hence the 'perfect' image is represented by

$$I(u',v') = \left(\frac{1}{2\pi}\right)^2 \sum_m \sum_n \sum_p \sum_q a(m,n)\,a^*(p,q)\,e^{i\{(m-p)u'+(n-q)v'\}}. \tag{19}$$

Comparison of (17) and (19) shows that some information regarding the object will be lost or falsified unless (when suitably normalized) $C(m,n;\,p,q) = 1$ for all values of m, n, p, q, for which the coefficients $a(m,n), a^*(p,q)$ are both other than zero. The influence of the coherence conditions in the object plane may be studied by giving suitable forms to $\gamma(x,y)$—the effective source defined in (3)—and the effects of aberrations and defect of focus of the optical system appear in the influence of $f(x,y)$ on the coefficients $C(m,n;\,p,q)$.

If the object is coherently illuminated we must place a point source at (x_0, y_0) of the effective source; (x_0, y_0) then determine the obliquity of the illuminating wave. For example, $x_0 = y_0 = 0$ denotes that the undiffracted light (zero-order spectrum) is along the direction of the axis, and $\sqrt{(x_0^2+y_0^2)} = 1$ denotes that the zero-order spectrum has the direction of the ray through the edge of the diaphragm. We write now

$$2\pi\gamma(x,y) = \delta(x-x_0, y-y_0),$$

where δ is again the Dirac delta function. Inserting this in (18) gives

$$C(m,n;\,p,q) = f(x_0+m, y_0+n) f^*(x_0+p, y_0+q), \tag{20}$$

and, if the optical system has a circular aperture of radius $\sqrt{(x^2+y^2)} = 1$, $C(m,n;\,p,q)$ will be zero when either, or both, of the conditions

$$\left.\begin{aligned} (x_0+m)^2+(y_0+n)^2 < 1, \\ (x_0+p)^2+(y_0+q)^2 < 1, \end{aligned}\right\} \tag{21}$$

is not fulfilled, since $f(x,y) = 0$ outside the unit circle. In other words, the Fourier spectra of frequencies m, n, p, q must fall within the lens aperture if they are to appear in the image. Within the aperture $|f(x,y)|$ is finite. Hence, if a structure is resolved with a perfectly focused system free from aberration, it will also be 'resolved' in the presence of aberrations and defects of focus, but generally with reduced (or reversed) contrast. In the absence of aberration and defect of focus $f(x,y) = 1$ for $x^2+y^2<1$. Hence, providing (21) are fulfilled, $C(m,n;\,p,q) = 1$ and the image is a perfect reproduction of the object. In coherent light we shall find that it is the size of aperture of the lens system which determines its resolving limit, whereas the aberrations and choice of focus determine the contrast of the resolved frequencies as they appear in the image.

Periodic line structures

If the structure is singly-periodic in the U-direction, the coefficients in the series for $E(u,v)$ and its complex conjugate must be restricted by the conditions $a(m,n) = 0$ $(n \neq 0)$, and $a^*(p,q) = 0$ $(q \neq 0)$. We then write

$$E(u) = \frac{1}{\sqrt{(2\pi)}} \sum_m a(m)\,\mathrm{e}^{\mathrm{i}mu}, \tag{22}$$

and in place of (16) we then find

$$I'(u') = \left(\frac{1}{2\pi}\right) \sum_m \sum_p C(m,p)\,a(m)\,a^*(p)\,\mathrm{e}^{\mathrm{i}(m-p)u'}, \tag{23}$$

where
$$C(m,p) = 2\pi \iint_{-\infty}^{+\infty} \gamma(x,y) f(x+m,y) f^*(x+p,y)\,\mathrm{d}y\,\mathrm{d}x. \tag{24}$$

If the effective illumination is a point source at $(x_0, 0)$, we find, in place of (20),

$$C(m,p) = f(x_0+m, 0) f^*(x_0+p, 0), \tag{25}$$

from which it is seen that $C(m,p) = 0$ when either of the conditions $|x_0+m|<1$, $|x_0+p|<1$ is not satisfied. The maximum values of $|m|$, $|p|$ which satisfy these conditions are equal to $1+|x_0|$. Since, according to (22), the length of period of the frequency m is given by $mu = 2\pi$, u being defined by (4), the limit of resolution occurs when

$$\xi = \frac{\lambda}{(1+\sigma) N \sin\alpha}, \tag{26}$$

where $\sigma = |x_0|$ is equal to the ratio $N_0 \sin\alpha_0 / N \sin\alpha$, α_0 being the angular distance of the effective point source from the axis, and N_0 being the refractive index of the space behind the object. (26) reduces to the well-known results of Abbé when $\sigma = 0$ and 1. Hence the Abbé formulae give true limits of resolution—at least in the case of coherently illuminated objects. The contrast of the image, however, is in general impaired or even falsified by the presence of aberrations and defective focusing.

Non-periodic structures

If the object is non-periodic it may be simpler to evaluate the integral (9) in the following manner, as opposed to the use of a Fourier integral in place of the series (16).

Consider first the integral (10), and denote by Φ, Ψ the expressions

$$\left.\begin{aligned}\Phi(u,v) &= E(-u,\,-v)\,F(u'+u,v'+v),\\ \Psi(u,v) &= G(u,v)\,H(u'-u,v'-v).\end{aligned}\right\} \tag{27}$$

In terms of these functions, (10) is equivalent to

$$I'(u',v') = \iiiint_{-\infty}^{+\infty} \Gamma(-u_1-u_2,\,-v_1-v_1)\,\Phi(u_1,v_1)\,\Psi(u_2,v_2)\,\mathrm{d}u_1\,\mathrm{d}u_2\,\mathrm{d}v_1\,\mathrm{d}v_2, \tag{28}$$

which, by virtue of the general theorem (14), may be transformed to give

$$I'(u',v') = 2\pi \iint_{-\infty}^{+\infty} \gamma(x,y)\,\phi(x,y)\,\psi(x,y)\,\mathrm{d}x\,\mathrm{d}y, \tag{29}$$

in which ϕ and ψ are the functions whose transforms are Φ and Ψ. These functions are given by

$$\phi(x,y) = \frac{1}{2\pi}\iint_{-\infty}^{+\infty} E(-u,\,-v)\,F(u'+u,v'+v)\,\mathrm{e}^{-\mathrm{i}(ux+vy)}\,\mathrm{d}u\,\mathrm{d}v,$$

$$\psi(x,y) = \frac{1}{2\pi}\iint_{-\infty}^{+\infty} G(u,v)\,H(u'-u,v'-v)\,\mathrm{e}^{-\mathrm{i}(ux+vy)}\,\mathrm{d}u\,\mathrm{d}v,$$

which, using the 'faltung' and 'shift' theorems for Fourier transforms, become

$$\left.\begin{aligned}\phi(x,y) &= \frac{1}{2\pi}\iint_{-\infty}^{+\infty} e(s-x,t-x)\,f(s,t)\,\mathrm{e}^{\mathrm{i}(su'+tv')}\,\mathrm{d}s\,\mathrm{d}t,\\ \psi(x,y) &= \frac{1}{2\pi}\iint_{-\infty}^{+\infty} g(s+x,t+y)\,h(s,t)\,\mathrm{e}^{\mathrm{i}(su'+tv')}\,\mathrm{d}s\,\mathrm{d}t,\end{aligned}\right\} \tag{30}$$

where, as before, e, f, g, h are the functions whose transforms are E, F, G, H.

To evaluate (9) we require to put $G(u,v) = E^*(u,v)$, and $H(u,v) = F^*(u,v)$. In place of the value in (30), we now find $\psi(x,y) = \phi^*(x,y)$, and (29) reduces to

$$I'(u',v') = 2\pi \iint_{-\infty}^{+\infty} \gamma(x,y)\,|\,\phi(x,y)\,|^2\,\mathrm{d}x\,\mathrm{d}y, \tag{31}$$

which, we shall see, leads in two simple cases to known results.

If the object is coherently and cophasally illuminated we must put

$$2\pi\gamma(x,y) = \delta(x,y).$$

(31) then gives $$I'(u',v') = |\,\phi(0,0)\,|^2$$

by virtue of the sifting property of the delta function. We may then expect the complex amplitude in the image to be found from the formula

$$E'(u',v') = \phi(0,0) = \frac{1}{2\pi}\iint_{-\infty}^{+\infty} e(s,t)f(s,t)\,\mathrm{e}^{\mathrm{i}(su'+tv')}\,\mathrm{d}s\,\mathrm{d}t, \tag{32}$$

a result which may be verified directly by means of an obvious application of the faltung theorem (see, for example, Duffieux 1946).

If the object is self-luminous (incoherence), we must write $\gamma(x,y) = 1$, and (31) reduces to

$$I'(u',v') = 2\pi \iint_{-\infty}^{+\infty} |\,\phi(x,y)\,|^2 \,dx\,dy$$

$$= 2\pi \iint_{-\infty}^{+\infty} |\,\Phi(u,v)\,|^2 \,du\,dv$$

by Parseval's theorem. Hence, recalling the definition (27),

$$I'(u',v') = 2\pi \iint_{-\infty}^{+\infty} |E(u,v)\,|^2\,|\,F(u'-u,\,v'-v)\,|^2 \,du\,dv, \tag{33}$$

the signs of u, v having been reversed. The two functions defined by

$$G(u,v) = |\,E(u,v)\,|^2,$$

$$H(u'-u,v'-v) = |\,F(u'-u,v'-v)\,|^2,$$

represent respectively the distributions of intensity in the object plane and in the diffraction image of a single point of the object. If these are substituted in (33) an obvious result follows. By use of the faltung theorem, (33) may then be transformed to give

$$I'(u',v') = 2\pi \iint_{-\infty}^{+\infty} g(x,y)\,h(x,y)\,e^{i(u'x+v'y)}\,dx\,dy. \tag{34}$$

Usually $h(x,y)$ is not known in simple form, and in consequence (34) is of restricted use. It is simpler to proceed using (30) and (31) with $\gamma(x,y) = 1$ for the whole (x,y) plane.

Duffieux, dealing with the two limiting cases of perfect cophasal coherence and complete incoherence, has called $f(x,y)$, $h(x,y)$ the transmission factors for the two cases respectively. Each Fourier frequency in the object function appears in the image distribution multiplied by the appropriate transmission factor. The same interpretation attaches to the coefficient $C(m,n;\,p,q)$, which appears, therefore, as a generalized transmission factor, from which $f(x,y)$, $h(x,y)$ derive as simple cases.

Non-periodic line structure

The structure of the object will now be supposed to vary only in the direction of u. In place of (27), we now write

$$\Phi(u,v) = E(-u)\,F(u'+u,v'+v). \tag{35}$$

The Fourier transform of unity is $\sqrt{(2\pi)}\,\delta(t)$. Hence the transform of $\Phi(u,v)$ will be given from (30) by

$$\phi(x,y) = \frac{1}{\sqrt{2\pi}} \iint_{-\infty}^{+\infty} e(s-x)\,\delta(t-y) f(s,t)\,e^{i(su'+tv')}\,ds\,dt,$$

or, because of the sifting property of $\delta(t-y)$,

$$\phi(x,y) = \frac{1}{\sqrt{(2\pi)}} e^{ivv'} \int_{-\infty}^{+\infty} e(s-x) f(s,y) e^{iu's} ds. \tag{36}$$

The intensity in the image is given by (31), $\phi(x,y)$ having now the form (36) in place of (30).

3. Images of coherent line structures

We shall first apply the formulae (23) and (24) to the simple object defined by the amplitude-transmission function

$$E(u) = \cos \omega u. \tag{37}$$

It is worth noting that physically the object so defined would need to have an intensity transmission $I(u) = \cos^2 \omega u$, together with $\frac{1}{2}\lambda$ phase-changing strips covering alternate periods of the structure.

The function (37) contains only two exponential frequencies, namely, $\pm\omega$. The two non-zero coefficients in (22) are $a(+\omega) = a(-\omega) = \sqrt{(\frac{1}{2}\pi)}$. Hence, in this case, (23) becomes

$$I'(u') = \tfrac{1}{4}\{C(\omega,\omega)+C(-\omega,-\omega)\} + \tfrac{1}{4}C(\omega,-\omega)\, e^{+i2\omega u'} + \tfrac{1}{4}C(-\omega,\omega)\, e^{-i2\omega u'}, \tag{38}$$

and, if the object is coherently illuminated at an obliquity (x_0, y_0), we write $2\pi\gamma(x,y) = \delta(x-x_0, y-y_0)$. The transmission factor (24) is then given by

$$C(m,p) = f(x_0+m, y_0) f^*(x_0+p, y_0), \tag{39}$$

providing the frequencies (m,p) are both resolved. This requires that neither $(x_0+m)^2$ nor $(x_0+p)^2$ exceed $1-y_0^2$. If these conditions are not satisfied $C(m,p) = 0$. Let $W(x,y)$ represent the wave aberration of the optical system, together with any defect of focus, then

$$\begin{aligned} f(x,y) &= e^{ikW(x,y)} \quad (x^2+y^2<1) \\ &= 0 \qquad\qquad (x^2+y^2>1), \end{aligned} \tag{40}$$

$k = 2\pi/\lambda$ being the propagation constant. The lens aperture is supposed to be of uniform transparency.

Substituting in (38) the values of $C(m,p)$ obtained by using (40) in (39), the image distribution is found after some reduction to be

$$I'(u') = \cos^2\{\omega u' + \tfrac{1}{2}k[W(x_0+\omega, y_0) - W(x_0-\omega, y_0)]\}.$$

A suitable transverse shift of focus will make $W(x_0+\omega, y_0) = W(x_0-\omega, y_0)$, and consequently

$$I'(u') = \cos^2 \omega u', \tag{41}$$

which is exactly the same as $I(u') = |E(u')|^2$. Thus, no matter what the aberration or plane of focus, the intensity distribution in the image is identical with that in the object—although the image may suffer a lateral displacement. This result is easy to understand if we recall the fact that only two diffraction spectra are formed by this structure.

The Fourier series representing the square-topped amplitude transmission of an ideal diffraction grating may be written

$$E(u) = \frac{1}{2} + \frac{2}{\pi} \Sigma \frac{1}{p} \cos p\omega u, \quad (p=1, 3, \ldots) \tag{42}$$

where ω is the fundamental frequency. To limit the analysis to essentials we shall consider a grating of the form

$$E(u) = \tfrac{1}{2}(1 + \cos \omega u). \tag{43}$$

The image will then roughly correspond to that obtained when only the zero and two first-order spectra of (42) take part in the formation of the image. The series (22) will now have three non-zero coefficients, and in place of (38) we now have nine terms of the series (23) to consider. (23) then becomes

$$\begin{aligned} I'(u') = {} & \tfrac{1}{4}C(0,0) + \tfrac{1}{8}C(\omega, \omega) + \tfrac{1}{8}C(-\omega, -\omega) \\ & + \tfrac{1}{8}\{C(0, -\omega) + C(\omega, 0)\} e^{i\omega u'} \\ & + \tfrac{1}{8}\{C(-\omega, 0) + C(0, \omega)\} e^{-i\omega u'} \\ & + \tfrac{1}{16}C(\omega, -\omega) e^{i2\omega u'} + \tfrac{1}{16}C(-\omega, \omega) e^{-i2\omega u'}, \end{aligned} \tag{44}$$

the coefficients being defined by (24).

With oblique coherent illumination the transmission factors will again be given by (39) and (40). The intensity distribution in the image is easily found from (44). It is

$$\begin{aligned} I'(u') = {} & \tfrac{1}{4} + \tfrac{1}{4}\cos\{\omega u' + kW(x_0 + \omega, y_0) - kW(x_0, y_0)\} \\ & + \tfrac{1}{4}\cos\{\omega u' + kW(x_0, y_0) - kW(x_0 - \omega, y_0)\} \\ & + \tfrac{1}{4}\cos^2\{\omega u' + \tfrac{1}{2}kW(x_0 + \omega, y_0) - \tfrac{1}{2}kW(x_0 - \omega, y_0)\}. \end{aligned}$$

By means of a change of radius of the image reference sphere, we may make $W(x_0, y_0) = 0$, and then a suitable transverse shift of focus will make

$$W(x_0 - \omega, y_0) = W(x_0 + \omega, y_0).$$

The above expression then becomes

$$I'(u') = \tfrac{1}{4} + \tfrac{1}{2}\cos kW(x_0 + \omega, y_0)\cos \omega u' + \tfrac{1}{4}\cos^2 \omega u'. \tag{45}$$

For comparison, the ideal image would have an intensity distribution given by the squared modulus of (43). This is

$$I(u') = \tfrac{1}{4} + \tfrac{1}{2}\cos \omega u' + \tfrac{1}{4}\cos^2 \omega u'. \tag{46}$$

The image differs from the object only by the presence of the factor $\cos kW(x_0 + \omega, y_0)$.

If now the focal plane is displaced along the axis, the resulting longitudinal focal shift term will produce an increment in $W(x_0 + \omega, y_0)$ proportional to the magnitude of the displacement. In consequence there will be a perfect image formed in the equally spaced planes for which $W(x_0 + \omega, y_0)$ is equal to an integral number of wavelengths. Hence the image of a simple periodic test-object formed by coherent light is independent of the aberration of the optical system. It follows that a coherently illuminated grating is of no value as a test-object. We shall see later that this conclusion is valid for any conditions of illumination and also for incoherent test-objects. I am indebted to Dr A. Maréchal for drawing my attention to this last point.

In figure 2 are shown curves of the intensity (45) for different values of $W(\omega)$—the effective source is taken to be at the point (0, 0), but this implies no loss of generality, providing the conditions following (39) are satisfied. It will be noted that the contrast is successively reversed as the plane of the focus is changed.

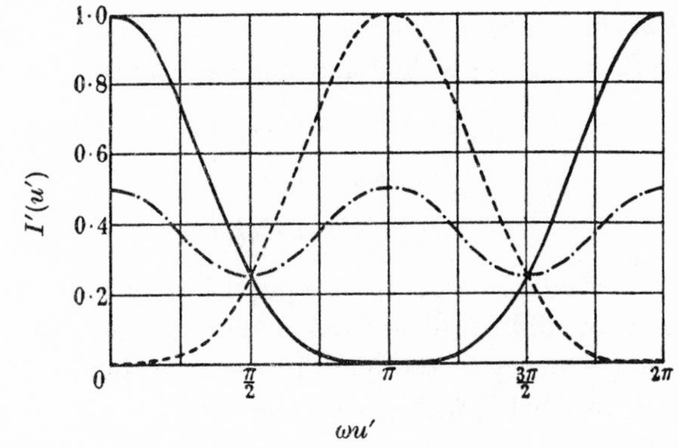

FIGURE 2. Images of the structure $E(u) = \frac{1}{2}(1+\cos \omega u)$. Different planes of focus and coherent illumination. $W(\omega) = 0, \pm 2\pi, \ldots$ (—); $\pm\frac{1}{2}\pi, \pm\frac{3}{2}\pi, \ldots$ (–·–); $\pm\pi, \pm 3\pi, \ldots$ (- - -).

The Rayleigh quarter wave-length limit for the tolerable aberration in an optical system ensures that $|W(x, y)| < \frac{1}{4}\lambda$ for all values of (x, y). In terms of the above result this amounts to saying that, at the optimum focus, none of the Fourier components of a structure appear falsified in the image by reversal of contrast.* On the other hand, the contrast of the components of different frequencies in the image will generally be reduced by the presence of aberrations, and some of them may be absent because $C(m, p) = 0$. Furthermore, a given component may be laterally displaced in the image relative to the other components.

A further conclusion may be noted in connexion with coherently illuminated objects. The coefficient $C(m, p)$ corresponding to a given term in (23) may be made equal to unity by an appropriate choice of the focal plane. Hence different planes of focus will select the finer or the coarser detail in the object. If the aberration of the optical system is zero all the resolved frequencies are, of course, perfectly reproduced in the focused image. The effect of a change of focus on $W(x_0+\omega, y_0)$ depends upon $(x_0+\omega)^2+y_0^2$, and will therefore be different for the different frequencies and will also depend on the obliquity (x_0, y_0) of the illuminating wave. For central illumination, when $x_0 = y_0 = 0$, the finer detail will be more susceptible to a change of focus than the coarser detail.

If all the terms in (44) are to appear in the image, the limit of resolution occurs when $\omega = 1$ $(\xi = \lambda/N \sin\alpha)$, and at this limit it is necessary to employ central coherent

* $|W(x, y)| < \frac{1}{4}\lambda$ implies that the $\lambda/4$ tolerance is allowed on both sides of the reference sphere. The stricter interpretation requires $|W(x, y)| < \frac{1}{8}\lambda$, $W(x, y)$ being referred to the optimum focus. In this case, of course, all the resolved components of the object appear in the image, and they have transmission factors not less than $1/\sqrt{2}$.

illumination for which $x_0 = y_0 = 0$. We shall see later that the contrast in the image is impaired, even for $\omega = 1$, when the object is self-luminous, although the limit of resolution in this case is $\omega = 2$. If the aperture of the illuminator is of radius $\sigma = 1$, the same limit of resolution will be found, and similarly the contrast, even for $\omega = 1$, is impaired.

By the use of oblique coherent illumination the limit of resolution may be increased with coherent light to the value $\omega = 2$, and it therefore is of interest to study the contrast in the image in this case.

We shall suppose that ω satisfies the following conditions

$$\left.\begin{aligned}(x_0+\omega)^2+y_0^2 &> 1-y_0^2,\\ x_0^2+y_0^2 &< 1,\\ (x_0-\omega)^2+y_0^2 &< 1-y_0^2,\end{aligned}\right\} \tag{47}$$

so that the spectrum $+\omega$ falls outside the lens pupil, whereas the zero order and $-\omega$ fall within it. The state of affairs is illustrated in figure 3, where the full circle of unit radius represents the periphery of the pupil in the (x, y) plane. If the expressions (39) and (40) are now recalled it will be seen that $C(m, p)$ will be zero if either, or both, of m and p is equal to $+\omega$, the conditions (47) being assumed to be satisfied. The remaining terms give

$$I'(u') = \tfrac{5}{8} + \tfrac{1}{4}\cos\{\omega u' + kW(x_0, y_0) - kW(x_0-\omega, y_0)\}. \tag{48}$$

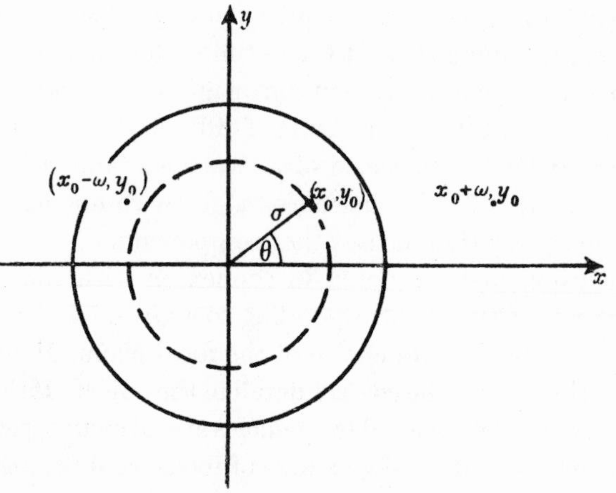

FIGURE 3. xy-plane; circular pupil and ring source.

By means of a transverse focal shift we may make $W(x_0, y_0) = W(x_0+\omega, y_0)$, and (48) becomes

$$I'(u') = \tfrac{5}{8} + \tfrac{1}{4}\cos \omega u', \tag{49}$$

which remains unaffected by a change of focal plane, except in so far as this makes necessary a different transverse focal shift. Hence the influence of aberration on the image will be to shift it laterally in the image plane. This shift will depend upon the frequency ω, the obliquity (x_0, y_0), and on the aberration and defect of focus. Thus a more complex structure will be blurred by the presence of aberration, and will appear to be 'focusable'.

The contrast in the image (49) is well marked (figure 6), and annular illumination has therefore great advantages for the study of fine detail, providing the peripheral parts of the lens are well corrected for aberration. However, it will be noted (figure 3) that some parts of a circular line source of radius (say) σ will be such that only the zero-order spectra fall within the pupil. These contribute only a constant term to (58) and in consequence have an adverse effect on the contrast in the image. If θ is the polar radius of a point on the source we require

$$\cos\theta < \frac{|\omega|-1}{\sigma} \tag{50}$$

to be satisfied if all points of the source are to contribute effectively to the image.

It would seem from the above to be a useful adjunct in central dark-ground illumination to employ, for the observation of line structures, a pair of 'shutters' capable of rotation about the optical axis and obscuring, by adjustment of their separation, those parts of the annulus not satisfying the condition (50). In practice the correct adjustment would be made using the observed contrast in the image as a criterion.

4. Images of self-luminous line structures

It is desirable to normalize the coefficients (24) in such a way that the uniform intensity in the image of an object of uniform intensity is itself unity. We have then to make $C(0,0) = 1$; and hence, in place of (24), we now write

$$C(m,p) = \frac{1}{A}\iint_{-\infty}^{+\infty} \gamma(x,y) f(x+m,y) f^*(x+p,y)\,\mathrm{d}x\,\mathrm{d}y, \tag{51}$$

where

$$A = \iint_{-\infty}^{+\infty} \gamma(x,y)|f(x,y)|^2\,\mathrm{d}x\,\mathrm{d}y \tag{52}$$

and $C(0,0) = 1$. It will be seen that, putting $\gamma(x,y) = 1$ over the area of the effective source and zero outside it, A is equal to the area in (x,y) units of the effective source or of the entrance pupil of the optical system, whichever is the less, assuming $|f(x,y)| = 1$ within the pupil.

It will be noted that the same normalization has been used above in connexion with coherent illumination. In the case of dark-ground observation $C(0,0) = 0$, since $f(x,y) = 0$ at all points for which $\gamma(x,y) \neq 0$. It is nevertheless best to retain effectively the same normalization by assuming $|f(x,y)| = 1$ in (52) for the whole area of the entrance pupil, including the opaque 'dark-ground' obstruction.

When the entrance pupil is a circle of radius $\sqrt{(x^2+y^2)} = 1$, the functions $f(x+m,y)$, $f^*(x+p,y)$ are zero outside circles of unit radius centred on the points $C_m(m,0)$ and $C_p(p,0)$ respectively. For self-luminous objects the effective source has $\gamma(x,y) = 1$ for the whole (x,y) plane. Hence in this case the integral (51) extends over the area common to the circles C_m and C_p. These circles are shown by the full lines in figure 4. The factor A is in this case equal to the area of the pupil, so that $A = \pi$.

If the object is illuminated by partially coherent light, the integrand in (51) will be equal to zero for all points exterior to the (now finite) region of the effective source, since for such points $\gamma(x, y) = 0$. Hence, unless the area common to C_m and C_p lies wholly within the region of the effective source, the domain of integration (51) will be further restricted as indicated by the dashed circle in figure 4. We shall return to this point later in connexion with the influence of the aperture of the illuminator on optical images.

For the moment we assume the object to be self-luminous, and the relevant function to describe it is the squared modulus of $E(u)$. This is, according to (22),

$$I(u) = \left(\frac{1}{2\pi}\right) \sum_m \sum_p a(m)\, a^*(p)\, e^{i(m-p)u},$$

which may be rewritten in the form

$$I(u) = \frac{1}{\sqrt{(2\pi)}} \sum_s b(2s)\, e^{i2su}, \tag{53}$$

where

$$b(2s) = \frac{1}{\sqrt{(2\pi)}} \sum_m a(m+s)\, a^*(m-s). \tag{54}$$

Since $I(u) = |E(u)|^2$, it is independent of the argument of $E(u)$, and consequently the coefficients $b(2S)$ also depend only on $|E(u)|$. We shall now show that for self-luminous objects the terms in (23) may also be grouped in a similar manner, and that the coefficients of the different frequencies $b(s)$ are 'transmitted' by the optical system independently of each other. This is in contrast to the case of coherent illumination, where different frequencies m and p may combine to give a separate term in the image intensity distribution.

Suppose then that the origin (figure 4) is moved to the point $(\frac{1}{2}(m+p), 0)$ in the (x, y) plane. (51) then becomes, for a self-luminous object,

$$C(m,p) = \frac{1}{\pi} \iint_{A(m,p)} f(x + \tfrac{1}{2}(m-p), 0) f^*(x - \tfrac{1}{2}(m-p), 0)\, dx\, dy, \tag{55}$$

where $A(m,p)$ denotes the area common to the circles C_m and C_p. Hence $C(m,p)$ will have the same value for all values of m and p for which $m-p = 2s$, where s is constant. Of course, s is subject to the condition $|s| < 1$, since for $|s| > 1$ the area $A(m,p)$ is equal to zero. We may therefore write

$$C(m+s, m-s) = C(s, -s) \tag{56}$$

and $C(m,m) = C(0,0) = 1$. With (56) in mind we may now group the terms of (23) in the following manner:

$$I'(u') = \frac{1}{\sqrt{(2\pi)}} \sum_s C(s,-s) \left\{ \frac{1}{\sqrt{(2\pi)}} \sum_m a(m+s)\, a^*(m-s) \right\} e^{i2su'} \tag{57}$$

or

$$I'(u') = \frac{1}{\sqrt{(2\pi)}} \sum_s D(s)\, b(s)\, e^{isu'}, \tag{58}$$

where
$$D(s) = \frac{1}{\pi} \iint\limits_{A(\frac{1}{2}s,\, -\frac{1}{2}s)} f(x+\tfrac{1}{2}s, 0)\, f^*(x-\tfrac{1}{2}s, 0)\, dx\, dy. \tag{59}$$

From this it is seen that $D(s)$ is the transmission factor for the frequency s of the object intensity function.

It will be observed that (56) is true for all values of m only when the object is self-luminous. If the effective source is of finite extent the region of integration $A(m, p)$ will be further limited, at least for some values of m and p, because $\gamma(x, y)$ may be equal to zero over part or all of $A(m, p)$. In these cases, $C(m+s, m-s) \neq C(s, -s)$, and the terms of the m-summation in (57) will therefore not all contain $C(s, s)$ as a factor. However, those terms for which (56) is true may still be grouped together, and regarded as an effectively self-luminous component of the object.

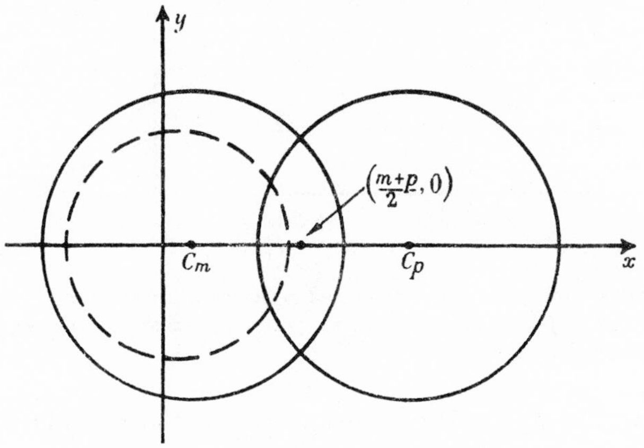

FIGURE 4. Region of integration for $C(m, p)$.

If the optical system is free from aberration and is perfectly focused, the function $f(x, y)$ is equal to unity inside the unit circle and zero outside. Hence, in this case the coefficients $D(s)$ in (58) are given, according to (59) by
$$D(s) = \frac{1}{\pi} A(\tfrac{1}{2}s, -\tfrac{1}{2}s), \tag{60}$$

$A(\frac{1}{2}s, -\frac{1}{2}s)$ being the area, in (x, y) units, of the region common to two circles of unit radius centred on $(\frac{1}{2}s, 0)$ and $(-\frac{1}{2}s, 0)$ respectively. This result is obviously valid for any shape of pupil, providing π is replaced by the area of the pupil. Thus, for a rectangular pupil with sides parallel with the x- and y-axes, and of unit length in the direction of x,
$$D(s) = \tfrac{1}{2}(2-s), \tag{61}$$
while, for a circular pupil,
$$D(s) = \frac{1}{\pi}(2\beta - \sin 2\beta) \quad (\beta = \cos^{-1}\tfrac{1}{2}s). \tag{62}$$

Hence, in contrast to a coherently illuminated object, the transmission coefficients decrease monotonically to zero with increasing frequency. The limit of resolution occurs when $s = 2$, that is, when
$$\xi = \frac{0{\cdot}5\lambda}{N \sin \alpha},$$

which agrees with the known result for this case. However, the contrast in the image is vanishingly small at this limit, since $D(s) \to 0$, whereas the contrast is unimpaired at the limit of resolution of a coherently illuminated object, although the finest resolvable structure is then twice as coarse as that denoted by the above limit.

Curves showing the contrast in the image of the terms of different frequency in the object are shown in figure 5 for coherent and self-luminous sources. Essentially the same results were obtained by Rayleigh (1896) for the special case of the image of a row of line sources formed by an optical system with a rectangular aperture.

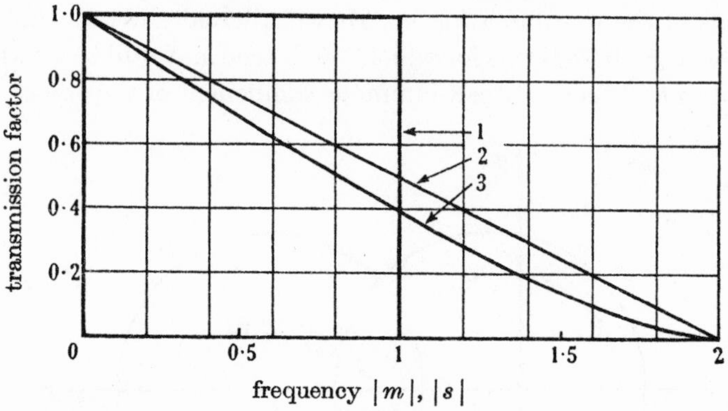

FIGURE 5. The transmission factors for central coherent illumination, $C(m, p)$, and for self-luminous objects, $D(s)$. Curve 1, $C(m, p)$, $|m| < 1$; curve 2, $D(s)$-rectangular aperture; curve 3, $D(s)$-circular aperture.

If a component of a self-luminous object is near to the limit of resolution the area $A(\frac{1}{2}s, -\frac{1}{2}s)$ will be very small, and we may write

$$D(s) = \frac{1}{\pi} A(\tfrac{1}{2}s, -\tfrac{1}{2}s) f(x_0 + \tfrac{1}{2}s, 0) f^*(x_0 - \tfrac{1}{2}s, 0),$$

where $(x_0, 0)$ is the mid-point of the area $A(\frac{1}{2}s, -\frac{1}{2}s)$. Comparison with (39) shows that the transmission factor is of exactly the same form as for a coherent source at $(x_0, 0)$, but the contrast is now $\frac{1}{\pi} A(\frac{1}{2}s, -\frac{1}{2}s)$. Hence the effect of aberration and defect of focus will be merely to shift the image laterally; they will have no effect on the contrast. Of course either of these can be so large that a significant variation of $f(x, y)$ occurs even over a small area, and the above conclusion is then valid only for objects which are on the limit of resolution, when the contrast is, in any case, vanishingly small.

It follows at once that measurements of 'resolving power' using a self-luminous grating give no information regarding the state of correction of aberration. This conclusion has to be modified if (say) a photographic plate is used to record the image, since the measured limit of resolution may then be only a fifth or so of the resolution of which the lens is inherently capable.

We shall now apply the above treatment to the structure defined by (43). The squared modulus $|E(u)|^2$ is given by

$$I(u) = \tfrac{3}{8} + \tfrac{1}{2} \cos \omega u + \tfrac{1}{8} \cos 2\omega u.$$

Since the transmission factor $D(s)$ is unchanged by a change of sign of s, the transmission factor for $\cos su$ is simply that for $e^{\pm isu}$. Hence the image formed, in the absence of aberration and defect of focus, is expressed by

$$I'(u') = \tfrac{3}{8} + \tfrac{1}{2}D(\omega)\cos\omega u' + \tfrac{1}{8}D(2\omega)\cos 2\omega u', \tag{63}$$

where $D(s)$ is given by (61) or (62) according as the pupil is rectangular or circular.

Intensity curves computed from (63) are shown in figure 6 for different values of ω. The dotted line shows the intensity in the image for the same structure formed by oblique coherent illumination. We may conclude that when $2 > |\omega| > 1$, oblique coherent illumination gives an image contrast superior to that obtained with a self-luminous object. If $|\omega| < 1$, the structure can be resolved with central coherent illumination, the contrast then being exactly reproduced. Hence it would seem that self-luminous objects give images which are inferior in contrast to properly controlled coherent illumination. We shall see later that the same conclusion holds true for coherent, or nearly coherent illumination, as against objects illuminated with a full cone of light.

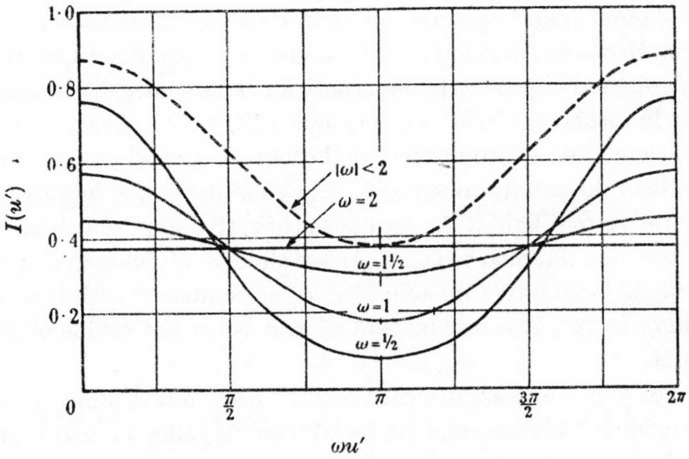

FIGURE 6. Images of the structure $E(u) = (1+\cos\omega u): W(x,y) = 0$, — self-luminous (circular aperture); --- oblique coherent illumination.

The above example is sufficient to show that the usual distinction between coherent and incoherent objects is over-simplified. Since in coherent light constant phase-differences exist, the image is influenced by the argument of the coherence factor. If $|\omega| > 1$, for example, the difference between the images formed with central and oblique coherent illumination amounts, in one case, to no resolution of the object, and, in the other case, to resolution equal to that with incoherent light but with superior contrast.

5. Partially coherent line structures

In connexion with equation (49) we have discussed the use of a ring source of radius σ, partly obscured to satisfy (50). If only the elements of the ring source at $(\pm x_0, 0)$ are used, the effective source is described by

$$\gamma(x,y) = \pi\{\delta(x-x_0, y) + \delta(x+x_0, y)\}, \tag{64}$$

π being a normalizing factor. The coherence factor for the object plane is thus

$$\Gamma(u_1 - u_2, v_1 - v_2) = \cos(u_1 - u_2)\,x_0, \tag{65}$$

and hence the illumination is partially coherent. It was found above that, in the absence of aberration, the image of the structure $E(u) = \frac{1}{2}(1+\cos\omega u)$ has the same intensity distribution in the case of partial coherence defined by (65) as for perfect coherence. Moreover, reference to the definitions (51) and (52) of the normalized coefficient $C(m,p)$ shows that this coefficient is equal to unity providing the region of the effective source lies wholly within the area common to the two circles C_m, C_p and that the aberration is zero. In this case also, the given component of the image appears exactly as if the object were illuminated coherently.

If, on the other hand, the area common to C_m, C_p lies wholly within that of the effective source, and the area of this latter is $\geqslant \pi$ (in the case of a circular pupil of radius $\sqrt{(x^2+y^2)} = 1$), the normalized coefficient $C(m,p)$ has the same value as for a self-luminous object.

It is thus apparent that, in some simple cases, certain components of the image of a partially coherent object will appear as in coherent illumination and others as self-luminous. However, in general, it does not seem possible to regard the image of a partially coherent object as the superposition of two images, the one in coherent and the other in incoherent light. For example, if the optical system shows aberration the coefficient $C(m,p)$ will not be equal to unity even when the effective source lies wholly within the area common to C_m, C_p; equally $C(m,p)$ has a value different from that for incoherent light if the area of the effective source, although containing the whole of the area common to C_m, C_p, is less than π. If this is in fact the case, the coefficient $C(m,p)$ is increased by a factor $1/\sigma^2$ as compared with incoherent image formation, since in (51) $A = \pi\sigma^2$ instead of π, σ being the radius of the (circular) effective source.

The above considerations are useful in special cases, and as indications of the way in which the mode of illumination may be adapted to bring out certain components in the image, but in cases of partially coherent objects it is necessary to consider the different components separately with the aid of diagrams such as figure 4.

The image of the structure $E(u) = \frac{1}{2}(1+\cos\omega u)$ has been studied in this way, assuming illumination by an effective source of radius σ, situated on the axis of the optical system. The intensity curves of the images for different values of σ are shown in figure 7 for $\omega = \frac{1}{2}$, 1 and $1\frac{1}{2}$. The structure corresponding to $\omega = 2$ is 'resolved' for $\sigma \geqslant 1$, but the contrast is vanishingly small. In all cases the image-forming system is assumed to be free from aberration, perfectly focused, and the radius of the pupil is $\sqrt{(x^2+y^2)} = 1$.

Using the type of consideration outlined above it is easy to show that, for $\omega = \frac{1}{2}$, the image is the same as in coherent light when $\sigma < 0{\cdot}50$. As σ is increased the contrast in the image decreases, but the influence of the mode of illumination on the image is not of great consequence practically. For larger values of ω, that is for finer structures, the influence of the mode of illumination is more marked. When $\omega = 1$, for example, the image is a perfect reproduction of the object for coherent illumination, but the contrast drops very rapidly with even a small increase in σ.

Curves are shown in figure 7 for $\sigma = 0{\cdot}25$, $1{\cdot}00$, and already with $\sigma = 0{\cdot}25$ the contrast is almost as small as that in the images for $\sigma = 1{\cdot}00$, and for incoherence (when $\sigma \to \infty$). When $\omega = 1\frac{1}{2}$ there is no resolution until σ exceeds $0{\cdot}50$. With increasing σ contrast appears in the image until $\sigma = 1$. Beyond this the effect of increasing σ is more to add a uniform illumination to $I'(u')$, thereby reducing the contrast.

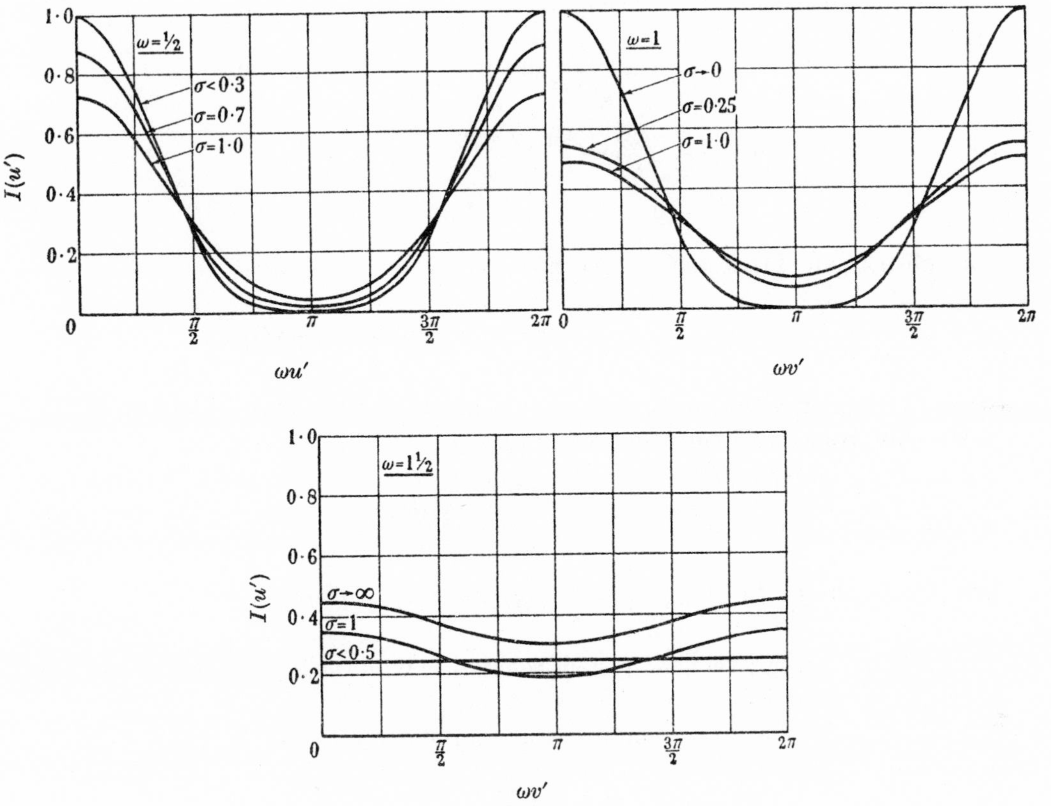

FIGURE 7. Influence of coherence on images of a periodic structure.

6. Images of a non-periodic structure

If the object consist of an isolated hole of negligible size in an opaque screen, the mode of illumination will obviously have no effect on the distribution of intensity in the image. In general, one will expect the influence of the coherence conditions in the object to be greater in those cases where a large number of points in the object contribute to the disturbance at any point in the image. As an illustration of the use of formulae (31) and (36) we shall consider the influence of the aperture of the illumination on the image of a transparent line of negligible width in an opaque screen.

The structure of the object may be specified by the expression

$$E(u) = \sqrt{(2\pi)}\,\delta(u), \tag{66}$$

whose transform is $e(s) = 1$. If the optical system has a circular pupil of radius $\sqrt{(x^2+y^2)} = 1$, and the aberration is zero, the complex transmission of the lens is

$$\left.\begin{aligned} f(x,y) &= 1, \quad x^2+y^2 \leqslant 1, \\ &= 0, \quad x^2+y^2 > 1, \end{aligned}\right\} \tag{67}$$

the image considered being that in the true focal plane. If (66) and (67) are substituted in (36) one obtains, for $|y| \leqslant 1$,

$$\phi(x,y) = \frac{1}{\sqrt{(2\pi)}} e^{iyv'} \int_{-\sqrt{(1-y^2)}}^{+\sqrt{(1-y^2)}} e^{iu's} \, ds,$$

or

$$\begin{aligned} \phi(x,y) &= \sqrt{\frac{2}{\pi}} e^{iyv'} \frac{\sin\{u'\sqrt{(1-y^2)}\}}{u'} \qquad (|y| \leqslant 1) \\ &= 0 \qquad (|y| > 1), \end{aligned}$$

and substitution of this result in (31) gives

$$I'(u') = \left(\frac{2}{u'}\right)^2 \int_{-\sigma}^{+\sigma} \left\{\int_{-\infty}^{+\infty} \gamma(x,y)\, dx\right\} \sin^2\{u'\sqrt{(1-y^2)}\}\, dy, \tag{68}$$

where the limit σ is the radius of the source or unity, whichever is the less.

We may write the integral in the bracket

$$\gamma_x(y) = \int_{-\infty}^{+\infty} \gamma(x,y)\, dx, \tag{69}$$

and (68) becomes $$I'(u') = \left(\frac{2}{u'}\right)^2 \int_{-\sigma}^{+\sigma} \gamma_x(y) \sin^2\{u'\sqrt{(1-y^2)}\}\, dy, \tag{70}$$

from which it appears that the system behaves as if all the light along the abscissa y were concentrated at the point $(0, y)$.

If the line is coherently illuminated at an obliquity y_0, the effective source is $\gamma_x(y) = \frac{1}{4}\delta(y-y_0)$, and

$$I'(u') = (1-y_0^2)\left[\frac{\sin\{u'\sqrt{(1-y_0^2)}\}}{u'\sqrt{(1-y_0^2)}}\right]^2 \tag{71}$$

from which it appears that the effect of the obliquity of illumination is to spread out the scale of the image by a factor $1/\sqrt{(1-y_0^2)}$, and to reduce the brightness. For central coherent illumination $y_0 = 0$, and (71) reduces to the known result.

If the effective source is uniform inside a circle of radius σ, the expression (69) gives $\gamma_x(y) = 2\sqrt{(\sigma^2-y^2)}$. When $\sigma < 1$, we write for (70)

$$I'(u') = \left(\frac{2}{u'}\right)^2 \int_{-\sigma}^{+\sigma} 2\sqrt{(\sigma^2-y^2)} \sin^2\{u'\sqrt{(1-y^2)}\}\, dy. \tag{72}$$

Putting $u' = 0$ we find $I'(0) = \pi\sigma^2(4-\sigma^2)$, and this we shall use to normalize the expression.

To evaluate (72) we put $y = \sin\theta$, $\sigma = \sin\beta$. Then

$$I'(u') = 2\left(\frac{2}{u'}\right)^2 \int_0^\beta \sqrt{(\sigma^2 - \sin^2\theta)}\{1-\cos(2u'\cos\theta)\}\cos\theta\, d\theta,$$

in which the term $\cos(2u'\cos\theta)$ may be expanded by Jacobi's series. The cosines of multiples of θ which occur as coefficients may then be expounded in powers of $\sin\theta$. Replacing $\sin\theta$ by y, one then finds

$$I'(u') = 2\left(\frac{2}{u'}\right)^2 \int_0^{\sigma} \sqrt{(\sigma^2-y^2)}\Bigg\{1-\cos 2u' + 2\sum_{n=2,4,\ldots}^{\infty} (-)^{\frac{1}{2}(n-2)} J_n(2u') \times \sum_{p=1}^{\frac{1}{2}n} (-)^p \frac{n^2(n^2-2^2)\ldots(n^2-\overline{2p-2}^{\,2})}{(2p)!}\, y^{2p}\Bigg\}\, dy,$$

and the substitution $y = \sigma\sin\psi$, then leads to integrals between 0 and $\frac{1}{2}\pi$ which can be written in terms of gamma functions. After some reduction one obtains

$$I'(u') = \left(\frac{4}{4-\sigma^2}\right)\Bigg\{\left(\frac{\sin u'}{u'}\right)^2 + \sum_{n=2,4,\ldots}^{\infty} (-)^{\frac{1}{2}(n-2)} \frac{J_n(2u')}{u'^2} \times \sum_{p=1}^{\frac{1}{2}n} (-)^p \frac{n^2(n^2-2^2)\ldots(n^2-\overline{2p-2}^{\,2})}{(p+1)!\,p!}\left(\frac{\sigma}{2}\right)^{2p}\Bigg\}, \qquad (73)$$

the above-mentioned normalizing factor having been introduced.

When $\sigma = 1$, the substitution $y = \sin\theta$ in (72) leads directly to the simple formula

$$I'(u') = \frac{2}{3u'^2}\{1 - J_0(2u') + J_2(2u')\}, \qquad (74)$$

which has again been normalized to make $I'(0) = 1$.

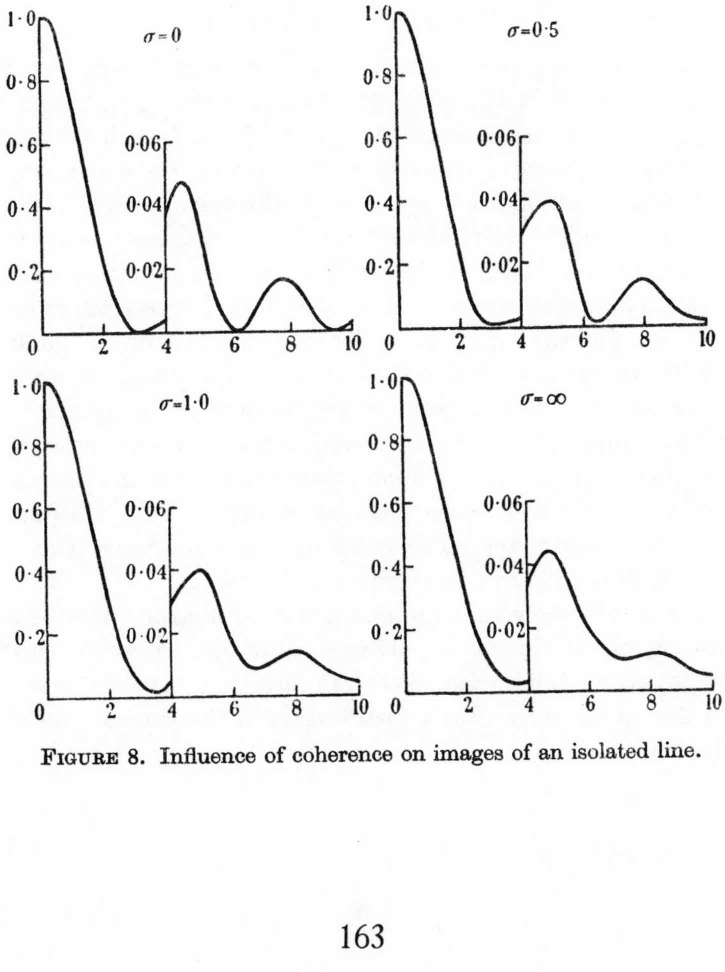

FIGURE 8. Influence of coherence on images of an isolated line.

If we let the effective source be a square of side 2σ, where $\sigma > 1$, the quantity $\gamma_x(y) =$ constant, and (70) may then be evaluated to give

$$I'(u') = \frac{3\pi}{8}\frac{H_1(2u')}{u'^2} \tag{75}$$

which is the known result in this case. One finds that a square effective source gives an image the same as that of a self-luminous line, providing $\sigma \geqslant 1$. $H_1(2u')$ is the Struve function of order unity (Gray & Mathews 1895). A proof of this formula using Fourier transforms has been given recently for this special case by Steel (1952).

For the case of self-luminous non-periodic structures a useful formula analogous to (58) may be obtained. The same argument is used as in that case, and one then proceeds to a limit obtaining the formula

$$I'(u') = \frac{1}{\sqrt{(2\pi)}}\int_{-\infty}^{+\infty} D(s)\, i(s)\, e^{isu'}\, ds, \tag{76}$$

where $D(s)$ is defined by (59) above. $i(s)$ is the inverse Fourier transform of the object intensity function $I(u)$.

In figure 8 are shown curves of the intensity distribution in the images of a transparent line illuminated by circular effective sources of radii $\sigma \to 0$, $\sigma = 0{\cdot}5$, $1{\cdot}0$, and $\sigma \to \infty$. It will be noted that, in this case, the influence of the coherence conditions on the image is not very great.

7. Physical interpretation

It is possible to give a physical interpretation of the preceding formulae. Corresponding to each point in the effective source a converging spherical wave of suitable amplitude and phase falls on the object. For each such wave a Fraunhofer diffraction process takes place between the object and the reference sphere in the object space. The various spectra appearing on this reference sphere appear at the corresponding points of the reference sphere in the image space, with their amplitudes and phases modified according to the complex transmission of the optical system. A second Fraunhofer transformation then takes place between the reference sphere in the image space and the image plane. The image is made up by the superposition of the intensity distributions due to the assembly of illuminating waves.

This interpretation is possible because, according to (3), any given conditions of coherence in the object plane may be reproduced by the superposition of a set of mutually incoherent waves having appropriate amplitudes and phases. If β is the angle of obliquity to the axis of the mean ray of a given wave, $\sin\beta$ must be given values from $-\infty$ to $+\infty$ to reproduce conditions of incoherence. This is physically meaningless, but it is valid as a mathematical device.

The above remarks show that the mathematical analysis may be regarded as describing a generalized Abbé theory of image formation. However, this interpretation is only physically valid when all the ray directions involved make very small angles with the optical axis. For a justification in the general case the analysis described herein is necessary.

I am indebted to a colleague, Dr A. Gordon, for checking most of the analysis, and for suggesting the alternative procedure described in the relations (27) to (31).

References

Duffieux, P. M. 1946 *L'intégrale de Fourier et ses applications à l'optique*. Besançon: privately printed.

Gray, A. & Mathews, G. B. 1895 *Bessel functions*, p. 202. London: Macmillan.

Hopkins, H. H. 1951 *Proc. Roy. Soc.* A, **208**, 263.

Rayleigh, Lord 1896 *Phil. Mag.* (5), **42**, 167.

Steel, W. H. 1952 *Rev. Opt.* (*theor. instrum.*), **31**, 7, p. 334.

Titchmarsh, E. C. 1937 *Theory of Fourier integrals*. Oxford University Press.

PAPER NO. 13

Reprinted from *American Journal of Physics*, Vol. 22, pp. 351–362, 1954.

Polarization and the Stokes Parameters

William H. McMaster
University of Virginia, Charlottesville, Virginia
(Received February 8, 1954)

The Stokes parameters have been found to offer a very convenient method for the description of polarization of both electromagnetic radiation and elementary particles. Their development is presented along with some applications to physical problems.

INTRODUCTION

In the study of both electromagnetic radiation and elementary particles the formalism of the Stokes parameters[1] offers a very convenient tool for the description of polarization. This method shows explicitly the direct analogy between polarization of photons and elementary particles and readily relates this polarization to the polarization of light studied in physical optics courses. It is the purpose of this article to organize into one reference the theory of the Stokes parameters.

In the development of the theory we shall use quantum-mechanical concepts; however, for the application of the Stokes parameters to the polarization of light, no knowledge of quantum mechanics is required. As an introduction we shall first discuss the application of the density matrix (or statistical matrix) to the description of polarization. The Stokes parameters are then defined in terms of elements of the density matrix.

THE DENSITY MATRIX

In quantum mechanics we know that an arbitrary wave equation can be expanded in any desired complete set of orthonormal eigenfunctions. That is,

$$\psi=\sum_i a_i\psi_i. \tag{1}$$

Then

$$|\psi|^*=\sum_{ij} a_i\psi_i a_j^*\psi_j^*=\sum_{ij} a_i a_j^*\psi_i\psi_j^*.$$

From the expansion coefficients we can form a matrix ρ by the rule

$$\rho_{ij}=a_i a_j^*. \tag{2}$$

This matrix is known as the density matrix[2] and has some very useful properties. First we note that $\rho_{ii}=a_i a_i^*$ gives the probability of finding the system in the state characterized by the eigenfunction ψ_i. If we consider the ψ function as being normalized then

$$\int\psi\psi^* d\tau=\sum_{ij} a_i a_j^*\int\psi_i\psi_j^* d\tau=\sum_i a_i a_i^*=\sum_i \rho_{ii}=1, \tag{3}$$

or

$$\mathrm{Tr}\rho=1.$$

---

[1] G. G. Stokes, Trans. Cambridge Phil. Soc. **9**, 399 (1852).

[2] R. C. Tolman, *The Principles of Statistical Mechanics* (Oxford University Press, New York, 1938), p. 325. See also, von Neumann, Nachr. Akad. Wiss. Göttingen, Math.-physik. Kl., p. 245 (1927).

If we make a measurement of some variable F in the system described by the function ψ, the result will be given by

$$\langle F\rangle=\int \psi^* F\psi d\tau=\sum_{ij}\int a_i^*\psi_i^* F a_j\psi_j d\tau,$$

$$\langle F\rangle=\sum_{ij} a_j a_i^* F_{ij}=\sum_{ij} F_{ij}\rho_{ji},$$

where the matrix F_{ij} is defined by the usual formula

$$F_{ij}=\int \psi_i^* F\psi_j d\tau;$$

but

$$\sum_j F_{ij}\rho_{ji}=(F\rho)_{ii}.$$

Therefore,

$$\langle F\rangle=\sum_i (F\rho)_{ii},$$

or

$$\langle F\rangle=\mathrm{Tr}(F\rho), \tag{4}$$

and since the matrices are all Hermitean, we also have

$$\langle F\rangle=\mathrm{Tr}(\rho F). \tag{5}$$

If we recall the classical use of density function $\rho(\mathbf{p},\mathbf{q})$, where the $\mathbf{p}$ and $\mathbf{q}$ are the momentum and position, respectively, which is normalized by

$$\int \rho d\tau=1, \tag{6}$$

and in terms of which the average value of a variable F is given by

$$\langle F\rangle=\int F\rho d\tau, \tag{7}$$

then we see immediately the similar role played by the density matrix in quantum mechanics from a comparison of Eqs. (3) and (5) with (6) and (7).

Polarization of electromagnetic radiation is usually described by the vibration of the electric vector. For a complete description, it can be shown[3] that a plane wave may be thought of as being composed of two physically independent (incoherent) beams of orthogonal polarization. That is, the electric vector may be analyzed by the equation

$$\mathbf{E}=a_1\mathbf{E}_1+a_2\mathbf{E}_2, \tag{8}$$

where $\mathbf{E}_1$ and $\mathbf{E}_2$ are two orthogonal unit vectors, and the a_i, which in general are complex, describe the amplitude and phase of the two vibrations. From the two expansion coefficients we can form a 2×2 density matrix.

Polarization of particles is described by the orientation of the spin. For the case of particles of spin 1/2, we need only two orthonormal wave functions to form a complete set describing the polarization of the particle.[4] Hence

$$\psi=a_1\Phi_1+a_2\Phi_2, \tag{9}$$

where Φ_1 is the eigenfunction for the spin quantum number $+1/2$ and Φ_2 is that for $-1/2$. Again the expansion coefficients form a 2×2 density matrix.

The similarity of Eqs. (8) and (9) immediately suggest that the description of polarization of electromagnetic radiation and of particles will be similar.[5] In fact we can use the same equation for the description of both[6] if we write

$$\psi=a_1\psi_1+a_2\psi_2, \tag{10}$$

where we must distinguish two cases.[7]

Case I. Electromagnetic Radiation or Photons

The ψ_1 and ψ_2 can either represent two orthogonal states of plane polarization or two states of circular polarization. Then $|a_1|^2$ and $|a_2|^2$ give the probabilities of detection of quanta by a detector sensitive only to states ψ_1 and ψ_2, respectively.

Case II. Particles of Spin 1/2

The wave functions ψ_1 and ψ_2 can represent either of two opposite spin orientations, parallel and antiparallel to the momentum. Again the $|a_i|^2$ give the probabilities of detection of particles specified by the eigenfunctions ψ_i.

[3] See, for example, L. Landau and E. Lifshitz, *The Classical Theory of Fields* (Addison-Wesley Press, Cambridge, 1951), pp. 126–129.

[4] See, for example, L. I. Schiff, *Quantum Mechanics* (McGraw-Hill Book Company, Inc., New York, 1949), p. 224.

[5] See also, P. Jordan, Z. Physik **44**, 292 (1927).

[6] D. L. Falkoff and J. E. MacDonald, J. Opt. Soc. Am. **41**, 861 (1951).

[7] See, for example, H. A. Tolhoek and S. R. DeGroot, Physica **17**, 1 (1951).

In both cases the density matrix

$$\rho=\begin{pmatrix} a_1a_1^* & a_1a_2^* \\ a_2a_1^* & a_2a_2^* \end{pmatrix} \tag{11}$$

completely characterizes the beam since we can obtain the intensities of the two polarization states from the diagonal elements while the off-diagonal elements furnish the relative phase.

Complete polarization can be represented by a single eigenfunction

$$\psi=a_1\psi_1 \quad \text{with} \quad \rho_+=\begin{pmatrix} 1 & 0 \\ 0 & 0 \end{pmatrix},$$

or (12)

$$\psi=a_2\psi_2 \quad \text{and} \quad \rho_-=\begin{pmatrix} 0 & 0 \\ 0 & 1 \end{pmatrix},$$

where the ψ_i refer to a state of pure polarization and we are considering a beam of unit intensity.

Since an unpolarized beam may be considered as the incoherent superposition of two polarized beams with equal intensity, to get the density matrix for an unpolarized beam we add those of the two polarized beams

$$\rho_u=1/2\begin{pmatrix} 1 & 0 \\ 0 & 1 \end{pmatrix}; \tag{13}$$

the 1/2 appears because of the normalization by Eq. (3).

This may be more apparent if we use the equation

$$\mathbf{E}=\cos\theta e^{i\Phi}\mathbf{E}_1+\sin\theta\mathbf{E}_1$$

to represent the beam. The density matrix is then

$$\begin{pmatrix} \cos^2\theta & \cos\theta\sin\theta e^{i\Phi} \\ \sin\theta\cos\theta e^{-i\Phi} & \sin^2\theta \end{pmatrix}.$$

For an unpolarized beam we must average over the angles θ and Φ. Hence,

$$\rho_u=\begin{pmatrix} \frac{1}{2} & 0 \\ 0 & \frac{1}{2} \end{pmatrix}=\frac{1}{2}\begin{pmatrix} 1 & 0 \\ 0 & 1 \end{pmatrix}.$$

In general, a beam will have an arbitrary degree of polarization and we can characterize such a beam by the incoherent superposition of an unpolarized beam and a totally polarized one. If the polarized portion is described by Eq. (9), then its contribution to the density matrix will be given by Eq. (11). The unpolarized portion will be characterized by Eq. (13) so that the density matrix for an arbitrary beam can be written in the form

$$\rho=U\begin{pmatrix} 1 & 0 \\ 0 & 1 \end{pmatrix}+P\begin{pmatrix} a_1a_1^* & a_1a_2^* \\ a_2a_1^* & a_2a_2^* \end{pmatrix}, \tag{14}$$

where P represents the degree of polarization. P is real and $0\leq P\leq 1$. We now note the following three cases:

(1) If $0<P<1$, then the beam is partially polarized.
(2) If $P=0$, then the beam is unpolarized.
(3) If $P=1$, then the beam is totally polarized.

For the case $P=0$, we know that

$$\rho=1/2\begin{pmatrix} 1 & 0 \\ 0 & 1 \end{pmatrix};$$

thus $U=1/2$ when $P=0$.

For the case $P=1$, the density matrix is given by Eq. (11) so that

$$U=0 \quad \text{when} \quad P=1.$$

These conditions are satisfied by $U=1/2 \times(1-P)$ so that for the general case, the density matrix is given by

$$\rho=1/2(1-P)\begin{pmatrix} 1 & 0 \\ 0 & 1 \end{pmatrix}+P\begin{pmatrix} a_1a_1^* & a_1a_2^* \\ a_2a_1^* & a_2a_2^* \end{pmatrix}. \tag{15}$$

By the proper choice of pure states of polarization ψ_i, the part of the density matrix representing total polarization can be written in one of the forms given by Eq. (12); therefore, we may write the general density matrix as

$$\rho=1/2(1-P)\begin{pmatrix} 1 & 0 \\ 0 & 1 \end{pmatrix}+P\begin{pmatrix} 1 & 0 \\ 0 & 0 \end{pmatrix},$$

or

$$\rho=1/2\begin{pmatrix} 1+P & 0 \\ 0 & 1-P \end{pmatrix}. \tag{16}$$

Hence any intensity measurement made in relation to these pure states will yield the eigenvalues

$$I_1=1/2(1+P),$$
$$I_2=1/2(1-P). \tag{17}$$

We note that the definition of P is independent of the choice of pure states and gives only the degree of mixture of polarized and unpolarized light. We shall now introduce a description which is dependent upon the type of polarization.

THE STOKES PARAMETERS FOR ELECTROMAGNETIC RADIATION

To determine experimentally the state of polarization of an arbitrary beam of electromagnetic radiation (photons) we must make a set of four measurements. These measurements can best be explained using the analogy of optics where Nicol prisms and quarter-wave plates are used. However, it must be remembered that these devices will not work for high-energy photons and particles, but that other analogous experiments must be used. These analogous experiments will be discussed in a later section. The most convenient set of four measurements are those that yield the following information:

(1) The intensity of the beam.

(2) The degree of plane polarization with respect to two arbitrary orthogonal axes.

(3) The degree of plane polarization with respect to a set of axes oriented at 45° to the right of the previous one.

(4) The degree of circular polarization.

In optics the second and third of these measurements can be made with a Nicol prism while the fourth requires the additional use of a quarter-wave plate.

Let us now consider these measurements in terms of Eq. (10):

$$\psi = a_1\psi_1 + a_2\psi_2.$$

Throughout this discussion we shall consider the beam to be normalized to unit intensity so that an intensity measurement will yield

$$I = a_1a_1^* + a_2a_2^* = 1. \qquad (18)$$

In terms of the states described by ψ_1 and ψ_2 we shall now define an orientation coefficient

$$P(\psi_1,\psi_2) = a_1a_1^* - a_2a_2^* = \rho_{11} - \rho_{22}, \qquad (19)$$

which gives the difference of the intensity measurements of the pure states defined by ψ_1 and ψ_2.

So far the ψ_i have not been chosen and can refer equally well to plane polarization or circular polarization states.

The general equation for polarization in terms of the $\mathbf{E}$ vector is

$$\mathbf{E} = b_1 \exp[i(\omega t + \delta_1)]\mathbf{E}_1 + b_2 \exp[i(\omega t + \delta_2)]\mathbf{E}_2, \qquad (20)$$

where the b_i are real, and the $\mathbf{E}_i$ are orthogonal unit vectors chosen arbitrarily in the plane orthogonal to the direction of propagation. From this equation we note the following cases:

(1) If the phase difference $\varphi = \delta_1 - \delta_2 = 0$ we have plane-polarized radiation.

(2) If $b_1 = b_2$ and $\varphi = \pm\pi/2$, we have right or left circular polarization depending upon whether the sign is $+$ or $-$, respectively.

(3) If $b_1 \neq b_2 \neq 0$ and $\varphi \neq 0$, we have elliptical polarization.

For the first polarization measurement we measure the intensities transmitted by a Nicol prism oriented in the two directions $\mathbf{E}_1$ and $\mathbf{E}_2$. Thus we have chosen

$$\psi_1 = \mathbf{E}_1 \quad \text{and} \quad \psi_2 = \mathbf{E}_2.$$

The first orientation coefficient will then be

$$P(\psi_1,\psi_2) = P_1 = a_1a_1^* - a_2a_2^*$$

or

$$P_1 = \rho_{11} - \rho_{22}. \qquad (21)$$

In terms of Eq. (20) this is $P_1 = b_1^2 - b_2^2$.

The second measurement is similar to the first but oriented at 45° to the right. With this choice of axes

$$\psi = a_1'\psi_1' + a_2'\psi_2',$$

where

$$\psi_1' = \mathbf{E}_1 \cos 45° + \mathbf{E}_2 \sin 45°,$$

$$\psi_2' = -\mathbf{E}_1 \sin 45° + \mathbf{E}_2 \cos 45°,$$

and

$$P(\psi_1',\psi_2') = P_2 = a_1'a_1'^* - a_2'a_2'^*,$$

or

$$P_2 = \rho_{11}' - \rho_{22}'.$$

Since we have the same beam,

$$\psi = a_1'\psi_1' + a_2'\psi_2' = a_1\psi_1 + a_2\psi_2.$$

From Eq. (22) we have

$$\psi_1 = (1/\sqrt{2})\psi_1' - (1/\sqrt{2})\psi_2',$$

$$\psi_2 = (1/\sqrt{2})\psi_1' + (1/\sqrt{2})\psi_2',$$

or

$$a_1'\psi_1' + a_2'\psi_2' = (1/\sqrt{2})a_1(\psi_1' - \psi_2') + (1/\sqrt{2})a_2(\psi_1' + \psi_2'),$$
$$= (1/\sqrt{2})(a_1 + a_2)\psi_1' + (1/\sqrt{2})(a_2 - a_1)\psi_2'.$$

Thus

$$a_1' = (1/\sqrt{2})(a_1 + a_2), \quad \text{and} \quad a_2' = (1/\sqrt{2})(a_2 - a_1).$$

From which, after a simple multiplication, we find

$$P_2 = \rho_{12} + \rho_{21}. \qquad (22)$$

In terms of Eq. (20) this is readily seen to be

$$P_2 = 2b_1b_2\cos(\delta_1 - \delta_2).$$

The third measurement is one for circularly polarized light. To make the measurement we insert a quarter-wave plate with its fast axis 45° to the right of $\mathbf{E}_1$ and make intensity measurements with the transmission axis of the Nicol prism oriented along $\mathbf{E}_1$ and $\mathbf{E}_2$. That is, we are making the choice

$$\psi_R = (1/\sqrt{2})e^{i\varphi}\psi_1 + (1/\sqrt{2})\psi_2,$$

where $\varphi = \frac{1}{4}2\pi$, or $e^{i\pi/2} = i$, and

$$\psi_L = (1/\sqrt{2})\psi_1 + (/1\sqrt{2})e^{i\varphi}\psi_2,$$

with

$$P(\psi_R, \psi_2) = \rho_{11}^c - \rho_{22}^c. \qquad (23)$$

Thus

$$\left.\begin{aligned} \psi_1 &= (1/\sqrt{2})(\psi_L - i\psi_R) \\ \psi_2 &= (1/\sqrt{2})(\psi_R - i\psi_L) \end{aligned}\right\}, \qquad (24)$$

then since $\psi = a_1\psi_1 + a_2\psi_2$, we have

$$\psi = (a_1/\sqrt{2})(\psi_L - i\psi_R) + (a_2/\sqrt{2})(\psi_R - \psi_L),$$
$$\psi = (1/\sqrt{2})(a_2 - ia_1)\psi_R + (1/\sqrt{2})(a_1 - ia_2)\psi_L$$
$$= a_1^c\psi_R + a_2^c\psi_L. \qquad (25)$$

Substituting the values of a_i^c from Eq. (25) in Eq. (23) we find

$$P(\psi_R, \psi_L) = P_3 = i(\rho_{21} - \rho_{12}). \qquad (26)$$

In terms of Eq. (20) this is $2b_1b_2\sin(\delta_1 - \delta_2)$. The appearance of the imaginary number i in P_3 is the mathematical expression of the fact that we must use a quarter-wave plate since a Nicol prism by itself is not sensitive to circular polarization.

As a result of these four measurements, we have a set of four quantities:

$$\left.\begin{aligned} I &= \rho_{11} + \rho_{22} \\ P_1 &= \rho_{11} - \rho_{22} \\ P_2 &= \rho_{12} + \rho_{21} \\ P_3 &= i(\rho_{21} - \rho_{12}) \end{aligned}\right\}, \qquad (27)$$

which are known as the Stokes parameters and completely characterize a beam of electromagnetic radiation. For pure states of polarization a measurement of the Stokes parameters can easily be seen to provide the following results:

$P_1 = +1$; plane polarization along $\mathbf{E}_1$
$P_1 = -1$; plane polarization along $\mathbf{E}_2$
$P_2 = \pm 1$; plane polarization at an angle of 45° to the right of $\mathbf{E}_1$ and $\mathbf{E}_2$, respectively,
$P_3 = +1$; right circular polarization
$P_3 = -1$; left circular polarization.

From the Stokes parameters we can readily construct the density matrix

$$\rho = 1/2\begin{pmatrix} 1 + P_1 & P_2 + iP_3 \\ P_2 - iP_3 & 1 - P_1 \end{pmatrix}. \qquad (28)$$

THE STOKES PARAMETERS OF AN ELECTRON

In this section we will follow the development given by Tolhoek and De Groot.[7] The general state of a beam of electrons can again be described by

$$\psi = a_1\psi_1 + a_2\psi_2, \qquad (10)$$

where ψ_1 represents a state of spin $+1/2$ in the positive z direction, ψ_2 represents a state of spin $-1/2$ in the positive z direction, and the spin direction is defined in the rest system of the electron. It can be shown[8] that an arbitrary spin direction in relation to the z axis is given by

$$a_2/a_1 = \tan\tfrac{1}{2}\theta e^{i\varphi}, \qquad (29)$$

where the angles θ and φ give the orientation of the spin as shown in Fig. 1. If an arbitrary phase

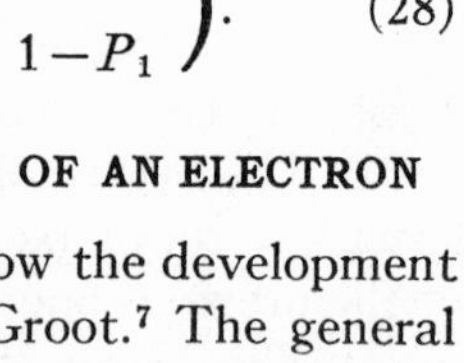

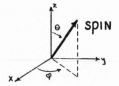

FIG. 1. Angles θ and ϕ give the spin orientation with respect to the axes.

[8] P. A. M. Dirac, *The Principles of Quantum Mechanics* (Oxford University Press, New York, 1930), p. 136.

is neglected, we can choose

$$a_1=\cos 1/2\theta;\quad a_2=\sin 1/2\theta e^{i\varphi}. \qquad (30)$$

Then the state of polarization in the x direction will be given by the values

$$\psi_1'\begin{cases}\theta=\pi/2\\ \varphi=0\end{cases};\quad \psi_2'\begin{cases}\theta=\pi/2\\ \varphi=\pi\end{cases}.$$

Therefore

$$\psi_1'=(1/\sqrt{2})(\psi_1+\psi_2)$$

and (31)

$$\psi_2'=(1/\sqrt{2})(\psi_1-\psi_2);$$

and for states of polarization in the y direction, designated by the functions ψ_1'' and ψ_2'', we have:

$$\psi_1''\begin{cases}\theta=\pi/2\\ \varphi=\pi/2\end{cases}\quad\text{and}\quad \psi_2''\begin{cases}\theta=\pi/2\\ \varphi=\pi/2\end{cases}.$$

Therefore

$$\psi_1''=(1/\sqrt{2})(\psi_1+i\psi_2);$$
$$\psi_2''=(1/\sqrt{2})(\psi_1-i\psi_2). \qquad (32)$$

Let us again consider a beam of electrons normalized to unit intensity so that

$$I=a_1a_1^*+a_2a_2^*=1.$$

If we investigate a beam of electrons by taking a measurement to determine the number of spins in the z direction with the value $+1/2$ (which yields $I_1=a_1a_1^*$) and another measurement for $-1/2$ (which yields $I_2=a_2a_2^*$), then we obtain the orientation coefficient.

$$P_1=a_1a_1^*-a_2a_2^*=\rho_{11}-\rho_{22}. \qquad (33)$$

To investigate a spin state at right angles to this one in the x direction, we have from Eq. (31),

$$\psi_1=(1/\sqrt{2})(\psi_1'+\psi_2');\psi_2=(1/\sqrt{2})(\psi_1'-\psi_2'). \qquad (34)$$

Upon substitution of Eq. (34) into Eq. (10),

$$\psi=(a_1/\sqrt{2})(\psi_1'+\psi_2')+(a_2/\sqrt{2})(\psi_1'-\psi_2')$$
$$=(1/\sqrt{2})(a_1+a_2)\psi_1+(1/\sqrt{2})(a_1-a_2)\psi_2.$$

Then

$$P_2=\rho_{11}'-\rho_{22}'=(1/2)(a_1+a_2)(a_1^*+a_2^*)$$
$$-(1/2)(a_1-a_2)(a_1^*-a_2^*)$$
$$=a_1a_2^*+a_2a_1^*.$$

Therefore,

$$P_2=\rho_{12}+\rho_{21}. \qquad (35)$$

TABLE I. Meanings of Stokes parameters.

Stokes parameter	Photon observation	Electron observation
I	Intensity	Intensity
P_1	Plane polarization	Transverse spin
P_2	Plane polarization at an angle of $\pi/4$ to the previous direction	Transverse spin at an angle of $\pi/2$ to the previous direction
P_3	Circular polarization	Longitudinal spin

For states polarized in the y direction, from Eq. (32):

$$\psi_1=(1/\sqrt{2})(\psi_1''+\psi_2'')$$

and

$$\psi_2=(1/\sqrt{2})(-i\psi_1''+i\psi_2'').$$

A similar calculation leads to

$$P_3=i(\rho_{12}-\rho_{21}). \qquad (36)$$

Thus the four measured quantities, again called the Stokes parameters, are given by

$$\left.\begin{aligned}I&=\rho_{11}+\rho_{22}\\ P_1&=\rho_{11}-\rho_{22}\\ P_2&=\rho_{12}+\rho_{21}\\ P_3&=i(\rho_{21}-\rho_{12})\end{aligned}\right\}, \qquad (37)$$

and we notice that Eqs. (37) are identical to Eqs. (27).

Scattering experiments are generally used to determine the Stokes parameters of electrons; Tolhoek and DeGroot[7] show in their article that scattering experiments are not sensitive to longitudinal spin, so that if we desire to make a measurement of longitudinal spin we must first change it to transverse spin. This reminds us of the fact that in optics a Nicol prism is not sensitive to circular polarization and that the circular polarization must be transformed to plane polarization by means of a quarter-wave plate before a measurement can be made. Thus if we choose single scattering as the electron polarization detector as being analogous to the Nicol prism, then we can attach the meanings to the Stokes parameters as shown in Table I.

This description for electrons can readily be extended to all elementary particles of spin $\pm 1/2$.

PROPERTIES OF THE STOKES PARAMETERS

In the application of the Stokes parameters to any problem it is customary to write them in the form of a four vector.

$$\begin{pmatrix} I \\ P_1 \\ P_2 \\ P_3 \end{pmatrix} = \begin{pmatrix} I \\ \mathbf{P} \end{pmatrix}. \tag{38}$$

As an example of its use, consider the following simple cases:

$\begin{pmatrix} 1 \\ 0 \\ 0 \\ 0 \end{pmatrix}$ represents an unpolarized beam;

$\begin{pmatrix} 1 \\ \pm 1 \\ 0 \\ 0 \end{pmatrix}$ and $\begin{pmatrix} 1 \\ 0 \\ \pm 1 \\ 0 \end{pmatrix}$ represent plane polarization or transverse spin;

$\begin{pmatrix} 1 \\ 0 \\ 0 \\ \pm 1 \end{pmatrix}$ represent circular polarization or longitudinal spin.

Since the Stokes parameters are dependent upon the choice of axes, there must exist a transformation matrix M which will relate the Stokes parameters in one coordinate system to another system. If we consider a second coordinate system rotated about the direction of propagation an angle θ to the right of the original coordinate system, then

$$\begin{pmatrix} I' \\ \mathbf{P}' \end{pmatrix} = M \begin{pmatrix} I \\ \mathbf{P} \end{pmatrix}, \tag{39}$$

where

$$M = \begin{pmatrix} 1 & 0 & 0 & 0 \\ 0 & \cos 2\theta & \sin 2\theta & 0 \\ 0 & -\sin 2\theta & \cos 2\theta & 0 \\ 0 & 0 & 0 & 1 \end{pmatrix}. \tag{40}$$

Thus if in the old system we had (I, P_1, P_2, P_3), then in the new system of coordinates, the Stokes parameters of the same beam would be

$$\begin{pmatrix} I \\ P_1 \cos 2\theta + P_2 \sin 2\theta \\ -P_1 \sin 2\theta + P_2 \cos 2\theta \\ P_3 \end{pmatrix}. \tag{41}$$

A rotation in the opposite direction changes the sign of the $\sin 2\theta$ terms.

Next let us consider a partially polarized beam and determine the probability of detecting a given polarization in it. First let us consider the probability of finding a photon or an electron in the states described by ψ_1 and ψ_2. For a normalized beam

$$I = a_1 a_1^* + a_2 a_2^* = 1 \quad \text{and} \quad P_1 = a_1 a_1^* - a_2 a_2^*.$$

Then by subtraction and addition we find, respectively,

$$a_1 a_1^* = (1/2)(1+P_1)$$

and (42)

$$a_2 a_2^* = (1/2)(1-P_1).$$

Since the state represented by ψ_1 is characterized by the Stokes parameters (1 1 0 0) and that of ψ_2 by (1 −1 0 0), we can get the same probability for detection in an arbitrary beam characterized by $(I\ P_1\ P_2\ P_3)$ from the simple vector multiplications

$$(1/2)(1\ \ 1\ \ 0\ \ 0) \begin{pmatrix} I \\ P_1 \\ P_2 \\ P_3 \end{pmatrix} = (1/2)(1+P_1);$$

and

$$(1/2)(1\ \ -1\ \ 0\ \ 0) \begin{pmatrix} I \\ P_1 \\ P_2 \\ P_3 \end{pmatrix} = (1/2)(1-P_1).$$

In a similar manner it can easily be shown that the probabilities of detection of polarization in the other two states of polarization are given by

$$(1/2)(1 \pm P_2) \quad \text{and} \quad (1/2)(1 \pm P_3). \tag{43}$$

Now it can readily be verified by a simple calculation that these results can be expressed in two forms which are useful in the application of the theory.

(1) Let w be the probability of detecting a photon or a particle characterized by the density matrix ρ' in an arbitrary beam characterized by the density matrix ρ. Then

$$w = \mathrm{Tr}(\rho\rho'), \tag{44}$$

where we note the similarity of Eq. (44) to Eq. (4).

(2) In terms of the Stokes parameters, let us determine the probability of detecting a photon or a particle characterized by the Stokes parameters $(1, \mathbf{Q})$ in an arbitrary beam characterized by the Stokes parameters $(I, \mathbf{P})$. For this we use an analyzer which will pass only states of polarization $(1, \mathbf{Q})$ then

$$w = (1/2)(1 + \mathbf{P} \cdot \mathbf{Q}) = (1/2)(1, \mathbf{Q}) \begin{pmatrix} I \\ \mathbf{P} \end{pmatrix}. \tag{45}$$

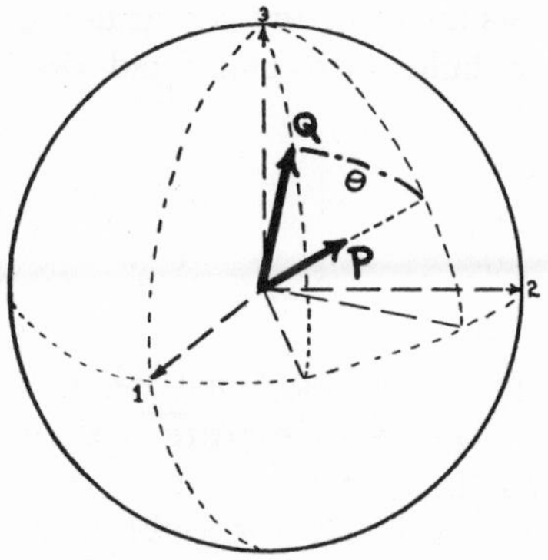

FIG. 2. Poincaré sphere representation of the orientation coefficients. **Q** is a unit vector representing the analyzer setting, while **P** is a vector ($0 \leq |P| \leq 1$) representing the state of polarization of the beam.

With this last notation we may map[9] the orientation coefficients on the Poincaré sphere of radius 1 as shown in Fig. 2. In this representation, the three components P_1, P_2, and P_3 are orthogonal. With respect to the polarization axes specified by **1**, **2**, **3** in the figure, the polarization of the beam will be given by a vector $\mathbf{P}(0 \leq |\mathbf{P}| \leq 1)$ oriented in some given direction. The different settings of an analyzer used to determine the polarization will be mapped by a unit vector **Q**. Then from the figure we see that $\mathbf{P}\cdot\mathbf{Q}$ is given by $P\cos\theta$.

The Stokes parameters also have the property that, if several independent beams are superposed incoherently, then the Stokes parameters of the resulting beam are just the sum of the parameters characterizing the individual beams.[10] That is, for the final beam, we have

$$\begin{aligned} I &= \sum_i I^{(i)}, \\ P_1 &= \sum_i P_1^{(i)}, \\ \times\quad P_2 &= \sum_i P_2^{(i)}, \\ P_3 &= \sum_i P_3^{(i)}. \end{aligned} \tag{46}$$

DESCRIPTION OF INTERACTIONS

If photons or particles undergo an interaction which is sensitive to polarization then, in general, the Stokes parameters of the initial beam will be transformed into a new set of parameters. The relations between these two Stokes vectors is given by a matrix T characteristic of the interaction. That is,

$$\begin{pmatrix} I \\ \mathbf{P} \end{pmatrix} = T\begin{pmatrix} I_0 \\ \mathbf{P}_0 \end{pmatrix}, \tag{47}$$

where T is a 4×4 array of numbers. Unfortunately, the interaction matrices have not been worked out for all polarization-sensitive interactions. In this section we shall discuss a few of those that have been worked out to provide an example of the ease of description in terms of the Stokes parameters.

When we pass a beam $(I, \mathbf{P})$ through a polarizer T and detect it with an analyzer $(1, \mathbf{Q})$, then the fractional intensity detected is given by

$$w = (1/2)(1,\mathbf{Q})T\begin{pmatrix} I \\ \mathbf{P} \end{pmatrix}. \tag{48}$$

Since polarization is first studied in the realm of optics, let us start with some examples applied to this field.

(1) A Nicol prism with its transmission axis along $\mathbf{E}_1$, i.e., one which will only pass light characterized by the Stokes parameters (1 1 0 0). In this case the interaction matrix is

$$T = (1/2)\begin{bmatrix} 1 & 1 & 0 & 0 \\ 1 & 1 & 0 & 0 \\ 0 & 0 & 0 & 0 \\ 0 & 0 & 0 & 0 \end{bmatrix}. \tag{49}$$

This matrix expresses the $\cos^2\Phi$ dependence on the orientation of the electric vector as can be seen from the following considerations. For the vector expression $\mathbf{E} = \cos\Phi\mathbf{E}_1 + \sin\Phi\mathbf{E}_2$ we have the Stokes vector

$$\begin{bmatrix} 1 \\ \cos^2\Phi - \sin^2\Phi \\ 2\sin\Phi\cos\Phi \\ 0 \end{bmatrix},$$

and hence

$$(1/2)\begin{bmatrix} 1 & 1 & 0 & 0 \\ 1 & 1 & 0 & 0 \\ 0 & 0 & 0 & 0 \\ 0 & 0 & 0 & 0 \end{bmatrix}\begin{bmatrix} 1 \\ \cos^2\Phi - \sin^2\Phi \\ 2\sin\Phi\cos\Phi \\ 0 \end{bmatrix} = (1/2)\begin{bmatrix} 1+\cos^2\Phi - \sin^2\Phi \\ 1+\cos^2\Phi - \sin^2\Phi \\ 0 \\ 0 \end{bmatrix} = \cos^2\Phi\begin{bmatrix} 1 \\ 1 \\ 0 \\ 0 \end{bmatrix}.$$

[9] U. Fano, J. Opt. Soc. Am. **39**, 859 (1949).
[10] S. Chandrasekhar, Astrophys. J. **105**, 424 (1947).

Now consider that we have two Nicol prisms. The first acts as a polarizer and has the interaction matrix T; the second we use as an analyzer which accepts polarization characterized by $(1, \mathbf{Q})$. Then the intensity accepted by the analyzer after the light has passed through the polarizer is given by Eq. (48).

If the initial beam is unpolarized and the transmission axes of the Nicol prisms are parallel, then

$$w=(1/2)(1\ \ 1\ \ 0\ \ 0)T\begin{pmatrix}1\\0\\0\\0\end{pmatrix}$$

$$=(1/2)(1\ \ 1\ \ 0\ \ 0)(1/2)\begin{pmatrix}1\\1\\0\\0\end{pmatrix}=1/2,$$

i.e., only half the intensity of the original beam is transmitted. For crossed Nicols

$$w=(1/2)(1\ \ -1\ \ 0\ \ 0)T\begin{pmatrix}1\\0\\0\\0\end{pmatrix}=0,$$

i.e., no light is passed. Other more complicated combinations can readily be worked out, where it must be remembered that the Stokes parameters of the initial beam must be given so that $P_1=1$ refers to plane polarization along the transmission axis of the Nicol prism.

The next two examples are among those worked out by Perrin[11] in his article on the scattering of light.

(2) A birefringent crystal which introduces a phase φ between the components of the vibration along two orthogonal axes. If we take the fast axis along $\mathbf{E}_1$, then

$$T=\begin{pmatrix}1&0&0&0\\0&1&0&0\\0&0&\cos\varphi&-\sin\varphi\\0&0&\sin\varphi&\cos\varphi\end{pmatrix}. \tag{50}$$

In particular, a quarter-wave plate with its fast axis along $\mathbf{E}_1$ is given by

$$T=\begin{pmatrix}1&0&0&0\\0&1&0&0\\0&0&0&-1\\0&0&1&0\end{pmatrix}. \tag{51}$$

[11] F. Perrin, J. Chem. Phys. 10, 415 (1942).

As an example, let us consider the conversion of circularly polarized light into plane polarized light.

$$T\begin{pmatrix}1\\0\\0\\1\end{pmatrix}=\begin{pmatrix}1\\0\\-1\\0\end{pmatrix},$$

which says that right circularly polarized light passing through a quarter-wave plate with its fast axis along $\mathbf{E}_1$ is converted into plane-polarized light oriented 45° to the right of $\mathbf{E}_2$. The inverse effect is given by

$$T\begin{pmatrix}1\\0\\1\\0\end{pmatrix}=\begin{pmatrix}1\\0\\0\\1\end{pmatrix},$$

where light polarized 45° to the right of $\mathbf{E}_1$ is converted into right circularly polarized light.

(3) A crystal exhibiting optical activity which rotates the plane of polarization an angle φ to the right has the interaction matrix

$$T=\begin{pmatrix}1&0&0&0\\0&\cos 2\varphi&-\sin 2\varphi&0\\0&0&0&0\\0&0&0&1\end{pmatrix}. \tag{52}$$

These examples demonstrate the simplicity of treating polarized light in the formalism of the Stokes parameters.

Next let us consider the polarization of gamma rays. In the Compton scattering of photons the Klein-Nishina formula shows that the cross section depends on the angle between the directions of polarization of the incident and scattered photon. Fano[9] has shown that the treatment of Compton scattering in terms of the Stokes parameters greatly facilitates the computation of the various cross sections. For our examples in this discussion we shall only discuss plane polarization effects, in which case the interaction matrix developed by Fano reduces to a 3×3 matrix expressing the usual Klein-Nishina cross section:

$$T=(1/2)(e^2/mc^2)(k/k_0)^2 \times\begin{pmatrix}1+\cos^2\theta+(k_0-k)(1-\cos\theta)&-\sin^2\theta&0\\-\sin^2\theta&1+\cos^2\theta&0\\0&0&2\cos\theta\end{pmatrix}, \tag{53}$$

where k_0 and k are the energy of the incident and scattered quanta, respectively, in units of mc^2, and θ is the angle of scattering between k_0 and k. With this matrix, positive values of the Stokes parameters have the following significance:

P_1 represents plane polarization in the plane of scattering.

P_2 represents plane polarization in a plane 45° to the right.

The complete 4×4 matrix is given in Fano's article. The fourth row and column, omitted here, contain terms dependent upon the spin orientation of the scattering electron. Using Fano's interaction matrix, the Compton cross section per unit solid angle is then given by

$$\frac{d\sigma}{d\Omega}=(1/2)(1,\mathbf{Q})T\begin{pmatrix} I \\ \mathbf{P} \end{pmatrix}. \tag{54}$$

As examples of its use we shall discuss three cases.

(1) Compton Scattering of Unpolarized Gamma Rays

As a result of Compton scattering the Stokes parameters of an unpolarized beam undergo the transformation

$$T\begin{pmatrix} 1 \\ 0 \\ 0 \end{pmatrix}\sim\begin{pmatrix} 1+\cos^2\theta+(k_0-k)(1-\cos\theta) \\ -\sin^2\theta \\ 0 \end{pmatrix}, \tag{55}$$

from which, since $P_1\sim-\sin\theta$, we see that as a result of Compton scattering the beam is partially polarized orthogonal to the plane of scattering. The degree of polarization is usually defined as

$$p=d\sigma_\perp/d\sigma_{\|},\quad \text{where}\quad d\sigma_\perp=(1/2)(1\ \ -1\ \ 0)T\begin{pmatrix} 1 \\ 0 \\ 0 \end{pmatrix},$$

$$d\sigma_{\|}=(1/2)(1\ \ 1\ \ 0)T\begin{pmatrix} 1 \\ 0 \\ 0 \end{pmatrix},$$

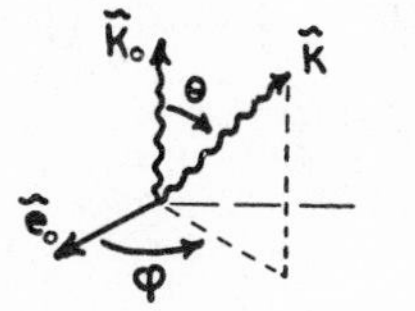

FIG. 3. Compton scattering of a polarized gamma-ray beam. $\mathbf{e}_0$ gives the plane of polarization of the incident beam $\mathbf{k}_0$, which is oriented at an angle ϕ to the plane of scattering. θ is the Compton scattering angle.

yielding the result

$$p=\frac{(k_0-k)(1-\cos\theta)+2}{(k_0-k)(1-\cos\theta)+2\cos^2\theta},$$

where the $\perp$ and $\|$ refer to plane polarization perpendicular to and parallel to the plane of scattering, respectively. Using the ordinary Klein-Nishina formula the derivation of this result requires much more effort and careful consideration of the angles involved. Thus in analogy to optics we see that the polarization of gamma rays by Compton scattering is very similar to the polarization of light by reflection from a dielectric.

(2) Compton Scattering of a Polarized Beam

For this discussion we shall use the geometry shown in Fig. 3. To use Fano's matrix we must recall that positive values of P_1 refer to plane polarization in the plane of scattering. The simplest representation of the initial beam is (1 1 0), but this is in a coordinate system rotated through an angle Φ to the right of the plane of scattering (looking in the direction $-\mathbf{k}_0$); therefore, we must rotate the coordinate system an angle Φ to the left using the matrix M given by Eq. (40) with the appropriate change of sign. Thus, using the plane of scattering as a reference plane, we have

$$M\begin{pmatrix} 1 \\ 1 \\ 0 \end{pmatrix}=\begin{pmatrix} 1 \\ \cos 2\Phi \\ \sin 2\Phi \end{pmatrix}=\begin{pmatrix} 1 \\ \cos^2\Phi-\sin^2\Phi \\ 2\sin\Phi\cos\Phi \end{pmatrix},$$

which yields the final result

$$d\sigma/d\Omega=(1/2)(1\ \ 0\ \ 0)T\begin{pmatrix} 1 \\ \cos^2\Phi-\sin^2\Phi \\ 2\sin\Phi\cos\Phi \end{pmatrix}$$

$$\sim k_0/k+k/k_0-2\sin^2\theta\cos^2\Phi,$$

the usual result, but found in a much simpler manner. Here we have used (1 0 0) as an analyzer since we are interested in the intensity of the scattered beam regardless of its polarization. This is just the Stokes vector characterizing an ordinary photon detector, such as a scintillation counter. Thus we see that the polarized gamma rays are preferentially scattered in a direction perpendicular to the electric vector. This allows

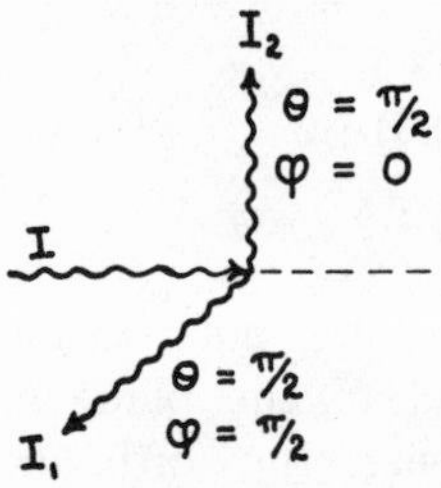

FIG. 4. Compton scattering as an analyzer for plane polarization such that for $P_1=1$ the beam is plane polarized in the (I, I_2) plane. The degree of polarization is given by

$$I_2/I_1=(p+R)/(pR+1),$$

where R is the ratio that would be obtained for a totally polarized beam.

us to use Compton scattering as an analyzer in a way analogous to the Nicol prism of optics.

To use Compton scattering as an analyzer for plane polarization, the two measurements shown in Fig. 4 are made. The partially polarized beam is Compton scattered and intensity measurements are made for a scattering angle of $\theta=\pi/2$ in the two directions I_2 and I_1, where I_1 is perpendicular to the plane of I and I_2. The degree of polarization p with respect to these two directions is found from $I_2/I_1=(p+R)/pR+1)$, where R is the ratio that would be obtained for a beam totally polarized in the (I, I_2) plane.

(3) The Double Compton Scattering Experiment, Shown in Fig. 5

The cross section for double scattering is given by

$$d\sigma/d\Omega=(1/2)(1\ \ 0\ \ 0)T_2T_1\begin{pmatrix}1\\0\\0\end{pmatrix}.$$

The unpolarized incident beam, $\mathbf{k}_0$, characterized by (1 0 0), will be partially polarized as a result of the first scattering. After this first scattering, the beam $\mathbf{k}_1$ will be characterized by

$$T_1\begin{pmatrix}1\\0\\0\end{pmatrix}\sim\begin{pmatrix}1+\cos^2\theta_1+(k_0-k_1)(1-\cos\theta_1)\\-\sin^2\theta_1\\0\end{pmatrix},$$

from Eq. (55). Since these Stokes parameters refer to the $(\mathbf{k}_0, \mathbf{k}_1)$ plane, in order to use Fano's matrix for the second scattering we must

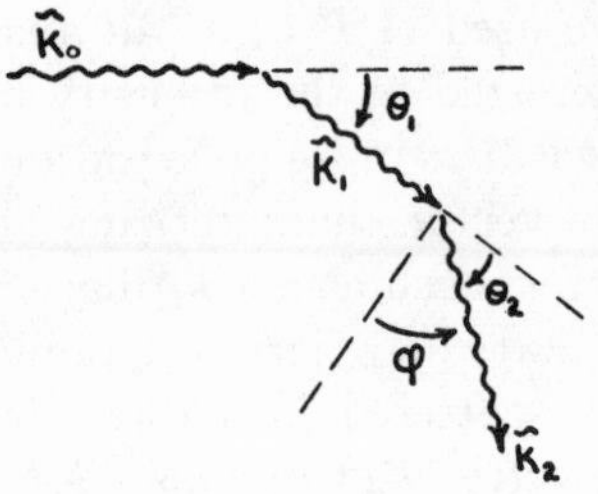

FIG. 5. Double Compton scattering of an unpolarized gamma-ray beam. φ is the angle between the $(\mathbf{k}_0, \mathbf{k}_1)$ plane and the $(\mathbf{k}_1, \mathbf{k}_2)$ plane.

transform by the matrix M these Stokes parameters to refer to the $(\mathbf{k}_1, \mathbf{k}_2)$ plane, which is at an angle Φ to the $(\mathbf{k}_1, \mathbf{k}_2)$ plane. Therefore,

$$d\sigma\sim(1/2)(1\ \ 0\ \ 0)T_2\times\begin{pmatrix}1+\cos^2\theta_1+(k_0-k_1)(1-\cos\theta_1)\\-\sin^2\theta_1(\cos^2\Phi-\sin^2\Phi)\\0\end{pmatrix},$$

which yields $d\sigma\sim\gamma_{01}\gamma_{12}-\gamma_{01}\sin^2\theta_2-\gamma_{12}\sin^2\theta_1+2\sin^2\theta_1\sin^2\theta_2\cos^2\Phi$, where $\gamma_{01}=k_1/k_0+k_0/k_1$; $\gamma_{12}=k_2/k_1+k_1/k_2$, which is the same result given by Wightman[12] using the density matrix.

Next let us consider the polarization of electrons as a result of a scattering experiment. If we write the wave function ψ for an electron in terms of orthogonal states of longitudinal spin

$$\psi=a_1\psi_1''+a_2\psi_2'',$$

then the spin direction of the electron is given by Eq. (29). In terms of these spin angles Mott[13]

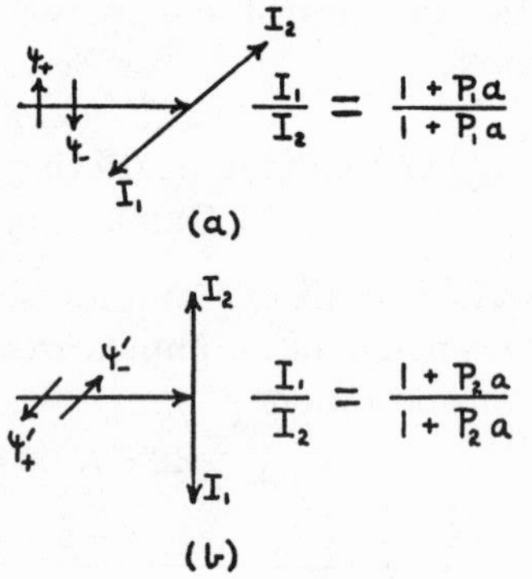

FIG. 6. Single scattering of electrons for the determination of the orientation coefficients with respect to the fundamental spin states shown by the small vectors on the incident beam. $a=D/\bar{I}$.

has shown that the scattered intensity is given by

$$I=\bar{I}-D\sin\theta\sin(\Phi-\pi/2)$$

for right angle scattering, where $\bar{I}$ and D are functions of the scattering angle. Tolhoek[7] has pointed out that the orientation coefficients of the electron beam can then be determined by the two right angle scattering experiments shown in Fig. (6). A measurement of the intensities I_1 and I_2 in each of the two positions shown yields the orientation coefficients P_1 and P_2, where positive values of P_i mean a spin orientation

[12] A. Wightman, Phys. Rev. **74**, 1813 (1948).
[13] N. F. Mott, Proc. Roy. Soc. (London) **A124**, 425 (1929), also N. F. Mott and H. S. W. Massey, *The Theory of Atomic Collisions* (Oxford University Press, New York, 1950), second edition, p. 76.

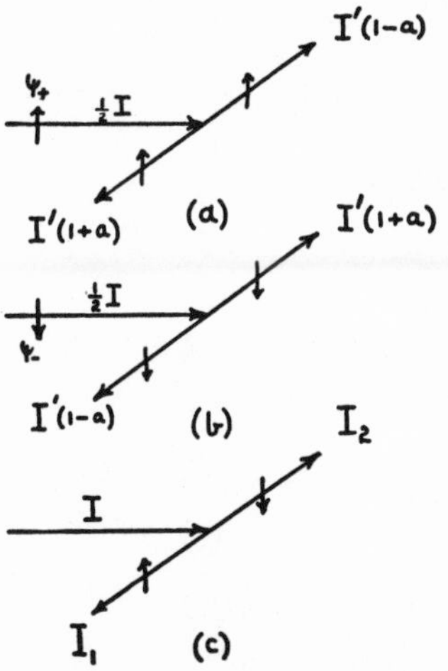

FIG. 7. In (a) and (b) we have an analysis of an unpolarized beam into two independent components of opposite spin. Each component will be scattered with a different intensity so that the resulting beams I_1 and I_2 of the unpolarized beam shown in (c) will be partially polarized. The arrows in the figure show the predominant spin direction of the partially polarized beams.

orthogonal to the plane of scattering as shown in the figure. As an example, for pure states of spin, ψ_+, i.e., $P_1=1$, we have

$$\frac{I_1}{I_2}=\frac{\bar{I}-D\sin\pi/2\,\sin(-\pi/2)}{\bar{I}-D\sin\pi/2\,\sin(\pi-\pi/2)}=\frac{\bar{I}+D}{\bar{I}-D}=\frac{1+a}{1-a}, \tag{56}$$

where $a=D/\bar{I}$. For a partially polarized beam then the measurement of I_1 and I_2 will yield

$$I_1/I_2=(1+Pa)/(1-Pa),$$

where the P's refer to transverse polarization. Now longitudinal polarization cannot be measured by the scattering experiment alone, just as circular polarization of a beam of light cannot be measured by a Nicol prism alone. The longitudinal polarization must be converted into transverse polarization just as circular polarization must be transformed into plane polarization by means of a quarter-wave plate. For electrons Tolhoek shows that this transformation may be obtained by the use of a transverse electric field, which changes the direction of motion of the electron, but leaves its spin orientation in space unchanged.

We have seen the analyzing properties of electron scattering experiments from the above discussion; now let us investigate the polarization of an unpolarized electron beam by single scattering. Since scattering is sensitive only to spin states orthogonal to the scattering plane we should expect that the scattered beam would be partially polarized orthogonal to the plane of scattering which is indeed the case. Now an unpolarized beam of electrons can be thought of as an incoherent superposition of two beams of opposite polarization. Hence, an unpolarized beam can be subdivided into the two components shown in Fig. 7(a) and (b). These two components will be scattered differently. Hence, the scattered beams I_1 and I_2 as shown in Fig. 7(c) will be partially polarized with the predominant spin direction as shown in the figure.

CONCLUSION

By the introduction of the Stokes parameters for the description of polarization, once an interaction matrix has been found, the mathematical description of polarization reduces to a very simple form. This method has the advantage that it shows the direct analogy of polarization of particles and high-energy photons to the polarization of light studied in optics courses.

The author wishes to express thanks to Professor F. L. Hereford, who interested the author in this subject, and Dr. Stephen Berko for their helpful suggestions in the preparation of this article. Thanks are also due to other members of the Rouss Physical Laboratory who proofread this article.

PAPER NO. 14

Reprinted from *Proceedings of the Royal Society*, Ser. A, Vol. 225, pp. 96–111, 1954.

A macroscopic theory of interference and diffraction of light from finite sources

I. Fields with a narrow spectral range

By E. Wolf

Department of Mathematical Physics, University of Edinburgh

(*Communicated by M. Born, F.R.S.—Received* 27 *January* 1954)

A macroscopic theory of interference and diffraction of light in stationary fields produced by finite sources which emit light within a finite spectral range is formulated. It is shown that a generalized Huygens principle may be obtained for such fields, which involves only observable quantities. The generalized Huygens principle expresses the *intensity* at a typical point of the field in terms of an integral taken twice independently over an arbitrary surface, the integral involving the intensity distribution over the surface and the values of a certain correlation factor, which is found to be the 'degree of coherence' previously introduced by Zernike. Next it is shown that under fairly general conditions, this correlation factor is essentially the normalized integral over the source of the Fourier (frequency) transform of the spectral intensity function of the source, and that it may be determined from simple interference experiments. Further, it is shown that in regions where geometrical optics is a valid approximation, the coherence factor itself then obeys a simple geometrical law of propagation. Several results on partially coherent fields, established previously by Van Cittert, Zernike, Hopkins and Rogers, follow as special cases from these theorems.

The results have a bearing on many optical problems and can also be applied in investigations concerned with other types of radiation.

1. Introduction

In the usual treatments of interference and diffraction of light, the source is assumed to be of vanishingly small dimensions (a point source), emitting strictly monochromate radiation. Such treatments correspond essentially to an idealized wave field created by a (classical) oscillator. Huygens's principle, in the extended formulation of Fresnel, may be regarded as an approximate propagation law for such fields.

In recent years remarkable advances have been made in practical optics in connexion with interference and diffraction of light and electrons; in particular, the phase-contrast method (Zernike 1934 *a*, *b*), the method of the coherent background (Zernike 1948) and the method of reconstructed wave-fronts (Gabor 1949, 1951) must be mentioned. These discoveries, as well as numerous problems in both theoretical and practical optics, make it highly desirable to extend the theory of interference and diffraction to fields produced by an actual source, i.e. a source of finite extension and one which emits light within a finite frequency range.

First steps towards formulating such a theory were made by Berek (1926 *a*, *b*, *c*, *d*), Van Cittert (1934, 1939), Zernike (1938) and Hopkins (1951, 1953). The experimental counterpart of these investigations dates back to Michelson† (1890, 1891 *a*, *b*, 1892, 1920).‡ In the theoretical papers just referred to correlation factors for light disturbances at two arbitrary points or at two instants of time were introduced and applied to particular problems, notably by Hopkins (1953). However, only a moderate progress was achieved in formulating the general laws relating to such fields and in fact the subject presents to-day a somewhat confused picture. This is mainly because each of the four authors introduced a formally different correlation factor, which in turn led to many disconnected results.

In the present investigation a systematic study is made of interference and diffraction in stationary§ optical fields produced by finite sources which emit light within a finite frequency range. Part I is mainly restricted to the case when the effective frequency range is sufficiently narrow. In § 2 it is shown that a Huygens principle may be formulated for such fields, which involves only observable quantities. In this generalized form, the Huygens principle expresses the *intensity* at an arbitrary point of the field in terms of an integral taken twice independently over an arbitrary surface, the integrand involving (1) the values of the intensity at all points of this surface and (2) a correlation factor, which turns out to be the complex form of the 'degree of coherence' introduced by Zernike. In this formulation the Huygens principle is subject to similar restrictions on its range of validity as encountered in connexion with its usual form, but a rigorous formulation is possible and will be given in part II of this investigation.

In § 3 the significance of the coherence factor is discussed, and it is shown that it may be determined from simple interference experiments. In § 4 it is shown that under fairly general conditions the coherence factor is essentially the normalized integral over the source of the Fourier (frequency) transform of the intensity function $j(\xi, \nu)$ of the source. This relation takes a particularly simple form when the frequency range of the radiation is sufficiently narrow and when some further simplifying conditions are satisfied. Under these restrictions several of the earlier

† See also Michelson & Pease (1921) and Pease (1931).

‡ A fuller historical survey is given in my article in *Vistas in astronomy* (Wolf 1954). In addition to the literature quoted there, reference to a discussion of a more abstract kind may be added: Wiener (1930), chapter III, § 9.

§ By a stationary field we mean here a field of which all observable properties are constant in time. This definition includes as special case the usual case of high frequency sinusoidal time dependence; or the field constituted by the steady flux of (polychromatic) radiation through an optical system. But it excludes fields for which the time average over a macroscopic time-interval of the flux of radiation depends on time.

results of Van Cittert (1934), Zernike (1938) and Hopkins (1951) are then shown to follow as special cases. It is also found that in regions where the approximations of geometrical optics hold, the coherence factor itself then obeys a simple geometrical law of propagation. In § 6 it is shown that when the source is small, the generalized Huygens principle may be expressed in a simple form in which the properties of the source and the transmission properties of the medium are completely separated.

The results have an immediate bearing on many optical problems and can also be applied in investigations concerned with other types of radiation.

2. A generalized Huygens's principle

We shall be concerned with stationary optical fields and begin by considering the propagation of a beam of natural, nearly monochromatic light from a finite source Σ. For reasons of convergence we assume that the radiation field exists only between the instants $t = -T$ and $t = +T$. It is easy to pass to the limit $T \to \infty$ subsequently.

Let $V(\mathbf{x}, t)$ denote the disturbance at a point specified by the position vector $\mathbf{x}$, at time t. We shall represent V in the form of a Fourier integral:

$$V(\mathbf{x}, t) = \int_{-\infty}^{+\infty} v(\mathbf{x}, \nu)\, e^{-2\pi i \nu t}\, d\nu. \tag{2·1}$$

Then
$$v(\mathbf{x}, \nu) = \int_{-T}^{T} V(\mathbf{x}, t)\, e^{2\pi i \nu t}\, dt. \tag{2·2}$$

Since the light is assumed to be almost monochromatic, $|v(\mathbf{x}, \nu)|$ will differ appreciably from zero only in a narrow frequency range $\nu_0 - \Delta\nu \leqslant \nu \leqslant \nu_0 + \Delta\nu$.

Let us take a surface $\mathscr{A}$ cutting across the beam and consider the intensity at a point $P(\mathbf{x})$ on that side of Σ towards which the light is advancing (figure 1).

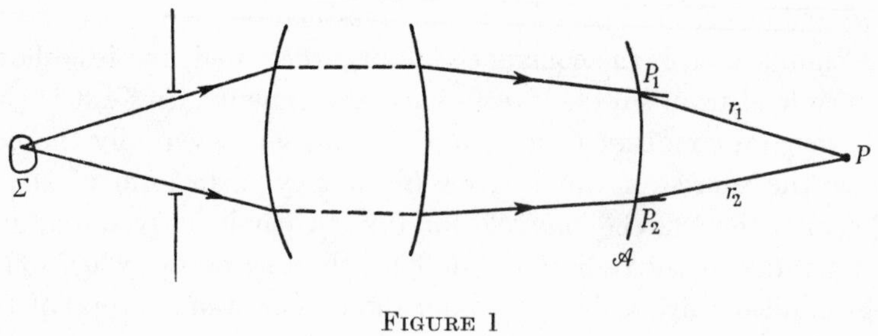

Figure 1

Each Fourier component of (2·1) represents a perfectly monochromatic wave, and therefore (under the usual restrictions on its range of validity) obeys Huygens's principle in the usual form

$$v(\mathbf{x}, \nu) = \int_{\mathscr{A}} v(\mathbf{x}_1, \nu) \frac{e^{ikr_1}}{r_1} \Lambda_1\, d\mathbf{x}_1. \tag{2·3}$$

Here $\mathbf{x}_1$ is the position vector of a typical point P_1 on the surface $\mathscr{A}$, r_1 is the distance from P_1 to P, $k = \frac{2\pi\nu}{c} = \frac{2\pi}{\lambda}$, c being the vacuum velocity of light and λ the wave-length. Further, Λ_1 is the usual inclination factor of Huygens's principle:

$$\Lambda_1 = \frac{\mathrm{i}}{2\lambda}[\cos\phi_1' - \cos\phi_1]; \tag{2·4}$$

the meaning of the angles ϕ_1 and ϕ_1' is shown in figure 2.

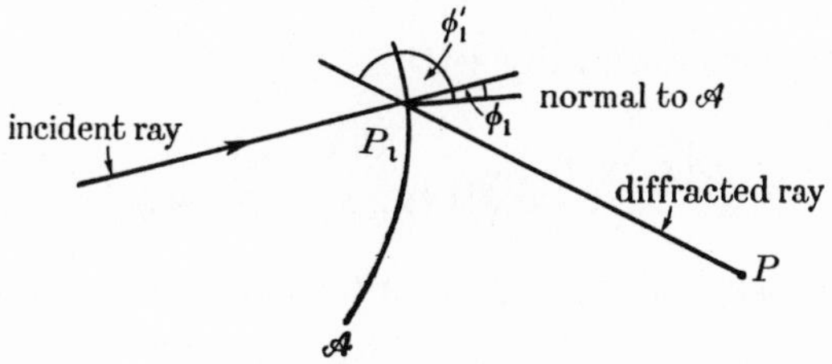

FIGURE 2

From (2·1) and (2·3) it follows that

$$V(\mathbf{x}, t) = \int_{-\infty}^{+\infty} \mathrm{d}\nu\, \mathrm{e}^{-2\pi\mathrm{i}\nu t} \int_{\mathscr{A}} v(\mathbf{x}_1, \nu) \frac{\mathrm{e}^{\mathrm{i}kr_1}}{r_1} \Lambda_1\, \mathrm{d}\mathbf{x}_1. \tag{2·5}$$

The intensity at P is then given by

$$I(\mathbf{x}) = \langle V(\mathbf{x}, t)\, V^*(\mathbf{x}, t)\rangle, \tag{2·6}$$

where asterisks denote the complex conjugate and the brackets denote average over the interval $-T \leqslant t \leqslant T$. Substituting from (2·5) into (2·6) we obtain

$$I(\mathbf{x}) = \Big\langle \int_{-\infty}^{+\infty}\int_{-\infty}^{+\infty} \mathrm{d}\nu\,\mathrm{d}\nu' \exp\{-2\pi\mathrm{i}(\nu-\nu')t\} \times \int_{\mathscr{A}}\int_{\mathscr{A}} v(\mathbf{x}_1, \nu)\, v^*(\mathbf{x}_2, \nu') \frac{\exp\{\mathrm{i}[kr_1 - k'r_2]\}}{r_1 r_2} \Lambda_1 \Lambda_2'^*\, \mathrm{d}\mathbf{x}_1\, \mathrm{d}\mathbf{x}_2 \Big\rangle, \tag{2·7}$$

where $k' = 2\pi\nu'/c$, $\Lambda' = \Lambda(k')$, and the points $P_1(\mathbf{x}_1)$ and $P_2(\mathbf{x}_2)$ explore the surface $\mathscr{A}$ independently.

Since the light is assumed to be nearly monochromatic and of frequency ν_0 the curve $v(\mathbf{x}, \nu)$, considered as a function of ν, will have a peak at $\nu = \nu_0$ and fall off rapidly on both sides, being practically zero outside the range $(\nu_0 - \Delta\nu, \nu_0 + \Delta\nu)$. Under these conditions we may replace both k and k' in the term

$$\exp\{\mathrm{i}[kr_1 - k'r_2]\}\, \Lambda_1 \Lambda_2'^*$$

by $k^{(0)} = 2\pi\nu_0/c$. (2·7) then becomes

$$I(\mathbf{x}) = \int_{\mathscr{A}}\int_{\mathscr{A}} \Gamma(\mathbf{x}_1, \mathbf{x}_2) \frac{\exp\{\mathrm{i}k^{(0)}(r_1 - r_2)\}}{r_1 r_2} \Lambda_1^{(0)} \Lambda_2^{(0)*}\, \mathrm{d}\mathbf{x}_1\, \mathrm{d}\mathbf{x}_2, \tag{2·8}$$

where

$$\Gamma(\mathbf{x}_1, \mathbf{x}_2) = \frac{1}{2T}\int_{-T}^{T} \mathrm{d}t \int_{-\infty}^{+\infty}\int_{-\infty}^{+\infty} v(\mathbf{x}_1, \nu)\, v^*(\mathbf{x}_2, \nu') \exp\{-2\pi\mathrm{i}(\nu-\nu')t\}\, \mathrm{d}\nu\, \mathrm{d}\nu'. \tag{2·9}$$

Since
$$\lim_{T\to\infty}\int_{-T}^{+T}\exp\{-2\pi i(\nu-\nu')t\}\,dt=\delta(\nu-\nu'),$$
where δ is the Dirac Delta function, it follows that, for sufficiently large T, (2·9) reduces to
$$\Gamma(\mathbf{x}_1,\mathbf{x}_2)=\frac{1}{2T}\int_{-\infty}^{+\infty}v(\mathbf{x}_1,\nu)\,v^*(\mathbf{x}_2,\nu)\,d\nu, \tag{2·10}$$
or, using the convolution (Faltung) theorem,
$$\Gamma(\mathbf{x}_1,\mathbf{x}_2)=\langle V(\mathbf{x}_1,t)\,V^*(\mathbf{x}_2,t)\rangle. \tag{2·11}$$
It will be useful to normalize Γ by setting
$$\gamma(\mathbf{x}_1,\mathbf{x}_2)=\frac{\Gamma(\mathbf{x}_1,\mathbf{x}_2)}{\sqrt{\{\Gamma(\mathbf{x}_1,\mathbf{x}_1)\,\Gamma(\mathbf{x}_2,\mathbf{x}_2)\}}}. \tag{2·12}$$
Now $\Gamma(\mathbf{x}_s,\mathbf{x}_s)$ $(s=1,2)$ is nothing but the intensity I_s at the point $\mathbf{x}_s$, so that (2·12) may be written as
$$\gamma(\mathbf{x}_1,\mathbf{x}_2)=\frac{\Gamma(\mathbf{x}_1,\mathbf{x}_2)}{\sqrt{(I_1 I_2)}}. \tag{2·13}$$
Substituting from (2·13) into (2·8), we finally obtain
$$I(\mathbf{x})=\int_{\mathscr{A}}\int_{\mathscr{A}}\sqrt{(I_1I_2)}\,\gamma_{12}\frac{\exp\{ik^{(0)}(r_1-r_2)\}}{r_1r_2}\Lambda_1^{(0)}\Lambda_2^{(0)*}\,d\mathbf{x}_1\,d\mathbf{x}_2, \tag{2·14}$$
where γ_{12} has been written for $\gamma(\mathbf{x}_1,\mathbf{x}_2)$.

Equation (2·14) may be regarded as a *generalized Huygens principle*. In the usual formulation (2·3), Huygens's principle applies only to strictly coherent radiation and expresses the (non-observable) *disturbance* at a point in the wave field as sum of contributions from each element $d\mathbf{x}_1$ of the primary wave (or, more generally, of an arbitrary surface). In the present formulation, the restriction of strict coherence is dropped and the *intensity* is calculated by summing over all products $d\mathbf{x}_1\,d\mathbf{x}_2$ of the surface, each contribution being weighted by the appropriate value of the correlation factor γ_{12}. It will be shown that in any particular case, this factor may be determined from simple experiments, and that under fairly general conditions it may also be calculated from the knowledge of the intensity function of the source and the optical transmission properties of the medium. Hence our generalized Huygens principle involves *observable quantities*† only.

3. Determination of the γ factor from experiment

The relations (2·13) and (2·11) are formally equivalent to relations (4) and (5) of Zernike's (1938) paper and show that $\Gamma(\mathbf{x}_1,\mathbf{x}_2)$ is the *mutual intensity* and $\gamma(\mathbf{x}_1,\mathbf{x}_2)$

† An earlier formulation of Huygens's principle in terms of observable quantities, due to Gabor (1952; Private communication), must also be mentioned: 'Set up a coherent radiation field and apply to it a small perturbation by introducing objects in the path of radiation which do not destroy the coherence. If the *absolute amplitudes* of the perturbed field are known in one cross-section, they are thereby determined in all cross-sections.' This formulation is, however, not sufficiently general, being restricted to strictly coherent radiation.

the complex degree of coherence,† two important concepts introduced by Zernike. The degree of coherence has been previously defined in a different way and under more restrictive conditions by Van Cittert (1934).

In general γ is complex. It is easily seen that its absolute value is less than unity. For one has, using the well-known modulus inequality for integrals,

$$\left|\int_{-T}^{T} V(\mathbf{x}_1, t)\, V^*(\mathbf{x}_2, t)\, \mathrm{d}t\right| \leqslant \left\{\int_{-T}^{T} |\, V(\mathbf{x}_1, t)\, V^*(\mathbf{x}_2, t)\,|\, \mathrm{d}t\right\}. \tag{3·1}$$

Moreover, by Schwarz's inequality,

$$\left\{\int_{-T}^{T} |\, V(\mathbf{x}_1, t)\, V^*(\mathbf{x}_2, t)\,|\, \mathrm{d}t\right\}^2 \leqslant \int_{-T}^{T} |\, V(\mathbf{x}_1, t)\,|^2\, \mathrm{d}t \int_{-T}^{T} |\, V(\mathbf{x}_2, t)\,|^2\, \mathrm{d}t. \tag{3·2}$$

From (3·1) and (3·2),

$$\left|\int_{-T}^{T} V(\mathbf{x}_1, t)\, V^*(\mathbf{x}_2, t)\, \mathrm{d}t\right| \leqslant \left\{\int_{-T}^{T} |\, V(\mathbf{x}_1, t)\,|^2\, \mathrm{d}t\right\}^{\frac{1}{2}} \left\{\int_{-T}^{T} |\, V(\mathbf{x}_2, t)\,|^2\, \mathrm{d}t\right\}^{\frac{1}{2}}, \tag{3·3}$$

or, in terms of Γ, $\qquad |\,\Gamma(\mathbf{x}_1, \mathbf{x}_2)\,| \leqslant \surd\{\Gamma(\mathbf{x}_1, \mathbf{x}_1)\,\Gamma(\mathbf{x}_2, \mathbf{x}_2)\}$,

whence

$$|\,\gamma(\mathbf{x}_1, \mathbf{x}_2)\,| \leqslant 1. \tag{3·4}$$

In order to see the significance of γ and also to confirm our earlier statement that it is an observable quantity, we shall apply our generalized Huygens principle to a simple interference experiment.

Figure 3 shows the arrangement. Light from a finite source Σ falls either directly or via an optical system on to a screen $\mathscr{A}$ which has small openings at P_1 and P_2. The resulting interference fringes are observed on a second screen $\mathscr{A}'$.

If $\mathrm{d}\mathscr{A}_1$ and $\mathrm{d}\mathscr{A}_2$ denote the areas of the openings at P_1 and P_2, our generalized Huygens principle (2·14) reduces to the following expression when integration is taken over the non-illuminated side of the screen $\mathscr{A}$:

$$\begin{aligned} I \sim I_1\gamma_{11}\left(\frac{1}{r_1}\right)^2 \Lambda_1\Lambda_1^*(\mathrm{d}\mathscr{A}_1)^2 + \surd(I_1 I_2)\,\gamma_{12}\frac{\exp\{ik(r_1-r_2)\}}{r_1 r_2}\Lambda_1\Lambda_2^*\,\mathrm{d}\mathscr{A}_1\mathrm{d}\mathscr{A}_2 \\ + I_2\gamma_{22}\left(\frac{1}{r_2}\right)^2 \Lambda_2\Lambda_2^*(\mathrm{d}\mathscr{A}_2)^2 + \surd(I_2 I_1)\,\gamma_{21}\frac{\exp\{ik(r_2-r_1)\}}{r_2 r_1}\Lambda_2\Lambda_1^*\,\mathrm{d}\mathscr{A}_2\mathrm{d}\mathscr{A}_1. \end{aligned} \tag{3·5}$$

The upper index zero has now been omitted on k, Λ_1 and Λ_2^*. Since $\gamma_{11} \equiv 1$ it follows that the first term gives precisely the value $I^{(1)}$ of the intensity which would be obtained at P if the opening at P_1 alone was open ($\mathrm{d}\mathscr{A}_2 = 0$), the third term having a similar interpretation:

$$\left.\begin{aligned} I^{(1)}(\mathbf{x}) &= \frac{I_1}{r_1^2}|\,\Lambda_1\,|^2\,(\mathrm{d}\mathscr{A}_1)^2, \\ I^{(2)}(\mathbf{x}) &= \frac{I_2}{r_2^2}|\,\Lambda_2\,|^2\,(\mathrm{d}\mathscr{A}_2)^2. \end{aligned}\right\} \tag{3·6}$$

† In the first part of this paper, Zernike defined the degree of coherence of two light vibrations as the visibility of the interference fringes that may be obtained from them under the best circumstances, i.e. when both intensities are made equal and only small path-differences introduced. The analytical definition given by equation (4) of his paper [equivalent to (2·12) above] is however more general and applies whether or not the two intensities are equal; nor is it restricted to small path differences.

The second and the fourth term in (3·5) are complex conjugates of each other. Hence using the identity

$$z+z^* = 2\mathscr{R}(z) \tag{3·7}$$

($\mathscr{R}$ denoting the real part), the sum of the two terms may be written as

$$2\sqrt{(I_1 I_2)}\frac{1}{r_1}\frac{1}{r_2}|\Lambda_1||\Lambda_2||\gamma_{12}|\cos[\arg\gamma_{12}+k(r_1-r_2)]\,\mathrm{d}\mathscr{A}_1\mathrm{d}\mathscr{A}_2$$
$$= 2\sqrt{(I^{(1)}I^{(2)})}\,|\gamma_{12}|\cos[\arg\gamma_{12}+k(r_1-r_2)], \tag{3·8}$$

and (3·5) then reduces to

$$I = I^{(1)}+I^{(2)}+2\sqrt{(I^{(1)}I^{(2)})}\,|\gamma_{12}|\cos[\arg\gamma_{12}+k(r_1-r_2)]. \tag{3·9}$$

We see that in the limiting cases $|\gamma_{12}| = 1$ and $|\gamma_{12}| = 0$ (3·9) reduces to the usual laws for the combination of *perfectly coherent* and *completely incoherent* disturbances. Hence (3·9) may be regarded as a *generalized interference law* in which the factor γ_{12} is a measure of the degree of correlation between the disturbances at P_1 and P_2. This law was derived previously in a different manner by Zernike

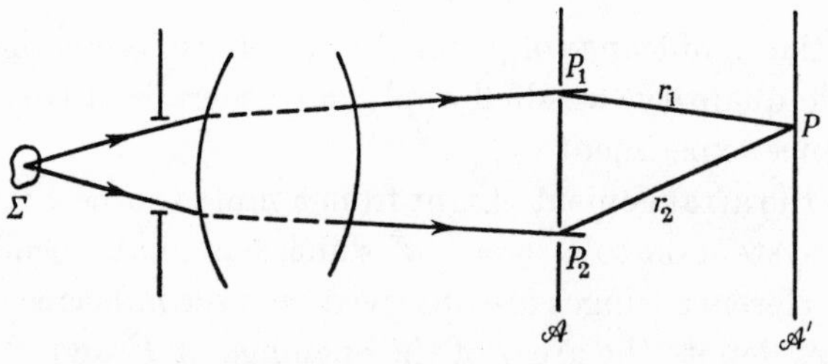

FIGURE 3. An interference experiment with a finite source.

(1938). It has also been derived under a more restrictive definition of the coherence factor by Hopkins (1951).

The generalized interference law (3·9) enables the calculation of $I(\mathbf{x})$ when $I^{(1)}$, $I^{(2)}$ and γ_{12} is known. Conversely when $I^{(1)}$, $I^{(2)}$ and I are known γ_{12} may be determined from measurements of intensities. One has only to measure $I^{(1)}$, $I^{(2)}$ and I for different values of r_1 and r_2 and solve the resulting equations obtained from (3·9) for $|\gamma_{12}|$ and $\arg\gamma_{12}$. Hence the coherence factor may be considered to be an *observable quantity*.

As pointed out by Zernike and Hopkins, the coherence factor is closely related to the *visibility*† and the *position* of the fringes. By (3·9) one has

$$\left.\begin{aligned} I_{\text{max.}} &= I^{(1)}+I^{(2)}+2\sqrt{(I^{(1)}I^{(2)})}\,|\gamma_{12}|, \\ I_{\text{min.}} &= I^{(1)}+I^{(2)}-2\sqrt{(I^{(1)}I^{(2)})}\,|\gamma_{12}|. \end{aligned}\right\} \tag{3·10}$$

† The visibility $\mathscr{V}(\mathbf{x})$ of interference fringes, a concept due to Michelson, is defined by

$$\mathscr{V} = \frac{I_{\text{max.}}-I_{\text{min.}}}{I_{\text{max.}}+I_{\text{min.}}},$$

where $I_{\text{max.}}$ is the maximum of the intensity at the centre of the brightest fringe near $P(\mathbf{x})$ and $I_{\text{min.}}$ is the intensity at the centre of the adjacent dark fringe.

Hence the visibility $\mathscr{V}$ is given by

$$\mathscr{V} = \frac{2}{\sqrt{\frac{I^{(1)}}{I^{(2)}}} + \sqrt{\frac{I^{(2)}}{I^{(1)}}}} |\gamma_{12}|. \tag{3·11}$$

Moreover, it is seen from (3·9) that $\arg \gamma_{12}$ appears formally as the 'phase difference' between the disturbances at the two points; it is equal (in suitable units) to the amount of lateral displacement of the fringe system from the position which it would occupy if $\arg \gamma_{12}$ was equal to zero. In practice one can often set $I^{(1)} = I^{(2)}$. (3·11) shows that the visibility is then simply equal to the absolute value of the coherence factor.

In the paper already referred to, Hopkins (1951) claimed that Zernike's results relating to partially coherent fields may be derived without any resort to statistical analysis. This claim is not justified; Hopkins's analysis, unlike Zernike's, is applicable only to the limiting case of vanishingly narrow frequency range. For a similar reason, a claim by Van Cittert (1939) as to the equivalence of his correlation factor with the degree of coherence of Zernike also does not appear to be justified.

For a full understanding of the coherence problem, it is obviously essential to take the finite frequency range of actual radiation into account. This may be done with the help of Fourier analysis as described in § 2. Alternatively, one may express the disturbance in the form

$$V(\mathbf{x}, t) = U(\mathbf{x}, t) \exp\{-2\pi i \nu_0 t\}, \tag{3·12}$$

where the complex amplitude U is now a function of both position and time. Since the light is assumed to be nearly monochromatic and of frequency ν_0, U will, however, at each point vary slowly (and irregularly) in comparison with the frequency ν_0, and will remain practically constant over an interval of time depending on the coherence length of the light.†

The path differences encountered in instrumental optics are, as a rule, small compared to the coherence length. Hopkins's theory will in these cases lead substantially to the same results as the present analysis. When large path differences are involved Hopkins's theory can, however, no longer be expected to be applicable.

The representation (3·12) was used by Zernike (1938) and also previously by Berek (1926 (*a*)–(*d*)). For a quantitative treatment the Fourier integral approach seems, however, to be more appropriate and has also the attractive feature that it brings the optical coherence problems within the scope of methods well established in connexion with other statistical problems encountered in physics.

Experimental investigations, using small path differences, were recently carried out by Baker (1953) and Arnulf, Dupuy & Flamant (1953). Good agreement with the existing theories was obtained.

† The coherence length may be defined as the maximum path difference between the interfering beams for which interference fringes may be obtained. The finite value of the coherence length arises mainly from (1) natural broadening of spectral lines due to the finite lifetime of atomic states, (2) broadening due to atomic collisions and (3) Doppler broadening due to the thermal motion of the atoms. For an excellent account of the coherence length see Born (1933).

Finally a paper by Duffieux (1953) may also be mentioned. It contains some criticisms of the earlier theories and discusses the connexion between the coherence factor and the transmission functions introduced by him and Lansraux in earlier investigations.

4. Expressions for the mutual intensity function

We shall now derive an explicit expression for the mutual intensity function Γ_{12} in terms of the intensity function of the source and the transmission function of the medium. No restriction on the frequency range will now be imposed.

We assume that the source Σ is a radiating element of a plane surface and divide it into elements $d\Sigma_1, d\Sigma_2, \ldots$, whose linear dimensions are small compared to the wave-length. If $V_m(\mathbf{x}, t)$ denotes the disturbance at the point $P(\mathbf{x})$ and time t due to the mth element of the source, then

$$V(\mathbf{x}, t) = \sum_m V_m(\mathbf{x}, t),$$

and (2·11) becomes

$$\begin{aligned}\Gamma_{12} &= \langle \sum_m V_m(\mathbf{x}_1, t) \sum_n V_n^*(\mathbf{x}_2, t)\rangle \\ &= \langle \sum_m V_m(\mathbf{x}_1, t)\, V_m^*(\mathbf{x}_2, t)\rangle + \langle \sum_{m \neq n} V_m(\mathbf{x}_1, t)\, V_n^*(\mathbf{x}_2, t)\rangle, \end{aligned} \tag{4·1}$$

where m and n run independently through all possible values. Now to a very good approximation, the disturbances from different elements of the source may be treated as statistically independent, i.e.

$$\langle V_m(\mathbf{x}_1, t)\, V_n^*(\mathbf{x}_2, t)\rangle = 0 \quad \text{when} \quad m \neq n,$$

so that the last term in (4·1) vanishes. If we also introduce the Fourier inverse $v_m(\mathbf{x}, \nu)$ of $V_m(\mathbf{x}, t)$

$$v_m(\mathbf{x}, \nu) = \int_{-T}^{T} V_m(\mathbf{x}, t)\, e^{2\pi i \nu t}\, dt,$$

and use the convolution theorem, we then obtain the following expression for Γ_{12}:

$$\Gamma_{12} = \langle \sum_m V_m(\mathbf{x}_1, t)\, V_m^*(\mathbf{x}_2, t)\rangle \tag{4·2a}$$

$$= \frac{1}{2T} \sum_m \int_{-\infty}^{+\infty} v_m(\mathbf{x}_1, \nu)\, v_m^*(\mathbf{x}_2, \nu)\, d\nu. \tag{4·2b}$$

Let
$$K(\mathbf{x}', \mathbf{x}'', \nu) = a(\mathbf{x}', \mathbf{x}'') \exp\{2\pi i (\nu/c) \mathscr{S}(\mathbf{x}', \mathbf{x}'')\} \tag{4·3}$$

(a and $\mathscr{S}$ real) be the *transmission function* of the medium, defined as the disturbance at $P''(\mathbf{x}'')$ due to a monochromatic point source of frequency ν and of unit strength† at $P'(\mathbf{x}')$. (Strictly a and $\mathscr{S}$ also depend on the frequency, but this dependence may here be neglected.) In particular, if the points are situated in a region where diffraction effects are not dominant, then the phase function $\mathscr{S}$ is simply the

† We define a source of unit strength as one which *in vacuo* would give rise to a disturbance of unit amplitude at a unit distance from it.

Hamilton point characteristic function of the medium, i.e. the optical length of the natural ray joining $\mathbf{x}'$ to $\mathbf{x}''$, and the amplitude a may be obtained from the usual conservation law of geometrical optics. *In vacuo*, for example,

$$K = \frac{\exp\{ik\,|\,\mathbf{x}'-\mathbf{x}''\,|\}}{|\,\mathbf{x}'-\mathbf{x}''\,|}. \tag{4·4}$$

If $\boldsymbol{\xi}_m$ denotes the position vector of the element $\mathrm{d}\Sigma_m$ of the source, then the disturbance at a point $P(\mathbf{x})$ due to a typical Fourier component $v_m(\mathbf{x},\nu)$ of $V_m(\mathbf{x},t)$ may be written as

$$v_m(\mathbf{x},\nu) = v_m(\boldsymbol{\xi}_m,\nu)\,K(\boldsymbol{\xi}_m,\mathbf{x},\nu). \tag{4·5}$$

Hence (4·2*b*) becomes

$$\Gamma_{12} = \frac{1}{2T}\sum_m \int_{-\infty}^{+\infty} |\,v_m(\boldsymbol{\xi}_m,\nu)\,|^2\,K(\boldsymbol{\xi}_m,\mathbf{x}_1,\nu)\,K^*(\boldsymbol{\xi}_m,\mathbf{x}_2,\nu)\,\mathrm{d}\nu$$

$$= \frac{1}{2T}\sum_m a(\boldsymbol{\xi}_m,\mathbf{x}_1)\,a(\boldsymbol{\xi}_m,\mathbf{x}_2)\int_{-\infty}^{+\infty} |\,v_m(\boldsymbol{\xi}_m,\nu)\,|^2 \exp\{2\pi i\nu\tau_{12}(\boldsymbol{\xi}_m)\}\,\mathrm{d}\nu, \tag{4·6}$$

where

$$\tau_{12}(\boldsymbol{\xi}_m) = \frac{1}{c}[\mathscr{S}(\boldsymbol{\xi}_m,\mathbf{x}_1) - \mathscr{S}(\boldsymbol{\xi}_m,\mathbf{x}_2)] \tag{4·7}$$

represents the difference in the time needed for light to travel to $\mathbf{x}_1$ and $\mathbf{x}_2$ from the element $\mathrm{d}\Sigma_m$ of the source. Applying the convolution theorem to (4·6) it follows that Γ may also be written in the form

$$\Gamma_{12} = \sum_m a(\boldsymbol{\xi}_m,\mathbf{x}_1)\,a(\boldsymbol{\xi}_m,\mathbf{x}_2)\,\langle V_m(\boldsymbol{\xi}_m,t-\tau_{12})\,V_m^*(\boldsymbol{\xi}_m,t)\rangle. \tag{4·8}$$

Instead of $|\,v_m(\boldsymbol{\xi}_m,\nu)\,|$ which is defined for only a discontinuous set of values $\boldsymbol{\xi}_m$, we may introduce a function $j(\boldsymbol{\xi},\nu)$ which represents the *intensity per unit area* of the source, *per unit frequency range*;† it is defined for the whole continuous set of $\boldsymbol{\xi}$ values. Absorbing the factor $1/2T$ in our definition of j, (4·6) then becomes

$$\Gamma_{12} = \int_\Sigma \mathrm{d}\boldsymbol{\xi}\,a(\boldsymbol{\xi},\mathbf{x}_1)\,a(\boldsymbol{\xi},\mathbf{x}_2)\int_{-\infty}^{+\infty} j(\boldsymbol{\xi},\nu)\exp\{2\pi i\nu\tau_{12}(\boldsymbol{\xi})\}\,\mathrm{d}\nu. \tag{4·9}$$

The amplitude factor $a(\boldsymbol{\xi},\mathbf{x}_s)$ $(s = 1, 2)$ of the transmission function will as a rule be a slowly varying function of $\boldsymbol{\xi}$. Also in most applications the linear dimensions of the source will be small compared to the distance from the source to $\mathbf{x}_s$. Hence in (4·8) and (4·9) we may as a rule replace $a(\boldsymbol{\xi},\mathbf{x}_s)$ by $a(0,\mathbf{x}_s)$ without introducing an appreciable error. Finally, normalizing Γ as before, we obtain

$$\gamma_{12} = \frac{\sum_m \langle V_m(\boldsymbol{\xi}_m,t-\tau_{12})\,V_m^*(\boldsymbol{\xi}_m,t)\rangle}{\sum_m \langle V_m(\boldsymbol{\xi}_m,t)\,V_m^*(\boldsymbol{\xi}_m,t)\rangle} \tag{4·10a}$$

$$= \frac{\int_\Sigma \mathrm{d}\boldsymbol{\xi}\int_{-\infty}^{+\infty} j(\boldsymbol{\xi},\nu)\exp\{2\pi i\nu\tau_{12}(\boldsymbol{\xi})\}\,\mathrm{d}\nu}{\int_\Sigma \mathrm{d}\boldsymbol{\xi}\int_{-\infty}^{+\infty} j(\boldsymbol{\xi},\nu)\,\mathrm{d}\nu}, \tag{4·10b}$$

i.e. *the coherence factor is essentially the normalized integral over the source of the Fourier (frequency) transform of the intensity function of the source.*

† We neglect here the variation of the intensity with direction. The effect of this variation can also be taken into account by introducing a more general intensity function $j(\boldsymbol{\xi},\mathbf{p},\nu)$, $\mathbf{p}$ being a directional variable.

5. Some approximate expressions for the coherence factor

We shall now consider the form which the expressions for γ take when the effective frequency range is sufficiently narrow. We shall show that they lead, in special cases, to several results obtained previously by Van Cittert (1934), Zernike (1938), Hopkins (1951) and Rogers (1953).

Let us assume that the effective frequency range is so narrow that ν may be replaced by ν_0 in the integral in (4·9). This will be permissible, if

$$|\Delta\nu\tau_{12}(\boldsymbol{\xi})| \ll 1,$$

or, since $\Delta\nu = -c\Delta\lambda/\lambda^2$ and $\tau_{12}(\boldsymbol{\xi}) = \frac{1}{c}[\mathscr{S}(\boldsymbol{\xi}, \mathbf{x}_1) - \mathscr{S}(\boldsymbol{\xi}, \mathbf{x}_2)]$, if

$$\frac{|\Delta\lambda|}{\lambda} \ll \frac{\lambda}{|\mathscr{S}(\boldsymbol{\xi}, \mathbf{x}_1) - \mathscr{S}(\boldsymbol{\xi}, \mathbf{x}_2)|}. \tag{5·1}$$

We also set

$$\int_{-\infty}^{\infty} j(\boldsymbol{\xi}, \nu)\, \mathrm{d}\nu = J(\boldsymbol{\xi}); \tag{5·2}$$

(4·9) then gives on normalizing

$$\gamma_{12} = \frac{1}{\sqrt{(I_1 I_2)}} \int_\Sigma J(\boldsymbol{\xi})\, a(\boldsymbol{\xi}, \mathbf{x}_1)\, a(\boldsymbol{\xi}, \mathbf{x}_2) \exp\{2\pi \mathrm{i} \nu_0 \tau_{12}(\boldsymbol{\xi})\}\, \mathrm{d}\boldsymbol{\xi}, \tag{5·3}$$

where

$$I_s = \Gamma_{ss} = \int_\Sigma J(\boldsymbol{\xi})\, a^2(\boldsymbol{\xi}, \mathbf{x}_s)\, \mathrm{d}\boldsymbol{\xi} \quad (s = 1, 2). \tag{5·3a}$$

In particular, *in vacuo* (air) one has (cf. (4·4))

$$a(\boldsymbol{\xi}, \mathbf{x}_s) = \frac{1}{R_s}, \quad \mathscr{S}(\boldsymbol{\xi}, \mathbf{x}_s) = R_s,$$

with

$$R_s = |\mathbf{x}_s - \boldsymbol{\xi}|.$$

(5·3) then reduces to

$$\gamma_{12} = \frac{1}{\sqrt{(I_1 I_2)}} \int_\Sigma \frac{J(\boldsymbol{\xi})}{R_1 R_2} \exp\{\mathrm{i}k^{(0)}(R_1 - R_2)\}\, \mathrm{d}\boldsymbol{\xi}, \tag{5·4}$$

where

$$I_s = \int_\Sigma \frac{J(\boldsymbol{\xi})}{R_s^2}\, \mathrm{d}\boldsymbol{\xi} \quad (s = 1, 2).$$

Integrals of the form (5·4) are well known in optics. They represent the complex amplitude at the point P_2 in a diffraction pattern around P_1, when diffraction takes place at an aperture which is identical in form with Σ, the amplitude in the diffracting aperture being proportional to $J(\boldsymbol{\xi})$. This expression for the coherence factor was first obtained by Zernike (1938) and later by Hopkins (1951).

If, as before, we neglect the variation of the amplitude factor with $\boldsymbol{\xi}$, we obtain from (5·3)

$$\gamma_{12} = \frac{\int_\Sigma J(\boldsymbol{\xi}) \exp\{2\pi \mathrm{i} \nu_0 \tau_{12}(\boldsymbol{\xi})\}\, \mathrm{d}\boldsymbol{\xi}}{\int_\Sigma J(\boldsymbol{\xi})\, \mathrm{d}\boldsymbol{\xi}}. \tag{5·5}$$

This relation may be further simplified if either the linear dimensions of the source, or the distance between P_1 and P_2 are small compared to the distance from the source to P_1 and P_2. In the first case one may expand $\tau_{12}(\boldsymbol{\xi})$ at a suitable point O ($\boldsymbol{\xi}=0$) of the source and neglect higher order terms:

$$\tau_{12}(\boldsymbol{\xi}) \sim \frac{1}{c}[\mathscr{S}(0,\mathbf{x}_1)-\mathscr{S}(0,\mathbf{x}_2)]+\frac{1}{c}\boldsymbol{\xi}\,.\,\frac{\partial}{\partial\boldsymbol{\xi}}[\mathscr{S}(\boldsymbol{\xi},\mathbf{x}_1)-\mathscr{S}(\boldsymbol{\xi},\mathbf{x}_2)]_{\boldsymbol{\xi}=0}. \tag{5·6}$$

Now by the fundamental property of the $\mathscr{S}$ function,

$$\frac{\partial}{\partial\boldsymbol{\xi}}[\mathscr{S}(\boldsymbol{\xi},\mathbf{x}_s)]_{\boldsymbol{\xi}=0} = -\mathbf{p}_s \quad (s=1,2), \tag{5·7}$$

where $\mathbf{p}_s$ denotes the ray vector† at O of the ray $\overrightarrow{OP_s}$. Hence

$$\tau_{12}(\boldsymbol{\xi}) \sim \frac{1}{c}[\mathscr{S}(0,\mathbf{x}_1)-\mathscr{S}(0,\mathbf{x}_2)]-\frac{1}{c}\boldsymbol{\xi}\,.\,(\mathbf{p}_1-\mathbf{p}_2), \tag{5·8}$$

and (5·5) reduces to

$$\gamma_{12} = \frac{\int_\Sigma J(\boldsymbol{\xi})\exp\{-\mathrm{i}k^{(0)}\boldsymbol{\xi}\,.\,(\mathbf{p}_1-\mathbf{p}_2)\}\,\mathrm{d}\boldsymbol{\xi}}{\int_\Sigma J(\boldsymbol{\xi})\,\mathrm{d}\boldsymbol{\xi}}\exp\{\mathrm{i}k^{(0)}[\mathscr{S}(0,\mathbf{x}_1)-\mathscr{S}(0,\mathbf{x}_2)]\}. \tag{5·9}$$

Hence *when the effective frequency range is sufficiently narrow and the source small enough, the coherence factor is equal to the product of the term*

$$\exp\{\mathrm{i}k^{(0)}[\mathscr{S}(0,\mathbf{x}_1)-\mathscr{S}(0,\mathbf{x}_2)]\}$$

and the complex amplitude in an associated Fraunhofer diffraction pattern.

Similar analysis may be used when the distance between P_1 and P_2 is small compared to the distance from the source to these points. In place of (5·6) we now have

$$\begin{aligned}\tau_{12}(\boldsymbol{\xi}) &\sim -\frac{1}{c}(\mathbf{x}_2-\mathbf{x}_1)\left[\frac{\partial\mathscr{S}(\boldsymbol{\xi},\mathbf{x})}{\partial\mathbf{x}}\right]_{\mathbf{x}=\mathbf{x}_1}\\ &= -\frac{1}{c}(\mathbf{x}_2-\mathbf{x}_1)\,.\,\mathbf{p}_1'(\boldsymbol{\xi}),\end{aligned} \tag{5·10}$$

where $\mathbf{p}_1'(\boldsymbol{\xi})$ is the ray vector at P_1 of the ray $\boldsymbol{\xi}\to\mathbf{x}_1$. (5·5) now reduces to

$$\gamma_{12} = \frac{\int_\Sigma J(\boldsymbol{\xi})\exp\{-\mathrm{i}k^{(0)}(\mathbf{x}_2-\mathbf{x}_1)\,.\,\mathbf{p}_1'(\boldsymbol{\xi})\}\,\mathrm{d}\boldsymbol{\xi}}{\int_\Sigma J(\boldsymbol{\xi})\,\mathrm{d}\boldsymbol{\xi}}. \tag{5·11}$$

In particular, assume that P_1 and P_2 are in a plane parallel to the source and illuminated directly by it and that the medium between the source and the plane is homogeneous and of refractive index $n=1$. If we choose as origin of the position vector the point P_1, then $\mathbf{p}_1'(\boldsymbol{\xi}) = -\boldsymbol{\xi}/|\boldsymbol{\xi}|$. Moreover, it will often be permissible to

† A ray vector $\mathbf{p}$ at a point P is defined by the relation $\mathbf{p}=n\mathbf{s}$, $\mathbf{s}$ being the unit vector along the ray at P and n the value of the refractive index at that point.

replace $|\boldsymbol{\xi}|$ by the distance d between the plane of the source and the plane containing P_1 and P_2. (5·11) then reduces to

$$\gamma_{12} = \frac{\int_\Sigma J(\boldsymbol{\xi}) \exp\{ik^{(0)}\mathbf{x}_2 . \boldsymbol{\xi}/d\}\, d\boldsymbol{\xi}}{\int_\Sigma J(\boldsymbol{\xi})\, d\boldsymbol{\xi}}; \qquad (5\cdot12)$$

hence *the coherence factor is again expressed in the form of a complex amplitude in an associated Fraunhofer diffraction pattern.* This result was first obtained, for a source of circular or rectangular form, under a somewhat different definition of the γ factor by Van Cittert (1934). It is of importance in the theory of the stellar interferometer (cf. Hopkins 1951).

From (5·9) one can easily derive a simple '*propagation law*' for the coherence factor, valid (subject to the restrictions mentioned) in regions where diffraction effects are not dominant.† Consider two pairs of points $P_1(\mathbf{x}_1)$, $P_2(\mathbf{x}_2)$ and $P_1'(\mathbf{x}_1')$, $P_2'(\mathbf{x}_2')$ in the field, such that P_1' lies on the ray from O to P_1 and P_2' lies on the ray from O to P_2 (figure 4). Then it immediately follows from (5·9) that

$$\gamma(\mathbf{x}_1', \mathbf{x}_2') = \gamma(\mathbf{x}_1, \mathbf{x}_2) \exp\{ik^{(0)}[\mathscr{S}(\mathbf{x}_1, \mathbf{x}_1') - \mathscr{S}(\mathbf{x}_2, \mathbf{x}_2')]\}. \qquad (5\cdot13)$$

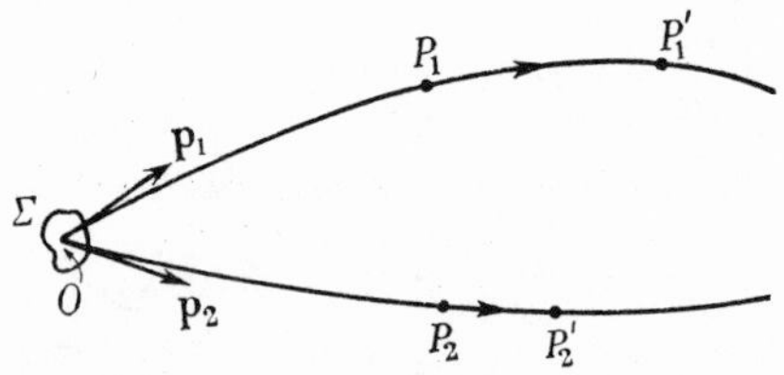

FIGURE 4. Propagation of 'coherence' in regions where geometrical optics is a valid approximation (small source and narrow frequency range assumed).

In particular, if the optical path $[P_1P_1']$ equals the optical path $[P_2P_2']$, e.g. when P_1 and P_2 are one wave-front and P_1' and P_2' on another wave-front, then $\mathscr{S}(\mathbf{x}_1, \mathbf{x}_1') = \mathscr{S}(\mathbf{x}_2, \mathbf{x}_2')$ and (5·13) reduces to

$$\gamma(\mathbf{x}_1', \mathbf{x}_2') = \gamma(\mathbf{x}_1, \mathbf{x}_2). \qquad (5\cdot14)$$

We may therefore say that *under the restrictions mentioned, the coherence is propagated in accordance with the laws of geometrical optics.*

It was shown by Hopkins (1951) that the absolute value of the coherence factor for two points in the entry pupil of an optical system and of the corresponding conjugate points in the exit pupil is the same. The result is seen to be an immediate consequence of the theorem just established.

It was also pointed out by Rogers (1953) that the degree of coherence is conserved in a beam of light from image plane to image plane in Gaussian systems. This result too is a special case of the law expressed by (5·13).

† A more general propagation law valid everywhere will be discussed in part II of this investigation.

6. The generalized Huygens principle in the special case of a small source

In this section we shall reformulate the generalized Huygens principle by making use of the relation which expresses the coherence factor in terms of the intensity function of the source, the source being assumed to be sufficiently small. Before doing this, we shall, however, generalize (2·14) by dropping the restriction that the medium between the surface of integration and the point of observation is homogeneous.

If the medium between $\mathscr{A}$ and P is heterogeneous or contains refracting or reflecting surfaces, then clearly the factor $\exp\{ik^{(0)}r_s)/r_s\}$ (with $s = 1, 2$) in (2·14) has to be replaced by the appropriate transmission function $K(\mathbf{x}_s, \mathbf{x}, \nu_0)$, giving

$$I(\mathbf{x}) = \int_{\mathscr{A}}\int_{\mathscr{A}} \sqrt{(I_1 I_2)}\, \gamma_{12} K(\mathbf{x}_1, \mathbf{x}, \nu_0)\, K^*(\mathbf{x}_2, \mathbf{x}, \nu_0)\, \Lambda_1^{(0)} \Lambda_2^{(0)*}\, d\mathbf{x}_1 d\mathbf{x}_2. \qquad (6\cdot1)$$

Assume that the conditions under which (5·9) was derived are satisfied. Then, from (5·3*a*),

$$I_s = \Gamma_{ss} = a^2(0, \mathbf{x}_s) \int_{\Sigma} J(\boldsymbol{\xi})\, d\boldsymbol{\xi} = a^2(0, \mathbf{x}_s)\, \mathscr{J}, \qquad (6\cdot2)$$

where

$$\mathscr{J} = \int_{\Sigma} J(\boldsymbol{\xi})\, d\boldsymbol{\xi} = \int_{\Sigma} d\boldsymbol{\xi} \int_{-\infty}^{+\infty} j(\boldsymbol{\xi}, \nu)\, d\nu. \qquad (6\cdot3)$$

(6·1) then becomes

$$I(\mathbf{x}) = \mathscr{J} \int_{\mathscr{A}}\int_{\mathscr{A}} \gamma_{12} a(0, \mathbf{x}_1)\, a(0, \mathbf{x}_2)\, K(\mathbf{x}_1, \mathbf{x}, \nu_0)\, K^*(\mathbf{x}_2, \mathbf{x}, \nu_0)\, \Lambda_1^{(0)} \Lambda_2^{(0)*}\, d\mathbf{x}_1 d\mathbf{x}_2. \qquad (6\cdot4)$$

Substituting from (5·9), setting

$$\frac{\int_{\Sigma} J(\boldsymbol{\xi}) \exp\{-ik^{(0)}\boldsymbol{\xi} . (\mathbf{p}_1 - \mathbf{p}_2)\}\, d\boldsymbol{\xi}}{\int_{\Sigma} J(\boldsymbol{\xi})\, d\boldsymbol{\xi}} = \sigma(\mathbf{p}_1 - \mathbf{p}_2), \qquad (6\cdot5)$$

and using (4·3) we obtain

$$I(\mathbf{x}) = \mathscr{J} \int_{\mathscr{A}}\int_{\mathscr{A}} \sigma(\mathbf{p}_1 - \mathbf{p}_2)\, K(0, \mathbf{x}_1, \nu_0)\, K^*(0, \mathbf{x}_2, \nu_0) \times K(\mathbf{x}_1, \mathbf{x}, \nu_0)\, K^*(\mathbf{x}_2, \mathbf{x}, \nu_0)\, \Lambda_1^{(0)} \Lambda_2^{(0)*}\, d\mathbf{x}_1 d\mathbf{x}_2. \qquad (6\cdot6)$$

In this formula, *the effect of the source and the transmission properties of the medium are completely separated.* The source is characterized by the factor $\sigma(\mathbf{p}_1 - \mathbf{p}_2)$ which is the normalized Fourier transform of $J(\boldsymbol{\xi})$, and the medium is characterized by the transmission function K which in practice can be calculated by methods well known in optical designing (e.g. from a ray trace). The ray vectors $\mathbf{p}_1$ and $\mathbf{p}_2$ occurring in the source factor σ and the position vectors $\mathbf{x}_1$ and $\mathbf{x}_2$ of points on the surface of integration are connected by the canonical relations (5·7). It is clear that the integration over the surface $\mathscr{A}$ may be replaced by integration over the solid angle which the entry pupil of the instrument subtends at the source. Also instead of integration *twice* over *one* surface (e.g. the exit pupil) one may reformulate (6·6) so as to involve *one* integration over each of *two* different surfaces (e.g. the entry and the exit pupils). Similar formulae were found by Hopkins (1951, 1953).

Within the accuracy of the present analysis (6·6) shows that *in a given medium, sources which have identical σ factors will give rise to the same intensity distribution.* This is probably the main reason why, in Gabor's method of reconstructed wave-fronts (Gabor 1949, 1951), one may use a different source for the reconstruction from that employed in taking the hologram. In practice, the intensity function will be often practically independent of the position of the radiating element, and it then follows that provided the geometrical shapes of the sources are the same the σ factors will be identical.

7. Concluding remarks

Our generalized Huygens principle makes it possible to obtain solutions to a variety of problems encountered in light optics, and with suitable modifications may also be applied in investigations concerned with other kinds of radiation (electron beams, X-rays, micro-waves). Some possible applications of our results to astronomical investigations and their relation to Michelson's pioneering researches on the application of interference methods to astronomy have already been briefly discussed elsewhere (Fürth & Finlay-Freundlich 1954; Wolf 1954).

It is well known that in applications to problems of image formation in optical systems with low numerical aperture, Huygens's principle gives results in excellent agreement with experiments, but it fails at higher apertures. Now (5·9) and (5·11) show that in systems with high aperture the γ factor may vary appreciably over the domain of integration. Consequently it may be expected that our generalized Huygens principle, which takes into account this variation, will have a wider range of validity than Huygens's principle in its usual form. Since in our formulation the Huygens principle involves observable quantities only, it should be possible to determine its range of validity from experiment.

It can be shown that in the limiting case as $|\gamma_{12}| \to 1$, (6·1) reduces to usual expressions for the intensity due to an ideal monochromatic point source. Thus (6·1) contains practically the whole elementary diffraction theory of image formation as a special case, and together with (4·9) leads, as we have seen, to many of the previously derived results concerning partially coherent fields. However, the limitations of our analysis must also be stressed, by summarizing the main assumptions under which these formulae were derived: The generalized Huygens principle has been established on the assumption that the usual conditions for the validity of Huygens's principle are satisfied and that the effective frequency range is sufficiently narrow; in the derivation of the expression (4·9) for the Γ factor it has been assumed that the effect of the medium is described with a sufficient accuracy by a transmission function of the form (4·3). In (6·6) it was assumed, in addition to both these conditions, that the source is sufficiently small. Some of these restrictions will be removed in part II of this investigation, where it will be shown that the coherence factors of Van Cittert, Zernike and Hopkins are special cases of a more general correlation function which rigorously obeys the wave equation and with the help of which precise propagation laws may be formulated.

In conclusion, I wish to thank Professor Max Born, F.R.S., for stimulating and helpful discussions, and Professor E. Finlay-Freundlich for valuable information concerning possible applications to astronomy. I am also indebted to Dr A. B. Bhatia and Mr G. Weeden for useful suggestions. Finally, I wish to thank Mr and Mrs R. M. Sillitto for having drawn my attention to some important aspects of Michelson's work.

Part of this work was carried out during the tenure of an Imperial Chemical Industries Research Fellowship and was also supported by a Research Grant from the Carnegie Trust for the Universities of Scotland, both of which are gratefully acknowledged.

References

Arnulf, A., Dupuy, O. & Flamant, F. 1953 *Rev. Opt.* (*théor. instrum.*), **32**, 529.
Baker, L. R. 1953 *Proc. Phys. Soc.* B, **66**, 975.
Berek, M. 1926*a* *Z. Phys.* **36**, 675.
Berek, M. 1926*b* *Z. Phys.* **36**, 824.
Berek, M. 1926*c* *Z. Phys.* **37**, 287.
Berek, M. 1926*d* *Z. Phys.* **40**, 420.
Born, M. 1933 *Optik*, §42, 132. Berlin: Springer.
Duffieux, P. 1953 *Rev. Opt.* (*théor. instrum.*), **32**, 129.
Fürth, R. & Finlay-Freundlich, E. 1954 Contribution in *Vistas in astronomy*. London: Pergamon Press.
Gabor, D. 1949 *Proc. Roy. Soc.* A, **197**, 454.
Gabor, D. 1951 *Proc. Phys. Soc.* B, **64**, 449.
Hopkins, H. H. 1951 *Proc. Roy. Soc.* A, **208**, 263.
Hopkins, H. H. 1953 *Proc. Roy. Soc.* A, **217**, 408.
Michelson, A. A. 1890 *Phil. Mag.* **30**, 1.
Michelson, A. A. 1891*a* *Phil. Mag.* **31**, 256.
Michelson, A. A. 1891*b* *Phil. Mag.* **31**, 338.
Michelson, A. A. 1892 *Phil. Mag.* **34**, 280.
Michelson, A. A. 1920 *Astrophys. J.* **51**, 257.
Michelson, A. A. & Pease, F. G. 1921 *Astrophys. J.* **53**, 249.
Pease, F. G. 1931 Contribution to *Ergebnisse der Exacten Naturwissenschaften*. Berlin: Springer, **10**, 84.
Rogers, G. L. 1953 *Nature, Lond.*, **172**, 118.
Van Cittert, P. H. 1934 *Physica*, **1**, 201.
Van Cittert, P. H. 1939 *Physica*, **6**, 1129.
Wiener, N. 1930 *Acta Math.* **55**, 182.
Wolf, E. 1954 Contribution in *Vistas in astronomy*. London: Pergamon Press.
Zernike, F. 1934*a* *Physica*, **1**, 689.
Zernike, F. 1934*b* *Mon. Not. R. Astr. Soc.* **94**, 377.
Zernike, F. 1938 *Physica*, **5**, 785.
Zernike, F. 1948 *Proc. Phys. Soc.* **61**, 158.

PAPER NO. 15

Reprinted from *Il Nuovo Cimento*, Vol. 12, pp. 884–888, 1954.

Optics in Terms of Observable Quantities.

E. Wolf

Department of Astronomy, The University, Manchester, England

(ricevuto il 26 Settembre 1954)

Summary — Space-time correlation functions are defined which express the correlation between components of the electromagnetic field vectors in stationary fields. These functions form sets of 3×3 matrices, the individual elements of which obey the wave equation. Unlike the field vectors which are not measurable at the high frequencies encountered in Optics our correlation functions may be determined with the help of standard optical instruments. The results enable a unified treatment of theories of partial coherence and partial polarization to be obtained, and suggest a formulation of a wide branch of Optics in terms of observable quantities only.

In all Optical experiments the only quantities which are observable are the averages of certain quadratic functions of the field components. It is therefore tempting to try to formulate the laws of Optical fields directly in terms of such quantities rather than in terms of the unmeasurable field vectors as has been customary in the past.

It was as early as 1852 that Stokes (1) showed that a nearly monochromatic (plane) light wave may be characterized at each point by four parameters which now bear his name (*). If

(1) $E_x = a_1(\boldsymbol{x}, t) \cos \{2\pi\nu_0 t - \alpha_1(\boldsymbol{x}, t)\}\,, \qquad E_y = a_2(\boldsymbol{x}, t) \cos \{2\pi\nu_0 t - \alpha_2(\boldsymbol{x}, t)\},$

(1) G. G. Stokes: *Trans. Camb. Phil. Soc.*, **9**, 399 (1852). Also his *Mathematical and Physical Papers* (Cambridge, 1901), vol. III, p. 233.

(*) Very good accounts of the Stoke's parameters may be found in Chandrasekhar (2) and Walker (3).

are the components of the electric vector of such a wave in two mutually orthogonal directions at right angles to the direction of propagation, the Stokes parameters are defined by

$$(2)\qquad \begin{cases} P=\langle a_1^2+a_2^2\rangle\,, & Q=\langle a_1^2-a_2^2\rangle\,,\\ U=\langle 2a_1a_2\cos(\alpha_1-\alpha_2)\rangle\,, & V=\langle 2a_1a_2\sin(\alpha_1-\alpha_2)\rangle\,,\end{cases}$$

the brackets $\langle....\rangle$ denoting time average. Then the intensity $I(\psi, \varepsilon)$ associated with vibrations in the direction which makes an angle ψ with the x-direction, when retardation ε is introduced between the two components, is given by (see CHANDRASEKHAR [2], p. 29)

$$(3)\qquad I(\psi,\varepsilon)=\tfrac{1}{2}[P+Q\cos 2\psi+(U\cos\varepsilon-V\sin\varepsilon)\sin 2\psi]\,.$$

By measuring I for different values of ψ and ε, the four parameters may be determined.

Let us now introduce in place of E_x and E_y, the associated complex vectors

$$(4)\qquad \hat{E}_x=a_1(\boldsymbol{x},t)\exp\{i[2\pi\nu_0 t-\alpha_1(\boldsymbol{x},t)]\},\quad \hat{E}_y=a_2(\boldsymbol{x},t)\exp\{i[2\pi\nu_0 t-\alpha_2(\boldsymbol{x},t)]\}.$$

Next we construct the four functions

$$(5)\qquad \mathcal{E}_{ij}=\langle\hat{E}_i(\boldsymbol{x},t)\hat{E}_j^*(\boldsymbol{x},t)\rangle,$$

where i and j can each take on the value x or y and asterisk denotes the complex conjugate. The knowledge of these four quantities is equivalent to the knowledge of the four Stokes parameters; in fact the Stokes parameters are simple linear combinations of the $\mathcal{E}_{ij}$'s. If (3) is expressed in terms of these quantities, one finds after a simple calculation that

$$(6)\qquad I(\psi,\varepsilon)=I_x(\psi)+I_y(\psi)+2\sqrt{I_x(\psi)}\sqrt{I_y(\psi)}\,|\gamma_{xy}|\cos[\arg\gamma_{xy}+\varepsilon]\,,$$

where

$$(7)\qquad I_x(\psi)=\mathcal{E}_{xx}\cos^2\psi\,,\qquad I_y(\psi)=\mathcal{E}_{yy}\sin^2\psi\,,$$

and

$$(8)\qquad \gamma_{xy}=\frac{\mathcal{E}_{xy}}{\sqrt{\mathcal{E}_{xx}}\sqrt{\mathcal{E}_{yy}}}\;;$$

---

(2) S. CHANDRASEKHAR: *Radiative Transfer* (Oxford, 1950).

(3) M. J. WALKER: *Amer. Journ. Phys.*, **22**, 170 (1954).

and the well-known inequality $P^2 \geqslant Q^2 + U^2 + V^2$ becomes simply $|\gamma_{xy}| \leqslant 1$. Equation (6) is formally identical with the *generalized interference law* derived in recent years in the theory of partially coherent scalar fields (ZERNIKE ([4]), HOPKINS ([5]), WOLF ([6])). In the special case when $\gamma = 0$, (6) reduces to the usual law for the combination of completely incoherent fields; when $|\gamma| = 1$ it reduces to the ordinary interference law for fields which are perfectly coherent. Partially coherent and partially polarized fields are characterized by the intermediate values of $|\gamma|$.

Now the Stokes parameters (or the 2 by 2 matrix whose elements are defined by (5)) give the intensity and express the correlation between the components of $\boldsymbol{E}$ at the *same* point in space and at the *same* instant of time. Moreover they are defined only for a plane wave whose effective frequency range is sufficiently narrow. In order to characterize a general stationary field (*), we introduce, by analogy with the scalar case (see WOLF ([7])), correlation functions between field components at different points in space and at different instants of time.

Let

$$E_i(\boldsymbol{x}, t) = \int_0^\infty a_{\nu i}(\boldsymbol{x}) \cos \{2\pi\nu t - \alpha_{\nu i}(\boldsymbol{x})\} \, d\nu \, , \qquad (i = x, \, y \text{ or } z) \tag{9}$$

be the Fourier representation of a typical field component over the time interval $-T \leqslant t \leqslant T$, $\boldsymbol{E}$ being formally assumed to be zero outside this range, and define the functions

$$\hat{E}_i(\boldsymbol{x}, t) = \int_0^\infty a_{\nu i}(\boldsymbol{x}) \exp \{i[2\pi\nu t - \alpha_{\nu i}(\boldsymbol{x})\} \, d\nu \, . \tag{10}$$

We now introduce a 3×3 correlation matrix $\mathcal{E}$ whose elements are

$$\mathcal{E}_{ij}(\boldsymbol{x}_1, \boldsymbol{x}_2, \tau) = \langle \hat{E}_i(\boldsymbol{x}_1, t+\tau) \hat{E}_j^*(\boldsymbol{x}_2, t) \rangle \, . \tag{11}$$

Similar considerations to those employed in connection with scalar fields of

---

([4]) F. ZERNIKE: *Physica*, **5**, 785 (1938).

([5]) H. H. HOPKINS: *Proc. Roy. Soc.*, A **208**, 263 (1951).

([6]) E. WOLF: *Proc. Roy. Soc.*, A **225**, 96 (1954).

(*) By a stationary field we mean here a field of which all *observable* properties are constant in time. This includes as a special case the usual case of high frequency periodic time-dependence; or the field constituted by the steady flux of polychromatic radiation through an optical system. But it excludes fields for which the time average over a macroscopic time interval of the flux of radiation depends on time.

([7]) E. WOLF: *Proc. Roy. Soc.*, in press.

arbitrary frequency range and with vector fields characterized by the Stokes parameters indicate, that the expressions for the electric energy density appropriate to various experimental conditions are simple functions of the elements of the $\mathcal{E}$-matrix; and moreover that all these elements may be determined from experiments by means of standard optical instruments.

Since the electric vector satisfies the wave equation, it can readily be shown that each element of the $\mathcal{E}$-matrix satisfies two equations

$$\left\{\begin{aligned} \nabla_1^2\mathcal{E}_{ij} &= \frac{1}{c^2}\frac{\partial^2\mathcal{E}_{ij}}{\partial\tau^2}\,, \\ \nabla_2^2\mathcal{E}_{ij} &= \frac{1}{c^2}\frac{\partial^2\mathcal{E}_{ij}}{\partial\tau^2}\,, \end{aligned}\right. \tag{12}$$

where ∇_1^2 and ∇_2^2 are the Laplacian operators with respect to the coordinates of $\boldsymbol{x}_1$ and $\boldsymbol{x}_2$, and c is the velocity of light in the vacuum. Thus not only the unmesurable field vectors, but also *the observable correlation functions* here introduced *obey rigorous propagation laws*. This result should prove particularly useful in connection with scattering problems.

In addition to the $\mathcal{E}$-matrix, one can introduce similar matrices involving components of the other field vectors ($\boldsymbol{H}$, $\boldsymbol{D}$ and $\boldsymbol{B}$) and also matrices involving mixed pairs like E_i and H_j. On account of Maxwell's equations, these matrices are related by a set of first order partial differential equations with respect to the variables $\boldsymbol{x}_1$, $\boldsymbol{x}_2$, and τ. In the analysis of all optical experiments $c\tau$ will play the part of an optical path difference. The actual time, like the frequency has been eliminated.

Unlike the $\mathcal{E}$-matrix, the $\mathcal{H}$ matrix is not likely to be of any interest in Optics, since no radiation detectors appear to be available at Optical wavelengths which would respond to the magnetic rather than the electric field. It may, however, prove useful in connection with applications to other types of partially coherent radiation, e.g. in Radio Astronomy. The matrix containing the mixed pairs should prove useful in experiments where the (averaged) flux of energy rather than the energy density is measured.

The matrices here introduced may be expected to play a role in Electromagnetic field theory which is somewhat analogous to that which the Density matrix of von NEUMAN ([8]) plays in Quantum Mechanics. An analogy between the Stokes parameters and the Density Matrix has been noted previously (PERRIN ([9]), FALKOFF and MACDONALD ([10]); see also FANO ([11])). It is, how

---

([8]) J. v. NEUMANN: *Gött. Nachr.*, 245 (1927).
([9]) F. PERRIN, *Journ. Chem. Phys.*, **10**, 415 (1942).
([10]) D. L. FALKOFF and J. E. MACDONALD: *Journ. Opt. Soc. Amer.*, **41**, 861 (1951).
([11]) U. FANO: *Phys. Rev.*, **93**, 121 (1954).

ever, evident that only by considering more general correlation functions, such as those here introduced, does one obtain an adequate tool for the study of propagation problems in a general stationary electromagnetic field.

A fuller discussion of the subject matter of this note will be published at a later date.

This work was carried out during the tenure of an Imperial Chemical Industries Research Fellowship and was also supported by a grant from the Carnegie Trust for the Universities of Scotland, both of which are gratefully acknowledged.

RIASSUNTO (*)

Si definiscono funzioni di correlazione spazio-tempo che esprimono la correlazione fra componenti dei vettori del campo elettromagnetico in campi stazionari. Queste funzioni formano gruppi di 3×3 matrici i cui elementi individuali soddisfano all'equazione d'onda. A differenza dei vettori di campo che, alle alte frequenze che intervengono in Ottica, non sono misurabili, le nostre funzioni di correlazione possono essere determinate con l'ausilio degli ordinari apparecchi ottici. I risultati consentono un trattamento unificato delle teorie della coerenza parziale e della polarizzazione parziale e suggeriscono una formulazione di un ampio settore dell'Ottica in termini di sole grandezze osservabili.

(*) *Traduzione a cura della Redazione.*

PAPER NO. 16

Reprinted from *Revue d'Optique*, Vol. 34, pp. 1–21, 1955.

LA NOTION DE COHÉRENCE EN OPTIQUE

par André BLANC-LAPIERRE et Pierre DUMONTET

Laboratoire de Physique théorique. Faculté des Sciences d'Alger

SOMMAIRE. — *L'effet de la diffraction sur la correspondance objet-image a été surtout étudié dans le cas d'un* objet cohérent *ou d'un* objet incohérent. *Récemment, certains auteurs ont considéré le cas d'une* cohérence partielle. *L'étude de la cohérence n'étant qu'une application de la* théorie de la corrélation, *il nous a paru utile de la développer systématiquement en nous plaçant dans le cadre de la* théorie des fonctions aléatoires. *Afin de préciser les notions dans un cadre général,* nous n'avons pas, dans les premiers chapitres, fait l'hypothèse du monochromatisme.

Nous avons consacré un chapitre aux sources approximativement monochromatiques. *Nous étudions ensuite la* relation entre la cohérence d'une source qui joue le rôle d'objet et celle de son image dans un dispositif optique. *Nous faisons cette étude dans le cadre du monochromatisme ; la généralisation à une radiation stationnaire quelconque ne présente pas de difficulté.*

Enfin, il nous a paru utile de dire quelques mots sur les propriétés statistiques de la variable lumineuse.

SUMMARY. — *The effect of diffraction on the relation object-image has been studied chiefly in the case of* coherent object *or of* incoherent object. *Recently, some authors have considered the case of* partial coherence. *The study of coherence being only an application of* correlation theory, *we thought useful to develop it systematically in the outline of the* random functions theory. *In the first chapters, we have not supposed monochromatic light, to extend the informations to the general case.*

One chapter is devoted to nearly monochromatic sources. *Then we study the* relation between the coherence of a source acting as object, and the coherence of its image given by an optical design. *This study is done with monochromatic light ; it is not difficult then to generalize to any stationary radiation.*

Finally, it seemed useful to say few words about the statistical properties of the luminous variable.

1. **Introduction.** — On a tendance, en optique physique, lorsqu'il s'agit de calculer la répartition d'intensité dans une image, à se placer dans l'un ou l'autre des deux cas extrêmes suivants : *l'objet est cohérent* ou *l'objet est incohérent.* Or, dans les exemples courants, on se trouve très souvent dans des cas intermédiaires, d'où la nécessité de définir avec précision une *cohérence partielle* afin de pouvoir étudier la correspondance objet-image sans se limiter aux cas de la cohérence ou de l'incohérence. C'est une question très importante et de nombreux auteurs, parmi lesquels on peut citer P. H. van Cittert [1], F. Zernike [2], H. H. Hop-

kins [3, 4], P. M. Duffieux [5, 6], E. Wolf [7], ont défini cette notion de cohérence partielle et analysé les propriétés qui lui sont rattachées. En général, leurs études sont limitées au cas d'une radiation approximativement monochromatique. Par ailleurs, il nous paraît évident que *la notion de cohérence partielle n'est autre chose que l'application à des problèmes d'optique de la théorie générale de la corrélation.* Son véritable cadre mathématique est celui de la théorie des *fonctions aléatoires.* C'est dans cet esprit que nous avons écrit cet article [1]. Nous pensons y avoir montré que les résultats généraux de la théorie des fonctions aléatoires permettent une étude complète et systématique des questions relatives à la cohérence. *Afin de bien préciser les notions dans un cadre général nous n'avons pas, dans les paragraphes* 2, 3 *et* 4, *fait l'hypothèse du monochromatisme.* Le paragraphe 5 est consacré aux *sources approximativement monochromatiques.* Dans le paragraphe 6, nous étudions *la relation entre la cohérence d'une source, qui joue le rôle d'objet, et celle de son image dans un dispositif optique.* Nous faisons cette étude dans le cadre du monochromatisme, la généralisation à une radiation stationnaire quelconque ne présentant pas de difficulté. A propos de l'étude de cette correspondance, nous discutons sommairement la validité de l'approximation qui consiste à traduire les propriétés d'un système optique par un *filtre linéaire* et nous étudions la transformation de la cohérence dans un filtre linéaire. Enfin, il nous a paru utile de dire quelques mots sur les propriétés statistiques de la variable lumineuse et sur les raisons générales qui conduisent à penser qu'elle obéit à une *loi de Laplace-Gauss* : c'est l'objet du paragraphe 7.

Naturellement, comme le lecteur s'en rendra compte, beaucoup de résultats contenus dans ce mémoire ont déjà été indiqués par les auteurs cités [2].

2. **Généralités.** — *Les notions de cohérence et d'incohérence sont inséparables de concepts de nature statistique.* Réfléchissons à une expérience d'interférences. Soient deux sources S_1 et S_2 fournissant respectivement deux vibrations $x_1(t)$ et $x_2(t)$. En un certain point, superposons $x_1(t)$ et $x_2(t-\tau)$ et étudions l'intensité résultante $I(t, \tau)$

$$I(t, \tau) = [x_1(t) + x_2(t-\tau)]^2. \tag{1}$$

En réalité, ce que nous percevons sera une certaine moyenne de $I(t, \tau)$ prise pendant un temps dont l'ordre de grandeur est égal à la constante de temps λ du dispositif d'observation. Par exemple, et pour fixer les idées, nous pourrons percevoir

$$\mathfrak{I}(t, \tau) = \frac{1}{\lambda}\int_{-\infty}^{t} I(\theta, \tau)\, \mathbf{e}^{\frac{-(t-\theta)}{\lambda}}\, d\theta. \tag{2}$$

(1) Pour les résultats de la théorie des fonctions aléatoires que nous utilisons, nous prions le lecteur de se reporter à l'ouvrage « Théorie des fonctions aléatoires » par A. Blanc-Lapierre & R. Fortet (Masson éditeur, Paris 1953). Les renvois à cet ouvrage sont indiqués par l'abréviation F. A.

(2) *Note ajoutée à la correction des épreuves.* Depuis que cet article a été écrit nous avons eu connaissance du très intéressant article de E. WOLF, A macroscopic theory of interference and diffraction of light from finite sources. I. Fields with a narrow spectral range, *Proc. Royal Soc.* [A], t. **225**, 1954, p. 96. Le lecteur notera les liens évidents entre certains résultats donnés ici et d'autres contenus dans cet article.

A cause des fluctuations, nous ne pouvons préciser complètement, à priori, les fonctions $x_1(t)$ et $x_2(t)$. Les phénomènes sont irreproductibles à *l'échelle microscopique.* Nous pouvons seulement définir et imposer des valeurs de *nature macroscopique* correspondant à des grandeurs moyennes ; $x_1(t)$, $x_2(t)$ et $\mathfrak{J}(t,\tau)$ *sont des fonctions aléatoires. Nous imaginons un très grand nombre d'expériences macroscopiquement identiques* : chacune d'elles correspond à ce que, dans la théorie des fonctions aléatoires, on appelle une *épreuve* $\mathcal{E}$. $x_1(t)$ ne dépend pas seulement du temps t, mais aussi de l'épreuve $\mathcal{E}$. Nous pourrions écrire, pour marquer ce fait, $x_1 = x_1(t, \mathcal{E})$. Si on fixe $\mathcal{E}$, x_1 est une fonction du temps au sens de l'analyse. Pour un instant t donné, x_1 dépend de l'épreuve $\mathcal{E}$. On suppose qu'une probabilité est définie sur l'ensemble des épreuves $\mathcal{E}$. Si on fixe t, x_1 est une *variable aléatoire.*

Nous supposerons que les sources considérées sont *macroscopiquement constantes* dans le temps ; les propriétés statistiques des fonctions aléatoires introduites seront invariantes pour tout changement de l'origine des temps. Nous avons affaire à des *fonctions aléatoires stationnaires* (F. A. p. 15, p. 111-114, p. 443-470).

La description macroscopique de notre phénomène d'interférences se limite à l'aspect moyen du phénomène. Elle se contentera de nous décrire le comportement de l'espérance mathématique $E\{\mathfrak{J}(t,\tau)\}$ qui, d'ailleurs, à cause du caractère stationnaire, ne dépendra que de τ. La figure d'interférence sera décrite par la fonction

$$J(\tau) = E\{\mathfrak{J}(t,\tau)\} = E\{x_1^2\} + E\{x_2^2\} + 2\,\gamma(\tau) \tag{3}$$

en posant

$$\gamma(\tau) = E\{x_1(t)\,x_2(t-\tau)\} \tag{4}$$

(γ ne dépend que de τ à cause du caractère stationnaire).

Les fonctions aléatoires stationnaires introduites dans les applications physiques présentent d'ailleurs toujours un *ergodisme* suffisant pour permettre de *remplacer les espérances mathématiques par des moyennes temporelles* (F. A. p. 465-470).

$$\begin{aligned}(5)\quad J(\tau) = \lim_{T\to\infty} \frac{1}{T}\int_0^T x_1^2(t)\,dt + \lim_{T\to\infty}\frac{1}{T}\int_0^T x_2^2(t)\,dt\ + \\ + \ 2\lim_{T\to\infty}\frac{1}{T}\int_0^T x_1(t)\,x_2(t-\tau)\,dt,\end{aligned}$$

où les intégrales sont calculées sur une épreuve quelconque $\mathcal{E}$.

Cet exemple simple montre que le cadre mathématique normal pour l'étude des problèmes de cohérence, qui sont intimement liés à des questions d'interférences, est celui de la théorie des fonctions aléatoires. C'est ce point de vue que nous allons développer systématiquement. *L'étude de la cohérence n'est qu'un cas particulier de l'étude de la corrélation.*

3. **Définition d'une source.** — Pour bien situer les notions à utiliser, plaçons-nous volontairement dans un cadre général : *nous ne supposerons pas nos sources*

monochromatiques. Par contre, *nous conserverons l'hypothèse de stationnarité dans le temps.* Nous donnerons des définitions valables pour des *sources primaires* (corps noir, gaz excité,...) ou pour des *sources secondaires* (objets ou surfaces éclairés par des sources primaires et jouant ou pouvant jouer le rôle de sources).

1. ***Définition des sources élémentaires.*** — Soit une source plane $(\alpha \times \beta)$. A un *petit* élément $d\alpha\, d\beta$ d'aire $d\varphi = d\alpha\, d\beta$ entourant un point M (fig. 1), associons la vibration élémentaire

$$ds\,[\alpha, \beta, t] = ds\,[M, t]. \tag{6}$$

ds est une valeur réelle, fonction aléatoire de l'élément $dM = d\alpha\, d\beta$ et du temps t. Il est possible qu'il existe une dérivée aléatoire $\sigma\,[M, t]$ telle que

$$ds\,[\alpha, \beta, t] \sim \sigma\,[\alpha, \beta, t]\, d\alpha\, d\beta = \sigma\,[M, t]\, dM. \tag{7}$$

L'existence de σ n'est pas obligatoire, En particulier, σ n'existe pas pour une source incohérente. Cependant nous admettrons, dans la suite, que σ existe et cela dans le seul but de ne pas alourdir les expressions mathématiques utilisées ; le cas d'incohérence pourra toujours être réintroduit dans nos calculs comme aspect limite, en utilisant la fonction δ (impulsion unité de Dirac).

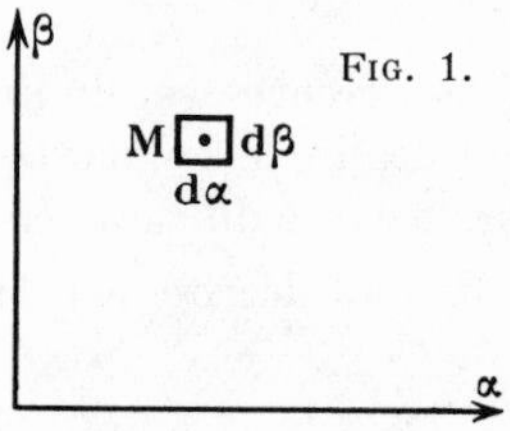

FIG. 1.

Les composantes constantes n'ont aucun intérêt en optique ; nous supposons donc que les fonctions aléatoires introduites ont une espérance mathématique identiquement nulle.

2. ***Analyse harmonique.*** — Il est utile d'introduire des représentations harmoniques de $\sigma\,[M, t]$ (F. A. p. 343-373).

a) *En faisant jouer à t un rôle privilégié,* on écrira

$$\sigma\,[M, t] = \int_{-\infty}^{+\infty} e^{2\pi i \nu t}\, d\Sigma_\nu\,[M, \nu] \quad , \tag{8}$$

où $\Sigma\,[M, \nu]$ est une fonction aléatoire de M et de ν et $d\Sigma_\nu$ son accroissement sur un intervalle $\Delta\nu$ de ν (M restant constant).

b) *Si on veut traiter en bloc* M *et* ν, on écrira

$$\sigma\,[M, t] = \int_{(\Omega)} \int_{(\nu)} e^{2\pi i(\nu t + \overrightarrow{\Omega}\,\overrightarrow{M})}\, d\Sigma'_{\Omega,\nu}\,[\Omega, \nu] \quad , \tag{9}$$

où $\Sigma'\,[\Omega, \nu]$ est une fonction aléatoire de $\overrightarrow{\Omega}$ et de ν définie par la somme de ses accroissements $d\Sigma'_{\Omega,\nu}\,[\Omega, \nu]$ sur les domaines élémentaires $d\,\Omega \times d\nu$.

$\overrightarrow{\Omega}$ est une *fréquence spatiale* (à deux dimensions Ω_1 et Ω_2 ; $d\Omega = d\Omega_1\, d\Omega_2$) ν est la *fréquence habituelle* (*couleur*).

σ étant réel, les accroissements $d\Sigma_\nu$ (*ou* $d\Sigma'_{\Omega,\nu}$) *correspondant à deux domaines en* ν (*ou en* $\Omega \times \nu$) *symétriques par rapport à l'origine, seront imaginaires conjugués.*

Examinons maintenant cette source à travers un dispositif optique quelconque D_1, *invariable dans le temps.* Ce *dispositif sera linéaire.* Pour connaître l'intensité à un instant et en un point quelconques dans l'espace-image lié à D_1, il suffit donc de connaître, pour deux points M_1 et M_2 et deux instants t_1 et t_2 *quelconques*, la valeur de

$$(10) \qquad E\left\{ \sigma[M_1, t_1]\, \sigma^*[M_2, t_2] \right\} = \Gamma[M_1, M_2, t_1, t_2] \text{ (}^3\text{)}$$

Ainsi. Γ permet le calcul de toutes les intensités. Γ *renferme tout ce que l'on peut dire au sujet de la cohérence.*

3. ***Propriétés diverses.*** — Les fonctions aléatoires de t correspondant aux divers points M sont *stationnaires en t dans leur ensemble. Si on se limite aux moyennes du second ordre*, cela se traduit par l'identité suivante valable pour M_1, M_2, t_1, t_2 et h quelconques :

$$(11) \qquad \Gamma[M_1, M_2, t_1, t_2] = \Gamma[M_1, M_2, t_1 + h, t_2 + h].$$

Introduisons $t_1 - t_2 = \tau$, la fonction Γ ne dépendra du temps que par τ et nous la noterons $\Gamma[M_1, M_2, \tau]$.

Si $M_1 = M_2$, la fonction $\Gamma[M_1, M_2, \tau]$ (symétrique en τ) est ce qu'on appelle la *fonction d'autocorrélation* de $\sigma[M_1, t]$; si $M_1 \neq M_2$, $\Gamma[M_1, M_2, \tau]$ est la *fonction de corrélation* de $\sigma[M_1, t_1]$ et de $\sigma[M_2, t_2]$.

Des résultats classiques dans la théorie des fonctions aléatoires stationnaires (voir F. A. p. 450-470) *nous permettent d'énoncer quelques propriétés utiles* :

A) La condition (11) est équivalente à la suivante

$$(12) \quad E\left\{ d\Sigma_\nu[M_1, \nu_1]\, d\Sigma^*_\nu[M_2, \nu_2] \right\} \equiv 0, \text{ pour } \nu_1 \neq \nu_2, M_1 \text{ et } M_2 \text{ quelconques.}$$

B) Si les conditions équivalentes (11) et (12) sont remplies, il existe une fonction $F[M_1, M_2, \nu]$ telle que

$$(13) \qquad \Gamma[M_1, M_2, \tau] = \int_{-\infty}^{+\infty} e^{2\pi i \nu \tau}\, dF_\nu[M_1, M_2, \nu],$$

l'accroissement $dF_\nu[M_1, M_2, \nu]$ de F sur l'intervalle ν, $\nu + d\nu$ étant défini par

$$(14) \qquad dF_\nu[M_1, M_2, \nu] = E\left\{ d\Sigma_\nu[M_1, \nu]\, d\Sigma^*_\nu[M_2, \nu] \right\}.$$

C) *Les fonctions* $F[M_1, M_2, \nu]$ *possèdent un ensemble de propriétés remarquables* :

a) *Si* $M_1 = M_2$ *les accroissements* dF_ν *sont tous réels et non négatifs* ; la fonction $F[M_1, M_1, \nu]$ est alors la *fonction de répartition spectrale de la puissance moyenne* $E\left\{ \sigma^2[M_1, t] \right\}$ *associée à* $\sigma[M_1, t]$.

L'accroissement

$$(15) \qquad \Delta F[M_1, M_1, a, b] = F[M_1, M_1, b] - F[M_1, M_1, a] \qquad (b \geqslant a)$$

représente la contribution de la bande de fréquences (a, b) à $E\left\{ \sigma^2[M_1, t] \right\}$. La

(3) σ est réel, mais nous introduisons son imaginaire conjugué σ* pour donner à nos formules l'aspect qu'elles conserveront lorsque nous utiliserons des valeurs complexes.

puissance moyenne $E\{\sigma^2[M_1, t]\}$ (indépendante de t à cause du caractère stationnaire) est reliée à F par la relation

$$(16) \qquad E\{\sigma^2[M_1, t]\} = \int_{-\infty}^{+\infty} dF_\nu[M_1, M_1, \nu].$$

$F[M_1, M_1, \nu]$ définit *le spectre* de $\sigma[M_1, t]$. Il est possible que $F[M_1, M_1, \nu]$ admette une dérivée $f[M_1, \nu]$ (naturellement jamais négative). Ce sera la densité spectrale de $\sigma[M_1, t]$. *La densité spectrale et la fonction d'autocorrélation associées à* $\sigma[M, t]$ *sont transformées de Fourier l'une de l'autre.*

b) *Si* $M \neq M_2$, *on peut seulement affirmer que la fonction* $F[M_1, M_2, \nu]$ *est à variation totale bornée sur* $-\infty < \nu < +\infty$.

c) *Les fonctions* $F[M_1, M_2, \nu]$ *considérées comme des fonctions de* ν *dépendant des paramètres* M_1 *et* M_2 *ne sont pas indépendantes.* En particulier, si $\Delta F[M_1, M_2, b, a]$ est la *variation totale* de F sur un intervalle a, b ($a < b$) quelconque (si $M_1 = M_2$, la variation et la variation totale se confondent), on peut établir que l'on a

$$(17) \qquad |\Delta F[M_1, M_2, b, a]|^2 \leqslant \Delta F[M_1, M_1, b, a]\, \Delta F[M_2, M_2, b, a].$$

*4. **Cas particulier des sources spatialement uniformes.**— Alors, la fonction* $\sigma[M, t]$ *qui, jusqu'ici, était seulement stationnaire en* t *est maintenant stationnaire en* t *et en* M. On doit avoir, pour M_1, M_2, t_1, t_2, M et h quelconques,

$$(18) \qquad \Gamma[M_1, M_2, t_1, t_2] = \Gamma[M_1 + M, M_2 + M, t_1 + h, t_2 + h].$$

On peut remplacer (18) par une autre condition équivalente faisant intervenir les composantes harmoniques de σ. Cette condition s'énonce ainsi : *sauf si on a simultanément* $\Omega_1 = \Omega_2$ *et* $\nu_1 = \nu_2$, *on a*

$$(19) \qquad E\{d\Sigma'_{\Omega,\nu}[\Omega_1, \nu_1]\; d\Sigma'^{*}_{\Omega,\nu}[\Omega_2, \nu_2]\} \equiv 0.$$

Posons alors

$$(20) \qquad M = M_1 - M_2 \quad \text{et} \quad dF'_{\Omega,\nu}[\Omega, \nu] = E\{|d\Sigma'_{\Omega,\nu}[\Omega, \nu]|^2\}.$$

On a

$$\Gamma[M_1, M_2, t_1, t_2] = \Gamma[M, \tau]$$

$$(21) \qquad = \iint_{\Omega \times \nu} e^{2\pi i[\nu\tau + \vec{\Omega}.\vec{M}]}\, dF'_{\Omega,\nu}[\Omega, \nu].$$

La fonction $F'[\Omega, \nu]$ ainsi introduite est *à accroissements non négatifs.* C'est la *fonction de répartition spectrale de la puissance moyenne* $E\{\sigma^2[M, t]\}$ *de* $\sigma[M, t]$ *considérée comme une fonction simultanée de* M *et de* t.

4. **Cohérence. Cohérence partielle. Incohérence.** — ***1. Définition de la cohérence.*** — La corrélation entre les différentes sources élémentaires est définie par $\Gamma[M_1, M_2, \tau]$. Les accroissements $d\Sigma_\nu[M, \nu]$ satisfont à la relation de non corrélation (12). Les équations (13) et (14) introduisent les fonctions spectrales $F[M_1, M_2, \nu]$.

A) *Soient deux points* M_1 *et* M_2 *particuliers* ; intéressons-nous aux sources $\sigma[M_1, t]\, dM_1$ et $\sigma[M_2, t]\, dM_2$ associées aux *petits* éléments dM_1 et dM_2 entourant M_1 et M_2. *Quand dirons-nous que ces deux sources sont cohérentes ?*

Pour répondre à cette question, comparons les figures d'interférences obtenues en faisant interférer

a) $\sigma\,[M_1, t]\,dM_1$ *avec lui-même* (*cas* I),
b) $\sigma\,[M_2, t]\,dM_2$ *avec lui-même* (*cas* II),
c) $\sigma\,[M_1, t]\,dM_1$ *avec* $\sigma\,[M_2, t]\,dM_2$ (*cas* III).

Le *schéma mathématique* de nos expériences est le suivant. A chaque point d'un domaine (ρ) de l'espace, nous associons une certaine *différence de marche l.* Organisons nos interférences dans un *milieu non dispersif* de sorte qu'à une valeur donnée de l correspondra un *retard* τ bien défini. Nous admettrons que des miroirs peuvent exister dans notre dispositif interférentiel ce qui pourrait remplacer σ par $-\sigma$. Soit $J(\tau)$ l'espérance mathématique $E\{\mathfrak{J}\} = E\{I\}$ [voir (1) et (2)] au point de retard τ. *Si nous choisissons des éléments* dM_1 *et* dM_2 *d'égales extensions*, la figure d'interférence sera caractérisée, au facteur dM_1^2 près, de la façon suivante :

Cas I (22 a) $J_{\text{I}}[\tau] = 2[\Gamma_{11}[0] \pm \Gamma_{11}[\tau]]$,
Cas II (22 b) $J_{\text{II}}[\tau] = 2[\Gamma_{22}[0] \pm \Gamma_{22}[\tau]]$,
Cas III (22 c) $J_{\text{III}}[\tau] = [\Gamma_{11}[0] + \Gamma_{22}[0] \pm 2\Gamma_{12}[\tau]]$,

où

(22 d) $\Gamma_{ij} = \Gamma\,[M_i, M_j, \tau]$.

Les signes $\pm$ correspondent au fait que, sur chacun des deux faisceaux qui interfèrent, on peut prendre σ ou $-\sigma$.

La figure d'interférences relative au signe + est complémentaire de celle qui correspond au signe — ; si nous ne voulons pas distinguer ces deux figures complémentaires et si, de plus, nous ne tenons pas à distinguer entre $J[\tau]$ et $K\,J[\tau]$ (K réel) qui ne diffèrent que par l'intensité totale, les figures d'interférences associées aux trois cas considérés seront respectivement caractérisées par

$$(23)\quad H_{\text{I}}[\tau] = \frac{|\Gamma_{11}[\tau]|}{\Gamma_{11}[0]}\ ;\ H_{\text{II}}[\tau] = \frac{|\Gamma_{22}[\tau]|}{\Gamma_{22}[0]}\ ;\ H_{\text{III}}[\tau] = 2\,\frac{|\Gamma_{12}[\tau]|}{\Gamma_{11}[0] + \Gamma_{22}[0]}.$$

Posons-nous alors la question suivante : *Nos trois figures d'interférences peuvent-elles être identiques, c'est-à-dire, peut-on avoir*

$$(24)\quad H_{\text{I}}[\tau] = H_{\text{II}}[\tau] = H_{\text{III}}[\tau + \tau_0]\ ?$$

Nous avons introduit une constante τ_0 dans H_{III} ; en effet, dans I et II, $\tau = 0$ définit un élément de symétrie ; cela n'est plus certain dans III. La condition (24) exprime alors que la figure d'interférences relative au cas III peut être rendue identique à l'une et l'autre des figures correspondant aux cas I et II par une translation τ_0 convenable.

$H[\tau]$ peut être interprété comme *facteur de contraste* ; l'intensité moyenne dans (P) est en effet égale à $J[\infty]$ et on a

$$(25)\quad H[\tau] = \frac{|J[\tau] - J[\infty]|}{J[\infty]}.$$

De l'inégalité de Schwarz et aussi des propriétés relatives aux moyennes arithmétiques et géométriques, on déduit

$$(26)\qquad \Gamma_{12}\,[\tau] \leqslant \sqrt{\Gamma_{11}\,[0]\,\Gamma_{22}\,[0]} \leqslant \frac{1}{2}\,[\Gamma_{11}\,[0] + \Gamma_{22}\,[0]]\,.$$

Il en résulte que l'on a toujours $0 \leqslant H\,[\tau] \leqslant 1$. Dans les cas I et II, la valeur 1 est effectivement atteinte pour $\tau = 0$. Dans III, la valeur 1 n'est pas nécessairement atteinte.

Si (24) *est vérifié, il existe au moins une valeur* τ_0 *pour laquelle on a* $H_{\text{III}} = 1$.

Cela suffit pour que (24) soit vérifié pour tout τ. En effet $H_{\text{III}}\,[\tau_0] = 1$ implique que, pour $\tau = \tau_0$, (26) se réduise à deux égalités. Cela entraîne

a) $\Gamma_{11}\,[0] = \Gamma_{22}\,[0]$, *les deux sources doivent donc avoir la même puissance* ;

b) (27) $\quad | E \{ \sigma[M_1, t]\ \sigma^*\,[M_2, t - \tau_0] \} |^2 = E \{ \sigma^2\,[M_1, t] \} E \{ \sigma^2(M_2, t) \}$,

d'où l'existence d'un nombre certain, positif ou négatif k, *tel que l'on ait presque sûrement*

$$(28)\qquad \sigma\,[M_1, t] = \mathrm{k}\,\sigma\,[M_2, t - \tau_0]\ .$$

Les deux sources devant avoir la même puissance, on aura $\mathrm{k} = \pm 1$. (24) est alors évidemment vérifié pour tout τ. Nous venons d'établir les résultats suivants :

1° *Si le facteur de contraste relatif à la figure* III *atteint la valeur* 1, *cela suffit à assurer l'identité des trois figures d'interférences.*

2° *Ceci se produira si on a, presque sûrement,*

$$(29)\qquad \sigma\,[M_1, t] = \pm\,\sigma\,[M_2, t - \tau_0],$$

où τ_0 *est une valeur non aléatoire.*

Remarque. — Des raisonnements analogues au précédent peuvent fournir *de façon rigoureuse* certains résultats sur l'aspect des figures d'interférences, que des considérations intuitives rendent naturels.

a) Les valeurs de τ pour lesquelles on a $H_{\text{I}}\,[\tau] = 1$ sont nécessairement des valeurs isolées (sinon $\sigma\,[M, t]$ serait presque sûrement indépendant de t et on aurait $H_{\text{I}} \equiv 1$).

b) Si on a $H_{\text{I}}(h) = 1$ pour $h \neq 0$, on a aussi $H_{\text{I}}\,(\mathrm{n}\,h) = 1$ pour n entier positif, négatif ou nul et σ est périodique en t de période h.

c) Si on a $H_{\text{I}}(h) = H_{\text{I}}(h') = 1$, h et h' sont commensurables car la fonction $H_{\text{I}}[\tau]$ ne peut avoir deux périodes incommensurables sans être constante.

En résumé, on n'aura $H_{\text{I}}(\tau) = 1$ pour plusieurs valeurs distinctes que si σ résulte de la superposition de composantes monochromatiques correspondant à des fréquences ν commensurables. La distinction entre commensurable et incommensurable n'a pas un grand sens expérimental et, *pratiquement*, $H_{\text{I}}(\tau)$ ne sera exactement égal à 1 pour plusieurs valeurs distinctes que si σ est monochromatique.

Si (29) est rempli, $E \{ \sigma^2\,[M, t] \}$, qui était indépendant de t, devient aussi indépendant de M. On peut éviter cette condition très restrictive en modifiant un peu notre question et en la formulant de la façon suivante :

Quelle est la condition nécessaire et suffisante pour que les trois figures d'interférences puissent être rendues identiques sous réserve que, grâce à un écran absorbant neutre, on ait ramené les sources à avoir la même intensité ? Autrement dit, on admet la possibilité d'avoir $E\{\sigma^2[M_1,t]\} \neq E\{\sigma^2[M_2,t]\}$, quitte à rendre les deux sources de même puissance grâce à un écran absorbant neutre. Il faut alors modifier notre énoncé comme suit :

La condition nécessaire et suffisante cherchée est qu'il existe deux nombres réels k *et* τ_0 *indépendants de* M *et de* t *tels que l'on ait presque sûrement*

$$\sigma[M_1, t] = k\,\sigma[M_2, t - \tau_0]. \tag{30}$$

S'il en est ainsi, nous dirons que les deux sources considérées sont cohérentes.

La condition (30) *peut être formulée de façon équivalente en faisant intervenir* $\Sigma[M_1, \nu]$ *et* $\Sigma[M_2, \nu]$. Il faut alors remplacer (30) par la relation suivante qui doit être vérifiée presque sûrement

$$d\Sigma_\nu[M_1, \nu] = k\,e^{-2\pi i \nu \tau_0}\, d\Sigma_\nu[M_2, \nu]. \tag{31}$$

On peut aussi donner la condition équivalente suivante : *il existe deux nombres réels* k *et* τ_0 *tels que l'on ait identiquement*

$$dF_\nu[M_1, M_2, \nu] = k\,e^{-2\pi i \nu \tau_0}\, dF_\nu[M_2, M_2, \nu] = \frac{1}{k}\, e^{-2\pi i \nu \tau_0}\, dF_\nu[M_1, M_1, \nu]. \tag{32}$$

Cela entraîne

$$\mu[M_1, M_2, \nu] = \frac{E\{d\Sigma_\nu[M_1, \nu]\, d\Sigma_\nu^*[M_2, \nu]\}}{\sqrt{E\{|d\Sigma_\nu[M_1, \nu]|^2\}\, E\{|d\Sigma_\nu[M_2, \nu]|^2\}}} = e^{-2\pi i \nu \tau_0}. \tag{33}$$

Mais, réciproquement, (33) n'entraîne pas nécessairement (32) *avec* k *indépendant de* ν.

Si on se reporte à l'équation

$$\Gamma_{12}[\tau] = \int_{-\infty}^{+\infty} e^{2\pi i \nu \tau}\, dF_\nu[M_1, M_2, \nu], \tag{34}$$

on peut considérer qu'elle exprime que la figure d'interférence $\Gamma_{12}[\tau]$ résulte de la superposition des *franges monochromatiques*

$$\text{partie réelle de } \{e^{2\pi i \nu \tau}\, dF_\nu[M_1, M_2, \nu]\}. \tag{35}$$

Ceci va nous conduire à des interprétations simples des conditions précédentes.

a) *Supposons que, pour* ν *quelconque, on ait* $|\mu[M_1, M_2, \nu]| = 1$; alors, pour chaque valeur de ν, il existe un nombre $k[\nu]$ et un nombre $\tau_0[\nu]$ réels tels que l'on ait presque sûrement

$$d\Sigma_\nu[M_1, \nu] = k[\nu]\, e^{-2\pi i \nu \tau_0[\nu]}\, d\Sigma_\nu[M_2, \nu]. \tag{36}$$

Si on isole une tranche $\Delta\nu$ *infiniment fine* et si on revient aux expériences (I), (II), (III) :

1° (I) et (II) donneront des franges à minima nuls ;

2°) si on a ramené les sources à avoir la même intensité, (III) fournira aussi des franges à minima nuls.

Mais si on utilise une tranche $\Delta\nu$ *finie*

1° (I) et (II) donneront des systèmes semblables,

2° (III) différera de (I) et (II).

Nos trois figures d'interférences ne seront identiques que pour $\Delta\nu \to 0$. Nous dirons alors qu'il y a *cohérence monochromatique* ; on remarquera que la condition $|\mu| = 1$ *signifie que, pour chaque fréquence, le coefficient de Hopkins est égal à l'unité.*

b) τ_0 *indépendant de* ν signifie que, pour $\tau = \tau_0$, les franges monochromatiques (35) ont, quel que soit ν, toutes une frange brillante ou toutes une frange sombre, ou encore que, pour $\tau = \tau_0$, on a une frange sombre (ou brillante) *achromatique.*

c) k *indépendant de* ν permet, en utilisant un filtre absorbant *neutre* d'opacité convenable, de ramener nos deux sources à avoir le même spectre.

On peut illustrer ce qui précède par la remarque suivante : Soient S une source primaire et S_1 et S_2 deux sources secondaires éclairées par S.

a) *Plaçons-nous dans le vide.* $(SS_1) - (SS_2)$ correspond à un retard bien défini et, entre S_1 et S_2, on aura la relation (29) [avec le signe + en l'absence de miroir].

b) *Plaçons sur* (SS_1) *un système absorbant neutre* dont nous supposons (sans nous préoccuper de savoir comment on le réalisera) qu'il n'introduit pas de variation du retard τ. Alors nous sommes dans le cas où (30), (31), (32) sont remplis, k et τ_0 ne dépendant pas de ν. Nous disons qu'il y a *cohérence.*

c) *Plaçons sur* (SS_1) *un système dispersif* (absorbant et déphasant). Alors k et τ_0 deviendront des fonctions de ν. $|\mu|$ sera encore égal à 1. Il y aura seulement *cohérence monochromatique.*

B) *Au lieu de considérer deux points* M_1 *et* M_2 *particuliers, considérons maintenant l'ensemble de tous les points* M *du plan source* $\alpha \times \beta$.

Nous dirons que l'ensemble des sources du plan $\alpha \times \beta$ est cohérent si ces sources sont cohérentes deux à deux. Par une généralisation immédiate de ce qui précède on obtient les énoncés suivants :

a) *Pour que l'ensemble des sources du plan* $\alpha \times \beta$ *présente la cohérence monochromatique il faut et il suffit que la fonction* σ *puisse se mettre sous la forme*

$$\sigma\,[M, t] = \int_{-\infty}^{+\infty} e^{2\pi i \nu t}\, K[M, \nu]\, dy\,(\nu), \tag{37}$$

où $K\,[M, \nu]$ *est une fonction certaine à valeurs complexes telle que*

$$K\,[M, \nu] = K^*\,[M, -\nu],$$

$y\,[\nu]$ *est une fonction aléatoire à accroissements non correlés satisfaisant* à

$$dy\,(\nu) = dy^*\,(-\nu)\ .$$

Toutes les fonctions de t, $\sigma[M, t]$ dérivent alors, par filtrage linéaire, d'une seule fonction aléatoire à savoir

$$Y\,[t] = \int_{-\infty}^{+\infty} e^{2\pi i \nu t}\, dy\,(\nu)$$

qui joue le rôle d'une *source unique* d'où dérivent toutes les *sources secondaires.*

b) *Pour que l'ensemble des sources du plan* $\alpha \times \beta$ *soit cohérent, il faut et il suffit que, en plus de la condition précédente, on ait*

$$K[M, \nu] = \mathcal{K}[M] \, e^{-2\pi i \nu \tau [M]},$$

où $\mathcal{K}[M]$ et $\tau[M]$ sont des fonctions réelles de M.

2. ***Définition de l'incohérence.*** — *Il y a incohérence si, pour deux éléments* dM_1 *et* dM_2 *disjoints quelconques, on n'observe aucun phénomène d'interférences pour le cas* III. Il en est ainsi lorsque, pour deux éléments disjoints dM_1 et dM_2 quelconques, on a

$$E\{ ds[M_1, t_1] . ds^*[M_2, t_2] \} \equiv 0. \tag{38}$$

Alors, on peut voir que $\sigma[M, t]$ qui se présente comme une dérivée aléatoire n'existe pas de sorte que, stricto sensu, *nos considérations précédentes faites en raisonnant sur* σ *excluent le cas de l'incohérence.* Cette difficulté n'existerait pas si on avait raisonné sur s définie comme l'intégrale des ds. En fait, cela aurait alourdi inutilement nos calculs et, d'ailleurs, l'introduction des fonctions de Dirac permet de traiter le cas de l'incohérence comme cas limite. La condition d'incohérence s'exprimera alors de la façon suivante :

$$\Gamma[M, M + m, \tau] = \delta[m] \, I(M, \tau). \tag{39}$$

3. ***Cohérence partielle.*** — Il y aura cohérence partielle toutes les fois qu'il n'y a ni cohérence ni incohérence.

5. **Remarques sur le problème de la cohérence pour les sources monochromatiques.** — *1.* ***Quasimonochromatisme et monochromatisme absolu.*** — *Supposons toujours la stationnarité dans le temps* et posons

$$\sigma[M, t] = \mathcal{R}[Z[M, t]] = \text{partie réelle de } Z, \tag{40}$$

où

$$Z[M, t] = 2\int_0^\infty e^{2\pi i \nu t} \, d\Sigma_\nu[M, \nu]. \tag{41}$$

$Z[M, t]$ est ce qu'on appelle le *signal analytique* (F. A. p. 411) associé à $\sigma[M, t]$. On voit sans peine que l'on a

$$\Gamma[M_1, M_2, \tau] = E\{ \sigma[M_1, t] \, \sigma^*[M_2, t - \tau] \} = \frac{1}{2} \mathcal{R}[C[M_1, M_2, \tau]], \tag{42}$$

où

$$C[M_1, M_2, \tau] = E\{ Z[M_1, t] \, Z^*[M_2, t - \tau] \} = \tag{43}$$
$$= 4\int_0^\infty e^{2\pi i \nu \tau} E\{ d\Sigma_\nu[M_1, \nu] \, d\Sigma_\nu^*[M_2, \nu] \}.$$

Si on tient compte du fait que les accroissements $d\Sigma_\nu[M, \nu]$ correspondant à deux domaines en ν symétriques par rapport à $\nu = 0$, sont imaginaires conjugués, on voit sans peine que les connaissances de Γ et de C sont équivalentes. *C peut donc être utilisé, comme Γ, pour décrire l'état de cohérence des sources* M_1 *et* M_2. *En définitive, au lieu de raisonner sur les expressions réelles* $\sigma[M_1, t]$, $\sigma[M_2, t]$,

$\Gamma[M_1, M_2, \tau]$, *on peut, de façon équivalente, utiliser les grandeurs complexes* $Z[M_1, t]$, $Z[M_2, t]$, $C[M_1, M_2, \tau]$.

a) *Supposons notre source quasimonochromatique.* Elle n'émet alors que des fréquences ν telles que $\nu_0 - \Delta\nu \leqslant |\nu| \leqslant \nu_0 + \Delta\nu$; $\Delta\nu \ll \nu_0$ et on a

$$(44) \qquad Z\,[M, t] = 2\int_{\nu_0-\Delta\nu}^{\nu_0+\Delta\nu} \mathbf{e}^{2\pi \mathbf{i}\nu t} d\Sigma_\nu\,[M, \nu] = A\,[M, t]\;\mathbf{e}^{2\pi \mathbf{i}\nu_0 t}.$$

$Z\,[M, t]$ se présente comme une onde d'apparence monochromatique de fréquence ν_0 et d'*amplitude complexe* $A\,[M, t]$ *très lentement variable en* M *et* t.

$$(45) \qquad C\,[M_1, M_2, \tau] = \mathbf{e}^{2\pi \mathbf{i}\nu_0\tau} E\left\{ A\,[M_1, t]\; A^*\,[M_2, t-\tau] \right\}.$$

Pour $|\tau| < \tau_1$ (τ_1 *est fixé*), *c'est-à-dire pour un ordre d'interférence borné*, on peut choisir $\Delta\nu$ assez petit pour avoir avec une approximation aussi bonne que l'on veut

$$(46) \qquad C\,[M_1, M_2, \tau] = \mathbf{e}^{2\pi \mathbf{i}\nu_0\tau}\; E\left\{ A\,[M_1, t]\; A^*\,[M_2, t] \right\}$$

et, en général, on aura des propriétés ergodiques suffisantes pour pouvoir écrire

$$(47) \qquad E\left\{ A\,[M_1, t]\; A^*\,[M_2, t] \right\} = \lim_{T\to\infty} \frac{1}{T}\int_0^T A\,[M_1, \theta]\; A^*\,[M_2, \theta]\; d\theta,$$

l'intégrale étant calculée sur une épreuve particulière $\mathcal{E}$ d'ailleurs quelconque. Naturellement (caractère stationnaire), t n'intervient pas dans le résultat ; en désignant par C' la valeur de $E\left\{ A\,[M_1, t]\, A^*[M_2, t]\right\}$, on a, à cette approximation.

$$C\,[M_1, M_2, \tau] = \mathbf{e}^{2\pi \mathbf{i}\nu_0\tau}\; C'\,[M_1, M_2].$$

La cohérence partielle entre M_1 *et* M_2 *est à cette approximation, complètement caractérisée par la fonction* C' *indépendante du temps qui peut être interprétée soit comme une espérance mathématique prise à un instant t quelconque, soit comme une moyenne temporelle prise sur une épreuve* $\mathcal{E}$ *quelconque.*

b) *Supposons maintenant la source parfaitement monochromatique.* Alors $\Delta\nu = 0$ et A ne dépend plus de t, la moyenne temporelle contenue dans (47) est égale à $A\,[M_1]\; A^*\,[M_2]$; c'est une *variable aléatoire*, fonction de l'épreuve mais non de t qui n'a aucune raison d'être égale à son espérance mathématique. (47) n'est plus valable et C' perd sa signification de moyenne temporelle pour ne conserver que celle d'une espérance mathématique.

Dans le cas du monochromatisme parfait, la notion de cohérence partielle échappe complètement à un observateur qui ne suit que ce qui se passe sur une épreuve particulière ; *cette notion n'a alors de sens que pour l'ensemble des épreuves.* En fait, cette remarque n'a qu'une portée mathématique ; *physiquement, il n'y a pas de monochromatisme absolu et nous appellerons source monochromatique une source quasimonochromatique correspondant à une valeur extrêmement faible de* $\Delta\nu/\nu_0$. Nous raisonnerons dans le cas du monochromatisme mais il ne faudra pas perdre de vue que ce cas n'est que la traduction approchée d'un quasimonochromatisme réel. Nous raisonnerons sur C' ce qui revient à supposer que A ne dépend pas de t.

2. ***Cohérence partielle pour des sources monochromatiques.*** — *La fréquence* ν_0 *étant fixée*, notre source S sera définie par la fonction aléatoire

$$A\,[M] = \int e^{2\pi i\,\vec{\Omega}\cdot\vec{M}}\, d\Sigma'_{\Omega}[\Omega] \quad . \tag{48}$$

Les propriétés relatives à la cohérence seront caractérisées par C' ou par les accroissements $d\Phi\,[\Omega_1, \Omega_2, \nu_0]$ (que nous noterons $d\Phi\,[\Omega_1, \Omega_2]$) définis par

$$d\Phi\,[\Omega_1, \Omega_2] = E\left\{ d\Sigma'_{\Omega}\,[\Omega_1, \nu_0]\; d\Sigma'^{*}_{\Omega}\,[\Omega_2, \nu_0] \right\} \quad . \tag{49}$$

Dans le cas d'une source spatialement uniforme, $A[M]$ *est stationnaire en* M. $C'[M_1, M_2]$ ne dépend alors que de $N = M_1 - M_2$ et $d\Phi\,[\Omega_1, \Omega_2]$ est identiquement nul pour $\Omega_1 \neq \Omega_2$. Le spectre $E\{\,|A|^2\,\}$ est caractérisé par les contributions non négatives $d\Phi\,[\Omega, \Omega]$ des domaines $d\Omega$. En général, ce spectre aura une densité $g\,[\Omega] = \Delta\Phi\,[\Omega, \Omega]/\Delta\Omega$: g et C' seront transformées de Fourier l'une de l'autre :

$$C'\,[N] = \int_{-\infty}^{+\infty} e^{2\pi i\,\vec{\Omega}\cdot\vec{N}}\, g\,[\Omega]\, d\Omega \quad . \tag{50}$$

Si on ne suppose pas la source spatialement uniforme, la discussion de la *cohérence* — qui se ramène ici à la *cohérence monochromatique* — se fait comme précédemment en utilisant le coefficient de Hopkins. *Dans le cas particulier d'une source spatialement uniforme*, cette discussion prend l'aspect suivant.

a) *Il y a cohérence* si $|\mu| = 1$, c'est-à-dire s'il existe un nombre réel certain φ (M) et une variable aléatoire Y tels que

$$A\,[M, \mathcal{E}] = e^{2\pi i \varphi[M]}\; Y\,[\mathcal{E}] \tag{51}$$

Si on suppose $E\{\,|A|^2\,\} = 1$ (ce qui ne restreint pas la généralité), on a

$$C'\,[N] = e^{2\pi i\,[\varphi[M+N] - \varphi[M]]} \tag{52}$$

D'ailleurs, à cause de la stationnarité, $\varphi\,[M + N] - \varphi\,[M]$ sera une fonction linéaire de N et on aura

$$\varphi\,[M] = \vec{\Omega}_0\,\vec{M} + \text{Cte} \quad .$$

De sorte que la condition de cohérence pourra s'exprimer de la façon suivante : il existe un vecteur $\vec{\Omega}_0$ tel que $C'\,[N] = \exp\,(2\pi i\,\vec{\Omega}_0\,\vec{N})$ ce qui équivaut à $g\,[\Omega] = \delta\,[\Omega_0 - \Omega]$. La condition $\Omega_0 = 0$ signifie que toutes les sources M sont en phase.

b) *Il y a incohérence si* $\mu = 0$ *ce qui est équivalent aux conditions*

$$C'\,[N] = \delta\,[N] \qquad \text{ou} \qquad g\,[\Omega] \equiv 1.$$

6. **Relation entre la cohérence d'une source et celle de son image.** — *1.* Soit une source (S) conforme à ce qui a été dit au paragraphe 3 (définie par l'ensemble des $ds\,[M, t]$ ou par $\sigma\,[M, t]$) et un plan $P_1\,[\alpha_1, \beta_1]$. Entre P et P_1, nous supposons l'existence d'un système optique D quelconque mais *invariable dans le temps*. Du fait de l'existence de (S), il existe une certaine vibration $\sigma_1\,[M_1, t]$ au point

M_1 de P_1 à l'intant t. Les propriétés du système optique D sont décrites par celles de la correspondance entre $\sigma[M, t]$ et σ_1 $[M_1, t]$ ou encore entre A [M] et A_1 $[M_1]$. Nous nous proposons d'étudier ici comment l'on passe de la fonction Γ relative à σ à la fonction Γ_1 relative à σ_1. *Nous nous bornerons à discuter la question dans le cadre du monochromatisme, les résultats s'étendant très facilement au cas général.*

2. ***Conséquences du caractère linéaire de la correspondance*** D. — *La correspondance entre A [M] et A_1 $[M_1]$ est toujours linéaire dans le sens suivant* : désignons par $\mathcal{D}$ $[A]$ la transformée A_1 de A, on a toujours

$$\mathcal{D}\,[A + A'] = \mathcal{D}\,[A] + \mathcal{D}\,[A'] . \tag{53}$$

Désignons alors par R $[M_1, M]$ la transformée de A lorsque A se réduit à une impulsion unité δ [M] en M. En d'autres termes, soit R $[M_1, M]$ la répartition d'amplitude complexe dans la tache *image* associée à M. La correspondance $\mathcal{D}$ est complètement caractérisée par la donnée de R $[M_1, M]$. On peut donc calculer la fonction C_1' relative à A_1 à partir de R et de C' relative à A. On obtient les relations suivantes :

$$A_1\,[M_1] = \int_M A\,[M]\,R\,[M_1, M]\,dM \tag{54}$$

et

$$C'[M_1, M_1'] = \iint_{M_2 \times M_2'} R[M_1, M_2]\,R^*[M_1', M_2']\,C[M_2, M_2']\,dM_2\,dM_2'. \tag{55}$$

Si la source primaire A (M) est incohérente, c'est-à-dire si l'on a

$$C(M, M'] = \delta\,[M - M']\,I\,(M),$$

on déduit de (55)

$$C'\,[M_1, M_1'] = \int_{M_2} I(M_2)\,R\,[M_1, M_2]^*\,R\,[M_1', M_2]\,dM_2, \tag{56}$$

de sorte que l'existence de taches images entraîne une certaine cohérence dans l'image A_1. Si A est spatialement uniforme, il n'en sera pas de même, en général, pour A_1.

3. ***L'approximation des filtres linéaires.*** — La correspondance $\mathcal{D}$ peut prendre un aspect particulièrement simple dans les circonstances suivantes. Il peut arriver que, peut-être au prix d'un déplacement et d'une homothétie dans P_1, on puisse rendre la correspondance entre A [M] et A_1 $[M_1]$ invariante par translation, c'est-à-dire que si $A_1\,[M_1] = \mathcal{D}\,[A\,[M]]$ on aura aussi

$$A_1\,[M_1 + M_0] = \mathcal{D}\,[A\,[M + M_0]] .$$

Il n'y a alors aucune raison de distinguer systématiquement par un indice un point courant de P_1 d'un point courant quelconque de P et nous dirons que $\mathcal{D}$ transforme A [M] en A_1 [M] et que la correspondance satisfait à la relation d'invariance

$$\mathcal{D}\,[A\,[M + M_0]] = A_1\,[M + M_0] . \tag{57}$$

Dans ces conditions (linéarité + invariance par translation), $\mathcal{D}$ constitue un filtre linéaire (F. A. p. 347-359). Physiquement, cela signifie que, si R $[M, M_0]$ est l'image de $\delta\,[M_0]$, R $[M, M_0]$ ne dépend que de $N = M - M_0$, c'est-à-dire que si on déplace M_0, la tache image subit la même translation sans que la fonction R [M — M_0] subisse la moindre modification ni en module (ce qui peut parfois ne pas être trop restrictif) ni en argument (ce qui sera en général beaucoup plus restrictif compte tenu de la petitesse de la longueur d'onde). *Une telle invariance ne peut évidemment être réalisée que dans une région centrale du champ qui peut d'ailleurs, si elle existe, être très restreinte.* Etant donnée la simplicité de calculs auxquels se prêtent les filtres linéaires, nous allons développer cette approximation : nous montrerons ensuite sur un exemple son caractère limité.

A) Cohérence et filtres linéaires. — La relation (54) devient

$$A_1 [\mathrm{M}] = \int_{(\mathrm{M}')} A [\mathrm{M}'] \, R [\mathrm{M} - \mathrm{M}'] \, d\mathrm{M}'. \tag{58}$$

R est la *réponse percussionnelle du filtre*. Il est utile d'introduire le *gain* $G [\Omega]$ du filtre ; il est défini par

$$G [\Omega] = \int_{\mathrm{M}} \mathbf{e}^{-2\pi i \vec{\Omega} . \vec{\mathrm{M}}} R [\mathrm{M}] \, d\mathrm{M}. \tag{59}$$

La correspondance entre A et A_1 s'exprime simplement si on utilise les $d\Sigma'_\Omega [\Omega]$; on a

$$d\Sigma'_{1\Omega} [\Omega] = G [\Omega] \, d\Sigma'_\Omega [\Omega]. \tag{60}$$

Le passage des moyennes du second ordre relatives à A aux moyennes du second ordre relatives à A_1 s'effectue simplement en utilisant (55) où R[M, M'] ne dépend plus que de M — M' ou bien, dans l'espace des $\vec{\Omega}$, en utilisant la relation suivante déduite de (60) :

$$d\Phi_1 [\Omega_1, \Omega_2] = G [\Omega_1] \, G^* [\Omega_2] \, d\Phi [\Omega_1, \Omega_2]. \tag{61}$$

D'où

$$C'_1 [\mathrm{M}_1, \mathrm{M}_2] = \int_{\Omega_1 \times \Omega_2} \mathbf{e}^{2\pi i [\vec{\mathrm{M}_1} \vec{\Omega_1} - \vec{\mathrm{M}_2} \vec{\Omega_2}]} d\Phi_1 [\Omega_1, \Omega_2]. \tag{62}$$

Cas particulier d'une source spatialement uniforme. — S'il existe une densité $g [\Omega]$ dans le spectre de $E \{ | A |^2 \}$, les deux relations précédentes prendront la forme suivante :

$$g_1 [\Omega] = | G [\Omega] |^2 g [\Omega], \tag{63}$$

$$C'_1 [\mathrm{N}] = \int_{\Omega} \mathbf{e}^{2\pi i \vec{\Omega} \vec{\mathrm{N}}} | G [\Omega] |^2 g [\Omega] \, d\Omega. \tag{64}$$

Il est très important de remarquer que, alors, l'argument du gain n'intervient pas dans l'expression de C'.

B) Etude d'un exemple particulier. — *Etudions les limites de l'approximation des filtres linéaires sur un exemple sans intérêt pratique mais se prêtant à des calculs extrêmement simples.* Plaçons-nous dans un cas unidimensionnel et supposons le système optique réduit à un miroir plan limité (fig. 2), par rapport auquel les plans (P) et (P_1) sont conjugués. La répartition des amplitudes dans (P_1) correspondant à une fente objet de coordonnée α sera, à un facteur constant près, donnée par

$$(65) \qquad A_1\,[\alpha_1] = \int_{-a}^{+a} e^{-\frac{2\pi i}{\lambda}(r-r_1)}\, dx.$$

Or, *en supposant les rayons pas trop inclinés sur l'axe*, on a

$$(66) \quad r - r_1 = \sqrt{d^2 + (x-\alpha)^2} - \sqrt{d^2 + (x-\alpha_1)^2} \simeq \frac{(x-\alpha)^2 - (x-\alpha_1)^2}{2\,d},$$

$$(67) \qquad r - r_1 \simeq -\frac{(\alpha-\alpha_1)\,x}{d} + \frac{(\alpha-\alpha_1)\,(\alpha+\alpha_1)}{2\,d}.$$

D'où

$$(68) \qquad A_1\,[\alpha_1] = e^{-\frac{2\pi i}{\lambda}\frac{(\alpha-\alpha_1)\,(\alpha+\alpha_1)}{2\,d}} \int_{-a}^{+a} e^{\frac{2\pi i}{\lambda}(\alpha-\alpha_1)\,x}\, dx.$$

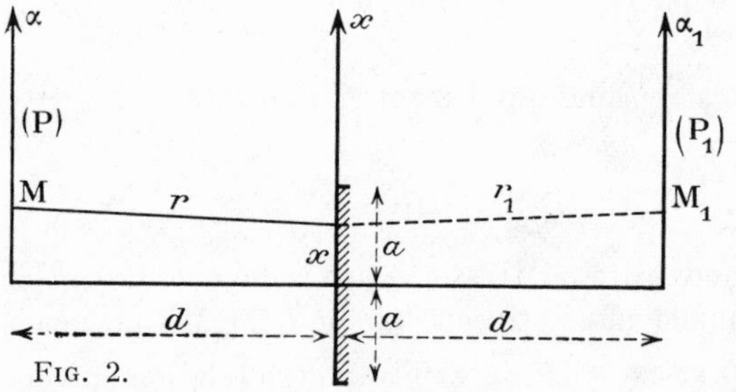

FIG. 2.

Il paraît raisonnable de penser que, pour que $A_1\,[\alpha_1]$ ne dépende que de $\alpha - \alpha_1$, il faut que l'on puisse admettre que l'exponentielle qui précède le signe $\int$ reste très voisine de 1 ou, tout au moins, qu'elle reste voisine de 1 pour des $|\alpha - \alpha_1|$ déjà assez grands par rapport à la demi-longueur conventionnelle $\varepsilon = \lambda d/2a$ de la tache de diffraction, par exemple, pour $|\alpha_1 - \alpha| = \mathrm{K}\lambda d/2a$ (avec peut-être K de l'ordre de 5). Admettons que nous poserons $e^{i\theta} \simeq 1$ si $\theta < 1/10$; on pourra alors éliminer l'exponentielle précédant le signe $\int$ si

$$\frac{2\pi}{\lambda}\ \mathrm{K}\ \frac{\lambda d}{2a}\ .\ 2\ \frac{\alpha_m}{2d}\ .\ 10 < 1$$

(où α_m est le maximum admis de α) ou encore si, en posant $\mathrm{K} = 5$, on a

$$\alpha_m < a/50\,\pi \simeq a/150.$$

Or la demi-largeur conventionnelle de la tache de diffraction vaut $\varepsilon = \lambda d/2a$; on ne pourra dire raisonnablement que la correspondance $A \to A_1$ définit un filtre linéaire dans le domaine $|\alpha| < \alpha_m$ que si ce domaine est nettement supérieur à ε. Cela n'est vérifié que si l'on a $\lambda d/2\,a \ll a/150$ ou encore $d \ll d_0$ avec $d_0 = (a/75)\,(a/\lambda)$; prenons $a = 1$ cm et $\lambda = 0{,}5\,\mu$, la distance d_0 ainsi définie vaudra environ 260 cm ; on ne pourra donc, raisonnablement, être sûr que l'on peut commencer à parler de filtre linéaire que, par exemple, pour $d = 30$ cm ; en particulier, pour $d = 10$ cm, on a $\alpha_m \simeq 60\,\mu$, $\varepsilon = 2{,}5\,\mu$, $\alpha_m = 24\,\varepsilon$.

Cet exemple montre qu'il n'est pas sûr que l'approximation constituée par les filtres linéaires soit très précise. Elle fournit cependant un schéma commode qui permet, avec un appareil mathématique simple, d'obtenir des indications intéressantes ; il est prudent de penser que, dans certains cas, cette méthode ne fournit peut-être que des renseignements semi-quantitatifs.

Remarques. — 1°) Si on admet que la représentation de la correspondance $\mathcal{O}$ par un filtre linéaire est légitime, la relation (68) en fournit immédiatement le gain : il vaut 1 pour les fréquences $|\Omega| < a/\lambda$ et 0 pour les autres. C'est là une propriété générale. Lorsque l'on utilise l'approximation des filtres dans un problème de diffraction par une ouverture, si $T[x, y]$ représente la *transparence de la pupille* $[T = 1$ dans la pupille et 0 à l'extérieur], pour un choix convenable des unités dans les espaces objet et image, le gain peut être mis sous la forme

$$G[\Omega_x, \Omega_y] = T[\lambda\,\Omega_x, \lambda\,\Omega_y].$$

La pupille étant limitée, il s'agit d'un filtre passe-bande.

2°) Si, *l'éclairage de* (P) *étant supposé incohérent*, on s'intéresse à la correspondance entre les intensités dans (P) et (P_1), on sera amené à associer à une source ponctuelle M une tache de diffraction dans laquelle la répartition des intensités sera décrite par $|A_1|^2$. L'exponentielle précédant le signe $\int$ s'éliminera d'elle-même et le dispositif se présentera comme un filtre linéaire en intensité sous la seule condition que l'approximation contenue dans (67) soit légitime, c'est-à-dire que les rayons soient peu inclinés sur l'axe. Pour l'utilisation en éclairage incohérent de filtres linéaires opérant sur l'intensité, nous renvoyons à P. M. Duffieux [5] et A. Blanc-Lapierre & M. Perrot [8, 9].

7. **Propriétés statistiques de la lumière et loi de Laplace-Gauss.** — Dans tout ce qui précède, nous ne nous sommes occupés que de *moyennes du second ordre*, c'est-à-dire essentiellement des propriétés de la fonction Γ. On peut être plus exigeant et vouloir préciser l'ensemble des propriétés statistiques de la variable lumineuse. Si nous désignons par $\varphi[\mathrm{m}, t]$ la variable lumineuse en m et à l'instant t (on supposera que le point m ne vient pas sur la source primaire elle-même) on peut se poser la question suivante :

Etant donnés N *points* $\mathrm{m}_1, \mathrm{m}_2, \ldots, \mathrm{m}_n, \ldots, \mathrm{m}_N$ *et* K *instants* $t_1, t_2, \ldots, t_k, \ldots, t_K$, *quelle est la fonction de distribution relative à l'ensemble des* N × K *variables aléatoires* $\varphi[\mathrm{m}_n, t_k]$?

Un raisonnement général, développé en appendice, permet d'affirmer, sous des hypothèses très larges, que cette fonction de distribution est une loi de Laplace-Gauss. *La fonction aléatoire* $\varphi[\mathrm{m}, t]$ *est une fonction aléatoire laplacienne* (F. A. p. 473). A cause des propriétés particulières des lois de Laplace-Gauss, toutes les propriétés statistiques sont alors complètement déterminées par la connaissance de la seule fonction

$$\gamma[\mathrm{m}_1, \mathrm{m}_2, t_1, t_2] = E\left\{\varphi(\mathrm{m}_1, t_1]\,\varphi[\mathrm{m}_2, t_2]\right\}.$$

APPENDICE

Propriétés statistiques de la lumière et loi de Laplace-Gauss

1. **Un argument d'ordre statistique.** — La question des propriétés statistiques de v [m, t] a été abordée, tout au moins dans le cas monochromatique, par divers auteurs qui s'attachent à démontrer le caractère laplacien (ou gaussien) de la variable lumineuse. Il ne semble pas que la valeur démonstrative des raisonnements faits à ce sujet soit supérieure à celle de la remarque suivante : *Soit une variable aléatoire X, somme d'un nombre* N *de variables aléatoires réelles indépendantes* x_i $[E \{ x_i \} \equiv 0]$ *qui apportent toutes des contributions du même ordre à X.* (*Cela exclut le cas où quelques-unes des variables* x_i *auraient des carrés moyens très supérieurs à la somme des autres carrés moyens, ce qui signifierait que, pratiquement, X ne dépend que d'un petit nombre de* x_i.) *Si* N *est grand, on peut dire que X obéit pratiquement à une loi de Laplace (ou de Gauss) d'espérance mathématique nulle et de carré moyen*

$$E \{ X^2 \} = \sum_{i=1}^{N} E \{ x_i^2 \}.$$

Ce résultat nous permet d'affirmer ce qui suit :

Si nous ne nous plaçons pas sur la source primaire elle-même, supposée assez étendue, et si le flux lumineux n'est pas trop faible, on pourra toujours considérer que v[m, t] résulte de la superposition d'un très grand nombre de contributions pratiquement indépendantes provenant de nombreux éléments de la source primaire qui est toujours incohérente à une certaine échelle encore assez fine par rapport à nos moyens d'observation. La remarque précédente permet d'affirmer le caractère laplacien de v [m, t]. D'une façon générale, si $m_1, \ldots, m_n, \ldots, m_N$ et $t_1, \ldots, t_k, \ldots, t_K$ sont des points et des instants quelconques et en nombre quelconque, les N $\times$ K variables aléatoires v $[m_n, t_k]$ formeront un ensemble de N $\times$ K variables aléatoires laplaciennes.

En fait, notre raisonnement ne nous a permis de conclure que moyennant deux restrictions :

a) *On ne se place pas sur la source primaire elle-même.* Notre raisonnement est essentiellement de nature statistique ; il ignore, dans une large mesure, les caractéristiques particulières des sources élémentaires et ne s'applique que *si* v *résulte de l'action de beaucoup de sources élémentaires indépendantes.* Il ne peut donc rien apporter sur les propriétés particulières d'une source élémentaire, celles-ci devant découler d'une étude directe de la dite source. Ce raisonnement élude donc le problème des propriétés de chaque source élémentaire.

b) *Le flux lumineux est assez intense et la source assez étendue.* Le raisonnement s'appliquera si on peut décomposer la source primaire en un très grand nombre N de sources élémentaires indépendantes (on suppose la source assez étendue) telles que chacune de ces sources indépendantes contribue effectivement à v de façon non négligeable, c'est-à-dire envoie à l'instant t une puissance non négligeable en m. C'est pour cela que la source doit être assez intense.

Il paraît vraisemblable qu'une analyse plus élaborée conduirait à ne retenir comme seule condition que la restriction relative au fait que, à l'instant t, le flux lumineux doit être assez dense près du point m *où est défini* φ.

2. **Remarques sur un modèle classique.** — *Reportons-nous à une image courante dans la théorie classique de l'émission et assimilons l'action d'une source lumineuse* à celle d'oscillateurs amortis relancés en des instants distribués au hasard. *Supposons, pour simplifier le raisonnement, la source macroscopiquement constante.* On pourra traduire cette image en associant à la source une vibration

$$\mathcal{A}\,[t] = \sigma\,[t]\,\mathrm{d\,M} = \Sigma_{t_j}\, R(t - t_j) \tag{1}$$

avec (⁴)

$$E\,\{\,\sigma\,\} \equiv 0\ . \tag{2}$$

(Les instants t_j sont répartis dans le temps suivant une distribution de Poisson de densité constante ρ et R représentera un train *d'oscillations amorties.*) *Assez loin de la source et dans un domaine relativement petit*, on pourra considérer que la variable lumineuse est égale à

$$\varphi\,[x, t] = \varphi\left[t - \frac{x}{c}\right] = K\,\sigma\left[t - \frac{x}{c}\right]\mathrm{d\,M}\quad ,$$

où x est la distance à la source, c la vitesse de propagation de la lumière dans le vide et K une constante fonction de la distance à la source du centre du petit domaine considéré, les ondes se comportant alors comme des ondes planes normales à Ox.

Les fonctions aléatoires introduites par (1) sont classiques dans les problèmes de bruit de fond (F. A. p. 142-190). Si θ représente la constante de temps liée à $R\ (t - t_j)$, on peut établir le résultat suivant.

La condition pour que $\varphi\ [x, t]$ *soit pratiquement laplacienne est que l'on ait*

$$\rho\,\theta \gg 1\ . \tag{3}$$

Cette condition signifie que, en moyenne, beaucoup de trains d'oscillations sont lancés pendant la durée θ, c'est-à-dire pendant le temps nécessaire à l'amortissement de chacun d'eux. Elle est certainement remplie si le flux lumineux est important.

On peut tirer des remarques intéressantes de la considération de l'énergie emmagasinée dans le volume du parallélipipède centré sur x_0 *limité en* x *par les plans* $x_0 - l$ *et* $x_0 + l$ *et de surface de base unité suivant* y O z. Cette énergie est, à l'instant t,

$$W[l, t] = K'^2 \int_{x_0-l}^{x_0+l} \varphi^2\,[x, t]\,\mathrm{d}x\ . \tag{4}$$

Son *espérance mathématique* vaut

$$a\,[l] = E\,\{\,W\,[l]\,\} = K'^2 c \int_{x_0-(l/c)}^{x_0+(l/c)} E\,\{\,\varphi^2(t)\,\}\,\mathrm{d}t\ . \tag{5}$$

(⁴) Si R était tel que l'on n'ait pas $E\,\{\,\sigma\,\} \equiv 0$, on retrancherait $E\,\{\,\sigma\,\}$ de σ puisque la composante constante $E\,\{\,\sigma\,\}$ n'a aucun intérêt en optique.

Son *écart quadratique moyen* δ [l] est donné par

(6) $\delta^2[l] = E\left\{[W - E\{W\}]^2\right\}$

$$= C^2 K'^4 \iint_{x_0-(l/c)}^{x_0+(l/c)} E\left\{[v^2(t) - E\{v^2\}]\;[v^2(t') - E\{v^2\}]\right\} dt\,dt'.$$

Certains résultats classiques de l'étude des fonctions aléatoires dérivées des processus de Poisson permettent le calcul de $a[l]$ et de $\delta^2[l]$. Nous nous bornerons ici à donner le résultat pour $l \gg c\,\theta$. On trouve alors

(7) $a\,[l] = 2\,K'^2\,l\,\rho\,\mu_1\,(0).$

(8) $\delta^2[l] = 2\,l\,c\,K'^4[\rho \int_{-\infty}^{+\infty} \mu_2(\theta)\,d\theta + 2\,\rho^2 \int_{-\infty}^{+\infty} \mu_1^2[\theta]\,d\theta],$

où

(9) $\mu_1[\theta] = \int_{-\infty}^{+\infty} R[t]\,R[t-\theta]\,dt$ et $\mu_2(\theta) = \int_{-\infty}^{+\infty} R^2[t]\,R^2[t-\theta]\,dt.$

Voyons ce que donne la formule précédente en prenant pour $R(t)$ un paquet d'ondes

(10) $R(t) = \int_{-\infty}^{+\infty} \cos 2\pi\nu t\; \mathbf{e}^{-\frac{(\nu-\nu_0)^2}{2\sigma^2}}\,d\nu = \sqrt{2\pi}\,\sigma \cos 2\pi\nu_0 t\; \mathbf{e}^{-2\pi^2 t^2 \sigma^2}.$

On trouve alors, pour $\sigma \ll \nu_0$ (ce que nous supposons),

(11) $\mu_1[0] \simeq \dfrac{\sigma\sqrt{\pi}}{2},\quad \int_{-\infty}^{+\infty} \mu_2(\theta)\,d\theta \simeq \sigma^2\,\dfrac{\pi}{4},\quad \int_{-\infty}^{+\infty} \mu_1^2[\theta]\,d\theta \simeq \dfrac{\sigma}{8}\sqrt{\dfrac{\pi}{2}},$

d'où

(12) $a\,[l] = K'^2\,l\,\rho\,\sigma\sqrt{\pi}\,,$

(13) $\delta^2\,[l] = 2\,l\,c\,K'^4\left\{\rho\,\sigma^2\,\dfrac{\pi}{4} + \dfrac{\sigma}{8}\,\rho^2\,\sqrt{2\pi}\right\}.$

Le carré moyen δ^2 est la somme de deux termes : l'un prépondérant pour les faibles valeurs de ρ, croît comme $a[l]$ lorsque ρ varie ; l'autre, prépondérant pour les grandes valeurs de ρ croît comme $a^2[l]$ lorsque ρ varie. Le premier terme existe seul si les trains d'ondes sont assez espacés pour ne pas se recouvrir, le second est lié aux interférences entre les divers trains d'ondes. On retrouve cette situation pour le carré moyen $\overline{\delta[W_\nu]^2}$ des fluctuations de l'énergie W_ν dans le *rayonnement isotherme de fréquence* ν (à $d\nu$ près) pour un élément de volume dV [10]. On a alors

(14) $\overline{\delta\,[W_\nu]^2} = \overline{(W_\nu - \overline{W_\nu})^2} = h\nu\,\overline{W_\nu} + \dfrac{(\overline{W_\nu})^2}{g_\nu}\quad,$

où g_ν est le nombre d'états distincts correspondant au domaine $dV \times d\nu$.

On peut donner à ce rapprochement une forme plus saisissante : à chaque train d'oscillations, on peut associer une *énergie moyenne*

(15) $$\varepsilon_0 = c\, K'^2 \int_{-\infty}^{+\infty} R^2(t)\, \mathrm{d}t = c\, K'^2\, \sigma \frac{\sqrt{\pi}}{2}$$

et un *étalement* (dans l'espace)

(16) $$\Delta x = \frac{c}{\sigma \sqrt{2\,\pi}}$$

(le coefficient $1/\sqrt{2\,\pi}$ est choisi pour que l'on ait $\Delta\nu . \Delta t = 1$).

En adoptant cette valeur de Δx, on peut dire que la longueur $2\,l$ du domaine considéré peut contenir un nombre de trains d'oscillations successifs égal à

(17) $$g'_i = \frac{2\,l}{\Delta x} = \frac{2\,l\,\sigma\sqrt{2\,\pi}}{c} \quad .$$

La relation (13) devient alors

(18) $$\delta^2\,[l] = \varepsilon_0\, a\,[l] + \frac{a^2\,[l]}{g'_i} \quad .$$

C'est exactement la formule (14) si on remplace g_i par g'_i et ε_0 par $h\nu$.

Il n'y a pas à insister sur cette analogie de formules et *il est bien certain que* (18) *ne saurait constituer une démonstration de* (14) ; cependant cette ressemblance suggère une certaine correspondance entre les termes des seconds membres des relations (14) et (18). Dans le modèle classique, le caractère laplacien correspond à la possibilité de négliger le terme en a devant celui en a^2 ; par suite de l'analogie signalée, il peut paraître raisonnable de penser que cette situation correspond à la possibilité de négliger le terme en $\overline{W_\nu}$ devant celui en $(\overline{W_\nu})^2$ ce qui est possible s'il y a une grande densité de photons dans le volume considéré.

Manuscrit reçu le 8 février 1954.

Références

[1] P. H. van Cittert, *Physica*, t. **1**, 1934, p. 201.
[2] F. Zernicke, *Physica*, t. **5**, 1938, p. 785.
[3] H. H. Hopkins, *Proc. Royal Soc.* [A], t. **208**, 1951, p. 263.
[4] H. H. Hopkins, *Proc. Royal Soc.* [A], t. **217**, 1953, p. 408.
[5] P. M. Duffieux, L'intégrale de Fourier et ses applications à l'optique, chez l'Auteur, Université de Besançon, 1946.
[6] P. M. Duffieux, *Rev. Opt.*, t. **32**, 1953, p. 129.
[7] E. Wolf, A macroscopic theory of interference and diffraction of light from finite sources, *Nature*, t. **172**, 1953, p. 535.
L'auteur indique qu'il publiera ailleurs ses résultats sous une forme plus développée ([5]).
[8] A. Blanc-Lapierre & M. Perrot, *C. R. Ac. Sc.*, t. **231**, 1950, p. 539.
[9] A. Blanc-Lapierre, *Bull. Soc. Fr. Electr.* [7], t. **2**, 1952, p. 481.
[10] F. Perrin, Mécanique statistique quantique, Gauthier-Villars, Paris, 1939, chap. XI, p. 93.

([5]) *Note ajoutée à la correction des épreuves.* Voir : E. Wolf, *Proc. Royal Soc.* [A], t. **225**, 1954, p. 96.

PAPER NO. 17

Reprinted from *Physical Review*, Vol. 99, pp. 1691–1700, 1955.

Photoelectric Mixing of Incoherent Light*

A. Theodore Forrester,† Richard A. Gudmundsen,‡ and Philip O. Johnson§
University of Southern California, Los Angeles, California
(Received April 20, 1955)

Beats have been obtained between incoherent light sources by mixing Zeeman components of a visible spectral line at a photosurface. Periodicity in emission was observed through the excitation of a 3-cm cavity. Because of incoherence between the spectral lines and incoherence between the beats from different photocathode areas, the signal-to-shot-noise ratio at the cavity is only 3×10^{-5} but the beats were modulated optically, while maintaining constant total intensity and our receiver was able to yield a signal-to-noise ratio of two at the indicator. The basic idea is that, in the photoelectric process, the emission probability for electrons is proportional to the square of the resultant electric field amplitude, implying an interference between light originating in independent sources. This is a point of view which does not appear to be tested in any other experiment involving quantum effects. The experiment also demonstrates that any time delay between photon absorption and electron release must be significantly less than 10^{-10} second.

I. INTRODUCTION

THE combination of two wave trains of slightly different frequencies is equivalent to a wave of the average frequency modulated by the difference frequency. This is evidenced in the phenomenon of acoustical beats and is responsible for the operation of superheterodyne radio receivers. The periodic variation in intensity which occurs at a fixed point on the image of a Michelson interferometer when one of the mirrors is moving may also be interpreted as beats between the light reflected from the stationary mirror and light which has had its frequency changed by reflection from the moving mirror.[1–3] However, the problem of beating incoherent light waves, i.e., light waves which originate in different sources, is quite different, and since the publication of the original suggestion,[4,5] it has been argued[6] that the observation of such beats is impossible. These arguments, when examined carefully, really provide reasons why beats between incoherent light sources are difficult, rather than impossible, to detect. Were optical lines much sharper than they are, or possible to produce, without great broadening, in much greater intensity than present techniques permit, beats between incoherent light waves would be easy to observe.

Following the publication of the original suggestion[4] for this experiment, Ruark,[7] calling attention to an

* Supported in part by the Office of Naval Research. Reproduction in whole or in part is permitted for any purpose of the United States Government.

† Now on leave at Westinghouse Research Laboratories, East Pittsburgh, Pennsylvania.
‡ Now at Hughes Aircraft Corporation, Culver City, California.
§ Now at North American Aviation, Downey, California.

[1] A. Righi, J. Physique 2, 437 (1883) describes an ingenious production of light beats. To demonstrate that light which has passed through a rotating Nichol prism may be resolved into two circularly polarized beams, one increased and one decreased in frequency with respect to the incident light, he performed an experiment, in its essence a double-slit interference experiment, in which one slit was illuminated by light of reduced frequency and the other slit by the increased frequency, both altered to be plane polarized in the same plane. The moving fringe pattern he observed, and interpreted as beats, is, in principle, no different from those produced by moving a mirror of a Michelson interferometer.

[2] E. Rüchardt, Optik 6, 238 (1950), raises an objection to this point of view based on a requirement for coherence over a beat period. His objection is adequately answered in reference 3.

[3] C. V. Fragstein, Optik **8**, 289 (1951).

[4] Forrester, Parkins, and Gerjuoy, Phys. Rev. **72**, 728 (1947).

[5] Gerjuoy, Forrester, and Parkins, Phys. Rev. **73**, 922 (1948).

[6] L. R. Griffith, Phys. Rev. **73**, 922 (1948).

[7] A. Ruark, Phys. Rev. **73**, 181 (1948).

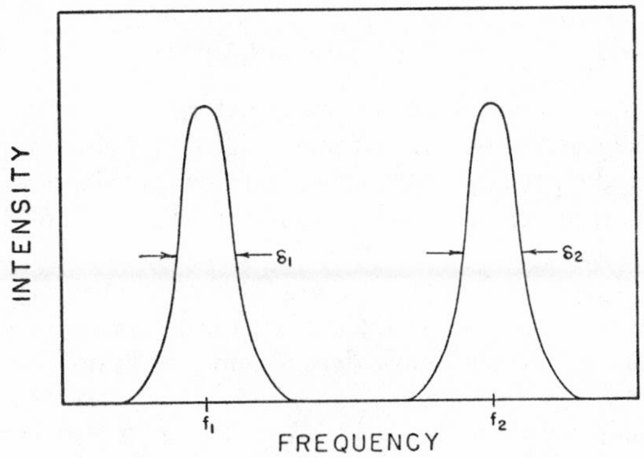

FIG. 1. Spectrum required for the production of sharp beats.

earlier paper,[8] pointed out that waves of different frequencies are not necessarily incoherent, a point of view with which we are in complete agreement. When the mirror of a Michelson interferometer is moved, or in Righi's experiment,[1] the two frequencies are thoroughly coherent. When a periodic modulation of a wave occurs, the sidebands generated in the process have phases which are related and are therefore coherent. We wish to make a special point of the fact that the lines we are mixing are regarded as completely incoherent with respect to each other; i.e., that the relative phase shift of the two lines is completely random and occurs at a rate limited only by the widths of the two lines.

Imagine a source of light emitting two spectral lines arising from two sets of atoms and therefore incoherent. The line widths δ_1 and δ_2, shown in Fig. 1, can arise in many ways, e.g., as a natural width or from Doppler broadening. Whatever the source of the broadening, this curve may be thought of as a plot of the squares of the amplitudes of the Fourier components of the electromagnetic field which is the light. The combination of these two lines should show beats between all of the Fourier components of one line and all of the components of the other (and also between components of a single line which can be ignored because they occur in a different frequency range). The variation in amplitude with time is shown in Fig. 2(a) where the effect of a line width which is not negligible is displayed by irregularities in the wave envelope. The beat frequencies will vary approximately from $f_2-f_1-\delta$ to $f_2-f_1+\delta$ but will nevertheless be a well-defined beat pattern providing that

$$f_2-f_1\gg\delta, \tag{1}$$

which is equivalent to stating that the coherence time $1/\delta$ should be long compared to the beat period $1/(f_2-f_1)$.

Line widths, in the visible region, for allowed transitions, run about 10^8 cps but because it is hard to avoid a great deal of broadening when high intensity is sought, 10^9 cps is a more reasonable figure to use in making estimates. The inequality (1) then requires that f_2-f_1 be of the order of 10^{10} cps or greater to produce a sharp beat frequency. Fortunately 10^{10} cps is a very convenient region of the electromagnetic spectrum in which to detect energy and is a frequency separation easily produced by the Zeeman effect.

A nonlinear device is required for the generation of the difference frequency. For light waves a solution is provided by the photoelectric effect, in which, over a small range of frequencies, the current is proportional to light intensity, or electric field amplitude squared. Indeed, the assumption that the probability of emission of an electron is proportional to the square of the total electric field intensity is really the fundamental assumption of this experiment. It implies an interference between light originating in different atoms, which many physicists find contradictory to their ideas about the nature of interference.

The validity of this approach to the photoelectric effect, rather than, for example, a squaring of the amplitudes of the Fourier components before addition (which would yield a time-independent current), although extremely basic, does not appear to be tested, at least with anything like this degree of directness, in any other experiment. Furthermore, it seems that in no other experiment involving a quantum effect as detector, is an effect observed which must be interpreted as interference between independently generated waves, including wave functions for, say, electrons.

It is worth noting that the generation of the beat frequency depends on what would ordinarily be called a one-quantum process. Any effect which required that photons act in pairs, or successively, to produce emission is so small compared even to the observed effect, as to be considered, at present, indiscernible.

Makinson[9] has made a calculation which is a more rigorous treatment of the photoelectric effect than we attempted. Since he describes the electric fields classically and permits the addition of waves of different frequencies, it is a calculation which we would expect

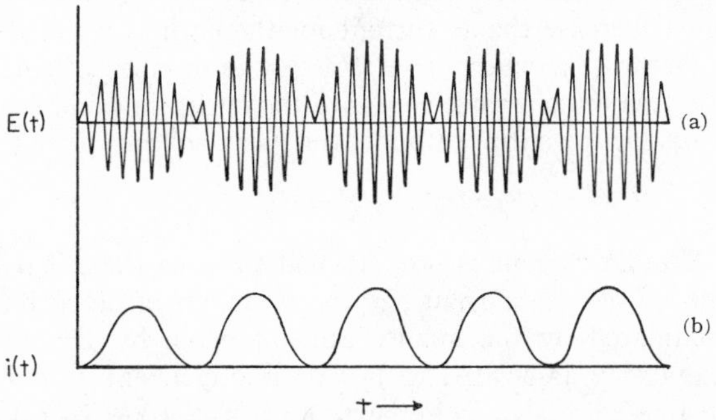

FIG. 2. Electric field variation (a) due to an admixture of two spectral lines and the accompanying photoelectric current (b). The irregularities show the effect of line widths which are not negligibly small.

[8] Breit, Ruark, and Brickwedde, Phil. Mag. 3, 1306 (1927).

[9] R. E. B. Makinson, Phys. Rev. 75, 1908 (1949).

to reduce to our own. However, in his application of his results to the problem of light beats he finds the phase of the beats to vary with the velocity of the emitted electrons which, he states, would result in a very small percentage modulation, whereas the starting point in our calculations is that the photocurrent from a sufficiently small area is 100 percent modulated, as shown in Fig. 2(b). Makinson's conclusion should be based on a demonstration that the variation in phase with velocity is of the order of π or greater. As far as we are aware, the magnitude of this phase variation has not been calculated.

II. ORDER OF MAGNITUDE CALCULATIONS

A. Comparison between Signal and Shot Noise

It is shot noise which provides the basic obstacle to the observation of the process as outlined and a very simple picture suffices for an estimation of the signal-to-shot noise ratio which can be expected. In Fig. 2(b) the photocurrent, resulting from a spectrum made up of two lines satisfying the criterion (1), is shown. Shot noise which would cause variations in current not mirroring those in the field intensity is neglected for the moment. Unfortunately, beats from widely separated areas of the cathode cannot be expected to be in phase, unless we use perfectly plane light. In a more realistic case we can expect the relative phase of each of the light waves and, therefore, the phase of the beats to be constant over an area of the order of the size of the diffraction pattern of a single point on the source,[10] i.e., λ^2/Ω, where Ω is the solid angular spread in the light at the photocathode. From this area the average peak value of the ac current, at the beat frequency, is equal to the average or dc current, so that the mean square value of the ac current from this elementary area is

$$\langle i_0^2\rangle = \tfrac{1}{2}i_{\mathrm{dc}}^2 = \tfrac{1}{2}(I\lambda^2/A\Omega)^2, \qquad (2)$$

where I is the total photocurrent from a cathode of total area A. If we consider the signal from each area of coherence (meaning an area of the size λ^2/Ω) to be randomly phased with respect to all others, then we must increase the ac current for the entire cathode by a factor $\sqrt{n}$, where $n = A\Omega/\lambda^2$ is the number of times the area of coherence goes into the entire cathode area. Thus for the entire cathode, the mean square current is

$$\langle i^2\rangle = \langle i_0^2\rangle A\Omega/\lambda^2 = I^2\lambda^2/2A\Omega. \qquad (3)$$

The dc current is proportional to n, so that, if n is very large, the current has the character of dc current modulated by the minute amount given by Eq. (3). Inevitably associated with a dc photocurrent I is the mean square shot noise current $\langle i_n^2\rangle = 2eI\Delta f$. To compare this with the signal, Δf should be taken as the frequency interval throughout which the signal is spread, approximately the line width δ times $\sqrt{2}$, leading to

$$\langle i^2\rangle/\langle i_n^2\rangle = \lambda^2 I/4\sqrt{2}e\delta A\Omega, \qquad (4)$$

an equation which we can use in determining the order of magnitude of the expected signal-to-noise in the emitted current. The experimental signal-to-noise ratio, aside from errors arising from the character of the approximations made in deriving Eq. (4), will be worse than this due to receiver noise. However when circuit noise is minimized and I made as large as inequality (1) allows, the shot noise does become the major source of noise.

Using the green line $\lambda = 5.461\times10^{-5}$ cm of Hg^{202}, and, after passing the light through an optical system containing a filter to remove all Hg lines but this one and a polaroid to select the appropriate components of the Zeeman spectra, we were able to get a photocurrent of 3.85×10^{-6} ampere from an $A\Omega$ of 0.17 cm^2-steradian, under conditions for which $\delta = 8\times10^8$ cps.[11] These numbers in Eq. (4) lead to an expected signal-to-noise ratio of approximately 10^{-4}.

B. Observability of this Effect

The detection of an effect so heavily overwhelmed by noise demands that a modulation be imposed on the signal in such a way that the noise remains unmodulated. In a technique which is now common[12] the modulated signal, after detection, is passed through a very narrow band amplifier, preferably phase selective, before registering on the indicator. A system of this kind gives a signal-to-noise ratio at the final indicator which is,[13] approximately,

$$S/N = \epsilon(\Delta_1 f/\Delta_2 f)^{\frac{1}{2}}, \qquad (5)$$

where ϵ is the input signal-to-noise ratio, $\Delta_1 f$ the band width of the input, and $\Delta_2 f$ the band width of the narrow-band amplifier.

Unfortunately, it is not feasible to choose $\Delta_1 f$ equal to the spectral line width, approximately 1000 megacycles per second. It was found difficult to make a very low-noise microwave receiver with a band width greater than 7 megacycles. Since we arrived at $\epsilon = 10^{-4}$ in Sec. IIA, we require here, for $S/N = 1$ that $\Delta_2 f = 0.07$ cps. Actually effects not taken into account in this rough calculation, such as the effect of extraneous light on the photocathode, worsen the situation so that, in practice, a response time of 250 seconds, implying $\Delta_2 f \approx 0.004$, was able to give a signal-to-noise ratio of only two at the final indicator. A great deal of patience is required to obtain data under these conditions and pains have to be taken to see that effects of amplifier drift do not give spurious signals.

[10] Discussed at greater length in Sec. IVB.

[11] Factors governing the choice of this spectral line, and conditions under which these numbers were obtained are discussed in Sec. IIID.

[12] R. H. Dicke, Rev. Sci. Instr. 17, 268 (1946).

[13] Easily seen to be an approximation to the exact Eq. (17), as shown following Eq. (22). A qualitative argument is presented by R. H. Dicke (reference 12).

The accurate quantitative calculation from which Eqs. (4) and (5) may be found as approximations, is deferred to Sec. IV so that it can be developed with specific reference to the circuitry presented in Sec. IIID.

III. EXPERIMENTAL EQUIPMENT AND PROCEDURES

A. Modulation Technique

While producing the modulation of the signal it is necessary that the shot noise remain unmodulated to the order of one part in 10^5. It is not possible, for example, to produce the desired modulation by modulating the magnetic field responsible for the line separation because the light intensity is too dependent on the magnetic field intensity. It did prove possible to modulate with constant total intensity by taking advantage of the difference in the polarization of the central and outer components of the Zeeman spectrum with the optics shown in the upper left of Fig. 3. Ignoring, for the moment, the optics to the left of the light source, light from the source passes through a rotating half-wave plate whose effect on a polarized beam of light is to cause a rotation of the plane of polarization at twice the rate at which the retardation plate rotates. The action on polarized light of the rotating $\lambda/2$ plate followed by a polaroid is to produce an interruption of the light at four times the rotation frequency. Viewed perpendicular to the magnetic field, the light is composed of the π components polarized parallel to the magnetic field, and the σ components, polarized normal to the magnetic field so that the effect of the rotating $\lambda/2$ plate plus polaroid B is to produce an alternation between the π and σ components at the photosurface. If the magnetic field is set so that beats between σ components occur at the frequency of the cavity, then this will result in a modulation of these beats.

If the total light from the source is unpolarized this modulation will occur with no variation in total light

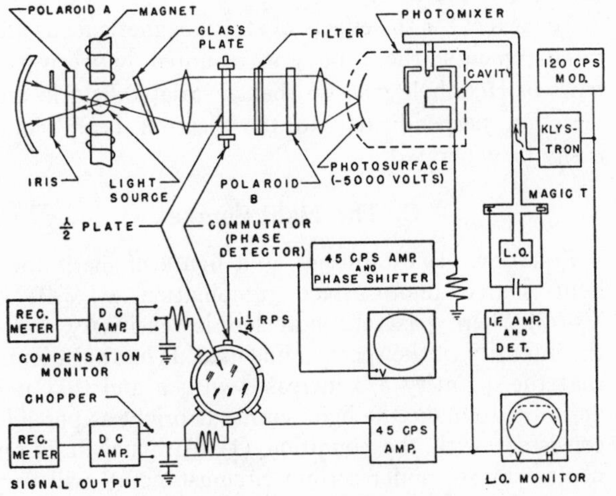

Fig. 3. Basic apparatus for the photoelectric mixing experiment.

intensity or photoelectric emission. However, it was found that our sources were never unpolarized to a degree even approaching one part in 10^5. Even in the absence of a magnetic field a light source is usually polarized by an amount of the order of 1 percent. In a magnetic field the polarization of our microwave excited electrodeless discharges was of the order of 10 percent,[14] or 10^4 times too large. An adjustable tipped glass plate which could compensate for 6–8 percent polarization was placed between the source and the rotating $\lambda/2$ plate as shown in Fig. 3. The remainder of the polarization compensation was accomplished by making use of the light leaving the source in the reverse direction. This light was polarized, adjusted in intensity and plane of polarization and sent back through the source so as to produce zero signal on the compensation monitoring oscilloscope and zero signal on the recording meter which is the output of the compensation monitoring circuit chain.

B. Photomixer

It is in the photoelectric mixing tube, or photomixer, shown in Fig. 4, that the beat frequency is generated. Because of the small signal, even under the most favorable conditions, it is important that the photosurface be as sensitive as possible. The other requirements for the tube can be quantitatively obtained from the calculations of Sec. IVA, but it is easy to understand qualitatively what they are. The cathode must have the largest possible area and see the largest solid angle so that the photocurrent can be large. The signal-to-shot-noise ratio is not affected by this increase in total current but it is important in order to get above detector noise. For the same reason it is important that the resonant cavity in which the current pulses are converted to electrical energy receive almost all of the current and have as high a Q and R (shunt resistance) as possible. The latter requirement calls for a large gap in the cavity which in turn requires high-energy electrons in order that transit time be held to less than half a period.

The obvious photosurface is the $SbCs_3$ photosurface because of its great sensitivity and because it can be laid down on the glass in a semitransparent layer having a very large light-gathering power. For electron focusing purposes the front surface of the tube was ground to the shape of a sphere of radius of curvature 2 in. Contact to the $\frac{1}{2}$-in. diameter cathode was made by means of a layer of platinum laid down on the glass as shown. The electrode following the cathode was a sheet of molybdenum formed to be a sphere concentric with the cathode.

[14] The large degree of polarization in a magnetic field has an interesting probable explanation. Electrons moving along the magnetic field are quickly swept out of the discharge. Electrons moving at right angles to H are confined to tight spirals. The plasma can be expected, then, to be made up predominantly of electrons whose velocity is normal to H. These can be expected to excite radiation polarized normal to H, the σ lines. These are observed to predominate.

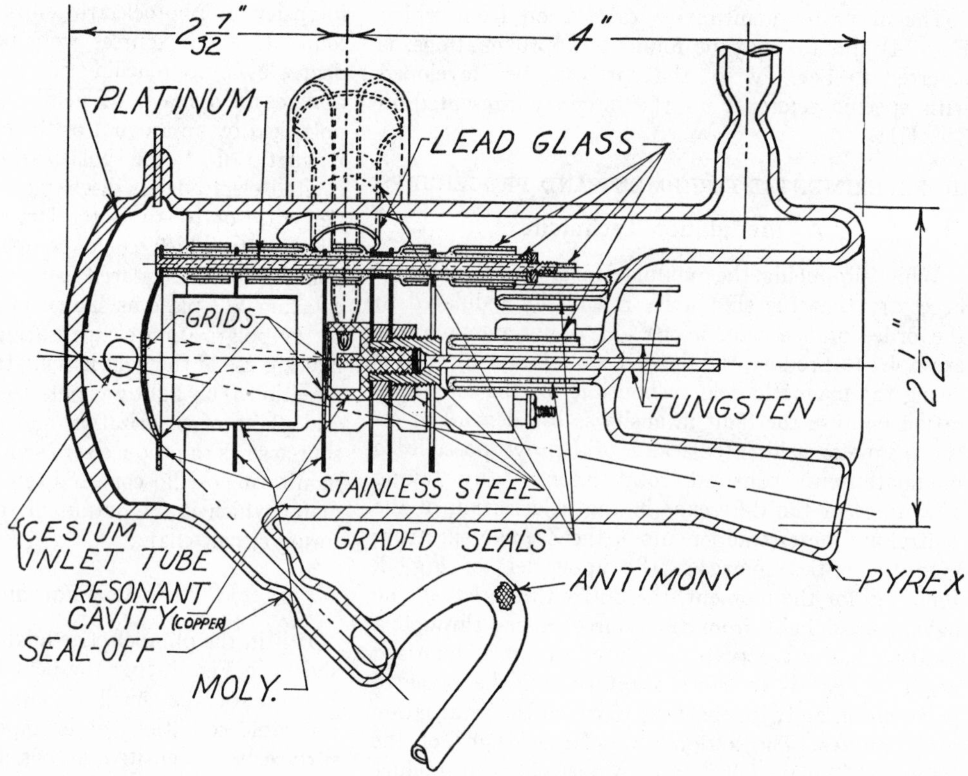

FIG. 4. Photoelectric mixing tube (photomixer).

A grid strung across the $\frac{3}{8}$-in. diameter hole was woven of 0.0005-in. diameter tungsten wire spaced 0.020 in. between wires. The purpose of the cylindrical electrode is to permit slight adjustments in focal length, and of the flat plate to shield the cavity from electrons which originate elsewhere than at the photosurface and to make possible a focusing prior to examination of the microwave power delivered to the cavity. The entire electrode structure except for the cathode is assembled, as shown in Fig. 4, on three columns of Pb glass beads held together with platinum tie rods, and carried on a single 0.100 in. diameter tungsten lead.

The requirement of high voltage, in order to minimize transit time, and the use of Cs in the photosurface manufacture, almost proved incompatible. After the admission of Cs it becomes impossible to hold high voltage across Pyrex. The Cs appears to form a conducting layer which does not leave the glass at temperatures up to 150°C, the highest temperature to which the tube could be subjected after the photosurface was laid down. However the use of lead glass[15] in strategic locations makes it possible to hold high voltages. Each lead-in had a graded seal and a $\frac{1}{2}$-in. length of lead glass and the support structure was made of lead glass beads held together by a tie rod made of platinum to match the coefficient of expansion of the glass.

Serious technical difficulties associated with the making of seals for bringing energy out of the cavity were eliminated by folding the envelope down into a hole in the side of the cavity, so that a pick-up loop on the end of a coaxial line could be inserted externally. The portion of the glass inside the cavity was a very thin-walled bubble and did not load the cavity measurably. A low-loss glass had been planned for this application but Pyrex was found satisfactory. Here, again, to prevent electrical leakage a lead glass shield had to be used. This method of withdrawing energy from the cavity not only eliminated the problem of difficult vacuum seals but provided the very great advantage of permitting external adjustment of the coupling to the cavity.

To correct for the effects of stray magnetic fields and misalignments, the tube was mounted between two pairs of Helmholz coils so that a variable field in any direction perpendicular to the electron beam could easily be generated.

C. The Light Source

Equation (4) establishes as a figure of merit for a light source photosurface combination $\lambda^2 I/\delta A\Omega$ or $\lambda^2 g\mathcal{I}/\delta$, where g is the photoelectric sensitivity and $\mathcal{I}$ the intensity of the spectral line used. It has been found that the quantity $\mathcal{I}/\delta$ increased with $\mathcal{I}$ and that it is desirable to make the light source as bright as possible, consistent with the condition (1). Although it is not necessarily so, under certain circumstances δ will vary as $1/\lambda$. In that case the figure of merit becomes $\lambda^3 g\mathcal{I}$.

[15] V. K. Zworykin and E. G. Ramberg, *Photoelectricity* (John Wiley and Sons, Inc., New York, 1949), p. 88.

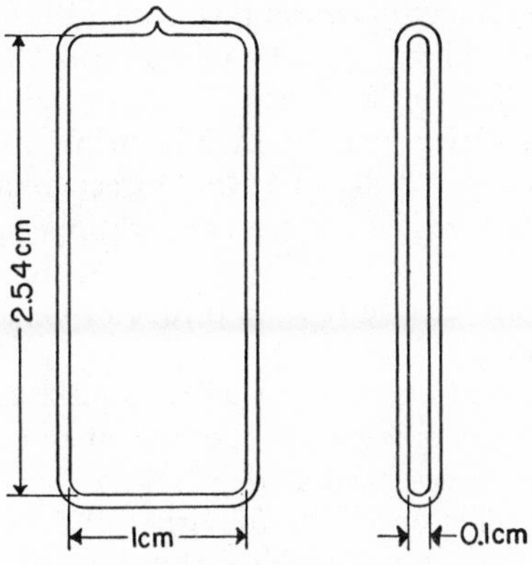

FIG. 5. Light source.

The semitransparent $SbCs_3$ photosurface has its maximum sensitivity at about 5000 A.[16] The factor $\lambda^2 g$ has its maximum at 5300 A and $\lambda^3 g$ at 5400 A. These figures immediately suggest the use of the Hg 5461 A line (see Fig. 5). In addition many other factors point to the use of this line. Mercury is an excellent choice because its large mass minimizes Doppler effect, its availability in separated isotopes makes it possible to eliminate hyperfine structure as a source of broadening, its sparse spectra contains very intense lines which are easily separated from each other and its convenient form and vapor pressure makes its use in discharge tubes extremely convenient. The green line not only occurs close to the optimum wavelength but is very brilliant and there is a great practical convenience in working with the wavelength which has become the standard for retardation plates, nonreflecting coatings, etc. There are two drawbacks to this line. The ease with which the line reverses prevents the use of deep columns at high intensity; and the Zeeman spectra for this line, shown in Fig. 6, are such that only 60 percent of the energy in the σ components is effective in producing beats. It is not inconceivable that a careful examination of the lines λ4046 or λ4358 might make them intrinsically somewhat better than λ5461 but the difficulty of separating the blue and violet lines with filters of sufficient transmissivity rules strongly in favor of the green line. The filter used was made up of 3 mm of Jena glass BG 20 and 2 mm of Jena Glass GG 11 "A" which passed 97 percent of the green line, excluding the effect of surface reflections, while effectively eliminating all of the other mercury lines. To minimize surface reflections the filter components and a polaroid were cemented together with Canada balsam.

In line with recent light source developments,[17–20] our light source was a microwave-excited electrodeless discharge using a Raytheon Microtherm CDM-2 with a maximum rating of 125 watts (but measuring only 100 watts) as a source of power. The advantages of exciting at 2450 Mc/sec were very great as compared to lower frequency excitation. With this supply we were able to get greater intensity and much more stable operation than with a 70-megacycle 2-kilowatt generator we had used previously; and the lifetimes of the tubes, limited by blackening of the walls of the quartz tubes, were much longer at the higher frequencies.[21]

The tubes were made of quartz and shaped as shown in Fig. 5, with light drawn off through the broad faces. The discharge was narrow in one dimension to minimize self-absorption and of large area to produce the maximum total amount of light. With this source we were able to get intensities of 6×10^{-4} watt per cm² per steradian[22] through an optical system containing the green filter and a polaroid under conditions for which the line width, using Hg^{202}, was approximately 10^9 cps. This was obtained with 100 percent of power to the source and cooling of the source by directing jets of air at the two 1-in. long edges. The cooling air was directed at the edges, rather than the faces so that the condensed mercury would not obscure the light.

It was possible, by cutting down on the cooling, to get intensities as high as 2.4×10^{-2} watt per cm² per steradian through the filter and polaroid, but under circumstances in which the line widths were too great for this experiment.

D. Circuits

The essential circuitry is shown in block form in Fig. 3. The power delivered to the cavity by the electrons is withdrawn by a coupling loop on the end of

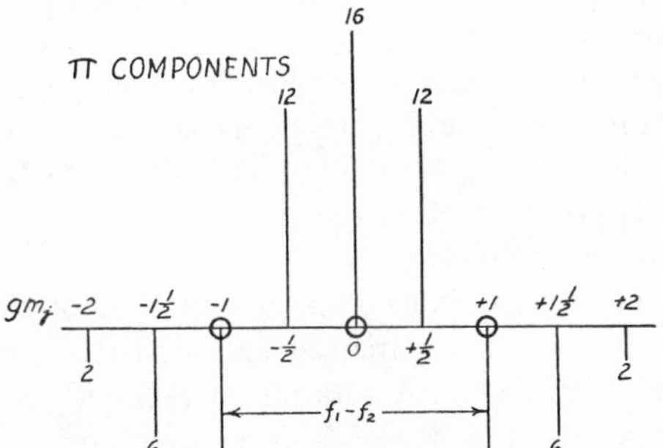

FIG. 6. Zeeman pattern for the 5461 A line of Hg, a $^3S_1 \rightarrow {}^3P_2$ transition. π lines, polarized with the magnetic field, are drawn above and σ lines below. Heights of lines represent relative intensities and the circles indicate the positions in the normal triplet.

[16] V. K. Zworykin and E. G. Ramberg (reference 15), p. 98.
[17] W. F. Meggers and F. O. Westfall, J. Research Natl. Bur. Standards 44, 447 (1950).
[18] Kerr and Des Lattes, Office of Naval Research Technical Report No. 10, Contract Nonr, 248, T. O. 8 (unpublished).
[19] Zelakoff, Wykoff, Aschenhand, and Loomis, J. Opt. Soc. Am. 39, 12 (1949).
[20] E. Jacobsen and G. R. Harrison, J. Opt. Soc. Am. 39, 1054 (1949).

[21] The light source is described much more fully in a separate paper being prepared for another journal.
[22] Intensities quoted are values estimated from the photocurrent from a 929 and the bandbook value of photosensitivity at λ5461.

a coaxial line coupling into a waveguide which feeds the power to a magic T. In the T this power is mixed with local oscillator power at a pair of crystals in a technique now common in microwave detection. To avoid the difficult transformation problems encountered in changing a balanced signal from the crystals to a one-sided input for the intermediate frequency (i.f.), the crystals are Microwave Associates IN23CMR, a matched pair of crystals of opposite polarity with respect to the cartridge. This retains the advantage of balanced detection, i.e., the cancellation of local oscillator noise, and simplifies the problem of an appropriate input transformer for the i.f. amplifier. Since the transformer is one of the most important components in achieving a low-noise, wide-band amplifier, this is an important feature of the detector. The i.f. amplifier is a 30-megacycle amplifier with a noise figure of 1.0 db and a band width of 7 Mc/sec with a gain of about 80 db. The i.f. detector output is fed into a 4-cycle wide 45-cps amplifier whose output was rectified by the phase detector which was a mechanical switch made up of four 90° segments, rotating with the half-wave plate which produced the signal modulation, and three brushes at 45° intervals connected as shown in Fig. 4. The phasing was adjusted to produce a maximum output for a signal of the proper phase by rotating polaroid B. The rectified signal was passed through a low-pass RC filter with an RC of 250 seconds, which sets the band pass of the 45-cps detecting circuitry at approximately 1/250 cps. The filter output is fed to a low frequency (0.1-cps) periodic chopper at the input of the balanced dc amplifier which eliminates dc amplifier drift as a source of spurious signal. The effect of the chopper is to discriminate between signals originating prior to the dc amplifier and dc amplifier drift in a manner which is made clear in the photographs of Fig. 7. The positive and negative portions of the chopper cycle are unequal in length to discriminate positive from negative signals. This is particularly important when working near noise level, as in trace 3, Fig. 7, in order to be able to observe, when frequent reversals occur, whether the trace is not more often in one direction than another.

The system has inherent in it a method of checking to see whether an apparent signal originates anywhere else in the circuitry. A rotation of polaroid B by 90° should reverse the sign of a signal which originates in the light source, but leave unaffected a signal originating between the photosurface and the chopper.

Because of klystron drift a device for maintaining the local oscillator frequency at 30 Mc/sec from the cavity frequency is necessary. A 120-cps frequency-modulated signal is reflected from the cavity as shown in Fig. 3. Far from the cavity resonance, the signal displayed on the oscilloscope trace mirrors the i.f. response, as shown by the dotted curve on the oscilloscope. When the local oscillator is properly set with respect to the cavity, the oscilloscope trace shows a dip whose width is characteristic of the cavity Q and whose depth is characteristic of the coupling to the cavity. This simple circuit, then, provides us with a guide to i.f. alignment, a monitor for holding the local oscillator at the correct frequency with respect to the cavity, and a guide for proper adjustment of the cavity coupling. The 120-cps signal and its harmonics are unable to pass through the narrow-band 45-cps amplifier.

The polarization compensation monitor chain is fed by the 45 cps component of the current to the resonant cavity, and is essentially identical to that portion of the main detector chain following the i.f. detector. However, it contains a phase shifter so that the two chains can be set to amplify exactly in phase.

IV. CALCULATIONS

A. Precise Calculations for an Arbitrary Area of Coherence

Because some interesting conclusions can be drawn from an accurate comparison of the observed signal with the theoretical signal strength, and because the detecting system is one which has a wide application, it seems pertinent to present an outline and the results of a precise calculation.

If two spectral lines of equal intensity and equal width, and with Gaussian shapes, are split into Fourier components which are added and squared, representing the action of the photosurface, that part of the current per unit area which arises from differences in frequencies in the two lines is

$$i=[(2\ln 2)/\pi]^{\frac{1}{4}}(i_{\mathrm{dc}}/\sqrt{\delta})\sum_k \exp[-(\ln 2)(\omega_0-\omega_k)^2/\delta^2]\times\cos(\omega_k t+\phi_k)(\Delta\omega)^{\frac{1}{2}}, \quad (6)$$

where ω_0 is the difference in frequency between line centers, δ is the line width, i_{dc} is the photocurrent per unit area, ϕ_k is a random phase and $\Delta\omega$ is the band width of a single component of the sum. (All frequencies are here expressed in radians/sec.) The current is proportional to cathode area only over a small area $\gamma\lambda^2/\Omega$, referred to as the area of coherence, where Ω is the solid angular spread in the light striking the cathode, λ is the wavelength, and γ is a numerical constant of the order of one.[23] Since the beats from separate areas are randomly phased, the rms beat current from the entire cathode is proportional to the square root of the number of such areas in the cathode surface. This leads to a mean square current, in the frequency range ω to $\omega+\Delta\omega$, including here the effect of the modulation at a frequency ω_M,

$$\langle I^2\rangle=\frac{1}{2}\left(\frac{\ln 2}{2\pi}\right)^{\frac{1}{2}}\frac{\alpha_2{}^2 I_c{}^2\gamma\lambda^2}{\delta\alpha_1 A\Omega}\times\exp\left[-\frac{2(\ln 2)(\omega_0-\omega)^2}{\delta^2}\right](\Delta\omega)\cos\omega_M t, \quad (7)$$

[23] This point is amplified in Sec. IVB.

where I_c is the dc current to the cavity, α_1 is the fraction of the photocathode area from which electrons enter the microwave cavity and α_2 is the fraction of the photoelectrons due to light from the two lines being mixed.

Accompanying this current is the unavoidable shot noise given by

$$\langle I_n^2\rangle = eI_c\Delta\omega/\pi, \quad (8)$$

where e is the electron charge. The currents given by Eqs. (7) and (8) deliver power to the cavity according to

$$P=\frac{\langle I^2\rangle R}{1+4Q^2(\omega-\omega_c)^2/\omega_c^2} \quad (9)$$

where R, Q, and ω_c are the shunt resistance, Q-factor and resonant frequency of the cavity, leading to a power per unit frequency range into the microwave receiver, including receiver noise referred to its input,

$$\frac{dP_1}{d\omega}=\frac{1}{1+4Q^2(\omega-\omega_c)^2/\omega_c^2}+K_F+\frac{K_B\cos\omega_M t}{1+4Q^2(\omega-\omega_c)^2/\omega_c^2}, \quad (10)$$

where

$$K_B=\frac{1}{2}\left(\frac{\pi\ln 2}{2}\right)^{\frac{1}{2}}\frac{\gamma\lambda^2}{e}\frac{I_c\alpha_2^2}{\delta\alpha_1 A\Omega} \quad (11)$$

is the signal-to-shot-noise ratio at the cavity and

$$K_F=FkT(1+\beta)/2eI_cR\beta \quad (12)$$

is the ratio of the contribution of receiver noise to that of shot noise at the central frequency. F is the noise figure of the microwave receiver and β, the coupling coefficient, is the ratio of power withdrawn from the cavity to power dissipated in the cavity. Q and R are both inversely proportional to $(1+\beta)$.

The effect of the superheterodyne detection and i.f. amplification is to lower all frequencies and multiply the power by $G_1(\omega)$, the i.f. power gain, leading to

$$\frac{dP_2}{d\omega}=G_1(\omega)\left\{\frac{1}{1+4Q^2(\omega-\omega_{\text{i.f.}})^2/\omega_c^2}+K_F+\frac{K_B\cos\omega_M t}{1+4Q^2(\omega-\omega_{\text{i.f.}})^2/\omega_c^2}\right\}. \quad (13)$$

An expression for the output of the detector[24] following the i.f. may be obtained by repeating the process which led to Eq. (6), i.e., representing the power P_2 as the square of the voltage expressed in terms of the Fourier components. If we retain only terms $\omega\approx\omega_M=2\pi\times 45$, we get for the input to the 45-cps amplifier:

$$V=2Y\textstyle\sum_j\cos(\omega_j t+\phi_j)(\Delta\omega)^{\frac{1}{2}}+K_BX\cos\omega_M t, \quad (14)$$

where

$$X=\int_0^\infty\frac{G_1(\omega)d\omega}{1+4Q^2(\omega-\omega_{\text{i.f.}})^2/\omega_c^2}, \quad (15)$$

and

$$Y=\left\{\int_0^\infty G_1^2(\omega)\left[K_F+\frac{1}{1+4Q^2(\omega-\omega_{\text{i.f.}})^2/\omega_c^2}\right]^2 d\omega\right\}^{\frac{1}{2}}. \quad (16)$$

The effect of the phase detector plus the low-pass filter is to give, finally, a signal-to-rms-noise ratio[25] at the indicator:

$$S/N=K_BX/2YZ, \quad (17)$$

where

$$Z=\left\{\int_0^\infty[G_2(\omega)/G_2(0)]^2 d\omega\right\}^{\frac{1}{2}}, \quad (18)$$

$G_2(\omega)$ being the voltage gain of the low-pass filter.

In going from Eq. (13) to Eq. (14) we assumed, above, a square-law detector. Our detector, as operated, is more nearly linear than square-law but, according to Selove,[26] this will not alter Eq. (17).

For the low-pass filter shown in Fig. 3,

$$S/N=(RC/2\pi)^{\frac{1}{2}}K_BX/Y, \quad (19)$$

where R is the parallel impedance of the filter resistance and dc amplifier input.

For a flat i.f. response over a band width B,

$$X=B(\tan^{-1}s)/s, \quad (20)$$

where $s=QB/\omega_c$ is the ratio of i.f. to cavity width, and

$$Y=B\left[\frac{\tan^{-1}s}{s}\left(\frac{1}{2}+2K_F\right)+\frac{1}{2}\frac{1}{1+s^2}+K_F^2\right]^{\frac{1}{2}}, \quad (21)$$

so that

$$\frac{S}{N}=\frac{K_B(RCB')^{\frac{1}{2}}(\tan^{-1}s)/s}{\left[\frac{\tan^{-1}s}{s}\left(\frac{1}{2}+2K_F\right)+\frac{1}{2}\frac{1}{1+s^2}+K_F^2\right]^{\frac{1}{2}}}, \quad (22)$$

where $B'=B/2\pi$ is the i.f. band width in cycles per second. For $K_F=0$ and $s\ll 1$, this reduces to $S/N=(RCB')^{\frac{1}{2}}K_B$ in confirmation of Eq. (5).

B. Area of Coherence

There remains to be calculated the constant γ, giving the ratio of the effective area of coherence to λ^2/Ω. It is plain that the area of coherence must be of the order of the size of a diffraction image from the following considerations: Over a diffraction pattern a large phase shift occurs but, since the diffraction pattern is essentially identical for two frequencies very close together, the relative phase shift and therefore the phase of the

[24] In ordinary microwave circuitry this would be the second detector. In this application it is the third, the first being the photosurface.

[25] This result is easily seen to be quite similar to that of R. H. Dicke (reference 12) as expressed in his Eq. (20).

[26] W. Selove, Rev. Sci. Instr. 25, 120 (1954).

beats remains constant over a diffraction pattern, if the two frequencies originated at a single source point. Of course, the area of a single diffraction pattern contains many overlapping patterns but it is still true that the beats will not be completely random until we get to two image points separated by the size of the diffraction image. The area of the central image is $3.67\lambda^2/\Omega$ so that the effective area of coherence is certainly less than this, the intensity in the secondary diffraction rings being very small.

Gerjuoy[27] has made a calculation of the signal strength based on an integration over the entire cathode using an expression for the field intensity which considers the light as spread uniformly over an area A with a solid angular spread Ω, at each point, and derived an expression for the intensity which an appropriate comparison with the derivation of the preceding section yields a value of $\gamma=1$.

Another, and very interesting, approach to γ may be made through a work of Landé.[28] He derives an expression for the number of degrees of freedom in a beam of light acting for a specified time, which can be converted, for a narrow band of frequencies, into an expression for the number of independent oscillators required to duplicate a light beam of given area and angular spread. If this number of oscillators is equated to the number of areas of coherence, we are again led to $\gamma=1$.

V. EXPERIMENTAL RESULTS

Several recording meter traces are shown in Fig. 7. The signal size is gotten from a comparison of trace 1 with trace 3, corresponding to a rotation of the polaroid B, Fig. 3, through 90°. The signal does not merely change sign in this process, as it should, indicating that some signal originates in the circuitry and this is borne out by trace 2 taken for zero light intensity. The source of the spurious signal apparently lies in a mechanical coupling between the rotating $\lambda/2$ plate and the microwave circuitry. At least it is true that pains taken to isolate these and to make the rotation as quiet as possible reduced this to the low level shown from a very much higher one. The signal obtained in this manner, divided by the rms deviation from the mean was taken for several magnetic field variations and is shown plotted in Fig. 8. The lengths of the lines drawn through each point are the probable error in the results and are less than the rms values of the deviations out of the dc amplifier because the trace was observed for a time long compared to RC for the low-pass filter. Because of darkening of the light source tubes, and other effects, it was not possible to duplicate the conditions for each point and the signal heights have all been corrected to 3.85 μa.

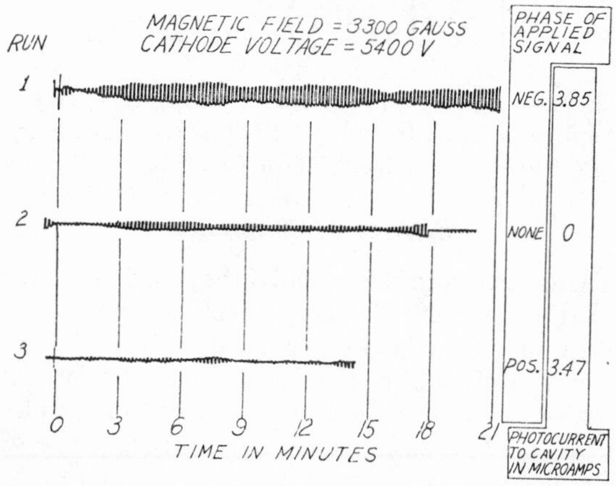

FIG. 7. Photograph of meter traces.

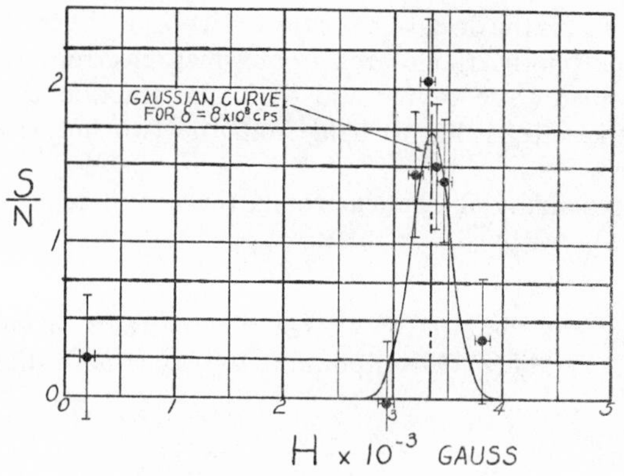

FIG. 8. Final signal-to-noise ratio *vs* Zeeman field. Cavity current $=3.85$ μa. Accelerator voltage$=5400$ v.

The relative contribution of circuit noise and shot noise is contained in Eq. (22) and a measurement of the variation in total noise with cavity current yielded, through this equation, a value of $K_F=0.206$ at $I_c=3.85$ μa. [This value in Eq. (12) yields a noise figure for the microwave receiver of approximately 8, which is consistent with that expected from the measured i.f. noise figures and checks the value roughly measured directly.]

For the calculation of K_B by Eq. (11), $\delta/2\pi$ was taken as 8×10^8 cps. This value is obtained from Fig. 7 if it is realized that the width, at half-maximum, of the best spectrum is $\sqrt{2}$ times the spectral line width. The fraction of photocurrent which entered the cavity was $\alpha_1=0.6$, and for the spectral line used (see Fig. 6) $\alpha_2=0.6$. The factor $A\Omega$ was approximately 0.17, leading to $K_B=2.63\times10^{-5}$.

When these values of K_B and K_F together with $RC=240$ sec, $B'=6.8\times10^6$ cps and $s=1$ are used, Eq. (23) leads to

$$(S/N)_{\text{calculated}}=0.83. \qquad (23)$$

This is one-half of the observed S/N shown in Fig. 8, close enough to be regarded as good agreement. Furthermore, a spatial variation in emission over the cathode

[27] E. Gerjuoy, University of Pittsburgh (unpublished).
[28] A. Landé, *Handbuch der Physik* (Verlag Julius Springer, Berlin, 1928), Vol. 20, p. 453.

area, due either to light source or photocathode inhomogeneity, will cause an S/N greater than that calculated on the basis of uniform emission and it is likely that the nonuniformity in emission was considerable.

VI. CONCLUSIONS

The agreement between calculations and observations is regarded as a verification of the basic premise discussed in the introduction, i.e., that photoelectric emission is proportional to the square of the total wave amplitude, implying an interference between independently generated light waves, and as a generalization, perhaps, the same kind of interference between other waves such as quantum mechanical wave functions. An alternative to this conclusion would be that there existed some coherence between components of a Zeeman spectrum. However, from a quantum-mechanical point of view two Zeeman lines originate in two distinct sets of atoms with different orientations in the magnetic field and are as independent as if they originated in different atomic species. Furthermore, it seems unlikely that an effect due to a completely unrelated cause would produce so nearly the same size signal as the one we calculated.

A secondary conclusion is that any delay between photon absorption and electron emission must be significantly less than 10^{-10} second, since a decay time equal to the beat period would decrease the signal by a factor of 6.4. This time is so short compared to the lifetimes of allowed transitions in the visible that it becomes significant. For example, the possibility that an electron could be raised to some metastable state from which it might decay to the free electron state is ruled out at least for the antimony-cesium photosurface used in this experiment.

ACKNOWLEDGMENTS

We wish to thank the Office of Naval Research for their financial support and loan of valuable equipment during a portion of this experimentation. We wish to thank Dr. Meggers and Dr. Westfall of the National Bureau of Standards for filling light-source tubes with Hg^{198}, even though in our final runs Hg^{202}, which was kindly loaned by the U. S. Atomic Energy Commission, was used. We are especially grateful to Mr. H. M. Joseph, now with Lear Inc., Research and Development Laboratories, who worked many hours with us, without remuneration, to help achieve a very low noise i.f. amplifier essential to the success of the experiment. Dr. Gordon Locher, of Western Radiation Laboratory, also supplied, without charge, considerable help in the manufacture of the photoelectric mixing tube. We wish to acknowledge also the important role of the extreme glassblowing skill of Mr. Greiner in the photomixer manufacture. We are grateful to Dr. A. J. Allen, Director of the University of Pittsburgh Radiation Laboratory and his staff for their gracious help in the preparation of the manuscript during the period when one of us, A. T. Forrester, was associated with that group.

Above all we wish to acknowledge that the idea was originally conceived by W. E. Parkins and A. T. Forrester and given credence through discussions and calculations in which E. Gerjuoy participated as well. We are also pleased to acknowledge the stimulation of numerous discussions with E. Gerjuoy throughout the progress of this work.

PAPER NO. 18

Reprinted from *Proceedings of the Royal Society*, Ser. A, Vol. 230, pp. 246–265, 1955.

A macroscopic theory of interference and diffraction of light from finite sources

II. Fields with a spectral range of arbitrary width

BY E. WOLF

The Physical Laboratories, University of Manchester

(*Communicated by M. Born, F.R.S.—Received* 29 *November* 1954)

The results of part I of this investigation are generalized to stationary fields with a spectral range of arbitrary width. For this purpose it is found necessary to introduce in place of the mutual intensity function of Zernike a more general correlation function

$$\hat{\Gamma}(\mathbf{x}_1, \mathbf{x}_2, \tau) = \langle \hat{V}(\mathbf{x}_1, t+\tau)\, \hat{V}^*(\mathbf{x}_2, t)\rangle,$$

which expresses the correlation between disturbances at any two given points $P_1(\mathbf{x}_1)$, $P_2(\mathbf{x}_2)$ in the field, the disturbance at P_1 being considered at a time τ later than at P_2. It is shown that $\hat{\Gamma}$ is an observable quantity. Expressions for $\hat{\Gamma}$ in terms of functions which specify the source and the transmission properties of the medium are derived.

Further, it is shown that *in vacuo* the correlation function obeys rigorously the two wave equations

$$\nabla_s^2 \hat{\Gamma} = \frac{1}{c^2}\frac{\partial^2 \hat{\Gamma}}{\partial \tau^2} \quad (s = 1, 2),$$

where ∇_s^2 is the Laplacian operator with respect to the co-ordinates (x_s, y_s, z_s) of $P_s(\mathbf{x}_s)$. Using this result, a formula is obtained which expresses rigorously the correlation between disturbances at P_1 and P_2 in terms of the values of the correlation and of its derivatives at all pairs of points on an arbitrary closed surface which surrounds P_1 and P_2. A special case of this formula ($P_2 = P_1$, $\tau = 0$) represents a rigorous formulation of the generalized Huygens principle, involving observable quantities only.

1. INTRODUCTION

In part I of this investigation (Wolf (1954*a*), to be referred to as I), interference and diffraction of light in stationary fields produced by finite sources which emit light within a small but finite spectral range were studied. Whilst the results obtained are of interest in connexion with a variety of problems, they are inadequate in the treatment of problems where the path differences between the interfering beams are sufficiently large (cf. appendix 2 below). The present paper removes this restriction, and extends the results to stationary fields of any spectral range.

In the investigation of I, the mutual intensity function

$$\hat{\Gamma}(\mathbf{x}_1, \mathbf{x}_2) = \langle \hat{V}(\mathbf{x}_1, t)\, \hat{V}^*(\mathbf{x}_2, t)\rangle \tag{1·1}$$

of Zernike, played an essential part. In (1·1), $\hat{V}(\mathbf{x}, t)$ represents the complex disturbance at a point specified by the position vector $\mathbf{x}$, at time t, asterisks denoting the complex conjugate and sharp brackets the time average. In order to extend the analysis to fields with a spectral range of arbitrary width, it is found necessary to introduce in place of (1·1) the more general correlation function

$$\hat{\Gamma}(\mathbf{x}_1, \mathbf{x}_2, \tau) = \langle \hat{V}(\mathbf{x}_1, t+\tau)\, \hat{V}^*(\mathbf{x}_2, t)\rangle. \tag{1·2}$$

With the help of this function which is shown to represent an observable quantity, the generalized Huygens principle derived in I and the generalized interference law (I, (3·9)) of Zernike & Hopkins are extended to the wider class of fields under consideration. Expressions for $\hat{\Gamma}(\mathbf{x}_1, \mathbf{x}_2, \tau)$ in terms of quantities which specify the source and the medium are also derived.

In § 7, it is shown that *in vacuo* the correlation function (1·2) obeys rigorously the wave equations

$$\left.\begin{aligned}\nabla_1^2 \hat{\Gamma} &= \frac{1}{c^2}\frac{\partial^2 \hat{\Gamma}}{\partial \tau^2},\\ \nabla_2^2 \hat{\Gamma} &= \frac{1}{c^2}\frac{\partial^2 \hat{\Gamma}}{\partial \tau^2},\end{aligned}\right\} \tag{1·3}$$

where ∇_1^2 and ∇_2^2 are the Laplacian operators with respect to the co-ordinates of the points specified by the position vectors $\mathbf{x}_1$ and $\mathbf{x}_2$ respectively.

In § 8, a formula is derived which expresses rigorously the correlation between the disturbances at two given points $P_1(\mathbf{x}_1)$ and $P_2(\mathbf{x}_2)$ in the field in terms of the correlation, and its derivatives, between the disturbances at all pairs of points on an arbitrary closed surface surrounding P_1 and P_2. A special case of this formula ($\mathbf{x}_1 = \mathbf{x}_2$, $\tau = 0$) represents a rigorous formulation of our generalized Huygens principle.

As in I, polarization effects are not considered in the present paper. They will be taken into account in part III of this investigation, where it will be shown that a natural generalization of our results leads to a unified treatment of partial coherence and partial polarization and to a formulation of a wide branch of optics in terms of observable quantities only.†

2. Preliminary: a complex representation of real, polychromatic fields

When dealing with a real, monochromatic (or nearly monochromatic) field, it is usual to employ a complex representation, the field variable being identified with the real part of an appropriate complex function.

† A brief preliminary account of some of these results will be found in Wolf (1954*b*).

In the present paper we shall be concerned with polychromatic fields, i.e. with fields which cover a finite spectral range. It will be convenient to use also in this case a complex representation, this being a natural extension of the representation used in connexion with monochromatic fields.

Let $F(t)$ be a real function, defined for all values of t $(-\infty < t < \infty)$ which possess a Fourier integral representation:

$$F(t) = \int_0^\infty (a_\nu \cos 2\pi\nu t + b_\nu \sin 2\pi\nu t)\, d\nu. \tag{2·1}$$

With F we associate the (generally complex) function $\hat{F}$, defined as

$$\hat{F}(t) = \int_0^\infty (a_\nu + ib_\nu)\, e^{-2\pi i\nu t} d\nu. \tag{2·2}$$

It is seen that

$$F(t) = \mathscr{R}\hat{F}(t), \tag{2·3}$$

where $\mathscr{R}$ denotes the real part.

$\hat{F}(t)$ will be referred to as the *half-range complex function* associated with the real function $F(t)$; it is characterized by the property that it may be represented by a Fourier integral which contains no terms of negative frequencies.† F defines $\hat{F}$ uniquely and vice versa.

Throughout this paper a real function and the associated half-range complex function will be denoted by the same symbol, the latter being distinguished by a circumflex. The use of half-range complex functions in place of real functions considerably shortens some of our calculations, and enables the resulting formulae to be expressed in a form which closely resembles those obtained in connexion with almost monochromatic fields in I.

3. A space-time correlation function of stationary fields

In order to extend the analysis of I to stationary fields the spectral range of which is arbitrarily wide, it is necessary, as will be seen below, to introduce, in place of the mutual intensity function (1·1) of Zernike, the more general correlation function‡

$$\hat{\Gamma}(\mathbf{x}_1, \mathbf{x}_2, \tau) = \langle \hat{V}(\mathbf{x}_1, t+\tau)\, \hat{V}^*(\mathbf{x}_2, t)\rangle, \tag{3·1}$$

† *Added in proof* 21 *March* 1955: Since this was written I find that the same complex representation of real fields was introduced previously by Gabor (1946, p. 432) in his interesting investigations in communication theory. Gabor also points out that the real and imaginary parts of such 'half-range complex functions' are Hilbert transforms of each other.

‡ The auto-correlation function, which in our notation would be written as $\hat{\Gamma}(\mathbf{x}, \mathbf{x}, \tau)$, was previously employed in optics by a number of authors, e.g. Wiener (1930), Van Cittert (1939) and Parke (1948). Wiener (1930, p. 119) points out that this function played a fundamental part in Schuster's theory of white light.

Added in proof 21 *March* 1955: In a very interesting paper which was published since this was written, Blanc-Lapierre & Dumontet (1955) applied the general theory of random functions to the optical coherence problems. In their treatment which leads to several new results some of which are closely related to ours, the cross-correlation function (3·2) plays also a basic role.

which expresses the correlation between disturbances at the points $P_1(\mathbf{x}_1)$ and $P_2(\mathbf{x}_2)$, the disturbance at $\mathbf{x}_1$ being considered at a time τ later than at $\mathbf{x}_2$. The sharp brackets denote the time average.†

By a straightforward calculation carried out fully in appendix 1, it may be shown that if $\hat{V}$ is a half-range complex function (in the sense defined in the previous section) the correlation function defined by (3·1) is likewise a half-range complex function, the knowledge of $\hat{\Gamma}(\mathbf{x}_1, \mathbf{x}_2, \tau)$ being equivalent to the knowledge of the real function

$$\Gamma(\mathbf{x}_1, \mathbf{x}_2, \tau) = \mathscr{R}\hat{\Gamma}(\mathbf{x}_1, \mathbf{x}_2, \tau) = 2\langle V(\mathbf{x}_1, t+\tau)\, V(\mathbf{x}_2, t)\rangle. \tag{3·2}$$

We shall need an expression for $\hat{\Gamma}$ in terms of the Fourier components of $\hat{V}$. Let

$$\left.\begin{aligned} \hat{V}(\mathbf{x}, t) &= \int_0^\infty v(\mathbf{x}, \nu)\, \mathrm{e}^{-2\pi i\nu t}\, d\nu; \\ \text{then}\qquad v(\mathbf{x}, \nu) &= \int_{-T}^{T} \hat{V}(\mathbf{x}, t)\, \mathrm{e}^{2\pi i\nu t}\, dt. \end{aligned}\right\} \tag{3·3}$$

It follows from (3·1), on using the convolution theorem, that

$$\hat{\Gamma}(\mathbf{x}_1, \mathbf{x}_2, \tau) = \lim_{T\to\infty} \frac{1}{2T} \int_0^\infty v(\mathbf{x}_1, \nu)\, v^*(\mathbf{x}_2, \nu)\, \mathrm{e}^{-2\pi i\nu\tau}\, d\nu. \tag{3·4}$$

We note a useful relation between $\hat{\Gamma}(\mathbf{x}_2, \mathbf{x}_1, -\tau)$ and $\hat{\Gamma}(\mathbf{x}_1, \mathbf{x}_2, \tau)$. From (3·4),

$$\hat{\Gamma}(\mathbf{x}_2, \mathbf{x}_1, -\tau) = \hat{\Gamma}^*(\mathbf{x}_1, \mathbf{x}_2, \tau). \tag{3·5}$$

We shall normalize $\hat{\Gamma}$ by setting

$$\hat{\gamma}(\mathbf{x}_1, \mathbf{x}_2, \tau) = \frac{\hat{\Gamma}(\mathbf{x}_1, \mathbf{x}_2, \tau)}{\sqrt{\hat{\Gamma}(\mathbf{x}_1, \mathbf{x}_1, 0)}\sqrt{\hat{\Gamma}(\mathbf{x}_2, \mathbf{x}_2, 0)}} = \frac{\hat{\Gamma}(\mathbf{x}_1, \mathbf{x}_2, \tau)}{\sqrt{I(\mathbf{x}_1)}\sqrt{I(\mathbf{x}_2)}}, \tag{3·6}$$

where

$$I(\mathbf{x}_s) = \hat{\Gamma}(\mathbf{x}_s, \mathbf{x}_s, 0) = \langle \hat{V}(\mathbf{x}_s, t)\, \hat{V}^*(\mathbf{x}_s, t)\rangle \quad (s = 1, 2) \tag{3·7}$$

is the intensity at the point $P_s(\mathbf{x}_s)$. On applying the modulus inequality for integrals and the Schwarz inequality to (3·4), one has, by a similar argument as in I, § 3,

$$|\hat{\gamma}(\mathbf{x}_1, \mathbf{x}_2, \tau)| \leqslant 1. \tag{3·8}$$

4. An approximate propagation law for $\hat{\Gamma}(\mathbf{x}_1, \mathbf{x}_2, \tau)$ and an extension of the generalized Huygens principle to fields with spectral range of arbitrary width

As in I we consider the propagation of a beam of light from a finite source Σ. The field will again be assumed to be stationary, but no restriction on the width of the spectral range will now be imposed.

† In I, the time average was taken over the finite range $-T \leqslant t \leqslant T$, but it is more convenient mathematically (although it is not more significant physically) to allow the range to become infinite by proceeding to the limit $T \to \infty$, as customary. We shall now understand the time average in this sense:

$$\hat{\Gamma}(\mathbf{x}_1, \mathbf{x}_2, \tau) = \lim_{T\to\infty} \frac{1}{2T} \int_{-T}^{T} \hat{V}(\mathbf{x}_1, t+\tau)\, \hat{V}^*(\mathbf{x}_2, t)\, dt.$$

The integration over the frequency range was taken in I from $-\infty$ to $+\infty$, but as explained in the preceding section we may set $v(\mathbf{x}, \nu) = 0$ for $\nu < 0$ and hence integrate over the positive range $0 \leqslant \nu < \infty$ only. The quantities which were denoted by $V(\mathbf{x}, t)$, $\Gamma(\mathbf{x}_1, \mathbf{x}_2)$ and $\gamma(\mathbf{x}_1, \mathbf{x}_2)$ in I would in the present notation be therefore written as $\hat{V}(\mathbf{x}, t)$, $\hat{\Gamma}(\mathbf{x}_1, \mathbf{x}_2)$ and $\hat{\gamma}(\mathbf{x}_1, \mathbf{x}_2)$.

Let $\mathscr{A}$ be a surface cutting across the beam (see figure 1) and let $P_1(\mathbf{x}_1)$ and $P_2(\mathbf{x}_2)$ be two points on that side of $\mathscr{A}$ towards which the light is advancing. We shall derive an expression for the correlation $\hat{\Gamma}(\mathbf{x}_1, \mathbf{x}_2, \tau)$ between the disturbances at P_1 and P_2, in terms of the values which $\hat{\Gamma}$ takes at all pairs of points on the surface $\mathscr{A}$.

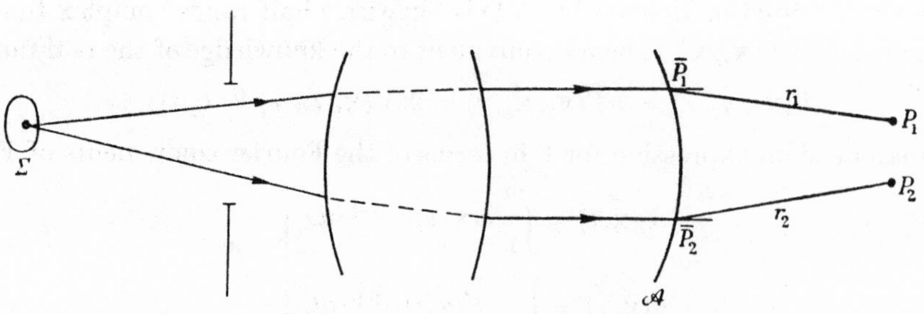

FIGURE 1. Derivation of a propagation law for $\hat{\Gamma}(\mathbf{x}_1, \mathbf{x}_2, \tau)$.

Each Fourier term $v(\mathbf{x}, \nu)\exp\{-\mathrm{i}\,2\pi\nu t\}$ of $\hat{V}$ represents a perfectly monochromatic wave. Hence the value of v at P_1 may be expressed (under the usual restrictions) in terms of the values which v takes at all points on the surface $\mathscr{A}$, by means of the ordinary Huygens principle:

$$v(\mathbf{x}_1, \nu) = \int_{\mathscr{A}} v(\bar{\mathbf{x}}_1, \nu)\frac{\mathrm{e}^{\mathrm{i}kr_1}}{r_1}\Lambda_1\,\mathrm{d}\bar{\mathbf{x}}_1. \tag{4·1}$$

Here $\bar{\mathbf{x}}_1$ is the position vector at a typical point $\bar{P}_1$ on the surface $\mathscr{A}$, r_1 is the distance from $\bar{P}_1$ to P_1 (see figure 1), $k = 2\pi\nu/c = 2\pi/\lambda$, c being the vacuum velocity of light and λ the wave-length. Λ denotes as before the usual inclination factor (see I, (2·4)).

In a similar way, the values $v(\mathbf{x}_2, \nu)$ at P_2 may be expressed in the form

$$v(\mathbf{x}_2, \nu) = \int_{\mathscr{A}} v(\bar{\mathbf{x}}_2, \nu)\frac{\mathrm{e}^{\mathrm{i}kr_2}}{r_2}\Lambda_2\,\mathrm{d}\bar{\mathbf{x}}_2. \tag{4·2}$$

From (4·1) and (4·2),

$$v(\mathbf{x}_1, \nu)\,v^*(\mathbf{x}_2, \nu) = \int_{\mathscr{A}}\int_{\mathscr{A}} v(\bar{\mathbf{x}}_1, \nu)\,v^*(\bar{\mathbf{x}}_2, \nu)\frac{\mathrm{e}^{\mathrm{i}k(r_1-r_2)}}{r_1 r_2}\Lambda_1\Lambda_2^*\,\mathrm{d}\bar{\mathbf{x}}_1\,\mathrm{d}\bar{\mathbf{x}}_2, \tag{4·3}$$

the points $\bar{P}_1(\bar{\mathbf{x}}_1)$ and $\bar{P}_2(\bar{\mathbf{x}}_2)$ exploring the surface $\mathscr{A}$ independently. Next we multiply (4·3) by $\frac{1}{2T}\exp\{-2\pi\mathrm{i}\nu\tau\}$, integrate over the frequency range and proceed to the limit $T\to\infty$. We then obtain the following expression for $\hat{\Gamma}(\mathbf{x}_1, \mathbf{x}_2, \tau)$:

$$\hat{\Gamma}(\mathbf{x}_1, \mathbf{x}_2, \tau) = \int_{\mathscr{A}}\int_{\mathscr{A}}\left\{\lim_{T\to\infty}\frac{1}{2T}\int_0^\infty v(\bar{\mathbf{x}}_1, \nu)\,v^*(\bar{\mathbf{x}}_2, \nu)\exp\left\{-2\pi\mathrm{i}\nu\left[\tau-\frac{r_1-r_2}{c}\right]\right\}\Lambda_1\Lambda_2^*\,\mathrm{d}\nu\right\}\frac{1}{r_1}\frac{1}{r_2}\,\mathrm{d}\bar{\mathbf{x}}_1\,\mathrm{d}\bar{\mathbf{x}}_2. \tag{4·4}$$

Now in the integral over the frequency range, the factors Λ_1 and Λ_2^* are well-behaved functions, depending on the frequency only through a multiplicate factor ν.

We shall take Λ_1 and Λ_2^* outside the ν integration in mean values† (denoted by $\bar{\Lambda}_1$ and $\bar{\Lambda}_2^*$). The remaining part of the frequency integral gives, in the limit $T\rightarrow\infty$, precisely $\hat{\Gamma}\left(\bar{\mathbf{x}}_1, \bar{\mathbf{x}}_2, \tau-\frac{r_1-r_2}{c}\right)$. Hence (4·4) reduces to the relatively simple law

$$\hat{\Gamma}(\mathbf{x}_1, \mathbf{x}_2, \tau) = \int_{\mathscr{A}}\int_{\mathscr{A}} \frac{\hat{\Gamma}\left(\bar{\mathbf{x}}_1, \bar{\mathbf{x}}_2, \tau-\frac{r_1-r_2}{c}\right)}{r_1 r_2} \bar{\Lambda}_1 \bar{\Lambda}_2^* \, d\bar{\mathbf{x}}_1 \, d\bar{\mathbf{x}}_2. \tag{4·5}$$

Equation (4·5) expresses $\hat{\Gamma}(\mathbf{x}_1, \mathbf{x}_2, \tau)$ in terms of the values which this function takes at all pairs of points $\bar{P}_1$ and $\bar{P}_2$ on the surface, the time argument for each pair having the values $\tau-(\tau_1-\tau_2)$, where $\tau_1 = r_1/c$ and $\tau_2 = r_2/c$ are the times needed for light to travel from $\bar{P}_1$ to P_1 and from $\bar{P}_2$ to P_2 respectively.

Of particular interest is the special case when the points P_1 and P_2 coincide and when, in addition, $\tau = 0$. Denoting the common point by $P(\mathbf{x})$, the left-hand side of (4·5) reduces to the intensity $I(\mathbf{x})$ at P, and, if (3·6) is also used, one obtains

$$I(\mathbf{x}) = \int_{\mathscr{A}}\int_{\mathscr{A}} \frac{\sqrt{I(\bar{\mathbf{x}}_1)}\sqrt{I(\bar{\mathbf{x}}_2)}}{r_1 r_2} \hat{\gamma}\left(\bar{\mathbf{x}}_1, \bar{\mathbf{x}}_2, \frac{r_2-r_1}{c}\right) \bar{\Lambda}_1 \bar{\Lambda}_2^* \, d\bar{\mathbf{x}}_1 \, d\bar{\mathbf{x}}_2, \tag{4·6}$$

$I(\bar{\mathbf{x}}_1)$ and $I(\bar{\mathbf{x}}_2)$ being the intensities at two typical points $\bar{P}_1$ and $\bar{P}_2$ of the surface $\mathscr{A}$.

In (4·6) $\hat{\gamma}$ may be replaced by $\gamma = \mathscr{R}\hat{\gamma}$, since the imaginary part of $\hat{\gamma}$ contributes nothing to the integral, as I and $\bar{\Lambda}_1\bar{\Lambda}_2^*$ are real. That the integral is actually real may be shown formally by verifying that it remains unchanged when its complex conjugate is taken. This result follows immediately on using (3·5) and interchanging the independent variables $\bar{\mathbf{x}}_1$ and $\bar{\mathbf{x}}_2$. In place of (4·6) we may therefore write

$$I(\mathbf{x}) = \int_{\mathscr{A}}\int_{\mathscr{A}} \frac{\sqrt{I(\bar{\mathbf{x}}_1)}\sqrt{I(\bar{\mathbf{x}}_2)}}{r_1 r_2} \gamma\left(\bar{\mathbf{x}}_1, \bar{\mathbf{x}}_2, \frac{r_2-r_1}{c}\right) \bar{\Lambda}_1 \bar{\Lambda}_2^* \, d\bar{\mathbf{x}}_1 \, d\bar{\mathbf{x}}_2. \tag{4·6a}$$

Equation (4·6*a*) (or (4·6)) may be regarded as a *generalized Huygens principle for stationary fields of an arbitrary spectral range*. It expresses the *intensity* at the point $P(\mathbf{x})$ in terms of the intensity distribution over an arbitrary surface $\mathscr{A}$, the contribution from each pair of elements of the surface being weighed by the appropriate value of the correlation factor $\gamma(\mathbf{x}_1, \mathbf{x}_2, \tau)$. This factor, which is a generalization of the degree of coherence of Zernike, may, like the latter, be determined from experiments (cf. § 5 below). It may also be calculated from the knowledge of an (observable) correlation function of the source and of the transmission properties of the medium (cf. § 6 below). Hence our extended formulation of the generalized Huygens principle involves again observable quantities only.

† This step of the analysis, though formally correct, is somewhat unsatisfactory, since the mean values will depend not only on the geometrical situation, but also on the form of v, as function of frequency. It should, however, be borne in mind that the inclination factor Λ of the ordinary Huygens principle represents only a rough approximation, and is, in most practical cases, simply replaced by a constant. The difficulty disappears in the rigorous formulation given in § 8 below.

In order to see more clearly the connexion between the formula just derived, and some of the formulae given previously, consider the special case when the light is practically monochromatic, of frequency ν_0. If we express $\hat{\Gamma}$ and $\hat{\gamma}$ in the form

$$\left.\begin{aligned}\hat{\Gamma}(\mathbf{x}_1, \mathbf{x}_2, \tau) &= G(\mathbf{x}_1, \mathbf{x}_2, \tau)\, \mathrm{e}^{-2\pi \mathrm{i}\nu_0\tau},\\ \hat{\gamma}(\mathbf{x}_1, \mathbf{x}_2, \tau) &= g(\mathbf{x}_1, \mathbf{x}_2, \tau)\, \mathrm{e}^{-2\pi \mathrm{i}\nu_0\tau},\end{aligned}\right\} \tag{4·7}$$

the quantities G and g, considered as functions of τ, will vary slowly compared with the variation of the exponential term. If sufficiently small path differences are involved, the variations of G and g, with τ, may be completely neglected, i.e. we may then write

$$\left.\begin{aligned}G \sim G(\mathbf{x}_1, \mathbf{x}_2) &\sim \hat{\Gamma}(\mathbf{x}_1, \mathbf{x}_2, 0),\\ g \sim g(\mathbf{x}_1, \mathbf{x}_2) &\sim \hat{\gamma}(\mathbf{x}_1, \mathbf{x}_2, 0).\end{aligned}\right\} \tag{4·8}$$

G is essentially Zernike's mutual intensity (denoted by $\Gamma(\mathbf{x}_1, \mathbf{x}_2)$ in I) and g his complex degree of coherence (denoted by $\gamma(\mathbf{x}_1, \mathbf{x}_2)$ in I). With this substitution, (4·5) reduces to Zernike's propagation law (Zernike 1938, equation (9)).†

$$G(\mathbf{x}_1, \mathbf{x}_2) \sim \int_{\mathscr{A}}\int_{\mathscr{A}} \frac{G(\overline{\mathbf{x}}_1, \overline{\mathbf{x}}_2)}{r_1 r_2}\, \mathrm{e}^{\mathrm{i}k^{(0)}(r_1 - r_2)}\, \Lambda_1^{(0)} \Lambda_2^{(0)*}\, \mathrm{d}\overline{\mathbf{x}}_1\, \mathrm{d}\overline{\mathbf{x}}_2, \tag{4·9}$$

where $k^{(0)} = 2\pi\nu_0/c$, $\Lambda^{(0)} = \Lambda(\nu_0)$; and with the same approximation (4·6) reduces to the more restricted formulation of the generalized Huygens principle given in I.

5. The generalized interference law and determination of the correlation function from experiments

5·1. *The case* $\mathbf{x}_1 \neq \mathbf{x}_2$

In order to determine the correlation functions from experiment we may use a procedure similar to that described in I in connexion with the determination of the less general correlation functions of Zernike.

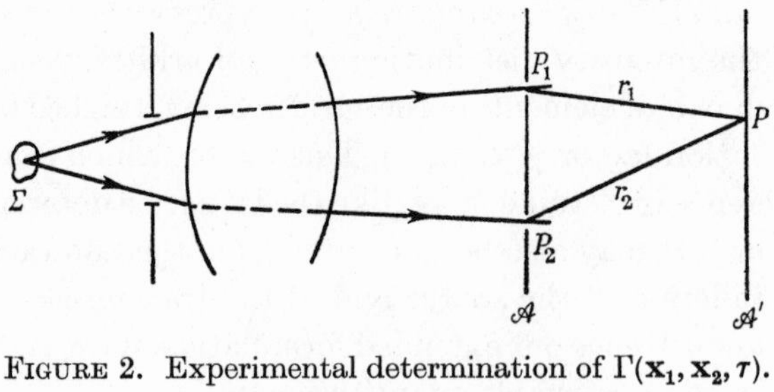

Figure 2. Experimental determination of $\Gamma(\mathbf{x}_1, \mathbf{x}_2, \tau)$.

We place a screen $\mathscr{A}$ across the field so as to pass through the points $P_1(\mathbf{x}_1)$ and $P_2(\mathbf{x}_2)$, small openings $\mathrm{d}\mathscr{A}_1$ and $\mathrm{d}\mathscr{A}_2$ being made at these points. The resulting intensity distribution is observed on a second screen $\mathscr{A}'$ (figure 2).

† Zernike actually neglected the variation of the inclination factor over the surface of integration, setting as usual, $\Lambda = \mathrm{i}/\lambda$.

Taking the integrals (4·6a) over the non-illuminated side of the screen $\mathscr{A}$, we obtain for the intensity $I(\mathbf{x})$ at the point $P(\mathbf{x})$ the expression

$$I(\mathbf{x}) \sim \frac{I(\mathbf{x}_1)}{r_1^2} |\tilde{\Lambda}_1|^2 (\mathrm{d}\mathscr{A}_1)^2 + \frac{I(\mathbf{x}_2)}{r_2^2} |\tilde{\Lambda}_2|^2 (\mathrm{d}\mathscr{A}_2)^2$$
$$+ \frac{\sqrt{I(\mathbf{x}_1)}\sqrt{I(\mathbf{x}_2)}}{r_1 r_2} \gamma\left(\mathbf{x}_1, \mathbf{x}_2, \frac{r_2 - r_1}{c}\right) \tilde{\Lambda}_1 \tilde{\Lambda}_2^* \mathrm{d}\mathscr{A}_1 \mathrm{d}\mathscr{A}_2$$
$$+ \frac{\sqrt{I(\mathbf{x}_2)}\sqrt{I(\mathbf{x}_1)}}{r_2 r_1} \gamma\left(\mathbf{x}_2, \mathbf{x}_1 \frac{r_1 - r_2}{c}\right) \tilde{\Lambda}_2 \tilde{\Lambda}_1^* \mathrm{d}\mathscr{A}_2 \mathrm{d}\mathscr{A}_1. \quad (5·1)$$

Now the first term on the right is precisely the intensity $I^{(1)}(\mathbf{x})$ which would be obtained at P if the openings at P_1 alone was open ($\mathrm{d}\mathscr{A}_2 = 0$):

$$I^{(1)}(\mathbf{x}) = \frac{I(\mathbf{x}_1)}{r_1^2} |\tilde{\Lambda}_1|^2 (\mathrm{d}\mathscr{A}_1)^2, \quad (5·2)$$

the second term

$$I^{(2)}(\mathbf{x}) = \frac{I(\mathbf{x}_2)}{r_2^2} |\tilde{\Lambda}_2|^2 (\mathrm{d}\mathscr{A}_2)^2 \quad (5·3)$$

having a similar interpretation. Also, on account of (3·5) and remembering that γ is the real part of $\hat{\gamma}$,

$$\gamma\left(\mathbf{x}_2, \mathbf{x}_1, \frac{r_1 - r_2}{c}\right) = \gamma\left(\mathbf{x}_1, \mathbf{x}_2, \frac{r_2 - r_1}{c}\right). \quad (5·4)$$

Further, since $\tilde{\Lambda}$ is purely imaginary, $\tilde{\Lambda}_2 \tilde{\Lambda}_1^* = \tilde{\Lambda}_1 \tilde{\Lambda}_2^* = |\tilde{\Lambda}_1||\tilde{\Lambda}_2|$. Hence (5·1) reduces to

$$I(\mathbf{x}) = I^{(1)}(\mathbf{x}) + I^{(2)}(\mathbf{x}) + 2\sqrt{I^{(1)}(\mathbf{x})}\sqrt{I^{(2)}(\mathbf{x})}\gamma\left(\mathbf{x}_1, \mathbf{x}_2, \frac{r_2 - r_1}{c}\right). \quad (5·5)$$

(5·5) represents *a general interference law for stationary fields.*

It is seen that in order to determine $\gamma(\mathbf{x}_1, \mathbf{x}_2, \tau)$ it is only necessary to take the distances r_1 and r_2 such that $\frac{r_2 - r_1}{c} = \tau$, and to measure the intensities $I(\mathbf{x})$, $I^{(1)}(\mathbf{x})$ and $I^{(2)}(\mathbf{x})$. γ is then given by

$$\gamma\left(\mathbf{x}_1, \mathbf{x}_2, \frac{r_2 - r_1}{c}\right) = \frac{I(\mathbf{x}) - I^{(1)}(\mathbf{x}) - I^{(2)}(\mathbf{x})}{2\sqrt{I^{(1)}(\mathbf{x})}\sqrt{I^{(2)}(\mathbf{x})}}. \quad (5·6)$$

If the value of Γ is also required, it is necessary, in addition, to measure the intensities $I(\mathbf{x}_1)$ and $I(\mathbf{x}_2)$ at the points P_1 and P_2. Then from (5·6) and (3·6),

$$\Gamma\left(\mathbf{x}_1, \mathbf{x}_2, \frac{r_2 - r_1}{c}\right) = \frac{1}{2}\sqrt{\frac{I(\mathbf{x}_1)}{I^{(1)}(\mathbf{x})}}\sqrt{\frac{I(\mathbf{x}_2)}{I^{(2)}(\mathbf{x})}}[I(\mathbf{x}) - I^{(1)}(\mathbf{x}) - I^{(2)}(\mathbf{x})]. \quad (5·7)$$

In the special case when the effective frequency range is sufficiently narrow (path differences small enough), (5·5) reduces, on substitution from (4·7), to the formula I, (3·9) of Zernike and Hopkins:

$$I(\mathbf{x}) = I^{(1)}(\mathbf{x}) + I^{(2)}(\mathbf{x}) + 2\sqrt{I^{(1)}(\mathbf{x})}\sqrt{I^{(2)}(\mathbf{x})}\,|g(\mathbf{x}_1, \mathbf{x}_2)|\cos[\arg g(\mathbf{x}_1, \mathbf{x}_2) + k^{(0)}(r_1 - r_2)]. \quad (5·8)$$

5·2. *The case* $\mathbf{x}_1 = \mathbf{x}_2$

It has been assumed so far that the two points P_1 and P_2 are distinct. We must now consider the special case when they coincide.

When $P_1 = P_2$, (3·1) and (3·4) reduce to

$$\hat{\Gamma}(\mathbf{x}_1, \mathbf{x}_1, \tau) = \langle \hat{V}(\mathbf{x}_1, t+\tau)\, \hat{V}^*(\mathbf{x}_1, t)\rangle = \lim_{T\to\infty} \frac{1}{2T}\int_0^\infty |\, v(\mathbf{x}_1, \nu)\,|^2\, \mathrm{e}^{-2\pi \mathrm{i}\nu\tau}\, \mathrm{d}\nu. \tag{5·9}$$

Hence $\hat{\Gamma}$ may now be determined by measuring the spectral intensity function $I(\mathbf{x}_1, \nu) = \lim\limits_{T\to\infty} \frac{1}{2T} |\, v(\mathbf{x}_1, \nu)\,|^2$ at P_1 and evaluating the Fourier integral. (If, in addition to $\mathbf{x}_1 = \mathbf{x}_2$, one also has $\tau = 0$, $\hat{\Gamma}$ reduces, of course, to the intensity at P_1.)

A more direct determination of $\hat{\Gamma}(\mathbf{x}_1, \mathbf{x}_2, \tau)$ appears to be possible, in principle, by the use of some interferometer based on the principle of division of amplitude (cf. Williams 1941).

Suppose that the beam of light is divided at the point $P_1(\mathbf{x}_1)$ into two beams (for example, in the Michelson interferometer), which then proceed via different paths and are reunited again at a point $P(\mathbf{x})$. Let the transmission functions (cf. I, § 4) of the two paths for strictly monochromatic radiation of frequency ν be

$$K(\mathbf{x}_1, \mathbf{x}, \nu) = |\, K_1\,|\, \mathrm{e}^{2\pi \mathrm{i}\nu\phi_1}, \quad K(\mathbf{x}_2, \mathbf{x}, \nu) = |\, K_2\,|\, \mathrm{e}^{2\pi \mathrm{i}\nu\phi_2}. \tag{5·10}$$

Then the Fourier component $v(\mathbf{x}, \nu)$ at P is related to that at P_1 by

$$v(\mathbf{x}, \nu) = (K_1\Lambda_1 + K_2\Lambda_2)\, v(\mathbf{x}_1, \nu)\, \delta\mathscr{A}. \tag{5·11}$$

where $\delta\mathscr{A}$ is the element of area around P_1 which reflects and transmit the incident light. Hence the intensity at P is given by

$$\begin{aligned} I(\mathbf{x}) &= \lim_{T\to\infty} \frac{1}{2T}\int_0^\infty |\, v(\mathbf{x}, \nu)\,|^2\, \mathrm{d}\nu \\ &= I^{(1)}(\mathbf{x}) + I^{(2)}(\mathbf{x}) + \mathscr{J}(\mathbf{x}), \end{aligned} \tag{5·12}$$

where

$$\left.\begin{aligned} I^{(1)}(\mathbf{x}) &= \lim_{T\to\infty} \frac{1}{2T}\int_0^\infty |\, v(\mathbf{x}_1, \nu)\,|^2\, |\, K_1\,|^2\, |\, \Lambda_1\,|^2 (\delta\mathscr{A})^2\, \mathrm{d}\nu, \\ I^{(2)}(\mathbf{x}) &= \lim_{T\to\infty} \frac{1}{2T}\int_0^\infty |\, v(\mathbf{x}_1, \nu)\,|^2\, |\, K_2\,|^2\, |\, \Lambda_2\,|^2 (\delta\mathscr{A})^2\, \mathrm{d}\nu, \\ \mathscr{J}(\mathbf{x}) &= \lim_{T\to\infty} \frac{1}{T}\int_0^\infty |\, v(\mathbf{x}_1, \nu)\,|^2\, |\, K_1\,|\, |\, K_2\,|\, |\, \Lambda_1\,|\, |\, \Lambda_2\,| (\delta\mathscr{A})^2 \cos 2\pi\nu(\phi_2 - \phi_1)\, \mathrm{d}\nu. \end{aligned}\right\} \tag{5·13}$$

In general, $|\, K_1\,|$, $|\, K_2\,|$, ϕ_1 and ϕ_2 will depend on the frequency, since owing to the refraction at the individual elements of the interferometer, light of different frequencies will proceed along slightly different paths. In many cases, however, this effect will be negligible; $|\, K_1\,|$ and $|\, K_2\,|$ may then be taken outside the integrals in (5·13) and $\mathscr{J}$ becomes, apart from a multiplicative factor, the Fourier cosine transform of the spectral intensity function, this being equal to Γ according to

(5·9). The expression (5·12) for the intensity may then be expressed in a form strictly analogous to (5·5):

$$I(\mathbf{x}) = I^{(1)}(\mathbf{x}) + I^{(2)}(\mathbf{x}) + 2\sqrt{I^{(1)}(\mathbf{x})}\sqrt{I^{(2)}(\mathbf{x})}\,\gamma\left(\mathbf{x}_1, \mathbf{x}_1, \frac{\phi_2 - \phi_1}{c}\right), \tag{5·14}$$

with

$$\left.\begin{aligned} I^{(1)}(\mathbf{x}) &= |K_1|^2 |\Lambda_1|^2 (\delta\mathscr{A})^2 I(\mathbf{x}_1), \\ I^{(2)}(\mathbf{x}) &= |K_2|^2 |\Lambda_2|^2 (\delta\mathscr{A})^2 I(\mathbf{x}_1), \\ I(\mathbf{x}_1) &= \lim_{T\to\infty} \frac{1}{2T}\int_0^\infty |v(\mathbf{x}_1, \nu|^2 \,\mathrm{d}\nu = \langle \hat{V}(\mathbf{x}_1, t)\, \hat{V}^*(\mathbf{x}_1, t)\rangle. \end{aligned}\right\} \tag{5·15}$$

The term $I^{(1)}(\mathbf{x})$ represents the intensity which is obtained at the point $P(\mathbf{x})$ if the second beam is excluded ($K_2 = 0$), the term $I^{(2)}(\mathbf{x})$ having a similar interpretation. Hence to measure $\gamma(\mathbf{x}_1, \mathbf{x}_1, \tau)$ it is only necessary (provided $|K_1|$, $|K_2|$, ϕ_1 and ϕ_2 may be treated as independent of ν) to set the interferometer so that $(\phi_2 - \phi_1)/c = \tau$. γ is then given by a formula analogous to (5·6):

$$\gamma\left(\mathbf{x}_1, \mathbf{x}_1, \frac{\phi_2 - \phi_1}{c}\right) = \frac{I(\mathbf{x}) - I^{(1)}(\mathbf{x}) - I^{(2)}(\mathbf{x})}{2\sqrt{I^{(1)}(\mathbf{x})}\sqrt{I^{(2)}(\mathbf{x})}}. \tag{5·16}$$

And Γ is given by

$$\Gamma\left(\mathbf{x}_1, \mathbf{x}_1, \frac{\phi_2 - \phi_1}{c}\right) = I(\mathbf{x}_1)\,\gamma\left(\mathbf{x}_1, \mathbf{x}_1, \frac{\phi_2 - \phi_1}{c}\right). \tag{5·17}$$

The investigations of Zernike (1938, 1948; see also I, §3) brought out the close connexion which exists between the visibility factor of Michelson and the *correlation* (characterized by $\hat{\gamma}(\mathbf{x}_1, \mathbf{x}_2, 0)$) of disturbances *at two points in the field, at the same instant of time*. These investigations interpreted, also from a new point of view, Michelson's method for the determination of the intensity of radiation across a radiating source from the measurements of the visibility of fringes, a method which in recent years has become of fundamental importance in radio astronomy (cf. Ryle 1950; Smith 1952). With the help of the preceding analysis, it will now be shown that there is a complementary relation between the visibility function and the *correlation* (characterized by $\hat{\gamma}(\mathbf{x}_1, \mathbf{x}_1, \tau)$) of disturbances *at two instants of time, at the same point in the field*. This result leads to a new interpretation of Michelson's well-known method (Michelson 1891, 1892) for the determination of the energy distribution in spectral lines from measurements of the visibility.†

Let ν_0 be the mean frequency of the spectral distribution, assumed now to be confined within a narrow frequency range $\nu_0 - \Delta\nu \leqslant \nu \leqslant \nu_0 + \Delta\nu$ and set

$$\hat{\gamma}(\mathbf{x}_1, \mathbf{x}_1, \tau) = h(\mathbf{x}_1, \tau)\,\mathrm{e}^{-2\pi i\nu_0\tau}. \tag{5·18}$$

Further, assume, as is usually the case, that $I^{(1)}(\mathbf{x}) \sim I^{(2)}(\mathbf{x})$. Equation (5·14) then becomes

$$I(\mathbf{x}) = 2I^{(1)}(\mathbf{x})\{1 + |h(\mathbf{x}_1, \tau)| \cos[\arg h(\mathbf{x}_1, \tau) - 2\pi\nu_0\tau]\}. \tag{5·19}$$

Since $\Delta\nu/\nu_0 \ll 1$, h, considered as a function of τ, will change very slowly in comparison with $\cos 2\pi\nu_0\tau$ and $\sin 2\pi\nu_0\tau$, so that the minima and maxima of the intensity are effectively given by τ values which satisfy the relation

$$\sin[\arg h(\mathbf{x}_1, \tau) - 2\pi\nu_0\tau] = 0. \tag{5·20}$$

† In this connexion see also the paper by Van Cittert (1939).

The corresponding maxima and minima have the values

$$\left.\begin{aligned} I_{\text{max.}} &= 2I^{(1)}(\mathbf{x})\,[1+|\,h(\mathbf{x}_1,\tau)\,|],\\ I_{\text{min.}} &= 2I^{(1)}(\mathbf{x})\,[1-|\,h(\mathbf{x}_1,\tau)\,|].\end{aligned}\right\} \tag{5·21}$$

The visibility $\mathscr{V}$ of the fringes is therefore

$$\mathscr{V} = \frac{I_{\text{max.}}-I_{\text{min.}}}{I_{\text{max.}}+I_{\text{min.}}} = |\,h(\mathbf{x}_1,\tau)\,| = |\,\hat{\gamma}(\mathbf{x}_1,\mathbf{x}_1,\tau)\,|, \tag{5·22}$$

i.e. the visibility is equal to the absolute value of the complex correlation factor $\hat{\gamma}(\mathbf{x}_1,\mathbf{x}_1,\tau)$. Hence according to (5·9), using the Fourier inversion formula, it is possible to obtain information about the intensity distribution in the spectrum from measurements of the visibility, provided that suitable assumptions about the associated phase $[\arg\hat{\gamma}(\mathbf{x}_1,\mathbf{x}_1,\tau)]$ are made.†

It is now seen that the two methods of Michelson correspond essentially to the two limiting cases $\tau\to 0$ and $\mathbf{x}_2\to\mathbf{x}_1$ of our theory.

6. Expressions for $\hat{\Gamma}$ in terms of quantities which specify the source and the medium

We shall now derive explicit expressions for the $\hat{\Gamma}$ factor in terms of quantities which specify the source and the transmission properties of the medium.

As in I we assume the source Σ to be a radiating plane area and divide it into elements $\delta\Sigma_1, \delta\Sigma_2, \ldots$, which are small in linear dimensions in comparison with the optical wave-lengths. Let

$$\hat{V}_m(\mathbf{x},t) = \int_0^\infty v_m(\mathbf{x},\nu)\,\mathrm{e}^{-2\pi\mathrm{i}\nu t}\,\mathrm{d}\nu \tag{6·1}$$

be the disturbance due to the mth element; the total disturbance $\hat{V}(\mathbf{x},t)$ is then given by

$$\hat{V}(\mathbf{x},t) = \sum_m \hat{V}_m(\mathbf{x},t). \tag{6·2}$$

Hence

$$\begin{aligned}\hat{\Gamma}(\mathbf{x}_1,\mathbf{x}_2,\tau) &= \langle \hat{V}(\mathbf{x}_1,t+\tau)\,\hat{V}^*(\mathbf{x}_2,t)\rangle\\ &= \sum_m\sum_n \hat{\Gamma}_{mn}(\mathbf{x}_1,\mathbf{x}_2,\tau),\end{aligned} \tag{6·3}$$

where

$$\begin{aligned}\hat{\Gamma}_{mn}(\mathbf{x}_1,\mathbf{x}_2,\tau) &= \langle \hat{V}_m(\mathbf{x}_1,t+\tau)\,\hat{V}^*_n(\mathbf{x}_2,t)\rangle\\ &= \lim_{T\to\infty}\frac{1}{2T}\int_0^\infty v_m(\mathbf{x}_1,\nu)\,v^*_n(\mathbf{x}_2,\nu)\,\mathrm{e}^{-2\pi\mathrm{i}\nu\tau}\,\mathrm{d}\nu.\end{aligned} \tag{6·4}$$

In most cases of practical interest (e.g. for a gas discharge or incandescent solid) it will be permissible to assume that the radiation from the different elements of the source is mutually incoherent, i.e. that for all values of $\mathbf{x}_1$, $\mathbf{x}_2$ and τ

$$\hat{\Gamma}_{mn}(\mathbf{x}_1,\mathbf{x}_2,\tau) = 0 \quad \text{when} \quad m \neq n. \tag{6·5}$$

† As is evident from (5·19), the phase may in principle be obtained from the measurement of the position of the fringes. This has been pointed out already by Rayleigh (1892) in an open letter to Michelson, in which he discussed the question of a complete determination of the intensity distribution from Michelson's experiments.

There are, however, important cases when this assumption does not hold. For example, if the source is not a 'natural' source, but is a secondary source obtained by imagining a source of natural light by a lens of a finite aperture, then on account of diffraction there will exist a finite degree of correlation in the plane of the secondary source at points which are sufficiently close to each other. Accordingly, we shall first consider the general case when $\hat{\Gamma}_{mn} \neq 0$.

Let $a_m(\nu)$ be the strength of the radiation from the element $\delta\Sigma_m$, at frequency ν, and let $K(\mathbf{x}', \mathbf{x}'', \nu)$ be the transmission function of the medium. Then†

$$v_m(\mathbf{x}, \nu) = a_m(\nu)\, K(\boldsymbol{\xi}_m, \mathbf{x}, \nu), \tag{6·6}$$

where $\boldsymbol{\xi}_m$ denotes the position vector of the mth element of the source. (6·3) then becomes

$$\hat{\Gamma}(\mathbf{x}_1, \mathbf{x}_2, \tau) = \sum_m \sum_n \int_0^\infty J_{mn}(\nu) L(\boldsymbol{\xi}_m, \boldsymbol{\xi}_n;\, \mathbf{x}_1, \mathbf{x}_2;\, \nu)\, \mathrm{e}^{-2\pi i \nu\tau}\, \mathrm{d}\nu, \tag{6·7}$$

with

$$\left.\begin{aligned} J_{mn}(\nu) &= \lim_{T\to\infty} \frac{1}{2T}[a_m(\nu)\, a_n^*(\nu)], \\ L(\boldsymbol{\xi}_m, \boldsymbol{\xi}_n;\, \mathbf{x}_1, \mathbf{x}_2;\, \nu) &= K(\boldsymbol{\xi}_m, \mathbf{x}_1, \nu)\, K^*(\boldsymbol{\xi}_n, \mathbf{x}_2, \nu). \end{aligned}\right\} \tag{6·8}$$

and (second line)

Let

$$\int_0^\infty J_{mn}(\nu)\, \mathrm{e}^{-2\pi i \nu u}\, \mathrm{d}\nu = \hat{\Gamma}_{mn}(u). \tag{6·9}$$

We also introduce the frequency transform of L:

$$\int_0^\infty L(\boldsymbol{\xi}_m, \boldsymbol{\xi}_n;\, \mathbf{x}_1, \mathbf{x}_2;\, \nu)\, \mathrm{e}^{-2\pi i \nu u}\, \mathrm{d}\nu = M(\boldsymbol{\xi}_m, \boldsymbol{\xi}_n;\, \mathbf{x}_1, \mathbf{x}_2;\, u). \tag{6·10}$$

The relation (6·7) then becomes, on using the convolution theorem,

$$\hat{\Gamma}(\mathbf{x}_1, \mathbf{x}_2, \tau) = \sum_m \sum_n \int_{-\infty}^{+\infty} \hat{\Gamma}_{mn}(u)\, M(\boldsymbol{\xi}_m, \boldsymbol{\xi}_n;\, \mathbf{x}_1, \mathbf{x}_2;\, \tau - u)\, \mathrm{d}u. \tag{6·11}$$

(6·11) expresses $\hat{\Gamma}(\mathbf{x}_1, \mathbf{x}_2, \tau)$ in terms of $\hat{\Gamma}_{mn}(u)$ and M. The former specifies the source and may be determined from experiments described in § 5. The latter specifies the medium and may be obtained from calculations based on a ray trace.

For incoherent sources (i.e. sources for which (6·5) holds), the double summation in (6·11) reduces to a single summation:

$$\hat{\Gamma}(\mathbf{x}_1, \mathbf{x}_2, \tau) = \sum_m \int_0^\infty J_{mn}(\nu)\, L(\boldsymbol{\xi}_m, \boldsymbol{\xi}_n;\, \mathbf{x}_1, \mathbf{x}_2;\, \nu)\, \mathrm{e}^{-2\pi i \nu\tau}\, \mathrm{d}\nu \tag{6·12}$$

$$= \sum_m \int_{-\infty}^{+\infty} \hat{\Gamma}_{mn}(u)\, M(\boldsymbol{\xi}_m, \boldsymbol{\xi}_n;\, \mathbf{x}_1, \mathbf{x}_2;\, \tau - u)\, \mathrm{d}u. \tag{6·13}$$

If the elements $\delta\Sigma_1, \delta\Sigma_2, \ldots$ are taken small enough, one may replace the summations by integrations over the source, provided that obvious modifications are made: One introduces in place of $\hat{\Gamma}_{mn}$ and J_{mn} which are defined only for the discrete set of $\boldsymbol{\xi}$ values, the functions $\hat{\Omega}$ and j defined for a continuous range of $\boldsymbol{\xi}$ and $\boldsymbol{\xi}'$, such that

$$\hat{\Gamma}_{mn}(u) = \hat{\Omega}(\boldsymbol{\xi}, \boldsymbol{\xi}', u)\, \mathrm{d}\Sigma\, \mathrm{d}\Sigma', \tag{6·14}$$

$$J_{mn}(\nu) = j(\boldsymbol{\xi}, \boldsymbol{\xi}', \nu)\, \mathrm{d}\Sigma\, \mathrm{d}\Sigma', \tag{6·15}$$

† We neglect here the variation of the source strength with direction.

it being assumed that $m \neq n$. $\hat{\Omega}$ and j are, on account of (6·9), Fourier transforms of each other. In place of (6·7) and (6·11) one then obtains

$$\hat{\Gamma}(\mathbf{x}_1, \mathbf{x}_2, \tau) = \int_{\Sigma}\int_{\Sigma} \mathrm{d}\boldsymbol{\xi}\,\mathrm{d}\boldsymbol{\xi}' \int_0^{\infty} j(\boldsymbol{\xi}, \boldsymbol{\xi}', \nu)\, L(\boldsymbol{\xi}, \boldsymbol{\xi}'; \mathbf{x}_1, \mathbf{x}_2; \nu)\, \mathrm{e}^{-2\pi i \nu\tau}\, \mathrm{d}\nu \tag{6·16}$$

$$= \int_{\Sigma}\int_{\Sigma} \mathrm{d}\boldsymbol{\xi}\,\mathrm{d}\boldsymbol{\xi}' \int_{-\infty}^{+\infty} \hat{\Omega}(\boldsymbol{\xi}, \boldsymbol{\xi}', u)\, M(\boldsymbol{\xi}, \boldsymbol{\xi}'; \mathbf{x}_1, \mathbf{x}_2; \tau - u)\, \mathrm{d}u, \tag{6·17}$$

where $\boldsymbol{\xi}$ and $\boldsymbol{\xi}'$ explore the surface Σ of the source independently.

In the case of an incoherent source, we set

$$\hat{\Gamma}_{mm}(u) = \hat{\Omega}(\boldsymbol{\xi}, u)\, \mathrm{d}\Sigma, \tag{6·18}$$

$$J_{mm}(u) = j(\boldsymbol{\xi}, \nu)\, \mathrm{d}\Sigma. \tag{6·19}$$

$j(\boldsymbol{\xi}, \nu)$ is nothing but the spectral intensity function of the source; it represents the intensity per unit area of the source per unit frequency range. In place of (6·12) and (6·13) one then obtains

$$\hat{\Gamma}(\mathbf{x}_1, \mathbf{x}_2, \tau) = \int_{\Sigma} \mathrm{d}\boldsymbol{\xi} \int_0^{\infty} j(\boldsymbol{\xi}, \nu)\, L(\boldsymbol{\xi}, \boldsymbol{\xi}; \mathbf{x}_1, \mathbf{x}_2; \nu) \mathrm{e}^{-2\pi i \nu\tau}\, \mathrm{d}\nu \tag{6·20}$$

$$= \int_{\Sigma} \mathrm{d}\boldsymbol{\xi} \int_{-\infty}^{+\infty} \hat{\Omega}(\boldsymbol{\xi}, u)\, M(\boldsymbol{\xi}, \boldsymbol{\xi}; \mathbf{x}_1, \mathbf{x}_2; \tau - u)\, \mathrm{d}u. \tag{6·21}$$

As an example, consider the case of an incoherent source in a homogeneous medium. The transmission function of a homogeneous medium is

$$K(\boldsymbol{\xi}, \mathbf{x}, \nu) = \frac{\exp\left\{2\pi i \frac{\nu}{c} |\mathbf{x} - \boldsymbol{\xi}|\right\}}{|\mathbf{x} - \boldsymbol{\xi}|}, \tag{6·22}$$

so that

$$L(\boldsymbol{\xi}, \boldsymbol{\xi}; \mathbf{x}_1, \mathbf{x}_2, \nu) = \frac{\exp\left\{2\pi i \frac{\nu}{c} (R_1 - R_2)\right\}}{R_1 R_2}, \tag{6·23}$$

with

$$R_1 = |\mathbf{x}_1 - \boldsymbol{\xi}|, \quad R_2 = |\mathbf{x}_2 - \boldsymbol{\xi}|. \tag{6·24}$$

(6·20) then gives the following expression for $\hat{\Gamma}(\mathbf{x}_1, \mathbf{x}_2, \tau)$:

$$\hat{\Gamma}(\mathbf{x}_1, \mathbf{x}_2, \tau) = \int_{\Sigma} \mathrm{d}\boldsymbol{\xi} \int_0^{\infty} \frac{j(\boldsymbol{\xi}, \nu)}{R_1 R_2} \exp\left\{-2\pi i \nu \left[\tau - \frac{R_1 - R_2}{c}\right]\right\} \mathrm{d}\nu \tag{6·25}$$

$$= \int_{\Sigma} \frac{\hat{\Omega}\left(\boldsymbol{\xi}, \tau - \dfrac{R_1 - R_2}{c}\right)}{R_1 R_2}\, \mathrm{d}\boldsymbol{\xi}. \tag{6·26}$$

7. Differential equations for the correlation function $\hat{\Gamma}$

In its usual form, Huygens's principle describes within a certain degree of accuracy the propagation of the light disturbance $\hat{V}(\mathbf{x}, t)$. As is well known, this principle may be regarded as an approximate formulation of a rigorous theorem due to Kirchhoff (see, for example, Baker & Copson 1950), this theorem being a consequence of the fact that $\hat{V}$ obeys rigorously the wave equation.

In the present investigation we have found a generalization of the Huygens principle which applies to the intensity rather than to the complex disturbance, and, more generally, we have found a kind of Huygens principle for the propagation of the correlation function $\hat{\Gamma}(\mathbf{x}_1, \mathbf{x}_2, \tau)$. These results suggest that $\hat{\Gamma}$ itself obeys certain differential equations and that our propagation laws are essentially some approximate formulations of the associated 'Kirchhoff's theorems'. We shall now show that this indeed is the case.

In vacuo, the complex disturbance $\hat{V}(\mathbf{x}, t)$ satisfies the wave equation

$$\nabla^2 \hat{V} - \frac{1}{c^2}\frac{\partial^2 \hat{V}}{\partial t^2} = 0. \tag{7·1}$$

Consequently each Fourier component $v(\mathbf{x}, \nu)$ obeys the equation

$$\nabla^2 v + \left(\frac{2\pi\nu}{c}\right)^2 v = 0. \tag{7·2}$$

Let

$$\nabla_1^2 \equiv \frac{\partial^2}{\partial x_1^2} + \frac{\partial^2}{\partial y_1^2} + \frac{\partial^2}{\partial z_1^2} \tag{7·3}$$

be the Laplacian operator with respect to the co-ordinates x_1, y_1, z_1 of the point $P_1(\mathbf{x}_1)$. It then follows from (3·4) and (7·2) that

$$\begin{aligned}\nabla_1^2 \hat{\Gamma}(\mathbf{x}_1, \mathbf{x}_2, \tau) &= \lim_{T\to\infty} \frac{1}{2T}\int_0^{+\infty} [\nabla_1^2 v(\mathbf{x}_1, \nu)]\, v^*(\mathbf{x}_2, \nu)\, \mathrm{e}^{-2\pi i\nu\tau}\, d\nu \\ &= -\frac{2\pi^2}{c^2}\lim_{T\to\infty}\frac{1}{T}\int_0^{\infty} \nu^2 v(\mathbf{x}_1, \nu)\, v^*(\mathbf{x}_2, \nu)\, \mathrm{e}^{-2\pi i\nu\tau}\, d\nu.\end{aligned} \tag{7·4}$$

Also from (3·4),

$$\frac{\partial^2}{\partial \tau^2}\hat{\Gamma}(\mathbf{x}_1, \mathbf{x}_2, \tau) = -2\pi^2 \lim_{T\to\infty}\frac{1}{T}\int_0^{\infty} \nu^2 v(\mathbf{x}_1, \nu)\, v^*(\mathbf{x}_2, \nu)\, \mathrm{e}^{-2\pi i\nu\tau}\, d\nu. \tag{7·5}$$

Comparison of (7·4) and (7·5) shows that

$$\nabla_1^2 \hat{\Gamma} - \frac{1}{c^2}\frac{\partial^2 \hat{\Gamma}}{\partial \tau^2} = 0. \tag{7·6}$$

Similarly, if ∇_2^2 denotes the Laplacian operator with respect to the co-ordinates x_2, y_2, z_2 of the point $P_2(\mathbf{x}_2)$, then

$$\nabla_2^2 \hat{\Gamma} - \frac{1}{c^2}\frac{\partial^2 \hat{\Gamma}}{\partial \tau^2} = 0. \tag{7·7}$$

Hence, in vacuo, *the correlation function* $\hat{\Gamma}(\mathbf{x}_1, \mathbf{x}_2, \tau)$ *obeys rigorously the two wave equations* (7·6) *and* (7·7).

Each of the two wave equations describes the variation of the correlation when one of the points (P_2 or P_1) is fixed whilst the other point as well as the parameter τ varies. It will be recalled that τ denotes a time difference; in all experiments it will play the part of the difference in the optical path (divided by c). The 'actual' time makes no appearance in our formulae. This is a most desirable aspect of the theory, since true time variations are not observed in optical fields.

8. A rigorous formulation of the propagation law for $\hat{\Gamma}$ and of the generalized Huygens principle

We are now in a position to derive a rigorous propagation law for $\hat{\Gamma}$ and also to formulate rigorously the generalized Huygens principle.

Let $P(\mathbf{x})$ and $A(\mathbf{a})$ be any two points in the field and let $\mathscr{A}$ be any closed surface surrounding P; the point A may be either inside or outside this surface.

If ∇^2 denotes the Laplacian operator with respect to the co-ordinates of $P(\mathbf{x})$, then, according to (7·6),

$$\nabla^2\hat{\Gamma}(\mathbf{x},\mathbf{a},\tau)-\frac{1}{c^2}\frac{\partial^2\hat{\Gamma}(\mathbf{x},\mathbf{a},\tau)}{\partial\tau^2}=0. \tag{8·1}$$

Hence, using Kirchhoff's integral formula (cf. Baker & Copson 1950, p. 37), we may express $\hat{\Gamma}(\mathbf{x},\mathbf{a},\tau)$ in the following form:

$$\hat{\Gamma}(\mathbf{x},\mathbf{a},\tau)=\frac{1}{4\pi}\int_{\mathscr{A}}\left\{f[\hat{\Gamma}]^-+g\left[\frac{\partial}{\partial\tau}\hat{\Gamma}\right]^-+h\left[\frac{\partial}{\partial n}\hat{\Gamma}\right]^-\right\}\mathrm{d}\bar{\mathbf{x}}, \tag{8·2}$$

where

$$f=\frac{\partial}{\partial n}\left(\frac{1}{r}\right),\quad g=-\frac{1}{cr}\frac{\partial r}{\partial n},\quad h=-\frac{1}{r}, \tag{8·3}$$

r being the distance from a typical point $\bar{P}(\bar{\mathbf{x}})$ on the surface to $P(\mathbf{x})$ (see figure 3a), $\partial/\partial n$ denoting differentiation along the inward normal to $\mathscr{A}$; and the brackets $[\ldots]^-$ denote retarded values, i.e. values obtained by replacing τ by $\tau-r/c$, e.g.

$$[\hat{\Gamma}]^-=\hat{\Gamma}(\bar{\mathbf{x}},\mathbf{a},\tau-r/c). \tag{8·4}$$

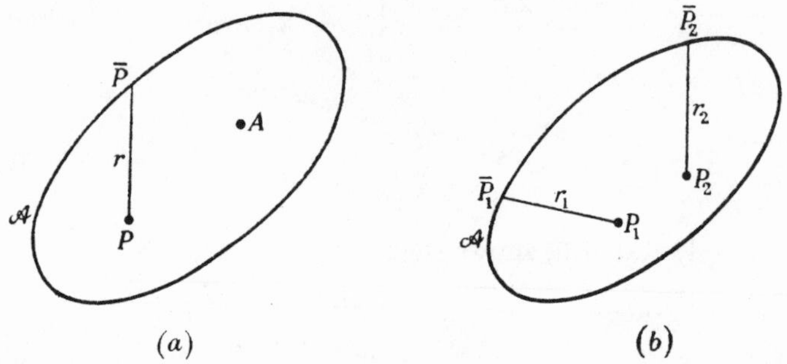

FIGURE 3. Illustrating Kirchhoff's integral theorem and the 'propagation law' for $\hat{\Gamma}$.

It may be shown by an argument similar to that used to derive Kirchhoff's formula that the integral (8·2) with advanced values ($\tau+r/c$ in place of $\tau-r/c$) also satisfies the wave equation, provided g is replaced by $-g$. If expressions for $\hat{V}$ in terms of retarded terms only are admitted, as it is reasonable to do on physical grounds, then Kirchhoff's integral for $\hat{\Gamma}(\mathbf{x},\mathbf{a},\tau)$ in terms of the retarded values of $\hat{\Gamma}$ clearly represents the physical solution. The situation is, however, different in the case of $\hat{\Gamma}(\mathbf{a},\mathbf{x},\tau)$. For if again only expressions in terms of retarded values are admitted for $\hat{V}$, Kirchhoff's integral for $\hat{\Gamma}(\mathbf{a},\mathbf{x},\tau)$ must involve terms like

$$\begin{aligned}f\langle\hat{V}(\mathbf{a},t+\tau)\,\hat{V}^*(\bar{\mathbf{x}},t-r/c)\rangle&=f\langle\hat{V}(\mathbf{a},t+\tau+r/c)\,\hat{V}^*(\bar{\mathbf{x}},t)\rangle\\&=f\hat{\Gamma}(\mathbf{a},\bar{\mathbf{x}},\tau+r/c),\end{aligned} \tag{8·5}$$

i.e. it must involve 'advanced' values of $\hat{\Gamma}$. The full formula for $\hat{\Gamma}(\mathbf{a}, \mathbf{x}, \tau)$ therefore is

$$\Gamma(\mathbf{a}, \mathbf{x}, \tau) = \frac{1}{4\pi}\int_{\mathscr{A}} \left\{ f[\hat{\Gamma}]^+ - g\left[\frac{\partial}{\partial \tau}\hat{\Gamma}\right]^+ + h\left[\frac{\partial}{\partial n}\hat{\Gamma}\right]^+ \right\} d\bar{\mathbf{x}}, \tag{8·6}$$

the brackets $[\dots]^+$ denoting the advanced values, e.g.

$$[\hat{\Gamma}]^+ = \hat{\Gamma}(\mathbf{a}, \bar{\mathbf{x}}, \tau + r/c). \tag{8·7}$$

We now apply (8·2) and (8·6) to derive an expression for $\hat{\Gamma}(\mathbf{x}_1, \mathbf{x}_2, \tau)$ in terms of an integral which involves the values of $\hat{\Gamma}$ and its derivatives at all pairs of points $\bar{P}_1$, $\bar{P}_2$ on an arbitrary closed surface $\mathscr{A}$ surrounding P_1 and P_2 (see figure 3*b*).

We set $\mathbf{x} = \mathbf{x}_1$, $\mathbf{a} = \mathbf{x}_2$, $\bar{\mathbf{x}} = \bar{\mathbf{x}}_1$ in (8·2). This gives

$$\Gamma(\mathbf{x}_1, \mathbf{x}_2, \tau) = \frac{1}{4\pi}\int_{\mathscr{A}} \left\{ f_1[\hat{\Gamma}]_1^- + g_1\left[\frac{\partial}{\partial \tau}\hat{\Gamma}\right]_1^- + h_1\left[\frac{\partial}{\partial n_1}\hat{\Gamma}\right]_1^- \right\} d\bar{\mathbf{x}}_1, \tag{8·8}$$

where the arguments in the brackets $[\dots]_1^-$ are $\bar{\mathbf{x}}_1$, $\mathbf{x}_2$, $\tau - r_1/c$, e.g.

$$[\hat{\Gamma}]_1^- = \hat{\Gamma}(\bar{\mathbf{x}}_1, \mathbf{x}_2, \tau - r_1/c); \tag{8·9}$$

r_1 being the distance from $\bar{P}_1$ to P_1, f_1, g_1 and h_1 the appropriate values of f, g and h $\partial/\partial n_1$ denoting differentiation at $\bar{P}_1$ along the inward normal to $\mathscr{A}$.

Next we express each of the retarded terms in (8·8) in terms of Kirchhoff's integral over the surface. Setting $\mathbf{a} = \bar{\mathbf{x}}_1$, $\mathbf{x} = \mathbf{x}_2$, $\bar{\mathbf{x}} = \bar{\mathbf{x}}_2$ in (8·6) and writing τ' in place of τ where τ' is arbitrary for the present, we obtain

$$\hat{\Gamma}(\bar{\mathbf{x}}_1, \mathbf{x}_2, \tau') = \frac{1}{4\pi}\int_{\mathscr{A}} \left\{ f_2[\hat{\Gamma}]_2^+ - g_2\left[\frac{\partial}{\partial \tau'}\hat{\Gamma}\right]_2^+ + h_2\left[\frac{\partial}{\partial n_2}\hat{\Gamma}\right]_2^+ \right\} d\bar{\mathbf{x}}_2, \tag{8·10}$$

where the arguments of the terms in the brackets $[\dots]_2^+$ are $\bar{\mathbf{x}}_1$, $\bar{\mathbf{x}}_2$, $\tau' + r_2/c$, e.g.

$$[\hat{\Gamma}]_2^+ = \hat{\Gamma}(\bar{\mathbf{x}}_1, \bar{\mathbf{x}}_2, \tau' + r_2/c), \tag{8·11}$$

the other symbols having a similar meaning as before. Differentiating (8·10) with respect to τ', we obtain

$$\frac{\partial}{\partial \tau'}\hat{\Gamma}(\bar{\mathbf{x}}_1, \mathbf{x}_2, \tau') = \frac{1}{4\pi}\int_{\mathscr{A}} \left\{ f_2\left[\frac{\partial}{\partial \tau'}\hat{\Gamma}\right]_2^+ - g_2\left[\frac{\partial^2}{\partial \tau'^2}\hat{\Gamma}\right]_2^+ + h_2\left[\frac{\partial^2}{\partial \tau' \partial n_2}\hat{\Gamma}\right]_2^+ \right\} d\bar{\mathbf{x}}_2. \tag{8·12}$$

Differentiation of (8·10) with respect to n_1 gives

$$\frac{\partial}{\partial n_1}\hat{\Gamma}(\bar{\mathbf{x}}_1, \mathbf{x}_2, \tau') = \frac{1}{4\pi}\int_{\mathscr{A}} \left\{ f_2\left[\frac{\partial}{\partial n_1}\hat{\Gamma}\right]_2^+ - g_2\left[\frac{\partial^2}{\partial n_1 \partial \tau'}\hat{\Gamma}\right]_2^+ + h_2\left[\frac{\partial^2}{\partial n_1 \partial n_2}\hat{\Gamma}\right]_2^+ \right\} d\bar{\mathbf{x}}_2. \tag{8·13}$$

Setting $\tau' = \tau - r_1/c$ in (8·10) to (8·13) and substituting into (8·8), we finally obtain

$$\begin{aligned}\hat{\Gamma}(\mathbf{x}_1, \mathbf{x}_2 \tau) = \frac{1}{(4\pi)^2}\int_{\mathscr{A}}\int_{\mathscr{A}} \Big\{ & f_1 f_2[\hat{\Gamma}] - f_1 g_2\left[\frac{\partial}{\partial \tau}\hat{\Gamma}\right] + f_1 h_2\left[\frac{\partial}{\partial n_2}\hat{\Gamma}\right] \\ & + g_1 f_2\left[\frac{\partial}{\partial \tau}\hat{\Gamma}\right] - g_1 g_2\left[\frac{\partial^2}{\partial \tau^2}\hat{\Gamma}\right] + g_1 h_2\left[\frac{\partial^2}{\partial \tau \partial n_2}\hat{\Gamma}\right] \\ & + h_1 f_2\left[\frac{\partial}{\partial n_1}\hat{\Gamma}\right] - h_1 g_2\left[\frac{\partial^2}{\partial n_1 \partial \tau}\hat{\Gamma}\right] + h_1 h_2\left[\frac{\partial^2}{\partial n_1 \partial n_2}\hat{\Gamma}\right]\Big\} d\bar{\mathbf{x}}_1 d\bar{\mathbf{x}}_2, \end{aligned} \tag{8·14}$$

where the arguments of the terms in the brackets are $\bar{\mathbf{x}}_1$, $\bar{\mathbf{x}}_2$, $\tau - \frac{r_1 - r_2}{c}$, e.g.

$$[\hat{\Gamma}] = \hat{\Gamma}\left(\bar{\mathbf{x}}_1, \bar{\mathbf{x}}_2, \tau - \frac{r_1 - r_2}{c}\right). \tag{8·15}$$

The formula (8·14) may be regarded as a rigorous formulation of the propagation law for $\hat{\Gamma}$. It expresses the correlation between disturbances at P_1 and P_2 in terms of the correlations and its derivatives at all pairs of points on an arbitrary closed surface surrounding P_1 and P_2.

In the special case when P_1 and P_2 coincide ($\mathbf{x}_1 = \mathbf{x}_2 = \mathbf{x}$ (say)) and when in addition $\tau = 0$, (8·14) reduces to, when we also substitute from (3·6),

$$\begin{aligned} I(\mathbf{x}) = \frac{1}{(4\pi)^2} \int_{\mathscr{A}} \int_{\mathscr{A}} & \sqrt{\bar{I}_1}\sqrt{\bar{I}_2}\left\{ f_1 f_2 [\hat{\gamma}] + (f_2 g_1 - f_1 g_2)\left[\frac{\partial}{\partial \tau}\hat{\gamma}\right] - g_1 g_2 \left[\frac{\partial^2}{\partial \tau^2}\hat{\gamma}\right]\right\} \\ & + \sqrt{\bar{I}_1}\left\{ f_1 h_2 \frac{\partial}{\partial n_2}(\sqrt{\bar{I}_2}[\hat{\gamma}]) + g_1 h_2 \frac{\partial}{\partial n_2}\left(\sqrt{\bar{I}_2}\left[\frac{\partial}{\partial \tau}\hat{\gamma}\right]\right)\right\} \\ & + \sqrt{\bar{I}_2}\left\{ f_2 h_1 \frac{\partial}{\partial n_1}(\sqrt{\bar{I}_1}[\hat{\gamma}]) - g_2 h_1 \frac{\partial}{\partial n_1}\left(\sqrt{\bar{I}_1}\left[\frac{\partial}{\partial \tau}\hat{\gamma}\right]\right)\right\} \\ & + h_1 h_2 \frac{\partial^2}{\partial n_1 \partial n_2}(\sqrt{\bar{I}_1}\sqrt{\bar{I}_2}[\hat{\gamma}])\, d\bar{\mathbf{x}}_1\, d\bar{\mathbf{x}}_2, \end{aligned} \tag{8·16}$$

where $\bar{I}_1 = I(\bar{\mathbf{x}}_1)$, $\bar{I}_2 = I(\bar{\mathbf{x}}_2)$ and $[\hat{\gamma}] = \hat{\gamma}\,(\bar{\mathbf{x}}_1, \bar{\mathbf{x}}_2, (r_2 - r_1)/c)$. By a similar argument to that used in connexion with (4·6), $\hat{\gamma}$ may be replaced in (8·16) by γ. (8·16) may be regarded as a rigorous formulation of the generalized Huygens principle.

Appendix 1. Cross-correlation between real functions in terms of the associated half-range complex functions

We shall establish the following theorem:

Let $F(t)$ and $G(t)$ be any two real functions, which possess Fourier integral representations and let $P(\tau)$ be the cross correlation function

$$P(\tau) = 2\int_{-\infty}^{+\infty} F(t+\tau)\, G(t)\, dt. \tag{A 1·1}$$

Then the half-range complex function $\hat{P}(\tau)$ associated with $P(\tau)$ is given by

$$\hat{P}(\tau) = \int_{-\infty}^{+\infty} \hat{F}(t+\tau)\, \hat{G}^*(t)\, dt, \tag{A 1·2}$$

where $\hat{F}$ and $\hat{G}$ are the half-range complex functions associated with F and G respectively.

Proof. Let a_ν and b_ν be the Fourier coefficients of F and c_ν and d_ν the Fourier coefficients of G:

$$F(t) = \int_0^\infty (a_\nu \cos 2\pi\nu t + b_\nu \sin 2\pi\nu t)\, d\nu, \tag{A 1·3}$$

$$G(t) = \int_0^\infty (c_\nu \cos 2\pi\nu t + d_\nu \sin 2\pi\nu t)\, d\nu. \tag{A 1·4}$$

In accordance with the definition of § 2, the associated half-range complex functions are then given by

$$\hat{F}(t) = \int_0^{\infty} (a_\nu + \mathrm{i}b_\nu)\, \mathrm{e}^{-2\pi \mathrm{i}\nu t}\, \mathrm{d}\nu, \tag{A 1·5}$$

$$\hat{G}(t) = \int_0^{\infty} (c_\nu + \mathrm{i}d_\nu)\, \mathrm{e}^{-2\pi \mathrm{i}\nu t}\, \mathrm{d}\nu. \tag{A 1·6}$$

Consider now the Fourier representations of $P(\tau)$ and $\hat{P}(\tau)$. Substitution from (A 1·4) into (A 1·1) gives

$$P(\tau) = 2\int_{-\infty}^{+\infty} \mathrm{d}t\, F(t+\tau) \int_0^{\infty} (c_\nu \cos 2\pi\nu t + d_\nu \sin 2\pi\nu t)\, \mathrm{d}\nu. \tag{A 1·7}$$

Now

$$\begin{aligned}\int_{-\infty}^{\infty} F(t+\tau) \cos 2\pi\nu t\, \mathrm{d}t &= \int_{-\infty}^{+\infty} F(u) \cos 2\pi\nu(u-\tau)\, \mathrm{d}u \\ &= \cos 2\pi\nu\tau \int_{-\infty}^{+\infty} F(u) \cos 2\pi\nu u\, \mathrm{d}u + \sin 2\pi\nu\tau \int_{-\infty}^{+\infty} F(u) \sin 2\pi\nu u\, \mathrm{d}u \\ &= \tfrac{1}{2}(a_\nu \cos 2\pi\nu\tau + b_\nu \sin 2\pi\nu\tau),\end{aligned} \tag{A 1·8}$$

where the Fourier inversion formula was used. Similarly

$$\int_{-\infty}^{\infty} F(t+\tau) \sin 2\pi\nu t\, \mathrm{d}t = \tfrac{1}{2}(b_\nu \cos 2\pi\nu\tau - a_\nu \sin 2\pi\nu\tau). \tag{A 1·9}$$

Hence, interchanging the order of integration in (A 1·7) and using the last two relations, one obtains

$$P(\tau) = \int_0^{\infty} \{[a_\nu c_\nu + b_\nu d_\nu] \cos 2\pi\nu\tau + [b_\nu c_\nu - a_\nu d_\nu] \sin 2\pi\nu\tau\}\, \mathrm{d}\nu. \tag{A 1·10}$$

The associated half-range complex function $\hat{P}(\tau)$ is therefore given by

$$\begin{aligned}\hat{P}(\tau) &= \int_0^{\infty} \{[a_\nu c_\nu + b_\nu d_\nu] + \mathrm{i}[b_\nu c_\nu - a_\nu d_\nu]\}\, \mathrm{e}^{-2\pi \mathrm{i}\nu\tau}\, \mathrm{d}\nu \\ &= \int_0^{\infty} (a_\nu + \mathrm{i}b_\nu)(c_\nu - \mathrm{i}d_\nu)\, \mathrm{e}^{-2\pi \mathrm{i}\nu\tau}\, \mathrm{d}\nu.\end{aligned} \tag{A 1·11}$$

Next consider the integral on the right of (A 1·2). Substitution from (A 1·6) gives

$$\int_{-\infty}^{\infty} \hat{F}(t+\tau)\hat{G}^*(t)\, \mathrm{d}t = \int_{-\infty}^{+\infty} \mathrm{d}t\, \hat{F}(t+\tau) \int_0^{\infty} (c_\nu - \mathrm{i}d_\nu)\, \mathrm{e}^{2\pi \mathrm{i}\nu t}\, \mathrm{d}\nu. \tag{A 1·12}$$

But

$$\begin{aligned}\int_{-\infty}^{\infty} \hat{F}(t+\tau)\, \mathrm{e}^{2\pi \mathrm{i}\nu t}\, \mathrm{d}t &= \int_{-\infty}^{+\infty} \hat{F}(u)\, \mathrm{e}^{2\pi \mathrm{i}\nu(u-\tau)}\, \mathrm{d}u \\ &= (a_\nu + \mathrm{i}b_\nu)\, \mathrm{e}^{-2\pi \mathrm{i}\nu\tau},\end{aligned} \tag{A 1·13}$$

where the Fourier inversion formula was used. Interchanging the order of integration in (A 1·12) and using (A 1·13), one obtains

$$\int_{-\infty}^{+\infty} \hat{F}(t+\tau)\,\hat{G}^*(t)\,\mathrm{d}t = \int_0^{\infty} (a_\nu + \mathrm{i}b_\nu)\,(c_\nu - \mathrm{i}d_\nu)\,\mathrm{e}^{-2\pi\mathrm{i}\nu\tau}\,\mathrm{d}\nu. \tag{A 1·14}$$

On comparing (A 1·14) with (A 1·11), the theorem follows.

Appendix 2. The range of validity of the restricted theory of paper I

We shall now investigate the range of validity of the restricted formulation of the generalized Huygens principle and of the generalized interference law of Zernike and Hopkins, as given in the preceding paper of this series.

The essential approximation which was made in the derivation of these formulae in I, was the replacement of the exponential term

$$\exp\{\mathrm{i}(kr_1 - k'r_2)\} \quad \text{by} \quad \exp\{\mathrm{i}k^{(0)}(r_1 - r_2)\}$$

in the expression† I, (2·7):

$$I(\mathbf{x}) = \frac{1}{2T}\int_{-T}^{T}\mathrm{d}t\int_0^\infty\int_0^\infty \mathrm{d}\nu\,\mathrm{d}\nu'\exp\{-2\pi\mathrm{i}(\nu-\nu')\,t\}$$
$$\times \int_{\mathscr{A}}\int_{\mathscr{A}} v(\mathbf{x}_1,\nu)\,v^*(\mathbf{x}_2,\nu')\,\frac{\exp\{\mathrm{i}(kr_1 - k'r_2)\}}{r_1 r_2}\,\Lambda_1\Lambda_2^*\,\mathrm{d}\mathbf{x}_1\,\mathrm{d}\mathbf{x}_2. \tag{A 2·1}$$

To see what restriction this implies, we note that

$$kr_1 - k'r_2 = k^{(0)}(r_1 - r_2) + (k - k^{(0)})\,(r_1 - r_2) + (k - k')\,r_2. \tag{A 2·2}$$

(A 2·1) may therefore be written as

$$I(\mathbf{x}) = \int_{\mathscr{A}}\int_{\mathscr{A}} G(\mathbf{x}_1,\mathbf{x}_2,r_2)\,\frac{\exp\{\mathrm{i}[k^{(0)}(r_1 - r_2) + (k - k^{(0)})\,(r_1 - r_2)]\}}{r_1 r_2}\,\Lambda_1\Lambda_2^*\,\mathrm{d}\mathbf{x}_1\,\mathrm{d}\mathbf{x}_2, \tag{A 2·3}$$

where

$$G(\mathbf{x}_1,\mathbf{x}_2,r_2) = \frac{1}{2T}\int_{-T}^{T}\mathrm{d}t\int_0^\infty\int_0^\infty v(\mathbf{x}_1,\nu)\,v^*(\mathbf{x}_2,\nu')\exp\{-2\pi\mathrm{i}(\nu-\nu')\,(t - r_2/c)\}\,\mathrm{d}\nu\,\mathrm{d}\nu'. \tag{A 2·4}$$

We now change the variable of integration in (A 2·4) from t to $t' = t - r_2/c$. Then, since

$$\lim_{T'\to\infty}\int_{-T'}^{T'}\exp\{-2\pi\mathrm{i}(\nu-\nu')\,t'\}\,\mathrm{d}t' = \delta(\nu-\nu'), \tag{A 2·5}$$

where δ is the Dirac delta function, it follows, that in the limit $T\to\infty$, G becomes independent of r_2 and is equal to

$$\Gamma(\mathbf{x}_1,\mathbf{x}_2) = \lim_{T\to\infty} G(\mathbf{x}_1,\mathbf{x}_2,r_2) = \lim_{T\to\infty}\frac{1}{2T}\int_0^\infty v(\mathbf{x}_1,\nu)v^*(\mathbf{x}_2,\nu)\mathrm{d}\nu \tag{A 2·6}$$

$$= \lim_{T\to\infty}\frac{1}{2T}\int_{-T}^{T} V(\mathbf{x}_1,t)\,V^*(\mathbf{x}_2,t)\,\mathrm{d}t. \tag{A 2·7}$$

† In I, the lower limits of the integrals over the frequency range were actually taken as $-\infty$. We replace them here by 0, since as explained in § 2 above, $v(\mathbf{x},\nu) = 0$ for $\nu < 0$.

Hence in the limit $T \to \infty$, (A 2·3) differs from the approximate expression I, (2·8) by the presence of the term $(k-k^{(0)})(r_1-r_2)$ in the exponential term and by the factor $\Lambda_1^{(0)}$, $\Lambda_2^{(0)*}$ in place of $\Lambda_1 \Lambda_2^*$. Setting $k-k^{(0)} = \Delta k$, $r_1 - r_2 = \Delta r$, and replacing the inclination factors by their mean values, it follows that I, (2·8) will be a valid approximation if

$$|\Delta k|\,|\Delta r| \ll 2\pi, \tag{A 2·8}$$

or, since $k = 2\pi/\lambda$, $\Delta k = -2\pi \Delta\lambda/\lambda^2$, and the condition becomes†

$$|\Delta r| \ll \frac{\lambda^2}{|\Delta\lambda|}. \tag{A 2·9}$$

Hence *the more restricted formulation of the generalized Huygens principle and of the generalized interference law given in paper I is applicable if the path differences between the interfering beams are small compared to* $\lambda^2/|\Delta\lambda|$.

The quantity $\lambda^2/|\Delta\lambda|$ has a simple physical interpretation: it represents (apart from a multiplicative factor of the order of unity), the coherence length of the radiation. A proof of this result will be found in Born (1933, p. 137) and Kahan (1952, p. 310).

This work was carried out during the tenure of an Imperial Chemical Industries Research Fellowship, the award of which, by the University of Manchester is gratefully acknowledged. It is also a pleasure to thank Dr F. D. Kahn for some helpful discussions.

References

Baker, B. B. & Copson, E. T. 1950 *The mathematical theory of Huygens's principle*, 2nd ed. Oxford: Clarendon Press.
Blanc-Lapierre, A. & Dumontet, P. 1955 *Rev. Opt. (théor. instrum.)*, **34**, 1.
Born, M. 1933 *Optik*. Berlin: Springer.
Gabor, D. 1946 *J. Inst. Elect. Engrs.* **93**, Part III, 429.
Kahan, T. 1952 *Suppl. Nuovo Cim.* **9**, 310.
Michelson, A. A. 1891 *Phil. Mag.* **31**, 338.
Michelson, A. A. 1892 *Phil. Mag.* **34**, 280.
Parke III, N. G. 1948 Ph.D. thesis, Massachusetts Institute of Technology. (See also *Tech. Rep. Electron., Mass. Inst. Tech.* no. 95 (1949).)
Rayleigh, Lord 1892 *Phil. Mag.* **34**, 407.
Ryle, M. 1950 *Rep. Progr. Phys.* **13**, 184.
Smith, F. G. 1952 *Mon. Not. R. Astr. Soc.* **112**, 497.
Van Cittert, P. H. 1939 *Physica*, **6**, 1129.
Williams, W. E. 1941 *Applications of interferometry*, 2nd ed. London: Methuen and Co.
Wiener, N. 1930 *Acta Math. Uppsala*, **55**, 117.
Wolf, E. 1954*a* *Proc. Roy. Soc.* A, **225**, 96.
Wolf, E. 1954*b* *Nuovo Cim.* **12**, 884.
Zernike, F. 1938 *Physica*, **5**, 785.
Zernike, F. 1948 *Proc. Phys. Soc.* **61**, 158.

† It is to be noted that this is also essentially the condition I, (5·1) under which the approximate expressions of Van Cittert, Zernike and Hopkins for the space-correlation factor were derived in the previous paper.

PAPER NO. 19

Reprinted from *Nature*, Vol. 177, pp. 27–29, 1956.

CORRELATION BETWEEN PHOTONS IN TWO COHERENT BEAMS OF LIGHT

By R. HANBURY BROWN

University of Manchester, Jodrell Bank Experimental Station

AND

R. Q. TWISS

Services Electronics Research Laboratory, Baldock

IN an earlier paper[1], we have described a new type of interferometer which has been used to measure the angular diameter of radio stars[2]. In this instrument the signals from two aerials A_1 and A_2 (Fig. 1*a*) are detected independently and the correlation between the low-frequency outputs of the detectors is recorded. The relative phases of the two radio signals are therefore lost, and only the correlation in their intensity fluctuations is measured; so that the principle differs radically from that of the familiar Michelson interferometer where the signals are combined before detection and where their relative phase must be preserved.

This new system was developed for use with very long base-lines, and experimentally it has proved to be largely free of the effects of ionospheric scintillation[2]. These advantages led us to suggest[1] that the principle might be applied to the measurement of the angular diameter of visual stars. Thus one could replace the two aerials by two mirrors M_1, M_2 (Fig. 1*b*) and the radio-frequency detectors by photo-electric cells C_1, C_2, and measure, as a function of the separation of the mirrors, the correlation between the fluctuations in the currents from the cells when illuminated by a star.

It is, of course, essential to the operation of such a system that the time of arrival of photons at the two photocathodes should be correlated when the light beams incident upon the two mirrors are coherent. However, so far as we know, this fundamental effect has never been directly observed with light, and indeed its very existence has been questioned. Furthermore, it was by no means certain that the correlation would be fully preserved in the process of photo-electric emission. For these reasons a laboratory experiment was carried out as described below.

The apparatus is shown in outline in Fig. 2. A light source was formed by a small rectangular aperture, 0·13 mm. × 0·15 mm. in cross-section, on which the image of a high-pressure mercury arc was focused. The 4358 A. line was isolated by a system of filters, and the beam was divided by the half-silvered mirror M to illuminate the cathodes of the photomultipliers C_1, C_2. The two cathodes were at a distance of 2·65 m. from the source and their areas were limited by identical rectangular apertures O_1, O_2, 9·0 mm. × 8·5 mm. in cross-section. (It can be shown that for this type of instrument the two cathodes need not be located at precisely equal distances from the source. In the present case their distances were adjusted to be roughly equal to an accuracy of about 1 cm.) In order that the degree of coherence of the two light beams might be varied at will, the photomultiplier C_1 was mounted on a horizontal slide which could be traversed normal to the incident light. The two cathode apertures, as viewed from the source, could thus be superimposed

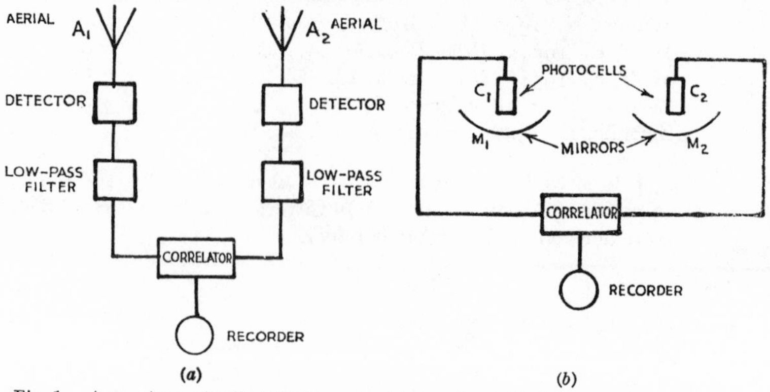

Fig. 1. A new type of radio interferometer (*a*), together with its analogue (*b*) at optical wave-lengths

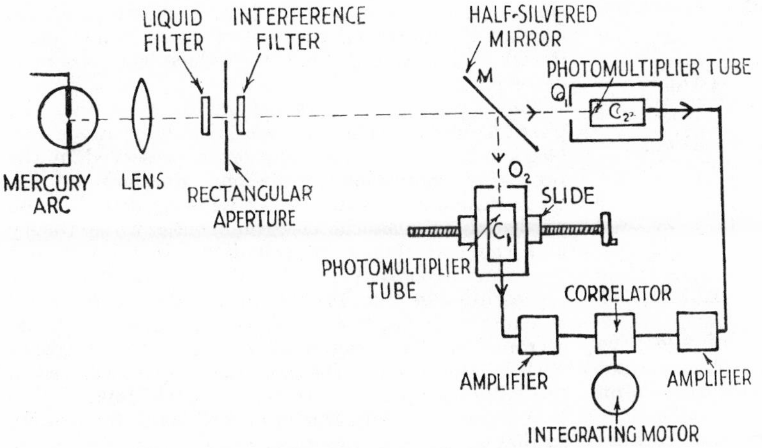

Fig. 2. Simplified diagram of the apparatus

or separated by any amount up to about three times their own width. The fluctuations in the output currents from the photomultipliers were amplified over the band 3–27 Mc./s. and multiplied together in a linear mixer. The average value of the product, which was recorded on the revolution counter of an integrating motor, gave a measure of the correlation in the fluctuations. To obtain a significant result it was necessary to integrate for periods of the order of one hour, so very great care had to be taken in the design of the electronic equipment to eliminate the effects of drift, of interference and of amplifier noise.

Assuming that the probability of emission of a photoelectron is proportional to the square of the amplitude of the incident light, one can use classical electromagnetic wave theory to calculate the correlation between the fluctuations in the current from the two cathodes. On this assumption it can be shown that, with the two cathodes superimposed, the correlation $S(0)$ is given by :

$$S(0) = A.T.b_v.\mathrm{f}\left(\frac{a_1\theta_1\pi}{\lambda_0}\right).\ \mathrm{f}\left(\frac{a_2\theta_2\pi}{\lambda_0}\right)\int\alpha^2(\nu).n_0{}^2(\nu).d\nu \quad (1)$$

It can also be shown that the associated root-mean-square fluctuations N are given by :

$$N = A.T.\frac{2m}{m-1}.\ b_v\,(b_vT)^{-\frac{1}{2}}\int\alpha(\nu).n_0(\nu).d\nu \quad (2)$$

where A is a constant of proportionality depending on the amplifier gain, etc. ; T is the time of observation ; $\alpha(\nu)$ is the quantum efficiency of the photocathodes at a frequency ν ; $n_0(\nu)$ is the number of quanta incident on a photocathode per second, per cycle bandwidth ; b_v is the bandwidth of the amplifiers ; $m/(m-1)$ is the familiar excess noise introduced by secondary multiplication ; a_1, a_2 are the horizontal and vertical dimensions of the photocathode apertures ; θ_1, θ_2 are the angular dimensions of the source as viewed from the photocathodes ; and λ_0 is the mean wave-length of the light. The integrals are taken over the complete optical spectrum and the phototubes are assumed to be identical. The factor $\mathrm{f}\left(\frac{a\theta\pi}{\lambda_0}\right)$ is determined by the dimensionless parameter η defined by

$$\eta = a\theta/\lambda_0 \quad (3)$$

which is a measure of the degree to which the light is coherent over a photocathode. When $\eta \ll 1$, as for a point source, $\mathrm{f}(\eta)$ is effectively unity ; however, in the laboratory experiment it proved convenient to make η_1, η_2 of the order of unity in order to increase the light incident on the cathodes and thereby improve the ratio of signal to noise. The corresponding values of $\mathrm{f}(\eta_1)$, $\mathrm{f}(\eta_2)$ were 0·62 and 0·69 respectively.

When the centres of the cathodes, as viewed from the source, are displaced horizontally by a distance d, the theoretical value of the correlation decreases in a manner dependent upon the dimensionless parameters, η_1 and d/a_1. In the simple case where $\eta_1 \ll 1$, which would apply to an experiment on a visual star, it can be shown that $S(d)$, the correlation as a function of d, is proportional to the square of the Fourier transform of the intensity distribution across the equivalent line source. However, when $\eta \gg 1$, as in the present experiment, the correlation is determined effectively by the apparent overlap of the cathodes and does not depend critically on the actual width of the source. For this reason no attempt was made in the present experiment to measure the apparent angular size of the source.

The initial observations were taken with the photocathodes effectively superimposed ($d = 0$) and with varying intensities of illumination. In all cases a positive correlation was observed which completely disappeared, as expected, when the separation of the photocathodes was large. In these first experiments the quantum efficiency of the photocathodes was too low to give a satisfactory ratio of signal to noise. However, when an improved type of photomultiplier became available with an appreciably higher quantum efficiency, it was possible to make a quantitative test of the theory.

A set of four runs, each of 90 min. duration, was made with the cathodes superimposed ($d = 0$), the counter readings being recorded at 5-min. intervals. From these readings an estimate was made of N_e, the root mean square deviation in the final reading $S(0)$ of the counter, and the observed values of $S_e(0)/N_e$ are shown in column 2 of Table 1. The results are given as a ratio in order to eliminate the factor A in equations (1) and (2), which is affected by changes in the gain of the equipment. For each run the factor

$$\frac{m-1}{m}\int\alpha^2(\nu)n_0{}^2(\nu)d\nu\Big/\int\alpha(\nu)n_0(\nu)d\nu$$

was determined from measurements of the spectrum of the incident light and of the d.c. current, gain and output noise of the photomultipliers ; the corresponding theoretical values of $S(0)/N$ are shown in the second column of Table 1. In a typical case, the photomultiplier gain was 3×10^5, the output current was 140 μamp., the quantum efficiency $\alpha(\nu_0)$ was of the order of 15 per cent and $n_0(\nu_0)$ was of the order of 3×10^{-3}. After each run a comparison run was taken with the centres of the photocathodes, as viewed from the source, separated by twice their width ($d = 2a$), in which position the theoretical

Table 1. COMPARISON BETWEEN THE THEORETICAL AND EXPERIMENTAL VALUES OF THE CORRELATION

	Cathodes superimposed ($d = 0$)		Cathodes separated ($d = 2a = 1\cdot 8$ cm.)	
	Experimental ratio of correlation to r.m.s. deviation $S_e(0)/N_e$	Theoretical ratio of correlation to r.m.s. deviation $S(0)/N$	Experimental ratio of correlation to r.m.s. deviation $S_e(d)/N_e$	Theoretical ratio of correlation to r.m.s. deviation $S(d)/N$
1	+ 7·4	+ 8·4	− 0·4	~ 0
2	+ 6·6	+ 8·0	+ 0·5	~ 0
3	+ 7·6	+ 8·4	+ 1·7	~ 0
4	+ 4·2	+ 5·2	− 0·3	~ 0

correlation is virtually zero. The ratio of $S_e(d)$, the counter reading after 90 minutes, to N_e, the root mean square deviation, is shown in the third column of Table 1.

The results shown in Table 1 confirm that correlation is observed when the cathodes are superimposed but not when they are widely separated. However, it may be noted that the correlations observed with $d=0$ are consistently lower than those predicted theoretically. The discrepancy may not be significant but, if it is real, it was possibly caused by defects in the optical system. In particular, the image of the arc showed striations due to imperfections in the glass bulb of the lamp ; this implies that unwanted differential phase-shifts were being introduced which would tend to reduce the observed correlation.

This experiment shows beyond question that the photons in two coherent beams of light are correlated, and that this correlation is preserved in the process of photoelectric emission. Furthermore, the quantitative results are in fair agreement with those predicted by classical electromagnetic wave theory and the correspondence principle. It follows that the fundamental principle of the interferometer represented in Fig. 1*b* is sound, and it is proposed to examine in further detail its application to visual astronomy. The basic mathematical theory together with a description of the electronic apparatus used in the laboratory experiment will be given later.

We thank the Director of Jodrell Bank for making available the necessary facilities, the Superintendent of the Services Electronics Research Laboratory for the loan of equipment, and Mr. J. Rodda, of the Ediswan Co., for the use of two experimental phototubes. One of us wishes to thank the Admiralty for permission to submit this communication for publication. [Oct. 5

[1] Hanbury Brown, R., and Twiss, R. Q., *Phil. Mag.*, **45**, 663 (1954).
[2] Jennison, R. C., and Das Gupta, M. K., *Phil. Mag.* (in the press).

PAPER NO. 20

Reprinted from *Nature*, Vol. 178, pp. 1046–1048, 1956.

A TEST OF A NEW TYPE OF STELLAR INTERFEROMETER ON SIRIUS

By R. HANBURY BROWN

Jodrell Bank Experimental Station, University of Manchester

AND

DR. R. Q. TWISS

Services Electronics Research Laboratory, Baldock

WE have recently described[1] a laboratory experiment which established that the time of arrival of photons in coherent beams of light is correlated, and we pointed out that this phenomenon might be utilized in an interferometer to measure the apparent angular diameter of bright visual stars.

The astronomical value of such an instrument, which might be called an 'intensity' interferometer, lies in its great potential resolving power, the maximum usable base-line being governed by the limitations of electronic rather than of optical technique. In particular, it should be possible to use it with base-lines of hundreds, if not thousands of feet, which are needed to resolve even the nearest of the *W*-, *O*- and *B*-type stars. It is for these stars that the measurements would be of particular interest since the theoretical estimates of their diameters are the most uncertain.

The first test of the new technique was made on Sirius (α Canis Majoris A), since this was the only star bright enough to give a workable signal-to-noise ratio with our preliminary equipment.

The basic equipment of the interferometer is shown schematically in Fig. 1. It consisted of two mirrors M_1, M_2, which focused light on to the cathodes of the photomultipliers P_1, P_2 and which were guided manually on to the star by means of an optical sight mounted on a remote-control column. The intensity fluctuations in the anode currents of the photomultipliers were amplified over the band 5–45 Mc./s., which excluded the scintillation frequencies, and a suitable delay was inserted into one or other of the amplifiers to compensate for the difference in the time of arrival of the light from the star at the two mirrors. The outputs from these amplifiers were multiplied together in a linear mixer and, after further amplification in a system where special precautions were taken to eliminate the effects of drift; the average value of the product was recorded on the revolution counter of an integrating motor. The readings of this counter gave a direct measure of the correlation between the intensity fluctuations in the light received at the two mirrors; however, the magnitude of the readings depended upon the gain of the equipment, and for this reason the r.m.s. value of the fluctuations at the input to the correlation motor was also recorded by a second motor. Since the readings of both revolution counters depend in the same manner upon the gain, it was possible to eliminate the effects of changes in amplification by expressing all results as the ratio of the integrated correlation to the r.m.s. fluctuations, or uncertainty in the final value. The same procedure was also followed in the laboratory experiment described in a previous communication[1].

There is no necessity in an 'intensity' interferometer to form a good optical image of the star. It is essential only that the mirrors should focus the light from the star on to a small area, so that the photocathodes may be stopped down by diaphragms to the point where the background light from the night sky is relatively insignificant. In the present case, the two mirrors were the reflectors of two standard searchlights, 156 cm. in diameter and 65 cm. in focal length, which focused the light into an area 8 mm. in diameter. However, for observations of Sirius, the circular diaphragms limiting the cathode areas of the photomultipliers (*R.C.A.* type 6342) were made as large as possible, namely, 2·5 cm. in diameter, thereby reducing the precision with which the mirrors had to be guided.

The first series of observations was made with the shortest possible base-line. The searchlights were placed north and south, 6·1 metres apart, and observations were made while Sirius was within 2 hr. of transit. Since the experiments were all carried out at Jodrell Bank, lat. 53° 14′ N., the elevation of the star varied between 15½° and 20°,

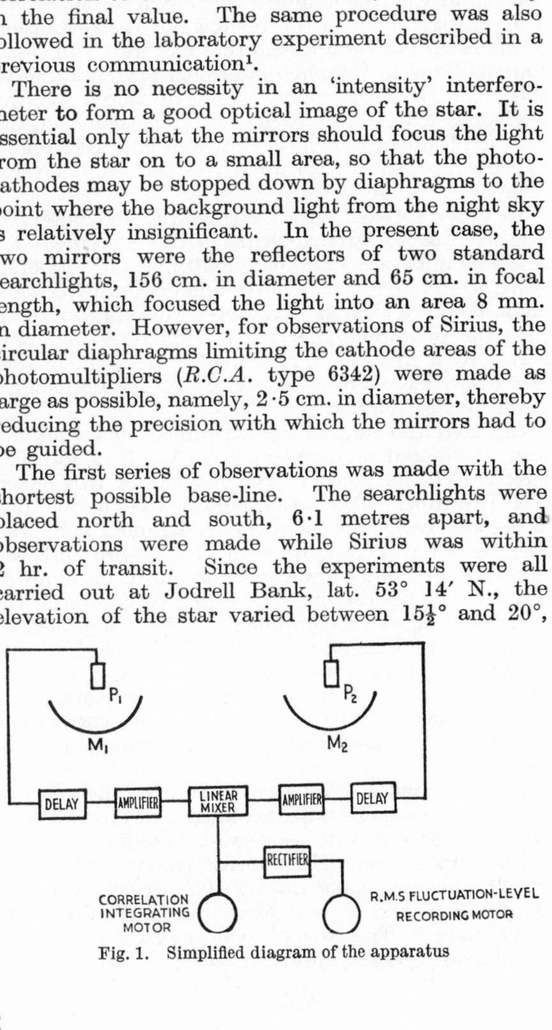

Fig. 1. Simplified diagram of the apparatus

Table 1. COMPARISON BETWEEN THEORETICAL AND OBSERVED CORRELATION

1. Base-line in metres	2·5 (N.S.)	5·54 (E.W.)	7·27 (E.W.)	9·20 (E.W.)
2. Observing time (min.)	345	285	280	170
3. Observed ratio of integrated correlation to r.m.s. deviation: $C(d)$	+8·50	+3·59	+2·65	+0·83
4. Theoretical ratio of integrated correlation to r.m.s. deviation, assuming star has an angular diameter of 0·0063″: $C(d)$	+9·35	+4·11	+2·89	+1·67
5. Theoretical ratio of integrated correlation to r.m.s. deviation, assuming star is a point source: $C(o)$	+10·15	+5.63	+5·06	+4·40
6. Theoretical normalized correlation coefficient for star of diameter 0·0063″: $\Gamma^2(d)$	0·92	0·73	0·57	0·38
7. Observed normalized correlation coefficient with associated probable errors: $\Gamma^2(d)$	0·84± 0·07	0·64± 0·12	0·52± 0·13	0·19± 0·15

and the average length of the base-line projected normal to the star was 2·5 metres; at this short distance Sirius should not be appreciably resolved.

Throughout the observations the average d.c. current in each photomultiplier was recorded every 5 min., together with the readings of the revolution counters on both the integrating motors. The small contributions to the photomultiplier currents due to the night-sky background were measured at the beginning and end of each run. The gains of the photomultipliers were also measured and were found to remain practically constant over periods of several hours.

In order to ensure that any correlation observed was not due to internal drifts in the equipment, or to coupling between the photomultipliers or amplifier systems, dummy runs of several hours duration were made before and after every observation; for these runs the photomultiplier in each mirror was illuminated by a small lamp mounted inside a detachable cap over the photocathode. In no case was any significant correlation observed.

In this initial stage of the experiment, observations were attempted on every night in the first and last quarters of the Moon in the months of November and December 1955; the period around the full moon was avoided because the background light was then too high. During these months a total observation time of 5 hr. 45 min. was obtained, an approximately equal period being lost due to failure of the searchlight control equipment. The experimental value for the integrated correlation $C(d)$ at the end of the observations is given in the line 3 of Table 1. The value of $C(d)$ is the ratio of the change in the reading of the counter on the correlation motor to the associated r.m.s. uncertainty in this reading.

In the second stage of the experiment the spacing between the mirrors was increased and observations were carried out with east–west base-lines of 5·6, 7·3 and 9·2 metres. These measurements were made on all possible nights during the period January–March 1956, and a total observing time of 12¼ hr. was obtained. The observed values of the integrated correlation $C(d)$ are shown in line 3 of Table 1.

As a final check that there was no significant contribution to the observed correlation from any other source of light in the sky, such as the Čerenkov component from cosmic rays[2], a series of observations was made with the mirrors close together and exposed to the night sky alone. No significant correlation was observed over a period of several hours.

The results have been used to derive an experimental value for the apparent angular diameter of Sirius. The four measured values of $C(d)$ were compared with theoretical values for uniformly illuminated disks of different angular sizes, and the best fit to the observations was found by minimizing the sum of the squares of the residuals weighted by the observational error at each point. In making this comparison, both the angular diameter of the disk and the value of $C(o)$, the correlation at zero base-line, were assumed to be unknown, and account was taken of the different light flux and observing time for each point. Thus the final experimental value for the diameter depends only on the relative values of $C(d)$ at the different base-lines, and rests on the assumption that these relative values are independent of systematic errors in the equipment or in the method of computing $C(d)$ for the models. The best fit to the observations was given by a disk of angular diameter 0·0068″ with a probable error of ± 0·0005″.

The angular diameter of Sirius, which is a star of spectral type *A*1 and photovisual magnitude −1·43, has never been measured directly; but if we assume that the star radiates like a uniform disk and that the effective black body temperature[3,4] and bolometric correction are 10,300° K. and −0·60, respectively, it can be shown that the apparent angular diameter is 0·0063″, a result not likely to be in error by more than 10 per cent. (In this calculation the effective temperature, bolometric magnitude and apparent angular diameter of the Sun were taken as 5,785° K., −26·95, and 1,919″, respectively.) Thus it follows that the experimental value for the angular diameter given above does not differ significantly from the value predicted from astrophysical theory.

A detailed comparison of the absolute values of the observed correlation with those expected theoretically has also been made, and the results are given in Table 1 and in Fig. 2. In making this com-

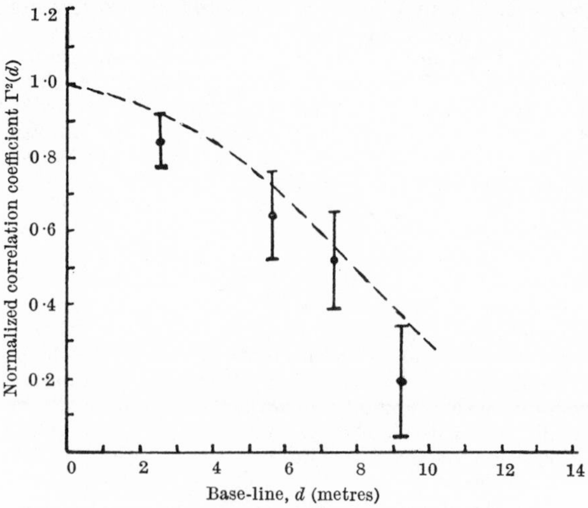

Fig. 2. Comparison between the values of the normalized correlation coefficient $\Gamma^2(d)$ observed from Sirius and the theoretical values for a star of angular diameter 0·0063″. The errors shown are the probable errors of the observations

parison, it is convenient to define a normalized correlation coefficient $\Gamma^2(d)$, which is independent of observing time, light flux and the characteristics of the equipment, where $\Gamma^2(d) = C(d)/C(o)$ and $C(d)$ is the correlation with a base-line of length d, and $C(o)$ is the correlation which would be observed with zero base-line under the same conditions of light flux and observing time. The theoretical values of $\Gamma^2(d)$ for a uniformly illuminated disk of diameter 0·0063″ are shown in line 6 of Table 1. For monochromatic radiation it is simple to evaluate $\Gamma^2(d)$, since it can be shown[5] that it is proportional to the square of the Fourier transform of the intensity distribution across the equivalent strip source ; however, in the present case, where the light band-width is large, the values of $\Gamma^2(d)$ were calculated by numerical integration.

The theoretical values of $C(o)$, given in line 5, were calculated for the conditions of light flux and observing time appropriate to each base-line by means of equations (1) and (2) of our previous communication[1] (though in the present experiment the r.m.s. fluctuations were smaller by a factor $1/\sqrt{2}$ than the value given in the previous paper, which refers to an alternative electronic technique). The most important quantities in this calculation are the gains and output currents of the photomultipliers and the band-widths of the amplifiers ; but it is also necessary to make a small correction for the combined spectral characteristics of the photocathodes, the atmospheric attenuation, the star and the mirrors. Finally, in line 4 of Table 1 the theoretical values of the correlation $C(d)$ are shown ; they were calculated from the theoretical values of $C(o)$ and $\Gamma^2(d)$ by means of the relation given above.

The correlation observed at the shortest base-line (2·5 m.) can be used as a rough test of the effects of atmospheric scintillation on the equipment, since the corresponding theoretical value depends only on well-known quantities and is almost independent of the angular diameter of the star. Throughout the observations Sirius was seen to be scintillating violently, although the corresponding fluctuations in the d.c. anode currents of the photomultiplier tubes, which were smoothed with a time constant of about 0·1 sec., were only of the order of ± 10 per cent, as might be expected with mirrors large by normal telescope standards. Nevertheless, the observed correlation $C(d) = + 8{\cdot}50$ does not differ significantly from the calculated value of + 9·35, and it follows that it cannot be greatly affected by scintillation.

The experimental values of $C(d)$ obtained at the four base-lines may be compared with the corresponding theoretical values $C(d)$ by means of lines 3 and 4 of Table 1. However, it is more convenient, since these values depend upon the different values of observing time and light flux at each base-line, to normalize the observed values of $C(d)$ by the corresponding values of $C(o)$, so as to give the normalized correlation coefficients $\Gamma^2(d)$ shown in line 7. In Fig. 2 these experimental values of $\Gamma^2(d)$ are shown together with their probable errors, and may be compared with the broken curve, which gives the theoretical values for a uniform disk of 0·0063″. It can be seen that both the relative and absolute values of $\Gamma^2(d)$ are in reasonable agreement with theory, and that within the rather wide limits of this preliminary test there is no significant difference between the correlation predicted and observed.

In assessing the potentialities of the technique described here, it is important to note that, although the measurements took five months to complete, the visibility was so poor that the total observing time was only 18 hr., while in this limited period additional absorption of 0·25–0·75 magnitudes due to haze or thin cloud was often present. If the observations had been made at a latitude where Sirius transits close to the zenith, the improved signal-to-noise ratio, due to decreased atmospheric absorption, would have made it possible to obtain the same data in a total observing time of about four hours.

Thus, despite their tentative nature, the results of this preliminary test show definitely that a practical stellar interferometer could be designed on the principles described above. Admittedly such an instrument would require the use of large mirrors. Judging from the results of this test experiment, where the peak quantum efficiency of the phototubes was about 16 per cent and the overall band-width of the amplifiers was about 38 Mc./s., one would need mirrors at least 3 metres in diameter to measure a star, near the zenith, with an apparent photographic magnitude + 1·5. Mirrors of at least 6 metres in diameter would be required to measure stars of mag. +3, and an increase in size would also be needed for stars at low elevation because of atmospheric absorption. However, the optical properties of such mirrors need be no better than those of searchlight reflectors, and their diameters could be decreased if the overall band-width of the photomultipliers and the electronic apparatus could be increased, or if photocathodes with higher quantum efficiencies become available. It must also be noted that the technique of using two mirrors, as described here, would probably be restricted to stars of spectral type earlier than G, since cooler stars of adequate apparent magnitude would be partially resolved by the individual mirrors.

The results of the present experiment also confirm the theoretical prediction[6] that an 'intensity' interferometer should be substantially unaffected by atmospheric scintillation. This expectation is also supported by experience with a radio 'intensity' interferometer[5,7,8] which proved to be virtually independent of ionospheric scintillation. It is also to be expected that the technique should be capable of giving an extremely high resolving power. Without further experience it is impossible to estimate the maximum practical length of the base-line ; however, it is to be expected that the resolving power could be at least one hundred times greater than the highest value so far employed in astronomy, and that almost any star of sufficient apparent magnitude could be resolved.

We thank the Director of Jodrell Bank for making available the necessary facilities, the Superintendent of the Services Electronics Research Laboratory for the loan of much of the equipment, and Dr. J. G. Davies for his assistance with setting up the searchlights. One of us (R. Q. T.) wishes to thank the Admiralty for permission to submit this communication for publication.

[1] Hanbury Brown, R., and Twiss, R. Q., *Nature*, **177**, 27 (1956).
[2] Galbraith, W., and Jelley, J. V., *J. Atmos. Terr. Phys.*, **6**, 250 (1955).
[3] Kuiper, G. P., *Astrophys. J.*, **88**, 429 (1938).
[4] Keenan, P. C., and Morgan, W. W., "Astrophysics", edit. J. A. Hynek (McGraw-Hill, New York, 1951).
[5] Hanbury Brown, R., and Twiss, R. Q., *Phil. Mag.*, **45**, 663 (1954).
[6] Twiss, R. Q., and Hanbury Brown, R. (in preparation).
[7] Hanbury Brown, R., Jennison, R. C., and Das Gupta, M. K., *Nature*, **170**, 1061 (1952).
[8] Jennison, R. C., and Das Gupta, M. K., *Phil. Mag.*, **1**, viii, 55 (1956).

PAPER NO. 21

Reprinted from *Proceedings of a Symposium on Astronomical Optics and Related Subjects*, edited by Z. Kopal (North-Holland Publishing Company, Amsterdam 1956), pp. 17–30.

LIGHT AND INFORMATION

D. GABOR

IMPERIAL COLLEGE, UNIVERSITY OF LONDON

It is just four years since I gave a talk of the same title in Edinburgh, as a Ritchie Lecture. To my regret, I still have not published it in its entirety, but I have repeated it so often, and the manuscript has been so widely circulated that it is now fairly well known among the rather few workers who are interested both in optics and in information theory. Several important papers have appeared in the meantime on this topic, of which I want to mention particularly the contributions by Blanc-Lapierre [1], and by Fellgett and Linfoot [2], which deal exhaustively with the Fourier-filter aspect of optical systems. Fortunately these authors, as well as P. M. Duffieux [3], are present at this Symposium, and this enables me to leave out the Fourier transformation theory, and to use the short time mainly for explaining certain methods and results which I have found only quite recently.

Communication theory has been widely publicised, but I am not so sure that it has been widely understood. It consists of two distinct parts. The first is the Structural Theory, concerned with the adequate representation of signals and noise in a communication channel, and their transformations. The principal result of the structural theory is the recognition that any physical system of communication or observation has only a *finite degree of freedom*. I will show later how this recognition enables us to obtain a very general and illuminating view of optical instruments and observation and to harness powerful mathematical methods into the service of optics.

The second part of communication theory is the Statistical Theory, in the main the creation of Claude E. Shannon and of Norbert Wiener, which is also often called, very misleadingly in my opinion, the Theory of Information. The central problem of this theory is the optimum utilisation of communication channels, given a certain ensemble of communication signs, and of noise. If these ensembles are given, one can define a certain quantity, in terms of the probabilities of the signals, which is called the Amount of Information; and one can correspondingly define the information handling capacity of the channel. There are two means of improving it; Coding and Matching. Coding is an agreed *language* between the transmitter and the receiver, matching is the translation of the signs of this language into the physical signals which go through the channel.

I want to emphasise very strongly how careful one must be when applying these most interesting and important concepts to research problems, such as astronomical optics. On the face of it, one would be inclined to say that the problems of research have nothing in common with the problems of communication engineering. When Mother Nature is talking to us, there is no question of a pre-arranged code; there is no question of finding out which of the communication signs, *all of which are supposed to be known beforehand*, Mother Nature has intended to send us.

Nevertheless, this would be going to far. We must remember that there are very different kinds of activities going under the name of "research". An astronomer might, for instance, wish to make a map of the stars in a certain part of the sky. He has then made up his mind that he is looking for point sources of light; only their whereabouts in the given region is unknown. His position is then not essentially different from that of the communication engineer, because as far as he is concerned the "signs" are point sources, hence the coding is known *ex hypothesi*. Though he has no influence on the coding, he is still free to use the teachings of communication theory as regards matching, in that small end-tract of the immensely long optical path which forms his observation system. I will give a few examples of this in my second contribution.

But if the research worker sets out to discover new things, this procedure is almost the opposite of reasonable. If one makes sure that one's observation system is most economical and efficient for observing things of a known kind, one has *ipso facto* almost made sure that anything new will slip through the net! If one wants to make new discoveries one must not specialise for known things; one must specialise for things which are *outside the code*. It is no reproach to Statistical Communication Theory that it cannot give advice in this art, except the one: "If you want to discover something new, keep away from my methods!"

GEOMETRICAL OPTICS

I propose to introduce the structural theory step by step, starting with geometrical optics, passing on to wave optics, and saying, on the way, a few things about quantum optics, which will take us to the borderline of the statistical theory.

The basic concept of geometrical optics is the *ray of light*. The rays in an optical system form a four-dimensional manifold; ∞^2 rays are crossing in every point of space. As Hamilton was the first to conceive, one can escape this confusion by adding two extra dimensions to the representation, corresponding to two angular coordinates. Fig. 1 illustrates this in the case of a one-dimensional imaging system. The extra coordinate in this case corresponds to the angle with the optic axis.

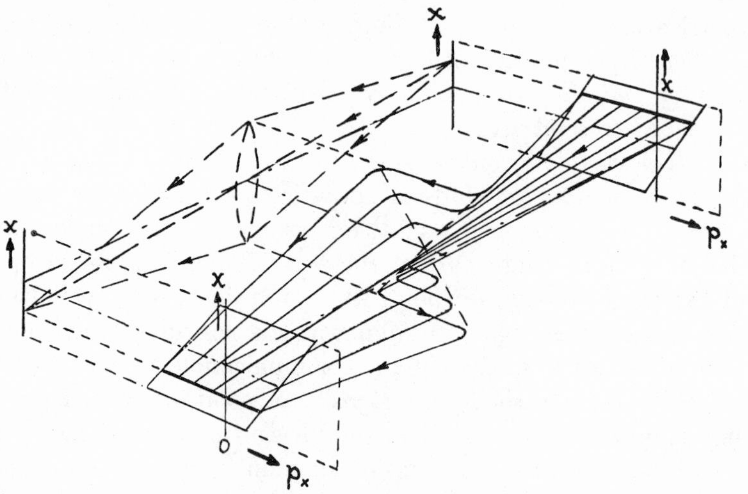

Fig. 1. One-dimensional imaging in configuration space and in phase-space

It is advantageous to make it equal or proportional to the sine of this angle. In Fig. 1 it is called p_x, as if it were $p_x = p \sin \theta$, the transversal momentum of light "corpuscles" with a total momentum p. The figure shows the rays which image a point. The rays retain constant momentum p_x until they reach the lens, which twists the ray-surface. One could similarly draw in the ray surfaces corresponding to other object points; they can never cross one another in phase space.

In geometrical optics any ray can carry a signal – an intensity – independently of the intensity of the neighbouring ray, however close, hence one could transmit through any optical channel a fourfold infinity of data, which evidently does not make physical sense. Yet, geometrical optics at least gives an important hint for removing this serious blemish. The hint is contained in the Smith-Lagrange theorem of the conservation of phase-space in optical transformations. This suggests very forcibly that the information is not carried by rays but by certain *tubes* of rays, whose cross section remains invariant

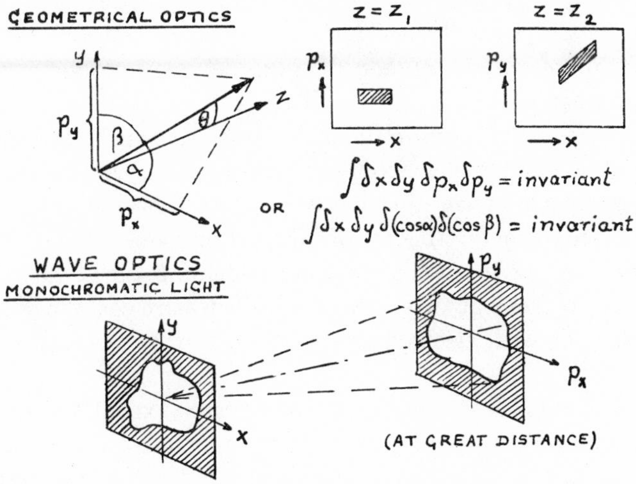

Fig. 2. The expansion theorem

along the optical path. But as geometrical optics does not know of any characteristic length, the cross-section of the information-carrying tubes remains entirely undetermined.

WAVE OPTICS. MONOCHROMATIC LIGHT

This uncertainty is at once removed in wave optics, which supplies the *wavelength* as a natural unit of optical lengths, and suggests that the elementary tube must have the cross section λ^2. This has been proved, or at any rate made plausible in special cases by McKay, the author, Blanc-Lapierre, Fellgett and Linfoot, but I propose to formulate it more exactly and generally in the form of the following *Expansion Theorem:*

"Assume that the object area, large compared with the square of the wavelength, is limited by a black screen. Assume also that there is a similar limitation in an aperture plane, at a great distance from the object plane. Then, in the domain limited by these two black screens, there exist N independent solutions of the wave equation

$$\nabla^2 u + \left(\frac{2\pi}{\lambda}\right)^2 u = 0;$$

that is to say solutions with $u = 0$ immediately behind the black screens, and N is

$$N = \frac{1}{h^2}\int \delta_x\delta_y\ \delta\ (\cos\alpha)\ \delta\ (\cos\beta) = \frac{1}{h^2}\int \delta_x\delta_y\ \delta p_x\delta p_y,\ \ (p = h/\lambda).$$

Any progressive wave through the object area and through the aperture can be expanded in terms of these N eigensolutions, with not more than N complex coefficients".

This is evidently an *optical* theorem, which is meaningful in much the same field as the Fresnel-Kirchhoff theorem. I have emphasized this by introducing the term "black screen", which has no meaning in the exact electromagnetic theory of diffraction. One cannot construct exact solutions of the wave equation which vanish at the surface of one screen, let alone at two screens. But as we are concerned here with optics and not with electromagnetic theory, this formulation will be quite sufficient.

This theorem has not been proved exactly except for rectangular and circular object areas and limiting apertures, and I have only made it plausible, (in 1951) for arbitrary but sufficiently large areas and apertures by taking "Gaussian beams" as the eigensolutions. But I do not see why we should worry about the lack of an exact proof any more then one does in Quantum Mechanics, where Dirac's Expansion Theorem has not been proved to this day.

Accepting the Expansion Theorem, we can now give a more exact interpretation to the "tubes of information". Assume that we have a set of N eigenfunctions u_k (x,y,z), and consider them in the object plane, $z = z_0$. It is well known that by linear combination one can form from these a set which is *orthogonal* in this plane; in physical language the energy fluxes in the object plane are *additive*. But if they are additive in one plane, they are also additive in all other planes of the optical system, otherwise we could *modulate* the energy of one beam by another, which, in the plane z_0 was completely independent of it. Hence the orthogonality of the elementary beams is an invariant property, and it is evident that we must replace the naive idea of "tubes of information" by a set of orthogonal, elementary beams.

The difficulty is only that though these beams keep their energies separate, they do not keep exactly to their own tubes. What is more, the better we separate them spatially in one cross section, the more they will overlap in others. Gaussian elementary beams have the smallest overspill, but they are not exactly orthogonal. One can, however, construct orthogonal beam systems with little more than the minimum overspill, and if we do not put too fine an edge on it, we can consider these as fairly close wave-mechanical substitutes for the "tubes of information" of the quasi-geometrical theory, with cross sections of λ^2.

It is very instructive to disregard this grain of salt, and to follow the adventures of elementary information tubes through an optical system, as illustrated in Fig. 3, which is again a one-dimensional example, like Fig. 1. The figures at the bottom show the lozenge-shaped x-p_x domains, which arise from an aperture at a finite distance, divided

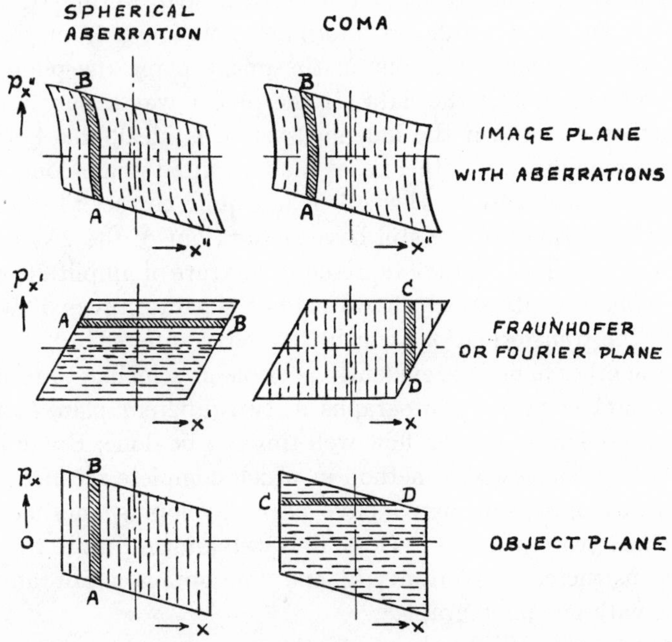

Fig. 3. Transformations of elementary areas in phase space

up into elementary areas in two different ways; at the left according to the position x, at the right according to p_x. In the rear focal plane of the imaging system, which is called the "Fraunhofer" or the "Fourier" plane, the strips have turned round by just 90 degrees; the positions are translated into angles, and *vice versa*. It is well known that this amounts to a Fourier representation of the object, but I will not dwell on this aspect of the subject.

In the image plane the vertical strips are turned upright again, but in the presence of geometrical aberrations these will not be straight. The figure at the left shows spherical aberration; the rays are displaced by $\Delta x \sim p^3{}_x$. The right figure illustrates coma; the displacement is $\Delta x = xp^2{}_x$.

In order to understand what is happening here, it is time to talk not only of the tubes, but of what is going through them, from plane to plane of the optical system. Each tube corresponds to one elementary beam or eigensolution of the wave equation for monochromatic light, and carries with it one complex datum, that is to say two real data. If we put a photographic plate in the way of the beam, this will analyse the x-p_x area into vertical strips, and will record only one datum; the energy flux through the vertical strip during the time of exposure. In other words no information was lost from the object plane to the image plane, but in the image plane the photographic plate loses one half of the data. If the object was a pure amplitude contrast object, and if the imaging system is ideal, this half is just what we require. In all other cases the unwanted phase-component of the contrast will intrude. Hence it is not quite correct to say that we have lost information, we still have N data out of the $2N$, but these are not what we want; they represent a mixture of amplitude contrast and of phase contrast which we cannot separate, even if we know exactly the transformations which they have suffered.

On the other hand it is evidently possible in principle to reconstruct all the data from *two* photographs in two different planes; it would take us too long to discuss how well this can be done. Bragg and Rogers [4] have suggested a method in which complete reconstruction is possible by optical means; I have been developing another method for the last years with W. P. Goss. It is well known that in the method of Reconstructed Wavefronts [5] one can obtain a fair amount of success with one photograph.

So far I have talked only of strictly monochromatic, that is to say absolutely coherent illumination. Partial coherence will be treated later, but it will be worth while to dwell for a moment on the case of completely incoherent illumination. In this case the strips A-B will be imaged, by the faulty objective, into curved and distorted strips, but, somewhat surprisingly, this is not a destructive transformation. Incoherent illumination can show up only the amplitude contrast, with N data, and if we know the aberrations, we have exactly N linear equations from which we can calculate the original figure. The question is only, with what accuracy? The problem may well be of considerable practical interest, especially in astronomy where the condition of complete incoherence is always fulfilled. It would be very tempting not to spend years figuring the mirrors of giant telescopes, but to correct the photographs by some electrical compensating filter, as

A. Blanc-Lapierre has proposed, or by re-photographing them through a compensating system. I have not looked very thoroughly into this problem of *"incoherent reconstruction"*, but it is easy to give an answer to the question of accuracy in the case of star images. Take first a corrected objective, and assume that N effective photons fall into the Airy figure. The mean square error in determining a coordinate x of the centre of gravity of the figure is of the order

$$\overline{\delta x^2} = d^2/\sqrt{N}$$

$d =$ inertial radius of the Airy figure.

If now one takes an uncorrected lens, which blows up the Airy figure M times, the r.m.s. error also increases M times. But this need not be a noticeable disadvantage. One does not bother in general to determine the centre of gravity of the Airy figure with an accuracy better than d, and if d is of the order of the Selwyn grain, one could not do it anyway. Hence the uncertainty in the position

$$d.M/\sqrt{N}$$

may still be smaller than d. With 100 effective photons in a star image one can afford blowing up the star image about 10 times, and yet reconstruct to an accuracy d. I think therefore that the correction of astronomical objectives by analogue methods deserves considerable attention.

"PRACTICALLY" MONOCHROMATIC LIGHT

After this excursion from absolute coherence to absolute incoherence we can attack the general problem of information transfer in the presence of partial coherence. Absolutely monochromatic light is of course an unrealistic abstraction; in reality we have at best what I propose to call *"practically"* monochromatic light.* This case arises, for instance, if one observes an object illuminated by a single spectral line, but from an extended source.

Partial coherence was first defined, as far as I known, by v. Laue [6] in 1907, later by van Cittert [7] and Zernike [8], and more recently the problem was treated with particular success by Hopkins [9], Blanc-Lapierre and Dumontet [10] and by Wolf [11]. I think the best start for a general theory is the *complex* coefficient of partial coherence, defined by H. H. Hopkins in 1951:

* This is what Blanc-Lapierre and Dumontet call "lumière quasimonochromatique".

Let P_1, P_2 be two points in a cross section of the optical tract, let us say a plane, which is illuminated by a source Σ. Let I_1, I_2 be the intensities at these points, and u_1, u_2 the complex light amplitudes. Then

$$\gamma_{12} = \frac{1}{\sqrt{I_1 I_2}} \int_{\Sigma} u_1 u_2^* \, d\sigma$$

is the coefficient of partial coherence.

It is important to note that γ_{12} can be determined experimentally by piercing two holes in the plane of P_1, P_2 and observing the Young interference figure behind them.

The general theory starts from the observation that

$$\gamma_{12} = \gamma_{21}^*.$$

Combined with the fact that in any plane we have only N elementary areas, (of which we can keep talking as "points",) this suggests forcibly to construct an *illumination matrix*

$$|I_{ik}| \equiv \begin{vmatrix} I_{11} & I_{12} & \cdots & I_{1N} \\ I_{21} & I_{22} & \cdots & I_{2N} \\ \cdots & \cdots & \cdots & \cdots \\ I_{N1} & I_{N2} & \cdots & I_{NN} \end{vmatrix}, \quad I_{ik} = \gamma_{ik} \sqrt{I_i I_k}.$$

This is a *Hermitian matrix* because

$$I_{ik} = I_{ki}^*.$$

The suffixes i, k which run from l to N, can be associated with the elementary areas in the plane of reference, (in any agreed order,) but just as well with the elementary waves, the eigensolutions of the monochromatic wave equation, as previously defined.

So far the matrix is nothing but a table of numbers, as we have defined it only in one plane. How is it transformed on its way through the optical system?

The answer is evident. The matrix terms I_{ik} are of the nature of the product of two complex amplitude components, $U_i U_k^*$, and in the special case of complete coherence the matrix can in fact be reduced to this form. We know that in an optical system the amplitudes suffer *linear* transformations. Hence if the illumination matrix has any physical significance, it must transform like the outer product of two *vectors*, whose components are complex conjugates.

At this point the explanations threaten to become too abstract, and it will be useful to supplement them by a geometrical picture.

Shannon [12] has introduced into communication theory the "signal space"; a Hilbert space of N dimensions, in which the coordinates are the expansion coefficients of the signal, in terms of the eigenfunctions. If one chooses these orthogonal, one has Cartesian coordinates in an N-dimensional space; squares of vectors represent energies.

We can use the same device, but with a modification; we must make all vector-components U_i complex, and introduce a *Hermitian* instead of Euclidian metric, that is to say define the energy of the vector U, as

$$\sum_{i=1}^{N} U_i U_i^*.$$

There is a very profound difference between the situation in communication theory and in optics, which forces us to introduce complex coordinates. Communication theory has been developed in the classical range of physical phenomena, in which a signal $s(t)$ is an observable function of time. On the other hand optics, with the exception of microwave optics, is entirely in the quantum range, in which it takes extraordinary large energies to make phases observable. (Gabor, 1950 [13].) We can only count photons, or observe energies. In other words in classical theory one postulates linear, amplitude-measuring instruments, while in optics we have only quadratic detectors, such as photographic plates. But though we cannot measure any of the complex vector-components U_i by themselves, we can infer them from interference experiments, up to a single unitary complex factor, common to all.

One could of course represent the complex vectors U_i in real terms by two vectors, one with the real and one with the imaginary components, but this would only conceal the fact that they are not observable. It is much better to adopt the complex representation. Though, strictly speaking one should not make a picture of it, there is no harm in a real model, and one can use such terms and concepts as "length", "angle", "rotation" with impunity, so long as one does not forget to translate them into Hermitian terms in the formulae.

With these reservations we can now state in geometrical language: All physical characteristics of practically monochromatic illumination can be represented by an *illumination ellipsoid* in the N-dimensional (complex) information space. All transformations of the illumination from plane to plane are (Hermitian) *rotations* of the ellipsoid in this space.

The principal axes of the ellipsoid are the square roots of the eigenvalues of the matrix, that is to say the solutions of the characteristic equation

$$\| I_{ik} - x^2\delta(i, k) \| = 0,$$

where $\delta(i, k)$ is Kronecker's unit tensor. We call the number of non-zero roots, that is to say of non-zero axes the *dimensionality* of the illumination. Fully coherent illumination * has the dimensionality one; the illuminating ellipsoid degenerates to a vector. Fully incoherent

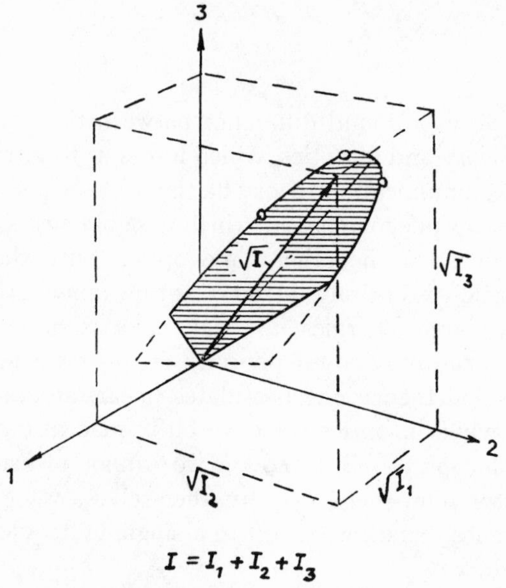

Fig. 4. The illumination ellipsoid in information space

illumination has the dimensionality N. A particularly interesting special case is if all N axes are equal. The illuminating ellipsoid now becomes a sphere, and it is seen that if incoherent illumination is even in one plane, it is even in all.

Fig. 4 illustrates the case of a two-dimensional ellipsoid in a three dimensional information space. It illustrates also the well known geometrical theorem that if one encloses an ellipsoid in a rectangular box, the square of the diagonal is equal to the sum of squares of the

* Rigorously full coherence obtains only in the case of a monochromatic point source, which is of course a non-physical abstraction. Practically we have full coherence if (a) the diameter of the source area is below the resolution limit of the optical system, (b) if the spectral width $\Delta\lambda$ is so small that $\lambda/\Delta\lambda$ is larger than the path differences in the optical system between interfering elementary beams. We can take (b) as the definition of "practical" monochromatism.

principal axes. This theorem corresponds to the conservation of energy along the optical path. The sum of the diagonal terms, the *trace* of the I-matrix remains an invariant in unitary Hermitian transformations, that is to say rotations.

Having now characterised the illumination, and its progress through the optical system, there is no difficulty in characterising the object. The object effects a linear transformation in the complex amplitudes, hence it must be a *linear Hermitian operator,* which can be again represented by a matrix, the *object matrix* Ω

$$\Omega \equiv |\,\omega_{ik}\,| \equiv \left|\begin{array}{cc} \omega_{11}\ \omega_{21} & \\ \omega_{21}\ \omega_{22} & \\ \cdots\cdots & \cdots\cdots \\ & \omega_{NN} \end{array}\right|, \quad (\omega_{ix} = \omega^*{}_{ki}).$$

This is understood to be referred to the same plane as te illumination matrix. The object operator transforms the illumination matrix into

$$|\, I^1{}_{i,j}\,| = \sum_{k}^{N}\sum_{m}^{N} \omega_{ik}\, I_{km}\, \omega_{mj},$$

which is also understood to be referred to the original plane *.

The general Hermitian linear operation consists of a deformation, that is to say a contraction or expansion in the direction of the axes, and a rotation. Let us exhibit these by decomposing Ω into the product of two operators:

$$\Omega = \left|\begin{array}{cc} \tau_1\ 0 & \\ 0\ \tau_2 & \\ \cdots\cdots & \cdots\cdots \\ & \tau_N \end{array}\right| \left|\begin{array}{cc} \pi_{11}\ \pi_{12} & \\ \pi_{21}\ \pi_{22} & \\ \cdots\cdots & \cdots\cdots \\ & \pi_{NN} \end{array}\right| .$$

Deformation *Rotation*

These are both Hermitian matrices, hence the diagonals must be real. Of a passive physical object we must expect that it can never increase the total energy, (and of course it cannot produce negative energies in the transmitted beam,) whatever the illumination, hence the "transmission coefficients τ must be all positive and smaller than or at most

* If the illumination is not monochromatic, we must form elementary monochromatic illumination matrices, transform each with its own operator, which is now a function of λ, and sum the intensities. This is of little interest from the point of view of information theory, as each distinguishable wavelength range $\Delta\lambda$ can be considered as an independent communication channel. In principle the distinguishable wavelength range is limited only by the observation time, which in optical problems is usually considered as unlimited.

equal to unity. The second operator, being unitary leaves the energy unaltered.

Note that this decomposition is *unique*. To prove this one has only to multiply the above equation from the right with the transpose, $\bar{\pi}$, of the π-matrix, which shows that π corresponds to the rotation which transforms Ω to a diagonal form. Now there are $N!$ ways for transforming a matrix to a diagonal form, corresponding tot the permutations, i.e. renumberings of the axes. But of these only one is of interest, as the "tubes of information" maintain their identity through the optical system. Thus the transmission coefficients τ are the eigenvalues of the object matrix, numbered in a certain definite way, and the π-matrix is also uniquely defined.

It is seen that the two operators into which we have decomposed the object operator possess physical reality, and this suggests considering the first as the *amplitude contrast matrix*, and the second as the *phase contrast matrix*.

It is of course not to be expected that these definitions will completely coincide with the current use of these terms in optics. For instance Hopkins [14], shows that a transparent object can show contrast on sharp focusing in an ordinary microscope, because part of the energy is diffracted outside the aperture. The new definition would interpret this as amplitude contrast. In fact, so long as one operates with one aperture, and if one knows nothing else about the object than what one sees, it is impossible to decide whether the energy has been absorbed in the object, or thrown outside the aperture. One can decide this only by prior knowledge, for instance by knowing the material of the object, or by varying the aperture and seeing whether the contrast vanishes or not.

I think this will be sufficient for establishing that Hermitian algebra is an adequate mathematical tool for a general theory of optical imaging. But having followed me so far, you may well ask whether anything new will come out of it? I think there is good hope for this. Something has already come out of it, which I for one have not expected, and this is that *the general optical object*, in an optical system with N complex degrees of freedom, *is characterised not by* $2N$ *but by* $\frac{1}{2}N\,(N+1)$ *real data,* and that moreover all these can be in principle determined from not more than N suitably designed experiments. I will omit the proof, which is based on illuminating the object in turn by N independent coherent beams of light. It is perhaps a little premature to say this, but I suspect that this might give the cue for the proper analysis of the confusing and not easily interpreted phenomena which one sees when looking into a *deep* object.

Another achievement which one may well expect from the general theory is an answer to the important practical question: "How far can one improve the accuracy in measuring parameters of interest by ignoring others?" In McKay's terminology: "How far can one concentrate metrons (photons) in the logons (degrees of freedom) which one wants to observe?" I hope to show in a second talk that there is indeed much to be gained in certain simple cases, but it is too early yet for a general answer.

Perhaps the most satisfactory feature of the theory is that it operates entirely with quantities which are in principle observable, in line with the valuable efforts of E. Wolf to rid optics of its metaphysical residues. Another is, that it is a good bridge to quantum optics. Indeed we have only to add the condition that the energies which appear in the diagonal of the I-matrix are integer multiples of a basic quantum to obtain a formulation of quantum optics, sufficient for most purposes. This is not surprising, seeing that the theory is modelled on quantum mechanics, in its form due to Born, Heisenberg and Dirac rather than that of Schrödinger. Optics was always considered as a good didactical preparation for Wave Mechanics, now it appears that Quantum Mechanics is not a bad preparation for optics.

REFERENCES

1. A. Blanc-Lapierre, Ann. de l'Institut Henri Poncaré, **13,** 245, 1953.
2. P. B. Fellgett ard E. H. Linfoot, Phil. Trans. Roy. Soc. A. No. 931, **247,** 369, 1955.
3. P. M. Duffieux, *L'integrale de Fourier et ses applications en optique,* Univ. de Besançon, 1946.
4. W. L. Bragg and G. L. Rogers, Nature **167,** 190, 951.
5. D. Gabor, Proc. Roy. Soc. A. **197,** 457, 1949; Proc. Phys. Soc. B. **23,** 449, 1951.
6. M. v. Laue, Ann. der Phys. **23,** 1, 1907.
7. H. van Cittert, Physica, **1,** 201, 1934.
8. F. Zernike, Physica, **5,** 785, 1938.
9. H. H. Hopkins, Proc. Roy. Soc. A. **208,** 263, 1951; **217,** 408, 1953.
10. A. Blanc-Lapierre and P. Dumontet, Revue d'Optique, **34,** 1, 1955.
11. E. Wolf, Proc. Roy. Soc. A. **225,** 96, 1954, **230,** 246, 1955.
12. C. E. Shannon, Proc. I.R.E. **37,** 10, 1949.
13. D. Gabor, Phil. Mag. (7), **41,** 1161, 1950.
14. H. H. Hopkins, Revue d'Optique, **31,** 142, 1952.

PAPER NO. 22

Reprinted from *Nature*, Vol. 178, pp. 1449–1450, 1956.

THE QUESTION OF CORRELATION BETWEEN PHOTONS IN COHERENT LIGHT RAYS

Brannen and Ferguson[1] have reported experimental results which they believe to be incompatible with the observation by Hanbury Brown and Twiss[2] of correlation in the fluctuations of two photoelectric currents evoked by coherent beams of light. Brannen and Ferguson suggest that the existence of such a correlation would call for a revision of quantum theory. It is the purpose of this communication to show that the results of the two investigations are not in conflict, the upper limit set by Brannen and Ferguson being in fact vastly greater than the effect to be expected under the conditions of their experiment. Moreover, the Brown–Twiss effect, far from requiring a revision of quantum mechanics, is an instructive illustration of its elementary principles. There is nothing in the argument below that is not implicit in the discussion of Brown and Twiss, but perhaps I may clarify matters by taking a different approach.

Consider first an experiment which is simpler in concept than either of those that have been performed, but which contains the essence of the problem. Let *one* beam of light fall on *one* photomultiplier, and examine the statistical fluctuations in the counting-rate. Let the source be nearly monochromatic and arrange the optics so that, as in the experiments already mentioned, the difference in the length of the two light-paths from a point A in the photocathode to two points B and C in the source remains constant, to within a small fraction of a wave-length, as A is moved over the photocathode surface. (This difference need not be small, nor need the path-lengths themselves remain constant.) Now it will be found, even with the steadiest source possible, that the fluctuations in the counting-rate are slightly greater than one would expect in a random sequence of independent events occurring at the same average rate. There is a tendency for the counts to 'clump'. From the quantum point of view this is not surprising. It is typical of fluctuations in a system of bosons. I shall show presently that this extra fluctuation in the single-channel rate necessarily implies the cross-correlation found by Brown and Twiss. But first I propose to examine its origin and calculate its magnitude.

Let P denote the square of the electric field in the light at the cathode surface in one polarization, averaged over a few cycles. P is substantially constant over the photocathode at any instant, but as time goes on it fluctuates in a manner determined by the spectrum of the disturbance, that is, by the 'line shape'. Supposing that the light contains frequencies around ν_0, we describe the line shape by the normalized spectral density $g(\nu-\nu_0)$. The width of the distribution g, whether it be set by circumstances in the source itself or by a filter, determines the rate at which P fluctuates. For our purpose, the stochastic behaviour of P can be described by the correlation function $\overline{P(t)\,P(t+\tau)}$, which is related in turn to $g(\nu-\nu_0)$ by[3]

$$\overline{P(t)P(t+\tau)} = \bar{P}^2(1+|\rho|^2),$$

$$\text{where } \rho = \int_{-\infty}^{\infty} g(x)\exp 2\pi i\tau x\, dx \qquad (1)$$

For the probability that a photoelectron will be ejected in time dt, we must write $\alpha P dt$, where α is constant throughout the experiment. It makes no difference whether we think of P as the square of an electric field-strength or as a photon probability density. (In this connexion the experiment of Forrester, Gudmundsen and Johnson[4] on the photoelectric mixing of incoherent light is interesting.) Assuming one polarization only, and one count for every photoelectron, we look at the number of counts n_T in a fixed interval T, and at the fluctuations in n_T over a sequence of such intervals. From the above relations, the following is readily derived:

$$\overline{n^2_T} - \overline{n}^2_T = \overline{n}_T(1 + \alpha\bar{P}\tau_0) \qquad (2)$$

$$\text{where } \tau_0 = \int_{-\infty}^{\infty} |\rho|^2 d\tau$$

and it has been assumed in deriving (2) that $T \gg \tau_0$. Now $\alpha\bar{P}$ is just the average counting-rate and τ_0, a correlation time determined by the light spectrum, is approximately the reciprocal of the spectral bandwidth $\Delta\nu$; in particular, if $\Delta\nu$ is the full width at half intensity of a Lorentzian density function, $\tau_0 = (\pi\Delta\nu)^{-1}$, while if $\Delta\nu$ is the width of a rectangular density function, $\tau_0 = \Delta\nu^{-1}$. We see that the fractional increase in mean-square fluctuation over the 'normal' amount is independent of T, and is about equal to the number of counts expected in an interval $1/\Delta\nu$. This number will ordinarily be very much smaller than one. The result, expressed in this way, does not depend on the counting efficiency.

If one insists on representing photons by wave packets and demands an explanation in those terms of the extra fluctuation, such an explanation can be given. But I shall have to use language which ought, as a rule, to be used warily. Think, then, of a stream of wave packets, each about $c/\Delta\nu$ long, in a random sequence. There is a certain probability that two such trains accidentally overlap. When this occurs they interfere and one may find (to speak rather loosely) four photons, or none, or something in between as a result. It is proper to speak of interference in this situation because the conditions of the experiment are just such as will ensure that these photons are in the same quantum state. To such interference one may ascribe the 'abnormal' density fluctuations in any assemblage of bosons.

Were we to carry out a similar experiment with a beam of electrons, we should, of course, find a slight suppression of the normal fluctuations instead of a slight enhancement; the accidentally overlapping wave trains are precisely the configurations excluded by the Pauli principle. Nor would we be entitled in that case to treat the wave function as a classical field.

Turning now to the split-beam experiment, let n_1 be the number of counts of one photomultiplier in an interval T, and let n_2 be the number of counts in the other in the same interval. As regards the fluctuations in n_1 alone, from interval to interval, we face the situation already analysed, except that we shall now assume both polarizations present. The fluctuations in orthogonal polarizations are independent, and we have, instead of (2),

$$\overline{\Delta n_1{}^2} = \overline{n_1{}^2} - \bar{n}_1{}^2 = \bar{n}_1(1 + \tfrac{1}{2}\bar{n}_1\tau_0/T) \qquad (3)$$

where n_1/T has been written for the average counting-rate in channel 1. A similar relation holds for n_2. Now if we should connect the two photomultiplier outputs together, we would clearly revert to a single-channel experiment with a count $n = n_1 + n_2$. We must then find:

$$\overline{\Delta n^2} = \bar{n}\,(1 + \tfrac{1}{2}\bar{n}\tau_0/T) \qquad (4)$$

But $\overline{\Delta n^2} = \overline{(\Delta n_1 + \Delta n_2)^2}$

$$= \bar{n}_1(1+\tfrac{1}{2}\bar{n}_1\tau_0/T) + \bar{n}_2(1+\tfrac{1}{2}\bar{n}_2\tau_0/T) + \overline{2\,\Delta n_1\Delta n_2} \qquad (5)$$

From (4) and (5) it follows that:

$$\overline{\Delta n_1\Delta n_2} = \tfrac{1}{2}\bar{n}_1{}^2\tau_0/T \qquad (6)$$

This is the positive cross-correlation effect of Brown and Twiss, although they express it in a slightly different way. It is merely another consequence of the 'clumping' of the photons. Note that if we had separated the branches by a polarizing filter, rather than a half-silvered mirror, the factor 1/2 would be lacking in (4), and (5) would have led to $\overline{\Delta n_1\Delta n_2} = 0$, which is as it should be.

If we were to split a beam of electrons by a non-polarizing mirror, allowing the beams to fall on separate electron multipliers, the outputs of the latter would show a negative cross-correlation. A split beam of classical particles would, of course, show zero cross-correlation. As usual in fluctuation phenomena, the behaviour of fermions and the behaviour of bosons deviate in opposite directions from that of classical particles. The Brown–Twiss effect is thus, from a *particle* point of view, a characteristic quantum effect.

It remains to show why Brannen and Ferguson did not find the effect. They looked for an increase in coincidence-rate over the 'normal' accidental rate, the latter being established by inserting a delay in one channel. Their single-channel rate was 5×10^4 counts per sec., their accidental coincidence rate about 20 per sec., and their resolving time about 10^{-8} sec. To analyse their experiment one may conveniently take the duration T of an interval of observation to be equal to the resolving time. One then finds that the coincidence-rate should be enhanced, in consequence of the cross-correlation, by the factor $(1 + \tau_0/2T)$. Unfortunately, Brannen and Ferguson do not specify their optical band-width; but it seems unlikely, judging from their description of their source, that it was much less than 10^{11} cycles/sec., which corresponds to a spread in wavelength of rather less than 1 A. at 4358 A. Adopting this figure for illustration, we have $\tau_0 = 10^{-11}$ sec., so that the expected fractional change in coincidence-rate is 0·0005. This is much less than the statistical uncertainty in the coincidence-rate in the Brannen and Ferguson experiment, which was about 0·01. Brown and Twiss did not count individual photoelectrons and coincidences, and were able to work with a primary photoelectric current some 10^4 times greater than that of Brannen and Ferguson. It ought to be possible to detect the correlation effect by the method of Brannen and Ferguson. Setting counting efficiency aside, the observing time required is proportional to the resolving time and inversely proportional to the square of the light flux per unit optical band-width. Without a substantial increase in the latter quantity, counting periods of the order of years would be needed to demonstrate the effect with the apparatus of Brannen and Ferguson. This only adds lustre to the notable achievement of Brown and Twiss.

E. M. PURCELL

Lyman Laboratory of Physics,
Harvard University,
Cambridge, Massachusetts.

[1] Brannen, E., and Ferguson, H. I. S., *Nature*, **178**, 481 (1956).
[2] Brown, H. R., and Twiss, R. Q., *Nature*, **177**, 27 (1956).
[3] Lawson, J. L., and Uhlenbeck, G. E., "Threshold Signals", p. 61 (McGraw-Hill, New York, 1950).
[4] Forrester, A. I., Gudmundsen, R. A., and Johnson, P. O., *Phys. Rev.*, **99**, 1691 (1955).

PAPER NO. 23a

Reprinted from *Proceedings of the Royal Society*, Ser. A, Vol. 242, pp. 300–324, 1957.

Interferometry of the intensity fluctuations in light
I. Basic theory: the correlation between photons in coherent beams of radiation

By R. Hanbury Brown

Jodrell Bank Experimental Station, University of Manchester

and R. Q. Twiss

Division of Radiophysics, C.S.I.R.O., Sydney, Australia

(*Communicated by A. C. B. Lovell, F.R.S.—Received* 15 *November* 1956—*Revised* 23 *May* 1957)

It is shown by a quantum-mechanical treatment that the emission times of photoelectrons at different points illuminated by a plane wave of light are partially correlated, and identical results are obtained by a classical theory in which the photocathode is regarded as a square-law detector of suitable conversion efficiency. It is argued that the phenomenon exemplifies the wave rather than the particle aspect of light and that it may most easily be interpreted as a correlation between the intensity fluctuations at different points on a wavefront which arise because of interference between different frequency components of the light.

From the point of view of the corpuscular picture the interpretation is much less straightforward but it is shown that the correlation is directly related to the so-called bunching of photons which arises because light quanta are mutually indistinguishable and obey Bose–Einstein statistics. However, it is stressed that the use of the photon concept before the light energy is actually detected is highly misleading since, in an interference experiment, the electromagnetic field behaves in a manner which cannot be explained in terms of classical particles.

The quantitative predictions of the theory have been confirmed by laboratory experiments and the phenomenon has been used, in an interferometer, to measure the apparent angular diameter of Sirius: these results, together with further applications to astronomy, will be discussed in detail in later papers.

It is shown that the classical and quantum treatments give identical results when applied to find the fluctuations in the photoemission current produced by a single light beam, and the connexion between these fluctuations and the correlation between photons in coherent beams is pointed out. The results given here are in full agreement with those obtained by Kahn from an analysis based on quantum statistics: however, they differ from those derived on thermodynamical grounds by Fellgett and by Clark Jones and the reasons for this discrepancy are discussed.

1. Introduction

In this paper, the first of a series on the interferometry of intensity fluctuations in light, we shall establish theoretically the underlying principle of the technique, which is that the times of emission of photoelectrons at different points illuminated by coherent beams of light are partially correlated. The chief application of this technique is to astronomy, and it has already been successfully tested in a measurement of the angular diameter of Sirius (Hanbury Brown & Twiss 1956*b*). The existence of a correlation between photons has been denied by some authors (Brannen & Ferguson 1956) who have stated, in our view wrongly, that it is contrary to the laws of quantum mechanics. The error appears to have arisen because of a too literal reliance on the corpuscular picture of light. As Bohr has pointed out, in his

Principle of Complementarity, a particular experiment can exemplify the wave or the particle aspect of light but not both; thus the interpretation is greatly simplified, and indeed is much more likely to be correct, if one confines oneself rigidly to the use of the appropriate language and talks of photons when the energy behaves like a classical particle but otherwise talks only of waves. In the present paper, as we shall show, we are dealing essentially with an interference phenomenon which can be interpreted, on the classical wave picture, as a correlation between intensity fluctuations due to beats between waves of different frequency; the concept of a photon need only be introduced at the stage where energy is extracted from the light beam in the process of photoemission.

This does not mean that one cannot interpret the phenomenon from the corpuscular point of view: one can, but only if one is prepared to endow the photons with properties very different from those of classical particles, and in practice the corpuscular picture is more of a hindrance than a help to an interpretation of the phenomenon. Indeed if photons did behave like independent classical particles, distinguishable one from another and obeying Boltzmann statistics, the correlation between them would be identically zero. However, photons are not independent since only states symmetrical between them can occur in nature; thus they obey Bose–Einstein statistics and must be regarded as mutually indistinguishable.

The connexion between the fact that photons are bosons and the existence of a correlation between light quanta may be illustrated by the familiar example of a cavity filled with thermal radiation. In this case, as is well known, the r.m.s. fluctuations in the number of photons in an elementary cell in phase space are greater than those predicted by the classical Boltzmann statistics; as Einstein (1909) pointed out, this excess noise is essentially a wave interference effect, but it can be interpreted in the corpuscular picture as the so-called 'bunching' of photons (Clark Jones 1953).

In principle this 'bunching' of photons could be measured directly if a single photocathode were illumined by a coherent beam of light, since the fluctuations in the photoemission current should be slightly greater than the pure noise fluctuations which would arise if the photoelectrons were emitted completely independently. In practice the difference between photon and shot noise, which we have called the excess photon noise, is too small to be detected conveniently with one photocathode (Fürth & MacDonald 1947), being swamped by effects such as space-charge smoothing in the photocell or fluctuations in the multiplication process in a photomultiplier. However, the 'bunching' can be measured with two separate phototubes, the cathodes of which lie in the same cell in phase space or, in other words, are illumined by coherent beams of light.† In this arrangement the shot noise currents, the space-charge smoothing effects and the multiplication noise in the two phototubes are uncorrelated, and thus the small correlation between the fluctuations in the two currents can be detected if the observations are carried out over a sufficiently long time. This correlation can only arise if there is a corresponding correlation in the

† The connexion between the extent, in real space, of an elementary cell in phase space and the volume over which a light beam may be regarded as coherent is not perhaps self-evident and it is therefore examined in appendix I.

time of emission of photoelectrons from the two cathodes, and it follows that this latter phenomenon is related to the fact that photons obey Bose–Einstein statistics. It is of course possible, by means of quantum statistics, to develop the theory given in this paper entirely in terms of the particle picture, as has been done by Kahn (1957); however, we have chosen an alternative approach which emphasizes that the correlation between photons is essentially an interference effect related to the wave picture rather than to the corpuscular aspect of light.

Experiments to measure directly the correlation in the arrival times of photons with coincidence counters have been carried out by Ádám, Jánossy & Varga (1955) and, with more sensitive equipment, by Brannen & Ferguson (1956), but with a negative result. However, as we have pointed out elsewhere (Hanbury Brown & Twiss 1956*c*), under the conditions of these experiments the expected correlation would have been far too small to be detected. We have carried out independently (Hanbury Brown & Twiss 1956*a*) a similar experiment in which we measured the correlation between fluctuations in the emission currents† of two phototubes, under conditions where the expected signal to noise ratio was of the order 10 to 1, and we have obtained a positive result in satisfactory quantitative agreement with theory. However, the detailed interpretation of this experiment will be left to a later paper of this series since the analysis is complicated by the fact that the light beam was not fully coherent over the surfaces of the photocathodes. In the present paper we shall consider only the idealized case of a plane wave of linearly polarized light in order to present the basic theory in the simplest form.

The phenomenon we are discussing is a general characteristic of an electromagnetic radiation field and will therefore occur not only at optical but also at radio wavelengths. The existence of the effect in the latter case has been demonstrated, implicitly, by experiments with an 'intensity' interferometer which has been used to measure the angular diameter of discrete radio sources (Hanbury Brown, Jennison & Das Gupta 1952). In these experiments energy was extracted from the electromagnetic field by two separate aerials, corresponding to the apertures of the phototubes in the optical experiment, and was then rectified by two square law detectors which correspond to the two photoelectric cathodes. The correlation between the fluctuations in the output currents of the two detectors was measured and was found to be equal to the theoretical value as calculated by classical electromagnetic theory.

The general theory of this radio interferometer has been given elsewhere (Hanbury Brown & Twiss 1954), but in the rather complex form required for practical applications to radio-astronomy. To bring out the connexion between the radio and the optical case we shall first develop a simple classical theory for the correlation between the intensity fluctuations at different points in space for the idealized case where the incident radiation field is a plane wave of radio frequency.

† The correlation was measured in this way, and not with a coincidence counter as in the experiments of Ádám *et al.* (1955), because the latter technique is not practical for the measurements on stars to which our work was primarily directed.

2. The classical theory of the intensity fluctuations in a plane electromagnetic wave

(a) *The intensity fluctuations in a plane wave*

Let us assume that the frequency components of the incident electromagnetic plane wave are confined to a limited region of the radio-frequency spectrum defined by

$$\nu_1 < \nu < \nu_2,$$

such that

$$\nu_1 > \nu_2 - \nu_1.$$

If the voltage induced by this radiation field in an aerial of aperture A is rectified in a square-law detector, then the low-frequency fluctuations in the output current of the detector can be expressed as a sum of the beats between the different radio-frequency components of the electromagnetic wave and correspond to the intensity fluctuations in the incident radiation. It is obvious that the amplitude and phase of these low-frequency fluctuations in the detector output current are the same at any point on the wavefront of a plane wave: so if signals are picked up by two separate aerials and rectified in separate square-law detectors, the low-frequency fluctuations in the two output currents will be perfectly correlated so long as the effects of shot noise in the detector current can be neglected. The fact that this correlation is equally to be expected, on a classical theory, at optical wavelengths appears to have been overlooked.

To develop this argument in a quantitative form, which will later be compared with the results obtained by a quantum theory, we proceed as follows. By a suitable choice of gauge a linearly polarized wave of electromagnetic radiation can be completely described by a vector potential $\mathfrak{A}$ with a single component perpendicular to the direction of propagation. If the observation is of duration T, this component can be represented by a Fourier series,

$$\mathfrak{A} = \sum_{r=1}^{\infty} q_r \exp\left[\frac{2\pi i r}{T}(t+\mathbf{k}.\mathbf{x})\right] + q_r^* \exp\left[-\frac{2\pi i r}{T}(t+\mathbf{k}.\mathbf{x})\right], \tag{2·1}$$

where q_r, q_r^* are quantities determining the amplitude and phase of the rth Fourier component, and the sign of $\mathbf{k}.\mathbf{x}$ is that appropriate to an inward travelling wave. In the present case we are assuming that q_r is zero except when $\nu_1 T < r < \nu_2 T$.

In a classical theory q_r is a complex number such that

$$q_r q_r^* = \left(\frac{p_r}{8\pi^2\nu_r^2 T}\bigg/\sqrt{\frac{\mu_0}{\epsilon_0}}\right) \tag{2·2}$$

where p_r/T is the power flow across unit area perpendicular to the direction of propagation associated with the rth Fourier component of frequency ν_r, where

$$\nu_r = r/T, \tag{2·3}$$

and where $(\mu_0/\epsilon_0)^{\frac{1}{2}}$ is the characteristic impedance of free space. If we define a quantity n_r by the equation

$$n_r h\nu_r = p_r, \tag{2·4}$$

then n_r/T may formally be identified with the average number of quanta of energy $h\nu_r$ crossing unit area in unit time, and we may put

$$q_r = \left(\frac{h}{8\pi^2\nu_r}\sqrt{\frac{\mu_0}{\epsilon_0}}\right)^{\frac{1}{2}} n_r^{\frac{1}{2}} \exp i\phi_r, \qquad (2\cdot5)$$

where ϕ_r is the phase of the rth Fourier component of the vector potential at the wavefront defined by

$$t + \mathbf{k}.\mathbf{x} = 0. \qquad (2\cdot6)$$

In the limiting case as $T \to \infty$, n_r is the average number of quanta per unit frequency bandwidth.

In what follows we shall assume that the phases of the different Fourier components are quite uncorrelated so that we may take the values of ϕ_r to be a set of independent random variables distributed with uniform probability over the range $0 < \phi_r < 2\pi$. This assumption is certainly valid as long as the radiation can be described by a stationary time series. Even when this is not the case, as when the electromagnetic energy is produced in bursts, the phases of the Fourier components of the radiation received by the observer will be effectively uncorrelated as long as the region of the source over which the intensity fluctuation is coherent is sufficiently small.

The voltage $V(t)$ produced across the input terminals to the square law detector by the vector potential defined by (2·1) will be of the form

$$V(t) = \sum_{r=1}^{\infty} \beta_r q_r \exp\left[\frac{2\pi i r}{T}(t + \mathbf{k}.\mathbf{x}_1)\right] + \beta_r^* q_r^* \exp\left[-\frac{2\pi i r}{T}(t + \mathbf{k}.\mathbf{x}_1)\right], \qquad (2\cdot7)$$

where $\mathbf{x}_1$ are the co-ordinates of the phase reference point and β_r is a complex quantity such that $\beta_r\beta_r^*$ is linearly proportional to the aerial aperture A and to the aerial efficiency at frequency r/T.

If we substitute from (2·5) and (2·7) in the equation

$$\mathrm{i}(t) = bV^2(t), \qquad (2\cdot8)$$

then the low-frequency components in the output current of the square-law detector are given by an expression of the general form

$$i(t) = eA\sum_{r=1}^{\infty}\frac{\alpha_r n_r}{T} + 2eA\sum_{r>s}^{\infty}\sum_{s=1}^{\infty}\left(\frac{\alpha_r\alpha_s n_r n_s}{T^2}\right)^{\frac{1}{2}}\cos\left[\frac{2\pi}{T}(r-s)(t+\mathbf{k}.\mathbf{x}_1) + \phi_r - \phi_s\right], \qquad (2\cdot9)$$

where e is the electronic charge and α_r is defined by the equation

$$\alpha_r = \frac{\beta_r\beta_r^*}{eA}\frac{h}{4\pi^2\nu_r}\sqrt{\left(\frac{\mu_0}{\epsilon_0}\right)}\,b. \qquad (2\cdot10)$$

This unconventional and somewhat clumsy symbolism has been adopted so that a direct comparison may be made with the quantum treatment of the optical case in which the symbol α_r will correspond to the photocathode quantum efficiency.

Let us now suppose that the a.c. fluctuations in the detector output are passed through a filter, with a frequency response $F(f)$ which does not transmit d.c. so that

$$F(0) = 0, \qquad (2\cdot11)$$

then $J(t)$ the output current of this filter may be written

$$J(t) = A \sum_{r>s}^{\infty} \sum_{s=1}^{\infty} \left(\frac{\alpha_r \alpha_s n_r n_s}{T^2}\right)^{\frac{1}{2}} \left\{ F\left(\frac{r-s}{T}\right) \exp \mathrm{i}\left[\frac{2\pi(r-s)}{T}(t+\mathbf{k}.\mathbf{x}_1)+(\phi_r-\phi_s)\right] + F^*\left(\frac{r-s}{T}\right) \exp -\mathrm{i}\left[\frac{2\pi(r-s)}{T}(t+\mathbf{k}.\mathbf{x}_1)+(\phi_r-\phi_s)\right]\right\}. \quad (2\cdot12)$$

This expression may be simplified if the filter bandwidth is so narrow that $\alpha_r n_r \simeq \alpha_s n_s$ for all values of r and s for which the frequency $(r-s)/T$ lies in the filter passband. In this case if we introduce two new indices l, m defined by

$$\tfrac{1}{2}(r+s) = l, \quad r-s = m, \quad (2\cdot13)$$

we have that

$$J(t) = A \sum_{m=1}^{M} \sum_{l=T\nu_1}^{T\nu_2} \frac{\alpha_l n_l}{T} \left\{ F\left(\frac{m}{T}\right) \exp \mathrm{i}\left[\frac{2\pi m}{T}(t+\mathbf{k}.\mathbf{x}_1)+(\phi_r-\phi_s)\right] + F^*\left(\frac{m}{T}\right) \exp -\mathrm{i}\left[\frac{2\pi m}{T}(t+\mathbf{k}.\mathbf{x}_1)+(\phi_r-\phi_s)\right]\right\}, \quad (2\cdot14)$$

where M/T, the highest beat frequency passed by the filter, is very much less than $\nu_2-\nu_1$, the bandwidth of the incident radiation, and where ϕ_r, ϕ_s are independent random variables distributed with uniform probability over the range

$$0<\phi_r<2\pi; \quad 0<\phi_s<2\pi.$$

This result will now be used to derive expressions for the correlation between intensity fluctuations at different points on the wavefront and for the mean square value of the intensity fluctuations at a single point.

(*b*) *The correlation between intensity fluctuations at different points on the wavefront*

Let us consider the case where the plane electromagnetic wave is incident on two aerials with apertures A_1, A_2 and phase reference points $\mathbf{x}_1$, $\mathbf{x}_2$ respectively. If $J_1(t)$, $J_2(t)$ are the a.c. output currents of the two low frequency filters, which we shall assume to have identical characteristics, the correlation $C(T_0)$ between these two currents, averaged over a time interval T_0, is given by

$$C(T_0) = \frac{1}{T_0}\int_0^{T_0} J_1(t-t_0)\, J_2(t)\, \mathrm{d}t, \quad (2\cdot15)$$

where
$$t_0 = \mathbf{k}.(\mathbf{x}_1-\mathbf{x}_2) \quad (2\cdot16)$$

is the difference in time between the arrival of the incident radiation at the two aerials, and where T_0 may have any value less than T.

For our present purposes the quantity of interest is $\bar{C}$, the ensemble average of $C(T_0)$ taken over an infinite number of independent time intervals each of length T_0, which is equal to the time average

$$\lim_{T_0\to\infty} C(T_0)$$

in the present case where the fluctuations are determined by a stationary time series.

For the classical radio case this calculation is very straightforward, since it is only necessary to average over the random radio-frequency phases. Terms in $C(T_0)$ which depend on ϕ_r, ϕ_s will average to zero and so we have immediately that

$$\bar{C} = 2e^2 A_1 A_2 \sum_{l=\nu_1 T}^{\nu_2 T} \frac{\alpha_l^2 n_l^2}{T} \sum_{m=1}^{M} \frac{|F(m/T)|^2}{T}, \tag{2·17}$$

which, as one would expect, is independent of T_0.

In the limiting case $T \to \infty$, so we may replace the sums in (2·17) by integrals on putting $1/T = \mathrm{d}\nu$ when we have that

$$\bar{C} = 2e^2 \int_{\nu_1}^{\nu_2} A_1 A_2 \alpha^2(\nu)\, n^2(\nu)\, \mathrm{d}\nu \int_0^\infty F^2(f)\, \mathrm{d}f. \tag{2·18}$$

(c) *The mean square value of the intensity fluctuations*

If, for the moment, we ignore the effects of shot noise in the current of the square-law detector, the mean square fluctuations $\overline{j_C^2}$ in the output current of the filter may be defined as the ensemble average of

$$\frac{1}{T_0} \int_0^{T_0} J^2(t)\, \mathrm{d}t,$$

where $J(t)$ is given by (2·14). Accordingly, $\overline{j_C^2}$ is given by (2·18) with $A_1 = A_2 = A$.

If a current I_0 flows in the detector circuit the mean square fluctuations are increased by the shot noise term $\overline{j_N^2}$ which, in the absence of space-charge smoothing (Rice 1944), is given by

$$\overline{j_N^2} = 2eI_0 \int_0^\infty |F(f)|^2\, \mathrm{d}f. \tag{2·19}$$

Now from (2·9) the incident radiation field increases the average current I_0 in the square-law detector by J_0, where

$$J_0 = eA \sum_{r=1}^{\infty} \alpha_r n_r / T, \tag{2·20}$$

so, in the limiting case, as $T \to \infty$,

$$J_0 = e \int_0^\infty A\alpha(\nu)\, n(\nu)\, \mathrm{d}\nu. \tag{2·21}$$

The total mean square fluctuations $\overline{J^2(t)}$ in the filtered output current of the square-law detector due to the incident radiation field are therefore given by

$$\overline{J^2(t)} = \overline{j_N^2} + \overline{j_C^2} = 2e^2 \left[\int_0^\infty A\alpha(\nu)\, n(\nu)\, \mathrm{d}\nu + \int_0^\infty A^2 \alpha^2(\nu)\, n^2(\nu)\, \mathrm{d}\nu \right] \int_0^\infty |F(f)|^2\, \mathrm{d}f, \tag{2·22}$$

since the noise currents j_N and j_C are uncorrelated.

The first term in (2·22) represents the *shot-noise* term due to the discrete nature of the particles carrying the detector current, while the second term, which is due to beats between the different Fourier components of the incident radiation field, may be called the *wave interaction* noise.

In a typical radio case $An(\nu)$, which is effectively equal to the number of quanta extracted from the radiation field by the aerial in unit time and unit bandwidth, is of the order of 10^5, while $\alpha(\nu)$, which is effectively equal to the average number of electrons transported from cathode to anode of the square-law detector by the incidence of a single photon, might be of the order of 10^6. Under these conditions $\overline{j_C^2}$ exceeds $\overline{j_N^2}$ by a factor of 10^{11}, so that the contribution of the latter is completely negligible.

However, at optical wavelengths $An(\nu)$ is of the order of 10^{-4} in a typical case, while $\alpha(\nu)$, the quantum efficiency of the photocathode, is of the order of 10^{-1}. Under these conditions everything is reversed and the classical theory would lead one to expect that $\overline{j_C^2}$ would be smaller than $\overline{j_N^2}$ by a factor $\sim 10^{-5}$. Admittedly it is not obvious that the quantitative predictions of a classical and determinist wave theory will be valid for the optical case, but it is shown below that indeed they are and that the wave interaction noise is simply another name for the excess photon noise, due on the corpuscular picture to the fact that photons obey Bose–Einstein statistics.

3. The fluctuations in the photoelectric emission due to a plane wave of light

(a) *The probability of photoelectric emission by a plane wave of light*

In order to calculate the correlation between the times of emission of photoelectrons at different points of a wavefront and to find the mean square fluctuations in the photoemission current from a given photocathode, we shall first obtain an expression for the probability of photoemission in terms of the observables of the incident beam of light.

In a quantum theory one must regard the quantities q_r, q_r^*, which occur in (2·1), as operators rather than as numbers, and the quantities n_r, $\exp \mathrm{i}\phi_r$, which correspond to the action and angle variables of the equivalent harmonic oscillator, are also operators satisfying commutation relations of the form (Heitler 1954),

$$n_r \exp(\mathrm{i}\phi_s) - \exp(\mathrm{i}\phi_s)\, n_r = \delta_{rs} \exp(\mathrm{i}\phi_s), \tag{3·1}$$

$$n_r \exp(-\mathrm{i}\phi_s) - \exp(-\mathrm{i}\phi_s)\, n_r = -\delta_{rs} \exp(-\mathrm{i}\phi_s), \tag{3·2}$$

where δ_{rs} is the familiar Kronecker symbol and (3·2) is the complex conjugate of (3·1).

In the standard treatment of the interaction between the matter and radiation fields, as given by Dirac (1947), one calculates the probability of a transition in which a photon is absorbed from a specific Fourier component of the radiation field so that the number of quanta associated with this component changes by unity. However, this procedure can clearly not be used to analyse an experiment in which we measure the correlation between the times of arrival of photons at different points of a wavefront, since, if the time of arrival of a photon is known to an accuracy Δt, the uncertainty, $\Delta E \equiv h\Delta\nu$, in the energy must satisfy the inequality

$$\Delta E \Delta t = h, \tag{3·3}$$

or

$$\Delta\nu \Delta t = 1. \tag{3·4}$$

If the particular Fourier component with which a specific photon is to be associated is known then $\Delta\nu = 1/T$, where T is the total observing time, and one has no knowledge whatever as to the actual moment, in the observation period, when the photon arrived.

It follows that the action and angle variables of the radiation field are not observables for the conditions under which one would look for a correlation between the arrival times of photons. As we have just shown in the classical analysis of the radio-frequency case the intensity fluctuations depend upon the beat frequencies between the different radio-frequency components of the incident radiation rather than upon the radio-frequency components themselves, while the correlation between the intensity fluctuations is determined by the amplitudes and relative phases of these beat frequencies.

When interpreting interference phenomena according to the corpuscular theory of radiation, it has been emphasized by Dirac (1947) that one must not talk of interference between two different photons, which never occurs, but rather of the interference of a photon with itself. This point was originally made for the case of spatial interference, as in an interferometer, but the arguments on which it is based are equally valid for temporal interference as in the phenomenon of a beat frequency. Accordingly, in the corpuscular theory, one must not interpret a beat frequency as an interference between photons of different energy, but rather as a phenomenon caused by the uncertainty in the energies of the individual photons which may be associated with either of the two Fourier components of the radiation field, the interference of which gives the beat frequency.

It follows that the observables appropriate to the measurement of a beat frequency are the *relative* phases of the two Fourier components and the *total* number of quanta associated with the two components. As is well known (Heitler 1954), these quantities can be measured simultaneously without violating the uncertainty principle since they are characterized by operators of the form

$$n_r + n_s \quad \text{and} \quad \exp i(\phi_r - \phi_s)$$

which commute. To prove this we have from (3·1) and (3·2) that

$$\begin{aligned}(n_r + n_s)\exp i(\phi_r - \phi_s) &= \exp i(\phi_r - \phi_s)n_r + \exp i(\phi_r - \phi_s) \\ &\quad + \exp i(\phi_r - \phi_s)n_s - \exp i(\phi_r - \phi_s) \\ &= \exp i(\phi_r - \phi_s)\,(n_r + n_s). \qquad (3{\cdot}5)\end{aligned}$$

For the specialized purposes of this paper, in which one is concerned simply with the fluctuations in the cathode currents of photocells, one may therefore discuss the interaction of the radiation field and the photocathode by a simplified theory in which the radiation field is characterized by a set of commuting operators and may therefore be treated classically. This procedure takes no specific account of the fact that the emission of a photoelectron reduces the total number of photons in the radiation field by one, but then there is no *a priori* knowledge of this number, still less of the actual distribution of these photons, with energy: all that is known, from a study of the light source, is the *average* number of photons arriving in unit time in unit bandwidth together with the fact that the fluctuations in the number of incident photons are controlled by Bose–Einstein statistics. It is this indeterminacy, basic to the existence of a correlation between photons, which makes it possible to use a classical treatment for the radiation and impossible to use the standard quantized field treatment of the photoelectric effect: the latter applies rigorously to an experiment where the energy of the incident photon and the momentum of the emitted electron can both be known to the maximum accuracy permitted by the uncertainty principle.

In what follows we shall assume that Ψ represents the total wave function for the electrons and ions forming the photocathode when acted upon by the incident radiation field. Then Ψ will be a solution of the Schrödinger equation

$$i\hbar\frac{\partial\Psi}{\partial t} = (H_0+H_1)\Psi, \tag{3·6}$$

where H_0 is the Hamiltonian for the matter field alone, and H_1, the interaction energy, is of the form

$$H_1 = \sum_l \rho_l \mathbf{v}_l \mathfrak{A}(\mathbf{x}_l), \tag{3·7}$$

where $\mathbf{v}_l$ is a dynamical variable describing the lth particle of charge ρ_l at the point $\mathbf{x}_l$, and $\mathfrak{A}(\mathbf{x}_l)$ is the vector potential acting on the lth particle.

From (2·1) and (2·5) the expression for H_1 may be written

$$H_1 = \sum_l \rho_l \mathbf{v}_l \sum_{r=1}^{\infty}\left(\frac{\hbar}{4\pi\nu_r T}\bigg/\sqrt{\frac{\mu_0}{\epsilon_0}}\right)^{\frac{1}{2}}\{n_r^{\frac{1}{2}}\exp(+i\phi_r)\exp(2\pi i\nu_r(t+\mathbf{k}.\mathbf{x}_l))$$
$$+\exp(-i\phi_r)\,n_r^{\frac{1}{2}}\exp(-2\pi i\nu_r(t+\mathbf{k}.\mathbf{x}_l))\} \tag{3·8}$$

If the wave function Ψ_0 satisfying the zero-order equation

$$i\hbar\frac{\partial\Psi_0}{\partial t} = H_0\Psi_0 \tag{3·9}$$

can be found, one can use a perturbation procedure to determine the first-order approximation to the exact solution from the equation

$$i\hbar\frac{\partial\Psi_1}{\partial t} - H_0\Psi_1 = H_1\Psi_0. \tag{3·10}$$

For our present purposes there is no need to derive a detailed theory for the photoelectric effect since quantitative data, such as the dependence of quantum efficiency on frequency, or the lower limit to the time delay of photoemission, can be taken from experiment. The important thing to note is that the interaction energy is linearly dependent upon the vector potential of the radiation field as, therefore, is the first-order perturbation in the wave function of the matter field. However, this is no longer the case in the second- and higher-order perturbation terms which describe processes in which several photons are simultaneously emitted or absorbed. Such processes are of two kinds. In the first, several photons are involved in the emission of a single photoelectron, but such events are very rare and can be ignored without significant error. In the second, two or more electrons are emitted in a process, in which each photoemission absorbs a single photon, which is clearly related to the problem of the coherent emission of photoelectrons. In a fully rigorous treatment one would have to use a higher-order perturbation theory to analyze this case but we shall make the simplifying assumption that the combined probability of obtaining two photoemissions in a very small time interval from areas A_1 A_2 of a photocathode is equal to the product of the probability of obtaining a photoemission from each area separately. Clearly this assumption is valid in the physically important case when the areas A_1 A_2 belong to quite separate photocathodes illumined by coherent light beams since the actual processes of photoemission in the two photocathodes are quite independent. The assumption will also be valid for a single photocathode as long as the fractional volume over which appreciable electronic interaction can take place inside the photocathode is very small compared with unity, a condition that will always be met in practice.

The solution for Ψ_1, corresponding to the *absorption* of a photon by an electron which is then emitted from the photocathode, is of the general form

$$\Psi_1 = \sum_l \sum_{r=1}^{\infty} \eta_{lr} \exp(-\mathrm{i}\phi_r) n_r^{\frac{1}{2}} \exp\{-2\pi \mathrm{i}\nu_r(t+\mathbf{k}:\mathbf{x}_l)\}, \tag{3·11}$$

where η_{lr} is a complex quantity involving the lth particle of the matter field and the rth component of the radiation field. The terms in H_1 proportional to $\exp\{2\pi \mathrm{i}\nu_r(t+\mathbf{k}.\mathbf{x}_l)\}$ do not contribute since they correspond to processes involving the *emission* of a photon by the particles of the photocathode.

The probability of a single photoemission in time $\mathrm{d}t$ is then proportional to

$$\mathrm{d}t \int \Psi_1^* \Psi_1 \,\mathrm{d}\tau,$$

where the integral is taken over the volume of the photocathode and a summation is made over all the particles of the matter field. If we assume that the photocathode of area A_1 is placed normal to the incident plane light wave so that $(\mathbf{x}_1)$ are the co-ordinates of the midpoint of the cathode, then

$$P(\mathbf{x}_1, t) \equiv \int \Psi_1^* \Psi_1 \,\mathrm{d}\tau \tag{3·12}$$

is given by an expression of the form

$$P(\mathbf{x}_1, t) = \sum_{r=1}^{\infty} A_1 \frac{\alpha_r n_r}{T} + 2A_1 \sum_{r>s}^{\infty} \sum_{s=1}^{\infty} \left(\frac{\alpha_r \alpha_s n_r n_s}{T^2}\right)^{\frac{1}{2}}$$
$$\times \cos(2\pi(\nu_r - \nu_s)(t+\mathbf{k}.\mathbf{x}_1) + (\phi_r - \phi_s) - (\theta_r - \theta_s)), \tag{3·13}$$

where θ_r/ν_r, θ_s/ν_s determine the delays in the emission of a photoelectron after the absorption of photons of energy $h\nu_r$, $h\nu_s$ respectively. As the delay in photoemission is known experimentally (Forrester, Gudmundsen & Johnson 1955) to be much less than 10^{-10} s while, because of the limited amplifier bandwidths and the spread in transit time through the photomultiplier tube, the beat frequencies which are significant in a practical case all lie below 10^8 or 10^9 c/s, we may put

$$\theta_r - \theta_s = 0$$

in (3·13) without introducing significant error.

The quantity α_r is simply the cathode quantum efficiency for a normally incident plane wave of frequency r/T and, as in the analysis for the radio case, we shall assume that α_r is a smoothly changing function of frequency effectively constant over the maximum beat frequency bandwidth that can arise in practice. It would be difficult to establish this assumption experimentally since the cathode quantum efficiency is normally measured with a light beam of bandwidth large compared to 10^8 c/s and any rapid changes in α_r with frequency would be smoothed out, but it is almost certainly valid in view of the appreciable energy spread of the electrons inside the photocathode. We shall also assume that the quantity n_r, which represents the average number of quanta of energy crossing unit area in unit time, is a smoothly varying function of frequency effectively constant over the beat frequency bandwidth. As before we now introduce two indices l, m defined by

$$\tfrac{1}{2}(r+s) = l, \quad r-s = m,$$

when $P(\mathbf{x}_1, t)$ is given by

$$P(\mathbf{x}_1, t) = \sum_{r=1}^{\infty} A_1 \frac{\alpha_r n_r}{T} + 2A_1 \sum_{m=1}^{M} \sum_{l=L_1}^{L_2} \frac{\alpha_l n_l}{T} \cos\left(\frac{2\pi m}{T}(t+\mathbf{k}.\mathbf{x}_1) + (\phi_r - \phi_s)\right), \tag{3·14}$$

a result that can be compared with (2·9).

(b) *The correlation between fluctuations in the emission currents of two separate photomultipliers*

Let us consider the case where a linearly polarized plane wave of light is normally incident on two separate photocathodes of areas A_1, A_2 centred at $\mathbf{x}_1$, $\mathbf{x}_2$ respectively. We assume that the photomultipliers are followed by bandpass filters with zero d.c. response, and we shall show that the average value of the correlation between the a.c. fluctuations in the output currents of these filters is identical with that derived above for the radio case by a classical deterministic theory.

As before, we take $J_1(t)$, $J_2(t)$ to be the a.c. output currents of the two filters and we must then calculate $\bar{C}$ the ensemble average of the integrated correlation defined by (2·15) and (2·16).

In this case the calculation is complicated by the necessity of averaging over the number and time of emission of the photoelectrons produced in time T as well as over the radio-frequency phases.

In the development of statistical theory a large number of methods have been evolved for analyzing problems of this kind. The one that we shall use is based on the so-called shot noise representation since, though not perhaps the most elegant procedure, it has the most direct physical interpretation in the present case. This procedure has been extensively studied by Rice (1944) and frequent use will be made of his results.

Following Rice we introduce a normalized probability function $p(t+\mathbf{k}.\mathbf{x}_1)$ such that

$$\int_0^T p(t+\mathbf{k}.\mathbf{x}_1) = 1, \tag{3·15}$$

which is related to $P(\mathbf{x}_1, t)$ by the equation

$$Np(t+\mathbf{k}.\mathbf{x}) = P(\mathbf{x}_1, t), \tag{3·16}$$

where

$$N = \sum_{r=1}^{\infty} A\alpha_r n_r = I_0/e \tag{3·17}$$

and I_0 is the average photoemission current.

We now consider a particular time interval of length T, in which exactly K electrons are emitted from one photocathode and exactly N electrons from the other, so that the output currents of the filters following the photomultipliers may be written

$$J_1(t) = \sum_{k=1}^{K} ef_k(t-t_k), \tag{3·18}$$

$$J_2(t) = \sum_{n=1}^{N} eg_n(t-t_n), \tag{3·19}$$

where $ef_k(t-t_k)$, $eg_n(t-t_n)$ are the effects produced in the first and second filters by electrons of charge e emitted at times t_k and t_n respectively.

Since the filters do not pass d.c.

$$\int_{-\infty}^{\infty} f_k(t-t_k)\,\mathrm{d}t = \int_{-\infty}^{\infty} g_n(t-t_n)\,\mathrm{d}t = 0 \tag{3·20}$$

and it will further be assumed that $f_k(t)$, $g_n(t)$ only differ appreciably from zero in an interval Δ which is negligibly small compared with T.

In a complete discussion it would be necessary to note that $f_k(t)$ will vary from one photoemission to another, since it will depend to some extent upon such things as the emission velocities of the photoelectrons and upon the number and momentum distribution of the secondary electrons produced at each stage of the photomultiplication process. For the present, however, we shall ignore these effects, which would considerably complicate the algebra without adding anything significant to a basic understanding of the phenomenon, though they are of real practical importance since they impose a lower limit to the resolving time of the electronic equipment.

Accordingly, we shall assume that

$$\left.\begin{aligned} f_k(t-t_k) &= f(t-t_k), \\ g_n(t-t_n) &= g(t-t_n). \end{aligned}\right\} \tag{3·21}$$

Finally, we shall limit ourselves to the idealized case

$$f(t) \equiv g(t), \tag{3·22}$$

though to begin with it will be more convenient to retain both symbols.

From (2·15), (3·18) and (3·19) we have that

$$C(T_0) = \frac{1}{T_0}\int_0^{T_0} \mathrm{d}t \sum_{k=1}^{K} \sum_{n=1}^{N} e^2 f(t-t_0-t_{1k})\, g(t-t_{2n}). \tag{3·23}$$

To find $\bar{C}$ we must average over the times of emission of the different photoelectrons, over the total number of photoelectrons emitted in a time interval T, and over the phases of the Fourier components of the radiation field so that

$$\bar{C} = \left\langle \frac{1}{T_0}\int_0^{T_0} \mathrm{d}t \sum_{K=1}^{\infty} \sum_{N=1}^{\infty} e^2 \rho_1(K)\,\rho_2(N) \sum_{k=1}^{K} \sum_{n=1}^{N} \int_0^T f(t-t_0-t_{1k})\, p_1(t_{1k}+\mathbf{k}.\mathbf{x}_1)\,\mathrm{d}t_{1k} \right.$$
$$\left. \times \int_0^T g(t-t_{2n})\, p_2(t_{2n}+\mathbf{k}.\mathbf{x}_2)\,\mathrm{d}t_{2n} \right\rangle_{\text{aver.}}, \tag{3·24}$$

where the angle brackets denote an averaging over the phases of the individual Fourier components of the radiation field. The quantities

$$p_1(t_1+\mathbf{k}.\mathbf{x}_1), \quad p_2(t_2+\mathbf{k}.\mathbf{x}_2)$$

are defined by (3·16), (3·17) and (3·14) and differ only in so far as the areas A_1, A_2 and the position vectors of the photocathodes are not identical. The quantities

$$\rho_1(K), \rho_2(N)$$

are the probabilities that exactly K and N photoelectrons are emitted in time T from the first and second photocathodes respectively. Rice (1944) has discussed the generalization of Campbell's theorem to the case where the probability of a fundamental event varies with time and it can easily be shown, along the lines of his analysis, that

$$\rho_1(K) = \frac{\bar{K}^K \exp(-\bar{K})}{K!}, \quad \rho_2(N) = \frac{\bar{N}^N \exp(-\bar{N})}{N!}, \tag{3·25}$$

where

$$\bar{K} = N_1 T, \quad \bar{N} = N_2 T, \tag{3·26}$$

but this result, which will be needed in the next section, is not essential to the present argument.

If we introduce new time variables τ_{1k}, τ_{2n} defined by

$$\tau_{1k} = t - t_0 - t_{1k}, \quad \tau_{2n} = t - t_{2n} \tag{3·27}$$

in place of $t_{1k} t_{2n}$ we see that

$$\bar{C} = \Big\langle \frac{1}{T}\int_0^{T_0} dt \sum_{K=1}^{\infty} \sum_{N=1}^{\infty} e^2 \rho_1(K)\rho_2(N) \sum_{k=1}^{K} \sum_{n=1}^{N} \int_{-(t-t_0)}^{T-(t-t_0)} f(\tau_{1k})\, p_1(t+\mathbf{k}.\mathbf{x}_1-\tau_{1k})\, d\tau_{1k}$$
$$\times \int_{-t}^{T-t} g(\tau_{2n})\, p_2(t+\mathbf{k}.\mathbf{x}_2-\tau_{2n})\, d\tau_{2n} \Big\rangle_{\text{aver.}}. \tag{3·28}$$

As long as $t < T-\Delta$ we may replace the integration limits over the variables τ_{1k}, τ_{2n} by $(-\infty, \infty)$ and since Δ/T is, *ex hypothesi*, negligibly small, the resulting error is also negligible.

From (3·16), (3·17) and (3·22) we get that

$$\bar{C} = \Big\langle \int_0^{T_0} \frac{dt}{T_0}\, e^2 \sum_{K=1}^{\infty} \sum_{N=1}^{\infty} K\rho_1(K)\, N\rho_2(N) \frac{A_2}{A_1} \Big\{ \int_0^{\infty} f(\tau_k)\, p_1(t+\mathbf{k}.\mathbf{x}_2-\tau_k)\, d\tau_k \Big\}^2 \Big\rangle_{\text{aver.}}. \tag{3·29}$$

From (3·20) only the time-dependent part will contribute to $\bar{C}$, while all the terms explicitly dependent upon the phases of the individual Fourier components of the incident light will average to zero.

Hence, since

$$\left.\begin{aligned} \sum_{K=1}^{\infty} K\rho_1(K) &= \bar{K} = N_1 T = A_1 T \sum_{r=1}^{\infty} \alpha_r n_r, \\ \sum_{N=1}^{\infty} N\rho_2(N) &= \bar{N} = N_2 T = A_2 T \sum_{r=1}^{\infty} \alpha_r n_r, \end{aligned}\right\} \tag{3·30}$$

we have from (3·16), (3·17) and (3·14) that

$$\bar{C} = 2e^2 \sum_{l=L_1}^{L_2} \frac{A_1 A_2 \alpha_l^2 n_l^2}{T} \sum_{m=1}^{M} \frac{|F(fm)|^2}{T}, \tag{3·31}$$

where

$$F(f) = \int_{-\infty}^{\infty} f(t) \exp[-2\pi i f t] \,dt \tag{3·32}$$

is the Fourier transform of $f(t)$ and satisfies the integral equation

$$f(t) = \int_{-\infty}^{\infty} F(f) \exp[2\pi i f t] df \tag{3·33}$$

as long as $\int_{-\infty}^{\infty} |f(t)| \,dt$ exists: a condition which certainly holds in the present case where $f(t)$ satisfies (3·7) and is zero outside $0 < t < \Delta$.

In the limiting case, as $T \to \infty$, (3·31) may be written

$$\bar{C} = 2e^2 A_1 A_2 \int_0^{\infty} \alpha^2(\nu)\, n^2(\nu)\, d\nu \int_0^{\infty} |F(f)|^2 df, \tag{3·34}$$

which is formally identical with the correlation for the radio case given by (2·18), although the physical interpretation of the symbols $\alpha(\nu)$, $F(f)$ is different in the two cases. Thus in the optical case $\alpha(\nu)$ is simply the quantum efficiency of the photocathode, while in the radio case it depends upon the aerial efficiency and the conversion characteristics of the square law detector; again, in the radio case $F(f)$ is simply the frequency characteristic of the filter, while in the optical case it also depends upon the frequency characteristic of the photomultiplier. However, these are minor points which do not alter the fundamental conclusion that the correlation can be found by a purely classical theory in which the photocathode is regarded as a square law detector of suitable conversion efficiency. It is this result which provides the basis for our claim that the correlation is essentially an interference effect exemplifying the wave rather than the corpuscular aspect of light.

(c) *The mean square fluctuations in the emission current of a phototube*

We have argued above that the excess photon noise and the correlation between photons in coherent beams are closely related, and to bring this out more explicitly we shall derive an expression for the mean square fluctuations in the output current of a bandpass filter following a phototube of cathode area A placed normal to an incident plane wave of light.

Let $J_K(t)$ be the output current in the case when exactly K photoelectrons are emitted in time T, then we may use the same shot noise representation in the previous section and write

$$J_K(t) = \sum_{k=1}^{K} e f(t - t_k), \tag{3·35}$$

so that

$$J_K^2(t) = \sum_{k=1}^{K} e^2 f^2(t - t_k) + 2 \sum_{k>n}^{K} \sum_{n=1}^{K} e^2 f(t - t_k) f(t - t_n). \tag{3·36}$$

The mean square fluctuations $\overline{J^2}(t)$ can then be found by averaging over the times of emission of different photoelectrons, over the total number of electrons emitted in time T and over the phases of the Fourier component of the incident light, so that we may write

$$\overline{J^2(t)} = \overline{j_N^2} + \overline{j_C^2}, \tag{3·37}$$

where

$$\overline{j_N^2} = \left\langle \sum_{K=1}^{\infty} e^2\rho(K) \int_0^T f^2(t-t_k)\, p_1(t+\mathbf{k}.\mathbf{x}_1)\, \mathrm{d}t_k \right\rangle_{\text{aver.}}, \tag{3·38}$$

$$\overline{j_C^2} = \left\langle 2 \sum_{k>n}^{K} \sum_{n=1}^{K} \int_0^T f(t-t_k)\, p_1(t+\mathbf{k}.\mathbf{x}_1)\, \mathrm{d}t_k \int_0^T f(t-t_n)\, p_1(t+\mathbf{k}.\mathbf{x}_1)\, \mathrm{d}t_n \right\rangle_{\text{aver.}}, \tag{3·39}$$

where $p(t+\mathbf{k}.\mathbf{x}_1)$ is defined by (3·14) and (3·16) and $\rho(K)$ is given by (3·25).

We shall now show that $\overline{j_N^2}$, the shot noise contribution to the mean square fluctuations, and $\overline{j_C^2}$ the wave interaction noise, are given by expressions formally identical with those derived in § 2 for the classical radio case.

Only the time *independent* part of $p(t+\mathbf{k}.\mathbf{x}_1)$ contributes to $\overline{j_N^2}$, since the time dependent part depends linearly on the random phases of the Fourier components of the incident light wave and averages to zero.

As long as $t_k < T-\Delta$ we may replace the limits of integration $(0, T)$ by $(-\infty, \infty)$ and, using Parseval's theorem in the form

$$\int_{-\infty}^{\infty} f^2(t)\, \mathrm{d}t = \int_{-\infty}^{\infty} |F^2(f)|\, \mathrm{d}f = 2\int_0^{\infty} |F^2(f)|\, \mathrm{d}f, \tag{3·40}$$

where $f(t)$, $F(\nu)$ are Fourier mates related by (3·32), (3·33), we get that

$$\overline{j_N^2} = 2e^2 A \int_0^{\infty} \alpha(\nu)\, n(\nu)\, \mathrm{d}\nu \int_0^{\infty} |F(f)|^2\, \mathrm{d}f \tag{3·41}$$

on substituting in (3·39) from (3·14), (3·16) and (3·25).

Since J_0, the average emission current of the photocell, is given by

$$J_0 = eA \int_0^{\infty} \alpha(\nu)\, n(\nu)\, \mathrm{d}\nu, \tag{3·42}$$

which is formally identical with (2·20), we see that j_N is indeed the shot noise current, for which

$$\overline{j_N^2} = 2eJ_0 \int_0^{\infty} |F(f)|^2\, \mathrm{d}f. \tag{3·43}$$

On the other hand, only the time *dependent* part of $p(t+\mathbf{k}.\mathbf{x}_1)$ contributes to $\overline{j_C^2}$, the contribution from the time-independent part being zero from (3·20). By a discussion along identical lines with that given in the previous section it can be shown that

$$\overline{j_C^2} = 2e^2 \int_0^{\infty} A^2\alpha^2(\nu)\, n^2(\nu)\, \mathrm{d}\nu \int_0^{\infty} |F(f)|^2\, \mathrm{d}f \sum_{K=1}^{\infty} \frac{K(K-1)\rho(K)}{K^2}. \tag{3·44}$$

But from (3·25) it follows immediately that

$$\sum_{K=1}^{\infty} \frac{K(K-1)\rho(K)}{K^2} = 1, \tag{3·45}$$

so that

$$\overline{j_C^2} = 2e^2 \int_0^{\infty} A^2\alpha^2(\nu)\, n^2(\nu)\, d\nu \int_0^{\infty} |F(f)|^2\, \mathrm{d}f \tag{3·46}$$

If (3·46) is compared with (3·34) it will be seen that the two expressions are identical in the special case $A_1 = A_2$, which establishes the close connexion between the excess photon noise in a coherent beam of light and the correlation between photons in two coherent light beams. It may also be seen that the expression given by (3·46) is identical with the second term in (2·21) which gives the wave interaction noise for the classical case, so that the excess photon noise due to the so-called 'bunching' of photons is the equivalent, in the corpuscular language, of the wave interaction noise of the undulatory picture. This identification is supported by the analysis of Kahn (1957) who obtains results identical with ours by a treatment based directly on the particle model of the incident light; but quite a different expression for the excess photon noise has been obtained by Fellgett (1949) and by Clark Jones (1953) who relied on thermodynamical arguments. In our view, however, thermodynamical considerations cannot be applied to the photoelectric effect and for this, and other reasons given in appendix II, we consider that their expression for the excess photon noise is wrong. If we may anticipate results which are to be given in a later paper, we may observe that this conclusion is supported by experimental measurements of the correlation between the fluctuations in separate phototubes. These results indicate that the ratio $\overline{j_C^2}/\overline{j_N^2}$ is approximately proportional to α, the quantum efficiency of the photocathodes, which is in accordance with the theory given above, but is incompatible with that of Fellgett (1949) in which $\overline{j_C^2}/\overline{j_N^2}$ should be independent of α.

Ideally, it would be desirable to confirm this conclusion by a direct measurement of the noise in the photoemission current, but this would be very difficult in practice since the excess photon noise is so small by comparison with the shot noise proper. Thus let us consider the case, appropriate to the laboratory experiment reported elsewhere (Hanbury Brown & Twiss 1956*a*), in which the light source is square in shape and subtends an angle θ^2 at the photocathode. Let us assume the idealized conditions in which the radiant energy is linearly polarized and concentrated into a narrow frequency band of rectangular shape centred at 4400 Å with an effective black-body temperature of 7000° K, and that the photocathode has a quantum efficiency of 20 % and a square aperture of width d. If the incident light is to be effectively a plane wave, the source must be so distant that it is not appreciably resolved by the photocathode and this sets an upper limit to the product θd given by the inequality $\theta d < 0{\cdot}2\lambda$ where $\lambda = 4400$ Å is the mean wavelength of the incident light. The number of quanta n incident on the photocathode in unit frequency bandwidth then obeys the inequality $n < 3{\cdot}7 \times 10^{-4}$ so that

$$\overline{j_C^2}/\overline{j_N^2} < \alpha n < 0{\cdot}74 \times 10^{-4}.$$

Since this is appreciably smaller than the uncertainties in the measurement of the shot noise proper, and since we have assumed conditions exceptionally favourable to the observation of the excess photon noise, we can conclude that the contribution of the latter to the total noise current in a phototube is quite negligible in a practical case.

(*d*) *The signal to noise ratio in a measurement of the correlation*

To complete the fundamental theory we shall calculate the signal to noise ratio in a measurement of the correlation. Thus if S is given by

$$S = \bar{C} = \left\langle \frac{1}{T_0}\int_0^{T_0} J_1(t-t_0)\, J_2(t)\,\mathrm{d}t \right\rangle_{\text{aver.}}, \tag{3·47}$$

and N is the r.m.s. fluctuation in $C(T)$ defined by

$$N^2 = \left\langle \left\{ \frac{1}{T_0} \int_0^{T_0} J_1(t-t_0)\, J_2(t)\, dt \right\}^2 \right\rangle_{\text{aver.}} - \bar{C}^2, \tag{3·48}$$

we shall calculate the ratio S/N.

As we have just seen, the contribution to N^2 due to the excess photon noise is negligible in comparison with the contribution from the shot noise proper, therefore in finding N we can assume that the fluctuations in the emission currents of the two photocathodes are due to independent shot noise currents. To this order of accuracy $J_1(t-t_0)$, $J_2(t)$ may be represented by the Fourier series

$$\left.\begin{aligned} J_1(t-t_0) &= \sum_{n=1}^{\infty} \gamma_n \cos\left(\frac{2\pi nt}{T} - \phi_n\right), \\ J_2(t) &= \sum_{n=1}^{\infty} \eta_m \cos\left(\frac{2\pi mt}{T} - \psi_m\right), \end{aligned}\right\} \tag{3·49}$$

where ϕ_n, ψ_m are independent random variables distributed with uniform probability over the range 0 to 2π.

If the photomultiplying process and the bandpass filters introduce no additional noise it follows immediately from (3·43), with $d\nu = 1/T$ that

$$\left.\begin{aligned} \tfrac{1}{2}j_N^2 &= 2eI_1 \frac{|F_1^2(fn)|}{T}, \\ \tfrac{1}{2}\eta_m^2 &= 2eI_2 \frac{|F_2^2(f_m)|}{T}, \end{aligned}\right\} \tag{3·50}$$

if the amplitude and phase response of the photomultiplier are included in $F(f)$. When the gain M of the photomultiplier is large, the noise introduced by the bandpass filters is normally negligible; but the number of secondary electrons emitted at a given stage of the photomultiplier is itself a fluctuating quantity, and it has been shown by Shockley & Pierce (1938) that this effect increases the output noise power by a term

$$\frac{M\mu - 1}{M(\mu-1)} \simeq \frac{\mu}{\mu-1}, \tag{3·51}$$

if $M \gg 1$, where μ is the secondary emission multiplication factor.

It is therefore more realistic to assume that

$$\left.\begin{aligned} \tfrac{1}{2}j_n^2 &= 2eI_1 \frac{\mu}{\mu-1} \frac{|F_1(f_n)|^2}{T}, \\ \tfrac{1}{2}\eta_m^2 &= 2eI_2 \frac{\mu}{\mu-1} \frac{|F_2(f_m)|^2}{T}. \end{aligned}\right\} \tag{3·52}$$

From (3·49) we have immediately that

$$N^2 = \left\langle \left| \frac{1}{T_0} \int_0^{T_0} dt \sum_{n,m=1}^{\infty} \gamma_n \eta_m \cos(2\pi f_n t - \phi_n) \cos(2\pi f_m t - \psi_m) \right|^2 \right\rangle_{\text{aver.}}, \tag{3·53}$$

where the angle brackets now mean that the expression contained within them is to be averaged over the random variables ϕ_n, ψ_m; since these phases are all mutually uncorrelated only terms independent of them contribute to N^2.

Integrating over time we get that

$$N^2 = \left\langle \left| \sum_{n,m=1}^{\infty} \frac{\gamma_n \eta_m}{2T_0} \left[\frac{\sin\{2\pi(f_n - f_m) T_0 - (\phi_n - \psi_m)\} + \sin(\phi_n - \psi_m)}{2\pi(f_n - f_m)} + \frac{\sin\{2\pi(f_n + f_m) T_0 - (\phi_n + \psi_m)\} + \sin(\phi_n + \psi_m)}{2\pi(f_n + f_m)} \right] \right|^2 \right\rangle_{\text{aver.}} . \quad (3\cdot54)$$

If we collect the terms independent of the random phases and proceed to the limit in which sums are replaced by integrals we get from (3·40) that

$$N^2 = \left(\frac{2e\sqrt{(I_1 I_2)}\,\mu}{\mu - 1} \right)^2 \frac{1}{T_0} \int_0^\infty |F_1(f_1)|^2 \, df_1 \int_0^\infty |F_2(f_2)|^2 \, df_2 \times \frac{1}{2} \left[\frac{\sin^2 \pi(f_1+f_2) T_0}{\pi^2 (f_1+f_2)^2} + \frac{\sin^2 \pi(f_1-f_2) T_0}{\pi^2 (f_1-f_2)^2} \right], \quad (3\cdot55)$$

a result very similar to that derived, for a somewhat different case, by Rice (1945). The contribution to N^2 from the term proportional to

$$\sin^2(\pi(f_1+f_2) T_0)/\pi^2(f_1+f_2)^2$$

is quite negligible in the practical case where the integration time $T_0 > 10^3$ s, and where the lowest frequency passed by the filter following the photomultiplier tube is $> 10^6$ c/s. Furthermore, with $T > 10^3$, then $F(f_1) \simeq F(f_2)$ for values of $f_1 - f_2$ for which $\sin^2(\pi(f_1-f_2)T_0)/\pi^2(f_1-f_2)^2$ differs significantly from zero, and in this case N^2 may be written in the simplified form

$$N^2 = \left(\frac{2e\sqrt{(I_1 I_2)}\,\mu}{\mu - 1} \right)^2 \frac{1}{T_0} \int_0^\infty |F_1(f) F_2(f)|^2 \, df \, \frac{1}{2\pi} \int_0^\infty \frac{\sin^2 X}{X^2} \, dX. \quad (3\cdot56)$$

The effective bandwidth of the bandpass filters, which we now assume to be identical, may be defined by the expression

$$b_v = \int_0^\infty |F(f)|^2 \, df / F^2_{\text{max.}}, \quad (3\cdot57)$$

where $$|F_1^2(f)| = |F_2^2(f)| = |F(f)|^2$$

and $F_{\text{max.}}$ is the maximum value of $F(f)$.

If we define a normalized spectral density coefficient η by the relation

$$\eta = \int_0^\infty |F^4(f)| \, df \Big/ F^2_{\text{max.}} \int_0^\infty |F^2(f)| \, df, \quad (3\cdot58)$$

then, since $$\frac{1}{2\pi} \int_{-\infty}^\infty \frac{\sin^2 X}{X^2} \, dX = \tfrac{1}{2}, \quad (3\cdot59)$$

we get from (2·20), (3·56) and (3·57) that†

$$N = \frac{\sqrt{2e^2\mu}}{\mu - 1}(A_1A_2)^{\frac{1}{2}}(b_v/T_0)^{\frac{1}{2}}\eta^{\frac{1}{2}}\,|\,F_{\text{max.}}|^2\int_0^\infty \alpha(\nu)\,n(\nu)\,d\nu. \tag{3·60}$$

From (3·34) and (3·60) we have that the signal to noise ratio S/N is given by

$$\left(\frac{S}{N}\right)_{\text{r.m.s.}} = \sqrt{2}\,(1-1/\mu)\,(A_1A_2)^{\frac{1}{2}}(b_v T_0)^{\frac{1}{2}}\eta^{-\frac{1}{2}}\int_0^\infty \alpha^2(\nu)\,n^2(\nu)\,d\nu\bigg/\int_0^\infty \alpha(\nu)\,n(\nu)\,d\nu, \tag{3·61}$$

which is independent of $|\,F_{\text{max.}}|^2$ and, therefore, of the gain of the photomultiplier tubes.

This result has been derived for the case where the incident light is linearly polarized. If the light is unpolarized, as is normally the case, the expression for N will be unaltered if the average number of quanta per cycle bandwidth is unaltered. However, the expression for S will be different since there is no correlation between quanta in different states of polarization, and we must decompose the incident beam into two independent components each with $\frac{1}{2}n$ quanta per cycle bandwidth. Since the correlation between coherent beams of polarized light is proportional to the square of the number of quanta per cycle bandwidth, the value for S must be reduced by a factor $\frac{1}{2}$, and the signal to noise ratio for the case of unpolarized light becomes

$$\left(\frac{S}{N}\right)_{\text{r.m.s.}} = (1/\sqrt{2})\,(1-1/\mu)\,(A_1A_2)^{\frac{1}{2}}(b_v T_0)^{\frac{1}{2}}\eta^{-\frac{1}{2}}\int_0^\infty \alpha^2(\nu)\,n^2(\nu)\,d\nu\bigg/\int_0^\infty \alpha(\nu)\,n(\nu)\,d\nu. \tag{3·62}$$

Because of the very small number of quanta received per unit bandwidth from even the hottest sources the signal to noise ratio will only be significant if one integrates for long times and also accepts the intensity fluctuations over the widest possible bandwidth. However, the signal to noise ratio is independent of the bandwidth of the incident light.

4. Discussion

In calculating the correlation between the emission times of photoelectrons at different points on the wavefront of a plane wave of light we have used a quantum theory in which the incident radiation field is treated classically. Such a procedure is justified theoretically by the fact that the relevant observables of the radiation field can be characterized by commuting operators, but it is opposed to one's natural tendency to regard a correlation between the emission times of photoelectrons as essentially a quantum phenomenon for which a classical treatment of the radiation field would only be valid in the limiting case where the number of incident quanta is very large. Accordingly, to make the argument more acceptable from a physical point of view we shall consider the analogous case of a diffraction grating illuminated by monochromatic light to produce an interference pattern on a screen.

In this latter case it is well known that the average distribution of light intensity over the screen can be found by a classical wave theory, even in the limiting case

† This expression for N is smaller by a factor $1/\sqrt{2}$ than that given in an earlier paper (Hanbury Brown & Twiss 1956*a*) which applies when the correlation is measured by a somewhat different technique.

where the light is so weak that one can count the arrival of individual photons. As Born (1945) points out in discussing an essentially similar experiment, things are in no way altered if the screen is replaced by a mosaic of photoelectric elements; the experiment still illustrates the wave aspect of light, since the particle aspect can only really be brought out by observations in which the position of a single quantum is measured at two successive instants of time. If such observations were introduced into the present experiment, the interference pattern would be destroyed.

From the point of view in which light is viewed as a photon stream the appearance of interference effects is closely related to Heisenberg's uncertainty principle; an accurate knowledge of the transverse point of impact of a photon involves a corresponding uncertainty in the transverse momentum, and therefore an uncertainty in the element of the grating from which the photon has come. If we perform an experiment, such as blackening out the rest of the grating to determine the transverse momentum of the photons, then the main interference pattern once more disappears.

This state of affairs is exactly paralleled by the correlation experiment, which is the subject of this paper, if one everywhere substitutes the concepts of time and energy for those of position and momentum. Thus the bandwidth of the light and the length of the observation time in the correlation experiment are the analogues of the width and number of lines per unit length of the grating. The interference pattern in time, the beat phenomenon of the correlation experiment, arises because of the uncertainty in the energy of the photon which produced a specific photoemission, and is the analogue of the interference pattern on the screen which arises because of the uncertainty in the transverse momentum of a photon reaching a specific point on the screen. Both phenomena are to be understood from the particle point of view as being due to an uncertainty in the behaviour of a single photon and not as due to interference between different quanta. Finally, it may be noted, that if one partly removes the uncertainty in the energy of the incident photons in the correlation experiment, by using a highly monochromatic source or by analyzing the light with a prism of very great resolving power, the higher beat frequency components will disappear, just as the analogous components of the interference pattern on the screen will disappear if the angular width of the grating is suitably reduced.

Accordingly, since the radiation field can be treated classically in the case of the diffraction grating, it is only to be expected that it can be treated classically in analyzing the correlation experiment.

In this paper we have considered the idealized case where the two photocathodes lie on the same wavefront of the incident light. However, the emission time of a photoelectron is uncertain within limits determinined in practice by the resolving time of the electronic equipment, so the observed correlation will not be affected as long as the difference in the time of arrival of a particular wavefront of the two photocathodes is small compared with this resolving time. This means that the position of the photocathodes need only be controlled to an accuracy determined by the bandwidth of the fluctuations rather than by the wavelength or bandwidth of the incident light.

To simplify the presentation, we have developed the fundamental theory for the case where this incident radiation is a plane wave. However, an arbitrary radiation field can be expressed as a sum of plane waves and, since the operators associated with the observables of one plane wave commute with all the operators associated with the observables of any other, one is equally justified in analyzing this general case by a theory in which the radiation field is treated classically; this will be done in a subsequent paper.

A quantum theory is needed to compute the probability of photoemission which, as we have shown, is proportional to the square of the amplitude of the incident light; but if this probability is known from experiment, one can calculate the correlation between the fluctuations in the photoemission currents at two separate phototubes by a fully classical theory in which the photocathodes are regarded as square law detectors of a suitable conversion efficiency. This emphasizes the fact that the theory is equally valid if the phototubes are replaced by true energy detectors such as bolometers or thermistors, though for reasons of signal to noise ratio these latter alternatives could not be used in a practical correlation experiment.

A purely classical theory can also be used to calculate the mean square fluctuations in the emission current of a single phototube. As we have shown these fluctuations can be represented as the sum of two terms, a *shot noise* term due to the discrete nature of the electrons carrying the photocurrent, and a term which we have called the *wave interaction noise* because in the classical theory it arises from the beats between the different Fourier components of the radiation field. The expressions for these terms are identical with those derived by Kahn (1957) in a treatment based directly on quantum statistics, and two conclusions can be drawn from this. First, that the shot noise is a consequence of the corpuscular nature of the electrons, it does not depend at all on the fact that the radiation field is also quantized; secondly, that the wave interaction noise is identical with the excess photon noise which is interpreted, in the language of the corpuscular theory, as due to the so-called 'bunching' of photons and which is essentially a consequence of the fact that light quanta obey Bose–Einstein statistics. This so-called 'bunching' is, of course, in no way dependent upon the actual mechanism by which the light energy was originally generated; still less does it imply that the photons must have been injected coherently into the radiation field. On the contrary, if one wishes to picture the electromagnetic field as a stream of photons, one has to imagine that the light quanta redistribute themselves over the wavefront, as the radiation field, which may be quite incoherent in origin, is focused and collimated into beams capable of mutual interference; thus the correlation between photons is determined solely by the energy distribution and coherence of the light reaching the photon detectors.

Appendix I. Coherent interference and the extent in real space of an elementary cell in phase space

When one is dealing with particles such as gas molecules the dimensions, in real space, of an elementary cell of volume L^3 in phase space are likely to be very small. Thus, in the extreme case of a hydrogen gas in which the uncertainties in the momenta are those appropriate to a thermal spread of $1°$ K, the dimensions, in real space, of the elementary cell are of the order of 10 Å cube and these will be correspondingly reduced for heavier gases or for larger uncertainties in the momenta of the individual molecules.

However, things are quite different in the case of light if the angular size of the source is very small. Thus, when light is received from a star, the volume in real space of an elementary cell can be many cubic metres, and we shall prove the general result that as long as two points are close enough together to permit virtually complete interference between the light rays reaching them, which implies that their separation is insufficient to resolve the star, then they lie in the same elementary cell in phase space.

For simplicity let us consider a very distant source of square angular aperture $\theta \times \theta$, where θ is very small, and two observing points on the earth with relative coordinates $(\Delta x, \Delta y, \Delta z)$ such that the light source lies on the z axis. Let us further suppose that the light is concentrated with a narrow frequency band of width $\Delta\nu$.

Then, since the volume of an elementary cell in phase space is h^3 we have that

$$\Delta p_x \Delta p_y \Delta p_z \Delta q_x \Delta q_y \Delta q_z = h^3, \tag{A 1}$$

where Δp, Δq are the uncertainties in the momentum and position respectively.

In the present case

$$\Delta p_x = \Delta p_y = \frac{h\nu}{c}\theta, \tag{A 2}$$

while if θ^2 is negligible compared with $\Delta\nu/\nu$

$$\Delta p_z = h\Delta\nu/c. \tag{A 3}$$

Substituting from (A 2) and (A 3) in (A 1) we get that

$$\theta^2 \frac{\Delta q_x \Delta q_y}{\lambda^2} \frac{\Delta q_z}{c} \Delta\nu = 1, \tag{A 4}$$

where $\lambda = c/\nu$.

Now if the interference fringes obtained from two coherent beams of light bandwidth $\Delta\nu$ are not to be appreciably weakened, the difference in the path length of the two beams must not exceed a wavelength of a frequency $\Delta\nu$ so that we must have

$$\Delta\nu\Delta z/c < 1. \tag{A 5}$$

Furthermore, if the transverse separation of the two points is to be so small that they do not resolve the source, one must certainly have

$$\theta\Delta x/\lambda < 1, \quad \theta\Delta y/\lambda < 1. \tag{A 6}$$

Combining the inequalities (A 5) and (A 6) with (A 4) one gets

$$\Delta x \Delta y \Delta z < \Delta q_x \Delta q_y \Delta q_z, \tag{A 7}$$

and since $\Delta q_x \Delta q_y \Delta q_z$ is the spatial volume of the elementary cell in phase space we have proved the required result.

The importance of this argument from the theoretical point of view is that it brings out the connexion between the wave and particle interpretations of the phenomenon intensity interference. Thus, on the classical picture one would expect the intensity fluctuations in the light at two different points in space to be correlated as long as the light rays reaching these two points were capable of mutual interference; while on the quantum picture one would expect a correlation between the arrival times of quanta at different points as long as these lie in the same cell in phase space, and the above discussion shows that if the latter condition is satisfied then so is the former.

Appendix II. On the excess photon noise of light detectors

In the text we derived an expression for the noise in a photoemission current from first principles, and noted that an identical result has been obtained by Kahn (1957) with the aid of quantum statistics. However, a quite different result has been given by Fellgett (1949) and also by Clark Jones (1953) from thermodynamical arguments, and in this appendix we give the reasons for rejecting their procedure.

The thermodynamical argument depends in the first place on an analysis of a thermal detector in thermal equilibrium with a blackbody enclosure at temperature T. The discussion given by Clark Jones is based on a general theorem by Callen & Welton (1951) which enables one to find the fluctuations associated with a linear dissipation process, and which represents a powerful generalization of Nyquist's theorem (1928) to cover any case where the underlying physical process can be characterized by a generalized impedance. The treatment by Fellgett is more specific in that the equivalent electrical circuit is explicitly derived for a given detector, but it is identical in essentials. Both writers assume that the fluctuations in the thermal detector output are equal to the energy fluctuations in the thermal radiation field, half being due to the absorbed and half to the emitted radiation. In the case of the photocell the emitted stream of radiation does not exist so, it is argued, the fluctuations in this case will be reduced by one-half.

Two objections to this treatment are immediately apparent. In the first place the theorem of Callen & Welton does not apply, since the dissipation process of a thermal detector is non-linear, the equivalent resistance being itself a function of the temperature. Admittedly if the temperature fluctuations are very small compared with T the error is also small, but then so is the contribution of the excess photon noise. If we consider the analogous case of a radio antenna in a blackbody of temperature T, then the voltage fluctuations across the output terminals of the antenna can be found from Nyquist's theorem by taking the radiation resistance of the antenna to be at temperature T. However, if a square-law detector were placed between the antenna and the output terminals, one could not use a generalized Nyquist's theorem to find the energy fluctuations in the incident field or the fluctuations across the output terminals of the square-law detector, since this would ignore the presence of beats between different components of the radiation field.

The second and more serious objection is that one cannot in general equate the fluctuations in the output of a thermal or photon detector to the energy fluctuations in the thermal radiation field; the principle of detailed balancing applies to the average flow of energy but not to the fluctuations themselves. It is essentially for this reason that the analysis for the thermal detector cannot be applied to the photocell and that the estimate of the excess photon noise given by Fellgett and Clark Jones is linearly proportional to the quantum efficiency rather than quadratically proportional as found by Kahn and by the present writers.

We thank Professor Rosenfeld for many helpful criticisms of this paper. We are also indebted to Dr Kahn for showing us his results before publication and for his valuable comments on our own approach. We also thank Dr Wolf and Professor le Couteur for their useful criticism.

References

Ádám, A., Jánossy, L. & Varga, P. 1955 *Acta Phys. Hung.* **4**, 301.
Born, M. 1945 *Atomic physics.* Glasgow: Blackie and Son.
Brannen, E. & Ferguson, H. I. S. 1956 *Nature, Lond.*, **178**, 481.
Callen, H. B. & Welton, T. A. 1951 *Phys. Rev.* **83**, 34.
Clark Jones, R. 1953 *Advances in electronics*, **5**. New York: Academic Press Inc.
Dirac, P. A. M. 1947 *The principles of quantum mechanics*, 3rd ed. Oxford: Clarendon Press.
Einstein, A. 1909 *Phys. Z.* **10**, 185.
Fellgett, P. B. 1949 *J. Opt. Soc. Amer.* **39**, 970.
Forrester, A. T., Gudmundsen, R. A. & Johnson, P. O. 1955 *Phys. Rev.* **99**, 1691.
Fürth, R. & MacDonald, D. K. C. 1947 *Nature, Lond.*, **159**, 608.
Hanbury Brown, R., Jennison, R. C. & Das Gupta, M. K. 1952 *Nature, Lond.*, **170**, 1061.
Hanbury Brown, R. & Twiss, R. Q. 1954 *Phil. Mag.* **45**, 663.
Hanbury Brown, R. & Twiss, R. Q. 1956*a* *Nature, Lond.*, **177**, 27.
Hanbury Brown, R. & Twiss, R. Q. 1956*b* *Nature, Lond.*, **178**, 1046.
Hanbury Brown, R. & Twiss, R. Q. 1956*c* *Nature, Lond.*, **178**, 1447.
Heitler, W. 1954 *The quantum theory of radiation*, 3rd ed. Oxford: Clarendon Press.
Jennison, R. C. & Das Gupta, M. K. 1956 *Phil. Mag.* (8), **1**, 55.
Kahn, F. 1957 In preparation.
Nyquist, H. 1928 *Phys. Rev.* **32**, 110.
Rice, S. O. 1944 *Bell. Syst. Tech. J.* **23**, 294.
Rice, S. O. 1945 *Bell. Syst. Tech. J.* **25**, 87.
Shockley, W. & Pierce, J. R. 1938 *Proc. Inst. Rad. Engrs*, **26**, 321.

PAPER NO. 23b

Reprinted from *Proceedings of the Royal Society*, Ser. A, Vol. 243, pp. 291–319, 1957.

Interferometry of the intensity fluctuations in light
II. An experimental test of the theory for partially coherent light

By R. Hanbury Brown

Jodrell Bank Experimental Station, University of Manchester

and R. Q. Twiss

Division of Radiophysics, C.S.I.R.O., Sydney, Australia

(*Communicated by A. C. B. Lovell, F.R.S.—Received* 20 *June* 1957)

A theoretical analysis is given of the correlation to be expected between the fluctuations in the outputs of two photoelectric detectors when these detectors are illuminated with partially coherent light. It is shown how this correlation depends upon the parameters of the equipment and upon the geometry of the experiment.

The correlation may be detected either by linear multiplication of the fluctuations in the two outputs or by a coincidence counter which counts the simultaneous arrival of photons at the detectors. The theory is given for both these techniques and it is shown that they are closely equivalent.

A laboratory test is described in which two photomultipliers were illuminated with partially coherent light and the correlation between the fluctuations in their outputs measured as a function of the degree of coherence. The results of this experiment are compared with the theory and it is shown that they agree within the limits of accuracy of the test; it is concluded that if there is any systematic error in the theory it is unlikely to exceed a few parts per cent.

1. Introduction

The basic principles of intensity interferometry, that is interferometry based on correlating the fluctuations of intensity in two beams of radiation, were presented in part I of this paper (Hanbury Brown & Twiss 1957). We showed there that the intensity fluctuations in two coherent beams are correlated and that, in the optical case, this correlation should be preserved in the process of photoemission so that the fluctuations in the anode currents of two phototubes should be partially coherent when they are illuminated by coherent beams of light.

We have already described (Hanbury Brown & Twiss 1956*a*) a preliminary test which established the existence of this correlation; however, this test was not designed to yield a precise quantitative result. It seemed desirable, especially in view of the considerable controversy about the principles involved, that a more exact check of the theory should be carried out, and so we have recently repeated the experiment with improved apparatus operating under more carefully controlled conditions. The results are reported below together with the necessary theoretical treatment of partially coherent light.

The principles of an intensity interferometer can be presented either in the classical terms of wave interference, or in terms of the relative times of arrival of photons. These two ways of looking at the phenomenon suggest two different experimental techniques for testing the correlation. The first technique, which may conveniently be regarded as illustrating the wave picture, consists in finding the correlation between the fluctuations in the anode currents of two light detectors by means of a linear multiplier which takes the time average of their cross-product. The second technique, which illustrates the corpuscular nature of light, makes use of a coincidence counter to detect individual events in which photoelectrons are emitted simultaneously from the cathodes of two light detectors.

At first sight, the second of these techniques appears more attractive since, by the use of coincidence counters, it should be possible to circumvent the major technical difficulty in the design of an extremely sensitive correlator, namely the elimination of random drifts in the apparatus. However, it is shown below that, with present-day counters and photomultipliers, the use of a coincidence counter demands a highly monochromatic and brilliant light source if significant results are to be obtained in a reasonable observing time. The apparatus used in the present experiment was also intended for use in a stellar interferometer, and it was not considered practicable to restrict the spectrum of the starlight to an extremely narrow band. It was therefore decided to employ a system based on a linear multiplier, since such a system can be used with light of arbitrarily broad bandwidth and significant correlation can be obtained in periods of a few minutes using a standard arc lamp as a source of light.

Although the present paper deals only with an experiment using a linear multiplier we also give in an appendix a short theoretical treatment of the results to be expected with a coincidence counter. This treatment demonstrates the equivalence of the two techniques and the results can be used to interpret the experimental data of other experimenters who have attempted to detect correlation using coincidence counters.

2. The correlation expected between the outputs of two photoelectric light detectors illuminated by partially coherent light

In this section we shall consider theoretically the behaviour of the system shown in simple outline in figure 1. Two phototubes $P_1 P_2$ are illuminated from a single source of light. The fluctuations in their anode currents are amplified by the amplifiers $B_1 B_2$ and are multiplied together by the linear multiplier C. The time average of the multiplier output is proportional to the correlation and is measured by the integrating device M_1.

The expressions given in part I of this paper for the average value of the correlation between the fluctuations in the anode currents of two separate phototubes were calculated for the idealized case in which the incident radiation was a plane wave. However, we also showed that the extension to the general case, where the light source and the apertures of the photocathodes are of arbitrary size, could be made entirely within the framework of a classical theory in which the photocathodes are regarded as square law detectors of suitable conversion efficiency.

2·1. *Light detectors with small aperture*

Under the conditions where the apertures of the individual photocathodes are too small to resolve the source appreciably, we can apply to the optical problem a similar analysis to that we have given elsewhere for a radio interferometer based on the same principle (Hanbury Brown & Twiss 1954), and this is given in the course of the general discussion in appendix A.

When the incident radiation is approximately monochromatic we show in appendix A that $\overline{C(d)}$, the average value of the correlation when the apertures of the radiation detectors are separated by a distance d, is related to $\overline{C(0)}$ the correlation observed with zero separation, by the equation

$$\overline{C(d)} = \Gamma^2(\nu_0, d)\,\overline{C(0)}, \qquad (2\cdot 1)$$

where ν_0 is the midband frequency of the incident radiation and where $\Gamma^2(\nu_0, d)$, the normalized *correlation factor*, is equal to the square of the modulus of the Fourier transform of the intensity across the equivalent line source. As we have pointed out in analyzing the radio case, this normalized *correlation factor* is related to the square

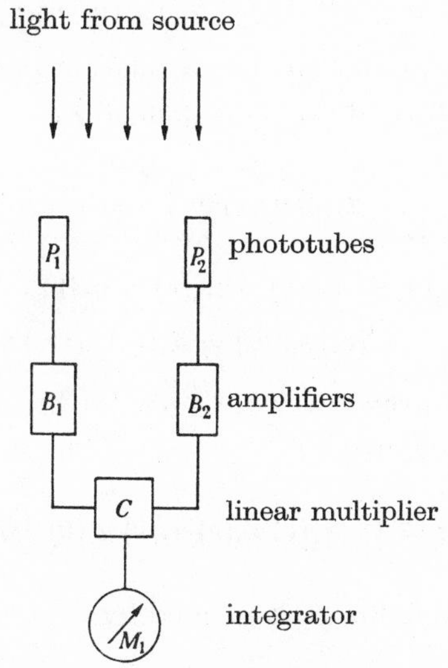

FIGURE 1. A simplified outline of an intensity interferometer.

of the visibility of the fringes that would be observed by a Michelson interferometer with the same baseline; it is also related to the square of the so-called *degree of coherence* (Zernike 1938) and the modulus of the *coherence phase factor* (Hopkins 1951).

Equation (2·1) is only valid when the fractional light bandwidth is very small and, when this is not the case, one must integrate over the light spectrum to find $\overline{C(d)}$. Thus, it was shown in part I that, if the incident light is a linearly polarized plane wave,

$$\overline{C(0)} = 2e^2 A_1 A_2 \int_0^\infty \alpha^2(\nu)\, n_0^2(\nu)\, \mathrm{d}\nu \int_0^\infty |\, F(f)\,|^2\, \mathrm{d}f, \qquad (2\cdot 2)$$

where A_1, A_2 are the areas of the photocathode apertures; $\alpha(\nu)$ is the cathode quantum efficiency; $n_0(\nu)$ is the number of quanta incident on unit area in unit bandwidth; $F(f)$ is the combined frequency response of the photomultipliers and of the amplifiers. This equation is equally valid when the light source is of finite angular size, provided that the photocathode apertures are themselves small, and hence we may write

$$\overline{C(d)} = 2e^2 A_1 A_2 \int_0^\infty \Gamma^2(\nu, d)\, \alpha^2(\nu)\, n_0^2(\nu)\, \mathrm{d}\nu \int_0^\infty |F(f)|^2 \mathrm{d}f. \tag{2·3}$$

The assumptions underlying equation (2·3) are, however, too idealized for it to be applicable to a practical case even when the apertures of the light detectors are very small. It is necessary to take account of the fact that the quantum efficiency and spectral response may be different at the two photocathodes, and that the light may not be linearly polarized; furthermore, the electrical frequency response of the photomultipliers and amplifiers may not be identical in the two channels and there will inevitably be a slight loss of correlation in the correlator system. To take account of these factors equation (2·3) may be written in the convenient form

$$\overline{C(d)} = e^2 \epsilon A_1 A_2 \beta_0 \overline{\Gamma^2(d)}\, \alpha^2(\nu_0)\, n_0^2(\nu_0)\, \sigma B_0 b_v\, |F_{\text{max.}}|^2, \tag{2·4}$$

where $(1-\epsilon)$ is the fraction of the correlation lost in the correlator circuits; B_0 is the effective bandwidth of the light defined by

$$B_0 = \left[\int_0^\infty \alpha_1(\nu)\, n_1(\nu)\, \mathrm{d}\nu \int_0^\infty \alpha_2(\nu)\, n_2(\nu)\, \mathrm{d}\nu\right]^{\frac{1}{2}} \Big/ \alpha(\nu_0)\, n_0(\nu_0) \tag{2·5}$$

and

$$\left.\begin{aligned} n_1(\nu) &= n_{1a}(\nu) + n_{1b}(\nu), \quad n_2(\nu) = n_{2a}(\nu) + n_{2b}(\nu), \\ \alpha^2(\nu_0)\, n_0^2(\nu_0) &= \alpha_1(\nu_0)\, \alpha_2(\nu_0)\, n_1(\nu_0)\, n_2(\nu_0) \end{aligned}\right\} \tag{2·6}$$

and the subscripts a, b refer to two orthogonal directions of polarization; where σ, the spectral density, is defined by

$$\sigma = \int_0^\infty \alpha_1(\nu)\, \alpha_2(\nu)\, n_1(\nu)\, n_2(\nu)\, \mathrm{d}\nu / B_0\, \alpha^2(\nu_0)\, n_0^2(\nu_0) \tag{2·7}$$

and the polarization factor β_0 is defined by

$$\beta_0 = 2[n_{1a}(\nu)\, n_{2a}(\nu) + n_{1b}(\nu)\, n_{2b}(\nu)]/n_1(\nu)\, n_2(\nu), \tag{2·8}$$

so that $\beta_0 = 1$ when $n_a(\nu) = n_b(\nu)$ as in the case of randomly polarized light. The mean value $\overline{\Gamma^2(d)}$ of the normalized correlation factor is defined by

$$\overline{\Gamma^2(d)} = \frac{\displaystyle\int_0^\infty \Gamma^2(\nu, d)\, \alpha_1(\nu)\, \alpha_2(\nu)\, [n_{1a}(\nu)\, n_{2a}(\nu) + n_{1b}(\nu)\, n_{2b}(\nu)]\, \mathrm{d}\nu}{\displaystyle\int_0^\infty \alpha_1(\nu)\, \alpha_2(\nu)\, [n_{1a}(\nu)\, n_{2a}(\nu) + n_{1b}(\nu)\, n_{2b}(\nu)]\, \mathrm{d}\nu} \tag{2·9}$$

and b_v the effective cross-correlation bandwidth of the amplifiers, is defined by

$$|F_{\text{max.}}|^2 b_v = \frac{1}{2} \int_0^\infty [F_1(f)\, F_2^*(f) + F_1^*(f)\, F_2(f)]\, \mathrm{d}f, \tag{2·10}$$

where $|F_{\text{max.}}|$ is the maximum value of $\frac{1}{2}[F_1(f)\, F_2^*(f) + F_1^*(f)\, F_2(f)]$.

We shall use equation (2·4) in a later part of this paper to interpret some observations on Sirius which have been briefly reported elsewhere (Hanbury Brown & Twiss 1956*b*). However, it cannot be applied to the experiment analyzed in the present paper in which the apertures of the light detectors were so large that they partially resolved the source. In this case the light cannot be regarded as coherent over the whole aperture of the light detectors and the correlation given by equation (2·4) must be reduced by the *partial coherence factor* $\bar{\Delta}$, where the bar denotes that the factor has been found by averaging over the frequency spectrum. The value of $\overline{\Gamma^2(d)}$, the normalized *correlation factor*, is now a function of both the angular size of the source and aperture of the light detectors.

2·2. *Light detectors with large apertures*

If no restrictions are placed upon the spectrum of the light or the intensity distribution over the source or the shape of the photocathodes, the general expression for the expected correlation is impracticably complex. However, for the purposes of the present paper we have greatly simplified the analysis by making the assumptions: (1) the intensity is uniform over the source of light, which is taken to be either circular or rectangular; (ii) the two photocathodes have identical rectangular apertures and the quantum efficiency is constant over each cathode; (iii) the quantum efficiency and the number of quanta received per unit bandwidth do not vary significantly

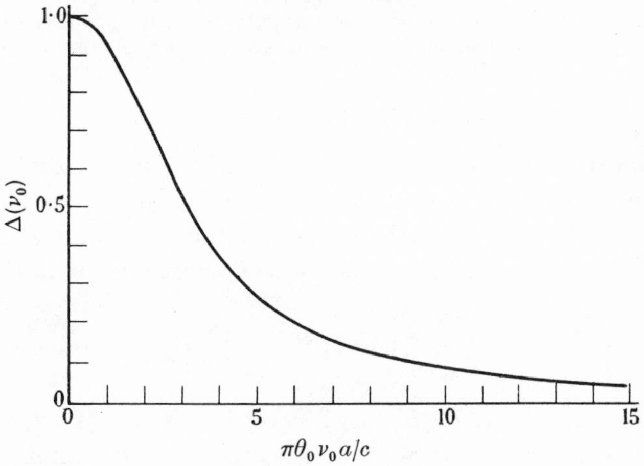

FIGURE 2. The variation of the partial coherence factor, $\Delta(\nu_0)$, with the parameter $\pi\theta_0\nu_0 a/c$, calculated for a circular source of angular diameter θ_0 viewed by two identical light detectors with square apertures $a \times a$.

over a frequency range equal to the bandwidth of the correlator. On the other hand, it will be assumed that the spectral width of the light is so narrow that both $\Gamma^2(\nu_0, d)$ and $\Delta(\nu_0)$ may be taken as constant over the light bandwidth, and that we may put

$$\overline{\Delta\Gamma^2(d)} \simeq \Delta(\nu_0)\,\Gamma^2(\nu_0, d).$$

The expressions for $\Gamma^2(\nu_0, d)$ and $\Delta(\nu_0)$, based on the assumptions stated above, have been derived in appendix A. The numerical values for the case of a circular source viewed by two square cathodes of identical size have been calculated from equations (A25) and (A26) by means of the Manchester University Electronic Computing Machine. Figure 2 shows the partial coherence factor $\Delta(\nu_0)$ as a function of $\pi\theta_0\nu_0 a/c$, where θ_0 is the apparent angular diameter of the source and a is the width of the photocathode apertures. Figure 3 shows the normalized correlation factor $\Gamma^2(\nu_0, d)$ as a function of $\pi\theta_0\nu_0 d/c$, where d is the cathode separation and $\pi\theta_0\nu_0 a/c$ has the value 3·06 which is appropriate to the present experiment.

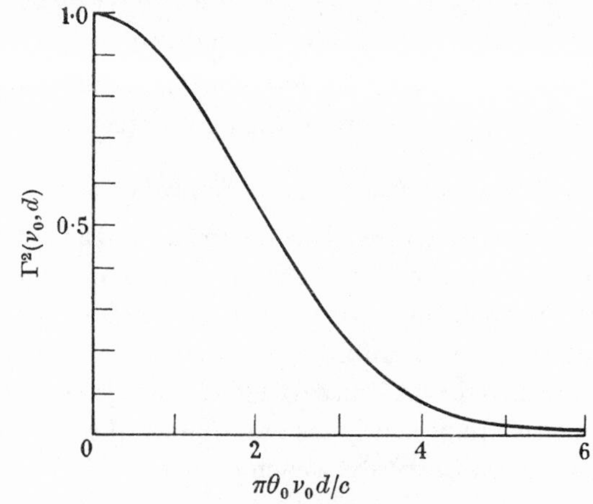

FIGURE 3. The variation of the normalized correlation factor $\Gamma^2(\nu_0, d)$ with the parameter $\pi\theta_0\nu_0 d/c$, calculated for a circular source of angular diameter θ_0 viewed by two identical light detectors with square apertures $a \times a$, where $\pi\theta_0\nu_0 a/c = 3{\cdot}06$.

2·3. *The expected signal to noise ratio*

We proved in part I that the r.m.s. uncertainty $N(T_0)$ in the correlation measured over an observing time T_0 depends, in a practical case, only upon the magnitude of the photoemission current, since the excess photon noise is negligible with any known light source. This enables us to write $N(T_0)$ immediately in the form

$$N(T_0) = \frac{e^2\mu}{\mu-1}(1+a)(1+\delta)(A_1A_2)^{\frac{1}{2}}\left(\frac{2b_v\eta}{T_0}\right)^{\frac{1}{2}} |F_{\text{max.}}|^2 \alpha(\nu_0)\, n_0(\nu_0)\, B_0, \qquad (2{\cdot}11)$$

which differs from equation (3·60) in part I by the two factors $(1+a)$ and $(1+\delta)$. The first factor has been introduced to allow for the presence of stray light, so that $(1+a)$ is the ratio of the total incident light to that from the source alone. The second factor has been introduced to allow for the presence of excess noise produced in the correlator, so that $(1+\delta)$ is the ratio of the total r.m.s. uncertainty in the reading of the integrating motor M_1 to the r.m.s. uncertainty due to fluctuations in the output of the photomultiplier tubes. In addition the quantity η, which is the normalized spectral density of the cross-correlation frequency response of the amplifiers following the phototubes, must be defined by

$$\eta = \int_0^\infty |F_1^2(f)\, F_2^2(f)|\, \mathrm{d}f/b_v\, |F_{\text{max.}}|^4 \qquad (2{\cdot}12)$$

in the practical case where the individual frequency responses of the two amplifiers are *not* identical, rather than by equation (3·58) of part I.

The only other quantity which has not previously been defined in the present paper is the factor $\mu/(\mu-1)$ which represents the excess noise introduced by the photomultiplier chain. It is known (Shockley & Pierce 1938) that μ is equal to the multiplication factor of the first stage, at least as long as this is not too great.

If we compare equations (2·4) and (2·11) it will be seen that both the correlation $\overline{C(d)}$ and the uncertainty, or noise, in the output are linearly proportional to $|F_{\text{max.}}|^2$. It follows that their ratio is independent of the gain of the equipment. We have therefore adopted the practice in the present papers of expressing both the observed and theoretical values of correlation as *signal to noise ratios* S/N. The theoretical value of this ratio, from equations (2·4) and (2·11) is

$$\frac{S}{N} = \frac{\overline{C(d)}}{N(T_0)} = \epsilon\beta_0 \frac{\mu-1}{\mu} \frac{(A_1A_2)^{\frac{1}{2}}}{(1+a)(1+\delta)} \alpha(\nu_0)\, n_0(\nu_0) \left(\frac{b_v T_0}{2\eta}\right)^{\frac{1}{2}} \sigma\Delta(\nu_0)\,\Gamma^2(\nu_0, d), \quad (2\cdot13)$$

where the factor $\Delta(\nu_0)\,\Gamma^2(\nu_0, d)$ has been substituted for $\Gamma^2(d)$ to allow for the finite size of the photocathode apertures.

To compare the theory with experiment we need to develop equation (2·13) to give the signal to noise ratio in the practical case where successive measurements may be made with different values of light flux. Let us suppose that, in the rth measurement, of duration T_r, the number of quanta incident in unit bandwidth at frequency ν_0 is $n_r(\nu_0)$. If the gain of the equipment has been kept the same for all the observations, the average value of the final reading of the integrating motor is proportional to $\Sigma n_r^2(\nu_0)T_r$; while the r.m.s. uncertainty in the value of a given reading can be found by adding the individual fluctuations incoherently, so that it is proportional to $(\Sigma n_r^2(\nu_0)\,T_r)^{\frac{1}{2}}$. The theoretical value of the signal to noise ratio, in a convenient form for comparison with experiment, may therefore be written,

$$\frac{S}{N} = \frac{\overline{C(d)}}{N(\Sigma T_r)} = \epsilon\beta_0 \frac{\mu-1}{\mu} \frac{(A_1A_2)^{\frac{1}{2}}}{(1+a)(1+\delta)} \alpha(\nu_0) \left(\frac{b_v}{2\eta}\right)^{\frac{1}{2}} \sigma\Delta(\nu_0, d)\, \frac{\sum\limits_{r=1}^{M} n_r^2(\nu_0)\, T_r}{\left[\sum\limits_{r=1}^{M} n_r^2(\nu_0)\, T_r\right]^{\frac{1}{2}}}. \quad (2\cdot14)$$

2·4. *The limiting signal to noise ratio for an arbitrarily large source*

In the limiting case where the source is completely resolved by an individual photocathode, the signal to noise ratio, with the cathodes optically superimposed, tends monotonically to a value determined simply by the effective black-body temperature of the source at the received wavelength, and this limit is independent of the source shape. To show this we consider first the case where the source is of arbitrary rectangular shape with an apparent angular width θ_1, θ_2. Then, in the limit,

$$A\theta_1\theta_2\nu_0^2/c^2 \to \infty \quad (2\cdot15)$$

and we have from equations (2·13, A23) that the signal to noise ratio tends to the limiting value

$$\lim_{\theta\to\infty} \left(\frac{S}{N}\right) = K_1 \frac{\alpha(\nu_0)\, n_0(\nu_0)\, c^2}{\nu_0^2\theta_1\theta_2}, \quad (2\cdot16)$$

where K_1 is a constant of proportionality given by

$$K_1 = \frac{\epsilon\beta_0\sigma}{(1+a)(1+\delta)} \frac{\mu-1}{\mu} \left(\frac{b_v T_0}{2\eta}\right)^{\frac{1}{2}}, \quad (2\cdot17)$$

which depends solely on the parameters of the electronic system and upon the spectral density of the light.

It may be noted that

$$Q_0 = \frac{2\gamma_0 n_0(\nu_0)}{\theta_1 \theta_2} \tag{2·18}$$

is the number of quanta of both polarizations emitted from an area of the source in unit bandwidth and unit solid angle. The factor 2 is included to allow for the fact that in practice the light beam must be split in order to superimpose the photocathodes, and the factor γ_0 takes account of any loss of light in the optical system. If the black-body temperature of the source at a frequency ν_0 is Θ_0, then

$$Q_0 = \frac{2\nu_0^2}{c^2}[\exp(h\nu_0/k\Theta_0) - 1]^{-1} \tag{2·19}$$

and substituting in equation (2·16) the maximum signal to noise ratio from a rectangular source, however large, is given by

$$\left(\frac{S}{N}\right)_{\text{max.}} = K_1 \frac{\alpha(\nu_0)}{\gamma_0}[\exp(h\nu_0/k\Theta_0) - 1]^{-1}. \tag{2·20}$$

This limit will clearly apply whatever the shape of the source, so long as it can be approximated by a series of rectangles or its area can be defined by a Riemann integral. It is interesting to note that under these conditions the signal to noise ratio depends upon the temperature but not upon the shape of the source; thus, effectively the equipment operates as pyrometer.

3. Description of the apparatus

3·1. *The optical equipment*

A simplified outline of the optical equipment is shown in figure 4. A secondary light source was formed by a circular pinhole 0·19 mm in diameter on which the image of a mercury arc was focused by a lens. The image of the arc in the plane of the pinhole was approximately 5 cm in length and its position was adjusted so that the pinhole lay in the relatively bright part of the arc close to one of the electrodes. The arc lamp, Mazda type ME/D 250 W was supplied by a direct current of approximately 4 A and the 4358 Å line of the mercury spectrum was isolated by means of a liquid filter with a transmission of 82 % at 4358 Å. The beam of light from the pinhole was divided by a semi-transparent mirror to illuminate the cathode of the photomultipliers P_1, P_2. The mirror surface was formed by evaporating pure aluminium on to glass, the reverse side being bloomed with cryolite to reduce unwanted internal reflexions. The area of each cathode exposed to the light was limited by a square aperture of 5 × 5 mm, and the distance from the pinhole to each cathode was adjusted to be 2·24 m with an accuracy of about 2 mm.

The photomultipliers were a matched pair, R.C.A. type 6342, with flat end-on cathodes and ten stages of multiplication. The photocathode surfaces had a maximum response at about 4000 Å. Tests at the National Physical Laboratory showed that the quantum efficiencies of the two cathodes, measured at 4000 Å, were 16·9 and 14·6 % and that the shapes of the spectral response curves were almost identical. The type 6342 photomultiplier has a small spread in electron transit time, particularly when the photocathode aperture is limited, and the effective bandwidth of the secondary emission amplification considerably exceeds the limit of 45 Mc/s set by the amplifiers in the correlator.

In order that the degree of coherence between the light on the two cathodes might be varied at will, one of the photomultipliers (P_2) was mounted on a horizontal slide which could be traversed normal to the incident light. Thus the cathode apertures, as viewed from the pinhole, could be superimposed or separated by any amount up to several times their width.

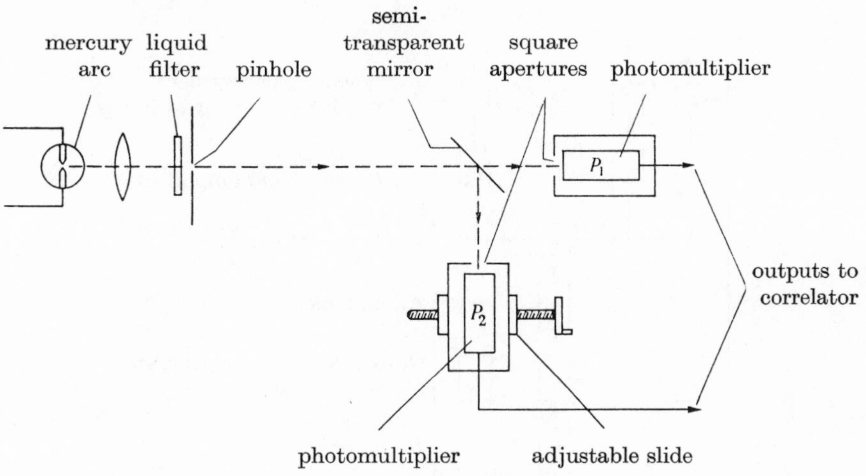

FIGURE 4. A simplified outline of the optical system.

The fluctuations in the anode currents of the photomultipliers were transmitted to the correlator through coaxial cables of equal length. In each case a simple high-pass filter was inserted between the anode and the input to the cable to remove the direct current component.

3·2. *The correlator*

A simplified diagram of the correlator is shown in figure 5. The cable from each of the photomultipliers was terminated in a matched load and the voltage fluctuations across this load were applied to one of the two input channels of the correlator. Both channels consisted of a phase-reversing switch followed by a wide-band amplifier. The switch (S_1) in channel 1 was electronic and reversed the phase of the input voltage 10 000 times per second in response to a 5 kc/s square wave from the generator G_1. It is essential to reduce amplitude modulation of the signal by this switch to an extremely low level in order to prevent spurious drifts in the equipment; for this reason the gain of the switch was equalized in both positions by means of an automatic balancing circuit comprising a detector, a selective 5 kc/s amplifier B_3 and a synchronous rectifier R_1. The phase-reversing switch S_2 in channel 2 consisted of a relay-operated coaxial switch which reversed the phase of the input every 10 s in response to a 0·05 c/s square wave from the generator G_2.

The wide-band amplifiers B_1, B_2 were identical in construction, their gain was substantially constant ($\pm$ 1db) from about 5 to 45 Mc/s and decreased rapidly outside this band. The outputs of these amplifiers were multiplied together in the multiplier C, which consisted of a balanced arrangement of two pentode valves with their anodes in push-pull. The output of the multiplier was then amplified by a high-gain selective amplifier B_4 tuned to 5 kc/s with a bandwidth of 70 c/s. The output of B_4 was applied to the synchronous rectifier R_2 which was of the conventional

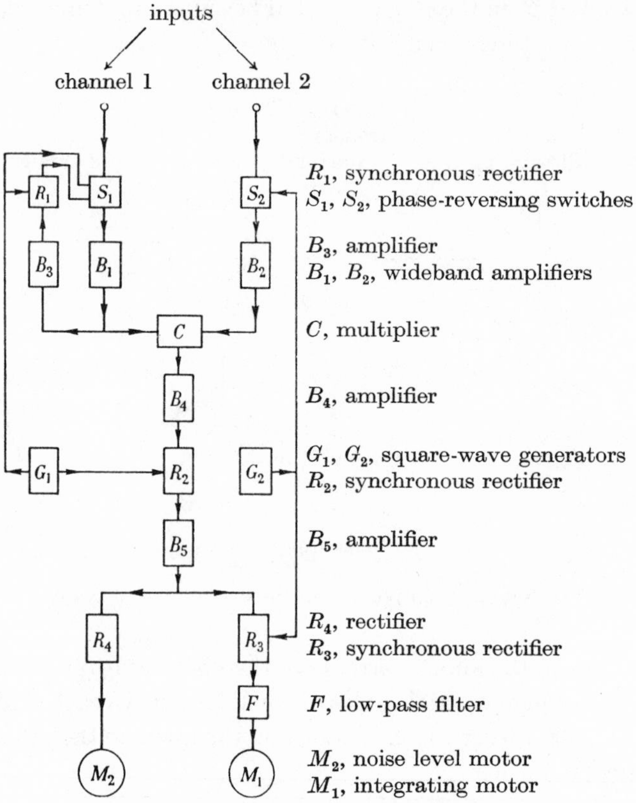

FIGURE 5. An outline of the correlator.

type using a ring of four diodes synchronized by the 5 kc/s switching wave generated by G_1. The rectifier R_2 was followed by the 0·05 c/s amplifier B_5 which was relatively broadband and passed frequencies from about 0·01 to 0·25 c/s. The final synchronous rectifier R_3 consists of a relay-operated switch which, in response to the 0·05 c/s square wave from G_2, periodically reversed the connexions between the output of the amplifier B_5 and the integrating motor M_1. A low-pass filter, containing only passive elements, was inserted between the output of R_3 and the motor to restrict the bandwidth of the signal to the range 0 to 0·01 c/s. The motor itself was a miniature integrating motor coupled through a reduction gear to a revolution counter; it was capable of rotation in either direction and tests showed the relation between speed and input voltage to be linear to better than 1 %. An additional integrating motor M_2 was provided to monitor the r.m.s. level of the output voltage from the amplifier B_5.

If the photomultipliers are illuminated with uncorrelated beams of light then the inputs to the correlator are mutually incoherent random noise voltages. Under these conditions the output of the multiplier is random noise with a spectral density which has a maximum around zero frequency and which decreases to zero at about 40 Mc/s. The corresponding output from the 5 kc/s amplifier B_4 is random noise centred about 5 kc/s with a bandwidth of 70 c/s. After passing through the synchronous rectifier R_2 the spectrum extends from 0 to 35 c/s, and after passing through amplifier B_5 and the second synchronous rectifier R_3 it is reduced to a band extending from about 0 to 0·25 c/s. The low-pass filter following the rectifier R_3 finally restricts the bandwidth to the range 0 to 0·01 c/s. Under the influence of this noise the motor spins in either direction at random and the reading of the revolution counter remains close to zero. However, if there is any correlation between the output voltage of the photomultipliers a 5 kc/s component appears at the anode of the multiplier; this component is coherent with the 5 kc/s switching wave and reverses in phase every 10 s in synchronism with the 0·05 c/s switching wave. After amplification by the selective amplifier B_4 the 5 kc/s component produces a 0·05 c/s square wave in the output of the synchronous rectifier R_2, which in turn is amplified by B_5 and rectified synchronously by R_3 to produce a direct current component in the voltage applied to the integrating motor M_1. Thus, when there is correlation between the input voltages, the integrating motor revolves more in one direction than the other and the reading on the revolution counter increases with time.

The principal difficulty in designing the correlator was to reduce the random drifts in the output to an acceptable value. It was desirable that any drift should be less than the r.m.s. deviation of the integrating motor M_1, due to noise alone, in a period of several hours. This requirement sets an unusually stringent limit to the tolerable level of any spurious signals in the correlator or to any drift in the synchronous rectifiers. For example a 5 kc/s signal, coherent with the switching wave frequency, will produce an output equal to the r.m.s. deviation of the output counter in 1 h if it is greater than 120 db below noise at the output of the multiplier. In a simple system employing a phase-reversing switch in only one channel, it is difficult to reduce random drifts to an acceptable value; however, by the use of two reversing switches and two synchronous rectifiers in cascade it was found possible to reduce the drift by several orders of magnitude without the use of precisely balanced circuits. It was also necessary to ensure that there was no electrical coupling between the inputs to the correlator, and that these circuits did not pick up signals from external sources. For this reason the photomultipliers were heavily screened and all their supply leads were thoroughly decoupled. Any coupling between the two channels which takes place after the phase-reversing switches does not give rise to spurious correlation, and therefore the switches were put as close as possible to the input terminals of the equipment. Apart from these precautions the equipment was mounted in enclosed racks to improve the screening and to help in stabilizing the temperature; all supplies to the equipment were stabilized.

Extensive tests of the correlator, using independent light sources to illuminate the photomultipliers, have shown that over a period of several hours the drift in the output is less than the r.m.s. uncertainty in the counter readings due to noise alone. However, a more detailed examination of the counter readings shows that over periods of a few minutes there are occasional deviations which are unexpectedly large, and it is believed that this effect is due to short-term drifts in the correlator. For the purposes of the present experiment these short-term drifts are unimportant since tests show that their average effect on the counter readings is not significant when readings are taken over periods of $\frac{1}{2}$ h or more.

4. Experimental procedure and results

4·1 *Calibration of the equipment*

The first step in calibrating the equipment was to measure the various parameters which are involved in the theoretical expression for the correlation in equation (2·14).

The combined spectral response of the arc lamp, lens, liquid filter and photocathodes was measured with a spectrograph with a resolving power of about 5 Å, and the result is shown in figure 6. The response has been plotted in terms of the frequency of the light and corresponds to the quantity $\alpha(\nu)\, n_0(\nu)/\alpha(\nu_0)\, n_0(\nu_0)$. From this curve the effective bandwidth B_0 and the normalized spectral density σ were found to be

$$\left.\begin{aligned} B_0 &= 0{\cdot}85 \times 10^{13}\,\text{c/s},\\ \sigma &= 0{\cdot}451, \end{aligned}\right\} \tag{4·1}$$

where B_0 is defined by equation (2·5), $\lambda_0 = c/\nu_0 = 4358$ Å, and σ is defined by equation (2·7).

The frequency response curves $|F_1^2(f)|$, $|F_2^2(f)|$ of the two amplifiers, B_1 and B_2 in figure 5 were measured directly with a signal generator. The cross-correlation frequency response $\frac{1}{2}[F_1(f)\,F_2^*(f) + F_1^*(f)\,F_2(f)]$ was measured by feeding a signal of variable frequency and constant amplitude into the inputs of the correlator in parallel and observing the output of the multiplier. From these results the bandwidth of the correlator b_v (equation (2·10)), and the spectral density factor η (equation (2·12)) were found to be

$$\left.\begin{aligned} b_v &= 38\,\text{Mc/s},\\ \eta &= 0{\cdot}98. \end{aligned}\right\} \tag{4·2}$$

The excess noise introduced by the correlator and by the stray light reaching the photocathodes is represented in equation (2·14) by the factors $(1+a)$ and $(1+\delta)$, respectively. Measurements showed that there was a small noise contribution from the correlator, mainly due to shot noise in the multiplier which was about 6 % of the total noise, but that stray light was negligible; the two factors therefore have the values

$$\left.\begin{aligned} 1+\delta &= 1{\cdot}06,\\ 1+a &= 1. \end{aligned}\right\} \tag{4·3}$$

It is also necessary, in order to evaluate equation (2·14) to know the value of $G\mu/(\mu-1)$ for each photomultiplier, where G is the overall current gain and μ is the gain of the first stage. As a preliminary test $G\mu/(\mu-1)$ was measured in two different ways. In the first method the current gains G and μ were measured directly by observing the ratio of the respective currents. In the second method the noise voltage across the anode load of each photomultiplier was compared with the noise generated across the same load by a temperature-limited tungsten-filament diode, this comparison being made at the output of the amplifiers B_1 and B_2 to ensure that the noise bandwidth was the same as that used in the actual tests. The quantity $G\mu/(\mu-1)$ for each photomultiplier was then calculated from the simple relation

$$G\mu/(\mu-1) = I_D/I_A, \tag{4·4}$$

where I_D, I_A are the anode currents of the diode and the photomultiplier, respectively, when their noise outputs are adjusted to be equal. The values obtained by the two methods described above agreed within the limits of experimental error.

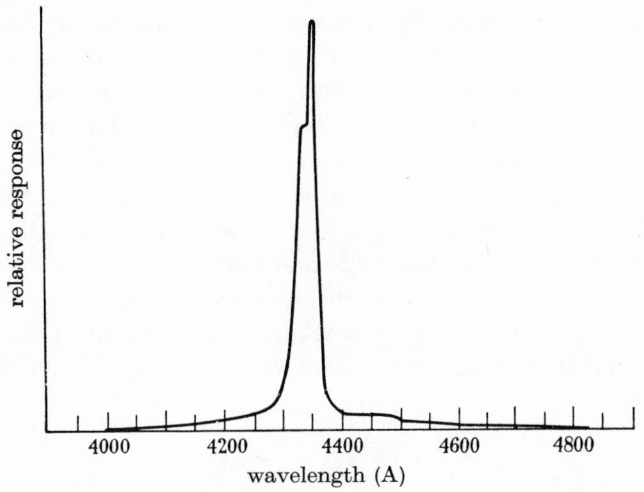

FIGURE 6. Combined spectral response of the arc lamp, optical system, filter and photocathodes.

It was therefore decided to use the second method, employing a noise diode, because it yields the value of $G\mu/(\mu-1)$ directly from a simple observation of the ratio of two currents, and the final theoretical value of the correlation is obtained without an independent measurement of the excess noise introduced by the multiplication process in the phototube. The measured values of $G\mu/(\mu-1)$, under the conditions of the experiment, were found to be $3{\cdot}43 \times 10^5$ and $4{\cdot}63 \times 10^5$ for the two photomultipliers, respectively.

Loss of correlation can occur both in the optical and in the electronic part of the equipment. In the present equipment the loss in the optical equipment arises because of polarization produced by the semi-transparent mirror and this is allowed for, in equation (2·14) by the factor β_0 defined by equation (2·8). For the optical system described here it was found that

$$\beta_0 = 0{\cdot}96. \tag{4·5}$$

The loss of correlation in the electronic equipment occurs almost entirely at the synchronous rectifiers. Ideally, the amplifiers before these rectifiers should pass all the harmonics of the square-wave switching waveform if there is to be no loss of sensitivity. However, in the present equipment it was necessary, in order to reduce the level at the input to the synchronous rectifier R_2 (figure 5) to a convenient value, to restrict the bandwidth of the amplifier B_4 to the fundamental of the phase-switching frequency of 5 kc/s. Under these conditions it can be shown (Dicke 1946) that the signal to noise ratio is reduced by the factor $\sqrt{8/\pi} \simeq 0{\cdot}90$. The frequency of the second phase-switch was only 0·05 c/s and it was possible to reduce the loss at the second synchronous rectifier R_3 to a factor 0·955. The overall loss of correlation in the electronic part of the equipment therefore reduced the signal to noise ratio by the factor

$$\epsilon = 0{\cdot}90 \times 0{\cdot}955 = 0{\cdot}86. \tag{4·6}$$

The final step in calibrating the equipment was to relate the r.m.s. fluctuations in the readings of the signal motor M_1 (figure 5) to the rate of revolution of the noise level motor M_2. This calibration was performed as follows. (*a*) Noise from two independent generators was fed into the inputs of the correlator and the output N_1, N_2 of the amplifiers B_1 and B_2 were adjusted so that the noise motor M_2 was turning at the arbitrarily chosen rate of 14 rev in 5 min. (*b*) The two noise generators were replaced by a single generator which was fed into both inputs in parallel, and the power output of this generator was adjusted to give levels N_1, N_2, as before, at the output of the amplifiers, the signal motor M_1 being disconnected to avoid overloading. (*c*) The power output of the single source was then reduced by a factor ρ^2 by means of a precision attenutator. The signal motor was reconnected and the change in output reading C_0 in a time of 5 min was recorded for various values of ρ^2.

It can be shown that the r.m.s. uncertainty $N(T)$ in the reading of the signal motor after a time T is

$$N(T) = \frac{C_0}{\rho^2}\left(\frac{1+\delta}{\epsilon}\right)\left(\frac{\eta}{2b_v T}\right)^{\frac{1}{2}}, \tag{4·7}$$

where δ, ϵ, η, b_v are parameters of the equipment defined previously. In the present case it was found that $C_0 = 131$ rev, $T = 5$ min, $\rho^2 = 1{\cdot}34 \times 10^4$, and substituting the appropriate numerical values for the other parameters, we get that,

$$N(5\,\text{min}) = N_0 = 14{\cdot}7 \text{ rev}, \tag{4·8}$$

where N_0 is defined as the r.m.s. uncertainty in the signal motor reading in a period of 5 min when the noise level motor is revolving at 14 rev in 5 min.

It is interesting to note that the factor $\dfrac{1+\delta}{\epsilon}\left(\dfrac{\eta}{2b_v T}\right)^{\frac{1}{2}}$ in equation (4·7) also appears in the theoretical expression for the signal to noise ratio equation (2·14). Therefore provided the measurement of N_0 is carried out by the method described here, a comparison of the theoretical and experimental signal to noise ratios is independent of the constants ϵ, δ, η and b_v.

4·2. *Experimental procedure*

The measurements were carried out as follows. The two photocathodes, as viewed from the light source, were superimposed by adjusting the position of the photomultiplier P_2 (figure 4). Readings were then taken every 5 min, for a total period of 4 h, of the revolution counters on the integrating motors M_1 and M_2 and also of the anode currents of the photomultipliers. The centres of the two photocathodes, as seen from the light source, were then separated by 1·25, 2·50, 3·75, 5·0 and 10·0 mm. In each of these positions readings were taken at 5 min intervals for about 30 min, the readings were then repeated with the cathodes separated by the same distances but in the opposition direction.

Throughout the experiment the gain of the amplifier B_4 (figure 5) was controlled to keep the output noise from the correlator approximately constant at a level such that the noise motor M_2 was recording 14 rev in 5 min. The gains of the two photomultipliers were measured before and after every run by comparison with a noise diode, as described in § 4·1. In practice the gains of the two photomultipliers, which were operated at an anode current of about 100 μA, remained constant throughout the experiment.

4·3. *Experimental results*

A marked correlation was observed in the first run with the cathodes superimposed; the total change in the reading of the integrating motor M_1 after 4 h was 1832 rev which, taking the value of N_0 given in equation (4·8), corresponds to an r.m.s. signal to noise ratio of about 18/1. This correlation was progressively reduced as the cathodes were separated until, when their centres were 10 mm apart, no significant correlation was observed.

Table 1. The experimental and theoretical correlation between the fluctuations in the outputs of two photoelectric detectors illuminated with partially coherent light

run no.	duration (h)	cathode separation (mm) *d*	observed correlation (r.m.s. signal to noise ratio) (S/N)	theoretical correlation (r.m.s. signal to noise ratio) (S/N)
1	4	0	+17·55	+17·10
2	1	1·25	+ 8·25	+ 8·51
3	1	2·50	+ 5·75	+ 6·33
4	1	3·75	+ 3·59	+ 4·19
5	1	5·00	+ 2·97	+ 2·22
6	1	10·00	+ 0·90	+ 0·13

The actual readings of the counters and the associated anode currents, etc., taken every 5 min, have not been reproduced here; instead the experimental results have been given in the more convenient form of r.m.s. signal to noise ratios, which are shown for each separation of the photocathodes in column 4 of table 1.

The experimental signal to noise ratios shown in table 1 were calculated from the original readings of the counters by the following method. Each 5 min interval was characterized by readings of the integrated correlation C_r recorded by the signal motor M_1, the noise level N_r recorded by the noise motor M_2, and the anode currents I_{1r}, I_{2r} of the photomultipliers. As they stood these results could not be added to give the final signal to noise ratios, because the gain of the correlator had been frequently altered during each run in an attempt to keep the noise level roughly constant and independent of the inevitable small changes in light flux from the arc lamp. However, it can be shown simply that the final signal to noise ratios, formed by combining the observations from M equal intervals, is independent of the correlator gain in each interval if the readings are weighted and added according to the formula

$$\frac{S}{N} = \frac{1}{N_0} \frac{\sum_{r=1}^{M} \frac{I_{1r} I_{2r}}{\bar{I}_1 \bar{I}_2} C_r \frac{\bar{N}}{N_r}}{\left[\sum_{r=1}^{M} \left(\frac{I_{1r} I_{2r}}{\bar{I}_1 \bar{I}_2}\right)\right]^{\frac{1}{2}}}, \tag{4·9}$$

where N_0 is the r.m.s. uncertainty in the correlation recorded in one interval for a standard noise level $\bar{N}$, and $\bar{I}_1$, $\bar{I}_2$ are averaged over all the intervals. The experimental signal to noise ratios shown in table 1 were therefore calculated for each position of the photocathodes by summing the individual readings taken every 5 min according to equation (4·9) using the experimental value of $N_0 = 14{\cdot}7$ rev given in equation (4·8).

5. Comparison between theory and experiment

The theoretical values of the expected correlation for each cathode separation were calculated as follows. For every 5 min interval the quantity

$$\frac{\mu - 1}{\mu} \alpha(\nu_0)\, n_r(\nu_0)\, (A_1 A_2)^{\frac{1}{2}}$$

was derived from the observed anode currents I_{1r}, I_{2r} of the photomultipliers by the relation

$$\frac{(I_{1r} I_{2r})^{\frac{1}{2}}}{eB_0\{G_1 G_2 \mu_1 \mu_2/(\mu_1 - 1)(\mu_2 - 1)\}^{\frac{1}{2}}} = \frac{\mu - 1}{\mu} \alpha(\nu_0)\, n_r(\nu_0)\, (A_1 A_2)^{\frac{1}{2}}, \tag{5·1}$$

using the values of B_0 and $G\mu/(\mu - 1)$ given in §4·3. In a typical 5 min interval

$$(I_{1r} I_{2r})^{\frac{1}{2}} = 104 \times 10^{-6}\,\text{A} \quad \text{and} \quad \frac{\mu - 1}{\mu} \alpha(\nu_0)\, n_r(\nu_0)\, (A_1 A_2)^{\frac{1}{2}} = 1{\cdot}92 \times 10^{-4}.$$

Since $(1 - 1/\mu)\,\alpha(\nu_0)$ had a value of about 0·12, a typical value for the number of quanta per second incident on each photocathode in unit bandwidth at the centre of the emission line was $1{\cdot}6 \times 10^{-3}$. Following equation (2·14), the results for each interval were then added together to give the theoretical signal to noise ratio, taking $T_r = 300$ s and assuming the values for the various parameters of the equipment given in §4. The partial coherence factor $\Delta(\nu_0)$ for the present equipment, where $\pi\theta_0 a\nu_0/c = \pi\theta_0 b\nu_0/c = 3{\cdot}06$, was calculated from equation (A 25) to be 0·52; the normalized correlation factor $\Gamma^2(\nu_0, d)$ was computed from equation (A 26) and the values are shown in column 6 of table 2. The final theoretical values for the correlation are shown in column 5 of table 1 where they may be compared with the experimental results in column 4.

The results have also been displayed in table 2 and in figure 7 in a form which is intended to show clearly how the correlation decreased with cathode spacing. To allow for the fact that the photomultiplier currents and the observation times were not the same for every cathode spacing, the observed value of the signal to noise ratio at each spacing has been normalized by the corresponding theoretical value calculated for zero cathode spacing, and for the appropriate values of incident light flux and observing time. Effectively, this procedure yields experimental values for the normalized correlation factor, and the results are shown in column 5 of table 2 where they can be compared directly with the theoretical values in column 6. The

TABLE 2. THE EXPERIMENTAL AND THEORETICAL VALUES FOR THE NORMALIZED CORRELATION FACTOR FOR DIFFERENT CATHODE SPACINGS

run no.	cathode separation (mm) d	observed correlation (r.m.s. signal to noise ratio) (S/N)	theoretical correlation assuming cathodes superimposed (r.m.s. signal to noise ratio) $(S/N)'$	experimental value of normalized correlation factor $\Gamma^2(\nu_0, d) = \frac{(S/N)}{(S/N)'}$	theoretical value of the normalized correlation factor $\Gamma^2(\nu_0, d)$
1	0	+17·55	+17·10	1·03 ± 0·04 (p.e.)	1·00
2	1·25	+ 8·25	+ 9·27	0·89 ± 0·07	0·928
3	2·50	+ 5·75	+ 8·85	0·65 ± 0·08	0·713
4	3·75	+ 3·59	+ 8·99	0·40 ± 0·07	0·461
5	5·00	+ 2·97	+ 9·00	0·33 ± 0·07	0·244
6	10·00	+ 0·90	+ 8·17	0·11 ± 0·08	0·015

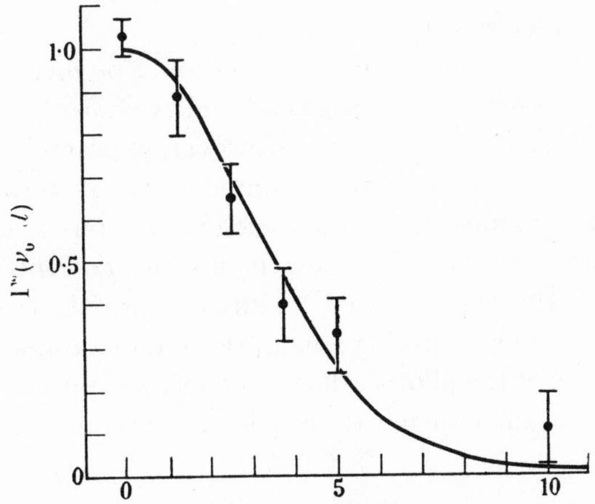

FIGURE 7. The experimental and theoretical values of the normalized correlation factor $\Gamma^2(\nu_0, d)$ for different values of separation between the photocathodes. The full line is the theoretical curve and the experimental results are plotted as points with their associated probable errors.

experimental results have also been plotted in figure 7 for comparison with the theoretical values shown as a solid curve.

A comparison between the theoretical and the experimental values of the correlation given in table 1 and also of the normalized correlation factor given in table 2 and figure 7 shows that, for all cathode spacings, the experimental results are in agreement with theory within the accuracy set by the statistical fluctuations in the measurements. It is true that the difference between theory and experiment is a little greater than the probable error in the case of the widest spacing, but this difference is not significant. The probable error is itself so small a fraction of the observed correlation that the probability that the agreement is due to chance fluctuation and that no real effect is being measured is negligibly small. The probability that the effect is due to some quite different cause, such as fluctuations in the source intensity, is rendered extremely unlikely by the closeness of the agreement between theory and experiment not only with the cathodes superimposed but also with five different cathode spacings.

There remains the possibility that the effect is real, but that its magnitude is not accurately predicted by our theory. This suggestion has been advanced by Fellgett (1957) who has argued on thermodynamical grounds that the excess photon noise, and therefore also the cross-correlation between the fluctuations in different phototubes, should be larger by a factor $\alpha^{-1}(\nu_0)$ than that which we have calculated. We have stated our theoretical objections to Fellgett's analysis in part I of this paper; in addition, his formula is quite incompatible with experimental results reported above, since it would require the correlation to have been about six times greater than that actually observed. In fact, significant disagreement with experiment would arise if the theoretical magnitude of the correlation at every cathode spacing, were increased or decreased† by more than about 12 or 10 %, respectively. However, it must be noted that a systematic error smaller than these limits could not be detected by the present measurements.

In principle, the accuracy of the experiment could be increased indefinitely by increasing the time of observation, since this would reduce the errors due to statistical fluctuations in the correlator output. However, in practice, it is to be expected that the accuracy would soon prove to be limited by errors in the calibration of the equipment or by errors in measurements of such parameters as the spectral density of the light, the cross-correlation frequency bandwidth and the incident light flux, which are probably of the order of 2 or 3 %; furthermore, the theoretical treatment was based on a number of simplifying assumptions, for example, that the quantum efficiency is constant over the photocathodes, and these approximations would also prevent any substantial increase in the precision of the comparison between theory and experiment.

6. Discussion

The experimental results given in the present paper confirm the results of our earlier test (Hanbury Brown & Twiss 1956*a*) and show that the fluctuations of intensity in two coherent beams of light are correlated. They also show that, for the general case of partially coherent light, the observed value of the correlation agrees, within the limits of accuracy of the measurements, with that calculated from a simple classical theory. If there is any systematic error in our calculation of the correlation, for example, due to some quantum effect which has been ignored, it is

† If the theoretical values of the correlation are all increased by the factor a, it can be shown that the best fit with the experimental data, found by minimizing the sum of the weighted squares of the residuals, occurs when $a = 1{\cdot}0137$. The standard deviation of a can be shown to be $0{\cdot}037$ and, if a significant disagreement is defined as three times the standard deviation, we get the result quoted above.

less than 10 %, since any greater error would have produced a significant disagreement with experiment.

Our earlier test has been criticized (Brannen & Ferguson 1956) on the grounds that the observed correlation might have been due to some effect which was modulating the intensity of the light source at frequencies within the passband of the correlator. This suggestion cannot, however, explain the observed decrease of correlation as the separation between the photocathodes is increased, which we have shown here to be in accordance with theory. Admittedly, it has also been suggested, by the same authors, that the decrease in correlation might have been due to light reaching the two photocathodes from different parts of the source as their separation was increased. However, in designing and adjusting the equipment used in both experiments we have taken considerable care to exclude this possibility, and even at the maximum separation of the photocathodes at least 95 % of the incident light came from regions of the source visible to both photocathodes.

The results of our preliminary demonstrations have also been criticized on the grounds that they do not agree with two other experiments which have failed to detect correlation. These experiments were carried out by Ádám, Jánossy & Varga (1955) and by Brannen & Ferguson (1956). In both cases an attempt was made to detect the correlation between the arrival times of photons in two coherent beams of light by means of a coincidence counter. Analyses of these experiments, which have been published elsewhere (Purcell 1956; Hanbury Brown & Twiss 1956*c*), show that they were both too insensitive, by several orders of magnitude, to detect any correlation. In appendix B of the present paper we show the equivalence of the techniques using a coincidence counter and a linear multiplier, but we have also shown that for practical reasons the use of a coincidence counter demands a highly monochromatic, as well as brilliant, source of light. Calculations show that, while it is not feasible to use a standard high-pressure mercury arc, it should be quite practicable to detect correlation with a coincidence counter using a low-pressure mercury isotope lamp. (Since the present paper was written a successful measurement of the correlation between photons using a low-pressure isotope lamp and a coincidence counter has been reported by Twiss, Little & Hanbury Brown (1957).)

It is possible that the principles described here will find practical application in the laboratory. For example, an intensity interferometer can be made to give an extremely high angular resolving power; alternatively, it might perhaps be applied to the measurement of the width and profile of extremely narrow spectral lines. An interesting property, which might have some practical use, was described in § 2·4, where it was shown that, when the source of light is completely resolved, the correlation is a function of the blackbody temperature of the source and effectively the equipment behaves as a pyrometer.

Although we have not considered any of these laboratory applications in detail, we have made a fairly thorough analysis of the application of an intensity interferometer to the measurement of the apparent angular sizes of the visible stars which is given in a later part of this paper. We have also reported briefly a test of the method on Sirius (Hanbury Brown & Twiss 1956*b*). A more detailed account of this work will also be given in a later part of this paper.

We thank the Director of Jodrell Bank Experimental Station for making available the necessary facilities, the Superintendent of the Services Electronic Research Laboratory for the loan of much of the equipment, Dr J. G. Davies for making the numerical computations shown in figures 2 and 3, and Dr A. Burawoy for developing the liquid optical filter.

Appendix A. The theoretical value of the correlation factor and partial coherence factor for partially coherent light

A 1. *The general formula for the correlation between partially coherent fields*

To calculate the correlation, in the general case, where the two light detectors are illuminated with partially coherent light, we shall follow a similar method to that used previously in analyzing a radio interferometer which operated on the same principle (Hanbury Brown & Twiss 1954).

To obtain a quantitative expression for the correlation factor $\Gamma^2(\nu_0, d)$ and the partial coherence factor $\Delta(\nu_0)$ in the case of the simple arrangement shown in figure 1, we shall consider a system of rectangular Cartesian co-ordinates such that the origin lies midway between the centres of the two light detectors, both of which lie in the x axis, and such that the z axis passes through the centre of the light source. We shall assume that the surface of the light source, distant R_0 from the plane containing the photocathodes, can be divided up into elementary areas $d\boldsymbol{\xi} = d\xi\, d\eta$ centred on the points (ξ, η, R_0). It has been shown by Kahn (1957) that the elementary area must be greater than $\lambda^2/2\pi$, where λ is the wavelength of the emitted radiation, but we shall assume here that the area of the source is so large that the error involved in replacing finite summations by integrals is negligible.

Consider the light emitted with a specific polarization from a particular elementary area in a time T. The vector potential at a point distant from the area may be represented by a Fourier series of the form

$$\sum_{r=0}^{\infty} h_r(\boldsymbol{\xi}) \cos\left[\frac{2\pi r}{T}\left(t - \frac{R(\boldsymbol{\xi}, \mathbf{x})}{c}\right) - \chi_r(\boldsymbol{\xi})\right], \tag{A 1}$$

where $\chi_r(\boldsymbol{\xi})$ is a random phase variable distributed with uniform probability between 0 and 2π such that

$$\overline{\chi_r(\boldsymbol{\xi})\, \chi_s(\boldsymbol{\xi}')} = \delta_{rs}\, \delta(\boldsymbol{\xi} - \boldsymbol{\xi}')$$

and $h_r^2(\boldsymbol{\xi})\, d\boldsymbol{\xi}$ is proportional to the number of quanta, incident on unit area at distance R_0, which are emitted with energy h_r/T from an elementary area $d\boldsymbol{\xi}$ of the source. In what follows we shall assume that the source is so far distant that its apparent angular size at the photocathodes is very small compared with unity.

If we assume that the photocathodes behave like square-law detectors with a conversion efficiency proportional to the quantum efficiency, it follows that the low-frequency fluctuations in one of the photoemission currents due to the intensity fluctuations in the incident light are proportional to

$$\int d\boldsymbol{\xi} . d\boldsymbol{\xi}' \sum_{r>s} \sum_{s=1}^{\infty} \int d\mathbf{x}\, \frac{2e(\alpha_{1r}\alpha_{1s} n_{1r} n_{1s})^{\frac{1}{2}}}{T}$$
$$\times \cos\left[\frac{2\pi(r-s)\,t}{T} - \frac{2\pi}{cT}\{rR(\boldsymbol{\xi}, \mathbf{x}) - sR(\boldsymbol{\xi}', \mathbf{x})\} - \{\chi_r(\boldsymbol{\xi}) - \chi_s(\boldsymbol{\xi}')\}\right],$$

where $n_r(\boldsymbol{\xi}) \sim h_r^2(\boldsymbol{\xi})$ is the number of quanta, emitted by unit area of the source, incident on unit area of the photocathode; e is the charge on the electron; $\boldsymbol{\xi}, \boldsymbol{\xi}'$ are the co-ordinates of arbitrary points on the surface of the source; $\mathbf{x} = (x, y, 0)$ are the co-ordinates of an arbitrary point on one of the photocathodes which we shall take to be rectangular in shape and defined by the inequalities

$$\left.\begin{aligned} -\tfrac{1}{2}b < y < \tfrac{1}{2}b, \quad & -\tfrac{1}{2}d - \tfrac{1}{2}a < x < -\tfrac{1}{2}d + \tfrac{1}{2}a, \\ -\tfrac{1}{2}b < y < \tfrac{1}{2}b, \quad & \tfrac{1}{2}d - \tfrac{1}{2}a < x < \tfrac{1}{2}d + \tfrac{1}{2}a, \end{aligned}\right\} \tag{A 2}$$

respectively.

These fluctuations are amplified by a secondary emission multiplier and an amplifier with a combined complex frequency response $F_1(f)$, and at the output of the amplifier they are proportional to

$$\iint d\xi\, d\xi' \sum_{r>s} \sum_{s=1}^{\infty} \int d\mathbf{x} \frac{2e(\alpha_{1r}\alpha_{1s}n_{1r}n_{1s})^{\frac{1}{2}}}{T} \times \mathscr{R}\left\{F_1\left(\frac{r-s}{T}\right) \exp \mathrm{i}\left[\frac{2\pi}{T}\left((r-s)\,t - \frac{rR(\boldsymbol{\xi},\mathbf{x}) - sR(\boldsymbol{\xi}',\mathbf{x})}{c}\right) - (\chi_r(\boldsymbol{\xi}) - \chi_s(\boldsymbol{\xi}'))\right]\right\}.$$

To the second order in R_0^{-1} we may write

$$\frac{R(\boldsymbol{\xi},\mathbf{x})}{R_0} = 1 - \frac{\mathbf{x}\,.\,\boldsymbol{\xi}}{R_0^2} + \frac{\boldsymbol{\xi}^2 + \mathbf{x}^2}{2R_0^2},$$

so that

$$\frac{rR(\boldsymbol{\xi},\mathbf{x}) - sR(\boldsymbol{\xi}',\mathbf{x})}{cT} = \frac{1}{cT}\left\{(r-s)\,R_0 - \mathbf{x}(r\boldsymbol{\xi} - s\boldsymbol{\xi}') + \frac{r\boldsymbol{\xi}^2 - s\boldsymbol{\xi}'^2}{2R_0} - \frac{(r-s)\,\mathbf{x}^2}{2R_0}\right\}. \tag{A 3}$$

In a practical case the last term in equation (A 3) is quite negligible for all values of r, s for which $F_1((r-s)/T)$ differs significantly from zero. Accordingly, the ensemble averaged correlation between the fluctuations in the outputs of the two amplifiers with complex response $F_1(f)$, $F_2(f)$ respectively is proportional to

$$\iiiint d\boldsymbol{\xi}\, d\boldsymbol{\xi}'\, d\mathbf{x}\, d\mathbf{x}' \sum_{r>s} \sum_{s=1}^{\infty} \frac{2e^2}{T} (\alpha_{1r}\alpha_{1s}n_{1r}n_{1s}\alpha_{2r}\alpha_{2s}n_{2r}n_{2s})^{\frac{1}{2}} \frac{1}{2}\left\{F_1\left(\frac{r-s}{T}\right) \times F_2^*\left(\frac{r-s}{T}\right) + F_1^*\left(\frac{r-s}{T}\right) F_2\left(\frac{r-s}{T}\right)\right\} \cos\left[\frac{2\pi}{cTR_0}\{r(\mathbf{x}-\mathbf{x}')\,\boldsymbol{\xi} - s(\mathbf{x}-\mathbf{x}')\,\boldsymbol{\xi}'\}\right],$$

where $\mathbf{x}$, $\mathbf{x}'$ are the co-ordinates of typical points on the first and second photocathodes, respectively. We shall assume that $\alpha_{1r}n_{1r} \simeq \alpha_{1s}n_{1s}$ and $\alpha_{2r}n_{2r} \simeq \alpha_{2s}n_{2s}$ for all values of r, s for which $F((r-s)/T)$ differs significantly from zero, and also that the angular size of the source is sufficiently small to ensure that,

$$\cos\left\{\frac{2\pi}{cTR_0}\left(r(\mathbf{x}-\mathbf{x}')\,\boldsymbol{\xi} - s(\mathbf{x}-\mathbf{x}')\,\boldsymbol{\xi}'\right)\right\} \simeq \cos\left\{\frac{2\pi(r+s)}{2cTR_0}(\mathbf{x}-\mathbf{x}')(\boldsymbol{\xi}-\boldsymbol{\xi}')\right\}. \tag{A 4}$$

In this case we can introduce new frequency variables f, ν such that

$$f = (r-s)/T, \quad \nu = (r+s)/2T. \tag{A 5}$$

Then in the limit as $T \to \infty$ we have that, with the cathodes at a spacing d, the ensemble average $\overline{C(d)}$ of the correlation is given by

$$\overline{C(d)} = 2e^2 \iiiint \mathrm{d}\boldsymbol{\xi}\,\mathrm{d}\boldsymbol{\xi}'\,\mathrm{d}\mathbf{x}\,\mathrm{d}\mathbf{x}' \int_0^\infty \mathrm{d}\nu\,\alpha_1(\nu)\,\alpha_2(\nu)\,[n_1(\nu,\boldsymbol{\xi})\,n_2(\nu,\boldsymbol{\xi})\,n_1(\nu,\boldsymbol{\xi}')\,n_2(\nu,\boldsymbol{\xi}')]^{\frac{1}{2}}$$
$$\times \cos\left\{\frac{2\pi\nu}{cR_0}(\mathbf{x}-\mathbf{x}')(\boldsymbol{\xi}-\boldsymbol{\xi}')\right\}\int_0^\infty \mathrm{d}f\,\tfrac{1}{2}(F_1(f)\,F_2^*(f)+F_1^*(f)\,F_2(f)), \quad \text{(A 6)}$$

where the integrals are taken over the surfaces of both photocathodes, twice over the surface of the source, over the frequency spectrum of the incident light and over the cross-correlation frequency response of the amplifiers following the photomultiplier tubes.

A 2. *The correlation when the apertures of the light detectors are small*

In the simple case where the apertures of the two photocathodes are too small to resolve the source appreciably and where the bandwidth of the light is small, we may replace $(\mathbf{x}-\mathbf{x}')(\boldsymbol{\xi}-\boldsymbol{\xi}')$ in equation (A 6) by $d(\xi-\xi')$ and write

$$\overline{C(d)} = \Gamma^2(\nu_0, d)\,\overline{C(0)}, \quad \text{(A 7)}$$

where ν_0 is the midband frequency of the light; $\overline{C(0)}$ is the correlation with zero spacing between the photocathodes; $\Gamma^2(\nu_0, d)$ is the normalized correlation factor. It follows from equation (A 6) that

$$\Gamma^2(\nu_0, d) = \frac{\iint \mathrm{d}\boldsymbol{\xi}\,\mathrm{d}\boldsymbol{\xi}'\,h^2(\boldsymbol{\xi})\,h^2(\boldsymbol{\xi}')\cos\left\{\frac{2\pi\nu_0}{cR_0}d(\xi-\xi')\right\}}{\iint \mathrm{d}\boldsymbol{\xi}\,\mathrm{d}\boldsymbol{\xi}'\,h^2(\boldsymbol{\xi})\,h^2(\boldsymbol{\xi}')}, \quad \text{(A 8)}$$

where $$h^2(\boldsymbol{\xi}) \sim [n_1(\nu_0,\boldsymbol{\xi})\,n_2(\nu_0,\boldsymbol{\xi}')]^{\frac{1}{2}}.$$

Alternatively, we may write

$$\Gamma^2(\nu_0, d) = H(\nu_0, d)\,H^*(\nu_0, d), \quad \text{(A 9)}$$

where $H(\nu_0, d)$ is defined by

$$H(\nu_0, d) = \int_{-\infty}^{\infty} \mathrm{d}\xi \exp\left(\frac{-2\pi \mathrm{i}\xi\nu_0 d}{cR_0}\right)\int_{-\infty}^{\infty}\mathrm{d}\eta\,h^2(\xi,\eta)\Big/\int_{-\infty}^{\infty}\mathrm{d}\xi\int_{-\infty}^{\infty}\mathrm{d}\eta\,h^2(\xi,\eta). \quad \text{(A 10)}$$

Now $$\overline{h^2(\xi)} = \int_{-\infty}^{\infty}\mathrm{d}\eta\,h^2(\xi,\eta)$$

is the intensity distribution over the equivalent line source projected parallel to the x axis and $H(\nu_0, d)$ is the normalized Fourier transform of this quantity. It follows, as stated in the text, that $\Gamma^2(\nu_0, d)$ is the square of the amplitude of this normalized Fourier transform.

A 3. *The correlation when the apertures of the light detectors are large*

A 3·1. *The general formula*

In the case where the aperture of the two photocathodes are so large that they appreciably resolve the source, the correlation with zero spacing is reduced by the *partial coherence factor* $\Delta(\nu_0)$ while $\Gamma^2(\nu_0, d)$, the *correlation factor* is now a function of the apertures of the photocathodes as well as of the angular size of the source.

If we assume (see § 2·2) that the light intensity may be taken as uniform over the source and that the quantum efficiency is constant over each photocathode, then from equation (A 6) the correlation is

$$\overline{C(d)} = 2e^2\iiiint \mathrm{d}\boldsymbol{\xi}\,\mathrm{d}\boldsymbol{\xi}'\mathrm{d}\mathbf{x}\,\mathrm{d}\mathbf{x}'\cos\left[\frac{2\pi\nu_0}{cR_0}(\mathbf{x}-\mathbf{x}')(\boldsymbol{\xi}-\boldsymbol{\xi}')\right]$$

$$\times\int_0^\infty \alpha_1(\nu)\,\alpha_2(\nu)\,n_1(\nu)\,n_2(\nu)\,\mathrm{d}\nu\int_0^\infty \tfrac{1}{2}\{F_1(f)\,F_2^*(f)+F_1^*(f)\,F_2(f)\}\,\mathrm{d}f \quad \text{(A 11)}$$

and it has been assumed that the light bandwidth is so narrow that

$$\cos\left[\frac{2\pi\nu}{cR_0}(\mathbf{x}-\mathbf{x}')(\boldsymbol{\xi}-\boldsymbol{\xi}')\right]$$

does not change significantly over the bandwidth for which $\alpha_1(\nu)\,\alpha_2(\nu)\,n_1(\nu)\,n_2(\nu)$ differs appreciably from zero.

From the definition of $\Delta(\nu_0)$ and $\Gamma^2(\nu_0, d)$ in § 2·2, it follows simply from equation (A 11) that

$$\Delta(\nu_0)\,\Gamma^2(\nu_0, d) = \frac{1}{\Omega_0^2 A_1 A_2}\iiiint \frac{\mathrm{d}\boldsymbol{\xi}\,\mathrm{d}\boldsymbol{\xi}'\,\mathrm{d}\mathbf{x}\,\mathrm{d}\mathbf{x}'}{R_0^4}\cos\left\{\frac{2\pi\nu_0(\mathbf{x}-\mathbf{x}')(\boldsymbol{\xi}-\boldsymbol{\xi}')}{cR_0}\right\}, \quad \text{(A 12)}$$

where the areas A_1, A_2 of the photocathodes and the solid angle subtended by the source at the photocathodes are given by

$$A_1 = \int \mathrm{d}\mathbf{x}, \quad A_2 = \int \mathrm{d}\mathbf{x}', \quad \Omega_0 = \int \frac{\mathrm{d}\boldsymbol{\xi}}{R_0^2} = \int \frac{\mathrm{d}\boldsymbol{\xi}'}{R_0^2}. \quad \text{(A 13)}$$

A 3·2. *A rectangular source viewed by two light detectors with rectangular apertures*

When the source is rectangular in shape with angular dimensions θ_1, θ_2, where $\theta_1\theta_2 = \Omega_0$, it is best to evaluate equation (A 12) by integrating initially over the variables $\mathbf{x}$, $\mathbf{x}'$ to get

$$\Delta(\nu_0)\Gamma^2(\nu_0, d) = \frac{1}{A_1A_2(\theta_1\theta_2)^2}\iint\frac{\mathrm{d}\eta\,\mathrm{d}\eta'}{R_0^2}\left[\frac{\sin\left(\pi(\eta-\eta')\,b\nu_0/cR_0\right)}{\pi(\eta-\eta')\,\nu_0/cR_0}\right]\iint\frac{\mathrm{d}\xi\,\mathrm{d}\xi'}{R_0^2}$$

$$\times\left[\frac{\sin\left(\pi(\xi-\xi')\,a\nu_0/cR_0\right)}{\pi(\xi-\xi')\,\nu_0/cR_0}\right]^2\cos\left[\frac{2\pi\nu_0 d(\xi-\xi')}{cR_0}\right], \quad \text{(A 14)}$$

where

$$A_1 = A_2 = ab. \quad \text{(A 15)}$$

We now introduce new variables defined by,

$$\left.\begin{aligned}\phi &= \frac{\pi a\nu_0}{cR_0}(\xi-\xi'), \quad \phi' = \frac{\pi a\nu_0}{cR_0}\left(\frac{\xi+\xi'}{2}\right),\\ \psi &= \frac{\pi b\nu_0}{cR_0}(\eta-\eta'), \quad \psi' = \frac{\pi b\nu_0}{cR_0}\left(\frac{\eta+\eta'}{2}\right),\end{aligned}\right\} \quad \text{(A 16)}$$

and integrate over ϕ', ψ' subject to the inequalities

$$\phi' < \left|\frac{\pi a\theta_1\nu_0}{c}-\phi\right|, \quad \psi' < \left|\frac{\pi b\theta_2\nu_0}{c}-\psi\right|. \quad \text{(A 17)}$$

We then get that

$$\Delta(\nu_0)\Gamma^2(\nu_0, d) = \left(\frac{c^2}{\pi^2\nu_0^2\theta_1\theta_2}\right)^2 \frac{1}{A_1A_2}$$
$$\times \int_0^{\Phi} \frac{2\sin^2\phi}{\phi^2}(\Phi-\phi)\cos\left(\frac{2d\phi}{a}\right)\mathrm{d}\phi \int_0^{\Psi}\frac{2\sin^2\psi}{\psi^2}(\Psi-\psi)\,\mathrm{d}\psi, \quad \text{(A 18)}$$

where θ_1, θ_2, the angular dimensions of the source, are defined by

$$\theta_1 = |\,\xi_1-\xi_2\,|_{\text{max.}}/R_0, \quad \theta_2 = |\,\eta_1-\eta_2\,|_{\text{max.}}/R_0, \quad \text{(A 19)}$$

and

$$\Phi = \frac{\pi a\theta_1\nu_0}{c}, \quad \Psi = \frac{\pi b\theta_2\nu_0}{c}. \quad \text{(A 20)}$$

Equation (A 18) involves integrals of the form

$$\mathfrak{J}(\Phi) = \int_0^{\Phi}\frac{2\sin^2\phi}{\phi^2}(\Phi-\phi)\,\mathrm{d}\phi \quad \text{(A 21)}$$

and, in the case where the source is not appreciably resolved by the individual photocathode apertures ($\Phi \ll 1$), we may write

$$\mathfrak{J}(\Phi) = \Phi^2 = \frac{\pi^2\theta_1^2a^2\nu_0^2}{c^2}, \quad \text{(A 22)}$$

so that in this case $\Delta(\nu_0) = 1$, as indeed it must be by definition.

For the opposite extremes, where the source is completely resolved by the individual photocathode apertures ($\Phi \to \infty$), then $\mathfrak{J}(\Phi) \to \pi\Phi$ and hence,

$$\Delta(\nu_0) \to \frac{c^2}{\nu_0^2\theta_1\theta_2}(A_1A_2)^{-\frac{1}{2}}. \quad \text{(A 23)}$$

For intermediate values of Φ, $\mathfrak{J}(\Phi)$ may be expressed conveniently in terms of tabulated functions

$$\mathfrak{J}(\Phi) = 2\Phi\,\mathrm{Si}\,(2\Phi) - (1-\cos 2\Phi) + \ln(\gamma\Phi) - \mathrm{Ci}\,(2\Phi), \quad \text{(A 24)}$$

where γ is Euler's constant given approximately by $\ln(\gamma) = 0{\cdot}5772$, and this result was used in computing the partial coherence factor for the case of the preliminary experiment that we have described elsewhere (Hanbury Brown & Twiss 1956*a*).

A 3·3. *A circular source viewed by two light detectors with rectangular apertures*

When the light source is circular the quantities $\Delta(\nu_0)$ and $\Gamma^2(\nu_0, d)$ can be derived more simply, since the distribution of intensity over the equivalent line source is independent of the direction of the line joining points on the two photocathodes. Thus, the correlation between the fluctuations in the currents emitted from points $(x_1, y_1, 0)$ and $(x_2, y_2, 0)$ on the two photocathodes is proportional to

$$\frac{4J_1^2\{(\pi\theta_0\nu_0/c)\,((x_1-x_2)^2+(y_1-y_2)^2)^{\frac{1}{2}}\}}{(\pi\theta_0\nu_0/c)^2\,((x_1-x_2)^2+(y_1-y_2)^2)},$$

where J_1 is a Bessel function of the first order, and θ_0 is the angular diameter of the source.

Thus from equation (A 12) the partial coherence factor $\Delta(\nu_0)$ is given by

$$\Delta(\nu_0) = \frac{1}{A_1 A_2}\int_{-\frac{1}{2}b}^{\frac{1}{2}b} dy_1 \int_{-\frac{1}{2}b}^{\frac{1}{2}b} dy_2 \int_{-\frac{1}{2}a}^{\frac{1}{2}a} dx_1 \int_{-\frac{1}{2}a}^{\frac{1}{2}a} dx_2 \frac{4J_1^2\{(\pi\theta_0\nu_0/c)((x_1-x_2)^2+(y_1-y_2)^2)^{\frac{1}{2}}\}}{(\pi\theta_0\nu_0/c)^2((x_1-x_2)^2+(y_1-y_2)^2)} \quad \text{(A 25)}$$

and the normalized correlation factor $\Gamma^2(\nu_0, d)$ is given by

$$\Gamma^2(\nu_0, d) = \frac{1}{A_1 A_2 \Delta(\nu_0)}\int_{-\frac{1}{2}b}^{\frac{1}{2}b} dy_1 \int_{-\frac{1}{2}b}^{\frac{1}{2}b} dy_2 \int_{-\frac{1}{2}(d+a)}^{-\frac{1}{2}(d-a)} dx_1 \int_{\frac{1}{2}(d-a)}^{\frac{1}{2}(d+a)} dx_2$$

$$\times \frac{4J_1^2\{(\pi\theta_0\nu_0/c)((x_1-x_2)^2+(y_1-y_2)^2)^{\frac{1}{2}}\}}{(\pi\theta_0\nu_0/c)^2\{(x_1-x_2)^2+(y_1-y_2)^2)\}}. \quad \text{(A 26)}$$

These last two results have been used to analyze the experiment described in the text.

Appendix B. The correlation between photons measured with a coincidence counter

We shall consider here an extremely simple system consisting of two photomultiplier tubes and a coincidence counter, and shall discuss briefly the idealized case in which a coincidence will be recorded if, and only if, two photoelectrons are emitted from the two photocathodes with a time difference less than the resolving time of the coincidence counter. We shall assume initially that the incident light is a linearly polarized plane wave; the extension to the general case of an arbitrarily polarized and partially coherent beam can then be made along identical lines to those developed in § 2 of the text.

B 1. *Mathematical theory*

If $P(t_n)\,dt$ is the probability that a photoelectron be emitted from a photocathode in the time interval

$$t_n \leqslant t < t_n + dt, \quad \text{(B 1)}$$

then we showed in equation (3·14) of part I that

$$P(t_n) = \sum_{r=0}^{\infty} \frac{\alpha_r n_r A}{T} + 2\sum_{r>s}\sum_{s=1}^{\infty} \frac{A}{T}(\alpha_r\alpha_s n_r n_s)^{\frac{1}{2}} \cos\left\{\frac{2\pi}{T}(r-s)t_n + \phi_r - \phi_s\right\}, \quad \text{(B 2)}$$

where T is an arbitrary time interval not less than the observation time of the experiment; α_r is the cathode quantum efficiency at frequency r/T; n_r is the number of linearly polarized quanta of frequency r/T incident in unit time on unit area of the photocathode; A is the area of the photocathode; ϕ_r, ϕ_s are independent random variables distributed with uniform probability over the range $0 \to 2\pi$.

It follows that $P(\tau_c, t_n)$, the probability that *one* electron be emitted in the time interval

$$t_n - \tau_c < t < t_n + \tau_c, \quad \text{(B 3)}$$

is given by

$$P(\tau_c, t_n) = \sum_{r=1}^{\infty} \frac{A}{T}\alpha_r n_r 2\tau_c + 2\sum_{r>s}\sum_{s=1}^{\infty} \frac{A}{T}(\alpha_r\alpha_s n_r n_s)^{\frac{1}{2}}$$

$$\times \frac{2\sin(2\pi(r-s)\tau_c/T)}{2\pi(r-s)/T} \cos\left[\frac{2\pi}{T}(r-s)t_n + \phi_r - \phi_s\right], \quad \text{(B 4)}$$

provided that τ_c is so small that one can neglect the probability that two or more electrons will be emitted in a time interval of duration $2\tau_c$. This last limitation is obviously essential to the use of a coincidence counter technique.

The ensemble average of the joint probability $P_1(\tau_c, t_n)$ that an election be emitted from one photocathode in the interval given by the inequality (B 1) while an electron is emitted from the other photocathode in the interval given by the inequality (B 3), may be written

$$\overline{P_1(\tau_c, t_n)\,P_2(t_n)\,\mathrm{d}t} = \mathrm{d}t\left\{\left(\sum_{r=1}^{\infty}\frac{A_1}{T}\alpha_{1r}n_{1r}\right)\left(\sum_{s=1}^{\infty}\frac{A_2}{T}\alpha_{2s}n_{2s}\right)2\tau_c + 4\sum_{r>s}\sum_{s=1}^{\infty}\frac{A_1A_2}{T^2}(\alpha_{1r}\alpha_{2r}n_{1r}n_{2r})\frac{\sin\left(2\pi(r-s)\,\tau_c/T\right)}{2\pi(r-s)/T}\right\}, \quad \text{(B 5)}$$

since all the terms which depend explicitly on ϕ_r, ϕ_s average to zero.

In the simple case where $A_1\alpha_1 n_1 = A_2\alpha_2 n_2 = A\alpha(\nu)\,n(\nu)$ the ensemble average $\overline{C(T_0)}$ of the number of coincidences in time T_0, is given by

$$\overline{C(T_0)} = 2\tau_c T_0\left(\int_0^{\infty} A\alpha(\nu)\,n(\nu)\,\mathrm{d}\nu\right)^2 + 4T_0\int_{\nu_s}^{\infty}\mathrm{d}\nu_r\int_0^{\infty}\mathrm{d}\nu_s\,A^2\alpha(\nu_r)\,\alpha(\nu_s)\,n(\nu_r)\,n(\nu_s)\frac{\sin\left(2\pi(\nu_r-\nu_s)\,\tau_c\right)}{2\pi(\nu_r-\nu_s)}, \quad \text{(B 6)}$$

where we have let $T\to\infty$.

If the light bandwidth is so large compared with $1/\tau_c$ the reciprocal resolving time of the coincidence counter, that $\alpha(\nu)\,n(\nu)$ does not vary appreciably over the frequency band for which $\sin\left(2\pi(\nu_r-\nu_s)\,\tau_c\right)/2\pi(\nu_r-\nu_s)$ differs significantly from zero, we have that

$$\overline{C(T_0)} = 2\tau_c T_0\left(\int_0^{\infty} A\alpha(\nu)\,n(\nu)\,\mathrm{d}\nu\right)^2 + T_0\int_0^{\infty} A^2\alpha^2(\nu)\,n^2(\nu)\,\mathrm{d}\nu, \quad \text{(B 7)}$$

which may be written

$$\overline{C(T_0)} = \overline{C_R(T_0)} + \overline{C_c(T_0)} = 2\tau_c T_0\,N_p^2 + \tau_0 T_0 N_p^2, \quad \text{(B 8)}$$

where N_p, the average number of photoelectrons emitted by either photocathode in unit time, is

$$N_p = \int_0^{\infty} A\alpha(\nu)\,n_0(\nu)\,\mathrm{d}\nu \quad \text{(B 9)}$$

and $c\tau_0$ is the 'coherence length' of the light (Born 1933) which, in the general case where the light bandwidth is of arbitrary shape, may be defined by

$$c\tau_0 = \frac{c\sigma}{B_0}, \quad \text{(B 10)}$$

where σ, and B_0 are defined by equations (2·7) and (2·5).

In equation (B 8) $C_R(T_0)$ is the average number of coincidences which would occur by chance if the emission of photoelectrons by the two photocathodes was completely random, and is given by

$$\overline{C_R(T_0)} = 2\tau_c T_0 N_p^2, \quad \text{(B 11)}$$

where it is assumed that $\tau_c \gg \tau_0$ and that $N_p \tau_c \ll 1$, while $\overline{C_c(T_0)}$ is the average number of excess coincidences due to the so-called 'bunching' of photons, and is given by

$$\overline{C_c(T_0)} = \tau_0 T_0 N_p^2. \tag{B 12}$$

If the incident light is randomly polarized and if N_0 is now the total average number of photoelectrons emitted in unit time from either photocathode, then the average number of excess coincidences is given by

$$\overline{C_c(T_0)} = \tfrac{1}{2}\tau_0 T_0 N_0^2. \tag{B 13}$$

since there is no correlation between quanta with mutually orthogonal polarizations. The random coincidences, on the other hand, depend only upon the total number of incident quanta so that,

$$\overline{C_R(T_0)} = 2\tau_c T_0 N_0^2 \tag{B 14}$$

as long as

$$\left.\begin{aligned} \tau_c &\gg \tau_0, \\ N_0 \tau_c &\ll 1. \end{aligned}\right\} \tag{B 15}$$

When the incident light is only partially coherent over the individual photocathode apertures and when the centres of the latter are separated by a distance d, the number of excess coincidences will be reduced by a factor $\overline{\Delta \Gamma^2(d)}$, where $\overline{\Delta}$ the *partial coherence factor* and $\overline{\Gamma^2(d)}$ the *normalized correlation factor* are defined in § 2. Thus, in this more general case, the correlation between the photons arriving at the two photocathodes increases the number of coincidences over the random rate by a factor

$$1 + \rho_c = 1 + \overline{\Delta \Gamma^2(d)} \frac{\tau_0}{4\tau_c}, \tag{B 16}$$

which is independent of the quantum efficiency and of the intensity of the incident light.

The random coincidences obey a Poisson distribution so that the r.m.s. fluctuation in their number is given by,

$$\{\langle C_R(T_0) - \overline{C_R(T_0)}\rangle^2_{\text{aver.}}\}^{\frac{1}{2}} = \overline{C_R(T_0)}^{\frac{1}{2}} = 2N_0(\tau_c T_0)^{\frac{1}{2}}. \tag{B 17}$$

Therefore the *signal to noise ratio*, defined as the ratio of the average number of excess coincidences to the r.m.s. fluctuation in the random coincidence rate, is

$$\frac{S}{N} = \frac{\overline{C_c(T_0)}}{\overline{C_R(T_0)}^{\frac{1}{2}}} = \frac{\overline{\Delta \Gamma^2(d)}\, N_0 \tau_0}{2} \left(\frac{T_0}{2\tau_c}\right)^{\frac{1}{2}}, \tag{B 18}$$

which from equations (2·3) (2·6) and (B 10) may be written in the equivalent form

$$\frac{S}{N} = \overline{\Delta \Gamma^2(d)}\, A \alpha(\nu_0)\, n_0(\nu_0)\, \sigma \left(\frac{T_0}{4\sqrt{2}\,\tau_c}\right)^{\frac{1}{2}}. \tag{B 19}$$

The results given above have been quoted elsewhere (Hanbury Brown & Twiss 1956*c*) for the special case $\overline{\Gamma^2(d)} = 1$; an exactly similar result has been derived by Purcell (1956) for the case $\overline{\Delta \Gamma^2(d)} = 1$ by an analysis based upon the auto-correlation function for the intensity fluctuations in the incident light. The present analysis, is, in effect, the Fourier transform dual to that of Purcell and it is simple to show that τ_0 as defined in (B 10) is identical with the τ_0 used by Purcell.

B 2. *Comparison with the alternative technique*

If equation (B 19) is compared with equation (2·13) which gives the signal to noise ratio for the linear multiplier technique, we see that the two expressions are the same if we put

$$\frac{b_v}{\eta} = \frac{1}{4\tau_c} \tag{B 20}$$

and assume that both equipments have an ideal performance

$$\text{(i.e.} \quad \epsilon\gamma_0(\mu-1)/\{\mu(1+a)(1+\delta)\} = 1).$$

Now b_v/η is roughly the response time of the amplifiers in the linear multiplier system, or alternatively is roughly the effective bandwidth of the coincidence counter. It is therefore clear that in the ideal case both techniques, coincidence counting of photoelectrons or linear multiplication of intensity fluctuations, give about the same signal to noise ratio provided that the effective bandwidth of their circuits, the spectral distribution of the incident light, and the primary photoemission currents are the same in both cases.

However, in a practical case there are considerable difficulties in meeting the condition that the primary photoemission currents should be the same for both techniques. Thus the need to satisfy the inequalities defined in equation (B 15) severely limits the maximum photoemission on current when the resolving time of coincidence counter is fixed, and in practice it is difficult to let N_0, the average number of photoelectrons emitted per second, rise much above 10^6 when one is using the coincidence-counter technique. As we have shown above the signal to noise ratio, which is by far the most important limitation of an 'intensity' interferometer, is determined by the number of incident quanta per cycle bandwidth rather than by the total incident light flux, and to ensure a workable signal to noise ratio, the average number of photoelectrons produced per second, by quanta in a frequency band of 1 c/s, must be of the order† of 10^{-5} or greater. This latter requirement can only be satisfied by a source of high equivalent temperature, especially if the source is to be negligibly resolved by the apertures of the photocathodes. For example, if $N_0 < 10^6$ this means that the light bandwidth must be less than 10^{11} c/s, which corresponds to a bandwidth of about 0·5 Å at a wavelength of 4000 Å.

In the laboratory these conditions can easily be met by using a low-pressure electrodeless isotope lamp. Thus a source described by Forrester, Gudmundsen & Johnson (1955) had a bandwidth of only $8 \cdot 10^8$ cycles, centred on the 5461 green line of ^{198}Hg, with an output flux of 0·004 W cm^{-2} sterad^{-1} which corresponds to an effective black-body temperature at the centre of the line of 6750° K.

However, in the case of a measurement on a star, where the incident light is spread over the whole visible spectrum, narrow bandwidths can only be obtained by means of interference filters (Ring 1956) which introduce appreciable attenuation and demand well collimated beams of light. For these and other technical reasons the coincidence-counting technique has not been seriously considered for astronomical applications.

† With practical values for the parameters of equation (2·14) and an amplifier bandwidth of 100 Mc/s one would get a signal to noise ratio of 3 to 1 in 1 h with $\sqrt{(A_1 A_2)}\,\alpha(\nu_0)\,n_0(\nu_0) = 10^{-5}$ and this is about the lowest sensitivity with which one could work comfortably.

B3. *The limiting case of arbitrarily small resolving time*

We have, so far, only considered the case in which τ_c, the resolving time of the coincidence counter is very much greater than τ_0, where $c\tau_0$ is the 'coherence length' of the incident light. However, if the light source is, for example, a low-pressure isotope lamp this condition is not necessarily valid and a brief discussion of the more general case is needed.

In the limiting case $\tau_c/\tau_0 \to 0$ it follows immediately from equation (B5) that

$$C(T_0) \to 2\tau_c T_0\left(\int_0^\infty A\alpha(\nu)\, n(\nu)\, d\nu\right)^2 + 2\tau_c T_0\left(\int_0^\infty A\alpha(\nu)\, n(\nu)\, d\nu\right)^2, \qquad \text{(B21)}$$

when

$$\overline{C_c(T_0)} \to \overline{C_R(T_0)} = 2\tau_c T_0 N_p^2 \qquad \text{(B22)}$$

in the idealized case of a linearly polarized plane wave of light. The average number of excess coincidences is therefore equal to the number of purely random coincidences which would arise from completely incoherent light beams, a result which is independent of the spectral distribution of the incident light.

In the general case, when τ_c/τ_0 is finite and non-zero, it can be shown from equation (B5) that

$$C_R(T_0) = \tau_0 T_0 N_p^2 \operatorname{erf}\left(\frac{\sqrt{\pi}\,\tau_c}{\tau_0}\right), \qquad \text{(B23)}$$

assuming that the spectral distribution of the incident light is Gaussian, as will be approximately the case when the line broadening is due to Doppler effect. The signal to noise ratio is therefore reduced by the factor

$$\operatorname{erf}(\sqrt{\pi}\,\tau_c/\tau_0),$$

which tends to zero as $\tau_c \to 0$, and this result emphasizes that the correlated photons are not in perfect time coincidence. There is a fundamental uncertainty in their arrival time which is of the order of the reciprocal of the bandwidth of the incident light, a result which can be deduced directly from the uncertainty principle.

References

Ádám, A, Jánossy, L. & Varga, P. 1955 *Acta Phys. Hungar.* **4**, 301.
Born, M. 1933 *Optik*. Berlin: Springer.
Brannen, E. & Ferguson, H. I. S. 1956 *Nature, Lond.* **178**, 481.
Dicke, R. H. 1946 *Rev. Sci. Instrum.* **17**, 268.
Fellgett, P. B. 1957 *Nature, Lond.* **179**, 956.
Forrester, A. T., Gudmundsen, R. A. & Johnson, P. O. 1955 *Phys. Rev.* **99**, 1691.
Hanbury Brown, R. & Twiss, R. Q. 1954 *Phil. Mag.* **45**, 663.
Hanbury Brown, R. & Twiss, R. Q. 1956*a* *Nature, Lond.* **177**, 27.
Hanbury Brown, R. & Twiss, R. Q. 1956*b* *Nature, Lond.* **178**, 1046.
Hanbury Brown, R. & Twiss, R. Q. 1956*c* *Nature, Lond.* **178**, 1447.
Hanbury Brown, R. & Twiss, R. Q. 1957 *Proc. Roy. Soc.* A, **242**, 300.
Hopkins, H. H. 1951 *Proc. Roy. Soc.* A, **208**, 263.
Kahn, F. D. *In preparation.*
Purcell, E. M. 1956 *Nature, Lond.* **178**, 1449.
Ring, J. 1956 *Astronomical optics* (ed. Z. Kopal). Amsterdam: North Holland Publ. Co.
Shockley, W. & Pierce, J. R. 1938 *Proc. Inst. Rad. Engn.* **26**, 321.
Twiss, R. Q., Little, A. G. & Hanbury Brown, R. 1957 *Nature, Lond.* **180**, 324.
Zernike, F. 1938 *Physica*, **5**, 785.

PAPER NO. 24

Reprinted from *Journal of the Optical Society of America*, Vol. 47, p. 976, 1957.

Intensity Matrix and Degree of Coherence

HIDEYA GAMO
Department of Physics, University of Tokyo, Tokyo, Japan
(Received May 20, 1957)

RECENTLY Dr. D. Gabor[1] in England and the writer[2] in Japan have quite independently reached the same conclusion, namely, that the intensity distribution of images can be described by using positive-definite Hermitian matrices under various conditions of illumination. This treatment is based on two important facts peculiar to optics; the one is that the only physical quantity directly observed is the intensity, and the other is that the information on an object to be derived from an observed image is dependent upon the nature of illumination, namely, coherence, partial coherence, or incoherence. Therefore, the matrix formalism may be useful for the information theory in optics, though it has not been applied practically. Since the procedure followed by the writer is somewhat different from D. Gabor's and some of the results obtained by the writer are not contained in D. Gabor's papers, the writer will explain here briefly the procedure and a few results.

We shall start from the well-known relation between the intensity of an image and the one of light source, because it suffices for two features mentioned above; namely,

$$I(x)=\int_\Sigma J(y)\,|U(y-x)|^2dy, \tag{1}$$

where $I(x)$ is the intensity of an image at point x, $J(y)$ the intensity of the source, and $U(y-x)$ the complex amplitude of waves at point x in the image produced by a point source having unit intensity at point y in the source; integration should be taken over the light-source Σ. The complex transmission function $U(y-x)$ in amplitude is, as in the following, expressed by using the transmission coefficient of the object $E(X)$, the amplitude of incident waves at the object $(Ay-X)$, and the transmission function of the pupil $u(X-x)$; that is,

$$U(y-x)=\int A(y-X)E(X)u(X-x)dX. \tag{2}$$

Since the Fourier transform of transmission function $u(X-x)$ is equal to the pupil function representing aberrations, and its frequency band width is limited by a given aperture, the function $u(X-x)$ can be expressed in series by using the sampling theorem[3]

$$u(X-x)=\sum_{n=-\infty}^{+\infty} u(X-n\pi/k\alpha)u_n(x) \tag{3}$$

where the sampling function $u_n(x)=\sin(k\alpha x-n\pi)/(k\alpha x-n\pi)$, $k=2\pi/\lambda$, λ the wavelength, α the aperture constant. Then putting Eqs. (2) and (3) into (1), we obtain finally

$$I(x)=\sum_{n=-\infty}^{+\infty}\sum_{m=-\infty}^{+\infty} u_n(X)A_{nm}u_m(x), \tag{4}$$

and

$$A_{nm}=\int\int \Gamma(X_1,X_2)E^*(X_1)E(X_2)u^*\times(X_1-n\pi/k\alpha)u(X_2-m\pi/k\alpha)dX_1dX_2, \tag{5}$$

where the mutual intensity $\Gamma(X_1,X_2)$ is equal to the phase-coherence factor[4] multiplied by the absolute values of amplitudes at points X_1 and X_2. The matrix whose n,m element is given by Eq. (5) is the *intensity matrix* now under consideration.

The intensity matrix stated above has some interesting properties. First, it is a positive, definite Hermitian matrix. To prove this, we note that the intensity expressed by Eq. (4) must always be positive and real. Second, its diagonal elements are given by the intensities at sampling points, and the trace or the sum of diagonal elements is equal to the total intensity integrated over the image. This is easily shown by considering the properties of sampling functions, namely, that each sampling function is unity at its own sampling point and zero at the other sampling points, and satisfies the orthogonal relation. Third, any intensity matrix can be diagonalized by a unitary transformation, following the well-known general theory of Hermitian matrices. Thus, the intensity given by Eq. (4) is expressed as

$$I(x)=\lambda_1|\Sigma_i S_{i1}u_1|^2+\lambda_2|\Sigma_i S_{i2}u_2|^2+\cdots, \tag{6}$$

where λ_n is the nth eigenvalue and S_{mn} is the mth component of the nth eigenvector of the given intensity matrix. These properties are stated also by D. Gabor in a somewhat different and more general manner.

Hereafter, we shall consider how the nature of illumination is reflected in an intensity matrix, since we find very little about it in Gabor's paper. In order to separate the effect of illumination on the intensity matrix from the effects of object and pupil, let us assume the transmission coefficient of an object is uniformly unity and the image is obtained by a pupil without either aberration or defect in focusing. In case of coherent illumination only the first term of the foregoing Eq. (6) remains and all the other terms are zero. In general, the secular equation of the coherent intensity matrix is easily solved, and its nonvanishing eigenvalue is equal to the total intensity of an image. On the other hand, all the terms in Eq. (6) remain in case of incoherent illumination. In other words, the rank of an incoherent intensity matrix is larger than the rank of any other coherent or partially coherent case. In case of partially coherent illumination, the term of the largest eigenvalue will play the leading part, for instance, when the phase-contrast method is applied. Thus, eigenvalues of an intensity matrix are evidently related to the nature of illumination.

Now let us introduce a new quantity having the same form as the entropy of statistical mechanics

$$D=-\Sigma_n(\lambda_n/I_0)\log(\lambda_n/I_0), \tag{7}$$

where λ_n is the nth eigenvalue, I_0 the trace of an intensity matrix. The quantity D is zero for the coherent illumination, and becomes $\log N$ for the incoherent illumination, where N is the number of the "degrees of freedom" of an image of area S, namely, $N=4\alpha^2S/\lambda^2$. The value of d for partially coherent illumination is positive and smaller than $\log N$, or $0\leqslant D\leqslant \log N=D_0$. Then, one minus the ratio of D to the maximum value D_0, that is, $d=(D_0-D)/D_0$ may be regarded as a measure of the "degree of coherence" of illumination, since it is unity for the coherent case and zero for perfectly incoherent case.

How the intensity matrix is transformed by changing the object or the aperture is another important problem, which will be discussed elsewhere.[5]

[1] D. Gabor, *Information Theory, Third London Symposium*, edited by C. Cherry (Butterworths Scientific Publications, London, 1956), pp. 26–33; D. Gabor, *Astronomical Optics*, edited by Z. Kopal (North-Holland Publishing Company, Amsterdam, 1956), pp. 17–30.

[2] H. Gamo, Kagaku (Tokyo) **26**, 470–471 (1956); H. Gamo, J. Appl. Phys. (Japan) **25**, 431–443 (1956), No. 11 Optics Issue.

[3] A. Blanc-Lapierre, Ann. inst. Henri Poincaré **13**, 283 (1953), No. 4; G. Toraldo di Francia, J. Opt. Soc. Am. **45**, 497 (1955).

[4] H. H. Hopkins, Proc. Roy. Soc. (London) **A208**, 263 (1951); **A218**, 408 (1953); E. Wolf, Proc. Roy. Soc. (London) **A225**, 96 (1954); **A230**, 246 (1954).

[5] H. Gamo, J. Phys. Soc. Japan (to be published).

PAPER NO. 25

Reprinted from *Nature*, Vol. 180, pp. 1035–1036, 1957.

TIME-CORRELATED PHOTONS

By G. A. REBKA, JUN.,* and PROF. R. V. POUND

Lyman Laboratory of Physics, Harvard University, Cambridge, Massachusetts

SEVERAL studies of the time correlation of photons in coherent light beams have recently been reported in this journal[1–8]. We have undertaken to demonstrate such correlation using pulse and coincidence techniques in contrast to the continuous-wave technique used by Brown and Twiss in their first study. Our experiment differs from their more recent one[9] in that counts from each detector are recorded. These are used for reducing each observation to a measurement of effective resolving time.

Following the formulation of Purcell[4], the number of coincidences N_c in the counting period T can be written

$$N_c = (N_1 N_2/T)\{2\tau_R + np\alpha\tau_0 f(\Delta)\}$$

The quantities N_1 and N_2 are the counts recorded from each detector, τ_R is the resolving time of the coincidence circuit, while τ_0 is the correlation time of the light and is related to the width and shape of the spectral line used, in the manner discussed by Purcell. The factor n accounts for the presence of pulses of other origin than the light beam, which are assumed random. In our experiment a typical value of n was 0·951. Similarly p accounts for polarization of the light beam. It would have a value of $\frac{1}{2}$ for completely unpolarized light, and had a value of 0·45 in our optical system. The degree of geometrical coherence resulting from the arrangement of the optical system is described by α, which has a maximum value of 1 for complete coherence. With the assumption of uniform sensitivity over the small photocathode areas used, a value of 0·55 was calculated for our optical system. Finally, $f(\Delta)$, where the time Δ is the difference in delay between the two channels, depends, in our limit where $2\tau_R \gg \tau_0$, mainly on the properties of the coincidence circuit and detectors. These are such that $f(0) \approx 1$, and $f(|\Delta| > \tau_R) \approx 0$. Were it possible to achieve a system with $2\tau_R \ll \tau_0$, $f(\Delta)$ would represent the square of the cosine transform of the line shape of the light source.

* National Science Foundation Predoctoral Fellow.

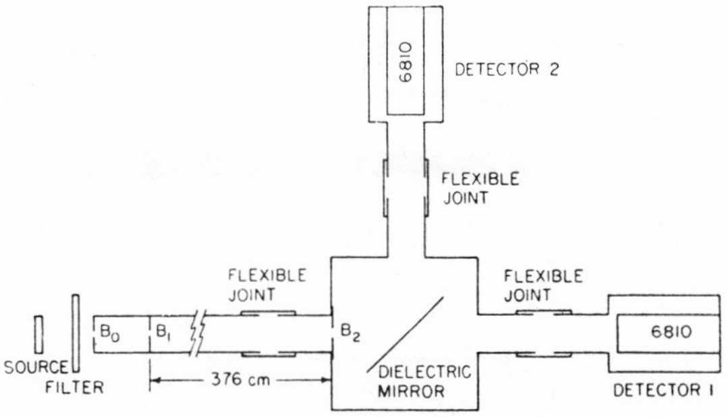

Fig. 1. A representation of the optical system. Baffles B_1 and B_2 define the geometrical coherence

Fig. 1 shows the arrangement of the optical system. A modified Baird–Atomic mercury-198 electrodeless discharge tube was driven by a 10-cm.-wave-length, continuous-wave magnetron delivering about 35 watts. Vapour from boiling nitrogen provided a coolant to prevent 'clean-up' of the mercury and to prevent broadening of the line in excess of Doppler effect[10]. Considerations of line width, intensity, and detector sensitivity led us to use the 4358-A. line, which was separated with a Baird–Atomic interference filter. A phase-coherent beam was selected from the light passing through the hole, of diameter 0·02 cm., in B_1 by the hole of diameter 0·84 cm. in B_2. The baffle B_2 was 376 cm. distant from B_1. The baffle B_0 reduced reflexion from the walls housing the optical system. The dielectric mirror split the light into two beams, each of which was incident on the photocathode of an R.C.A. 6810 photomultiplier tube. Definition of the beam prior to the mirror in our system, in contrast to that of Brown and Twiss[1,9], assures geometrical coherence at the two detectors without critical adjustment. Consequently we were not able to vary the correlation by changing the geometrical coherence.

The photomultiplier tubes were cooled throughout the experiment with solid carbon dioxide, and electrostatic shields at cathode potential were used to minimize dark current. The tubes were operated with an anode–cathode potential of 2,450 volts, and with the focusing grid and first dynode both 450 volts above the photocathode. Single photoelectrons produced an average pulse at the last dynode of 1/3 m.amp. and 15 × 10^{-9} sec. duration. Continuous application of the high voltage brought the dark current down after several days from an initially higher value to a steady state value equivalent to about 750 photoelectrons per second.

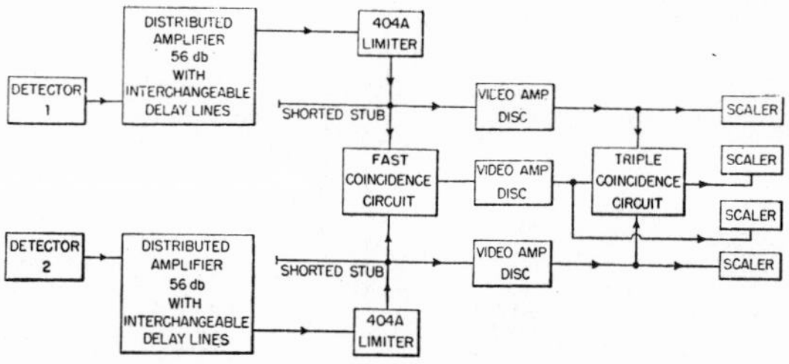

Fig. 2. Block diagram of the electronic system

A block diagram of our electronic system is shown in Fig. 2. It represents a modification of the fast–slow system often used in coincidence spectrometry[11] in that the slow pulse-height-selecting channels were applied to the limited fast-channel pulses. Limitation of the counting speed set by the scalers and discriminators in the slow singles-channels made necessary paralysis of the limiter for a period of 4×10^{-6} sec. to eliminate coincidences between counts not resolvable in the slow channels. This resulted in a 25 per cent loss in the singles-channels for a typical counting rate of 4×10^4 counts per second. The minimum resolving time, which was limited by transit-time effects in the photomultipliers, was established by replacing B_2 by a light-pulse source composed of liquid scintillator and cobalt-60. The intensity of each light pulse was such that few events detected from each photomultiplier corresponded to more than one photoelectron. With this source, a curve of coincidence counts versus time-delay difference indicated a coincidence efficiency corresponding to $f(0) \geqslant 0{\cdot}85$ for $\tau_R \geqslant 6 \times 10^{-9}$ sec.

With the parameters of the light source, optical system, and electronic system thus established, the correlated, relative to the random, coincidences were expected to be between 1 and 2 per cent. With the additional limitation in counting rate, at least 5 hr. of counting sample would be required to distinguish the correlation. By alternating the delay Δ between 0 and $6\tau_R$ every 100 sec., the effect of drift of the resolving time during this long counting period was minimized.

The combination of all measurements made with a resolving time of about 6×10^{-9} sec. yields

$$np \propto f(0)\tau_0 = (0{\cdot}173 \pm 0{\cdot}018) \times 10^{-9} \text{ sec.}$$

The error accounts only for the statistical accuracy of the sample. The experimental value of

$$\tau_0 = (0{\cdot}81 \pm 0{\cdot}16) \times 10^{-9} \text{ sec.}$$

is less accurately known since measurements and estimates of the other factors contribute to the uncertainty. For a line having a Doppler-broadened shape, this value of τ_0 would correspond to a full width at half height of 820 ± 160 Mc./s. The corresponding width for a mercury source at 300° K. would be 600 Mc./s.

Other measurements were made with a resolving time of about 9×10^{-9} sec., and these gave:

$$np \propto f(0)\tau_0 = (0{\cdot}214 \pm 0{\cdot}033) \times 10^{-9} \text{ sec.}$$

and

$$\tau_0 = (0{\cdot}90 \pm 0{\cdot}22) \times 10^{-9} \text{ sec.}$$

This result indicates that our estimates of the coincidence efficiency for $\tau_R = 6 \times 10^{-9}$ sec. are not significantly in error.

A check for any systematic error inherent in changing the delay was made by repeating the measurement with the light source replaced by a direct-current mercury arc which had a broad line. No significant effect was found within the statistical accuracy of the sample.

The authors wish to thank Prof. Francis Bitter, Mr. Adrian C. Melissinos, and the staff of the Research Laboratory of Electronics, Massachusetts Institute of Technology for the loan of their specially designed mercury-198 isotope lamp.

This work was supported in part by the joint programme of the Office of Naval Research and the U.S. Atomic Energy Commission.

[1] Hanbury Brown R., and Twiss, R. Q., *Nature*, **177**, 27 (1956).
[2] Brannen, E., and Ferguson, H. I. S., *Nature*, **178**, 481 (1956).
[3] Hanbury Brown, R., and Twiss, R. Q., *Nature*, **178**, 1447 (1956).
[4] Purcell, E. M., *Nature*, **178**, 1449 (1956).
[5] Hanbury Brown, R., and Twiss, R. Q., *Nature*, **178**, 1046 (1956).
[6] Fellgett, P. B., *Nature*, **179**, 956 (1957).
[7] Sillitto, R. M., *Nature*, **179**, 1127 (1957).
[8] Hanbury Brown, R., and Twiss, R. Q., *Nature*, **179**, 1128 (1957).
[9] Hanbury Brown, R., and Twiss, R. Q., *Proc. Roy. Soc.* (in the press).
[10] Meggers, W. F., and Kessler, K. G., *J. Opt. Soc. Amer.*, **40**, 737 (1950).
[11] Wertheim, G. K., Thesis, Harvard University (1955) (unpublished).

PAPER NO. 26

Reprinted from *Nature*, Vol. 180, pp. 324–326, 1957.

CORRELATION BETWEEN PHOTONS, IN COHERENT BEAMS OF LIGHT, DETECTED BY A COINCIDENCE COUNTING TECHNIQUE

By R. Q. TWISS and A. G. LITTLE
Division of Radiophysics, Commonwealth Scientific and Industrial Research Organization, Sydney

AND

R. HANBURY BROWN
University of Manchester Jodrell Bank Experimental Station

THERE has recently been a controversy[1–4] in the columns of *Nature* on the existence of a correlation between photons in coherent beams of light. In the initial communication[1], by Hanbury Brown and Twiss, an experiment was described in which light from a pinhole source was split, by a semi-transparent mirror, into two separate beams which then illuminated the cathodes of two separate phototubes. As long as the two light beams were partially coherent a correlation was observed between the fluctuations in the anode currents of the two phototubes and it was argued that this proved that the arrival times of light quanta at different points illuminated by coherent beams of light were correlated.

This experiment was criticized by Brannen and Ferguson[2], who had themselves carried out a somewhat different test, in which the simultaneous emission of photoelectrons at the two photocathodes were recorded by a coincidence counter, and in which no significant difference was observed in the number of counts with coherent or incoherent light. Furthermore, these authors suggested that the results obtained by Hanbury Brown and Twiss were due, not to a true correlation between the arrival times of quanta, but to some other cause such as an intensity fluctuation in the light source.

In reply it was pointed out by Purcell[3] and by Hanbury Brown and Twiss[4] that there was no real disagreement between the results of these two tests, since the experiment of Brannen and Ferguson[2], and the similar experiment of Adám, Jánossy and Varga[5], were far too insensitive to detect the effect. In addition, the reality of the phenomenon has been confirmed by observations[6] on Sirius; also the experiment of Hanbury Brown and Twiss has been repeated under precisely controlled conditions, when the observed correlation, with various degrees of optical superimposition of the photocathodes, agreed within a few per cent with the predictions of theory (unpublished work).

However, to make assurance doubly sure, we have repeated the experiment of Brannen and Ferguson, under conditions which permit the detection of the effect, and have obtained a positive result, in reasonable agreement with theory, as described below.

If a correlation between photons is to be detected in a convenient time by the coincidence counter technique it can be shown, not only that the light source must be brilliant, but also that the light reaching the photocathodes must be confined to a very narrow band. This immediately suggests the use of an isotope lamp and, in the present experiment, the light source was an electrodeless radio-frequency discharge in mercury-198 vapour, cooled by a forced air draught, to which the radio-frequency oscillator delivered $1\frac{1}{2}$ watts of power at a frequency of 800 Mc./s. In other respects the experimental arrangement was very similar to that of Brannen and Ferguson[2], employing $1P21$ phototubes, the coincidence circuit being that of Bell, Graham and Petch[7], with a resolution time of $3 \cdot 5 \times 10^{-9}$ sec.

The light source was limited by a pinhole 0·360 mm. in diameter; the photocathodes, at a distance of 1·25 metres, were limited by square apertures, 2 mm. × 2 mm., and the 5461 A. line was isolated by an optical filter. Both phototubes were mounted on movable slides so that, as seen by the source, they could be optically superimposed or separated by a distance of 5 mm. transverse to the line of sight; at the latter distance the pinhole source was completely resolved so that the incident light beams were effectively uncorrelated. The photon counts could also be uncorrelated by inserting a length of cable in one channel of the coincidence counter circuit with a delay of 15 mμsec., about four times the resolution time of the coincidence circuit.

The measurements were carried out as follows.

The photocathodes were optically superimposed for two minutes and the number of coincidences n_{1r} observed in this interval was recorded. One or other of the photocathodes was then moved to the displaced position, and n_{2r}, the number of coincidences in a two-minute period, was also recorded. This procedure was repeated ten times in a run in which the total observation time was 40 min. and the difference

$$\Delta n_c = \sum_{r=1}^{10} (n_{1r} - n_{2r}) \tag{1}$$

was measured. In order to eliminate the effect of any difference in the light intensity reaching the two different positions of the photocathode, an identical comparison run was then made with the extra cable inserted in one arm of the coincidence circuit and the difference

$$\Delta n'_c = \sum_{r=1}^{10} (n'_{1r} - n'_{2r}) \tag{2}$$

between the coincidences observed with the photocathodes superimposed and displaced was recorded. The quantity $N_c = (\Delta n_c - \Delta n'_c)$ was then taken as a measure of the number of correlated photoelectrons emitted in coincidence during a 20-min. interval.

A total of six such tests, with an overall observation time of 480 min., were performed, and the results are shown in Table 1. Combining all the measurements, we conclude that the ratio ρ of the correlated to the random counts is given by

$$\rho_{\text{exp.}} = 0{\cdot}0193 \pm 0{\cdot}0017 \text{ (probable error)} \tag{3}$$

The probability that this result is a random noise fluctuation is quite negligible, being less than 1 in 10^{15}; in addition, the possibility that it was significantly affected by drifts was eliminated by stabilizing the light intensity and the power supplies to the phototubes, with the result that any such drifts in the counting rate in a 40-min. run were too small, compared with the random fluctuation, to be detected.

A seventh run was then taken with the isotope light source replaced by a tungsten filament lamp, under which conditions no significant correlation is to be expected. As may be seen from Table 1, no significant correlation was in fact observed.

The experimental results have been compared with theory as follows. It has been shown by Purcell[3] that the theoretical ratio ρ of the correlated to the random coincidences is given by

$$\rho_{\text{theor.}} = \frac{\tau_0}{4\tau_c} \tag{4}$$

in the idealized case of a point source and unpolarized light. In this expression τ_c is the resolving time of the coincidence counter, and $c\tau_0$ is the 'coherence length' of the light when τ_0, which is approximately equal to the reciprocal light bandwidth, is defined by

$$\tau_0 = \int_{-\infty}^{\infty} | g^2(t) | \, dt \tag{5}$$

and $g(t)$, the Fourier transform of $f(\nu - \nu_0)$, the normalized spectral density of the incident light centred on frequency ν_0, is defined by

$$g(t) = \int_{-\infty}^{\infty} f(\nu) \exp (2\pi i\nu t) d\nu \tag{6}$$

The resolving time τ_c of the coincidence counter was measured to be $3{\cdot}5 \times 10^{-9}$ sec. under the conditions of the present experiment. The value of τ_0 for the mercury isotope lamp was measured by observing the visibility of fringes in a Michelson interferometer as a function of mirror spacing, and was found to be $0{\cdot}73 \times 10^{-9}$ sec.; this is somewhat

Table 1. RATIO OF CORRELATED TO RANDOM COINCIDENCES

Run No.	Observed number of coincidences in 20 min.* (N_T)	Observed number of correlated coincidences in 20 min. (N_c)	Percentage ratio of correlated to total coincidences ($\rho \times 100$)	Ratio of correlated coincidences to r.m.s. uncertainty in random coincidences† ($N_c/2(N_T)^{1/2}$)
1	136,941	2,986	2·18	4·03
2	130,067	2,190	1·68	3·04
3	137,435	2,369	1·72	3·2
4	100,732	2,185	2·18	3·43
5	97,258	1,864	1·93	3·0
6	94,739	1,761	1·86	2·8
7	81,576	185	0·22	0·33‡

* The average number of photoelectrons emitted in a 20 min. interval was of the order of 10^8.

† In deriving the root-mean-square uncertainty in the number of random coincidences it was assumed that they obeyed a Poisson distribution. The root-mean-square uncertainty for any one run is twice the square root of the number of coincidences observed in 20 min. since, as explained in the text, each run consisted of four 20 min. observation periods.

‡ Run No. 7 was taken with a broad bandwidth light source, a tungsten filament lamp, for which no significant correlation was expected.

smaller than the ideal value to be expected from a lamp of this kind with an atomic temperature of $\sim 300°$ K., but the decrease is caused by self-absorption at the centre of the line.

In applying equation (4) to the present equipment it is necessary to reduce the theoretical value of ρ by a factor $\Delta(\nu_0)\gamma$; where $\Delta(\nu_0)$ is the 'partial coherence' factor which allows for the fact that the source is of finite angular size and is therefore partially resolved by the individual photocathodes, and γ takes into account the polarization introduced by the semi-transparent mirror and the false counts due to dark current in the phototubes. The value of $\Delta(\nu_0)$ was calculated to be 0·475, and γ was measured to be 0·86.

With the numerical values given above, the theoretical value for ρ is

$$\rho_{\text{theor.}} = \Delta(\nu_0)\gamma\frac{\tau_0}{4\tau_c} = 0\cdot0207 \quad (7)$$

The error in this value is unlikely to exceed $\pm\ 0\cdot002$ and is principally caused by uncertainties in the value of τ_c, which in turn depends, in a complex manner, upon the amplitude distribution of the pulses from the phototubes.

A comparison of equations (3) and (7) shows that the experimental and theoretical values for the fraction of photoelectrons which are correlated and emitted in time coincidence are in satisfactory agreement, the discrepancy being less than the probable error in the measurement or the uncertainty in the theoretical value. Thus the present experiment confirms the conclusions drawn from the earlier test[1], and shows in a very direct manner that the arrival times of photons at different points are correlated when these points are illuminated by coherent beams of light.

We should like to thank Prof. R. E. Bell for very helpful advice on the operating conditions of the coincidence circuit, and the Division of Electrotechnology of the Commonwealth Scientific and Industrial Research Organization for the use of its high-speed counter. A detailed discussion of this and allied experiments will be submitted for publication in the *Australian Journal of Physics*.

[1] Hanbury Brown, R., and Twiss, R. Q., *Nature*, **177**, 27 (1956).

[2] Brannen, E., and Ferguson, H. I. S., *Nature*, **178**, 481 (1956).

[3] Purcell, E. M., *Nature*, **178**, 1449 (1956).

[4] Hanbury Brown, R., and Twiss, R. Q., *Nature*, **178**, 1447 (1956).

[5] Adám, A., Jánossy, L., and Varga, P., *Acta Hungarica*, **4**, 301 (1955).

[6] Hanbury Brown, R., and Twiss, R. Q., *Nature*, **178**, 1046 (1956).

[7] Bell, R. E., Graham, R. L., and Petch, H. E., *Canad. J. Phys.*, **30**, 35 (1952).

PAPER NO. 27

Reprinted from *Journal of the Optical Society of America*, Vol. 48, pp. 136–137, 1958.

Transformation of Intensity Matrix by the Transmission of a Pupil

HIDEYA GAMŌ
Department of Physics, University of Tokyo, Tokyo, Japan
(Received August 15, 1957)

RECENTLY in the course of researches concerning the information theory in optics, a positive-definite Hermitian matrix which can be used to describe the intensity distribution of images, or the intensity matrix, has been derived independently by Dr. D. Gabor[1] and the writer.[2] I should like to communicate one of the results on the intensity matrix which has been obtained by the writer. Assuming that the secondary image is obtained by transmitting through a given pupil the waves of an image obtained by another pupil as illustrated in Fig. 1, the problem to be discussed here is how the intensity matrix for the secondary image can be derived from the one of the primary images. For the sake of simplicity we shall take merely the one-dimensional image-forming system where the magnification is unity.

In order to facilitate the following discussion, first let us summarize briefly the definition of intensity matrix and its relation to the intensity distribution of an image, and omit its derivation which is stated elsewhere.[3] The intensity distribution of an image obtained by a pupil of numerical aperture α can be described as the following Hermitian form, that is,

$$I(x;\alpha)=[\phi(x;\alpha), A\phi(x;\alpha)]. \qquad (1)$$

Here A is the intensity matrix having n, m element

$$A_{nm}=\int\int\Gamma(X_1,X_2)E^*(X_1)E(X_2)u^*\times(X_1-n\pi/k\alpha)u(X_2-m\pi/k\alpha)dX_1dX_2, \qquad (2)$$

where $\Gamma(X_1, X_2)$ is the mutual intensity of illumination, or the phase-coherence factor multiplied by the amplitudes of waves at points X_1, X_2 in the object plane; $E(X)$ is the complex transmission coefficient of an object; $u(X-n\pi/k\alpha)$ is the transmission function of a given pupil, that is, the complex amplitude of waves at the nth sampling point $n\pi/k\alpha$ in the image plane, transmitted from the wave having unit amplitude at a point X and vanishing elsewhere in the object plane; $k=2\pi/\lambda$; and λ is the wavelength. $\phi(x;\alpha)$ in Eq. (1) is a vector whose nth component is given by the nth sampling function[4] for an image by a pupil of numerical aperture α

$$\phi_n(x;\alpha)=\frac{\sin(k\alpha x-n\pi)}{(k\alpha x-n\pi)}. \qquad (3)$$

We will now consider the intensity matrix of the image obtained by the combined system of two pupils mentioned above. Its matrix element is expressed in the same form as Eq. (2), except that the transmission function of the first pupil is replaced by the one of the combined system and the numerical aperture α is replaced by β of the second pupil. The transmission function $\bar{u}$ of the combined system, of course, is given by the convolution integral of transmission functions u and u' of the first and second pupils, respectively; that is,

$$\bar{u}(X-n\pi/k\beta)=\int_{-\infty}^{+\infty}u(X-\xi)u'(\xi-n\pi/k\beta)d\xi. \qquad (4)$$

Since the Fourier transform of the transmission function $u(x-\xi)$,

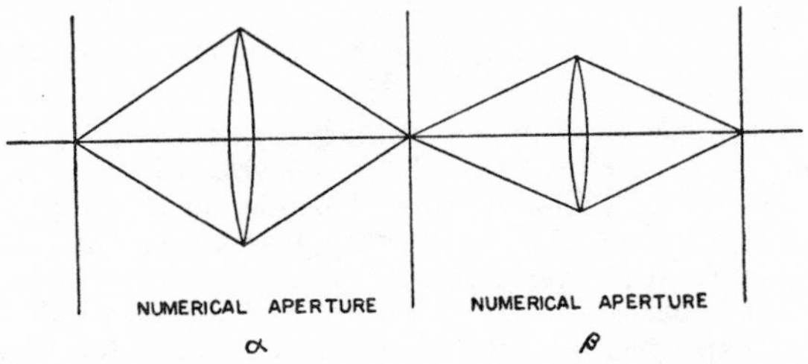

FIG. 1. Image waves obtained from one pupil are transmitted through a second pupil for the formation of a secondary image.

or the pupil function of the first pupil, vanishes outside the range 2α, the transmission function u can be expanded as a series by means of the sampling theorem[4] for the image by a pupil of numerical aperture α; namely,

$$u(X-\xi)=\sum_{n=-\infty}^{+\infty}u(X-n\pi/k\alpha)\frac{\pi}{k\alpha}u(\xi-n\pi/k\alpha). \qquad (5)$$

Here $(\pi/k\alpha)u(\xi-n\pi/k\alpha)$ is the nth sampling function given by Eq. (3). Putting the above equation into (4), the transmission function of the combined system $\bar{u}$ can be expressed as

$$\bar{u}(X-n\pi/k\beta)=\sum_{n'=-\infty}^{+\infty}u(X-n'\pi/k\alpha)T_{nn'}, \qquad (6)$$

where

$$T_{nn'}=\frac{\pi}{k\alpha}\int_{-\infty}^{+\infty}u(\xi-n\pi/k\alpha)u'(\xi-n'\pi/k\beta)d\xi. \qquad (7)$$

Inserting Eqs. (6), and (7) into the matrix element B_{nm} of the combined system and considering that the n, m element of the intensity matrix for the primary image is given by Eq. (2), we obtain finally

$$B_{nm}=\Sigma_k\ \Sigma_l\ T_{kn}{}^*A_{kl}T_{lm} \qquad (8)$$

According to the matrix representation it can be expressed as

$$B=T'^*AT, \qquad (9)$$

where T may be termed the "transmission matrix" of the second pupil whose matrix elements are given by Eq. (7), and T'^* is the matrix having n, m element equal to the complex conjugate of the m, n element of the matrix T, or the conjugate transposed matrix of T. The intensity matrix for the primary image, therefore, is transformed into another matrix for the secondary image according to the rule above mentioned.

The element of the transmission matrix mentioned above can easily be obtained, because in general the integral (7) is related closely to the sampling coefficient, which can be derived by using the well-known orthogonal relations[5] of sampling functions. When the numerical aperture β of the second pupil is smaller than the one of the first pupil α, the n, n' element of the transmission matrix $T_{nn'}$ can be given by the above-mentioned sampling coefficient at point $n\pi/k\alpha$ of the transmission function $u'(x-n'\pi/k\beta)$; that is,

$$T_{nn'}=(\pi/k\alpha)u'(n\pi/k\alpha-n'\pi/k\beta)\cdot(\alpha\geqslant\beta). \tag{10}$$

Accordingly, the $n-n'$ element of the above transmission matrix will represent how much complex amplitude of waves can be produced at the n'th sampling point of the secondary image plane by transmitting through a given pupil the wave having unit amplitude at the nth sampling point and vanishing elsewhere in the primary image plane. The transmission matrix for a given pupil without both aberration and defect of focusing is obtained as a special case of Eq. (10); that is,

$$S_{nm}{}^{(1)}=(\beta/\alpha)\sin(n\beta/\alpha-n')\pi/(n\beta/\alpha-n')\pi\cdot(\alpha\geqslant\beta). \tag{11}$$

On the other hand, when the numerical aperture α of the first pupil is smaller than the one of the second pupil β, the process of its transmission can be divided into two parts for the sake of convenience. The first process is that the intensity matrix for the primary image is rewritten by means of the sampling theorem for a pupil of numerical aperture β instead of α, and it is described by a transmission matrix $S^{(2)}$ having n, m element

$$S_{nm}{}^{(2)}=\sin(m\alpha/\beta-n)\pi/(m\alpha/\beta-n)\pi, \tag{12}$$

which is also derived by using the definition of sampling coefficient. The second process is that the intensity matrix newly expressed is transformed by the transmission of a given pupil of numerical aperture β, and it is described by the transmission matrix T which is the special case $\alpha=\beta$ of Eq. (10). Consequently, the transmission matrix in question can be given by the multiplication of these matrices $S^{(2)}$ and T.

Let us now consider how the intensity distribution of a secondary image mentioned above can be described by using the transmission matrix obtained above and the intensity matrix for the primary image. By inserting the matrix B of Eq. (9) in place of A and putting β in place of α in Eq. (1), the intensity distribution of a secondary image is given by

$$I(x;\beta)=[\phi(x;\beta),\ T'^{*}AT\phi(X;\beta)]. \tag{13}$$

Because of the property of Hermitian forms, it can be expressed as

$$I(x;\beta)=[\psi(x;\alpha),\ A\psi(x;\alpha)], \tag{14}$$

where

$$\psi(x;\alpha)=T\phi(x;\beta). \tag{15}$$

Thus, another picture of the optical transmission through a pupil is obtained; namely, that the intensity matrix A is invariant by the optical transmission, and the vector $\phi(x;\alpha)$ for the primary image, on the other hand, is replaced by another vector ψ given by Eq. (15). According to Eqs. (11) and (12), the vectors $\phi(x;\alpha)$ and $\phi(x;\beta)$ can be connected as follows; namely,

$$\begin{aligned}\phi(x;\beta)&=S^{(1)}\phi(x;\alpha), \quad \text{for} \quad \alpha\leqslant\beta,\\ &\quad S^{(2)}\phi(x;\alpha), \quad \text{for} \quad \alpha>\beta.\end{aligned} \tag{16}$$

Consequently, when preferring the latter picture, that the intensity matrix for the primary image is invariant by the optical transmission of a given pupil, we can consider that the vector $\phi(x;\alpha)$ for the primary image is transformed into another vector for the secondary image according to the following rule:

$$\psi(x;\alpha)=TS^{(i)}\phi(x;\alpha).\ \ i=1, \text{ or } 2 \text{ for } \alpha\leqslant\beta \text{ or } \alpha>\beta. \tag{17}$$

Now that elements of the transmission matrix mentioned above are, in general, closely related to the transmission function of a given pupil as in Eq. (7), the transmission matrix may be used as a quantity representing the performance of the image-forming system. The full paper will be published elsewhere.[6]

[1] D. Gabor, *Information Theory*, edited by C. Cherry (Butterworths Scientific Publications, London, 1956), pp. 26–33; D. Gabor, *Astronomical Optics*, edited by Z. Kopal (North-Holland Publishing Company, Amsterdam, 1956), pp. 17–30.

[2] Hideya Gamo, Kagaku (Tokyo) **26**, 470–471 (1956); J. Appl. Phys. (Japan) **25**, 431–443 (1956), No. **11**, Optics Issue.

[3] Reference 2 and Hideya Gamo, J. Opt. Soc. Am. **47**, 976 (1957).

[4] A. Blanc-Lapierre, Ann. inst. Henri Poincaré **13**, 283 (1953); Toraldo di Francia, J. Opt. Soc. Am. **45**, 497 (1955).

[5] C. E. Shannon, Proc. Inst. Radio Engrs. **37**, 10–21 (1949); S. Goldman, *Information Theory* (Prentice-Hall, Inc., Englewood Cliffs, New Jersey, 1953), p. 80.

[6] Hideya Gamo, J. Appl. Phys. (Japan) **26**, 414–421 (1957); J. Phys. Soc. Japan (to be published).

PAPER NO. 28

Reprinted from *Proceedings of the Physical Society*, Vol. 72, pp. 1037–1048, 1958.

Fluctuations of Photon Beams and their Correlations

By L. MANDEL

Department of Physics, Instrument Technology, Imperial College, London

MS. received 3rd March 1958, *and in revised form 28th April* 1958

Abstract. The distribution of counts from a photoelectric detector illuminated by light of bandwidth $\Delta\nu_0$ is analysed by associating the photons with Gaussian random waves. This is shown to lead to a full statistical description of the counts. It is shown that the number n_T in a time interval $T \ll 1/\Delta\nu_0$ obeys pure Bose–Einstein statistics, and that the fluctuations in longer intervals $T \gg 1/\Delta\nu_0$ are simply the density fluctuations of a boson assembly in a phase space of $\sim \Delta\nu_0 T$ cells. The correlation coefficient ρ of the fluctuations of counts from two detectors illuminated by partially coherent beams is found to be proportional to the local time average of the square of the coherence function $\langle\gamma_{12}{}^2\rangle$. The correlation is shown to depend on the degeneracy of the beams in such a way that $\rho \to 2\langle\gamma_{12}{}^2\rangle$ for highly degenerate beams. The results are all consistent with those obtained by Hanbury Brown and Twiss in 1957.

§ 1. Introduction

Hanbury Brown and Twiss (1956, 1958) and Twiss, Little and Hanbury Brown (1957) have described experiments in which they detected correlation between the arrival times of photons in two coherent light beams, but not in incoherent beams. It was first shown by Purcell (1956) that such correlation could be understood in terms of the non-classical fluctuations of photons.

The fluctuations in light beams and the resulting correlations have since been discussed by several authors. Jánossy (1957) and Wolf (1957) have examined the problem classically in terms of waves, while Hanbury Brown and Twiss (1957) have shown that identical results follow both from a classical analysis and a quantum analysis in terms of photons.

Here it is proposed to discuss the problem from the point of view of Purcell, in terms of the number of photons arriving in a certain time interval and to show that expressions for the variances and correlations are simply derivable by considering a photon beam as statistically associated with Gaussian random waves. This assumption, which is also implicit in Purcell's discussion and supported by the results of Hanbury Brown and Twiss (1957), is sufficient for a full statistical description of the fluctuations. It will be shown that it is consistent with the uncertainty principle, that it leads to the Bose–Einstein statistics in the appropriate case and that the fluctuations in short and long time intervals are the density fluctuations of an assembly of bosons in one and several cells of phase space respectively (Fürth 1928 a, b). The correlation between fluctuations in partially coherent beams will be seen to depend on the degeneracy of the beams, so that the effect should basically be regarded as a 'wave effect', as has already been suggested by Hanbury Brown and Twiss (1957).

The results obtained are consistent with those of Purcell (1956) and Hanbury Brown and Twiss (1957) but not with those of Fellgett (1949) and Clark Jones (1953).

§ 2. Fluctuations in an Homogeneous Beam

Consider a beam of light falling on some photoelectric detector, where n_T photoelectrons are ejected in a certain time interval T. Only the photoelectrons and not the photons are, of course, observable and our discussion must therefore be confined to the statistical behaviour of the photoelectrons. Although it is tempting to associate the ejection of a photoelectron with the arrival of a photon, this picture becomes inadmissible by the uncertainty principle for time intervals shorter than the reciprocal frequency spread of the light. We shall suppose that the light comes from a Gaussian random source (Jánossy 1957) emitting a narrow spectral line centred on the frequency ν_0, and that the line shape is describable by the normalized spectral density $\phi(\nu)$ with

$$\int_{-\infty}^{\infty} \phi(\nu)\,d\nu = 1 \text{ and } \phi(\nu)=\phi(-\nu).$$

While the shape of $\phi(\nu)$ is arbitrary we shall assume that its effective width $\Delta\nu_0$ defined by

$$\Delta\nu_0{}^2 = 2\int_0^{\infty} (\nu-\nu_0)^2\,\phi(\nu)\,d\nu$$

is small compared with ν_0. We shall suppose for the moment that the light is plane polarized, and denote the instantaneous amplitude by $y(t)$ and the corresponding intensity, i.e. the square of $y(t)$ averaged over a few cycles, by $P(t)$. Thus

$$P(t) = \frac{\nu_0}{r}\int_{t-r/2\nu_0}^{t+r/2\nu_0} Q(t')\,dt', \qquad \ldots\ldots(1)$$

where $Q(t)=y^2(t)$ and where r is a small integer. We shall denote this form of average as a 'local' average and write

$$\frac{\nu_0}{r}\int_{t-r/2\nu_0}^{t+r/2\nu_0} Q(t')\,dt' = \langle Q(t)\rangle.$$

Thus, from (1), $P(t)=\langle Q(t)\rangle$.

The restriction to a 'homogeneous' beam ensures that there are no large phase differences between different elements of the beam. More specifically, it is assumed that the phase difference between any two elements is much less than $2\pi\nu_0/\Delta\nu_0$, so that their intensity cross-correlation function decreases with increasing delay.

We shall now associate photons with the Gaussian random wave $y(t)$, by defining a probability that a photoelectron is ejected in a short time interval between t and $t+dt$. If we consider first order transitions only significant, in which one photon gives rise to one photoelectron, then this probability will be given by $\alpha P(t)\,dt$, where α is the quantum sensitivity of the photoelectric detector, assumed constant over the narrow frequency range $\Delta\nu_0$. The observable $P(t)$ provides the only link between the wave and the particle descriptions of the beam.

The fluctuations of the number of particles n_T therefore have two causes. There are first of all the fluctuations of the wave intensity $P(t)$, determined by the spectral line shape and there is the stochastic association of particles with the wave intensity. This two-fold source of the fluctuations results in the departure from classical statistics, as we shall show.

Let $p_n(t, T)$ denote the probability that n photoelectrons are ejected in the interval between t and $t+T$. This probability is therefore itself a stochastic function of time. In particular, from the definition,

$$p_1(t, dt) = \alpha P(t)\, dt \qquad \ldots\ldots(2)$$

and the expectation value of n in the interval t to $t+T$ will be

$$\alpha \int_t^{t+T} P(t')\, dt'.$$

The condition (2) leads from first principles in the usual way to the Poisson distribution in n:

$$p_n(t, T) = \frac{1}{n!}\left[\alpha \int_t^{t+T} P(t')\, dt'\right]^n \exp\left[-\alpha \int_t^{t+T} P(t')\, dt'\right]. \qquad \ldots\ldots(3)$$

$p_n(t, T)$ is not, however, a distribution that can be found experimentally. For, when $P(t)$ fluctuates at random, the given interval t to $t+T$ is unique and only the ensemble averages, which are equal to the time averages for a stationary process, are observable. But the operation of averaging $p_n(t, T)$ over time when $P(t)$ is fluctuating will not, in general, result in another Poisson distribution. This departure of the observed distribution from the classical form can therefore be seen to be a general consequence of the association of particles with fluctuating waves. We may say that the fluctuations of the waves lead to the non-classical fluctuations of the quanta.

The limiting case $T \to \infty$ is exceptional, for, in that case,

$$\int_t^{t+T} P(t')\, dt'$$

is practically independent of t. The operation of averaging over time does not therefore substantially alter $p_n(t, T)$, which remains Poissonian.

In order to determine the mean number of counts in a fixed interval T we have to use equation (3) and average over n and t. If the time average is denoted by a bar we find:

$$\overline{n_T} = \sum_{n=0}^{\infty} \overline{n p_n(t, T)}$$

and, from the properties of the Poisson distribution,

$$= \overline{\alpha \int_t^{t+T} P(t')\, dt'}$$

$$= \alpha \bar{P} T. \qquad \ldots\ldots(4)$$

Also

$$\overline{n_T{}^2} = \sum_{n=0}^{\infty} \overline{n^2 p_n(t, T)}$$

and, again using the well-known properties of the Poisson distribution,

$$= \overline{\alpha \int_t^{t+T} P(t')\, dt'} + \overline{\left[\alpha \int_t^{t+T} P(t')\, dt'\right]^2}$$

$$= \overline{n_T} + \alpha^2 \int_0^T \int_0^T R_p(y-x)\, dy\, dx,$$

where $R_P(\tau)$ is the autocorrelation function of $P(t)$.

Instead of integrating over the xy plane we can convert the double integral into a single integral by putting $\tau=y-x$ (e.g. see Rice 1945). Thus:

$$\overline{n_T^2}=\overline{n_T}+2\alpha^2\int_0^T (T-\tau)\,R_P(\tau)\,d\tau, \qquad \ldots\ldots(5)$$

where we have made use of the symmetry of $R_P(\tau)$. Before proceeding further we shall examine the relation between $R_P(\tau)$ and the normalized spectral density $\phi(\nu)$ of the light.

If $R_y(\tau)$ is the autocorrelation function of $y(t)$ and $R_y(\tau)/\overline{y^2}=\gamma(\tau)$, then it is well known (e.g. Rice 1944) that $\gamma(\tau)$, the normalized autocorrelation function, and $\phi(\nu)$, the normalized spectral density, are related by a Fourier transformation. Thus

$$\gamma(\tau)=\int_{-\infty}^{\infty}\phi(\nu)\exp(2\pi i\nu\tau)\,d\nu \qquad \ldots\ldots(6)$$

in which $\phi(\nu)=\phi(-\nu)$.

We also have the result (Lawson and Uhlenbeck 1950) that, for a Gaussian random process,

$$R_Q(\tau)=\bar{Q}^2[1+2\gamma^2(\tau)], \qquad \ldots\ldots(7)$$

where $R_Q(\tau)$ is the autocorrelation function of $Q(\tau)$. Now, from the definition of $P(t)$,

$$\begin{aligned} R_P(\tau)&=\overline{\langle Q(t+\tau)\rangle\langle Q(t)\rangle}\\ &=\left(\frac{\nu_0}{r}\right)^2\int_{-r/2\nu_0}^{r/2\nu_0}\int_{-r/2\nu_0}^{r/2\nu_0}\overline{Q(t+\tau+x)\,Q(t+y)}\,dx\,dy\\ &=\langle R_Q(\tau)\rangle, \qquad \ldots\ldots(8)\end{aligned}$$

whence $\quad R_P(\tau)=\bar{P}^2\,[1+2\langle\gamma^2(\tau)\rangle]$, since $\bar{Q}=\bar{P}$. $\qquad \ldots\ldots(9)$

It is shown in the Appendix that $\langle\gamma^2(\tau)\rangle$ is a slowly varying function of τ, which does not change much in an interval short compared with $1/\Delta\nu_0$, i.e. short compared with the characteristic coherence time of the light (e.g. Forrester 1956). In particular, $\langle\gamma^2(0)\rangle=\frac{1}{2}$. $R_P(\tau)$ is therefore also a slowly varying function and, like $\langle\gamma^2(\tau)\rangle$, it is appreciably different from zero only in a range of a few times $1/\Delta\nu_0$. We shall make use of these properties of $\langle\gamma^2(\tau)\rangle$ in evaluating the mean square fluctuations of n_T.

§ 3. The Mean Squared Variation of n_T

On introducing (9) into (5) we obtain

$$\begin{aligned}\overline{\Delta n_T^2}&=\overline{n_T^2}-\overline{n_T}^2\\ &=\overline{n_T}+4\alpha^2\bar{P}^2\int_0^T (T-\tau)\langle\gamma^2(\tau)\rangle\,d\tau. \qquad \ldots\ldots(10)\end{aligned}$$

Since the integrand $(T-\tau)\langle\gamma^2(\tau)\rangle$ is never negative, it follows at once that the fluctuations of n_T are greater than predicted by the classical particle statistics. This result, which is of course characteristic of the so-called bunching of bosons, is here seen to follow directly from associating photons with Gaussian random waves.

If we denote the integral in (10) by $\frac{1}{4}T\xi$, where ξ has the dimension of time, we can write

$$\overline{\Delta n_T^2}=\overline{n_T}\{1+\overline{n_T}(\xi/T)\}. \qquad \ldots\ldots(11)$$

Since $\langle\gamma^2(\tau)\rangle\leqslant\frac{1}{2}$ as shown in the Appendix, it follows from the definition of ξ that $\xi\leqslant T$ and

$$\overline{\Delta n_T^2}\leqslant\overline{n_T}\{1+\overline{n_T}\}.$$

The relation (11) holds generally but, by making use of the properties of $\langle\gamma^2(\tau)\rangle$ derived in the Appendix, we can immediately evaluate ξ in two limiting cases.

Case (a): $\Delta\nu_0 T \ll 1$.

This condition would be extremely difficult to satisfy experimentally. Thus, the narrowest line width obtainable from a ^{198}Hg discharge corresponds to a $\Delta\nu_0$ of the order of 6×10^8 c/s (von Klüber 1958), so that, even with this source, we should be dealing with time intervals T less than about 10^{-9} sec. Nevertheless, the result for this case is interesting.

Since $\langle\gamma^2(\tau)\rangle$ does not depart much from $\frac{1}{2}$ in an interval $T \ll 1/\Delta\nu_0$,

$$\tfrac{1}{4}T\xi \simeq \int_0^T \tfrac{1}{2}(T-\tau)\,d\tau = \tfrac{1}{4}T^2,$$

so that $\xi = T$ and

$$\overline{\Delta n_T^{\,2}} = \overline{n_T}\{1+\overline{n_T}\}. \qquad \ldots\ldots(12)$$

This is the well-known formula for the fluctuations of the occupation numbers of a single cell in phase space for an assembly of bosons. The photoelectrons in the interval T therefore obey 'pure' Bose–Einstein statistics. The reason for this can be seen at once if we examine the size of the elementary cell in phase space. In the direction of the beam this extends over a distance $\mathbf{c}\Delta t$, where $\Delta t \sim 1/\Delta\nu_0$. Thus, the photons in an interval $T \ll 1/\Delta\nu_0$ as above, i.e. much shorter than the so-called coherence time of the light (e.g. Forrester 1956), occupy the same cell in phase space. By the uncertainty principle they are therefore intrinsically indistinguishable and n_T obeys pure Bose–Einstein statistics. We shall see in the next section that, when $\Delta\nu_0 T \ll 1$, the complete probability distribution follows very simply from equation (3).

The departure from the classical particle statistics is expressible by the 'degeneracy' factor $1+\overline{n_T}$ which is a measure of the extent to which photons share the same cell in phase space. The excess fluctuations correspond to what Hanbury Brown and Twiss (1957) called the wave interaction noise, although these authors did not consider the case $\Delta\nu_0 T \ll 1$. The degeneracy is also indicative of whether the wave or the particle properties of the beam predominate. In the visible region of the spectrum $\overline{n_T}$, and therefore the degeneracy, are normally small and the non-classical behaviour of a single photon beam is difficult to observe. Finally we note from (12) that the percentage fluctuation defined by

$$(\overline{\Delta n_T^{\,2}})^{1/2}/\overline{n_T} = (1+1/\overline{n_T})^{1/2}$$

is always greater than one, even at high intensities. This feature is again characteristic of a boson assembly and quite different from the behaviour of classical particles, for which the percentage fluctuation tends to zero at high intensities.

Case (b): $\Delta\nu_0 T \gg 1$.

This is the condition assumed to hold in the analyses of both Purcell (1956) and Hanbury Brown and Twiss (1957). It is of course almost invariably satisfied in experiments as shown above, but corresponds to a slightly more complicated situation.

Since $\langle\gamma^2(\tau)\rangle$ extends appreciably only over an interval of the order a few times $1/\Delta\nu_0$, the integral in (10) can be written

$$\tfrac{1}{4}T\xi \simeq \int_0^\infty T\langle\gamma^2(\tau)\rangle\,d\tau = \tfrac{1}{4}T\int_{-\infty}^{\infty} 2\langle\gamma^2(\tau)\rangle\,d\tau.$$

Although the exact value of ξ therefore depends on the shape of the spectral line, we can see that it will be of the order $1/\Delta\nu_0$, since $\langle\gamma^2(\tau)\rangle$ changes only slowly and $\langle\gamma^2(0)\rangle = \frac{1}{2}$. We can therefore write

$$\xi = \kappa/\Delta\nu_0$$

where κ is a number depending on the spectral density $\phi(\nu)$ but of the order 1. Equation (11) therefore becomes

$$\overline{\Delta n_T^2} = \overline{n_T}\{1 + \kappa\overline{n_T}/\Delta\nu_0 T\}. \qquad \ldots\ldots(13)$$

This is the relation first obtained by Purcell (1956). It can also be seen to be equivalent to that of Hanbury Brown and Twiss (1957), when the integral over time defining ξ is converted to one over frequency.

Since $\gamma(\tau)$ and $\phi(\nu)$ are a Fourier transform pair, we have from Parseval's theorem

$$\int_{-\infty}^{\infty} \gamma^2(\tau)\,d\tau = \int_{-\infty}^{\infty} \phi^2(\nu)\,d\nu,$$

so that

$$\xi = 4\int_0^{\infty} \phi^2(\nu)\,d\nu = \kappa/\Delta\nu_0.$$

From (13) we note that the degeneracy factor is smaller than before for the same $\overline{n_T}$. The reason for this can again be seen if we remember that we are now dealing with a volume of phase space containing roughly $\Delta\nu_0 T = s$ cells.† The mean number of photons per cell is therefore $\overline{n_T}/s = \overline{m}$ and (13) can be written

$$\overline{\Delta n_T^2} = \overline{n_T}(1 + \kappa\,\overline{m}) = \overline{n_T}(1 + \kappa s\,\overline{m}^2/\overline{n_T}).$$

This is the expression in the conventional form for the density fluctuations in a larger volume of phase space (Fürth, 1928 a, b). The degeneracy is less because we are dealing with a time interval in which not all the photons are intrinsically indistinguishable. The departure from the classical statistics is again due to those photons which share one cell and therefore would become important only at very great intensities.

The excess fluctuations given by (13) correspond to the wave interaction noise of Hanbury Brown and Twiss (1957). The results do not agree with those of Fellgett (1949, 1957) and Clark Jones (1953) obtained from thermodynamic arguments, who quote larger values. A likely reason for the discrepancy has already been given by Hanbury Brown and Twiss, namely that these authors equate the fluctuations of the detector output to the energy fluctuations of the light, whereas the two are stochastically connected. In any case, the formula of Fellgett is not consistent with that of a boson assembly in a phase space of $\Delta\nu_0 T$ cells.

It is interesting to note that the percentage fluctuation

$$(\overline{\Delta n_T^2})^{1/2}/\overline{n_T} = (1/\overline{n_T} + \kappa/\Delta\nu_0 T)^{1/2}$$

could be either greater or less than one, depending on $\overline{n_T}$, since n_T no longer obeys the pure Bose–Einstein statistics.

† By restricting the discussion to what we have called homogeneous beams, we ensure that the spacial extent of the beam over the photo-detector does not include more than one cell.

§ 4. The Distribution Function of n_T when $\Delta\nu_0 T \ll 1$

So far we have been concerned only with the variance of n_T, but, when $\Delta\nu_0 T \ll 1$, it is not difficult to derive the full distribution. This will be given by the time or ensemble average of $p_n(t, T)$ in equation (3). Now, since $P(t)$ does not vary much in an interval $T \ll 1/\Delta\nu_0$, it follows that

$$\alpha \int_t^{t+T} P(t')\,dt' \simeq \alpha P(t) T.$$

We therefore find from (3),

$$\overline{p_n(t, T)} = \frac{1}{n!}\overline{(\alpha PT)^n \exp(-\alpha PT)}. \qquad \ldots\ldots(14)$$

If the probability distribution of P is known, we can evaluate $\overline{p_n(t, T)}$ by averaging over the ensemble. Since $y(t)$ is a narrow band Gaussian random variable, the local average

$$\langle y^2(t)\rangle = P(t) = \tfrac{1}{2}W^2(t),$$

where $W(t)$ is the envelope of $y(t)$. Now the probability density of $W(t)$ has been shown by Rice (1944) to be of the form:

$$(W/\bar{P})\exp(-W^2/2\bar{P}).$$

Hence, by transforming from W to P, we arrive at the probability density $p'(P)$ of P. Thus:

$$p'(P)\,dP = (1/\bar{P})\exp(-P/\bar{P})\,dP.$$

We can now evaluate (14) by averaging over the ensemble and we obtain

$$\overline{p_n(t, T)} = \frac{1}{\bar{P}n!}\int_0^\infty (\alpha PT)\exp(-\alpha PT - P/\bar{P})\,dP.$$

The integral is the well-known factorial function integral and leads to

$$\overline{p_n(t, T)} = \{(1+\alpha\bar{P}T)(1+1/\alpha\bar{P}T)^n\}^{-1}.$$

Since $\alpha\bar{P}T = \overline{n_T}$ from (4), we can write this as

$$\overline{p_n(t, T)} = (1-w)w^n, \qquad \ldots\ldots(15)$$

where $w = \{1+1/\overline{n_T}\}^{-1}$. This is the Bose–Einstein distribution function in standard form for the numbers in a single cell of phase space. The distribution can be seen to arise naturally from the association of photons with Gaussian random waves, when $\Delta\nu_0 T \ll 1$.

§ 5. Correlation between Fluctuations in Two Beams

Following the experimental work of Hanbury Brown and Twiss (1956, 1958) and Twiss, Little and Hanbury Brown (1957), the correlation between the fluctuations of two at least partially coherent beams has been studied theoretically by Purcell (1956), Wolf (1957), Jánossy (1957) and Hanbury Brown and Twiss (1957, 1958). The latter authors, in particular, have examined the problem in some detail and shown that the correlation between two beams and the 'excess' photon fluctuations of a single beam are very closely related.

Here it is proposed to show that similar relations are derivable very simply from equation (3) and that the correlation coefficient is very small unless the beams are substantially degenerate. We shall also find that, in sufficiently short time intervals, the correlation is independent of the spectral distribution.

We shall suppose that the two beams have the same spectral density, but not necessarily equal intensities $\overline{P_1}$ and $\overline{P_2}$ and that the degree of coherence is describable by the cross-correlation function $J_{12}(\tau)=\overline{y_1(t+\tau)y_2(t)}$ used by Wolf (1955). The normalized function $J_{12}(\tau)/(\overline{P_1}\,\overline{P_2})^{1/2}$ will be denoted by $\gamma_{12}(\tau)$. Wolf (1955) has shown that $\gamma_{12}(\tau)$ is an observable which is related to the visibility of the fringe system obtained from the superposition of the two beams. In the limit as the beams tend to complete coherence, $\gamma_{12}(\tau)\to\gamma(\tau)$.

From equation (3) we find

$$\overline{n_1 n_2}=\sum_{n_1=0}^{\infty}\sum_{n_2=0}^{\infty}\overline{n_1 n_2 p_{n_1}(t,T)p_{n_2}(t,T)} \qquad \ldots\ldots(16)$$

which reduces to

$$\alpha^2\int_0^T\int_0^T R_{P_1P_2}(y-x)\,dy\,dx, \qquad \ldots\ldots(17)$$

where $R_{P_1P_2}(\tau)$ is the cross-correlation function of the two intensities.

By using the result of Wolf (1957) that

$$R_{Q_1Q_2}=\overline{Q_1}\,\overline{Q_2}\,[1+2\gamma_{12}^2(\tau)] \qquad \ldots\ldots(18)$$

and taking local averages as before we arrive at

$$R_{P_1P_2}=\overline{P_1}\,\overline{P_2}\,[1+2\langle\gamma_{12}^2(\tau)\rangle], \qquad \ldots\ldots(19)$$

where $\langle\gamma_{12}^2(\tau)\rangle$ is a slowly varying function of τ, as shown in the Appendix.

By substituting in (17) and transforming to a single integral over $\tau=y-x$ as before, we obtain for the cross-correlation function of the fluctuations

$$\overline{\Delta n_1\,\Delta n_2}=4\alpha^2\,\overline{P_1}\,\overline{P_2}\int_0^T(T-\tau)\,\langle\gamma_{12}^2(\tau)\rangle\,d\tau. \qquad \ldots\ldots(20)$$

The value of the integral appears to depend on the detailed form of $\langle\gamma_{12}^2(\tau)\rangle$. We can, however, simplify it appreciably if any path difference between the two partially coherent beams at the two photoelectric detectors is rather less than the coherence length $\mathbf{c}\Delta\nu_0$; in other words, if $\langle\gamma_{12}^2(\tau)\rangle$ is a decreasing function of τ.

Under these conditions P_1 and P_2 will be related by an expression of the form:

$$P_2(t)=aP_2'(t)+bP_1(t),$$

where a and b are positive numbers and $P_2'(t)$ is a function with the same spectral density as $P_1(t)$ but uncorrelated with it. This leads to

$$R_{P_1P_2}(\tau)=\overline{P_1}\,\overline{P_2}+b[R_{P_1}(\tau)-\overline{P_1}^2],$$

so that

$$\langle\gamma_{12}^2(\tau)\rangle=\tfrac{1}{2}\,\frac{R_{P_1P_2}(\tau)-\overline{P_1}\,\overline{P_2}}{\overline{P_1P_2}-\overline{P_1}\,\overline{P_2}}\,\frac{\overline{P_1P_2}-\overline{P_1}\,\overline{P_2}}{\overline{P_1}\,\overline{P_2}}$$

$$=2\langle\gamma^2(\tau)\rangle\langle\gamma_{12}^2(0)\rangle. \qquad \ldots\ldots(21)$$

It follows that, for beams with small path difference, only the zero-time cross-correlation enters into the equations, as has already been pointed out by Wolf (1955). An expression similar to (21) has also been derived by Hanbury Brown and Twiss (1957) from more detailed considerations.

Under these conditions also the integral in equation (20) has the same form as that encountered earlier in equation (10) and we can immediately write down the two limiting solutions.

For $\Delta\nu_0 T \ll 1$,

$$\overline{\Delta n_1 \Delta n_2} = 2\bar{n}_1 \bar{n}_2 \langle \gamma_{12}^2(0) \rangle \qquad \text{......(22)}$$

and for $\Delta\nu_0 T \gg 1$,

$$\overline{\Delta n_1 \Delta n_2} = 2\bar{n}_1 \bar{n}_2 (\kappa/\Delta\nu_0 T) \langle \gamma_{12}^2(0) \rangle . \qquad \text{......(23)}$$

Finally, when the two beams are unpolarized, these correlations are halved. Since the fluctuations in two normal planes of polarization are independent, we may associate uncorrelated numbers of photons n_1' and n_1'' with each polarization, such that $\overline{n_1'} = \overline{n_1''} = \frac{1}{2}\bar{n}_1$ and $\overline{\Delta n_1 \Delta n_2} = 2\overline{\Delta n_1' \Delta n_2'}$.

The result for $\Delta\nu_0 T \gg 1$ is again equivalent to that obtained by Hanbury Brown and Twiss (1957), although expressed in a slightly different form. The case $\Delta\nu_0 T \ll 1$ was not considered by these authors and is unlikely to be of practical importance. But it is significant that the correlation in this case is independent of the spectral line shape. From (22) and (23) it is clear that $\overline{\Delta n_1 \Delta n_2}$ is directly proportional to the degeneracy of the beams, i.e. to the number of photons occupying the same cell in phase space.

§ 6. Superposition of Counts

When the counts n_1 and n_2 of two detectors illuminated by two similar partially coherent plane polarized beams are superposed, we obtained another variate whose degeneracy is in general intermediate between the plane polarized and the unpolarized case.

Thus if

$$n = n_1 + n_2,$$

$$\overline{\Delta n^2} = \overline{\Delta n_1^2} + \overline{\Delta n_2^2} + 2\overline{\Delta n_1 \Delta n_2}$$

and, by using the results of (12), (13), (22) and (23), we obtain

$$\overline{\Delta n^2} = \bar{n} \left\{ 1 + \bar{n} - \frac{2\bar{n}_1 \bar{n}_2}{\bar{n}} [1 - 2\langle \gamma_{12}^2(0) \rangle] \right\} \qquad \text{......(24)}$$

for $\Delta\nu_0 T \ll 1$, or

$$\overline{\Delta n^2} = \bar{n} \left\{ 1 + \bar{n} \left(\frac{\kappa}{\Delta\nu_0 T} \right) - \frac{2\bar{n}_1 \bar{n}_2}{\bar{n}} \left(\frac{\kappa}{\Delta\nu_0 T} \right) [1 - 2\langle \gamma_{12}^2(0) \rangle] \right\} \text{ for } \Delta\nu_0 T \gg 1. \qquad \text{......(25)}$$

Since $\langle \gamma_{12}^2(0) \rangle \leqslant \frac{1}{2}$, we see that the degeneracy of the distribution of n is generally less than that of n_1 and n_2. The difference depends on the normalized coherence factor $\langle \gamma_{12}^2(0) \rangle$. It follows at once that this factor also measures the extent to which the two beams share cells in phase space. In particular, for completely coherent beams, the statistical behaviour of n is identical with that of n_1 and n_2, since all the cells are shared. This has already been pointed out by Hanbury Brown and Twiss (1957).

§ 7. Discussion

It has been shown that a full statistical description of the fluctuations in photon beams is possible by associating the photons with Gaussian random waves, which are describable by their spectral density $\phi(\nu)$. The validity of this approach has been questioned by Fellgett (1957), but it is seen to lead to completely consistent results. While the results are all concerned with the number of counts in a definite time interval T, they can immediately be applied to continuous fluctuation and correlation measurements. For all such measurements are limited by a certain resolving time T, which takes the place of the standard interval.

As shown by equation (15), the number n_T in an interval $T \ll 1/\Delta\nu_0$ obeys the pure Bose–Einstein distribution, as we should expect for a boson assembly in a single cell of phase space. In longer intervals the fluctuations of n_T correspond to the boson density fluctuations in a phase space containing $\sim \Delta\nu_0 T$ cells. From the equations (12), (13), (22) and (23) it is clear that the correlation between fluctuations depends essentially on the non-classical fluctuations of the photons, as has already been shown by Purcell (1956) and Hanbury Brown and Twiss (1957) and on the degree of coherence between the beams. These, in turn, depend on the extent of the cells in phase space in real space and time.

The degree of coherence is therefore derivable from correlation measurements (Hanbury Brown and Twiss 1956) as well as from interference experiments (Wolf 1955), although the former always yield the local average $\langle \gamma_{12}^2(\tau) \rangle$. This is of course to be expected, since $\gamma_{12}(\tau)$ contains phase information about the beams which is incompatible with the detection of single photons. Even with a perfect detector having $\alpha = 100\%$ the counts must not be too closely identified with the wave intensity.

There is another important difference between the time dependent correlation effect and the interference effect. Since the correlation depends essentially on two or more photons sharing cells in phase space, it depends on the degeneracy and therefore, unlike the interference effect, varies with the intensity of the beams. This becomes most obvious if we examine the normalized correlation coefficient

$$\rho = \overline{\Delta n_1 \Delta n_2} / (\overline{\Delta n_1^2}\, \overline{\Delta n_2^2})^{1/2}.$$

We then find

$$\rho = \frac{2\bar{n}_1\bar{n}_2(\kappa/\Delta\nu_0 T)\langle \gamma_{12}^2(0) \rangle}{\{\bar{n}_1\bar{n}_2(1+\bar{n}_1\, \kappa/\Delta\nu_0 T)(1+n_2\, \kappa/\Delta\nu_0 T)\}^{1/2}}$$

if $\Delta\nu_0 T \gg 1$, or

$$\rho = \frac{2\bar{n}_1\bar{n}_2\langle \gamma_{12}^2(0) \rangle}{\{\bar{n}_1\bar{n}_2(1+\bar{n}_1)(1+\bar{n}_2)\}^{1/2}}$$

if $\Delta\nu_0 T \ll 1$.

It can be seen that, although $\rho \propto \langle \gamma_{12}^2(0) \rangle$, it is very much less than $\langle \gamma_{12}^2(0) \rangle$ under conditions of low degeneracy, but tends to $2\langle \gamma_{12}^2(0) \rangle$ at high degeneracy. In particular, for highly degenerate completely coherent beams $\rho = 1$. The correlation is therefore appreciable only when the wave properties, as distinct from the particle properties, of the beam become evident. This confirms the view of Hanbury Brown and Twiss (1957) that the effect should be regarded basically as a wave effect and shows that it will be more difficult to detect in an experiment with light than with radio waves (Hanbury Brown and Twiss 1954). It is a considerable credit to these authors that they were able to detect the effect while working with a degeneracy of the order of a few times 10^{-3}.

Acknowledgment

The author is grateful to Dr. R. Fürth for some valuable discussions.

REFERENCES

CLARK-JONES, R., 1953, *Advances in Electronics* 5 (New York : Academic Press).
FELLGETT, P. B., 1949, *J. Opt. Soc. Amer.*, **39,** 970; 1957, *Nature, Lond.*, **179,** 956.
FORRESTER, A. T., 1956, *Amer. J. Phys.*, **24,** 192.
FÜRTH, R., 1928 a, *Z. Phys.*, **48,** 323; 1928 b, *Ibid.*, **50,** 310.
HANBURY BROWN, R., and TWISS, R. Q., 1954, *Phil. Mag.*, **45,** 663; 1956, *Nature, Lond.*, **177,** 27; 1957, *Proc. Roy. Soc.* A, **242,** 300; 1958, *Ibid.*, **243,** 291.
JÁNOSSY, L., 1957, *Nuovo Cim.*, **6,** 111.
VON KLÜBER, H., 1958, *Nature, Lond.*, **181,** 1007.
LAWSON, J. L., and UHLENBECK, G. E., 1950, *Threshold Signals* (New York: McGraw-Hill).
PURCELL, E. M., 1956, *Nature, Lond.*, **178,** 1449.
RICE, S. O., 1944, *Bell Syst. Tech. J.*, **23,** 282; 1945, *Ibid.*, **24,** 46.
TWISS, R. Q., LITTLE, A. G., and HANBURY BROWN, R., 1957, *Nature, Lond.*, **180,** 324.
WOLF, E., 1955, *Proc. Roy. Soc.* A, **230,** 246; 1957, *Phil. Mag.*, **2,** 351.

APPENDIX

THE BEHAVIOUR OF $\langle\gamma_{12}^2(\tau)\rangle$

If $\phi_{12}(\nu)$ is the normalized cross power spectrum of the two beams, then

$$\gamma_{12}(\tau)=\int_{-\infty}^{\infty}\phi_{12}(\nu)\exp(2\pi i\nu\tau)\,d\nu.$$

Since $\phi_{12}(-\nu)=\phi_{12}^*(\nu)$, this can be written in the form

$$\gamma_{12}(\tau)=\int_0^{\infty}[\phi_{12}(\nu)\exp(2\pi i\nu\tau)+\phi_{12}^*(\nu)\exp(-2\pi i\nu\tau)]\,d\nu.$$

If we now introduce the substitution $\nu=\nu'+\nu_0$ and denote

$$\int_{-\nu_0}^{\infty}\phi_{12}(\nu_0+\nu')\exp(2\pi i\nu'\tau)\;d\nu'$$

by $V(\tau)$, the equation becomes

$$\gamma_{12}(\tau)=V(\tau)\exp(2\pi i\nu_0\tau)+V^*(\tau)\exp(-2\pi i\nu_0\tau). \quad \ldots\ldots(\text{A }1)$$

Now by hypothesis, $\phi_{12}(\nu_0+\nu')$ will be appreciably different from zero only for small ν' i.e. for $|\nu'|\lesssim\Delta\nu_0$. It follows from the integral defining $V(\tau)$ that this function will not change significantly in an interval $\Delta\tau\ll 1/\Delta\nu_0$. Thus $V(\tau)$ is a slowly varying function (i.e. compared with $y(t)$). From (A 1)

$$\gamma_{12}^2(\tau)=V^2(\tau)\exp(4\pi i\nu_0\tau)+V^{*2}(\tau)\exp(-4\pi i\nu_0\tau)+2\,|V(\tau)|^2. \quad \ldots\ldots(\text{A }2)$$

If we now calculate the local average, it follows at once from the properties of $V(\tau)$ that

$$\langle\gamma_{12}^2(\tau)\rangle=2\,|V(\tau)|^2, \quad \ldots\ldots(\text{A }3)$$

so that $\langle\gamma_{12}^2(\tau)\rangle$ is also a slowly varying function of τ. In particular, for $\tau\ll 1/\Delta\nu_0$,

$$\langle\gamma_{12}^2(\tau)\rangle=\langle\gamma_{12}^2(0)\rangle.$$

If the two beams are completely coherent so that $\phi_{12}(\nu)=\phi(\nu)$,

$$V(0)=\int_0^{\infty}\phi(\nu)\,d\nu=\tfrac{1}{2}.$$

Hence

$$\langle\gamma^2(0)\rangle=\tfrac{1}{2}. \quad \ldots\ldots(\text{A }4)$$

Further, from the integral defining $V(\tau)$, we see that, for $\tau\gg 1/\Delta\nu_0$, the integrand contains a rapidly oscillating factor $\exp(2\pi i\nu'\tau)$. It follows that $V(\tau)$—and therefore $\langle\gamma_{12}^2(\tau)\rangle$—is small for $\tau\gg 1/\Delta\nu_0$.

Finally from

$$V(\tau)=\exp(-2\pi i\nu_0\tau)\int_0^\infty \phi_{12}(\nu)\exp(2\pi i\nu\tau)\,d\nu$$

we have,

$$|V(\tau)|=\left|\int_0^\infty \phi_{12}(\nu)\exp(2\pi i\nu\tau)\,d\nu\right| \leqslant \int_0^\infty |\phi_{12}(\nu)|\,d\nu. \quad \ldots\ldots(A\,5)$$

Hence, using (A 3) and (A 5) we see that, for a single coherent beam,

$$\langle\gamma^2(\tau)\rangle \leqslant \langle\gamma^2(0)\rangle = \tfrac{1}{2}.$$

PAPER NO. 29

Reprinted from *Proceedings of the Physical Society*, Vol. 74, pp. 233–243, 1959.

Fluctuations of Photon Beams: The Distribution of the Photo-Electrons

By L. MANDEL

Department of Physics, Instrument Technology, Imperial College, London, S.W.7

MS. received 16*th October* 1958, *in revised form* 1*st April* 1959

Abstract. The probability distribution $p(n, T)$ of the number of counts n from a photoelectric detector illuminated by coherent light for a time T is studied, by associating photons stochastically with Gaussian random waves. The cumulants of the distribution are derived and it is shown to be of the expected form for a boson assembly in a limited volume of phase space. The distribution depends strongly on the degeneracy of the light beam. It approaches the Poisson form for classical particles at low degeneracies and the distribution characteristic of classical waves at high degeneracies. The analysis leads, incidentally, to an expression for the extent of the unit cell of phase space in the direction of the beam. It is argued that this should be adopted as the measure of coherence length.

§ 1. Introduction

Purcell (1956) has shown that the fluctuations of the number of photons n detected in a light beam can be calculated by associating the photons stochastically with Gaussian random waves of the appropriate spectral distribution. The validity of this approach has been questioned by Fellgett (1957), but it leads to results which are confirmed by experiment (Hanbury Brown and Twiss 1957) and consistent with Bose–Einstein statistics (Mandel 1958).

While the analysis in the last paper was, with one exception, confined to the calculation of the second moments of the fluctuations, here it is proposed to use the same approach to investigate the probability distribution of n. With the help of an approximation it will be shown that the distribution is of the form to be expected for a boson assembly in a finite volume of phase space. The results show that the form of the distribution is largely determined by the degree of degeneracy of the light beam and give a picture of the transition from the classical particle to the classical wave statistics.

In anticipation of some later considerations, it will be useful first to digress briefly, in order to derive generally the probability $p_s(n)$ of finding a state in which n bosons are distributed among s equal cells of phase space. We shall have occasion to refer to it later.

§ 2. The Probability $p_s(n)$ of finding n Bosons distributed among s Equal Cells of Phase Space

Let there be n_i particles in the ith cell, where $i=1, 2, \ldots. s$, so that

$$n=\sum_{i=1}^{s} n_i \qquad \ldots\ldots(1)$$

and

$$\bar{n}=s\bar{n}_i, \qquad \ldots\ldots(2)$$

since it is assumed that $\bar{n}_1=\bar{n}_2=\bar{n}_3$ etc.

If $p'(n_i)$ is the probability distribution of the n_i then it is well known (e.g. Fürth 1928) that

$$p'(n_i)=\frac{1}{(1+\bar{n}_i)(1+1/\bar{n}_i)^{n_i}} \text{ for } n_i \geqslant 0 \text{ and, using (2),}$$

$$=\frac{1}{(1+\bar{n}/s)(1+s/\bar{n})^{n_i}}. \qquad \ldots\ldots (3)$$

The probability of realizing any particular distribution of the n particles, i.e. any microstate, is therefore

$$\prod_{i=1}^{s} p'(n_i)=\frac{1}{(1+\bar{n}/s)^s(1+s/\bar{n})^n}. \qquad \ldots\ldots (4)$$

Since this is independent of the n_i's all distributions are equally likely. $p_s(n)$ is therefore obtained from the above by multiplying by the number of different realizable distributions, which is given by the well-known combinatorial expression

$$\frac{(s+n-1)!}{n!\,(s-1)!}.$$

The required probability $p_s(n)$ is then

$$\left.\begin{aligned} p_s(n)&=\frac{1}{(1+\bar{n}/s)^s(1+s/\bar{n})^n}\,\frac{(s+n-1)!}{n!\,(s-1)!}\\ &=\frac{(\bar{n}/s)^n}{(1+\bar{n}/s)^{n+s}}\,\frac{1}{(s-1)B(n+1,s-1)}, \end{aligned}\right\} \qquad \ldots\ldots (5)$$

where $B(a, b)$ is the beta-function defined by

$$B(a,b)=\int_0^\infty \frac{x^{a-1}}{(1+x)^{a+b}}\,dx=\frac{\Gamma(a)\Gamma(b)}{\Gamma(a+b)}.$$

When $s=1$ the product $(s-1)\,B\,(n+1,\,s-1)$ is to be interpreted as the limit $s\to 1$. By writing (5) in the form

$$p_s(n)=\frac{1}{(1+\bar{n}/s)^s(1+s/\bar{n})^n}\,\frac{(s-1)(s)(s+1)\ldots(s+n-1)}{(s-1)n!} \qquad \ldots\ldots (6)$$

we can immediately confirm that

$$\sum_{n=0}^{\infty} p_s(n)=\frac{1}{(1+\bar{n}/s)^s}\left[1-\frac{1}{1+s/\bar{n}}\right]^{-s}$$

$$=1, \text{ as required.}$$

We shall later have occasion to refer to these results, which are quite general. But we now return to consider the special case of a light beam falling on a photo-electric detector.

§ 3. The Probability Distribution of the Photo-Electrons

As before (Mandel 1958) we shall represent the narrow, coherent, plane polarized beam of light by a real Gaussian random variate $y(t)$ with normalized spectral density $\phi(\nu)$ and assume that the r.m.s. spectral width $\Delta\nu$ of $\phi(\nu)$ is small compared with the mean frequency ν_0. When this beam falls on the photoelectric surface, the probability that a photo-electron is ejected in the short interval between t and $t+dt$ will be $\alpha P(t)dt$, where α is a constant representing the quantum sensitivity of the photo-cathode and $P(t)$ is the classical beam intensity, i.e. $P(t)$ is the local average of $y^2(t)$ defined by

$$P(t) \equiv \langle y^2(t) \rangle \equiv \frac{\nu_0}{r} \int_{t-r/2\nu_0}^{t+r/2\nu_0} y^2(t')\, dt' \; ;$$

r is a small integer.

Now let $p(n, t, T)$ denote the probability that n photo-electrons are detected in the interval between t and $t+T$. From the stochastic association of the photons with the classical wave intensity, we can write

$$p(1, t, dt) = \alpha P(t)\, dt. \qquad \ldots\ldots(7)$$

It then follows from first principles that the expectation value of n in the interval t and $t+T$ is

$$\alpha \int_t^{t+T} P(t')\, dt'$$

and that $p(n, t, T)$ is a Poisson distribution in n:

$$p(n, t, T) = \frac{1}{n!}\left[\alpha \int_t^{t+T} P(t') dt'\right]^n \exp\left[-\alpha \int_t^{t+T} P(t')\, dt'\right]. \qquad \ldots\ldots(8)$$

However, as has already been pointed out (Mandel 1958) this distribution is not observable and only the time average $\overline{p(n, t, T)} \equiv p(n, T)$ has any physical significance. But, if $y(t)$ is a stationary random process, we may regard $p(n, t, T)$ as a function of

$$\int_t^{t+T} P(t')\, dt' \equiv E$$

and average over the ensemble of E.

Unfortunately, the exact probability distribution $\mathscr{P}(E)\, dE$ of E—which is related to that of

$$\int_t^{t+T} y^2(t')\, dt' \equiv F$$

(cf. Mandel, 1959†)—is not, in general, expressible explicitly (cf. Rice 1945, Kac and Siegert 1947, Siegert 1954). Rice (1945) has suggested a simple approximation to the distribution of F, which we shall take up later. But first we shall examine some properties of $p(n, t)$ that follow directly and quite generally from (8).

The moment generating function $M_p(x)$ of $p(n, T)$ is

$$M_p(x) = \sum_{n=0}^{\infty} \exp(xn) \overline{p(n, t, T)}.$$

By using the well-known properties of the Poisson distribution this can be written:

$$\begin{aligned} M_p(x) &= \overline{\exp[\alpha E(e^x - 1)]} \\ &= M_{\mathscr{P}}[\alpha(e^x - 1)], \qquad \ldots\ldots(9) \end{aligned}$$

where the bar now denotes the ensemble average over E and $M_{\mathscr{P}}(x)$ is the moment generating function of the distribution $\mathscr{P}(E)\, dE$. The corresponding expression for the cumulant generating functions $C_p(x)$ and $C_{\mathscr{P}}(x)$ is therefore

$$\begin{aligned} C_p(x) &= C_{\mathscr{P}}[\alpha(e^x - 1)] \\ &= \sum_{i=1}^{\infty} \frac{\alpha^i (e^x - 1)^i}{i!} \kappa_i \qquad \ldots\ldots(10\,a) \end{aligned}$$

† Paper read at Physical Society conference on Fluctuation Phenomena and Stochastic Processes, March 1959, to be published.

where κ_i is the ith cumulant of the distribution $\mathscr{P}(E)\,dE$. If K_i is the corresponding cumulant of the required distribution $p(n, T)$, we have, by comparing coefficients of powers of x in (10a):

$$\left.\begin{aligned} &K_1 = \alpha\kappa_1 \\ &K_2 = \alpha\kappa_1 + \alpha^2\kappa_2 \\ &K_3 = \alpha\kappa_1 + 3\alpha^2\kappa_2 + \alpha^3\kappa_3 \\ &K_4 = \alpha\kappa_1 + 7\alpha^2\kappa_2 + 6\alpha^3\kappa_3 + \alpha^4\kappa_4 \\ &\text{etc.} \end{aligned}\right\} \qquad \ldots\ldots(10\,b)$$

Now the cumulants κ_i of the distribution $\mathscr{P}(F)\,dF$ of F—which is similar to the distribution $P(E)\,dE$ of E when $T \gg 1/\nu_0$ or when $T = r/\nu_0$, where r is a small integer (cf. Mandel 1959)—have been obtained by Slepian (1958). From his results it follows that, for a stationary Gaussian random process $y(t)$, κ_i is expressible as a multiple convolution integral in terms of the normalized autocorrelation function

$$\gamma(\tau) \equiv \overline{y(t+\tau)y(t)}/\bar{P} \quad \text{of } y(t).$$

Thus:

$$\kappa_i = 2^{i-1}(i-1)!\,\bar{P}^i \int\int_{-T/2}^{+T/2} \int \gamma(t_1-t_2)\gamma(t_2-t_3)\ldots\gamma(t_i-t_1)\,dt_1\,dt_2\ldots dt_i$$

$$\text{for } i \geqslant 2. \qquad \ldots\ldots(11)$$

With the help of (11), the cumulants K_i of $p(n, T)$ then follow directly from (10). In principle, the distribution $p(n, T)$ is therefore determined.

Two limiting cases of $p(n, T)$ are of particular interest: (a) the mean intensity $\bar{P}$ is very small, corresponding to a non-degenerate photon beam; (b) the mean intensity $\bar{P}$ is very great, corresponding to a highly degenerate photon beam (cf. Mandel 1958).

Since $P(t)$ in (8) always occurs in conjunction with α, it is mathematically convenient to explore these two limiting cases by allowing $\alpha \to 0$ under (a) and $\alpha \to \infty$ under (b).

When $\alpha \to 0$, it follows at once from (10) that

$$K_i \to \alpha\kappa_1 = \alpha\bar{E} = \alpha\bar{P}T = \bar{n}, \qquad \ldots\ldots(12)$$

so that $p(n, T)$ is a Poisson distribution in n with mean $\bar{n}$. Thus, a weak, non-degenerate photon beam behaves like a beam of classical particles.

In the second case, when $\alpha \to \infty$, we see from (10) that

$$K_i \to \alpha^i\kappa_i, \qquad \ldots\ldots(13)$$

so that the distribution of n tends to become identical with that of αE. Now n is the quantal measure of the total energy absorbed in time T, while αE is the classical measure. It follows that the quantal and classical fluctuations of highly degenerate beams become identical. This result is, of course, to be expected from the correspondence principle.,

We see therefore that, as we pass from very weak to very intense light beams we move from the statistics of classical particles to the statistics of classical waves. But it is unfortunately true that the absence of an explicit expression for the distribution $p(n, T)$ prevents us from gaining very much insight into the nature of this transition. For this reason it is useful to explore the form of $p(n, T)$ with the aid of an explicit approximation to the distribution $\mathscr{P}(E)\,dE$, which arises from a suggestion first made by Rice (1945).

§ 4. The Approximate Form of the Probability Distribution $p(n, T)$

Rice appears to have been the first to investigate the distribution of F and to calculate some of the low order moments. He found (Rice 1945) that

$$\left.\begin{aligned} \bar{F} &= \bar{P}T \\ \text{and} \qquad \overline{\Delta^2 F} &= \overline{F^2} - \bar{F}^2 \\ &= 4\bar{P}^2 \int_0^T (T-\tau)\gamma^2(\tau)\,d\tau. \end{aligned}\right\} \qquad \ldots\ldots(14\,a)$$

It is readily shown that the corresponding expressions for E are

$$\left.\begin{aligned} \bar{E} &= \bar{P}T \\ \text{and} \qquad \overline{\Delta^2 E} &= 4\bar{P}^2 \int_0^T (T-\tau)\langle\gamma^2(\tau)\rangle\,d\tau \\ &= \bar{P}^2 T\xi. \end{aligned}\right\} \qquad \ldots\ldots(14\,b)$$

The latter integral was already encountered in the calculation of $\overline{\Delta^2 n}$ and denoted by $\frac{1}{4}T\xi$ (Mandel 1958), where ξ has the dimension of time and $\xi \gg T$. We shall continue to use this notation.

Rice now suggests that, as a first approximation, (aF) is a gamma-variate with parameter k, i.e.

$$P(F)\,dF \simeq \frac{(aF)^{k-1}\exp(-aF)}{\Gamma(k)}\,d(\ F). \qquad \ldots\ldots(15\,a)$$

where a and k are to be determined so as to satisfy (14 a). We shall now make the assumption that E approximately obeys a similar distribution, namely

$$P(E)\,dE \simeq \frac{(aE)^{k-1}\exp(-aE)}{\Gamma(k)}\,d(aE), \qquad \ldots\ldots(15\,b)$$

in which a and k are determined to satisfy (14 b). How this distribution can arise and some of the approximations involved in the derivation have already been discussed (Mandel 1959). Loosely speaking, (15 b) arises when the wave $y(t)$ is pictured as a serial superposition of a certain number of wave packets. The following properties of the distribution should be noted:

(1) By definition, it gives the first two moments of E correctly, whence it follows from (10 b), that the first two moments of n will also be correct (cf. Mandel 1958).

(2) It is correct for short time intervals $T \ll 1/\Delta\nu$.

(3) It is correct for time intervals $T \gg 1/\Delta\nu$ when the spectral distribution $\phi(\nu)$ is rectangular. This is shown in the Appendix, where it is proved that the rth cumulant of the distribution of E will, in general, be in error by the factor

$$\left[\int_{-\infty}^{\infty} \phi^2(\nu)\,d\nu\right]^{r-1} \Big/ \int_{-\infty}^{\infty} \phi^r(\nu)\,d\nu.$$

(4) It tends to the correct limit as the time interval $T \to \infty$.

From equations (14 b) and the properties of the gamma-distribution (15 b) we

have
$$\left.\begin{aligned}\bar{E}&=\bar{P}T \quad =k/a\\ \overline{\Delta^2 E}&=\bar{P}^2T\xi=k/a^2,\end{aligned}\right\} \qquad \ldots\ldots(16)$$
and

so that
$$\left.\begin{aligned}a&=1/\bar{P}\xi\\ k&=T/\xi.\end{aligned}\right\} \qquad \ldots\ldots(17)$$
and

We can now evaluate $p(n, T)$ by averaging (8) over the ensemble of E. Thus

$$\begin{aligned}p(n,T)&=\frac{1}{n!\Gamma(T/\xi)}\int_0^\infty(\alpha E)^n\left(\frac{E}{\bar{P}\xi}\right)^{T/\xi-1}\exp\left[-E\left(\alpha+\frac{1}{\bar{P}\xi}\right)\right]d\left(\frac{E}{\bar{P}\xi}\right)\\ &=\frac{(\alpha\bar{P}\xi)^n}{(1+\alpha\bar{P}\xi)^{n+T/\xi}}\frac{\Gamma(n+T/\xi)}{n!\Gamma(T/\xi)}.\end{aligned}$$

Since $\alpha\bar{P}T=\bar{n}$, this can be written in the form

$$\left.\begin{aligned}p(n,T)&=\frac{1}{(1+\bar{n}\xi/T)^{T/\xi}(1+T/\bar{n}\xi)^n}\frac{\Gamma(n+T/\xi)}{n!\,\Gamma(T/\xi)}\\ &=\frac{(\bar{n}\xi/T)^n}{(1+\bar{n}\xi/T)^{n+T/\xi}}\frac{1}{(T/\xi-1)B(n+1,T/\xi-1)}.\end{aligned}\right\} \qquad \ldots\ldots(18)$$

This is the observable probability distribution of the number of counts in an interval T. If we now compare the distributions (18) and (5) it will be seen that they are identical if T/ξ is integral and equal to s. In other words, $p(n, T)$ is the distribution of a boson assembly in a phase space of T/ξ cells.

Of course, while $p_s(n)$ in (5) was defined only for integral values of s, $p(n, T)$ is a continuous function of T/ξ. It can be regarded as a generalization, when one of the cells is not necessarily fully populated on the average. The distinction disappears for long intervals T, when T/ξ becomes a large number. Since $\xi \leqslant T$, we see that the number of cells $T/\xi \geqslant 1$ and the shortest interval T defines one cell. This is in accordance with the interpretation of T/ξ.

As has been shown (Mandel 1958), ξ becomes constant and independent of T for intervals long compared with $1/\Delta\nu$. The asymptotic value ξ_∞ is, by Parseval's theorem,

$$\xi_\infty=4\int_0^\infty\gamma^2(\tau)\,d\tau=4\int_0^\infty\phi^2(\nu)\,d\nu. \qquad \ldots\ldots(19)$$

It is therefore reasonable to regard $1/\xi_\infty$ as the average number of cells per unit time interval, or $c\xi_\infty$ as the average length of a cell associated with the light beam in the direction of motion. Since $c\xi_\infty$ is also the path difference within which interference effects can be observed, there are good reasons for calling ξ_∞ the coherence time of the light. As has been shown (Mandel 1958), ξ_∞ is of the order $1/\Delta\nu$.

We shall return to a discussion of these points. But first we must examine some properties of the distribution (18).

§ 5. Some Properties of the Distribution (18)

Inspection of equation (18) shows that the parameter $(\bar{n}\xi/T)$—the average number of particles per unit cell, or the degeneracy—plays a key role in determining the form of the distribution. If we regard $p(n, T) \equiv p(n, T, \bar{n}\xi/T)$ as a continuous function of the degeneracy parameter $\bar{n}\xi/T$ (with ξ/T constant) which may range from zero to infinity, then $(T/\xi-1)p(n, T, \bar{n}\xi/T)$ is a beta-prime distribution (cf. Kenney and Keeping 1951) of $\bar{n}\xi/T$ with parameters $n+1$ and $T/\xi-1$. The case $T/\xi=1$ must be regarded as a limiting case.

When the degeneracy is very small, $p(n, T)$ simplifies very considerably. By writing (18) in the form (6) it can be seen at once that, for $T/\xi \gg \bar{n}$,

$$p(n, T) \simeq \frac{\exp(-\bar{n})}{(T/\bar{n}\xi)^n} \frac{(T/\xi)^n}{n!} = \frac{\bar{n}^n}{n!} \exp(-\bar{n}), \qquad \ldots\ldots (20)$$

which is the classical Poisson distribution. This confirms the result obtained more generally in § 3.

This situation will generally apply when stellar sources are being studied. The light from these sources is always so weak that $\bar{n}\xi/T \ll 1$ and the degeneracy is unlikely to be detected in measurements on a single beam. The situation is, of course, improved when correlation measurements are undertaken on two or more coherent beams (Hanbury Brown and Twiss 1956), since these measurements single out the degenerate photons (Mandel 1958). Even so it is unlikely that any faint stars could be studied in this way.

When the degeneracy is very great, $p(n, T)$ approaches another limit. If $\bar{n}$ and n are large we can expand $n!$ and $\Gamma(n + T/\xi)$ in (18) by Stirling's theorem and so obtain

$$p(n, T) \simeq \left(\frac{T}{\bar{n}\xi}\right)^{T/\xi} \frac{\exp(-nT/\bar{n}\xi) n^{T/\xi - 1}}{\Gamma(T/\xi)}.$$

By introducing the almost continuous variate $z = n/\bar{n}$ we can express this in the form

$$p(z, T)\, dz = \frac{(zT/\xi)^{T/\xi - 1} \exp(-zT/\xi)}{\Gamma(T/\xi)} d(zT/\xi). \qquad \ldots\ldots (21)$$

Thus, for highly degenerate beams, zT/ξ becomes a gamma variate with parameter T/ξ and behaves like the variate $kE/\bar{E}$ of equation (15 *b*).

This result was, of course, to be expected from the more general considerations of § 3 and directly from the correspondence principle. In the limit of high degeneracy $n/\bar{n} \simeq E/\bar{E}$, since we are dealing with light of limited spectral extent and the distribution of z must tend to that of $E/\bar{E}$. Some examples of the function $p(z, T)$ are shown in the figure for different values of the parameter T/ξ. It should be stressed that the point $z = 0$ does not satisfy the condition $n \gg 1$.

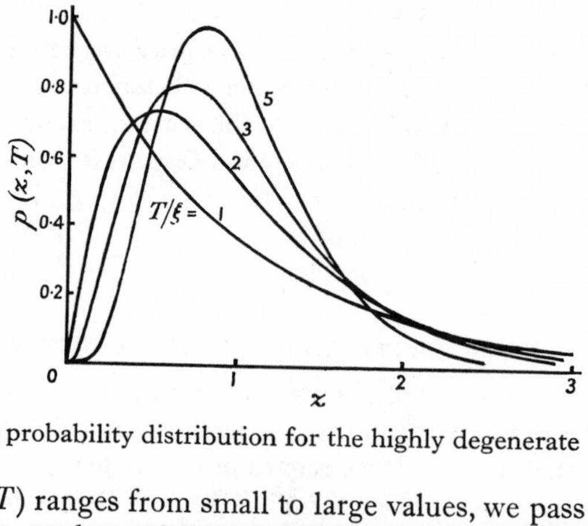

The probability distribution for the highly degenerate case.

Thus, as $(\bar{n}\xi/T)$ ranges from small to large values, we pass from the statistics of classical particles to the statistics of classical waves. Although the forms of the distributions (20) and (21) are generally quite different, they have a common limit. Thus, it is well known that the Poisson distribution for large $\bar{n}$ and the gamma distribution for large parameter T/ξ, both tend to the Gaussian form. However the similarity of these two limits disappears if we examine the variances.

The moment generating function $M(x, T)$ of $p(n, T)$ is easily found by writing (18) in the form (6). Thus

$$M(x,T)=\sum_{n=0}^{\infty} p(n,T)\exp(xn)$$
$$=\frac{1}{(1+\bar{n}\xi/T)^{T/\xi}}\left[1-\frac{\exp x}{1+T/\bar{n}\xi}\right]^{-T/\xi}$$
$$=[1+(\bar{n}\xi/T)(1-\exp x)]^{-T/\xi}. \qquad \ldots\ldots(22)$$

From this we can immediately obtain the second moment :

$$\overline{n^2}=\bar{n}+\bar{n}^2(1+\xi/T)$$

and

$$\overline{\Delta^2 n}=\bar{n}(1+\bar{n}\xi/T), \qquad \ldots\ldots(23)$$

which is the result found previously (Purcell, 1956, Mandel 1958). Thus for non-degenerate beams $\overline{\Delta^2 n}=\bar{n}$ and for highly degenerate beams $\overline{\Delta^2 n}=\bar{n}^2\xi/T$. The two variances therefore differ by the degeneracy factor, even though the distributions may be of similar form.

§ 6. Coherence Time and Degrees of Freedom

As has been shown in § 4, there are good reasons for adopting the quantity ξ_∞ defined by equation (19) as the measure of coherence time. This definition makes the coherence length in space and the size of the unit cell in phase space complementary concepts in the wave and particle descriptions of the beam. Since ξ_∞ is the interval within which the photons are intrinsically indistinguishable, we see that the idea of coherence, in general, implies a phase relationship in terms of classical waves and intrinsic indistinguishability in terms of photons.

Our definition ξ_∞ is based on the same distribution, viz. $\gamma^2(\tau)$, as the measure of coherence time $\Delta\tau_W$ recently proposed by Wolf (1958) where :

$$(\Delta\tau_W)^2=\int_0^\infty \tau^2\gamma^2(\tau)\,d\tau\Big/\int_0^\infty \gamma^2(\tau)\,d\tau. \qquad \ldots\ldots(24)$$

ξ_∞ and $\Delta\tau_W$ are both measures of the width of the $\gamma^2(\tau)$ distribution, but, whereas $\Delta\tau_W$ is chosen a little arbitrarily so as to satisfy an inequality, ξ_∞ arises naturally from consideration of the fluctuations.

It is useful to have some idea of the relation between them. If the spectral density for positive ν has the form of a narrow Gaussian distribution with mean ν_0 and r.m.s. width $\Delta\nu$, then $\gamma(\tau)=\cos(2\pi\nu_0\tau)\exp(-2\pi^2\Delta\nu^2\tau^2)$. It follows that

$$\xi_\infty=1/(2\sqrt{\pi}\Delta\nu)$$

and

$$\Delta\tau_W=1/(2\sqrt{2}\pi\Delta\nu),$$

so that $\xi_\infty/\Delta\tau_W=\sqrt{(2\pi)}$ in this case. Since the spectral distributions encountered in practice are often nearly Gaussian, the order of magnitude relation

$$(\xi_\infty\Delta\nu)^2\simeq 1/4\pi \qquad \ldots\ldots(25)$$

will usually hold.

We have seen that, for long intervals, T/ξ gives the mean number of independent cells of phase space contained within T and hence the number of independent data or statistical degrees of freedom of the energy distribution within T. This number should be compared with the upper limit fT given by the sampling theorem (Gabor 1946, Shannon 1949) for the intensity of a wave whose normalized spectral density $\phi(\nu)$ vanishes outside a frequency interval f. We find that, for any shape of the narrow distribution $\phi(\nu)$, $\xi_\infty \geqslant 1/f$, so that $T/\xi_\infty \leqslant fT$. The number of independent data determined by the coherence time ξ_∞ therefore has an upper bound given by the sampling theorem. These independent data describing the energy distribution of the beam within T are just the quantum numbers $n_i(i=1$ to $T/\xi_\infty)$ which were used to define the state of the radiation field in § 2 (cf. also Gabor 1950)†. In general, T/ξ_∞ will be of the order $\Delta\nu T$. For example, for the Gaussian spectral density, $T/\xi_\infty = 2\sqrt{\pi}\Delta\nu T$.

Although the definition of ξ_∞ is based on $\gamma^2(\tau)$, it is clear from (16) that ξ_∞/T is simply the percentage variance of the classical energy E in a long wave train and we can write

$$\lim_{T\to\infty} (\xi_\infty/T) = \overline{\Delta^2 E}/\bar{E}^2. \qquad \ldots\ldots (26)$$

This equation also follows from (23) by the correspondence principle, since $n \propto E$ for highly degenerate beams.

It is interesting to note that (26) is the relation to be expected if the long wave train of duration T is pictured as a random assembly of similar elementary wave packets of duration ξ_∞. Since $P(t)$ varies only slowly within the interval ξ_∞ (Mandel 1958), the picture is not altogether unsound. These wave packets in real space must not be confused with the wave packets in phase space which are occasionally used as representations of photons (cf. Sillito 1957). In the first case the number of wave packets within the interval T is quite independent of the intensity. Thus ξ_∞ can be defined and interpreted in purely classical, as well as quantum statistical terms.

We have seen that the quantity $\bar{n}\xi/T$, the mean number of photons per unit cell or the degeneracy, plays a crucial role in determining the statistical behaviour of n. When $\bar{n}\xi/T$ is large, the distribution of n or of energy tends to the classical form for Gaussian waves. Conversely, when $\bar{n}\xi/T$ is small $p(n, T)$ tends to the classical distribution for particles. Although most of the phase cells are then empty, the classical waves associated with the beam do not change their general form and, on the average, consist of just as many wave packets as before. If we interpret the variate $y(t)$ as the magnitude of the electric vector associated with the light beam, then the continuing rapid fluctuations of $y(t)$, even when few photons are present, can be regarded as a consequence of the uncertainty relations connecting the electric field strength and the number of photons. For the operators representing the field strength and the photon number do not commute (cf. Heitler 1954) and the product of the uncertainties in the field strength and in number is therefore finite. As the degeneracy gets less and less, the classical representation of the beam therefore becomes increasingly inaccurate.

† In a recent paper on beam fluctuations, Kahn (1958) used the sampling theorem to expand the classical wave amplitude in a series of sinc functions centred at the sampling points. The degrees of freedom or ' modes ' are the expansion coefficients.

References

Fellgett, P. B., 1957, *Nature, Lond.*, **179**, 956.
Fürth, R., 1928, *Z. Phys.*, **48**, 323.
Gabor, D., 1946, *J. Instn Elect. Engrs*, **93**, Pt. IV, 429.
—— 1950, *Phil. Mag.*, **41**, 1161.

HANBURY BROWN, R., and TWISS, R. Q., 1956, *Nature, Lond.*, **178,** 1046.
—— 1957, *Proc. Roy. Soc.* A, **243,** 291.
HEITLER, W., 1954, *The Quantum Theory of Radiation*, 3rd Edn (Oxford : Clarendon Press).
KAC, M., and SIEGERT, A. J. F., 1947, *J. Appl. Phys.* **18,** 383.
KAHN, F. D., 1958, *Opt. Acta*, **5,** 93.
KENNEY, J. F., and KEEPING, E. S., 1951, *Mathematics of Statistics II*, 2nd Edn (Amsterdam: Van Nostrand).
MANDEL, L., 1958, *Proc. Phys. Soc.*, **72,** 1037.
—— 1959, see footnote in § 3.
PURCELL, E. M., 1956, *Nature, Lond.*, **178,** 1449.
RICE, S. O., 1945, *Bell Syst. Tech. J.*, **24,** 46.
SHANNON, C. E., 1949, *Proc. Inst. Rad. Engrs*, **37,** 10.
SIEGERT, A. J. F., 1954, *Trans. Inst. Rad. Engrs, PGIT*-3, 4.
SILLITO, R. M., 1957, *Nature, Lond.*, **179,** 1127.
SLEPIAN, D., 1958, *Bell. Syst. Tech. J.*, **37,** 163.
WOLF, E., 1958. *Proc. Phys. Soc.*, **71,** 257.

APPENDIX

THE CUMULANTS OF E WHEN $T \gg 1/\Delta\nu$

It is well known that, for a narrow band process, $\gamma(\tau)$ is expressible in the form

$$\gamma(\tau) = V(\tau)\exp(2\pi i\nu_0\tau) + V^*(\tau)\exp(-2\pi i\nu_0\tau),$$

where $V(\tau)$ is a slowly varying, decreasing function, which becomes very small for $\tau \gg 1/\Delta\nu$ (cf. Mandel 1958). It follows that the integrand in (11) will be greatest when $t_1 \simeq t_2 \simeq \ldots \simeq t_r$ etc. and very small when $t_1 - t_2 \gg 1/\Delta\nu$, etc. If $T \gg 1/\Delta\nu$ and if we introduce the substitutions

$$\begin{aligned} t_r - t_1 &= x_r \\ t_{r-1} - t_r &= x_{r-1} \\ &\cdots\cdots\cdots\cdots \\ t_2 - t_3 &= x_2 \end{aligned}$$

into the multiple integral (11), we can write

$$\kappa_r \simeq 2^{r-1}(r-1)!\,\bar{P}^r \int_{-T/2}^{T/2} dt_1 \int\int_{-\infty}^{\infty} \ldots\ldots \int \gamma(x_2 + x_3 + \ldots x_r) \times \gamma(x_2)\gamma(x_3)\ldots\ldots\gamma(x_r)\,dx_2\,dx_3\ldots\ldots dx_r. \qquad \ldots\ldots(\text{A }1)$$

Now, from the Wiener–Khintchine theorem,

$$\int_{-\infty}^{\infty} \gamma(x_2 + x_3 + \ldots\ldots x_r)\gamma(x_2)\,dx_2 = \int_{-\infty}^{\infty} \phi^2(\nu)\exp[2\pi i\nu(x_3 + x_4 + \ldots x_r)]\,d\nu, \qquad \ldots\ldots(\text{A }2)$$

since $\phi(\nu)$ is the Fourier transform of $\gamma(\tau)$. When (A 2) is introduced into (A 1), all the remaining variables become separable and we obtain

$$\kappa_r \simeq 2^{r-1}(r-1)!\,\bar{P}^r T \int_{-\infty}^{\infty} \phi^r(\nu)\,d\nu. \qquad \ldots\ldots(\text{A }3)$$

Next consider the approximate distribution (15 *b*) with *a* and *k* given by (17). From the properties of the γ-variate the *r*th cumulant is given by

$$\kappa_r = (r-1)!\,\bar{P}^r T \xi^{r-1} \qquad \ldots\ldots (\text{A}\,4)$$

But, as has been shown (Mandel 1958), for $T \gg 1/\Delta\nu$,

$$\xi \simeq 2\int_{-\infty}^{\infty} \phi^2(\nu)\,d\nu,$$

so that (A 4) becomes

$$\kappa_r \simeq 2^{r-1}(r-1)!\,\bar{P}^r T\left[\int_{-\infty}^{\infty} \phi^2(\nu)\,d\nu\right]^{r-1}. \qquad \ldots\ldots (\text{A}\,5)$$

Comparison of (A 3) and (A 5) shows that the second cumulant is always correct, but that the higher ones are in error by the factor

$$\left[\int_{-\infty}^{\infty} \phi^2(\nu)\,d\nu\right]^{r-1} \Big/ \int_{-\infty}^{\infty} \phi^r(\nu)\,d\nu.$$

This factor is unity for a rectangular spectrum of the form

$$\phi(\nu) = 1/2\Delta\nu \text{ for } \nu_0 - \tfrac{1}{2}\Delta\nu < \nu < \nu_0 + \tfrac{1}{2}\Delta\nu,$$
$$= 0, \quad \text{for any other } \nu \geqslant 0,$$

and $\phi(-\nu) = \phi(\nu)$.

If we transform the variable E to the standard form

$$(E - \bar{E})/(\overline{\Delta^2 E})^{1/2} = (E - \bar{P}T)/\bar{P}(T\xi)^{1/2},$$

the first cumulant of the new distribution is zero and the higher cumulants are multiplied by $[\bar{P}(T\xi)^{1/2}]^{-r}$. Inspection of (A 3) and (A 5) now shows that, as $T \to \infty$, the cumulants of higher order than the second tend to zero. Since the second cumulants given by (A 3) and (A 5) are, in any case, equal it follows that the limit of the approximate distribution (15 *b*) is Gaussian with the correct mean and variance.

PAPER NO. 30

Reprinted from *Journal of the Optical Society of America*, Vol. 49, pp. 787–793, 1959.

On the Propagation of Mutual Coherence

George B. Parrent, Jr.*
Electromagnetic Radiation Laboratory, Air Force Cambridge Research Center, Bedford, Massachusetts
(Received December 1, 1958)

Section 1 presents a brief history of the development of coherence theory and discusses some of the advantages of Wolf's general formulation of the theory in terms of Gabor's analytic signals.

Section 2 contains an analysis of the limiting cases of coherence and incoherence, showing for these extremes the form of the mutual coherence function in a quasi-monochromatic field. In particular it is shown that in such a field *coherence* is characterized by a mutual coherence function which, apart from a simple periodic factor, is expressible as the product of a wave function with its complex conjugate, each factor depending on the coordinates of one point only. It is also shown that an incoherent field cannot exist in free space.

The mutual coherence function obeys rigorously two wave equations, and in Sec. 3 these equations are solved with the help of appropriate Green's functions to find the field produced by a general plane polychromatic source. The solution is simplified by the quasi-monochromatic approximation; and the limiting cases are discussed in detail showing that a coherent source always gives rise to a coherent field, and that a well known theorem of Van Cittert and Zernike represents an approximation to the incoherent limit of the quasi-monochromatic solution.

1. INTRODUCTION

1.1 History

In optics and several other fields employing electromagnetic wave theory, for example radar, radio astronomy, and scatter communications, many of the significant problems reduce ultimately to the determination of the correlation between the field disturbances at two different points. Historically, workers in visible optics first became concerned with this type of problem. The need for a mathematical framework to bridge the gap between coherence and incoherence (i.e., to treat partially coherent fields) was seen early; and at around the turn of the century von Laue[1] in a

* The research discussed here was conducted during the author's stay at the Physical Laboratories, University of Manchester, England.

[1] M. von Laue, Ann. Physik **23**, 1 (1907); also Handbuch der Experimentalphysik **XVIII**, 277 (1928).

paper treating the thermodynamics of optical fields introduced a measure of the "degree of coherence." This paper, however, apparently received little attention for it was some twenty years before the next paper on the subject appeared. Berek[2] then introduced another measure of the "degree of coherence" (the so-called degree of consonance), and while his paper revealed a keen insight into the physical aspects of the problem his results did not completely agree with experiment.[3] Some years later Van Cittert[4] introduced the "korrelation" (as a measure of the degree of coherence) and found an expression for it at points on a screen illuminated by an extended incoherent source. If the averages appearing in Van Cittert's paper are interpreted as ensemble averages and attention is limited to nearly monochromatic fields his "komplex korrelation" can, by invoking the ergodic hypothesis, be identified with the complex degree of coherence of current theory. Zernike[5] introduced another measure of the degree of coherence in terms of the time averaged product of the disturbances at the two points in question (the zero ordinate of the cross correlation function); and stressed the fact that the degree of coherence could be identified with the visibility of the fringes obtained by allowing the disturbances to interfere in an appropriate manner. Moreover he showed that under suitable conditions the mutual intensity (to be defined later) is propagated according to a relatively simple law. Hopkins[6] introduced yet another definition of the degree of coherence in terms of an integral over the primary source, assumed to be incoherent. His formulation is useful for calculations when applicable (incoherent, nearly monochromatic sources and small path differences).

Until the work of Blanc-Lapierre and Dumontet[7] and Wolf[8] none of the mathematical frameworks which had been introduced were applicable to treating any but the nearly monochromatic fields. Blanc-Lapierre and Dumontet defined the degree of coherence in terms of the cross correlation of the two real functions describing the disturbances at the two points. While this work was mathematically rigorous, for physical reasons it appears more desirable to formulate the problem in terms of complex functions. On the other hand, Wolf defined it in terms of a complex cross correlation function and provided a means of treating polychromatic fields with the physically more desirable representation in terms of analytic signals. Some of the advantages of this representation are discussed in the next section.

[2] M. Berek, Z. Physik **36**, 675 (1926).
[3] Dr. C. Lakeman, in J. Th. Groosmuller, *Physica* (Gravenhage, the Netherlands, 1928), Vol. 8, p. 199.
[4] P. H. van Cittert, Physica **1**, 201 (1934).
[5] F. Zernike, Physica **5**, 785 (1938).
[6] H. H. Hopkins, Proc. Roy. Soc. (London) **A208**, 263 (1951).
[7] A. Blanc-Lapierre, and P. Dumontet, Rev. opt. **34**, 1 (1955).
[8] E. Wolf, Proc. Roy. Soc. (London) **A230**, 246 (1955).

1.2 Analytic Signal and Some Definitions

Fundamental to the understanding of the latest developments in coherence theory is the analytic signal introduced by Gabor[9] for problems of communication theory. The need for the analytic signal has arisen from two sources, mathematical convenience and the physical requirements of the problem. Mathematically it is of course more convenient to deal with complex exponentials than with real sine or cosine functions. Physically the requirements are somewhat more subtle. First, since the basic quantity, a component of the electric vector, is real only half of its Fourier spectrum contains significant information since the spectrum of negative frequencies can be simply derived from that of positive frequencies. Second, it is desirable that the mutual coherence function, to be defined later, be related to the envelope of the intensity curve.

There are, of course, an infinite number of ways to select a complex representation and gain the advantages of complex exponentials over cosines, but in order that the definition of mutual coherence has a meaning a particular selection must be made. A confusion arises in earlier work over just this point. Reference is made to real and imaginary parts without specifying the type of complex representation used. Using the analytic signal as the complex representation has several advantages: there is a particularly simple relation between real and imaginary parts; the function, (regarded as a function of a complex variable), is analytic in half of the complex plane; the analytic signal can be expressed as a linear transform of the real signal with which it is associated; and it represents a natural generalization of the method of associating a complex function with the real monochromatic wave.

As mentioned earlier the central problem in coherence theory is the determination of the cross correlation between the disturbances at two points. The significant functions will be defined here in the manner introduced by Wolf and his terminology and notation (as given in reference 12) will be used throughout the paper. Let $V^r(t)$ be a real scalar time function (e.g., a Cartesian component of the electric vector) possessing a Fourier representation,

$$V^r(t)=\int_0^\infty a(\nu)\cos[\phi(\nu)-2\pi\nu t]d\nu. \qquad (1.2.1)$$

Associate with $V^r(t)$ the function $V^i(t)$ formed by changing the phase of each spectral component by $\pi/2$, i.e.,

$$V^i(t)=\int_0^\infty a(\nu)\sin[\phi(\nu)-2\pi\nu t]d\nu. \qquad (1.2.2)$$

[9] D. Gabor, J. Inst. Elec. Engrs. (London) Pt. III, **93**, 429 (1946).

The analytic signal, $V(t)$, associated with $V^r(t)$ is defined as

$$V(t)=V^r(t)+iV^i(t). \quad (1.2.3)$$

$V(t)$ can be written in terms of the full range spectrum of $V^r(t)$ as follows: let

$$V^r(t)=\int_{-\infty}^{\infty} v(\nu)e^{-2\pi i\nu t}d\nu, \quad (1.2.4)$$

where $v(\nu)=\frac{1}{2}a(\nu)e^{i\phi(\nu)}$ $(\nu>0)$, and $v^*(\nu)=v(-\nu)$. Then the analytic signal, $V(t)$, is given by

$$V(t)=\int_0^{\infty} v(\nu)e^{-2\pi i\nu t}d\nu. \quad (1.2.5)$$

It can be shown[10] that with the functions so defined $V^r(t)$ and $V^i(t)$ are Hilbert transforms, i.e.,

$$V_i(t)=\frac{P}{\pi}\int_{-\infty}^{\infty}\frac{V^r(t')}{t'-t}dt',$$

and

$$V^r(t)=-\frac{P}{\pi}\int_{-\infty}^{\infty}\frac{V^i(t')}{t'-t}dt', \quad (1.2.6)$$

where P denotes Cauchy's principle value at $t'=t$. In terms of the improper function,

$$\delta_-(t'-t)=\int_0^{\infty} e^{-2\pi i\nu(t'-t)}=\delta(t'-t)-\frac{iP}{\pi(t'-t)}, \quad (1.2.7)$$

$V(t)$ can be written as a linear transform of $V^r(t)$ in the form

$$V(t)=\int_{-\infty}^{\infty} V^r(t')\delta_-(t'-t)dt'. \quad (1.2.8)$$

In terms of these functions the basic quantities related to the concept of partial coherence may be defined unambigously. The "mutual coherence function" is defined as the complex cross correlation between the analytic signal representation of the real field at two points P_1 and P_2, i.e.,

$$\Gamma_{12}(\tau)=\langle V_1(t+\tau)V_2^*(t)\rangle, \quad (1.2.9)$$

where the sharp brackets denote time average; and the complex degree of coherence, $\gamma_{12}(\tau)$ is defined as

$$\gamma_{12}(\tau)=\frac{\Gamma_{12}(\tau)}{[\Gamma_{11}(0)\Gamma_{22}(0)]^{\frac{1}{2}}}=\frac{\Gamma_{12}(\tau)}{(I_1)^{\frac{1}{2}}(I_2)^{\frac{1}{2}}}. \quad (1.2.10)$$

Here I_s is the intensity at $P_s(S=1,2)$. It can be readily shown that $|\gamma_{12}(\tau)|$ can be identified with the visibility of the fringes obtained by allowing $V_1^r(t)$ and $V_2^r(t)$ to interfere after adjusting the intensities to equal values. By appealing to Schwarz inequality it is also easily shown that $0\leq|\gamma_{12}(\tau)|\leq 1$.

Since $V^r(t)$ represents a scalar wave we have in vacuum

$$\nabla^2V^r(t)=(1/c^2)[\partial^2V^r(t)/\partial t^2]. \quad (1.2.11)$$

If we multiply both sides of Eq. (1.2.11) by $\delta_-(t'-t)$ after setting $t=t'$ and integrate over t' we obtain

$$\nabla^2V(t)=(1/c^2)[\partial^2V(t)/\partial t^2], \quad (1.2.12)$$

where the theorem,

$$\frac{\partial^2V^i(t)}{\partial t^2}=\frac{P}{\pi}\int_{-\infty}^{\infty}\frac{\partial^2V^r(t')/\partial t'^2]}{t'-t}dt', \quad (1.2.13)$$

was used.

The mutual coherence function is propagated in vacuum according to two wave equations. These equations can be obtained by differentiating Eq. (1.2.9) with respect to P_1 and P_2 separately and then formally interchanging the order of differentiation and integration. Thus

$$\nabla_s^2\Gamma_{12}(\tau)=(1/c^2)[\partial^2\Gamma_{12}(\tau)/\partial\tau^2], \quad (s=1,2), \quad (1.2.14)$$

where ∇_s^2 and the Laplacian operator in the coordinates of P_s. These differential equations were first obtained by Wolf[8] and for an alternate derivation the reader is referred to this work. In Sec. 3 of his paper a rigorous solution of Eq. (1.2.14) is presented for the field produced by a plane polychromatic source.

If the ambiguity as to type of complex representation is removed from the earlier definitions, it can be shown that, when they are applicable, the quantities introduced are all simply expressible in terms of the mutual coherence function. The demonstrations of the various relations will form the subject of a separate paper.[11]

2. LIMITING CASES $|\gamma_{12}(\tau)|=1$ AND $|\gamma_{12}(\tau)|=0$

The extreme cases of $|\gamma_{12}(\tau)|=1$ and $|\gamma_{12}(\tau)|=0$ correspond to complete coherence and complete incoherence, respectively. It is of practical interest to examine some implications of our formulas for these limiting cases and in particular to examine the form which the *mutual coherence function*, $\Gamma_{12}(\tau)$, and the *mutual intensity*, $\Gamma_{12}(0)$, take.

We shall discuss in detail only a quasi-monochromatic field, i.e., one for which the effective spectral range, $\Delta\nu$, is small compared to the mean frequency, $\bar{\nu}$, i.e., such that

$$\Delta\nu/\bar{\nu}\ll 1. \quad (2.1)$$

It is known[12] that in this case, provided the time delay, τ, is small compared to the coherence time, $1/\Delta\nu$, the mutual coherence function is of the form

$$\Gamma_{12}(\tau)\cong\Gamma_{12}(0)e^{-2\pi i\bar{\nu}\tau}\quad\left(|\tau|\ll\frac{1}{\Delta\nu}\right); \quad (2.2)$$

[10] E. C. Titschmarch, *Introduction to the Theory of Fourier Integrals* (Clarendon Press, Oxford, 1948), second edition.

[11] G. B. Parrent and E. Wolf (to be published).

[12] M. Born and E. Wolf, *Principles of Optics* (Pergamon Press, London, 1959).

and we obtain from Eq. (1.2.12)

$$\nabla_s^2\Gamma_{12}(0)+\bar{k}^2\Gamma_{12}(0)=0, \tag{2.3}$$

where $\bar{k}$ is the mean wave number, $\bar{k}=2\pi\bar{\nu}/c$. It follows that for a coherent field

$$|\gamma_{12}(\tau)|=\frac{|\Gamma_{12}(\tau)|}{[\Gamma_{11}(0)\Gamma_{22}(0)]^{\frac{1}{2}}}=1. \tag{2.4}$$

Hence the mutual coherence function for this case must be of the form

$$\Gamma_{12}(\tau)=A(P_1)B(P_2)e^{i(\phi_{12}-2\pi\bar{\nu}\tau)}, \tag{2.5}$$

where $A(P_1)$ and $B(P_2)$ equal to $[\Gamma_{11}(0)]^{\frac{1}{2}}$ and $[\Gamma_{22}(0)]^{\frac{1}{2}}$, respectively, are functions of the coordinates of one point only. We shall now show that $\phi(P_1,P_2)$ also splits into two terms.

We have from Eqs. (2.5) and (2.3) the equation

$$(\nabla_1^2A_1/A_1)-(\nabla_1\phi_{12})^2+\bar{k}^2 +i[\nabla_1^2\phi_{12}+2(\nabla_1\phi_{12}\cdot\nabla_1A_1)/A_1]=0, \tag{2.6}$$

with a similar expression involving ∇_2 and B. Equating the real and imaginary parts separately to zero, we obtain the two equations

$$\nabla_1^2\phi_{12}=-[(2\nabla_1\phi_{12}\cdot\nabla_1A_1)/A_1], \tag{2.7}$$

and

$$(\nabla_1\phi_{12})^2=(\nabla_1^2A_1/A_1)+\bar{k}^2. \tag{2.8}$$

The right-hand side of Eq. (2.8) is a function of the coordinates of P_1 only and we can write Eq. (2.8) in the form

$$(\nabla_1\phi_{12})^2=f^2(x_1,y_1,z_1). \tag{2.9}$$

On the left side of Eq. (2.9) the coordinates of P_2 [contained in $\phi_{12}(P_1,P_2)$] may be regarded as parameters. Equation (2.9) is therefore of the form of the eikonal equation of geometrical optics or the Hamilton-Jacobi equation of dynamics. The general solution of this equation is well known and is

$$\phi_{12}=\phi^0(p,q,x_2,y_2,z_2)+\int_{P_{10}}^{P_1}f(x_1,y_1,z_1)ds, \tag{2.10}$$

where ϕ^0 is the value of the solution on a surface over which the solution has a constant value (independent of x_1,y_1,z_1), and the integral is taken along the extremal of the variational problem

$$\delta\int_{P_{10}}^{P_1}f(x_1,y_1,z_1)ds=0; \tag{2.11}$$

and P_1^0 is a typical point on the surface. Here p and q are two free parameters which may be regarded as characterizing the orientation of the surface ϕ^0.

Equation (2.10) shows that $\phi(P_1,P_2)$ is of the form

$$\phi_{12}=\alpha(P_1)+\beta(P_2), \tag{2.12}$$

where α and β depend only on the coordinates of P_1 and P_2, respectively. Hence Eq. (2.5) becomes

$$\Gamma_{12}(\tau)=[A_1e^{i\alpha_1}][B_2\,e^{i\beta_2}]e^{-2\pi i\bar{\nu}\tau}\left(|\tau|\ll\frac{1}{\Delta\nu}\right). \tag{2.13}$$

From the definition of $\Gamma_{12}(\tau)$ it is clear that $\Gamma_{12}(-\tau)=\Gamma_{21}^*(\tau)$. Applied to Eq. (2.13) this implies that

$$A(P_1)=B(P_1),$$

and

$$\beta(P_1)=-\alpha(P_1)\pm2n\pi, \tag{2.14}$$

where n is any integer. If we now set

$$U=A\,e^{i\alpha}. \tag{2.15}$$

Equation (2.13) becomes finally

$$\Gamma_{12}(\tau)=U(P_1)U^*(P_2)e^{-2\pi i\bar{\nu}\tau}\quad[|\tau|\ll(1/\Delta\nu)]. \tag{2.16}$$

Thus *in the quasi-monochromatic approximation a coherent field is completely characterized by a complex wave function U which depends on the coordinates of one point only, and in terms of it the mutual coherence function is given by Eq. (2.16).*

Conversely, *when $\Gamma_{12}(\tau)$ is of the form of Eq. (2.16) the field is completely coherent*; this is evident on substituting this form into the expression for $\gamma_{12}(\tau)$.

We shall now show that an incoherent quasi-monochromatic field cannot exist in free space. Incoherence is characterized by $|\gamma_{12}(0)|=0$ which implies that $\Gamma(P_1,P_2,0)=0$. However, it follows from the definition of $\Gamma(P_1,P_2,\tau)$ that

$$\Gamma(P_1,P_1,0)=I(P_1). \tag{2.17}$$

From these two considerations it follows that for an incoherent field the mutual intensity function is of the form

$$\Gamma(P_1,P_2,0)=\left\{\begin{matrix}0 & P_1\neq P_2\\ I(P_1) & P_1=P_2\end{matrix}\right\}. \tag{2.18}$$

Let Σ be a closed surface in the space, V, throughout which the field is assumed to be incoherent [i.e., Eq. (2.18) is satisfied throughout V]. The mutual intensity at all pairs of points, P_1 and P_2, interior to Σ is given by (see Sec. 3)

$$\Gamma(P_1,P_2,0)=\int_\Sigma\Gamma(S_1,S_2,0)F(r_1,r_2)dS_1dS_2, \tag{2.19}$$

where S_1 and S_2 are points on the surface Σ; $F(r_1,r_2)=(\partial\mathcal{G}_1/\partial n)(\partial\mathcal{G}_2/\partial n)$; $\mathcal{G}_1$ and $\mathcal{G}_2$ are Green's functions satisfying the two Helmholtz equations and vanishing on Σ; r_s is the distance from S_s to P_s.

The integral in Eq. (2.19) can be interpreted as a volume integral in a four space, $v^{(4)}$ but the integrand is nonzero only in a subspace, $v^{(2)}$ of $v^{(4)}$. Moreover, since the intensity is everywhere finite, the integral is identically zero for all P_1 and P_2 ($P_1=P_2$ included). This result is, however, contrary to the hypothesis that Eq. (2.18) holds throughout V. Since Eq. (2.19) is a direct consequence of the Helmholtz equations which must be satisfied by $\Gamma_{12}(0)$ in free space, we conclude that *an incoherent field cannot exist in free space.*

We may, however, define an incoherent *source* as one for which the mutual intensity is of the form (cf., Blanc-Lapierre and Dumontet[7]),

$$\Gamma(S_1,S_2,0)=I(S_2)\delta(S_2-S_1), \qquad (2.20)$$

for all pairs of source points (S_1 and S_2), where $\delta(S_2-S_1)$ is the Dirac delta function.

3. PROPAGATION OF MUTUAL COHERENCE WAVES

In this section we shall determine the mutual coherence function for a field created by a plane polychromatic source. In Fig. 1, which serves to define the coordinates, σ is the plane containing the extended polychromatic source with a known distribution of mutual coherence. P_1 and P_2 are points in the field and S_1 and S_2 are points in the source. P_1' and P_2' are the mirror images of P_1 and P_2 with respect to the plane σ.

As shown in Sec. 1 the propagation of the mutual coherence in vacuum is governed by the two wave equations

$$\nabla_s^2\Gamma_{12}(\tau)=(1/c^2)[\partial^2\Gamma_{12}(\tau)/\partial\tau^2] \quad (s=1,2). \qquad (3.1)$$

We assume that $\Gamma_{12}(\tau)$ is known for all pairs of points S_1 and S_2 in the plane of σ. Let $G_{12}(\nu)$ be the Fourier spectrum of $\Gamma_{12}(\tau)$, since $\Gamma_{12}(\tau)$ is an analytic signal it contains only positive frequencies, i.e.,

$$\Gamma_{12}(\tau)=\int_0^\infty G_{12}(\nu)e^{-2\pi i\nu\tau}d\nu; \qquad (3.2)$$

and by the Fourier inversion theorem

$$G_{12}(\nu)=\int_{-\infty}^{\infty}\Gamma_{12}(\tau)e^{-2\pi i\nu\tau}d\tau. \qquad (3.3)$$

Substituting Eq. (3.2) into Eq. (3.1) and interchanging the order of integration and differentiation we obtain

$$\int_0^\infty [\nabla_s^2+k^2(\nu)]G_{12}(\nu)e^{-2\pi i\nu\tau}d\nu=0. \qquad (3.4)$$

Since Eq. (3.4) must hold for all τ we have

$$[\nabla_s^2+k^2(\nu)]G_{12}(\nu)=0, \qquad (3.5)$$

where $k^2(\nu)=(2\pi\nu/c)$. Thus each spectral component of $\Gamma_{12}(\tau)$ satisfies two scalar Helmholtz equations. Equations (3.5) can be formally solved by employing Green's functions. To this end we integrate first over the coordinates of S_1 and obtain†

$$G(P_1,S_2,\nu)=-(1/4\pi)\int_\sigma G(S_1,S_2,\nu)(\partial\mathcal{G}_1/\partial n)d\sigma_1, \qquad (3.6)$$

† $G_{12}(\nu)$ and $\Gamma_{12}(\tau)$ will be written as $G(P_1,P_2,\nu)$ and $\Gamma(P_1,P_2,\tau)$ when necessary to stress the space dependence.

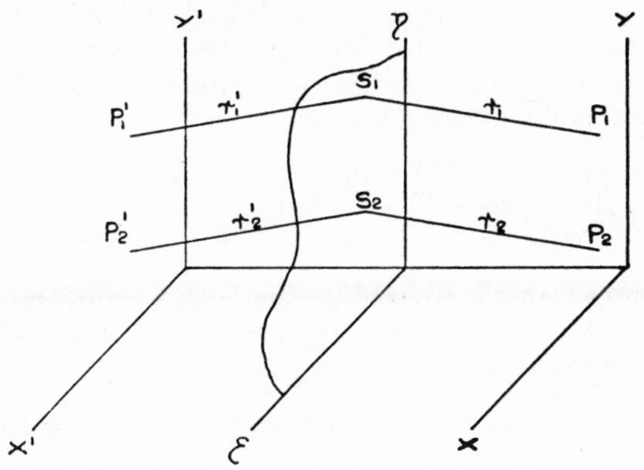

FIG. 1.

where $\mathcal{G}_1$ is a Green function satisfying the equation

$$[\nabla_1^2+k^2]\mathcal{G}_1=0, \qquad (3.7)$$

with boundary condition

$$\mathcal{G}_1(S)=0, \quad S\subset\sigma. \qquad (3.8)$$

Equation (3.6) provides the boundary condition for the solution of the second Helmholtz equation and employing the same theorem we obtain

$$G(P_1,P_2,\nu)=-(1/4\pi)\int_\sigma G(P_1,S_2,\nu)(\partial\mathcal{G}_2/\partial n)d\sigma, \qquad (3.9)$$

by integrating over the coordinates of S_2, where $\mathcal{G}_2$ is a second Green function satisfying the same conditions as $\mathcal{G}_1$. Substituting Eq. (3.6) into Eq. (3.9) yields

$$G(P_1,P_2,\nu)=\frac{1}{(4\pi)^2}\int_\sigma\int_\sigma G(S_1,S_2,\nu)\frac{\partial\mathcal{G}_1}{\partial n}\frac{\partial\mathcal{G}_2}{\partial n}d\sigma_1 d\sigma_2. \qquad (3.10)$$

In order to determine the exact form of $\mathcal{G}_1$ and $\mathcal{G}_2$ we must impose the radiation condition on $G(P_1,P_2,\nu)$. By applying the scalar radiation condition to $v_1(\nu)$ and $v_2^*(\nu)$ and appealing to the definition of $G(P_1,P_2,\nu)$ in terms of these functions it can be shown that $G(P_1,P_2,\nu)$ must behave asymptotically as

$$G(P_1,P_2,\nu)\rightarrow f(\theta_1,\theta_2,\phi_1,\phi_2)\frac{e^{ik(r_1-r_2)}}{r_1r_2}, \qquad (3.11)$$

where θ_s and ϕ_s are the polar angular coordinates of P_s. To satisfy this condition and those defining $\mathcal{G}_1$ and $\mathcal{G}_2$ the required Green functions are

$$\mathcal{G}_1=\frac{e^{ikr_1}}{r_1}-\frac{e^{ikr'_1}}{r'_1}$$

and

$$\mathcal{G}_2=\frac{e^{-ikr_2}}{r_2}-\frac{e^{-ikr'_2}}{r'_2} \qquad (3.12)$$

where r_1, r_2, r_1', and r_2' are defined in Fig. 1. That these are the required Green functions can be seen by direct substitution. Before substituting into Eq. (3.10) we obtain the normal derivatives as

$$\frac{\partial \mathcal{G}_1}{\partial n}=(ikr_1-1)\frac{e^{ikr_1}}{r_1}\frac{\partial r_1}{\partial n}+(1-ikr'_1)\frac{e^{ikr'_1}}{r'^2_1}\frac{\partial r'_1}{\partial n}. \quad (3.13)$$

Noting that $r_1|_\sigma=r_1'|_\sigma$ and that $(\partial r_1/\partial n)=z=-(\partial r_1'/\partial n)$, and setting $\cos\theta_s=(z_s/r_s)$ Eq. (3.13) can be written as

$$\partial\mathcal{G}_1/\partial n=2(ikr_1-1)\cos\theta_1(e^{ikr_1}/r_1), \quad (3.14)$$

and similarly

$$\partial\mathcal{G}_2/\partial n=-2(ikr_2+1)\cos\theta_2(e^{-ikr_2}/r_2).$$

On substituting Eq. (3.14) into Eq. (3.10) we obtain

$$G(P_1,P_2,\nu)=-\frac{1}{(2\pi)^2}\int_\sigma\int_\sigma G(S_1,S_2,\nu)$$
$$\times(ikr_2+1)(ikr_1-1)\cos\theta_1\cos\theta_2\frac{e^{ik(r_1-r_2)}}{r_1r_2}d\sigma_1d\sigma_2. \quad (3.15)$$

Equation (3.15) is the contribution from a single spectral component ν. The complete solution is obtained from Eq. (3.15) by integrating over the whole spectrum, i.e., taking the Fourier transform,

$$\Gamma_{12}(\tau)=\int_0^\infty\frac{1}{(2\pi)^2}\int_\sigma\int_\sigma G(S_1,S_2,\nu)(1+ikr_2)$$
$$\times(1-ikr_1)\cos\theta_1\cos\theta_2\frac{e^{ik(r_1-r_2)}}{r_1r_2}d\sigma_1d\sigma_2e^{-2\pi i\nu\tau}d\nu. \quad (3.16)$$

Since $d\sigma_1$ and $d\sigma_2$ and $d\nu$ are independent we invert the order of integration and obtain

$$\Gamma_{12}(\tau)=\frac{1}{(2\pi)^2}\int_\sigma\int_\sigma\frac{\cos\theta_1\cos\theta_2}{r_1r_2}$$
$$\times\int_0^\infty G(S_1,S_2,\nu)(1+ikr_2)(1-ikr_1)$$
$$\times\exp\left[-2\pi i\nu\left(\tau-\frac{r_1-r_2}{c}\right)\right]d\nu d\sigma_1d\sigma_2. \quad (3.17)$$

Expanding the inner integral we obtain

$$\int_0^\infty G(S_1,S_2,\nu)\exp\left[-2\pi i\left(\tau-\frac{r_1-r_2}{c}\right)\right]d\nu+\frac{i(r_2-r_1)}{c}$$
$$\times\int_0^\infty 2\pi\nu G(S_1,S_2,\nu)\exp\left[-2\pi i\nu\left(\tau-\frac{r_1-r_2}{c}\right)\right]d\nu$$
$$+\frac{r_1r_2}{c^2}\int_0^\infty \nu^2 4\pi^2 G(S_1,S_2,\nu)\exp\left[-2\pi i\nu\left(\tau-\frac{r_1-r_2}{c}\right)\right]d\nu.$$

For any function $f(t)$ and its half-range Fourier transform $g(\nu)$ we have

$$\frac{\partial^n f(t)}{\partial t^n}=\int_0^\infty(-2\pi i\nu)^n g(\nu)\exp(-2\pi i\nu t)d\nu. \quad (3.18)$$

Using Eq. (3.18), the integrals in the foregoing can be evaluated yielding

$$\Gamma\left(S_1,S_2,\tau-\frac{r_1-r_2}{c}\right)+\frac{r_1-r_2}{c}\frac{\partial}{\partial\tau}\Gamma\left(S_1,S_2,\tau-\frac{r_1-r_2}{c}\right)$$
$$-\frac{r_1r_2}{c^2}\frac{\partial^2}{\partial\tau^2}\Gamma\left(S_1,S_2,\tau-\frac{r_1-r_2}{c}\right), \quad (3.19)$$

and the final solution is

$$\Gamma(P_1,P_2,\tau)=\frac{1}{(2\pi)^2}\int_\sigma\int_\sigma\frac{\cos\theta_1\cos\theta_2}{r_1r_2}$$
$$\times\left[1+\frac{r_1-r_2}{c}\frac{\partial}{\partial\tau}-\frac{r_1r_2}{c}\frac{\partial^2}{\partial\tau^2}\right]$$
$$\times\Gamma\left(S_1,S_2,\tau-\frac{r_1-r_2}{c}\right)d\sigma_1d\sigma_2. \quad (3.20)$$

Equation (3.20) is the general solution for the mutual coherence in a field illuminated by a plane polychromatic source and expresses $\Gamma(P_1,P_2,\tau)$ in terms of its value on the source.

As mentioned earlier the most interesting case from a practical point of view is the quasi-monochromatic field. We shall therefore discuss in some detail the limiting forms of Eq. (3.20) for such fields. Under the justifying conditions the quasi-monochromatic approximation may be expressed in the form (cf., Sec. 2)

$$\Gamma_{12}(\tau)\simeq\Gamma_{12}(0)\exp(-2\pi i\bar{\nu}\tau)\quad\left(|\tau|\ll\frac{1}{\Delta\nu}\right); \quad (3.21)$$

substituting Eq. (3.21) into Eq. (3.20) yields

$$\Gamma_{12}(\tau)\simeq\frac{\exp(-2\pi i\bar{\nu}\tau)}{(2\pi)^2}\int_\sigma\int_\sigma(1-ikr_1)(1+ikr_2)$$
$$\times\cos\theta_1\cos\theta_2\Gamma(S_1,S_2,0)r_1r_2e^{ik(r_1-r_2)}d\sigma_1d\sigma_2. \quad (3.22)$$

To examine the incoherent limit we take $\Gamma(S_1,S_2,\tau)$ on the source to be of the form

$$\Gamma_{12}(\tau)\simeq I(S_2)\delta(S_2-S_1)e^{-2\pi i\bar{\nu}\tau}\quad\left(|\tau|\ll\frac{1}{\Delta\nu}\right). \quad (3.23)$$

Substituting Eq. (3.23) into Eq. (3.22) and integrating

over $d\sigma$, we obtain

$$\Gamma_{12}(\tau)\simeq\frac{\exp(-2\pi i\bar{\nu}\tau)}{(2\pi)^2}\int_\sigma I(S)(1-ikr_1)(1+ikr_2)$$

$$\times\cos\theta_1\cos\theta_2\frac{e^{ik(r_1-r_2)}}{r_1r_2}d\sigma\quad\left(|\tau|\ll\frac{1}{\Delta\nu}\right),\quad(3.24)$$

where r_1 and r_2 are now interpreted as the distance from a typical point, S, in the source to the field points P_1 and P_2, respectively, and $\cos\theta_s=z/r_s$ $(s=1,2)$. Equation (3.24) expresses the mutual coherence for sufficiently small τ in terms of the intensity distribution across the source. If Eq. (3.24) is evaluated at $\tau=0$ we obtain

$$\Gamma_{12}(0)\simeq\frac{1}{(2\pi)^2}\int_\sigma I(S)(1-ikr_1)(1+ikr_2)$$

$$\times\cos\theta_1\cos\theta_2\frac{e^{ik(r_1-r_2)}}{r_1r_2}d\sigma.\quad(3.25)$$

Equation (3.25) expresses the field created by an incoherent plane source in terms of the intensity distribution across the source. If attention is limited to the field on a plane parallel to the source plane and the obliquity neglected Eq. (3.25) reduces to the well-known theorem of Van Cittert and Zernike.

In most applications one is interested in the form of Eq. (3.25) in the Fraunhofer region. Under the usual approximations characterizing the far field, the right-hand side of Eq. (3.25) reduces to a Fourier transform of the intensity distribution. If the intensity distribution is suitably normalized Eq. (3.25) becomes

$$\gamma_{12}(0)=\int_\sigma I(\xi,\eta)\exp[ik(p\xi+q\eta)]d\xi d\eta,\quad(3.26)$$

where $p=(x_1/R_1)-(x_2/R_2)$, $q=(y_1/R_1)-(y_2/R_2)$, and $R_s=(x_s^2+y_s^2+z_s^2)^{\frac{1}{2}}$.

We next examine briefly the field due to a coherent quasi-monochromatic source. In this case the source distribution will be given by (cf., Sec. 2)

$$\Gamma(S_1,S_2,\tau)\simeq u_s(S_1)u_s^*(S_2)\,e^{-2\pi i\bar{\nu}\tau}\quad\left(|\tau|\ll\frac{1}{\Delta\nu}\right).\quad(3.27)$$

Substituting Eq. (3.27) into Eq. (3.22) yields

$$\Gamma(P_1,P_2,\tau)\simeq u(P_1)u^*(P_2)\,e^{-2\pi i\bar{\nu}\tau}\quad\left(|\tau|\ll\frac{1}{\Delta\nu}\right),\quad(3.28)$$

where

$$u(P_1)\simeq\frac{1}{2\pi}\int_\sigma u_s(S_1)(1-ikr_1)\cos\theta_1\frac{e^{ikr_1}}{r_1}d\sigma.\quad(3.29)$$

From Eq. (3.28) and the theorems established in Sec. 2 it is clear that *a coherent quasi-monochromatic source produces a coherent field.*

ACKNOWLEDGMENT

The author wishes to express his gratitude to Dr. E. Wolf both for suggesting the problems discussed in this paper and for helpful discussions as the work progressed.

PAPER NO. 31

Reprinted from *Il Nuovo Cimento*, Vol. 13, pp. 1165–1181, 1959.

Coherence Properties of Partially Polarized Electromagnetic Radiation (*).

E. Wolf (**)

Department of Theoretical Physics - University of Manchester

(ricevuto il 4 Giugno 1959)

Summary. — This paper is concerned with the analysis of partial polarization from the standpoint of coherence theory. After observing that the usual analytic definition of the Stokes parameters of a quasi-monochromatic wave are not unique, a simple experiment is analysed, which brings out clearly the observable parameters of a quasi-monochromatic light wave. The analysis leads to a unique coherency matrix and to a unique set of Stokes parameters, the latter being associated with the representation of the coherency matrix in terms of Pauli's spin matrices. In this analysis the concept of Gabor's analytic signal proves to be basic. The degree of coherence between the electric vibrations in any two mutually orthogonal directions of propagation of the wave depends in general on the choice of the two orthogonal directions. It is shown that its maximum value is equal to the degree of polarization of the wave. It is also shown that the degree of polarization may be determined in a new way from relatively simple experiments which involve a compensator and a polarizer, and that this determination is analogous to the determination of the degree of coherence from Young's interference experiment.

1. – Introduction.

Although it is generally known that there is an intimate connection between partial polarization and partial coherence (see, for example, Wiener (1), p. 187), a systematic analysis of the properties of partially polarized radiation from

(*) The research described in this paper has been partially sponsored by the Air Force Cambridge Research Center of the Air Research and Development Command, United States Air Force, through its European Office, under Contract No. AF 61(052)-169.

(**) Now at the Institute of Optics, University of Rochester, Rochester, N.Y.

(1) N. Wiener: *Acta Math.*, **55**, § 9 (1930).

the standpoint of coherence theory does not appear to have been made so far. There are several reasons why it seems desirable to carry out such an analysis. The numerous investigations made in recent years in connection with partially coherent light (ZERNIKE (2), HOPKINS (3), WOLF (4a,b,c), BLANC-LAPIERRE and DUMONTET (5)) have clearly shown that there is still a great deal to be learnt about the statistical properties of high frequency electromagnetic radiation. Moreover, as we shall briefly indicate, the usual treatments of partial polarization are not entirely satisfactory.

The most systematic treatments of partial polarization utilize the concept of Stokes parameters (introduced by G. G. STOKES in 1852 (6)) which are usually defined as follows: Consider a plane, quasi-monochromatic electromagnetic wave and let the components of the electric vector in two mutually perpendicular directions at right angles to the direction of propagation of the wave be represented in the form

$$(1.1)\qquad \begin{cases} E_x(t) = A_1(t)[\cos \Phi_1(t) - \bar{\omega}t]\,, \\ E_y(t) = A_2(t)[\cos \Phi_2(t) - \bar{\omega}t]\,, \end{cases}$$

where $\bar{\omega}$ denotes the mean frequency and t the time. The Stokes parameters are the four quantities

$$(1.2)\qquad \begin{cases} s_0 = \langle A_1^2\rangle + \langle A_2^2\rangle\,, \\ s_1 = \langle A_1^2\rangle - \langle A_2^2\rangle\,, \\ s_2 = 2\langle A_1 A_2 \cos(\Phi_1 - \Phi_2)\rangle\,, \\ s_3 = 2\langle A_1 A_2 \sin(\Phi_1 - \Phi_2)\rangle\,, \end{cases}$$

where the sharp brackets denote time average. Although to-day there exists an extensive literature in which the Stokes parameters play a central role (see, for example CHANDRASEKHAR (7)), it does not appear to have been noticed that the above relations do not define the parameters uniquely. For in (1.1) only $E_x(t)$ and $E_y(t)$ can be regarded as uniquely associated with the wave, whereas the A's and Φ's may evidently be chosen in many ways, leading to different sets of Stokes parameters. Only a careful analysis of experiment may be expected to lead to a unique set. Such an analysis is carried out in

(2) F. ZERNIKE: *Physica*, **5**, 785 (1938).

(3) H. H. HOPKINS: *Proc. Roy. Soc.*, A **208**, 263 (1951).

(4) E. WOLF: (*a*) *Proc. Roy. Soc.*, A **225**, 96 (1954); (*b*) *Nuovo Cimento*, **12**, 884 (1954); (*c*) *Proc. Roy. Soc.*, A **230**, 96 (1955).

(5) A. BLANC-LAPIERRE and P. DUMONTET: *Revue d'Optique*, **34**, 1 (1955).

(6) G. G. STOKES: *Trans. Camb. Phil. Soc.*, **9**, 399 (1852); also his *Mathematical and Physical Papers*, vol. **3** (Cambridge, 1901), p. 233.

(7) S. CHANDRASEKHAR: *Radiative Transfer* (Oxford, 1950), § 15.

Section **2** of this paper and shows that the unique set obtained is intimately related to the appropriate « degree of coherence » of the electric vibrations in the two orthogonal directions.

Another unsatisfactory feature of the usual treatments of partial polarization has been clearly pointed out in an interesting recent paper by PANCHARANTHAM ([8], expecially p. 399-340): A partially polarized beam is usually described in terms of incoherent superposition of polarized and unpolarized beams and the interference phenomena arising from the superposition of these beams are analysed by using the concept of coherence and incoherence alone. However, the decomposition may be carried out in many different ways and it is by no means evident that the different decompositions will always lead to identical results. In any case this approach masks completely the *invariant characteristics* of the different representations. These unsatisfactory features can only be expected to be removed by the introduction of intermediate states (partial coherence).

In the present paper the basic properties of a quasi-monochromatic partially polarized electromagnetic wave are discussed from the standpoint of coherency theory and the invariant characteristics of such a wave are clearly brought out. The basic tool used for this purpose is the coherency matrix introduced in a previous paper (WOLF ([4b])), specialized to the problem in question; however, unlike in the previous paper, the coherency matrix is introduced here from the analysis of a simple experiment.

2. – The coherency matrix of a plane, quasi-monochromatic electromagnetic wave.

Consider a plane, quasi-monochromatic wave and let $E_x^{(r)}(t)$ and $E_y^{(r)}(t)$ represent (*) the components of the electric vector $\boldsymbol{E}^{(r)}$ at a typical point in the wave field in two mutually orthogonal directions at right angles to the directions of propagation of the wave. We assume that $\boldsymbol{E}^{(r)}$ may be represented as a Fourier integral and write

$$(2.1)\qquad \left\{\begin{aligned} E_x^{(r)}(t) &= \int_0^\infty a_1(\omega)\cos[\varphi_1(\omega)-\omega t]\,\mathrm{d}\omega\,,\\ E_y^{(r)}(t) &= \int_0^\infty a_2(\omega)\cos[\varphi_2(\omega)-\omega t]\,\mathrm{d}\omega\,. \end{aligned}\right.$$

([8]) S. PANCHARATNAM: *Proc. Ind. Acad. Sci.*, A **44**, 398 (1956),

(*) The superscript « r » is introduced because the (real) electric wave function will shortly be regarded as the real part of a suitably chosen complex wave function.

Since the wave is assumed to be quasi-monochromatic, the spectral amplitudes $a_1(\omega)$ and $a_2(\omega)$ will be appreciable only in a narrow range

$$\bar{\omega} - \tfrac{1}{2}\Delta\omega \leqslant \omega \leqslant \bar{\omega} + \tfrac{1}{2}\Delta\omega \,, \tag{2.2}$$

where $\Delta\omega$ is small compared with the mean frequency $\bar{\omega}$.

Suppose that the wave is passed through a device (compensator) which introduces retardations in $E_x^{(r)}$ and $E_y^{(r)}$. Let $\varepsilon_1(\omega)$ and $\varepsilon_2(\omega)$ be the phase delays in the Fourier components of frequency ω of $E_x^{(r)}$ and $E_y^{(r)}$ respectively. The electric wave emerging from the compensator has the components

$$\left\{\begin{aligned} \mathscr{E}_x^{(r)}(t) &= \int_0^\infty a_1(\omega)\cos[\varphi_1(\omega) - \varepsilon_1(\omega) - \omega t]\,\mathrm{d}\omega \,, \\ \mathscr{E}_y^{(r)}(t) &= \int_0^\infty a_2(\omega)\cos[\varphi_2(\omega) - \varepsilon_2(\omega) - \omega t]\,\mathrm{d}\omega \,, \end{aligned}\right. \tag{2.3}$$

it being assumed that reflection and absorption losses are negligible. If we use the identity $\cos[A - \varepsilon] = \cos A \cos\varepsilon + \sin A \sin\varepsilon$ in (2.3) and assume that for any two frequencies ω' and ω'' in the range (2.2)

$$|\varepsilon_1(\omega') - \varepsilon_1(\omega'')| \ll 2\pi \,, \qquad |\varepsilon_2(\omega') - \varepsilon_2(\omega'')| \ll 2\pi \,, \tag{2.4}$$

(2.3) may be re-written as

$$\left\{\begin{aligned} \mathscr{E}_x^{(r)} &= E_x^{(r)}(t)\cos\bar{\varepsilon}_1 + E_x^{(i)}(t)\sin\bar{\varepsilon}_1 \,, \\ \mathscr{E}_y^{(r)} &= E_y^{(r)}(t)\cos\bar{\varepsilon}_2 + E_y^{(i)}(t)\sin\bar{\varepsilon}_2 \,, \end{aligned}\right. \tag{2.5}$$

where $\bar{\varepsilon}_1 = \varepsilon_1(\bar{\omega})$, $\bar{\varepsilon}_2 = \varepsilon_2(\bar{\omega})$ and $E_x^{(i)}$ and $E_y^{(i)}$ are the Fourier integrals *conjugate* to $E_x^{(r)}$ and $E_y^{(r)}$ respectively, *i.e.*

$$\left\{\begin{aligned} E_x^{(i)} &= \int_0^\infty a_1(\omega)\sin[\varphi_1(\omega) - \omega t]\,\mathrm{d}\omega \,, \\ E_y^{(i)} &= \int_0^\infty a_2(\omega)\sin[\varphi_2(\omega) - \omega t]\,\mathrm{d}\omega \,. \end{aligned}\right. \tag{2.6}$$

As is well known (cf. TITCHMARSH [9]), any two conjugate functions are related

---

([9]) E. C. TITCHMARSH: *Introduction to the Theory of Fourier Integrals* (Oxford, 1948), 2nd ed., chap. V.

by Hilbert's reciprocity relations, *i.e.* by relations of the form

$$(2.7)\qquad E_x^{(i)}(t) = \frac{P}{\pi}\int_{-\infty}^{\infty}\frac{E_x^{(r)}(t')}{t'-t}\,\mathrm{d}t'\,, \qquad E_x^{(r)}(t) = -\frac{P}{\pi}\int_{-\infty}^{\infty}\frac{E_x^{(i)}(t')}{t'-t}\,\mathrm{d}t'\,,$$

P denoting the Cauchy principal value at $t'=t$.

Suppose now that the wave emerging from the compensator is sent through a polarizer, which only transmits the component which makes an angle θ with the x-direction. This component is given by

$$(2.8)\qquad \mathscr{E}^{(r)}(t;\,\theta,\,\varepsilon_1,\,\varepsilon_2) = \mathscr{E}_x^{(r)}(t)\cos\theta + \mathscr{E}_y^{(r)}(t)\sin\theta\,,$$

so that the intensity of the light emerging from the polarizer is

$$(2.9)\qquad \begin{aligned} I(\theta_1,\,\varepsilon_1,\,\varepsilon_2) &= 2\langle\mathscr{E}^{(r)^2}(t,\,\theta,\,\varepsilon_1,\,\varepsilon_2)\rangle \\ &= 2\langle\mathscr{E}_x^{(r)^2}\rangle\cos^2\theta + 2\langle\mathscr{E}_y^{(r)^2}\rangle\sin^2\theta + 4\langle\mathscr{E}_x^{(r)}\mathscr{E}_y^{(r)}\rangle\cos\theta\sin\theta\,, \end{aligned}$$

where sharp brackets denote time average. Here the wave field was assumed to be stationary (*), so that the intensity I is independent of the time instant at which the average is taken and the factor 2 on the right of the first equation of (2.9) was introduced to simplify later calculations. Next we substitute from (2.5) into (2.9) and use the following relations which may be readily proved from the properties of Hilbert transforms (**):

$$(2.10)\qquad \left\{\begin{aligned} &\langle E_x^{(r)^2}\rangle = \langle E_x^{(i)^2}\rangle\,, \qquad \langle E_y^{(r)^2}\rangle = \langle E_y^{(i)^2}\rangle\,, \\ &\langle E_x^{(r)}E_y^{(r)}\rangle = \langle E_x^{(i)}E_y^{(i)}\rangle\,, \\ &\langle E_x^{(r)}E_y^{(i)}\rangle = -\langle E_x^{(i)}E_y^{(r)}\rangle\,, \\ &\langle E_x^{(r)}E_x^{(i)}\rangle = \langle E_y^{(i)}E_y^{(r)}\rangle = 0\,. \end{aligned}\right.$$

(*) Stationarity in the strict sense of the theory of random functions would imply that the field vectors are not square integrable and hence we would not be justified in using Fourier integral analysis. This difficulty may be avoided in the usual way by assuming that the field exists only for a finite time interval $-T\leqslant t\leqslant T$ and proceeding to the limit $T\to\infty$ at the end of the calculations. The final results are the same whether or not this refinement is made.

(**) The formulae (2.10) are valid quite generally. When the field is quasi-monochromatic as here assumed, they may be proved in a very simple way by the following

Further, if we set

$$(2.11) \qquad \delta = \bar{\varepsilon}_1 - \bar{\varepsilon}_2 ,$$

and write $I(\theta, \delta)$ in place of $I(\theta, \bar{\varepsilon}_1, \bar{\varepsilon}_2)$ [since $\bar{\varepsilon}_1$ and $\bar{\varepsilon}_2$ enter the expression for the intensity only through their difference] we obtain the following expression for the time averaged intensity:

$$(2.12) \qquad I(\theta, \delta) = 2\langle E_x^{(r)^2}\rangle \cos^2\theta + 2\langle E_y^{(r)^2}\rangle \sin^2\theta + \\ + 4\cos\theta \sin\theta\{\cos\delta\langle E_x^{(r)}E_y^{(r)}\rangle - \sin\delta\langle E_x^{(r)}E_y^{(i)}\rangle\} .$$

The formulae (2.12) may be expressed in a more convenient form, by using in place of the real wave functions the associated analytic signals of GABOR ([11]),

---

argument due to BRACEWELL [([10]), p. 102]. We set

$$E_x(t) = E_x^{(r)}(t) + iE_x^{(i)}(t) = A_1(t) \exp[-i\bar{\omega}t] ,$$

$$E_y(t) = E_y^{(r)}(t) + iE_y^{(i)}(t) = A_2(t) \exp[-i\bar{\omega}t] .$$

Then, if the field is quasi-monochromatic, the (generally complex) quantities A_1 and A_2 will vary slowly with t in comparison with the periodic term, and we have, for example,

$$\langle E_x^{(r)^2}\rangle = \left\langle\left(\frac{A_1 \exp[-i\omega t] + A_1^* \exp[i\omega t]}{2}\right)^2\right\rangle = \\ = \frac{1}{4}\langle A_1^2 \exp[-2i\bar{\omega}t]\rangle + \frac{1}{2}\langle A_1A_1^*\rangle + \frac{1}{4}\langle A_1^{*2} \exp[2i\bar{\omega}t]\rangle .$$

The first and the last term on the right vanish because of the rapidly varying terms $\exp[-2i\bar{\omega}t]$ and $\exp[2i\bar{\omega}t]$, so that

$$\langle E_x^{(r)^2}\rangle = \tfrac{1}{2}\langle A_1A_1^*\rangle .$$

Similarly

$$\langle E_x^{(i)^2}\rangle = \left\langle\left(\frac{A_1 \exp[-i\bar{\omega}t] - A_1^* \exp[i\bar{\omega}t]}{2i}\right)^2\right\rangle = \frac{1}{2}\langle A_1A_1^*\rangle .$$

Comparison of the last two formulae gives the first relation in (2.10); the other relations may be proved in a similar way.

([10]) R. N. BRACEWELL: *Proc. I.R.E.*, **46**, 97 (1958).

([11]) D. GABOR: *Journ. Inst. Elect. Engrs.*, **93**, part III, 429 (1946).

i.e. by using in place of $E_x^{(r)}$ and $E_y^{(r)}$, the functions (*)

$$(2.13)\quad \begin{cases} E_x(t) = E_x^{(r)}(t) + i\,E_x^{(i)}(t) = \int_0^\infty a_1(\omega)\exp\left[i[\varphi_1(\omega) - \overline{\omega}t]\right] d\omega\,, \\ E_y(t) = E_y^{(r)}(t) + i\,E_y^{(i)}(t) = \int_0^\infty a_2(\omega)\exp\left[i[\varphi_2(\omega) - \omega t]\right] d\omega\,. \end{cases}$$

Using (2.10) we have the relations

$$(2.14a)\qquad \langle E_x E_x^* \rangle = 2\langle E_x^{(r)2} \rangle = 2\langle E_x^{(i)2} \rangle\,,$$

$$(2.14b)\qquad \langle E_x E_y^* \rangle = 2\langle E_x^{(r)} E_y^{(r)} \rangle - 2i\langle E_x^{(i)} E_y^{(i)} \rangle\,,$$

etc. With the help of (2.14) the formula (2.12) becomes

$$(2.15)\qquad I(\theta, \delta) = J_{xx}\cos^2\theta + J_{yy}\sin^2\theta + J_{xy}\cos\theta\sin\theta\exp[-i\delta] + \\ + J_{yx}\sin\theta\cos\theta\exp[i\delta]\,,$$

where the J's are the elements of the *coherency matrix*

$$(2.16)\qquad \boldsymbol{J} = \begin{bmatrix} J_{xx} & J_{xy} \\ J_{yx} & J_{yy} \end{bmatrix} = \begin{bmatrix} \langle E_x E_x^* \rangle & \langle E_x E_y^* \rangle \\ \langle E_y E_x^* \rangle & \langle E_y E_y^* \rangle \end{bmatrix}.$$

The formula (2.15) expresses in a compact form the intensity of the wave after transmission through the compensator (which introduces a phase delay δ) and the polarizer (oriented so as to transmit the component which makes an angle θ with the x-axis) in terms of the coherency matrix $\boldsymbol{J}$ which characterizes the incident wave.

Since $J_{yx} = J_{xy}^*$ the coherency matrix is *Hermitian*. Its trace represents the intensity of the incident wave,

$$(2.17)\qquad \operatorname{Tr}\boldsymbol{J} = J_{xx} + J_{yy} = \langle E_x E_x^* \rangle + \langle E_y E_y^* \rangle = 2\langle E_x^{(r)2} \rangle + 2\langle E_y^{(r)2} \rangle\,,$$

and its non-diagonal elements express the correlation between the x and y-components of the complex vector $\boldsymbol{E}$. Further it follows from Schwarz' inequality

(*) An analytic signal is a complex function characterized by the property that its Fourier integral contains no spectral components of positive (or negative) frequencies. This fact alone implies that the real and imaginary parts of the signal are conjugate functions and hence Hilbert transforms of each other.

for integrals that $|J_{xy}| \leqslant \sqrt{J_{xx}}\sqrt{J_{yy}}$, $|J_{yx}| \leqslant \sqrt{J_{yy}}\sqrt{J_{xx}}$; hence, since $J_{yx} = J_{xy}^*$,

$$|\boldsymbol{J}| = J_{xx}J_{yy} - J_{xy}J_{yx} \geqslant 0 , \tag{2.18}$$

i.e. the discriminant of the coherency matrix is non-negative.

Let $A_1(t)$, $A_2(t)$ be the amplitudes and $\Psi_1(t)$ and $\Psi_2(t)$ the phases of $E_x(t)$ and $E_y(t)$ respectively, *i.e.*

$$E_x(t) = A_1(t) \exp[i\Psi_1(t)] , \qquad E_y(t) = A_2(t) \exp[i\Psi_2(t)] . \tag{2.19}$$

Then, from (2.13),

$$\begin{cases} E_x^{(r)}(t) = A_1(t) \cos[\Psi_1(t)] , & E_x^{(i)}(t) = A_1(t) \sin[\Psi_1(t)] , \\ E_y^{(r)}(t) = A_2(t) \cos[\Psi_2(t)] , & E_y^{(i)}(t) = A_2(t) \sin[\Psi_2(t)] , \end{cases} \tag{2.20}$$

and, if we introduce quantities $\Phi_1(t)$ and $\Phi_2(t)$ by the relations

$$\Psi_1(t) = \Phi_1(t) - \bar{\omega}t , \qquad \Psi_2(t) = \Phi_2(t) - \bar{\omega}t , \tag{2.21}$$

where $\bar{\omega}$ is the mean frequency, the components $E_x^{(r)}$, $E_y^{(r)}$ of the (real) electric vector are represented by expressions of the form (1.1), but the representation is now *unique*. In terms of the A's and Ψ's the elements of the coherency matrix are

$$\begin{cases} J_{xx} = \langle A_1^2 \rangle , \\ J_{yy} = \langle A_2^2 \rangle , \\ J_{xy} = \langle A_1 A_2 \exp[i(\Psi_1 - \Psi_2)] \rangle , \\ J_{yx} = \langle A_1 A_2 \exp[-i(\Psi_1 - \Psi_2)] \rangle . \end{cases} \tag{2.22}$$

We may now introduce a set of Stokes parameters by the relations

$$\begin{cases} s_0 = \langle A_1^2 \rangle + \langle A_2^2 \rangle & = J_{xx} + J_{yy} , \\ s_1 = \langle A_1^2 \rangle - \langle A_2^2 \rangle & = J_{xx} - J_{yy} , \\ s_2 = 2\langle A_1 A_2 \cos(\Psi_1 - \Psi_2) \rangle & = J_{xy} + J_{yx} , \\ s_3 = 2\langle A_1 A_2 \sin(\Psi_1 - \Psi_2) \rangle & = i(J_{yx} - J_{xy}) . \end{cases} \tag{2.23}$$

We see that this set of Stokes parameters is unique and that uniqueness has been achieved with the help of analytic signals, the introduction of which was suggested by the appearance of conjugate functions in the analysis of our experiment.

The relation between the Stokes parameters and the coherency matrix may also be expressed in the form

$$\boldsymbol{J} = \tfrac{1}{2}\sum_{i=0}^{3} s_i \boldsymbol{\sigma}_i\,, \tag{2.24}$$

where $\boldsymbol{\sigma}_0$ is the unit matrix

$$\boldsymbol{\sigma}_0 = \begin{bmatrix} 1 & 0 \\ 0 & 1 \end{bmatrix},$$

and $\boldsymbol{\sigma}_1$, $\boldsymbol{\sigma}_2$, $\boldsymbol{\sigma}_3$, are the Pauli spin matrices

$$\boldsymbol{\sigma}_1 = \begin{bmatrix} 1 & 0 \\ 0 & -1 \end{bmatrix}, \qquad \boldsymbol{\sigma}_2 = \begin{bmatrix} 0 & 1 \\ 1 & 0 \end{bmatrix}, \qquad \boldsymbol{\sigma}_3 = \begin{bmatrix} 0 & i \\ -i & 0 \end{bmatrix}. \tag{2.26}$$

The connection between a coherency matrix, Stokes parameters and Pauli's spin matrices has been noted previously (FANO (12)); however, as already mentioned, the non-uniqueness of the usual analytic definition of the Stokes parameters appears to have escaped attention.

Finally we mention that the coherency matrix (2.16) was introduced in an earlier paper (WOLF (4b)) from more formal considerations. Our present analysis shows that this matrix appears in a natural way from the analysis of a simple experiment.

3. – Some consequences of the basic intensity formula.

To see the physical significance of the intensity formula (2.15) we re-write it in a somewhat different form. We set

$$\frac{J_{xy}}{\sqrt{J_{xx}}\sqrt{J_{yy}}} = \mu_{xy} = |\mu_{xy}| \exp[i\beta_{xy}]\,. \tag{3.1}$$

---

(12) U. FANO: *Phys. Rev.*, **93**, 121 (1954).

It follows from (2.18) that

$$|\mu_{xy}| \leqslant 1 .$$

By analogy with the theory of partially coherent scalar fields we may call μ_{xy} *the complex degree of coherence* of the electric vibrations in the x and y directions. It absolute value $|\mu_{xy}|$ is a measure of the degree of correlation of the vibrations and its phase represents their « effective phase difference ».

If we substitute from (3.1) into the intensity formula (2.15) and use the relation $J_{yx} = J^*_{xy}$ we obtain the following expression for the intensity:

$$I(\theta, \delta) = J_{xx} \cos^2 \theta + J_{yy} \sin^2 \theta + 2\sqrt{J_{xx}}\sqrt{J_{yy}} \cos \theta \sin \theta \, |\mu_{xy}| \cos (\beta_{xy} - \delta) . \tag{3.3}$$

This expression is formally identical with the basic interference law of partially coherent fields [WOLF ([4a]), p. 102, THOMPSON and WOLF ([13]), p. 896]. It shows that the intensity $I(\theta, \delta)$ may be regarded as arising from the interference of two beams of intensities:

$$I^{(1)} = J_{xx} \cos^2 \theta , \qquad I^{(2)} = J_{yy} \sin^2 \theta , \tag{3.4}$$

and with complex degree of coherence μ_{xy}, after a phase difference δ has been introduced between them.

Returning to (2.15) we see that the elements of the coherency matrix of a quasi-monochromatic plane wave may be determined from very simple experiments. It is only necessary to measure the intensity for several different values of θ (orientation of polarizer) and δ (delay introduced by a compensator), and solve the corresponding relations obtained from (2.15). Let $\{\theta, \delta\}$, denote the measurement corresponding to a particular pair θ, δ. A convenient set of measurements is the following:

$$\{0°, 0\} , \quad \{45°, 0\} , \quad \{90°, 0\} , \quad \{135°, 0\} , \quad \left\{45°, \frac{\pi}{2}\right\} , \quad \left\{135°, \frac{\pi}{2}\right\} . \tag{3.5}$$

It follows from (2.15) that, in terms of the intensities determined from these six measurements, the elements of the coherency matrix are given by

$$\left\{\begin{aligned} J_{xx} &= I(0°, 0) , \\ J_{yy} &= I(90°, 0) , \\ J_{xy} &= \frac{1}{2}\{I(45°, 0) - I(135°, 0)\} + \frac{1}{2} i \left\{I\left(45°, \frac{\pi}{2}\right) - I\left(135°, \frac{\pi}{2}\right)\right\} , \\ J_{yx} &= \frac{1}{2}\{I(45°, 0) - I(135°, 0)\} - \frac{1}{2} i \left\{I\left(45°, \frac{\pi}{2}\right) - I\left(135°, \frac{\pi}{2}\right)\right\} . \end{aligned}\right. \tag{3.6}$$

([13]) B. J. THOMPSON and E. WOLF: *Journ. Opt. Soc. Amer.*, **47**, 895 (1957).

Thus we see that the elements of the coherency matrix represent ***measurable*** physical quantities.

In the theory of partially coherent scalar fields, the concept of Michelson's visibility (*) of fringes plays a central role (cf. ZERNIKE ([2])). We will now derive an expression for a quantity defined in a similar way and we shall see later that this quantity has a simple physical meaning.

It follows from (2.15) by a straightforward calculation, that the maxima and minima of the intensity (with respect to both θ and δ) are

$$(3.7)\qquad \left\{\begin{aligned} I_{\text{max}} &= \frac{1}{2}(J_{xx}+J_{yy})\left[1+\sqrt{1-\frac{4|\boldsymbol{J}|}{(J_{xx}+J_{yy})^2}}\right],\\ I_{\text{min}} &= \frac{1}{2}(J_{xx}+J_{yy})\left[1-\sqrt{1-\frac{4|\boldsymbol{J}|}{(J_{xx}+J_{yy})^2}}\right].\end{aligned}\right.$$

Hence

$$(3.8)\qquad \frac{I_{\text{max}}-I_{\text{min}}}{I_{\text{max}}+I_{\text{min}}}=\sqrt{1-\frac{4|\boldsymbol{J}|}{(J_{xx}+J_{yy})^2}}\,.$$

Now if the x and y axes are rotated about the direction of propagation of the wave, the coherency matrix will change. There are, however, two invariants for such rotations, namely the discriminant $|\boldsymbol{J}|$ and the trace $\text{Tr}\boldsymbol{J}=J_{xx}+J_{yy}$ of the matrix. Since on the right hand side of (3.8) the elements of $\boldsymbol{J}$ enter only in these combinations, it follows that the expression is invariant with respect to rotations of the axes and hence may be expected to have a physical significance. We shall see shortly (eq. (5.14) below) that it represents the degree of polarization of the wave.

4. – Coherency matrices of natural and of monochromatic radiation.

Light which is most frequently encountered in nature has the property that the intensity of its components in any direction perpendicular to the direction of propagation is the same; and, moreover, the intensity is not af-

---

(*) The visibility $\mathcal{V}$ of fringes at a point P in the fringe pattern is defined by the formula

$$\mathcal{V}=\frac{I_{\text{max}}-I_{\text{min}}}{I_{\text{max}}+I_{\text{min}}},$$

where I_{max} and I_{min} are the maximum and minimum intensities in the immediate neighbourhood of P.

fected by any previous retardation of one of the rectangular components relative to the other, into which the light may have been resolved. In other words

$$I(\theta,\delta) = \text{constant} \tag{4.1}$$

for all values of θ and δ. Such light is called *natural light*; and we may define « natural » electromagnetic radiation of any other spectral range in a strictly similar way.

It is evident from (3.3) that $I(\theta,\delta)$ is independent of δ and θ, if and only if

$$|\mu_{xy}| = 0\,, \qquad \text{and} \qquad J_{xx} = J_{yy}\,. \tag{4.2}$$

The first condition implies that the electric vibrations in the x and y directions are mutually incoherent. According to (3.1) and the relation $J_{yx} = J^*_{yx}$, (4.2) may also be written as

$$J_{xy} = J_{yx} = 0\,, \qquad J_{xx} = J_{yy}\,, \tag{4.3}$$

and it follows that *the coherency matrix of natural radiation* of intensity $J_{xx} + J_{yy} = I$ is

$$\frac{1}{2} I \begin{bmatrix} 1 & 0 \\ 0 & 1 \end{bmatrix}. \tag{4.4}$$

Next let us consider the coherency matrix of monochromatic radiation. In this case the amplitudes A_1 and A_2 and the phases Ψ_1 and Ψ_2 in (2.22) are independent of time and the coherency matrix has the form

$$\begin{bmatrix} A_1^2 & A_1 A_2 \exp[i(\Psi_1 - \Psi_2)] \\ A_1 A_2 \exp[-i(\Psi_1 - \Psi_2)] & A_2^2 \end{bmatrix}. \tag{4.5}$$

We see that in this case

$$|\boldsymbol{J}| = J_{xx} J_{yy} - J_{xy} J_{yx} = 0\,, \tag{4.6}$$

i.e. the discriminant of the coherency matrix is zero. The complex degree of coherence now is

$$\mu_{xy} = \frac{J_{xy}}{\sqrt{J_{xx}}\sqrt{J_{yy}}} = \exp[(\Psi_1 - \Psi_2)\,, \tag{4.7}$$

i.e. its absolute value is unity (complete coherence) and its phase is equal to the difference between the phases of the two components.

5. – The degree of polarization.

Before deriving an expression for the degree of polarization in terms of the coherency matrix we shall establish a simple theorem relating to the coherency matrix of a wave resulting from the superposition of a number of mutually independent waves.

Consider N mutually independent quasi-monochromatic waves propagated in the same direction (z-say) and let $E_x^{(n)}$, $E_y^{(n)}$ ($n = 1, 2, \ldots, N$) be the analytic signals associated with the components of the electric vibrations of the n-th wave in the directions of the x and y-axes. The components of the resulting wave then are

$$E_x = \sum_{n=1}^{N} E_x^{(n)}, \qquad E_y = \sum_{n=1}^{N} E_y^{(n)}, \tag{5.1}$$

so that the elements of the coherency matrix are

$$\left\{\begin{aligned} J_{kl} &= \langle E_k E_l^* \rangle = \sum_{n=1}^{N} \sum_{m=1}^{N} \langle E_k^{(n)} E_l^{(m)*} \rangle , \\ &= \sum_{n=1}^{N} \langle E_k^{(n)} E_l^{(n)*} \rangle + \sum_{n \neq m} \langle E_k^{(n)} E_l^{(m)*} \rangle . \end{aligned}\right. \tag{5.2}$$

Since the waves are assumed to be independent, each term under the last summation sign is zero, and it follows that

$$J_{kl} = \sum_{n=1}^{N} J_{kl}^{(n)} , \tag{5.3}$$

where $J_{kl}^{(n)} = \langle E_k^{(n)} E^{(n)*} \rangle$ are the elements of the coherency matrix of the n-th wave. The formula (5.3) shows that the coherency matrix of a wave resulting from the superposition of a number of independent waves is the sum of the coherency matrices of the individual waves.

To find an expression for the degree of polarization of a wave, we first represent the wave as a superposition of a wave of natural radiation and a wave of monochromatic radiation, independent of the former. Let $\mathbf{J}$ be the coherency matrix of the given wave and let $\mathbf{J}^{(1)}$ and $\mathbf{J}^{(2)}$ be the coherency matrices of the two independent waves into which we decompose it. Then according to (4.4) and (4.6) $\mathbf{J}^{(1)}$ and $\mathbf{J}^{(2)}$ must be of the form

$$\mathbf{J}^{(1)} = \begin{bmatrix} A & 0 \\ 0 & A \end{bmatrix}, \qquad \mathbf{J}^{(2)} = \begin{bmatrix} B & D \\ D^* & C \end{bmatrix}, \tag{5.4}$$

where $A \geqslant 0$, $B \geqslant 0$, $C \geqslant 0$ and

$$BC - DD^* = 0 . \tag{5.5}$$

In order to show that such a decomposition is possible we must determine quantities A, B, C, D, subject to the above conditions, such that the given coherency matrix $\mathbf{J} = [J_{ik}]$ is equal to the sum of two matrices of the form (5.4),

$$\mathbf{J} = \mathbf{J}^{(1)} + \mathbf{J}^{(2)} . \tag{5.6}$$

The relation (5.6) implies that

$$\begin{cases} J_{xx} = A + B , & J_{xy} = D , \\ J_{yx} = D^* , & J_{yy} = A + C . \end{cases} \tag{5.7}$$

On substituting for B, C, D and D^* from (5.7) into (5.5) we find that (*)

$$A = \tfrac{1}{2}(J_{xx} + J_{yy}) \pm \tfrac{1}{2}\sqrt{(J_{xx} + J_{yy})^2 - 4|\mathbf{J}|}. \tag{5.8}$$

Since $J_{yx} = J^*_{yx}$ the product $J_{xy}J^*_{yx}$ is non-negative, and it follows from (2.18) that

$$|\mathbf{J}| \leqslant J_{xx}J_{yy} \leqslant \tfrac{1}{4}(J_{xx} + J_{yy})^2 , \tag{5.9}$$

so that both the roots (5.8) are non-negative. Consider first the solution with the negative sign in front of the square root. We then have from (5.7),

$$\begin{cases} A = \tfrac{1}{2}(J_{xx} + J_{yy}) - \tfrac{1}{2}\sqrt{(J_{xx} + J_{yy})^2 - 4|\mathbf{J}|} , \\ B = \tfrac{1}{2}(J_{xx} - J_{yy}) + \tfrac{1}{2}\sqrt{(J_{xx} + J_{yy})^2 - 4|\mathbf{J}|} , \\ C = \tfrac{1}{2}(J_{yy} - J_{xx}) + \tfrac{1}{2}\sqrt{(J_{xx} + J_{yy})^2 - 4|\mathbf{J}|} , \\ D = J_{xy} , \\ D^* = J_{yx} . \end{cases} \tag{5.10}$$

Now

$$\sqrt{(J_{xx} + J_{yy})^2 - 4|\mathbf{J}|} = \sqrt{(J_{xx} - J_{yy})^2 + 4J_{xy}J_{yx}} \geqslant |J_{xx} - J_{yy}| . \tag{5.11}$$

Hence B and C are also non-negative as required. The other root given by (5.8) (with the positive sign in front of the square root) leads to negative values

(*) A is seen to be a characteristic root (eigenvalue) of the coherency matrix $\mathbf{J}$.

of B and C and must therefore be rejected. We have thus obtained a unique decomposition of the required kind.

The total intensity of the wave is

$$I_{\text{tot}} = \operatorname{Tr} \boldsymbol{J} = J_{xx} + J_{yy} \; ; \tag{5.12}$$

and the intensity of the monochromatic (and hence *polarized*) part is

$$I_{\text{pol}} = \operatorname{Tr} \boldsymbol{J}^{(2)} = B + C = \sqrt{(J_{xx} + J_{yy})^2 - 4|\boldsymbol{J}|} \, . \tag{5.13}$$

Hence the *degree of polarization* P of the original wave is

$$P = \frac{I_{\text{pol}}}{I_{\text{tot}}} = \sqrt{1 - \frac{4|\boldsymbol{J}|}{(J_{xx} + J_{yy})^2}} \, . \tag{5.14}$$

Since this expression involves only the two rotational invariants of the coherency matrix $\boldsymbol{J}$, the degree of polarization is independent of the particular choice of the x and y axes, as might have been expected.

Comparison of (5.14) with (3.8) shows that the quantity $(I_{\text{max}} - I_{\text{min}})/(I_{\text{max}} + I_{\text{min}})$ is precisely the degree of polarization P of the wave.

Unlike the degree of polarization, the degree of coherence depends on the choice of the x and y directions. We shall not investigate in detail the changes in the degree of coherence as the x, y axes are rotated; we shall only consider an extreme case, which is of special physical interest.

If in the expression (5.14) for the degree of polarization P we write out in full the discriminant $\boldsymbol{J}$, and use the expression (3.1) for the degree of coherence μ_{xy} we find that the following relation holds between P and $|\mu_{xy}|$:

$$1 - P^2 = \frac{J_{xx} J_{yy}}{[\frac{1}{2}(J_{xx} + J_{yy})]^2} [1 - |\mu_{xy}|^2] \, . \tag{5.15}$$

Since the geometric mean of any two positive numbers cannot exceed their arithmetic mean it follows that $1 - P^2 \leqslant 1 - |\mu_{xy}|^2$, *i.e.*

$$P \geqslant |\mu_{xy}| \, . \tag{5.16}$$

The equality sign in (5.16) will hold if an only if $J_{xx} = J_{yy}$, *i.e.* if the (time averaged) intensities in the two orthogonal directions are equal. We shall now show that a pair of directions always exists for which this is the case.

Suppose that we take a new pair of orthogonal directions x', y' perpendicular to the direction of propagation of the wave and let φ be the angle between x and x'. The components E_x, E_y, of the electric vector (in the complex representations (2.13)) in the new directions are

$$(5.17)\qquad \begin{cases} E_{x'} = \quad E_x \cos\varphi + E_y \sin\varphi\,, \\ E_{y'} = -E_x \sin\varphi + E_y \cos\varphi\,. \end{cases}$$

Hence the elements of the transformed coherency matrix $\mathsf{J}' = [J_{k'l'}] = [\langle E_{k'} E^*_{l'} \rangle]$ are

$$(5.18)\qquad \begin{cases} J_{x'x'} = J_{xx}c^2 + J_{yy}s^2 + (J_{xy} + J_{yx})cs\,, \\ J_{y'y'} = J_{xx}s^2 + J_{yy}c^2 - (J_{xy} + J_{yx})cs\,, \\ J_{x'y'} = (J_{yy} - J_{xx})cs + J_{xy}c^2 - J_{yx}s^2\,, \\ J_{y'x'} = (J_{yy} - J_{xx})cs + J_{yx}c^2 - J_{xy}s^2\,, \end{cases}$$

where

$$(5.19)\qquad c = \cos\varphi\,, \qquad s = \sin\varphi\,.$$

The intensities in the x' and y' directions will be equal (*i.e.* $J_{x'x'} = J_{y'y'}$) if

$$J_{xx}c^2 + J_{yy}s^2 + (J_{xy} + J_{yx})cs = J_{xx}s^2 + J_{yy}c^2 - (J_{xy} + J_{yx})cs\,.$$

Solving this equation for φ we obtain

$$(5.20)\qquad \operatorname{tg} 2\varphi = \frac{J_{yy} - J_{xx}}{J_{xy} + J_{yx}}\,.$$

Since $J_{yx} = J^*_{xy}$ and J_{xx} and J_{yy} are real, this equation always has a real root. Thus *there always exists a pair of directions for which the two intensities are equal. For this pair of directions the degree of coherence* $|\mu_{xy}|$ *of the electric vibrations has its maximum value and this value is equal to the degree of polarization of the wave.*

This special pair of directions has a simple geometrical significance. If, as in (5.6) we represent the wave as incoherent mixture of a wave of natural radiation and a wave of monochromatic (and therefore completely polarized) radiation, the angle χ which the major axes of the vibrational ellipse of the polarized portion makes with the x-direction is given by (see CHANDRASEKHAR ([7]), p. 33, eq. (180))

$$(5.21)\qquad \operatorname{tg} 2\chi = \frac{s_2}{s_1} = \frac{J_{xy} + J_{yx}}{J_{xx} - J_{yy}}\,.$$

It follows from (5.20) and (5.21) that $(\operatorname{tg} 2\chi)\cdot(\operatorname{tg} 2\varphi) = -1$ so that $\chi - \varphi = 45°$ or $135°$. This implies that *the directions for which $P = |\mu_{xy}|$ are the bisectors of the principal directions (directions of the major and minor axes) of the vibrational ellipse of the polarized portion of the wave.*

It is evident from the foregoing discussion that the introduction of coherence concepts into the theory of partial polarization leads to a clearer understanding of the behaviour of partially polarized radiation and suggests new ways for the measurement of its degree of polarization.

RIASSUNTO (*)

In questo articolo si fa l'analisi della polarizzazione parziale dal punto di vista della teoria della coerenza. Dopo aver rilevato che la usuale definizione analitica dei parametri di Stokes per un'onda quasi monocromatica non è univoca, si prende in esame un semplice esperimento il quale introduce, in maniera chiara, i parametri osservabili per un'onda luminosa quasi monocromatica. L'analisi conduce ad un'unica matrice di coerenza e ad un unico gruppo di parametri di Stokes; quest'ultimo è associato alla rappresentazione della matrice di coerenza in funzione delle matrici di spin di Pauli. Tale analisi mostra la fondamentale importanza del concetto di segnale analitico di Gabor. Il grado di coerenza fra le vibrazioni elettriche in due direzioni qualunque, reciprocamente ortogonali, di propagazione dell'onda dipende, in generale, dalla scelta delle due direzioni ortogonali. Si dimostra che il valore massimo di esso uguaglia il grado di polarizzazione dell'onda. Si dimostra altresì che il grado di polarizzazione può essere dedotto, seguendo una via nuova, da esperimenti relativamente semplici che richiedono l'uso di un compensatore e di un polarizzatore, e che tale determinazione è analoga a quella del grado di coerenza ricavata dall'esperimento di interferenza di Young.

(*) *Traduzione a cura della Redazione.*

PAPER NO. 32

Reprinted from *Il Nuovo Cimento*, Vol. 18, pp. 347–356, 1960.

Coherence Properties of Blackbody Radiation.

R. C. Bourret

Hughes Research Laboratories, A Division of Hughes Aircraft Company - Malibu, Cal.

(ricevuto il 12 Luglio 1960)

Summary. — An explicit calculation of the space and time correlation functions of the electric field components in blackbody radiation is given. These functions are determined by the statistics of a photon gas since the Planck distribution law is essentially the power spectrum of the field fluctuations. The spatial coherence functions are developed in close analogy with the theory of isotropic turbulence of an incompressible fluid.

1. – Introduction.

In recognition of the operationalism of much of contemporary physics it has lately been stressed, notably by Wolf (1), that the true observables of optics are not the fluctuating microfields themselves but various quadratic averages of them which are purveyed to us by detectors. The observables of greatest interest at the present time are the coherence functions of the electric fields, *i.e.*,

$$\langle E_i(r_1, t_1) E_j(r_2, t_2)\rangle ,$$

where the brackets signify an average over time. The canonical detectors for the determination of these quantities are polarizers, phase-plates and interferometers. Generally such coherence functions depend upon the properties of their sources (natural linewidth, polarization, line broadening, etc.) and upon the geometrical relations between source and detector. There is one case, however, in which these coherence functions depend solely upon the intrinsic statistical properties of the photons themselves (*i.e.*, Bose-Einstein

(1) E. Wolf: *Nuovo Cimento*, **12**, 884 (1954).

statistics) and the relative co-ordinates $|\boldsymbol{r}_2 - \boldsymbol{r}_2|$ and $|t_1 - t_2|$ of the coherence functions, namely, the case of blackbody radiation. We believe that this special case is of sufficient interest to warrent the explicit evaluation of the spatial and temporal coherence functions, which are developed in the following pages.

Consider the radiation field enclosed in a cavity with whose perfectly absorbing walls it is in thermal equilibrium. Its energy density (in Gaussian units) is

$$U = \frac{1}{8\pi}(\boldsymbol{E}\cdot\boldsymbol{D} + \boldsymbol{H}\cdot\boldsymbol{B}) . \tag{1}$$

We assume the cavity to be a vacuum in which (apart from dimensions)

$$D = E , \qquad B = H , \tag{2}$$

and

$$\operatorname{div} \boldsymbol{H} = \operatorname{div} \boldsymbol{E} = 0 . \tag{3}$$

These last relations will be required later. We assume, moreover, that the radiation is isotropic and homogeneous, and that

$$\langle E^2 \rangle = \langle H^2 \rangle . \tag{4}$$

The bracket symbol indicates an ensemble average at the same position and time in a large number of equivalent cavities. We shall require the ergodic hypothesis, too, since we shall need to regard the ensemble average as equal to the time and to the space average of fluctuating field quantities. The isotropicity of the electric field may be stated

$$\langle E_i E_j \rangle = \langle E_i^2 \rangle \delta_{ij} \qquad \text{(no summation).} \tag{5}$$

With (2) and (1) equation (5) becomes

$$\langle E_i E_j \rangle = \frac{4\pi}{3} \delta_{ij} U . \tag{6}$$

2. – Coherence functions and the wave-equation.

The quantities we wish to derive are the cross-correlation functions of the electric field components, namely,

$$\langle E_i(\boldsymbol{r}_1, t_1) E_j(\boldsymbol{r}_2, t_2) \rangle . \tag{7}$$

If now, stationarity in space and in time is assumed, these quantities depend only upon the spatial and temporal differences $\boldsymbol{r} = \boldsymbol{r}_2 - \boldsymbol{r}_1$ and $t = t_2 - t_1$. Making this assumption, we define the quantities of interest by

$$\mathscr{E}_{ij}(\boldsymbol{r}, t) = \langle E_i(\boldsymbol{r}_1, t_1) E_j(\boldsymbol{r}_2, t_2) \rangle . \tag{8}$$

WOLF (1) has described quantities similar to our $\mathscr{E}_{ij}(\boldsymbol{r}, t)$ which presuppose only temporal stationarity of the field fluctuations. Thus he writes

$$\mathscr{E}_{ij}(\boldsymbol{r}_1, \boldsymbol{r}_2, t) = \langle E_1(\boldsymbol{r}_1, t_1) E_j(\boldsymbol{r}_2, t_2) \rangle . \tag{9}$$

He then observes that these functions satisfy the two wave equations

$$\frac{1}{c^2}\frac{\partial^2 \mathscr{E}_{ij}}{\partial t^2} = \nabla_1^2 \mathscr{E}_{ij} , \tag{10a}$$

$$\frac{1}{c^2}\frac{\partial^2 \mathscr{E}_{ij}}{\partial t^2} = \nabla_2^2 \mathscr{E}_{ij} . \tag{10b}$$

Our stipulation of spatial stationarity enables us to write, in place of equations (10a) and (10b), the single equation

$$\frac{1}{c^2}\frac{\partial^2 \mathscr{E}_{ij}}{\partial t^2} = \nabla^2 \mathscr{E}_{ij} . \tag{11}$$

This may be shown in an elementary way from the definition $\mathscr{E}_{ij}$ and the assumption that the electric field components individually satisfy a wave equation of the form of equation (11). From the nature of stationary and isotropic fluctuations, it follows that the $\mathscr{E}_{ij}$ must be even functions of the arguments $\boldsymbol{r}$ and t; thus, a general even solution of (11) may be composed from the elementary solutions

$$\exp\,[i\,\boldsymbol{k}\cdot\boldsymbol{r}] \cos kct \tag{12}$$

by the integral

$$\mathscr{E}_{ij}(\boldsymbol{r}, t) = \int \exp\,[i\boldsymbol{k}\cdot\boldsymbol{r}] \cos kct f_{ij}(\boldsymbol{k})\, \mathrm{d}^3 k , \tag{13}$$

providing that

$$f_{ij}(\boldsymbol{k}) = f_{ij}(-\boldsymbol{k}) . \tag{14}$$

We now examine the case of simultaneous correlations (*i.e.*, $t = 0$). Equation (13) becomes

$$\mathscr{E}_{ij}(\boldsymbol{r}, 0) = \int \exp [i\boldsymbol{k}\cdot\boldsymbol{r}] f_{ij}(\boldsymbol{k})\, \mathrm{d}^3k \,. \tag{15}$$

3. – Boundary conditions and the Planck radiation law.

The functions represented are the spatial cross-correlation functions of the vector field components E_i. They are analogous to the velocity correlation field studied in the theory of homogeneous turbulence. In particular, they are analogous to the correlation functions of velocity in isotropic turbulence. The further restriction to the turbulence of an incompressible fluid, guaranteed by the condition

$$\operatorname{div} \boldsymbol{v} = 0 \,, \tag{16}$$

has its exact equivalent in our vacuum condition, equation (3):

$$\operatorname{div} \boldsymbol{E} = 0 \,. \tag{17}$$

Under these conditions of incompressibility (here, equation (17)) and isotropicity (equation (5)) it has been shown (by BATCHELOR (2), ROBERTSON (3), and KAMPÉ DE FÉRIET (4)) that $f_{ij}(\boldsymbol{k})$ takes the special form

$$f_{ij}(\boldsymbol{k}) = A(k)(k^2\,\delta_{ij} - k_i k_j) \,. \tag{18}$$

We have now only to determine the scalar function $A(k)$. This can be done as follows: Substitute (18) into equation (13) for the special case in which $\boldsymbol{r} = 0$. There results, after an integration over solid angle,

$$\mathscr{E}_{ij}(0, t) = \left(\frac{8\pi}{3}\,\delta_{ij}\right)\int_0^\infty k^4 A(k) \cos kct\, \mathrm{d}k \,. \tag{19}$$

Now according to the Wiener-Khintchine theorem the time cross-correlation functions (in our case $\mathscr{E}_{ij}(0, t)$) if stationary random functions are derivable

(2) G. K. BATCHELOR: *The Theory of Homogeneous Turbulence* (Cambridge, 1953).

(3) H. P. ROBERTSON: *Proc. Camb. Phil. Soc.*, **36**, 209 (1940).

(4) J. KAMPÉ DE FÉRIET: *Introduction to the statistical theory of turbulence-correlation and spectrum*, Lecture Series no. 8, Institute for Fluid Dynamics and Applied Mathematics, University of Maryland (1951).

from their cross-spectral densities by the formula

$$\mathscr{E}_{ij}(0, t) = \int_0^\infty \Phi_{ij}(k) \cos kct \, \mathrm{d}k \,. \tag{20}$$

It is thus easily seen, by comparison with (19), that $\Phi_{ij}(k)$ and $A(k)$ are related by

$$\Phi_{ij}(k) = \frac{8\pi}{3} \delta_{ij} k^4 A(k) \,. \tag{21}$$

The cross-spectral densities are simply related to the Planck distribution function for blackbody radiation since the total energy density, U, is the summation of the energy densities at all wave numbers. Thus

$$U = \frac{4\hbar c}{3\pi} \int_0^\infty \frac{k^3 \, \mathrm{d}k}{\exp[\alpha k] - 1} \,, \qquad \alpha = \frac{\hbar c}{kT} \,. \tag{22}$$

Equations (20) and (6) give

$$\mathscr{E}_{ij}(0, 0) = \langle E_i E_j \rangle = \frac{4\pi}{3} \delta_{ij} U = \int_0^\infty \Phi_{ij}(k) \, \mathrm{d}k \,. \tag{23}$$

Hence we obtain for $\Phi_{ij}(k)$ the expression

$$\Phi_{ij}(k) = \left(\frac{4}{3}\right)^2 \hbar c \delta_{ij} \frac{k^3}{\exp[\alpha k] - 1} \,. \tag{24}$$

Using the relation (21) we obtain for $A(k)$

$$kA(k) = \frac{2\hbar c}{3\pi} \frac{1}{\exp[\alpha k] - 1} \,. \tag{25}$$

The coherence matrix $\mathscr{E}_{ij}(\boldsymbol{r}, t)$ can now be obtained by the straightforward use of (13). It will be more instructive, however, to calculate the *time* and the *space* coherence functions separately, and to examine only special forms of the latter called the *lateral* and the *longitudinal* correlations (to be defined later). The time correlation functions are obtained from equation (19), and they prove to be

$$\mathscr{E}_{ij}(0, t) = -\left(\frac{4}{3}\right)^2 \hbar c \delta_{ij} \frac{1}{c^3} \frac{\partial^3}{\partial t^3} \int_0^\infty \frac{\sin kct \, \mathrm{d}k}{\exp[\alpha k] - 1} \,, \tag{26}$$

$$= -\delta_{ij} \frac{8}{9} \hbar c \left(\frac{\pi}{\alpha}\right)^4 L'''(\tau) \,, \qquad \tau = \frac{\pi c}{\alpha} t \,, \tag{27}$$

where

(28) $$L(\tau) = \text{ctgh}\ \tau - \frac{1}{\tau}\,(^*)\,.$$

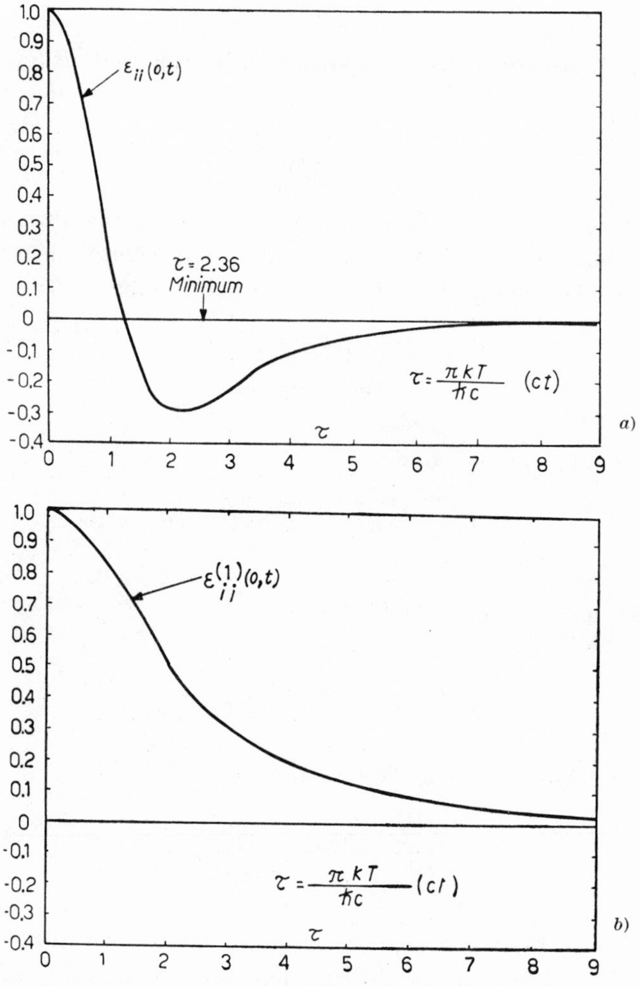

Fig. 1.

This coherence function (**), conveniently normalized, is displayed in Fig. 1*a*. For purposes of comparison, the corresponding coherence function for the (rather artificial) case of a one-dimensional pencil (one degree of freedom in

---

(*) Called the Langevin function in statistical mechanics.

(**) The author has just learned of some unpublished studies by G. HULTGREN of the California Institute of Tecnology in which this result, equation (27), is derived.

wave number) of blackbody radiation is easily shown to be

$$\mathscr{E}_{ii}^{(1)}(0, t) \approx L'(\tau)\,, \tag{29}$$

and this function is presented in Fig. 1*b*.

4. – Spatial coherence.

Because of the complications due to the vector character of the fields, the spatial coherence functions may be more easily interpreted if attention is confined to the *longitudinal* and the *lateral* case, *i.e.*, the cases in which the spatial separation is parallel and perpendicular, respectively, to the direction of a typical autocorrelated field component. The definitions are illustrated in Fig. 2. For brevity, we shall avail ourselves of some of the general formulae of the kinematics of isotropic turbulence of incompressible fluids. One such result is an expression for the (simultaneous) longitudinal correlation function in terms of $A(k)$. It is given (see *e.g.*, PAI (⁵), page 228) by

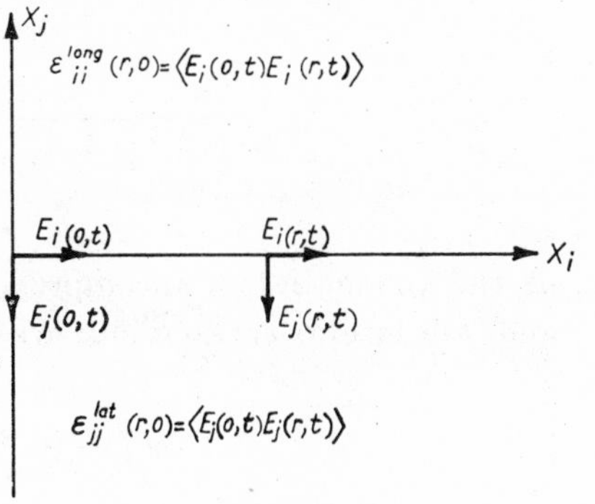

Fig. 2.

$$\mathscr{E}_{ii}^{\text{long}}(r, 0) = \frac{8\pi}{r^3}\int_0^\infty k^2 A(k)\left[\frac{\sin kr}{kr} - \cos kr\right] \mathrm{d}k\,. \tag{30}$$

It is convenient to rewrite this in the form

$$\mathscr{E}_{ii}^{\text{long}}(r, 0) = \frac{8\pi}{r^3}\left(1 - r\frac{\partial}{\partial r}\right)\int_0^\infty k A(k) \sin kr\, \mathrm{d}k\,. \tag{31}$$

Use of result (25) in this gives

$$\mathscr{E}_{ii}^{\text{long}}(r, 0) = \frac{8\hbar c\pi}{3\alpha}\frac{1}{r^3}\left(1 - r\frac{\partial}{\partial r}\right) L\left(\frac{\pi}{\alpha} r\right), \tag{32}$$

(⁵) SHIH-I PAI: *Viscous Flow Theory* - II: *Turbulent Flow* (New York, 1957), p. 228.

where $L(x)$ is the Langevin function defined previously. This longitudinal coherence function is shown in Fig. 3. It is a standard result (*) of the theory

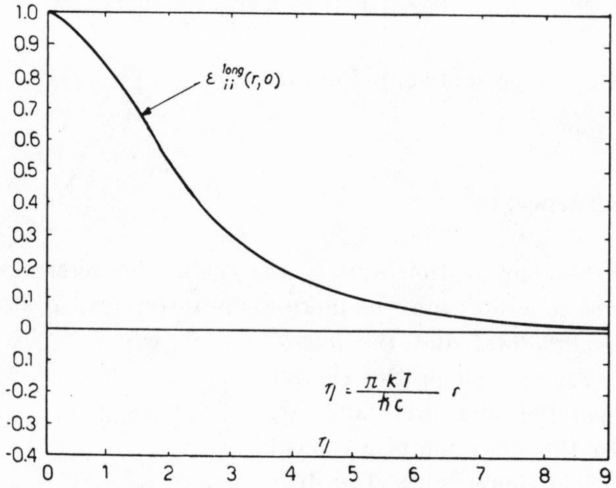

Fig. 3.

of the kinematics of incompressible isotropic turbulence that the longitudinal and the lateral correlations are related by

$$\mathscr{E}_{jj}^{\mathrm{lat}}(r,\,0) = \left(1 + \frac{r}{2}\frac{\partial}{\partial r}\right)\mathscr{E}_{ii}^{\mathrm{long}}(r,\,0)\,. \tag{33}$$

The lateral coherence function is shown in Fig. 4.

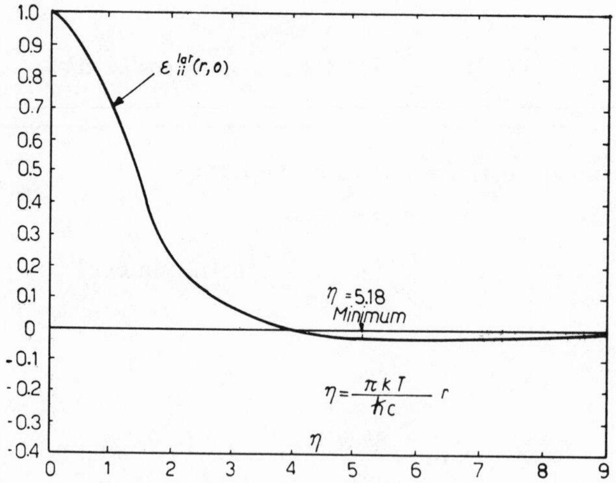

Fig. 4.

(*) Cf., *e.g.*, Pai: *op. cit.*, p. 193.

If it is desired to include the time (difference) dependence into the longitudinal or lateral coherence functions, one may simply replace $A(k)$ by $A(k)\cos kct$ in equation (29). By virtue of an elementary property of *sine* transforms there results

$$\mathscr{E}_{ii}^{\text{long}}(r, t) = \frac{1}{2r}[(r + ct)\mathscr{E}_{ii}^{\text{long}}(r + ct, 0) + (r - ct)\,\mathscr{E}_{ii}^{\text{long}}(r - ct, 0)] \tag{34}$$

and similarly for the lateral coherence function.

The results obtained here for the electric field are applicable without change to the magnetic field in view of the assumption $\langle E^2\rangle = \langle H^2\rangle$ and since $\operatorname{div}\boldsymbol{H} = 0$ is always satisfied.

If coherence functions for the vector potential are developed, cross-correlations between electric and magnetic fields may readily be expressed. This would be, in fact, a more satisfactory procedure from a general point of view, but the additional information obtained from it is of little value. One point however, seems worthy of mention, namely, that the time coherence of the vector potential is identical to that obtained for the electric field in the fictitious example of one-dimensional cavity radiation, Fig. 1*b*. This can be seen as follows: define

$$\mathscr{A}_{ij}(0, t) = \langle A_i(0, t_1)\, A_j(0, t_2)\rangle\,, \qquad t = t_2 - t_1\,. \tag{35}$$

Differentiating and using the stationarity of the bracketed function, we find that

$$\mathscr{A}''_{ij}(0, t) = -\,\langle A'_i(0, t_1)\, A'_j(0, t_2)\rangle\,. \tag{36}$$

We are free to choose Coulomb gauge (in vacuo) *i.e.*, $\operatorname{div}\boldsymbol{A} = 0$ and $\varphi = 0$, so that

$$E_i(0, t) = -\frac{1}{c}A'_i(0, t)\,. \tag{38}$$

Substituting this into (36) gives

$$\mathscr{A}''_{ij}(0, t) = -\,c^2\langle E_i(0, t_1)\, E_j(0, t_2)\rangle = -\,c^2\mathscr{E}_{ij}(0, t)\,. \tag{39}$$

Referring to equation (27) shows finally that

$$\mathscr{A}_{ij}(0, t) = \delta_{ij}\,\frac{8}{9}\,\hbar c^3\left(\frac{\pi}{\alpha}\right)^4 L'(\tau) \approx \mathscr{E}^{(1)}_{ij}(0, t)\,. \tag{40}$$

The vanishing divergence of $\boldsymbol{A}$ in this gauge makes possible the same development of spatial coherence functions for $\boldsymbol{A}$ as was employed previously for $\boldsymbol{E}$ (or $\boldsymbol{H}$), but we shall not develop the potential coherence functions further.

APPENDIX

Explicit formulas for coherence functions shown in the figures (subscript is figure number):

$$f_{1a}(x) = 15 \operatorname{cosech}^2 x\,[3 \operatorname{ctgh}^2 x - 1] - \frac{45}{x^4},$$

$$f_{1b}(x) = -3 \operatorname{cosech}^2 x + \frac{3}{x^2},$$

$$f_3(x) = \frac{45}{2x^3}\left[\operatorname{ctgh} x + x \operatorname{cosech}^2 x - \frac{2}{x}\right],$$

$$f_4(x) = \frac{45}{2x}\left[\frac{2}{x^3} - \operatorname{cosech}^2 x \cdot \operatorname{ctgh} x - \frac{1}{2x} \operatorname{cosech}^2 x - \frac{1}{2x^2} \operatorname{ctgh} x\right].$$

Since all these functions take the form of small differences of large numbers near $x = 0$, it is convenient to employ the following series expansions around the origin:

$$f_{1a}(x) = 1 + \frac{15}{2}\sum_{n=3}^{\infty}(-1)^n \frac{2^{2n} B_{2n-1}}{(2n)(2n-4)!}\, x^{2n-4},$$

$$f_{1b}(x) = 1 + 3\sum_{n=2}^{\infty}(-1)^{n-1} \frac{2^{2n} B_{2n-1}}{(2n)(2n-2)!}\, x^{2n-2},$$

$$f_3(x) = 1 + \frac{45}{2}\sum_{n=3}^{\infty}(-1)^n \frac{2^{2n}\, 2(n-1) B_{2n-1}}{(2n)!}\, x^{2n-4},$$

$$f_4(x) = 1 + \frac{45}{2}\sum_{n=3}^{\infty}(-1)^n \frac{2^{2n}\, 2(n-1)^2 B_{2n-1}}{(2n)!}\, x^{2n-4}.$$

The B_n are the Bernoulli numbers.

RIASSUNTO (*)

Si presenta un calcolo esplicito delle funzioni di correlazione dello spazio e del tempo delle componenti del campo elettrico nella radiazione del corpo nero. Queste funzioni sono determinate dalla statistica di un gas fotonico poichè la legge di distribuzione di Planck è essenzialmente lo spettro di potenza delle fluttuazioni di campo. Si sviluppano le funzioni spaziali di coerenza in stretta analogia con la teoria della turbolenza isotropica di un fluido incompressibile.

(*) *Traduzione a cura della Redazione.*

PAPER NO. 33

Reprinted from *Il Nuovo Cimento*, Vol. 15, pp. 370–388, 1960.

On the Matrix Formulation of the Theory of Partial Polarization in Terms of Observables.

G. B. PARRENT, Jr. (*) and P. ROMAN

Department of Theoretical Physics, The University - Manchester

(ricevuto il 19 Ottobre 1959)

Summary. — The coherency matrix of a quasi-monochromatic plane wave is deduced from a matrix representation of the analytic signal associated with the electric field. It is shown that if the radiation passes through a physical device, such as a compensator, absorber, rotator, or polarizer, the effect of this interaction can be fully described in terms of appropriately choosen operators which transform directly the coherency matrix. The complex degree of coherence is defined in terms of the operators mentioned above, and from this is deduced an expression characterizing the degree of polarization. It is shown that the quantity deduced in this manner is identical with that obtained from the more conventional definition. An experiment is described which can serve to measure the components of the correlation matrix.

1. – Introduction.

The study of partial polarization has a great deal in common with the study of partial coherence, since both are necessitated by the statistical nature of « natural » radiation. Thus for example a rigorously monochromatic beam of electromagnetic radiation would be completely coherent and completely polarized; and the behaviour of such a beam is simply determined. However, if one considers a more realistic model of a beam of light, *e.g.* quasi-monochro-

(*) Permanent address: Electromagnetic Radiation Lab., Electronics Research Directorate, Air Force Cambridge Research Center, L. G. Hanscom, A. F. B. Bedford, Mass., U.S.A.

matic (which results from the superposition of a large number of randomly timed statistically independent pulses with the same central frequency), the problem of describing its behaviour is essentially different. What was a completely determined problem becomes a statistical one in which even the concepts of amplitude and phase become ambiguous, requiring redefinition.

It is only when dealing with such statistical radiation that the terms partial coherence and partial polarization have any meaning.

In view of this genetic relationship between these two aspects of the study of natural light one would expect the techniques and concepts which prove important for one to be helpful in understanding the other. Consequently, the recent emphasis on the formulation of coherence theory in terms of measurable quantities (correlation functions), and the success which this approach has enjoyed, suggest that the more general problems of partial polarization might also be better understood if they were formulated in a similar way.

In fact, considerable progress with this approach has already been made through the introduction by WOLF (1) of the correlation tensor. The matrix of this tensor, the correlation matrix, is the appropriate entity for the description of partially polarized beams. Its physical significance was, however, left somewhat obscure. In a later paper WOLF (2) identified the elements of the coherency matrix $\mathcal{J}$ with the factors occurring in the expression for the intensity in a beam after transmission through a compensator and a polarizer. In his paper ref. (1) he pointed out that $\mathcal{J}$ is formally equivalent to the density matrix in the study of scattering phenomena.

In spite of the attention focussed upon the coherency matrix, the uniformity which this entity introduces has gone largely unnoticed. The formulation of the subject as introduced here stresses the analogy between this field of research and modern scattering theory.

We shall use here a matrix representation of the electric field and derive formulae for the solution of the problem arising in the description of partially polarized fields and their interaction with physical devices. We mention that in the last decade several attempts have been made to describe in terms of matrices the action of various physical devices on radiation passed through them (see, for example, CLARK JONES (3) or WESTFOLD (4), where also further literature is quoted). Our approach will be, however, very different from these and also more general. In particular, we shall avoid the use of the « Stokes vector »

$$\mathbf{s} = (s_0, s_1, s_2, s_3) ,$$

---

(1) E. WOLF: *Nuovo Cimento*, **12**, 884 (1954).
(2) E. WOLF: *Nuovo Cimento*, **13**, 1165 (1959).
(3) R. CLARK JONES: *Journ. Opt. Soc. Am.*, **46**, 126 (1956).
(4) K. C. WESTFOLD: *Journ. Opt. Soc. Am.*, **49**, 717 (1959).

which, in fact, has not the transformation property of a four-vector. Furthermore, we shall not need 4×4 matrices to characterize the interactions, but only 2×2 matrices. An attractive feature of the present approach is that *one obtains matrix operators characteristic of the physical devices (interactions) which operate directly on the coherency matrix.* Further, the physical characteristic to be measured (*e.g.* the intensity I) is always given by an expression of the form

$$F = \mathrm{Sp}\,[\mathscr{F}\mathscr{I}], \tag{1}$$

where F is the characteristic to be measured, $\mathscr{I}$ is the coherency matrix, $\mathscr{F}$ is an operator describing the experiment (the device), and Sp indicates the trace. In (1) *the operator* $\mathscr{F}$ *depends only on the physical devices and* $\mathscr{I}$ *depends only on the measurable properties of the beam.* That is, $\mathscr{I}$ describes the state of the field.

2. – Scope, definitions and notation.

Throughout this discussion we shall limit our attention to quasi-monochromatic fields. In essence this approximation requires that the spectral width of the radiation is negligible compared to the mean frequency, *i.e.*

$$\Delta\nu \ll \bar{\nu}\,. \tag{2a}$$

The introduction of this approximation enables us to evaluate frequency-dependent quantities at the mean frequency, but also limits the validity of the development to phenomena involving relatively small path differences Δl, *i.e.* the theory is valid when

$$\Delta l \ll c/\Delta\nu\,, \tag{2b}$$

where c is the velocity of light. For a full discussion of the consequences of this approximation the reader is referred to BORN and WOLF (5). Further, we limit our attention to plane waves in which case the electric vector has two components, perpendicular to the direction of propagation. While these approximations are appropriate for many problems of interest, the analysis given here may be extended to polychromatic non-planar waves. This generalization will be treated in a later communication.

(5) M. BORN and E. WOLF: *Principles of Optics* (London, 1959), p. 502.

The real electric field will be denoted by the column matrix

$$\mathscr{E}^r(\boldsymbol{x}, t) = \begin{pmatrix} E_x^r(\boldsymbol{x}, t) \\ E_y^r(\boldsymbol{x}, t) \end{pmatrix}. \tag{3}$$

Here, as throughout this paper, script letters denote matrices; bold face type denotes vectors; and a superscript r denotes a real function of a real variable.

Our analysis will not deal directly with $\mathscr{E}^r$ but rather with the associated analytic signal representation of the field. The advantage of this representation for problems involving correlation functions become apparent from the investigations of various authors ([6-8]); therefore we shall simply review briefly the method of obtaining the analytic signal from the real field.

It is assumed (see p. 1169 of ref. ([2])) that the field possesses a Fourier transform; thus

$$E_x^r(\boldsymbol{x}, t) = \int_0^\infty a(\boldsymbol{x}, \nu) \sin [\varphi(\boldsymbol{x}, \nu) - 2\pi\nu t]\, d\nu .$$

We associate with E_x^r its « conjugate function » $E_x^i(\boldsymbol{x}, t)$ defined by

$$E_x^i(\boldsymbol{x}, t) = \int_0^\infty a(\boldsymbol{x}, \nu) \cos [\varphi(\boldsymbol{x}, \nu) - 2\pi\nu t]\, d\nu .$$

It is easily shown that E_x^i is the Hilbert transform of E_x^r, *i.e.*

$$E_x^i(\boldsymbol{x}, t) = \frac{1}{\pi} P \int_{-\infty}^{+\infty} \frac{E_x^r(\boldsymbol{x}, t')}{t' - t}\, dt' ,$$

where P denotes Cauchy's principal value. The analytic signal E_x is then defined as

$$E_x(\boldsymbol{x}, t) = E_x^r(\boldsymbol{x}, t) + iE_x^i(\boldsymbol{x}, t) .$$

Similar considerations hold for E_y.

([6]) Cf. ref. ([5]), p. 492.
([7]) G. B. Parrent jr: *Journ. Opt. Soc. Am.*, **49**, 787 (1959).
([8]) P. Roman and E. Wolf: *Ann. Phys.*, in press.

In terms of these functions our representation becomes

$$\mathscr{E}(\boldsymbol{x}, t) = \begin{pmatrix} E_x(\boldsymbol{x}, t) \\ E_y(\boldsymbol{x}, t) \end{pmatrix} = \mathscr{E}^r(\boldsymbol{x}, t) + i\mathscr{E}^i(\boldsymbol{x}, t) . \tag{4}$$

In this representation the intensity I at the point $\boldsymbol{x}$ may be expressed as

$$I = \mathrm{Sp}\,\langle \mathscr{E} \times \mathscr{E}^\dagger \rangle . \tag{5}$$

Here $\mathscr{E}^\dagger$ is the Hermitian conjugate of $\mathscr{E}$, *i.e.* the row matrix

$$\mathscr{E}^\dagger = \tilde{\mathscr{E}}^* = [E_x^* \; E_y^*];$$

further, $\langle ... \rangle$ indicates the time average, and $\times$ denotes the Kronecker product of matrices. That this definition is equivalent to that normally given, may be seen by writing (5) in full, when we obtain

$$I = \langle \boldsymbol{E} \cdot \boldsymbol{E}^* \rangle = 2\langle (\boldsymbol{E}^r)^2 \rangle .$$

(In the last step some simple properties of Hilbert transforms have been utilized.)

We introduce now the coherency matriw $\mathscr{J}$ by the definition

$$\mathscr{J} = \langle \mathscr{E} \times \mathscr{E}^\dagger \rangle , \tag{6}$$

which is clearly Hermitian:

$$\mathscr{J}^\dagger = \mathscr{J} .$$

$\mathscr{J}$ is in fact the matrix of the correlation tensor introduced by WOLF [1] who proved that the elements of this matrix are observables of the radiation field. In terms of $\mathscr{J}$ the intensity (5) becomes simply

$$I = \mathrm{Sp}\,\mathscr{J} . \tag{7}$$

One further remark about notation. At one stage in the analysis the Pauli matrices will be used. Since our representation differs slightly from that normally used, we shall summarize it at this point. These matrices obey the algebra

$$\begin{cases} \sigma_\alpha \sigma_\beta = - i\sigma_\gamma , & (\alpha, \beta, \gamma) = (1, 2, 3) \text{ and cycl.}; \\ (\sigma_i)^2 = \sigma_0 , & (i = 0, 1, 2, 3) , \\ \sigma_i \sigma_0 = \sigma_0 \sigma_i = \sigma_i . & \end{cases} \tag{8}$$

It follows from (8) that

$$\mathrm{Sp}\,(\sigma_i\sigma_j) = 2\delta_{ij}\,, \qquad (i,\, j = 0, 1, 2, 3)\,. \tag{9}$$

The matrices may be represented by putting

$$\sigma_0 = \begin{pmatrix} 1 & 0 \\ 0 & 1 \end{pmatrix}, \quad \sigma_1 = \begin{pmatrix} 1 & 0 \\ 0 & -1 \end{pmatrix}, \quad \sigma_2 = \begin{pmatrix} 0 & 1 \\ 1 & 0 \end{pmatrix}, \quad \sigma_3 = \begin{pmatrix} 0 & i \\ -i & 0 \end{pmatrix}. \tag{10}$$

In the following sections we shall obtain transformation equations for $\mathscr{I}$ corresponding to the passage of partially polarized radiation through the various physical devices which are of interest in the study of such fields. In Sections **8** and **9** we discuss the various parameters used to specify the state of such statistical radiation and their relation to the coherency matrix. The relation between $\mathscr{I}$ and the density matrix will be briefly discussed in the concluding remarks.

3. – Compensator.

The compensator is a device which introduces a phase change ε_x in the x-component and ε_y in the y-component in each spectral component of the field vector. This results in a relative phase difference $\delta = \varepsilon_x - \varepsilon_y$ between the x- and y-components. Thus the compensator produces a selective rotation of the electric field in Fourier space. In the general case δ is of course a function of frequency; however, consistent with the quasi-monochromatic approximation we take the phase shift for each spectral component to be equal to that for the mean frequency. The compensator may, therefore, be represented or characterized by a unitary rotation matrix

$$\mathscr{C} = \begin{pmatrix} \exp\,[\frac{i}{2}\,\delta] & 0 \\ 0 & \exp\,[-\frac{i}{2}\,\delta] \end{pmatrix}. \tag{11}$$

If $\mathscr{E}$ is the field incident on the compensator, the emergent field $\mathscr{E}_C$ may be written as

$$\mathscr{E}_C = \mathscr{C}\mathscr{E}\,; \tag{12}$$

and the coherency matrix for the emergent field is given, according to the definition (6), by

$$\mathscr{I}_C = \langle \mathscr{E}_C \times \mathscr{E}_C^\dagger \rangle = \langle \mathscr{C}\mathscr{E} \times \mathscr{E}^\dagger \mathscr{C}^\dagger \rangle\,. \tag{13}$$

Hence we may write (*)

$$\mathscr{I}_C = \mathscr{C}\langle \mathscr{E} \times \mathscr{E}^\dagger \rangle \mathscr{C}^\dagger ; \tag{14}$$

and substituting from (6) we obtain, using also the unitary nature of $\mathscr{C}$,

$$\mathscr{I}_C = \mathscr{C}\mathscr{I}\mathscr{C}^\dagger = \mathscr{C}\mathscr{I}\mathscr{C}^{-1} . \tag{15}$$

Thus the effect of the compensator may be characterized by a rotation operator in Fourier space acting directly on the coherency matrix $\mathscr{I}$, without recourse to the unmeasurable quantities of the $\mathscr{E}$ field itself.

To compute the intensity I_C at a point in the emergent beam we take, according to (7), the trace on both sides of (15); and since the argument of the trace may be cyclically permuted, we obtain

$$I_C = I .$$

That the intensity is unchanged under the passage of the beam through a compensator, could have been anticipated from the fact that a rotation leaves the trace invariant and absorption and reflection were ignored. However, absorption is easily incorporated into the scheme as will be shown in the next section.

4. – Absorption.

Absorption is characterized by a decrease in field strength and may therefore be represented by a matrix of the form

$$\mathscr{A} = \begin{pmatrix} \exp[-\frac{1}{2}\eta_x] & 0 \\ 0 & \exp[-\frac{1}{2}\eta_y] \end{pmatrix} . \tag{16}$$

Here η_x and η_y are the absorption coefficients for the x- and y-components respectively, evaluated at the mean frequency $\bar{\nu}$. The effect of absorption is to produce a field $\mathscr{E}_A$ given by

$$\mathscr{E}_A = \mathscr{A}\mathscr{E} , \tag{17}$$

(*) The associative property used in obtaining (14) is of course not permissible in general; indeed in the general case this operation is meaningless. However, when it is possible to perform such an operation, as in the present and in the following cases, the associative property is valid as may be easily verified.

and the coherency matrix becomes

$$\mathscr{I}_A = \mathscr{A}\mathscr{I}\mathscr{A}^{\dagger}\,. \tag{18}$$

The reduced intensity is clearly

$$I_A = \mathrm{Sp}\,[\mathscr{A}\mathscr{I}\mathscr{A}^{\dagger}] = \mathrm{Sp}\,[\mathscr{A}^2\mathscr{I}]\,. \tag{19}$$

Note that the measurable intensity is given by an expression of the form (1).

5. – Rotator.

Various materials and physical devices produce a rotation of the electric vector, *i.e.* they rotate the plane of polarization. Such a device is termed a rotator and may clearly be characterized by a rotation operator

$$\mathscr{R}(\alpha) = \begin{pmatrix} \cos\alpha & \sin\alpha \\ -\sin\alpha & \cos\alpha \end{pmatrix}, \tag{20}$$

where α is the angle through which the field is rotated. Note that $\mathscr{R}$ is a real antisymmetric unimodular unitary matrix.

The emergent field $\mathscr{E}_R$ following a rotator is given by

$$\mathscr{E}_R = \mathscr{R}(\alpha)\mathscr{E}\,, \tag{21}$$

and the coherency matrix becomes

$$\mathscr{I}_R = \mathscr{R}(\alpha)\mathscr{I}\mathscr{R}^{\dagger}(\alpha) = \mathscr{R}(\alpha)\mathscr{I}\mathscr{R}^{-1}(\alpha) = \mathscr{R}(\alpha)\mathscr{I}\mathscr{R}(-\alpha)\,. \tag{22}$$

Thus the action of the rotator may also be represented as an operation directly on the coherency matrix which contains only observable quantities. The intensity of course remains unchanged.

6. – Polarizer.

The last device to be considered in this discussion before turning to general considerations is a polarizer, such as a Nichol prism, which passes only a particular component of the field, say the component making an angle θ with the x-direction. That is, the polarizer takes the projection of the $\mathscr{E}$ field on the direction θ. The polarizer may thus be characterized by a projection operator $\mathscr{P}_+(\theta)$. Projection operators are singular and satisfy the idempotency condition

$$\mathscr{P}_+(\theta)\mathscr{P}_+(\theta) = \mathscr{P}_+(\theta)\,. \tag{23}$$

Associated with every projection operator $\mathscr{P}_+(\theta)$ is an orthogonal proejction operator $\mathscr{P}_-(\theta)$ representing a projection on the direction orthogonal to θ and satisfying the conditions

$$\begin{cases} \mathscr{P}_-(\theta)\mathscr{P}_-(\theta) = \mathscr{P}_-(\theta)\,, \\ \mathscr{P}_+(\theta)\mathscr{P}_-(\theta) = \mathscr{P}_-(\theta)\mathscr{P}_+(\theta) = 0\,, \\ \mathscr{P}_+(\theta) + \mathscr{P}_-(\theta) = 1\,. \end{cases} \tag{24}$$

Since the polarizer takes a projection of the $\mathscr{E}$ field, it may be represented by the operator

$$\mathscr{P}_+(\theta) = \begin{pmatrix} \cos^2\theta & \sin\theta\cos\theta \\ \sin\theta\cos\theta & \sin^2\theta \end{pmatrix}. \tag{25}$$

The associated operator $\mathscr{P}_-(\theta)$ may be written as

$$\mathscr{P}_-(\theta) = \mathscr{P}_+\left(\theta + \frac{\pi}{2}\right) = \begin{pmatrix} \sin^2\theta & -\sin\theta\cos\theta \\ -\sin\theta\cos\theta & \cos^2\theta \end{pmatrix}. \tag{26}$$

That these Hermitian operators satisfy the above conditions (23) and (24) is readily verified.

The field $\mathscr{E}_P$ emerging from the polarizer is given as

$$\mathscr{E}_P = \mathscr{P}_+(\theta)\mathscr{E}\,; \tag{27}$$

and the coherency matrix becomes

$$\mathscr{J}_P = \mathscr{P}_+\mathscr{J}\mathscr{P}_+^\dagger = \mathscr{P}_+\mathscr{J}\mathscr{P}_+\,, \tag{28}$$

since $\mathscr{P}_+$ is Hermitian. Thus the effect of the polarizer is also represented by an operation directly on the observable entity $\mathscr{J}$. Unlike the devices previously considered, however, the polarizer does not leave the intensity unaltered. The intensity is given by

$$I_P = \mathrm{Sp}\mathscr{J}_P = \mathrm{Sp}[\mathscr{P}_+\mathscr{J}\mathscr{P}_+] = \mathrm{Sp}[\mathscr{P}_+\mathscr{J}]\,, \tag{29}$$

where the property (23) of $\mathscr{P}_+$ was used. The reduced intensity is again an expression of the form (1).

For further application we note here that, as follows directly from (25) and (20),

$$\mathscr{R}^\dagger(\alpha)\,\mathscr{P}_+(\theta)\mathscr{R}(\alpha) = \mathscr{P}_+(\theta+\alpha)\,, \tag{30}$$

and, in particular, using also (26),

$$\mathscr{R}^\dagger\left(-\frac{\pi}{2}\right)\mathscr{P}_+(0)\mathscr{R}\left(-\frac{\pi}{2}\right) = \mathscr{P}_+\left(-\frac{\pi}{2}\right) = \mathscr{P}_-(0)\,. \tag{30a}$$

7. – Cascaded systems.

The result of cascading the various devices discussed thus far is of course simply obtained by the consecutive application of the respective operators. One special case of such a cascaded system of particular interest is the compensator followed by a polarizer. Computing the intensity at a point in the beam emerging from such an arrangement we obtain

$$\begin{aligned} I_K(\theta,\delta) &= \mathrm{Sp}[\mathscr{P}_+(\theta)\mathscr{C}(\delta)\mathscr{I}\mathscr{C}^\dagger(\delta)\,\mathscr{P}_+(\theta)] = \mathrm{Sp}[\mathscr{P}_+^2(\theta)\mathscr{C}(\delta)\mathscr{I}\mathscr{C}^\dagger(\delta)] = \\ &= \mathrm{Sp}[\mathscr{C}^\dagger(\delta)\,\mathscr{P}_+(\theta)\mathscr{C}(\delta)\mathscr{I}] = \mathrm{Sp}[\mathscr{K}(\theta,\delta)\mathscr{I}]\,, \end{aligned} \tag{31}$$

where we have set

$$\mathscr{K} = \mathscr{C}^\dagger\mathscr{P}_+\mathscr{C} = \mathscr{C}^{-1}\mathscr{P}_+\mathscr{C}\,. \tag{32}$$

The matrix $\mathscr{K}$ which, incidentally, arises from $\mathscr{P}_+$ through a rotation in Fourier space, is useful enough to be written out in full here for later reference. We find

$$\mathscr{K}(\theta,\delta) = \begin{pmatrix} \cos^2\theta & \exp[\frac{i}{2}\delta]\sin\theta\cos\theta \\ \exp[-\frac{i}{2}\delta]\sin\theta\cos\theta & \sin^2\theta \end{pmatrix}. \tag{33}$$

Note that (31) is also of the form (1), and that the effect of the combination is represented by an operation directly on $\mathscr{I}$. If (31) is written in detail, we obtain the familiar result for such an arrangement, as discussed by WOLF [2]:

$$I_K(\theta,\delta) = J_{xx}\cos^2\theta + J_{yy}\sin^2\theta + (J_{xy}\exp[-i\delta] + J_{yx}\exp[i\delta])\sin\theta\cos\theta\,. \tag{34}$$

Here J_{xx} etc., are the components of the coherency matrix.

8. – The state of the field.

The state of polarization of an electromagnetic field has been described is several ways: in terms of Stokes parameters, the degree of coherence between the x- and y-components of the field, and the degree of polarization. We shall discuss each of these modes of description in terms of the coherency matrix. Most of the relationships discussed in this section were pointed out by WOLF ([2]). They are, however, obtained here from a quite different point of view which is being stressed as physically more meaningful in this discussion.

The customary set of Stokes parameters has been recently the subject of considerable discussion and it was pointed out by WOLF ([2]) that as usually defined they are not unique. This ambiguity was, however, removed by the introduction of the analytic signal representation. It was also shown recently by ROMAN ([9]) that the concept of Stokes parameters can be extended to the case of non-planar waves.

While it is the opinion of the authors that the elements of $\mathscr{I}$ are physically a more meaningful set of parameters, we will digress briefly and discuss the relation between $\mathscr{I}$ and the Stokes parameters.

Anticipating the discussion in Section **10**, we note that the coherency matrix is formally identical to the density matrix, and therefore the expansion of the density matrix given by FANO ([10]) may be used, *i.e.* we may set

$$\mathscr{I} = \tfrac{1}{2}\sum_{k=0}^{3} s_k \sigma_k \,, \tag{35}$$

where the σ_k are the Pauli matrices (10) and the s_k are the Stokes parameters. In keeping with the theme of this paper, we wish to express the Stokes parameters as derivable from $\mathscr{I}$, since they are observables of the field. Multiplying (35) by σ_i and taking the trace we obtain, using (9), the solution

$$s_i = \mathrm{Sp}[\sigma_i \mathscr{I}] \,, \qquad (i = 0, 1, 2, 3) \,. \tag{36}$$

Note that the Stokes parameters, as all measurables of the field, are obtained from an expression of the form (1).

Another parameter often used in the description of partially polarized fields is the degree of polarization, which may be defined as the ratio of the intensity of the completely polarized part of the radiation to the total intensity. It was shown by WOLF ([2]) that an unambiguous meaning can be given to the

(9) P. ROMAN: *Nuovo Cimento*, **13**, 974 (1959).
(10) U. FANO: *Phys. Rev.*, **93**, 121 (1954).

terms « completely polarized part » and « unpolarized part » and a unique expression for the degree of polarization can be obtained in terms of the invariants of $\mathscr{I}$.

The situation is analogous to that which existed in the closely related theory of partial coherence. Here the ratio of the intensity of the « coherent part » of the radiation to the total intensity was proposed as a definition of the degree of coherence. However, a much clearer understanding of the physical situation was obtained by defining the degree of coherence in terms of a cross-correlation function (cfr. WOLF [11]). Following this lead from scalar coherence theory we shall propose here a new definition of the degree of polarization which will be seen to have the advantage that it involves only *directly* observable quantities.

Our definition is suggested by the following considerations. A strictly monochromatic field is completely polarized and completely coherent. Further, as pointed out by WOLF [1,2], the intensity formula (34) is formally identical with the expression for the intensity resulting from the superposition of two partially coherent beams. Now, the significant entity for describing interference phenomena involving partially coherent radiation is the normalised cross-correlation of the interfering disturbances, *i.e.* the degree of coherence. Thus our considerations suggest that *a significant quantity for the specification of the state of a partially polarized field is the degree of coherence between the x- and y- components of the field.* This quantity μ_{xy} may then be defined as the normalized cross-correlation of the x- and y-components of the field.

The x-component of the field is of course given by $\mathscr{P}_+(0)\mathscr{E}$, and the y-component *which is rotated as to interfere with the x-component* can be expressed as $\mathscr{P}_+(0)\mathscr{R}(-\pi/2)\mathscr{E}$. Since, as a natural extension of the intensity formula (5), we can express the cross-correlation of two fields as

$$K = \mathrm{Sp}\,\langle \mathscr{E}_1 \times \mathscr{E}_2^\dagger \rangle\,,$$

the normalized cross-correlation between the aforementioned two field components may be written in the form

$$\text{(37)}\qquad \mu_{xy} = \\ = \frac{\mathrm{Sp}\,\langle [\mathscr{P}_+(0)\mathscr{E}] \times [\mathscr{P}_+(0)\mathscr{R}(-\pi/2)\mathscr{E}]^\dagger \rangle}{\{\mathrm{Sp}\,\langle [\mathscr{P}_+(0)\mathscr{E}] \times [\mathscr{P}_+(0)\mathscr{E}]^\dagger \rangle \cdot \mathrm{Sp}\,\langle [\mathscr{P}_+(0)\mathscr{R}(-\pi/2)\mathscr{E}] \times [\mathscr{P}_+(0)\mathscr{R}(-\pi/2)\mathscr{E}]^\dagger \rangle\}^{\frac{1}{2}}}\,.$$

Here the normalizing denominator represents the square root of the product of the intensities associated with the two interfering fields.

---

(11) E. WOLF: *Proc. Roy. Soc.*, A **230**, 96 (1955).

Permuting cyclically the arguments of the traces, utilizing the properties (23), (24) and (30a) of the projection operators, the unitarity of $\mathscr{R}$ and the Hermiticity of $\mathscr{P}$, and applying the « associative law » that was used first in the derivation of (14), this expression can be considerably simplified.

We obtain, in view of (6)

$$\mu_{xy} = \frac{\mathrm{Sp}\,[\mathscr{R}(\pi/2)\,\mathscr{P}_+(0)\mathscr{J}\,]}{\{\mathrm{Sp}\,[\mathscr{P}_+(0)\mathscr{J}\,]\cdot\mathrm{Sp}\,[\mathscr{P}_-(0)\mathscr{J}\,]\}^{\frac{1}{2}}}\,. \tag{38}$$

By the application of the Schwarz inequality it may be shown that the nodulus of μ_{xy} is bounded by zero and one and these extreme values are characteristic of incoherence and coherence respectively.

Since, apart from $\mathscr{J}$, the matrices in (38) are real, we may write

$$|\mu_{xy}|^2 = \frac{\mathrm{Sp}\,[\mathscr{R}(\pi/2)\,\mathscr{P}_+(0)\mathscr{J}\,]\cdot\mathrm{Sp}\,[\mathscr{R}(\pi/2)\,\mathscr{P}_+(0)\,\mathscr{J}^*]}{\mathrm{Sp}\,[\mathscr{P}_+(0)\mathscr{J}\,]\cdot\mathrm{Sp}\,[\mathscr{P}_-(0)\mathscr{J}\,]}\,. \tag{39}$$

Our expression for μ_{xy} is somewhat arbitrary, because for a different selection of the x- and y-directions we would of course obtain a different value. Clearly μ_{xy} is not yet a satisfactory measure for the state of polarization. However, we shall now show that there always exists a coordinate frame for which a maximum value of $|\mu_{xy}|^2$ is obtained.

The value of $|\mu_{xy}|^2$ referred to a coordinate system (XY) making an angle θ with the original one may be obtained from (39) by simply noting that in the new system the projection operator $\mathscr{P}(0)$ must be replaced by $\mathscr{P}(\theta)$. Thus

$$|\mu_{xy}(\theta)|^2 = \frac{\mathrm{Sp}\,[\mathscr{R}(\pi/2)\,\mathscr{P}_+(\theta)\mathscr{J}\,]\cdot\mathrm{Sp}\,[\mathscr{R}(\pi/2)\,\mathscr{P}_+(\theta)\mathscr{J}^*]}{\mathrm{Sp}\,[\mathscr{P}_+(\theta)\mathscr{J}\,]\cdot\mathrm{Sp}\,[\mathscr{P}_-(\theta)\mathscr{J}\,]}\,. \tag{40}$$

The straightforward but rather lengthy maximization of (40) with respect to θ yields the following condition on θ:

$$\mathrm{tg}\,2\theta_m = \frac{J_{yy} - J_{xx}}{J_{xy} + J_{yx}}\,, \tag{41}$$

where J_{xy} etc. are the elements of the coherency matrix with respect to the original choice of axes. Since $\mathscr{J}$ is Hermitian, it follows from (41) that there always exists a *real* θ such $|\mu_{xy}|$ is maximal.

Before proceeding we point out that the angle θ_m which maximises $|\mu_{xy}|$ is also the angle through which the coordinates must be rotated in order that the intensities associated with the x- and y-components are equal. This is easily shown by simply demanding that

$$\mathrm{Sp}\,[\mathscr{P}_+(\theta)\mathscr{I}] = \mathrm{Sp}\,[\mathscr{P}_-(\theta)\mathscr{I}]\,,$$

and solving for θ.

We are now in a position to define the degree of polarization of the beam. *The degree of polarization is defined as the maximum value of the modulus of the degree of coherence,* $|\mu_{xy}|$, maximized with respect to θ. According to this definition we obtain the degree of polarization, P, by substituting the value of θ_m given by (41) into (40). We find after some calculation that

$$P = \sqrt{1 - \frac{4\det\mathscr{I}}{(\mathrm{Sp}\,\mathscr{I})^2}}\,. \tag{42}$$

Since $\det\mathscr{I}$ and $\mathrm{Sp}\,\mathscr{I}$ are invariants, it is clear that P as given by (42) is independent of the choice of axes.

We note that equ. (42) is the result derived by WOLF [2] starting from the definition

$$P = \frac{I_{\mathrm{pol}}}{I_{\mathrm{tot}}}\,. \tag{43}$$

Since it will be of help in deriving an experimental scheme for measuring $\mathscr{I}$ (see Section **9**), we will digress now briefly to consider the determination of P from the definition (43)

Since $\mathscr{I}$ is Hermitian, it is possible to diagonalize it with a unitary matrix $\mathscr{D}$, *i.e.* there exists a $\mathscr{D}$ such that $\mathscr{D}^\dagger = \mathscr{D}^{-1}$ and

$$\mathscr{D}\mathscr{I}\mathscr{D}^{-1} = \begin{pmatrix} \lambda_+ & 0 \\ 0 & \lambda_- \end{pmatrix}, \tag{44}$$

where the eiginvalues are given by

$$\lambda_\pm = \frac{1}{2}\,\mathrm{Sp}\,\mathscr{I}\left\{1 \pm \sqrt{1 - \frac{4\det\mathscr{I}}{(\mathrm{Sp}\,\mathscr{I})^2}}\right\}. \tag{45}$$

Physically (44) implies that there exists a frame of reference (coordinate system and relative phase difference) such that the cross-correlation terms vanish. Thus the operator $\mathscr{D}$ must be of the form

$$\mathscr{D}(\alpha, \delta) = \mathscr{R}(\alpha)\mathscr{C}(\delta)\,. \tag{46}$$

The appropriate diagonalizing angles α and δ are readily found to be given by

$$\exp[2i\delta] = J_{yx}/J_{xy}\,, \tag{47a}$$

and

$$\operatorname{tg} 2\alpha = \frac{1}{\cos\delta}\frac{J_{xy}+J_{yx}}{J_{xx}-J_{yy}} = \frac{J_{xy}+J_{yx}}{J_{xx}-J_{yy}}\sqrt{\frac{2}{1+\operatorname{Re}(J_{yx}/J_{xy})}}\,. \tag{47b}$$

Proceeding, we may now write

$$\begin{pmatrix}\lambda_+ & 0\\ 0 & \lambda_-\end{pmatrix} = \lambda_-\begin{pmatrix}1 & 0\\ 0 & 1\end{pmatrix} + (\lambda_+ - \lambda_-)\begin{pmatrix}1 & 0\\ 0 & 0\end{pmatrix}. \tag{48}$$

The first term on the right of (48) represents evidently a completely unpolarized beam and the second a completely polarized beam. Furthermore, this decomposition is unique. Substituting from (48) into the definition (43) we obtain

$$P = \frac{\lambda_+ - \lambda_-}{\lambda_+ + \lambda_-}\,; \tag{49}$$

and substituting from (45) into (49) we obtain the same expression (42) for P as derived above from our definition. Hence the equivalence of the two definiions is established.

9. – Measurement of $\mathscr{J}$.

In order to give a simple and direct physical interpretation to each of the four measurements involved in the determination of $\mathscr{J}$, we consider first a decomposition of $\mathscr{J}$.

Setting

$$J_{xy} = \beta + i\gamma\,, \qquad \text{hence} \qquad J_{yx} = \beta - i\gamma\,,$$

we may separate $\mathscr{J}$ into a real and imaginary part as

$$\mathscr{J} = \begin{pmatrix}J_{xx} & \beta\\ \beta & J_{yy}\end{pmatrix} + i\gamma\begin{pmatrix}0 & 1\\ -1 & 0\end{pmatrix} \equiv \mathscr{J}^r + \gamma\sigma_3\,.$$

We note that

$$\operatorname{Sp}[\mathscr{P}_+(\theta)\mathscr{R}(\alpha)\sigma_3\mathscr{R}^\dagger(\alpha)] = 0$$

for *all* θ and α. In view of the discussion in the previous sections this implies that *the imaginary part of* $\mathscr{J}$ *does not contribute to the intensity unless a compensator is involved in the device through which the radiation has passed.* We may therefore describe the interaction of a partially polarized beam with a polarizer and a rotator solely in terms of the real (and symmetric) matrix $\mathscr{J}^r$.

Thus the intensity in a beam passed by a polarizer is given by

$$I_P(\theta) = \mathrm{Sp}\,[\mathscr{P}_+(\theta)\mathscr{J}^r]\,. \tag{51}$$

Since $\mathscr{J}^r$ is symmetric, it follows from (47*a*) and (46) that the matrix $\mathscr{D}$ which diagonalizes $\mathscr{J}^r$ is a pure rotation operator $\mathscr{R}$. Hence, denoting the eigenvalues of $\mathscr{J}^r$ by $\lambda^r_\pm$, we may write

$$\mathscr{R}(\alpha_d)\mathscr{J}^r\mathscr{R}^{-1}(\alpha_d) = \begin{pmatrix} \lambda^r_+ & 0 \\ 0 & \lambda^r_- \end{pmatrix}, \tag{52}$$

where the diagonalizing angle α_d is, according to (47*b*), determined by the condition

$$\mathrm{tg}\, 2\alpha_d = \frac{2\beta}{J_{xx} - J_{yy}}\,. \tag{53}$$

Using (52) and cyclically permuting the argument of the trace, (51) may be written as

$$I_P(\theta) = \mathrm{Sp}\left\{\mathscr{R}(\alpha_d)\mathscr{P}_+(\theta)\mathscr{R}^+(\alpha_d)\begin{pmatrix} \lambda^r_+ & 0 \\ 0 & \lambda^r_- \end{pmatrix}\right\} = \mathrm{Sp}\left\{\mathscr{P}_+(\theta-\alpha_d)\begin{pmatrix} \lambda^r_+ & 0 \\ 0 & \lambda^r_- \end{pmatrix}\right\}, \tag{54}$$

where use was made of (30). Taking now in particular $\theta = \alpha_d$, we obtain

$$I_P(\alpha_d) = \lambda^r_+\,, \tag{55a}$$

and similarly, for $\theta = \alpha_d + \pi/2$ we get

$$I_P\left(\alpha_d + \frac{\pi}{2}\right) = \lambda^r_-\,. \tag{55b}$$

We now notice that *the same angle* α_d *which diagonalizes* $\mathscr{J}^r$ *makes also the intensity* $I_P(\theta)$ *an extremum.* This can be shown simply by differentiating (51) with respect to θ. We obtain then the condition

$$\mathrm{tg}\, 2\theta_e = \frac{2\beta}{J_{xx} - J_{yy}}\,, \tag{56}$$

which is indeed identical with (53). Whether this extremum is a maximum or a minimum, depends on the sign of β. Combining this result with (55*a*, *b*) we can set

$$(57)\qquad \begin{cases} \lambda^r_+ = I_{e_1}, \\ \lambda^r_- = I_{e_2}; \end{cases}$$

where I_{e_1} and I_{e_2} denote the first and second extremal intensities (a maximum and a minimum if $\beta > 0$, and the opposite way round if $\beta < 0$). These occur at an angle $\theta_{e_1} = \alpha_d$ and $\theta_{e_2} = \alpha_d + \pi/2$ respectively, where α_d is given by (53).

The physical significance of the above mathematical considerations is obvious. To measure the elements J_{xx}, J_{yy}, and $\beta = \operatorname{Re} J_{xy}$ of $\mathscr{J}$, we use a polarizer and find first the smallest angle α_d for which we obtain an extremal intensity I_{e_1}. We measure this intensity and then turn the polarizer to the setting $\alpha_d + \pi/2$ and measure the new extremal intensity I_{e_2}. By (57) the two intensities give us directly the two eigen-values $\lambda^r_\pm$; then, knowing also the angle α_d we compute

$$\mathscr{J}^r = \begin{pmatrix} J_{xx} & \beta \\ \beta & J_{yy} \end{pmatrix} = \mathscr{R}^{-1}(\alpha_d) \begin{pmatrix} I_{e_1} & 0 \\ 0 & I_{e_2} \end{pmatrix} \mathscr{R}(\alpha_d)\,,$$

where $\mathscr{R}$ is given by (20).

If we like, we can now compute the first three Stokes parameters. According to (36) and (10)

$$(58)\qquad \begin{cases} s_0 = J_{xx} + J_{yy}\,, \\ s_1 = J_{xx} - J_{yy}\,, \\ s_2 = 2\beta\,. \end{cases}$$

It follows that if one is interested only in experiments which do not involve compensators, then these three parameters specify the beam completely.

For a complete description we also require, however, the fourth parameter $\gamma = \operatorname{Im} J_{xy}$. This can be obtained by an additional measurement involving a compensator and a polarizer. The intensity is then given by (31). Using a half-wave plate for which $\delta = \pi$ and setting the polarizer at an angle $\theta = \pi/4$ with respect to the x-axis, we obtain, using (33) and the notation of equ. (50)

$$I_K = \tfrac{1}{2}(J_{xx} + J_{yy}) + \gamma\,;$$

hence

$$(59)\qquad \gamma = I_K - \tfrac{1}{2} I_{\text{inc}}\,,$$

where $I_{\text{inc}} \equiv J_{xx} + J_{yy} = \text{Sp}\,\mathscr{J}$ is the incident intensity. Since J_{xx} and J_{yy} have been measured already in the previous experiment, the sole determination of the intensity I_K suffices to determine γ. As a matter of fact, settings other than $\theta = \pi/4$, $\delta = \pi$ could serve our purpose just as well. Knowing γ, the fourth Stokes parameter can be expressed as

$$s_3 = 2\gamma \,. \tag{60}$$

10. – Concluding remarks.

In the above discussion we have seen that a suitable and comprehensive description of the properties and the interactions of statistical radiation can be obtained in terms of the coherency matrix $\mathscr{J}$. It is interesting to note that both the definition and the properties of $\mathscr{J}$ resemble very much those of the statistical density matrix ϱ introduced for the characterization of statistical mixtures of quantum mechanical systems by VON NEUMANN (*). The application of the density matrix in the description of phenomena connected with electron- and photon-polarization has been discussed in great detail by TOLHOEK ([12]).

The main resemblance between the use of ϱ and of $\mathscr{J}$ is reflected in the fact that the expectation value of a physical observable is given in terms of the density matrix by

$$F = \langle \mathscr{F} \rangle = \text{Sp}\,[\mathscr{F}\varrho] \,,$$

while the observable value of a physical characteristic connected with a measurement is given in the present formalism by the analagous expression (1). In particular, as one finds easily by combining (15), (18), (21), (27), the « operator of the intensity » is given, in view of (5) and (1), by the operator

$$\mathscr{J} = \mathscr{A}^2(\eta)\,\mathscr{K}(\theta + \alpha, \delta) \,, \tag{61a}$$

where use was made of (30) and the notation (32). Similarly, the « operator of the Stokes parameter s_i » is, according to (36),

$$\mathscr{S}_i = \sigma_i \,. \tag{61b}$$

(*) For a detailed account see, for example, R. C. TOLMAN: *The principles of statistical mechanics* (Oxford, 1938), p. 327.

([12]) H. A. TOLHOEK: *Rev. Mod. Phys.*, **28**, 277 (1956).

Furthermore, specifying the statistical ensemble in a new representation of variables, *i.e.*, physically, performing a certain measurement on the system by letting it interact with some device, amounts to a similarity transformation on ϱ:

$$\varrho \to \varrho' = \mathcal{O}\varrho\mathcal{O}^{-1} = \mathcal{O}\varrho\mathcal{O}^{\dagger} \,. \tag{62a}$$

On the other hand, we have seen in Sect. **3** and **6** that the interaction of the statistical beam amounts to a transformation

$$\mathcal{J} \to \mathcal{J}' = \mathcal{O}\mathcal{J}\mathcal{O}^{\dagger} \,. \tag{62b}$$

There are, however, some important differences between the two formalisms. For one thing, ϱ is defined in terms of ensemble-averages, while the definition of $\mathcal{J}$ involves time-averages. Even though it is true that — by virtue of the ergodic hypothesis — the two averaging procedures are normally equivalent, this equivalence will break down in the limiting case of a strictly monochromatic beam.

Another difference is that the transforming operator in (62*a*) is always unitary, but in (62*b*) not necessarily so. For example, the operator $\mathcal{P}_{\pm}$ characterizing a polarizer is not. This circumstance arises from the fact that in our discussions we have confined our attention to describing the emergent beam only, while the complete specification of the interaction should involve also the non-transmitted (reflected or absorbed) component of the radiation as well.

RIASSUNTO (*)

La matrice di coerenza di un'onda piana quasi-monocromatica viene dedotta da una rappresentazione matriciale del segnale analitico associato al campo elettrico. Si mostra che se la radiazione passa attraverso un apparecchio fisico, quale un compensatore, un assorbitore, un rotatore o un polarizzatore, l'effetto di questa interazione può essere completamente descritto in termini di operatori opportunamente scelti che trasformano direttamente la matrice di coerenza. Il grado complesso di coerenza viene definito in termini degli operatori suddetti, e da questo viene dedotta un'espressione che caratterizza il grado di polarizzazione. Si mostra che la quantità così dedotta è uguale a quella ottenuta con la definizione più usuale. Si descrive un esperimento che può servire a misurare le componenti della matrice di correlazione.

(*) *Traduzione a cura della Redazione.*

PAPER NO. 34

Reprinted from *Il Nuovo Cimento*, Vol. 17, pp. 462–476, 1960.

Correlation Theory of Stationary Electromagnetic Fields.

PART I – The Basic Field Equations (*).

P. ROMAN

Department of Theoretical Physics, University of Manchester - Manchester

E. WOLF

Institute of Optics, University of Rochester - Rochester, N. Y.

(ricevuto il 22 Aprile 1960)

Summary. — In recent papers a class of second order correlation tensors was introduced and certain differential equations which govern their propagation were postulated. These correlation tensors, which may be regarded as natural generalizations of functions used in the analysis of partially coherent optical wavefields, characterize the correlations which exist between the electromagnetic field vectors at any two points in the field, at any two instants of time. In the present paper a derivation of the basic differential equations is presented; and it is shown that the two sets into which the equations naturally split are not independent, but in fact follow from each other as a consequence of certain symmetry properties which the correlation tensors exhibit.

1. – Introduction.

The researches carried out in recent years in the field of coherence theory of optics have indicated the possibility of formulating rigorous laws for the propagation of second order correlation functions of stationary electromagnetic fields. These functions (in general 3×3 Cartesian tensors) characterize the

(*) This research was supported in part by the United States Air Force under Contract no. AF 49(638)-602, monitored by the AF Office of Scientific Research of the Air Research and Development Command.

correlations between the electromagnetic field vectors at any two points in the field, at any two instants of time.

The correlation tensors were introduced in connection with an attempt to formulate a large branch of optics in terms of measurable quantities (1). The reason for attempting such a formulation is the fact, that in the great majority of optical experiments the quantities which are measured are not the field vectors themselves. These vectors vary so rapidly that no detector can record their detail behaviour (*). One can usually only measure the time averages of certain quadratic functions (*e.g.* the electric energy density) of the field components. Unlike the field vectors themselves, the correlation tensors are closely related to quantities that can be measured in the majority of optical experiments.

The correlation tensors were used, in specialized form, in the analysis of several problems. In reference (2) and (3) the elementary theory of partial polarization (for quasi-monochromatic plane waves) was reformulated and brought into a form strictly analogous to that employed in connection with the scalar theory of partially coherent optical fields. The formulation was found to be intimately connected with a well known representation of partially polarized radiation in terms of Stokes parameters. A generalization of the Stokes parameters to waves that are not necessarily plane, was recently carried out within the framework of correlation analysis by P. Roman (4). More recently it was shown (5) that the effect of the interaction of a partially polarized quasi-monochromatic plane wave with certain commonly used optical devices may be described as a transformation of the correlation tensor, the transformation involving an operator which is characteristic of the device.

In one of the earlier papers (Wolf (6)), differential equations which govern the propagation of the correlation tensors were postulated (in dyadic form). In the present paper they are derived from Maxwell's equations by the direct application of tensor analysis and it is shown that the two sets of equations into which they naturally split up are not independent. In fact, each set follows from the other by the use of certain symmetry properties which the correlation tensors exhibit. Some invariance properties of these equations are

---

(1) E. Wolf: *Nuovo Cimento*, **12**, 884 (1954).

(*) It may be recalled that whilst the optical periods are of the order of 10^{-14} s, the resolving time of the best photomultipliers which are at present available is of the order of 10^{-9} s.

(2) M. Born and E. Wolf: *Principles of Optics* (London and New York, 1959),

(3) E. Wolf: *Nuovo Cimento*, **13**, 1165 (1959).

(4) P. Roman: *Nuovo Cimento*, **13**, 974 (1959).

(5) G. B. Parrent and P. Roman: *Nuovo Cimento*, **15**, 370 (1960).

(6) E. Wolf: Contribution in *Proc. Symposium on Astronom. Optics* (Amsterdam, 1956), p. 177.

noted. Finally the wave equations for the correlation tensors are derived from the basic set of equations.

In Part II (ROMAN and WOLF [7]), a number of conservation laws involving the correlation tensors will be established.

2. – The correlation tensors.

We consider a stationary electromagnetic field in vacuum. Let $\boldsymbol{E}^{(r)}(\boldsymbol{x}, t)$ and $\boldsymbol{H}^{(r)}(\boldsymbol{x}, t)$ represent the electric and magnetic field vectors at a point specified by position vector $\boldsymbol{x}$, at time t. These vectors are real, but it is useful to employ a complex representation which may be introduced as follows:

Let $E_j^{(r)}$ $(j = 1, 2, 3)$ be the Cartesian components of $\boldsymbol{E}^{(r)}$ with respect to a fixed system of rectangular axes and let (*)

$$(2.1)\qquad \begin{cases} E_j^{(r)}(\boldsymbol{x}, t; T) = E_j^{(r)}(\boldsymbol{x}, t) & \text{when } |t| \leqslant T, \\ \qquad\qquad\quad = 0 & \text{when } |t| > T. \end{cases}$$

We develop $E_j^{(r)}$ into a real Fourier integral with respect to t, *i.e.* we write

$$(2.2)\qquad E_j^{(r)}(\boldsymbol{x}, t; T) = \int_0^\infty e_j(\boldsymbol{x}, \omega; T) \cos\left[\varphi_j(\boldsymbol{x}, \omega; T) - \omega t\right] d\omega\,.$$

With $E_j^{(r)}$ we now associate the complex vector E_j obtained by replacing in the integrand the cosine function by the exponential function:

$$(2.3)\qquad E_j(\boldsymbol{x}, t; T) = \int_0^\infty e_j(\boldsymbol{x}, \omega; T) \exp\left[i\left[\varphi_j(\boldsymbol{x}, \omega; T) - \omega t\right]\right] d\omega\,.$$

Then

$$(2.4)\qquad E_j(\boldsymbol{x}, t; T) = E_j^{(r)}(\boldsymbol{x}, t; T) + iE_j^{(i)}(\boldsymbol{x}, t; T)\,,$$

---

(7) P. ROMAN and E. WOLF: *Nuovo Cimento*, **17**, 477 (1960).

(*) The truncated functions $E_j^{(r)}(\boldsymbol{x}, t; T)$ are introduced for reasons well known in the theory of random processes: since the field is assumed to be stationary, $E_j^{(r)}(\boldsymbol{x}, t)$ is not square integrable and hence does not necessarily possess a Fourier representation. On the other hand the truncated functions $E_j^{(r)}(\boldsymbol{x}, t; T)$ are amenable to Fourier analysis and yield results of physical interest when one proceeds to the limit $T \to \infty$ in the appropriate formulae.

where

$$E_j^{(i)}(\boldsymbol{x}, t; T) = \int_0^\infty e_j(\boldsymbol{x}, \omega; T) \sin[\varphi_j(\boldsymbol{x}, \omega; T) - \omega t]\, d\omega \,. \tag{2.5}$$

It is seen that $E_j^{(r)}$ and $E_j^{(i)}$ are a pair of « conjugate functions » and, as is well known (cf. reference (8)), they are related by Hilbert's reciprocity relations,

$$\begin{cases} E_j^{(i)}(\boldsymbol{x}, t; T) = \dfrac{P}{\pi}\displaystyle\int_{-\infty}^{\infty} \dfrac{E_j^{(r)}(\boldsymbol{x}, t'; T)}{t'-t}\, dt' \,, \\[2ex] E_j^{(r)}(\boldsymbol{x}, t; T) = -\dfrac{P}{\pi}\displaystyle\int_{-\infty}^{\infty} \dfrac{E_j^{(i)}(\boldsymbol{x}, t'; T)}{t'-t}\, dt' \,, \end{cases} \tag{2.6}$$

where P denotes the Cauchy principal value at $t' = t$.

In a strictly similar manner we also associate a complex vector with the (truncated) magnetic field:

$$H_j(\boldsymbol{x}, t; T) = H_j^{(r)}(\boldsymbol{x}, t; T) + iH_j^{(i)}(\boldsymbol{x}, t; T)\,, \qquad (j = 1, 2, 3), \tag{2.7}$$

where $H_j^{(i)}$ is the conjugate function associated with the real magnetic field $H_j^{(r)}$.

A complex function such as E_j or H_j, the real and imaginary parts of which are conjugate functions is known as an *analytic signal* (*).

The correlation tensors referred to in the introduction may now be defined by formulae of the form

$$\mathscr{E}_{jk}(\boldsymbol{x}_1, \boldsymbol{x}_2, \tau) = \operatorname*{Lim}_{T\to\infty} \frac{1}{2T}\int_{-\infty}^{\infty} E_j(\boldsymbol{x}_1, t+\tau; T)\, E_k^*(\boldsymbol{x}_2, t; T)\, dt\,, \tag{2.8}$$

where the asterisk denotes the complex conjugate. Denoting the time averaging operation on the right hand side by sharp brackets, we have in all the following four correlation tensors:

$$\begin{cases} \mathscr{E}_{jk}(\boldsymbol{x}_1, \boldsymbol{x}_2, \tau) = \langle E_j(\boldsymbol{x}_1, t+\tau)\, E_k^*(\boldsymbol{x}_2, t)\rangle \,, \\ \mathscr{H}_{jk}(\boldsymbol{x}_1, \boldsymbol{x}_2, \tau) = \langle H_j(\boldsymbol{x}_1, t+\tau)\, H_k^*(\boldsymbol{x}_2, t)\rangle \,, \\ \mathscr{G}_{jk}(\boldsymbol{x}_1, \boldsymbol{x}_2, \tau) = \langle E_j(\boldsymbol{x}_1, t+\tau)\, H_k^*(\boldsymbol{x}_2, t)\rangle \,, \\ \tilde{\mathscr{G}}_{jk}(\boldsymbol{x}_1, \boldsymbol{x}_2, \tau) = \langle H_j(\boldsymbol{x}_1, t+\tau)\, E_k^*(\boldsymbol{x}_2, t)\rangle \,. \end{cases} \tag{2.9}$$

(8) E. C. TITCHMARSH: *Introduction to the Theory of Fourier Integrals*, 2nd Ed. (Oxford, 1948), chap. 5.

(*) References to papers dealing with properties of analytic signals will be found in BORN and WOLF (2), Section **10**·2.

The physical significance of these tensors is clear: they represent the correlations which exist between the field vectors at two typical points ($\boldsymbol{x}_1$ and $\boldsymbol{x}_2$), at time instants separated by a time interval τ.

Although the correlation tensors are in general complex, they are very closely related to the real correlation tensors associated with the physical (real) fields. To see this we separate the real and imaginary parts in each of the four tensors (2.9) and denote them by superscripts (r) and (i) respectively, *i.e.* we write

$$\mathscr{E}_{jk} = \mathscr{E}_{jk}^{(r)} + i\mathscr{E}_{jk}^{(i)}, \tag{2.10}$$

etc., where $\mathscr{E}_{jk}^{(r)}$ and $\mathscr{E}_{jk}^{(i)}$ are real. It then follows from formulae (VII) and (VI) established in the Appendix that

$$\left\{\begin{aligned} \mathscr{E}_{jk}^{(r)} &= 2\langle E_j^{(r)}(\boldsymbol{x}_1, t+\tau)\, E_k^{(r)}(\boldsymbol{x}_2, t)\rangle \\ &= 2\langle E_j^{(i)}(\boldsymbol{x}_1, t+\tau)\, E_k^{(i)}(\boldsymbol{x}_2, t)\rangle, \\ \mathscr{E}_{jk}^{(i)} &= 2\langle E_j^{(i)}(\boldsymbol{x}_1, t+\tau)\, E_k^{(r)}(\boldsymbol{x}_2, t)\rangle \\ &= -2\langle E_j^{(r)}(\boldsymbol{x}_1, t+\tau)\, E_k^{(i)}(\boldsymbol{x}_2, t)\rangle, \end{aligned}\right. \tag{2.11}$$

with similar expressions for the real and imaginary parts of the other correlation tensors. In particular we see that the real part of each correlation tensor is twice the correlation tensor formed by the components of the real field vectors. Moreover it is not difficult to show that $\mathscr{E}$ itself is an analytic signal, so that $\mathscr{E}_{jk}^{(r)}$ and $\mathscr{E}_{jk}^{(i)}$ bear the same relationship to each other as does $E_j^{(r)}$ and $E_j^{(i)}$, *i.e.* they form a pair of conjugate functions:

$$\left\{\begin{aligned} \mathscr{E}_{jk}^{(i)}(\boldsymbol{x}_1, \boldsymbol{x}_2, \tau) &= \frac{P}{\pi}\int_{-\infty}^{\infty} \frac{\mathscr{E}_{jk}^{(r)}(\boldsymbol{x}_1, \boldsymbol{x}_2, \tau')}{\tau'-\tau}\, d\tau', \\ \mathscr{E}_{jk}^{(r)}(\boldsymbol{x}_1, \boldsymbol{x}_2, \tau) &= -\frac{P}{\pi}\int_{-\infty}^{\infty} \frac{\mathscr{E}_{jk}^{(i)}(\boldsymbol{x}_1, \boldsymbol{x}_2, \tau')}{\tau'-\tau}\, d\tau'. \end{aligned}\right. \tag{2.12}$$

(Strictly similar result holds, of course, for the real and imaginary parts of $\mathscr{H}$, $\mathscr{G}$ and $\tilde{\mathscr{G}}$). The proof of (2.12) is identical with that given in connection with the optical scalar theory of partial coherence (cf. BORN and WOLF [2]), § **10**·3.2).

We note that the time averaged electric and magnetic energy densities, denoted by W_{el} and W_{mag} respectively, can easily be expressed in terms of the correlation tensors. We have

(2.13) $$\langle W_{el}\rangle = \frac{1}{8\pi}\langle \boldsymbol{E}^{(r)2}\rangle\,,$$

where $\boldsymbol{E}^{(r)} = \boldsymbol{E}^{(r)}(\boldsymbol{x}, t)$. Now according to (2.4),

(2.14) $$\langle \boldsymbol{E}\cdot\boldsymbol{E}^*\rangle = \langle \boldsymbol{E}^{(r)2}\rangle + \langle \boldsymbol{E}^{(i)2}\rangle + 2i\langle \boldsymbol{E}^{(r)}\cdot\boldsymbol{E}^{(i)}\rangle\,.$$

The first two terms on the right are, according to (2.11), equal to each other, and the last term vanishes since all the other terms are real. Hence $\langle \boldsymbol{E}\cdot\boldsymbol{E}^*\rangle = 2\langle \boldsymbol{E}^{(r)2}\rangle$ and (2.13) becomes

(2.15) $$\langle W_{el}\rangle = \frac{1}{16\pi}\langle \boldsymbol{E}\cdot\boldsymbol{E}^*\rangle = \frac{1}{16\pi}\,\mathrm{Sp}\,\mathscr{E}(\boldsymbol{x}, \boldsymbol{x}, 0)\,,$$

where Sp denotes the spur (trace). Similarly

(2.16) $$\langle W_{m}\rangle = \frac{1}{16\pi}\,\mathrm{Sp}\,\mathscr{H}(\boldsymbol{x}, \boldsymbol{x}, 0)\,.$$

Next consider the time average of the energy flow (Poynting vector) $\boldsymbol{S}$:

(2.17) $$\langle \boldsymbol{S}\rangle = \frac{c}{4\pi}\langle \boldsymbol{E}^{(r)}\wedge\boldsymbol{H}^{(r)}\rangle\,.$$

Now if $\mathscr{R}$ denotes the real part, we have

$$\begin{aligned}\mathscr{R}\langle \boldsymbol{E}\wedge\boldsymbol{H}^*\rangle &= \mathscr{R}\langle(\boldsymbol{E}^{(r)} + i\boldsymbol{E}^{(i)})\wedge(\boldsymbol{H}^{(r)} - i\boldsymbol{H}^{(i)})\rangle\\ &= 2\langle \boldsymbol{E}^{(r)}\wedge\boldsymbol{H}^{(r)}\rangle\,,\end{aligned}$$

if relations of the type $\langle E_j^{(r)}H_k^{(r)}\rangle = \langle E_j^{(i)}H_k^{(i)}\rangle$, which follow from formula (VI) of the Appendix, are used. Hence (2.17) may be written as

(2.18) $$\langle \boldsymbol{S}\rangle = \frac{c}{8\pi}\mathscr{R}\langle \boldsymbol{E}\wedge\boldsymbol{H}^*\rangle\,,$$

or, taking the components,

(2.19) $$\left|\begin{aligned}\langle S_j\rangle &= \frac{c}{8\pi}\mathscr{R}\{\langle E_k H_l^* - E_l H_k^*\rangle\}\\ &= \frac{c}{8\pi}\mathscr{R}\{\mathscr{G}_{kl}(\boldsymbol{x}, \boldsymbol{x}, 0) - \mathscr{G}_{lk}(\boldsymbol{x}, \boldsymbol{x}, 0)\}\\ &= \frac{c}{8\pi}\mathscr{R}\{\mathscr{G}_{kl}(\boldsymbol{x}, \boldsymbol{x}, 0) - \widetilde{\mathscr{G}}_{kl}(\boldsymbol{x}, \boldsymbol{x}, 0)\}\,,\end{aligned}\right.$$

with $(j, k, l) = (1, 2, 3)$ or cycl.

3. – The basic differential equations.

Since the electric and magnetic vectors are related by Maxwell's equations, the correlation tensors are evidently not independent of each other. We shall now derive the relations which exist between them.

Let ε_{jkl} be the completely antisymmetric unit tensor of Levi-Civita, *i.e.* ε_{jkl} is $+1$ or -1 according as the subscripts (j, k, l) are an even or an odd permutation of $(1, 2, 3)$ and $\varepsilon_{jkl}=0$ when two suffices are equal. The components of the vector product of any two vectors $\boldsymbol{V}$ and $\boldsymbol{W}$ may then be expressed in the form

$$(3.1)\qquad (\boldsymbol{V}\wedge\boldsymbol{W})_j=\varepsilon_{jkl}V_k W_l\,.$$

In particular, $\boldsymbol{V}$ may represent the operator $\nabla\equiv\partial/\partial x_k\equiv\partial_k$. Then

$$(3.2)\qquad (\nabla\wedge\boldsymbol{W})_j=\varepsilon_{jkl}\,\partial_k W_l\,.$$

Since we shall be dealing with functions which depend on the position of two points $\boldsymbol{x}_1$, $\boldsymbol{x}_2$, we must distinguish between two differential operators acting on space co-ordinates. We shall use superscript 1 or 2 according whether the operator acts on the co-ordinates of $\boldsymbol{x}_1$, or $\boldsymbol{x}_2$, *i.e.*

$$(3.3)\qquad \partial_k^1\equiv\frac{\partial}{\partial x_1^k}\,,\qquad \partial_k^2\equiv\frac{\partial}{\partial x_2^k}\qquad (k=1,2,3).$$

Now Maxwell's equations for the electromagnetic field in free space may be expressed in the form (*)

$$(3.4)\qquad \varepsilon_{jkl}\,\partial_k^1 E_l(\boldsymbol{x}_1,t_1)+\frac{1}{c}\frac{\partial}{\partial t_1}H_j(\boldsymbol{x}_1,t_1)=0\,,$$

$$(3.5)\qquad \varepsilon_{jkl}\,\partial_k^1 H_l(\boldsymbol{x}_1,t_1)-\frac{1}{c}\frac{\partial}{\partial t_1}E_j(\boldsymbol{x}_1,t_1)=0\,,$$

(*) It is assumed here that not only the real fields $\boldsymbol{E}^{(r)}$, $\boldsymbol{H}^{(r)}$ but also the associated complex fields $\boldsymbol{E}=\boldsymbol{E}^{(r)}+i\boldsymbol{E}^{(i)}$, $\boldsymbol{H}=\boldsymbol{H}^{(r)}+i\boldsymbol{H}^{(i)}$ where $\boldsymbol{E}^{(i)}$ and $\boldsymbol{H}^{(i)}$ are functions conjugate to $\boldsymbol{E}^{(r)}$ and $\boldsymbol{H}^{(r)}$ respectively, satisfy Maxwell's equations. That this is so may be readily shown by taking the Hilbert transform of Maxwell's equations for the real fields $\boldsymbol{E}^{(r)}$, $\boldsymbol{H}^{(r)}$ and using the fact that if two functions are Hilbert transforms of each other so are their derivatives. One then finds that $\boldsymbol{E}^{(i)}$, $\boldsymbol{H}^{(i)}$ and hence the complex fields $\boldsymbol{E}$, $\boldsymbol{H}$ also obey Maxwell's equations.

and the divergence conditions may be written as

$$\partial_j^1 E_j(\boldsymbol{x}_1, t_1) = 0 , \tag{3.6}$$

$$\partial_j^1 H_j(\boldsymbol{x}_1, t_1) = 0 . \tag{3.7}$$

We multiply (3.4) by $E_m^*(\boldsymbol{x}_2, t_2)$ and obtain the relation

$$\varepsilon_{jkl}\, \partial_k^1 E_l(\boldsymbol{x}_1, t_1)\, E_m^*(\boldsymbol{x}_2, t_2) + \frac{1}{c}\frac{\partial}{\partial t_1} H_j(\boldsymbol{x}_1, t_1)\, E_m^*(\boldsymbol{x}_2, t_2) = 0 . \tag{3.8}$$

Let $t_1 = t_2 + \tau$ and keep t_2 fixed. Then $\partial/\partial t_1 = \partial/\partial\tau$ and (3.8) becomes, on taking the time average with respect to t_2, and finally writing t in place of t_2:

$$\varepsilon_{jkl}\, \partial_k^1 \langle E_l(\boldsymbol{x}_1, t+\tau)\, E_m^*(\boldsymbol{x}_2, t)\rangle + \frac{1}{c}\frac{\partial}{\partial\tau} \langle H_j(\boldsymbol{x}_1, t+\tau)\, E_m^*(\boldsymbol{x}_2, t)\rangle = 0 ,$$

or, in terms of the tensors $\mathscr{E}$ and $\widetilde{\mathscr{G}}$ defined in (2.9),

$$\varepsilon_{jkl}\, \partial_k^1 \mathscr{E}_{lm} + \frac{1}{c}\frac{\partial}{\partial\tau} \widetilde{\mathscr{G}}_{jm} = 0 . \tag{3.9a}$$

In a similar way, if we multiply (3.4) by $H_m^*(\boldsymbol{x}_2, t_2)$ and apply the same procedure, we obtain the equation

$$\varepsilon_{jkl}\, \partial_k^1 \mathscr{G}_{lm} + \frac{1}{c}\frac{\partial}{\partial\tau} \mathscr{H}_{jm} = 0 . \tag{3.10a}$$

From (3.5) we obtain in the same way

$$\varepsilon_{jkl}\, \partial_k^1 \widetilde{\mathscr{G}}_{lm} - \frac{1}{c}\frac{\partial}{\partial\tau} \mathscr{E}_{jm} = 0 , \tag{3.11a}$$

and

$$\varepsilon_{ikl}\, \partial_k^1 \mathscr{H}_{lm} - \frac{1}{c}\frac{\partial}{\partial\tau} \mathscr{G}_{jm} = 0 . \tag{3.12a}$$

In a similar manner, the divergence conditions (3.6) and (3.7) yield the equations

$$\partial_j^1 \mathscr{E}_{jk} = 0 , \tag{3.13a}$$

$$\partial_j^1 \mathscr{G}_{jk} = 0 , \tag{3.14a}$$

$$\partial_j^1 \widetilde{\mathscr{G}}_{jk} = 0 , \tag{3.15a}$$

$$\partial_j^1 \mathscr{H}_{jk} = 0 . \tag{3.16a}$$

There is another set of equations which involve the operator ∂_k^2. It may be conveniently obtained by starting from equations which are the complex conjugates of the « complex » Maxwell equations (3.4) and (3.5), after replacing $\boldsymbol{x}_1$, t_1 and ∂_k^1 by $\boldsymbol{x}_2$, t_2 and ∂_k^2 respectively. If these equations are then multiplied by $E_m(\boldsymbol{x}_1, t_1)$ or $H_m(\boldsymbol{x}_1, t_1)$ and time averaging is performed as before, one obtains:

(3.9b) $$\varepsilon_{jkl}\,\partial_k^2\,\mathscr{E}_{ml} - \frac{1}{c}\frac{\partial}{\partial\tau}\mathscr{G}_{mj} = 0\,,$$

(3.10b) $$\varepsilon_{jkl}\,\partial_k^2\tilde{\mathscr{G}}_{ml} - \frac{1}{c}\frac{\partial}{\partial\tau}\mathscr{H}_{mj} = 0\,,$$

(3.11b) $$\varepsilon_{jkl}\,\partial_k^2\mathscr{G}_{ml} + \frac{1}{c}\frac{\partial}{\partial\tau}\mathscr{E}_{mj} = 0\,,$$

(3.12b) $$\varepsilon_{jkl}\,\partial_k^2\mathscr{H}_{ml} + \frac{1}{c}\frac{\partial}{\partial\tau}\mathscr{G}_{mj} = 0\,.$$

From the divergence conditions (3.6) and (3.7) we find in a similar way that

(3.13b) $$\partial_j^2\,\mathscr{E}_{kj} = 0\,,$$

(3.14b) $$\partial_j^2\,\tilde{\mathscr{G}}_{kj} = 0\,,$$

(3.15b) $$\partial_j^2\,\mathscr{G}_{kj} = 0\,,$$

(3.16b) $$\partial_j^2\mathscr{H}_{kj} = 0\,.$$

The differential equations (3.9*a*)–(3.16*a*) and (3.9*b*)–(3.16*b*) were postulated (in dyadic notation) by WOLF (6). Before we investigate some consequences of these equations, we note that each of the two sets of equations is invariant under the transformation

(3.17) $$\mathscr{E} \rightleftarrows -\mathscr{H}\,, \qquad \mathscr{G} \rightleftarrows \tilde{\mathscr{G}}\,.$$

Moreover the set (*a*) is also invariant under the transformation

(3.18) $$\mathscr{E} \rightleftarrows \mathscr{G}\,, \qquad \mathscr{H} \rightleftarrows \tilde{\mathscr{G}}\,,$$

and the set (*b*) under the transformation

(3.19) $$\mathscr{E} \rightleftarrows \tilde{\mathscr{G}}\,, \qquad \mathscr{H} \rightleftarrows \mathscr{G}\,.$$

4. – Equivalence of the sets of eqs. (*a*) and (*b*).

It will now be shown that the sets of equations (*a*) and (*b*) are not independent but follow from each other as a consequence of certain « symmetry properties » which the correlation tensors exhibit.

We have

$$(4.1)\qquad \left\{\begin{aligned} \mathscr{E}^*_{jk}(\boldsymbol{x}_1, \boldsymbol{x}_2, \tau) &= \langle E^*_j(\boldsymbol{x}_1, t+\tau)\, E_k(\boldsymbol{x}_2, t)\rangle \\ &= \langle E^*_j(\boldsymbol{x}_1, t)\, E_k(\boldsymbol{x}_2, t-\tau)\rangle \\ &= \mathscr{E}_{kj}(\boldsymbol{x}_2, \boldsymbol{x}_1, -\tau)\,. \end{aligned}\right.$$

On going from the first to the second line we have made use of the fact that the field is stationary. In a strictly similar manner

$$(4.2)\qquad \mathscr{H}^*_{jk}(\boldsymbol{x}_1, \boldsymbol{x}_2, \tau) = \mathscr{H}_{kj}(\boldsymbol{x}_2, \boldsymbol{x}_1, -\tau)\,,$$

whilst the remaining two correlation tensors are found to be related by the formula (*)

$$(4.3)\qquad \tilde{\mathscr{G}}_{jk}(\boldsymbol{x}_1, \boldsymbol{x}_2, \tau) = \mathscr{G}^*_{kj}(\boldsymbol{x}_2, \boldsymbol{x}_1, -\tau)\,.$$

We are again dealing with functions of two spacial arguments (and of the time difference τ) and must now distinguish between two « ∂ » operators, according to whether each operates on the *first* or the *second* spacial argument. We distinguish these by writing $\partial^{(1)}$ in the first case and $\partial^{(2)}$ in the second case. Thus if $F \equiv F(\boldsymbol{x}_1, \boldsymbol{x}_2, \tau)$ then

$$\partial^{(1)}_k = \frac{\partial F}{\partial x^k_1} = \partial^1_k \qquad (k = 1, 2, 3);$$

but if $G = G(\boldsymbol{x}_2, \boldsymbol{x}_1, \tau)$ then

$$\partial^{(1)}_k = \frac{\partial G}{\partial x^k_2} = \partial^2_k \qquad (k = 1, 2, 3),$$

etc. It then follows from (4.1) that

$$(4.4)\qquad \partial^{(1)}_l\, \mathscr{E}^*_{jk}(\boldsymbol{x}_1, \boldsymbol{x}_2, \tau) = \partial^{(2)}_l\, \mathscr{E}_{kj}(\boldsymbol{x}_2, \boldsymbol{x}_1, -\tau)\,.$$

We also have from (4.1)

$$(4.5)\qquad \frac{\partial}{\partial \tau}\, \mathscr{E}^*_{jk}(\boldsymbol{x}_1, \boldsymbol{x}_2, \tau) = \frac{\partial}{\partial \tau}\, \mathscr{E}_{kj}(\boldsymbol{x}_2, \boldsymbol{x}_1, -\tau)\,.$$

(*) There is a misprint in the corresponding formula (eq. (18), p. 183) in ref. (6).

From (4.2) one obtains, of course, strictly similar relations involving the tensor $\mathscr{H}$, whilst the corresponding relations involving $\widetilde{\mathscr{G}}$ and $\mathscr{G}$, deduced from (4.3), are

$$(4.6) \qquad \partial_l^{(1)} \, \widetilde{\mathscr{G}}_{jk}^*(\boldsymbol{x}_1, \boldsymbol{x}_2, \tau) = \partial_l^{(2)} \, \mathscr{G}_{kj}(\boldsymbol{x}_2, \boldsymbol{x}_1, -\tau) \,,$$

and

$$(4.7) \qquad \frac{\partial}{\partial \tau} \, \widetilde{\mathscr{G}}_{jk}^*(\boldsymbol{x}_1, \boldsymbol{x}_2, \tau) = \frac{\partial}{\partial \tau} \, \mathscr{G}_{kj}(\boldsymbol{x}_2, \boldsymbol{x}_1, -\tau) \,.$$

Let us now take the complex conjugate of (3.9a):

$$(4.8) \qquad \varepsilon_{jkl} \, \partial_k^{(1)} \, \mathscr{E}_{lm}^*(\boldsymbol{x}_1, \boldsymbol{x}_2, \tau) + \frac{1}{c} \frac{\partial}{\partial \tau} \, \widetilde{\mathscr{G}}_{jm}^*(\boldsymbol{x}_1, \boldsymbol{x}_2, \tau) = 0 \,.$$

Using the relation (4.4) and (4.7) this equation transforms into

$$(4.9) \qquad \varepsilon_{jkl} \, \partial_k^{(2)} \, \mathscr{E}_{ml}(\boldsymbol{x}_2, \boldsymbol{x}_1, -\tau) + \frac{1}{c} \frac{\partial}{\partial \tau} \, \mathscr{G}_{mj}(\boldsymbol{x}_2, \boldsymbol{x}_1, -\tau) = 0 \,.$$

If next we use the fact that

$$\frac{\partial}{\partial \tau} \, \mathscr{G}_{mj}(\boldsymbol{x}_2, \boldsymbol{x}_1, -\tau) = -\frac{\partial}{\partial(-\tau)} \, \mathscr{G}_{mj}(\boldsymbol{x}_2, \boldsymbol{x}_1, -\tau)$$

and set $\boldsymbol{x}_2 = \boldsymbol{u}_1$, $\boldsymbol{x}_1 = \boldsymbol{u}_2$, $-\tau = \varphi$, (4.9) becomes

$$(4.10) \qquad \varepsilon_{jkl} \, \partial_k^{(2)} \mathscr{E}_{ml}(\boldsymbol{u}_1, \boldsymbol{u}_2, \varphi) - \frac{1}{c} \frac{\partial}{\partial \varphi} \, \mathscr{G}_{mj}(\boldsymbol{u}_1, \boldsymbol{u}_2, \varphi) = 0 \,,$$

and this is just equation (3.9b) in another notation.

A similar procedure applied to equations (3.10a), (3.11a) and (3.12a) transforms them into equations (3.10b), (3.11b) and (3.12b) respectively.

Next consider the « divergence equations » (3.13a)–(3.16a). We have from (3.13a), with the help of (4.4)

$$(4.11) \qquad \partial_j^{(2)} \mathscr{E}_{kj}(\boldsymbol{x}_2, \boldsymbol{x}_1, -\tau) = 0 \,.$$

If we substitute $\boldsymbol{x}_2 = \boldsymbol{u}_1$, $\boldsymbol{x}_1 = \boldsymbol{u}_2$, $-\tau = \varphi$, this equation reduces to

$$(4.12) \qquad \partial_j^{(2)} \mathscr{E}_{kj}(\boldsymbol{u}_1, \boldsymbol{u}_2, \varphi) = 0$$

and this is equation (3.13b). In the same way one derives from the equations (3.14a), (3.15a) and (3.16a) the equations (3.14b), (3.15b) and (3.16b).

We have now shown that the equations (3.9*b*)–(3.16*b*) follow directly from the equations (3.9*a*)–(3.16*a*). Conversely it may be shown that the « *a* » equations follow from the « *b* » equations. We may thus restrict further discussions to equations of one of these two sets only. We select the « *a* » set and will from now on as a rule omit the superscript 1 on ∂_j^1. Thus ∂_j will denote differentiation with respect to the j-th component ($j=1, 2, 3$) of the spatial argument $\boldsymbol{x}_1$. In case of ambiguity we will, however, retain the superscript.

In analogy with Maxwell's equations one may also show that the « divergence equations » (3.13*a*)–(3.16*a*) follow from the main equations in the sense that if they hold for a given value $\tau=\tau_0$, they will hold for all values of τ. To see this we apply the operators $\partial_j^1 \equiv \partial_j$ to (3.11*a*) and find that

$$\varepsilon_{jkl}\,\partial_j\,\partial_k\,\widetilde{\mathscr{G}}_{lm} - \frac{1}{c}\,\frac{\partial}{\partial\tau}\,\partial_j\,\mathscr{E}_{jm} = 0\,.$$

Since ε_{jkl} is antisymmetric in the indices j, k, the first term, which involves the symmetric operator $\partial_j\partial_k$ vanishes and it follows that

$$\partial_j\,\mathscr{E}_{jm} = \text{constant}\,. \tag{4.13}$$

If this constant is zero when $\tau=\tau_0$, then the expression on the left is zero for all τ, in agreement with (3,13*a*). The same holds, of course, for the quantities $\partial_j\,\mathscr{G}_{jm}$, $\partial_j\,\widetilde{\mathscr{G}}_{jm}$ and $\partial_j\mathscr{H}_{jm}$, in agreement with (3.14*a*)–(3.16*a*).

5. – Wave equations.

We shall now show that each correlation tensor satisfies the wave equation.

On applying the operators $(1/c)\;\partial/\partial\tau$ to (3.11*a*), we have

$$\varepsilon_{jkl}\,\partial_k\,\frac{1}{c}\,\frac{\partial}{\partial\tau}\,\widetilde{\mathscr{G}}_{lm} = \frac{1}{c^2}\,\frac{\partial^2}{\partial\tau^2}\,\mathscr{E}_{jm}\,. \tag{5.1}$$

Next we substitute on the left hand side from (3.9*a*) and obtain

$$-\,\varepsilon_{jkl}\,\varepsilon_{lab}\,\partial_k\,\partial_a\,\mathscr{E}_{bm} = \frac{1}{c^2}\,\frac{\partial^2}{\partial\tau^2}\,\mathscr{E}_{jm}\,. \tag{5.2}$$

Now there is the identity (see, for example JEFFREYS [9], p. 73)

$$\varepsilon_{jkl}\,\varepsilon_{lab} = \varepsilon_{jkl}\,\varepsilon_{abl} = \delta_{ja}\,\delta_{kb} - \delta_{jb}\,\delta_{ka}\,, \tag{5.3}$$

---

[9] H. JEFFREYS and B. S. JEFFREYS: *Methods of Mathematical Physics*, 2nd Ed. (Cambridge, 1950).

where δ is the Kronecker symbol. Hence (5.3) becomes

$$-\partial_b \partial_j \mathscr{E}_{bm} + \nabla^2 \mathscr{E}_{jm} = \frac{1}{c^2} \frac{\partial^2}{\partial \tau^2} \mathscr{E}_{jm} , \tag{5.4}$$

where $\nabla^2 \equiv \partial_k \partial_k$ $(k = 1, 2, 3)$ is the wave operator. Now according to (3.13*a*), the first term on the left of (5.4) vanishes, so that one finally obtains the equation

$$\nabla^2 \mathscr{E}_{jm} = \frac{1}{c^2} \frac{\partial^2}{\partial \tau^2} \mathscr{E}_{jm} . \tag{5.5}$$

Hence the $\mathscr{E}$-tensor, and similarly also each of the tensors $\mathscr{H}$, $\mathscr{G}$ and $\widetilde{\mathscr{G}}$, satisfies the homogeneous wave equation, it being understood here that the wave operator acts on the first spacial argument. However, if the « symmetry relations » (4.4) and (4.5) are used, one also obtains homogeneous wave equations in which the wave operator acts on the second spacial argument. (Alternatively they may be derived from the « discarded » set (3.9*b*)–(3.16*b*)).

The wave equations for the correlation tensors were previously given by Wolf ([1]), who derived them by using the fact that each field vector itself satisfies the wave equation. Here we have shown that the wave equations for the correlation tensors follow directly from the basic set (3.9*a*)–(3.16*a*).

In Part II of this investigation a number of conservation laws which relate to the correlation tensors will be established.

Appendix

Some theorems on analytic signals.

In this Appendix we shall establish some theorems needed in the main text, which relate to averages involving products of analytic signals.

Let $F^{(r)}(t)$ and $G^{(r)}(t)$ be two real stationary fields and consider the correlation function

$$\langle F(t+\tau)\, G(t)\rangle = \lim_{T\to\infty} \frac{1}{2T} \int_{-\infty}^{\infty} F(t+\tau;\, T)\, G(t;\, T)\, \mathrm{d}t , \tag{A.1}$$

where $F(t;\, T)$ and $G(t;\, T)$ are the analytic signals associated with the truncated functions $F^{(r)}(t;\, T)$ and $G^{(r)}(t;\, T)$ (cf. (2.1)). To evaluate (A.1) we develop the functions on the right hand side of (A.1) into Fourier integrals (*):

$$(A.2)\qquad \begin{cases} F(t;\, T) = \displaystyle\int_0^{\infty} f(\omega;\, T) \exp\left[-\, i\omega t\right] d\omega\,, \\ G(t;\, T) = \displaystyle\int_0^{\infty} g(\omega;\, T) \exp\left[-\, i\omega t\right] d\omega\,. \end{cases}$$

From (A.1) and (A.2)

$$(A.3)\qquad \langle F(t+\tau)\, G(t)\rangle =$$

$$= \lim_{T\to\infty} \frac{1}{2T} \int_{-\infty}^{\infty} dt \int_0^{\infty}\!\!\int_0^{\infty} f(\omega;\, T)\, g(\omega';\, T) \exp\left[-\, i\omega\tau\right] \exp\left[-\, i(\omega+\omega')t\right] d\omega\, d\omega'\,.$$

Changing the order of the integrations and using the fact that

$$\int_{-\infty}^{\infty} \exp\left[-\, i(\omega+\omega')t\right] dt = 2\pi\delta(\omega+\omega')\,,$$

where δ is the Dirac delta function, it immediately follows that

$$(A.4)\qquad \langle F(t+\tau)\, G(t)\rangle = 0\,.$$

Let

$$(A.5)\qquad F = F^{(r)} + iF^{(i)}\,, \qquad G = G^{(r)} + iG^{(i)}\,,$$

($F^{(r)}$, $F^{(i)}$, $G^{(r)}$, $G^{(i)}$ all real). It then follows on substitution from (A.5) into A.4) and on equating real and imaginary parts, that

$$(A.6)\qquad \begin{cases} \langle F^{(r)}(t+\tau)\, G^{(r)}(t)\rangle = \langle F^{(i)}(t+\tau)\, G^{(i)}(t)\rangle\,, \\ \langle F^{(i)}(t+\tau)\, G^{(r)}(t)\rangle = -\,\langle F^{(r)}(t+\tau)\, G^{(i)}(t)\rangle\,. \end{cases}$$

(*) We assume here that F and G contain components corresponding to positive frequencies only, but strictly similar analysis and results hold when each contains spectral components corresponding only to negative frequencies. However, the results are not valid when F and G are analytic signal of « opposite kinds », *i.e.* when one contains a spectrum of positive frequencies, the other a spectrum of negative frequencies.

Finally we note a useful formula relating to the correlation function $\langle F(t+\tau)G^*(t)\rangle$. We have immediately, on using (A.5) and (A.6) that

$$\langle F(t+\tau)G^*(t)\rangle = 2\langle F^{(r)}(t+\tau)G^{(r)}(t) + 2i\langle F^{(i)}(t+\tau)G^{(r)}(t)\rangle \,. \tag{A.7}$$

RIASSUNTO (*)

In scritti recenti è stata introdotta una classe di tensori di correlazione di secondo ordine e sono state postulate alcune equazioni differenziali che regolano la loro propagazione. Questi tensori di correlazione, che possono essere considerati generalizzazioni naturali di funzioni usate nell'analisi dei campi d'onda ottici parzialmente coerenti, caratterizzano le correlazioni esistenti fra i vettori del campo elettromagnetico in qualsiasi coppia di punti del campo, e in qualsiasi due istanti. In questo scritto si presenta una deduzione delle equazioni differenziali fondamentali; e si mostra che i due gruppi nei quali le equazioni si suddividono naturalmente non sono indipendenti, ma in effetti derivano uno dall'altro come conseguenze di alcune proprietà di simmetria possedute dai tensori di correlazione.

(*) *Traduzione a cura della Redazione.*

PAPER NO. 35

Reprinted from *Il Nuovo Cimento,* Vol. 17, pp. 477–490, 1960.

Correlation Theory of Stationary Electromagnetic Fields.

PART II. – Conservation Laws (*).

P. ROMAN

Department of Theoretical Physics, University of Manchester - Manchester

E. WOLF

Institute of Optics, University of Rochester - Rochester, N. Y.

(ricevuto il 22 Aprile 1960)

Summary. — Two new second order space-time correlation tensors $\mathscr{W}_{jk}(\boldsymbol{x}_1, \boldsymbol{x}_2, \tau)$ and $\mathscr{S}_{jk}(\boldsymbol{x}_1, \boldsymbol{x}_2, \tau)$ are introduced, which are simple linear combinations of the correlation tensors discussed in Part I of this investigation ($\boldsymbol{x}_1$, $\boldsymbol{x}_2$ are position vectors of two points and τ a time delay). These new tensors are intimately related to certain generalizations of the (time averaged) energy density and the energy flow vector. Differential equations which $\mathscr{W}_{jk}$ and $\mathscr{S}_{jk}$ satisfy in free space are derived, and from them four new conservation laws are deduced. In the limit $\boldsymbol{x}_1 \to \boldsymbol{x}_2$, $\tau \to 0$ two of these laws reduce to the usual laws (in time averaged form) for the conservation of the energy and the momentum in an electromagnetic field. The other two laws reduce only to trivial identities in this limit, so that they have no analogy in the framework of the usual theory.

1. – Introduction.

In Part I of this investigation (ROMAN and WOLF (1), to be referred to as I), differential equations were derived, which govern the propagation of a

(*) This research was supported in part by the United States Air Force under Contract no. AF 49(638)-602, monitored by the AF Office of Scientific Research of the Air Research and Development Command.

(1) P. ROMAN and E. WOLF: *Nuovo Cimento*, **17**, 461 (1960).

class of certain second-order correlation tensors associated with stationary electromagnetic fields in free space. The tensors describe the correlation between the electromagnetic field vectors at any two points $(\boldsymbol{x}_1, \boldsymbol{x}_2)$ in space, at two instants of time separated by a time interval τ, and they contain information about the coherence as well as the polarization properties of the field. Since the correlation tensors represent averages taken over time intervals which are long compared with the mean periods of the vibrations of the field, they are of particular interest for optics, where, because of the great rapidity of the vibrations, only such averages and not the instantaneous values can be determined from experiment.

Second order correlation functions are, of course, not adequate to characterize completely all types of stationary electromagnetic fields. However, in many cases of practical importance, and certainly in most cases encountered in optics, the conditions for the validity of the central limit theorem of the theory of probability are effectively satisfied. One may then assume that the underlying probability distributions are Gaussian and in such cases second order correlations completely characterize the statistical behaviour of the field. It is, therefore, of some practical interest to investigate the consequences of the « second-order » theory, the mathematical structure of which was laid down in Part I.

In the present paper two new second-order correlation tensors $\mathscr{W}_{jk}(\boldsymbol{x}_1, \boldsymbol{x}_2, \tau)$ and $\mathscr{S}_{jk}(\boldsymbol{x}_1, \boldsymbol{x}_2, \tau)$ are introduced, which are simple linear combinations of the correlation tensors discussed in I. These new tensors are intimately related to certain generalizations of the (time averaged) energy density and the energy flow vector. With the help of the results of I, differential equations satisfied by $\mathscr{W}_{jk}$ and $\mathscr{S}_{jk}$ can be immediately written down and from them new types of conservation laws are derived. Some of these new laws are shown to be generalizations of the usual laws (in time averaged form) for the conservation of the energy and the momentum in an electromagnetic field. The other laws are found to have no analogy within the framework of the usual theory, which, unlike the present one deals only with averages evaluated for the same field point $(\boldsymbol{x}_1 = \boldsymbol{x}_2)$, at the same instant of time $(\tau = 0)$.

2. – The energy coherence tensor, the flow coherence tensor and the associated tensorial conservation laws.

Consider a stationary electromagnetic field and let $E_j(\boldsymbol{x}, t)$, $H_j(\boldsymbol{x}, t)$ $(j=1, 2, 3)$ be the Cartesian components of the complex analytic signals associated with the (real) electric and magnetic fields, at a point specified by the position vector $\boldsymbol{x}$, at time t. The four correlation tensors employed in Part I are then defined by the formulae

$$
(2.1)\qquad \begin{cases}
\mathscr{E}_{jk}(\boldsymbol{x}_1, \boldsymbol{x}_2, \tau) = \langle E_j(\boldsymbol{x}_1, t+\tau)\, E_k^*(\boldsymbol{x}_2, t)\rangle\,, \\
\mathscr{H}_{jk}(\boldsymbol{x}_1, \boldsymbol{x}_2, \tau) = \langle H_j(\boldsymbol{x}_1, t+\tau)\, H_k^*(\boldsymbol{x}_2, t)\rangle\,, \\
\mathscr{G}_{jk}(\boldsymbol{x}_1, \boldsymbol{x}_2, \tau) = \langle E_j(\boldsymbol{x}_1, t+\tau)\, H_k^*(\boldsymbol{x}_2, t)\rangle\,, \\
\widetilde{\mathscr{G}}_{jk}(\boldsymbol{x}_1, \boldsymbol{x}_2, \tau) = \langle H_j(\boldsymbol{x}_1, t+\tau)\, E_k^*(\boldsymbol{x}_2, t)\rangle\,,
\end{cases}
$$

$(j, k = 1, 2, 3)$, where asterisk denotes the complex conjugate and sharp bracket denote the time average.

It was shown that in vacuo, the four correlation tensors are related by the following set of differential equations:

$$(2.2)\qquad \varepsilon_{jkl}\,\partial_k\,\mathscr{E}_{lm} + \frac{1}{c}\frac{\partial}{\partial\tau}\widetilde{\mathscr{G}}_{jm} = 0\,,$$

$$(2.3)\qquad \varepsilon_{jkl}\,\partial_k\,\mathscr{G}_{lm} + \frac{1}{c}\frac{\partial}{\partial\tau}\mathscr{H}_{jm} = 0\,,$$

$$(2.4)\qquad \varepsilon_{jkl}\,\partial_k\,\widetilde{\mathscr{G}}_{lm} - \frac{1}{c}\frac{\partial}{\partial\tau}\mathscr{E}_{jm} = 0\,,$$

$$(2.5)\qquad \varepsilon_{jkl}\,\partial_k\,\mathscr{H}_{lm} - \frac{1}{c}\frac{\partial}{\partial\tau}\mathscr{G}_{jm} = 0\,,$$

$(j, m = 1, 2, 3)$. The correlation tensors also satisfy the subsidiary conditions

$$(2.6)\qquad \partial_j\,\mathscr{E}_{jk} = 0\,,$$

$$(2.7)\qquad \partial_j\,\mathscr{G}_{jk} = 0\,,$$

$$(2.8)\qquad \partial_j\,\widetilde{\mathscr{G}}_{jk} = 0\,,$$

$$(2.9)\qquad \partial_j\,\mathscr{H}_{jk} = 0\,,$$

$(k = 1, 2, 3)$. The symbol ε_{jkl} in equations (2.2)–(2.5) is the completely antisymmetric unit tensor of Levi-Civita ($\varepsilon_{jkl} = \pm 1$ according as j, k, l are even or odd permutations of $1, 2, 3$ and $\varepsilon_{jkl} = 0$ when two suffices are equal) and the symbol $\partial_k \equiv \partial_k^1$ denotes the differential operator $\partial/\partial x_1^k$. There is a second set of differential equations which the correlation tensor satisfies. These equations involve the differential operator $\partial_k^2 = \partial/\partial x_2^k$ and may be deduced from equations (2.2)–(2.9) by making the substitutions

$$(2.10)\qquad \partial_k \equiv \partial_k^1 \to \partial_k^2\,, \qquad \mathscr{G}_{lm} \rightleftarrows -\widetilde{\mathscr{G}}_{lm}\,;$$

this step utilizes the symmetry properties which the correlation tensors possess.

It was also shown in I that the time averaged energy density $\langle W\rangle$ and the time averaged energy flow vector $\langle \boldsymbol{S}\rangle$ are given by

$$(2.11)\qquad \langle W\rangle = \frac{1}{16\pi}\,\mathrm{Sp}\{\mathscr{E}(\boldsymbol{x}, \boldsymbol{x}, 0) + \mathscr{H}(\boldsymbol{x}, \boldsymbol{x}, 0)\},$$

$$(2.12)\qquad \langle S_l\rangle = \frac{c}{8\pi}\,\mathscr{R}\{\mathscr{G}_{jk}(\boldsymbol{x}, \boldsymbol{x}, 0) - \widetilde{\mathscr{G}}_{jk}(\boldsymbol{x}, \boldsymbol{x}, 0)\}\,,$$

where Sp denotes the spur (trace), $(j, k, l) = (1, 2, 3)$ or cycl., c is the vacuum velocity of light and $\mathscr{R}$ denotes the real part.

The (time averaged) energy density and energy flow vector are quadratic functions of the field components taken at the same point ($\boldsymbol{x}$) in space and at the same instant of time ($\tau = 0$). To generalize these quantities within the framework of our correlation analysis, one must evidently replace the arguments $(\boldsymbol{x}, \boldsymbol{x}, 0)$ by $(\boldsymbol{x}_1, \boldsymbol{x}_2, \tau)$ in (2.11) and (2.12). This leads to the introduction of two new tensors, *viz.*

$$(2.13)\qquad \left\{\begin{aligned} \mathscr{W}_{jk}(\boldsymbol{x}_1, \boldsymbol{x}_2, \tau) &= \mathscr{E}_{jk}(\boldsymbol{x}_1, \boldsymbol{x}_2, \tau) + \mathscr{H}_{jk}(\boldsymbol{x}_1, \boldsymbol{x}_2, \tau)\\ &= \langle \boldsymbol{E}_j(\boldsymbol{x}_1, t+\tau)\,\boldsymbol{E}_k^*(\boldsymbol{x}_2, t)\rangle + \langle \boldsymbol{H}_j(\boldsymbol{x}_1, t+\tau)\,\boldsymbol{H}_k^*(\boldsymbol{x}_2, t)\rangle\,,\end{aligned}\right.$$

$$(2.14)\qquad \left\{\begin{aligned} \mathscr{S}_{jk}(\boldsymbol{x}_1, \boldsymbol{x}_2, \tau) &= \mathscr{G}_{jk}(\boldsymbol{x}_1, \boldsymbol{x}_2, \tau) - \widetilde{\mathscr{G}}_{jk}(\boldsymbol{x}_1, \boldsymbol{x}_2, \tau)\\ &= \langle \boldsymbol{E}_j(\boldsymbol{x}_1, t+\tau)\,\boldsymbol{H}_k^*(\boldsymbol{x}_2, t) - \langle \boldsymbol{H}_j(\boldsymbol{x}_1, t+\tau)\,\boldsymbol{E}_k^*(\boldsymbol{x}_2, t)\rangle\,.\end{aligned}\right.$$

The tensor $\mathscr{W}_{jk}$ will be called the *energy coherence tensor* and $\mathscr{S}_{jk}$ the *flow coherence tensor*. These tensors may be regarded as generalizations of the *mutual coherence function* $\Gamma(\boldsymbol{x}_1, \boldsymbol{x}_2, \tau) = \langle V(\boldsymbol{x}_1, t+\tau)V^*(\boldsymbol{x}_2, t)\rangle$ which plays a central role in the scalar theory of partial coherence (cf. BORN and WOLF (2)).

If as in I superscripts r and i denote real and imaginary parts, one has

(2) M. BORN and E. WOLF: *Principles of Optics* (London and New York, 1959), chap. X.

from (2.13) and from the relations (2.11) of I (*)

$$(2.15)\quad \begin{cases} \mathscr{W}^{(r)}_{jk} = 2\{\langle E^{(r)}_j(\boldsymbol{x}_1, t+\tau)\, E^{(r)}_k(\boldsymbol{x}_2, t)\rangle + \langle \boldsymbol{H}^{(r)}_j(\boldsymbol{x}_1, t+\tau)\, H^{(r)}_k(\boldsymbol{x}_2, t)\rangle\}, \\ \mathscr{W}^{(i)}_{jk} = 2\{\langle E^{(i)}_j(\boldsymbol{x}_1, t+\tau)\, E^{(r)}_k(\boldsymbol{x}_2, t)\rangle + \langle \boldsymbol{H}^{(i)}_j(\boldsymbol{x}_1, t+\tau)\, H^{(r)}_k(\boldsymbol{x}_2, t)\rangle\}. \end{cases}$$

In a similar way it follows that

$$(2.16)\quad \begin{cases} \mathscr{S}^{(r)}_{jk} = 2\{\langle E^{(r)}_j(\boldsymbol{x}_1, t+\tau)\, H^{(r)}_k(\boldsymbol{x}_2, t)\rangle - \langle H^{(r)}_j(\boldsymbol{x}_1, t+\tau)\, E^{(r)}_k(\boldsymbol{x}_2, t)\rangle\}, \\ \mathscr{S}^{(i)}_{jk} = 2\{\langle E^{(i)}_j(\boldsymbol{x}_1, t+\tau)\, H^{(r)}_k(\boldsymbol{x}_2, t)\rangle - \langle H^{(i)}_j(\boldsymbol{x}_1, t+\tau)\, E^{(r)}_k(\boldsymbol{x}_2, t)\rangle\}. \end{cases}$$

Moreover, because of the relations of the form given by (2.12) of I it follows that $\mathscr{W}^{(r)}$ and $\mathscr{W}^{(i)}$, and also $\mathscr{S}^{(r)}$ and $\mathscr{S}^{(i)}$ form conjugate pairs *i.e.* they are related by Hilbert's reciprocity formulae:

$$(2.17)\quad \begin{cases} \mathscr{W}^{(i)}_{jk}(\boldsymbol{x}_1, \boldsymbol{x}_2, \tau) = \dfrac{P}{\pi}\displaystyle\int_{-\infty}^{\infty} \frac{\mathscr{W}^{(r)}_{jk}(\boldsymbol{x}_1, \boldsymbol{x}_2, \tau')}{\tau' - \tau}\, d\tau', \\ \mathscr{W}^{(r)}_{jk}(\boldsymbol{x}_1, \boldsymbol{x}_2, \tau) = -\dfrac{P}{\pi}\displaystyle\int_{-\infty}^{\infty} \frac{\mathscr{W}^{(i)}_{jk}(\boldsymbol{x}_1, \boldsymbol{x}_2, \tau')}{\tau' - \tau}\, d\tau', \end{cases}$$

$$(2.18)\quad \begin{cases} \mathscr{S}^{(i)}_{jk}(\boldsymbol{x}_1, \boldsymbol{x}_2, \tau) = \dfrac{P}{\pi}\displaystyle\int_{-\infty}^{\infty} \frac{\mathscr{S}^{(r)}_{jk}(\boldsymbol{x}_1, \boldsymbol{x}_2, \tau')}{\tau' - \tau}\, d\tau', \\ \mathscr{S}^{(r)}_{jk}(\boldsymbol{x}_1, \boldsymbol{x}_2, \tau) = -\dfrac{P}{\pi}\displaystyle\int_{-\infty}^{\infty} \frac{\mathscr{S}^{(i)}_{jk}(\boldsymbol{x}_1, \boldsymbol{x}_2, \tau')}{\tau' - \tau}\, d\tau'. \end{cases}$$

It follows from (2.11) and (2.13) that the time averaged energy density is given by

$$(2.19)\quad \langle W\rangle = \frac{1}{16\pi}\, \mathrm{Sp}\, \mathscr{W}(\boldsymbol{x}, \boldsymbol{x}, 0),$$

and from (2.12) and (2.14) one obtains for the components of the time averaged energy flow vector the expressions

$$(2.20)\quad \langle S_l\rangle = \frac{c}{8\pi}\, \mathscr{R}\mathscr{S}_{jk}(\boldsymbol{x}, \boldsymbol{x}, 0).$$

(*) Here the *first* of each of the expression for $\mathscr{E}^{(r)}_{jk}$, $\mathscr{E}^{(i)}_{jk}$ etc. given by (2.11) of I were used. Alternative but completely equivalent formulae for $\mathscr{W}^{(r)}_{jk}$ etc. may be obtained by using the second expression of each of the eq. (2.11).

Because the electric and magnetic vectors are related by Maxwell's equations, the correlation tensors $\mathscr{W}$ and $\mathscr{S}$ are also coupled by differential equations. To find these equations one only has to add (2.2) and (2.5) and use the defining equations (2.13) and (2.14). One then obtains

$$\varepsilon_{jkl}\,\partial_k\,\mathscr{W}_{lm} - \frac{1}{c}\frac{\partial}{\partial\tau}\,\mathscr{S}_{jm} = 0\,, \tag{2.21}$$

($j, m = 1, 2, 3$). Further, on substracting (2.4) from (2.3) and using (2.13) and (2.14) one obtains the equation

$$\varepsilon_{jkl}\,\partial_k\,\mathscr{S}_{lm} + \frac{1}{c}\frac{\partial}{\partial\tau}\,\mathscr{W}_{jm} = 0\,. \tag{2.22}$$

Moreover, from (2.6) and (2.9) it follows that

$$\partial_j \mathscr{W}_{jk} = 0\,. \tag{2.23}$$

Finally from (2.7) and (2.8) one has

$$\partial_j \mathscr{S}_{jk} = 0\,. \tag{2.24}$$

The set of equations (2.21)–(2.24) is formally analogous to Maxwell's equations for vacuo and is invariant under the transformation

$$\mathscr{W} \to \mathscr{S}\,, \qquad \mathscr{S} \to -\mathscr{W}\,. \tag{2.25}$$

If the symmetry relations of I, Section **4** are used, one obtains from (2.21)–(2.24) a second set of differential equations, involving in place of the differential operator $\partial_k \equiv \partial/\partial x_1^k$ the differential operator $\partial/\partial x_2^k$ taken with respect to the co-ordinates of the spatial argument $\boldsymbol{x}_2$. Formally the second set may be obtained from equations (2.21)–(2.24) on making the substitution

$$\partial_k \equiv \partial_k^1 \to \partial_k^2\,, \qquad \mathscr{S} \to -\mathscr{S}\,. \tag{2.26}$$

These equations may, of course, also be obtained directly from the equations (3.9*b*)–(3.16*b*) of Part I.

We have now obtained from the original set of equations (2.2)–(2.9) which couple the correlation tensors $\mathscr{E}$, $\mathscr{H}$, $\mathscr{G}$ and $\tilde{\mathscr{G}}$ the set (2.21)–(2.24) which couples the correlation tensors $\mathscr{W}$ and $\mathscr{S}$. The equations (2.2)–(2.9) are, however, more general, since on transition to (2.21)–(2.24) part of the information which these equations contain was lost when taking the linear combinations.

On using the same procedure as was employed in the derivation of eq. (5.5) of Part I, one finds from (2.21)–(2.24) that $\mathscr{W}$ and $\mathscr{S}$ satisfy the homogeneous wave equation:

$$\nabla^2 \mathscr{W}_{jk} = \frac{1}{c^2}\frac{\partial^2}{\partial\tau^2}\mathscr{W}_{jk}\,, \tag{2.27}$$

$$\nabla^2 \mathscr{S}_{jk} = \frac{1}{c^2}\frac{\partial^2}{\partial\tau^2}\mathscr{S}_{jk}\,. \tag{2.28}$$

These equations may, of course, also be derived by taking the appropriata linear combinations of the previously derived equations for $\mathscr{E}$, $\mathscr{H}$, $\mathscr{G}$ and $\tilde{\mathscr{G}}$.

The equations (2.21) and (2.22), just like the original equations (2.2)–(2.5), have the form of *tensorial* conservation laws. For example, according to (2.21) $\mathscr{S}_{jm}$ is a conserved quantity, $\varepsilon_{jkl}\mathscr{W}_{lm}$ being the associated flow. Another class of conservation laws may readily be derived, some of which may be regarded as generalizations of the familiar conservation laws (in their time averaged form) for the energy and the momentum of the electromagnetic field. This class of conservation laws will be derived in the next sections.

3. – Scalar conservation laws.

On taking the trace of (2.21) one obtains

$$\left\{\begin{aligned} &\varepsilon_{jkl}\,\partial_k \mathscr{W}_{lj} - \frac{1}{c}\frac{\partial}{\partial\tau}\mathscr{S}_{jj} = 0\,,\\ &\partial_k^1(-\,\varepsilon_{klj}\mathscr{W}_{lj}) + \frac{1}{c}\frac{\partial}{\partial\tau}\,\mathrm{Sp}\,\mathscr{S} = 0\,. \end{aligned}\right. \tag{3.1}$$

Here the superscript 1 has been attached to ∂_k to stress that the differential operator is to be taken with respect to the co-ordinates of the spatial argument $\boldsymbol{x}_1$.

In a strictly similar way one obtains from (2.22) (or more simply on transforming (3.1) in accordance to (2.25)),

$$\partial_k^1(\varepsilon_{klj}\mathscr{S}_{lj}) + \frac{1}{c}\frac{\partial}{\partial\tau}\,\mathrm{Sp}\,\mathscr{W} = 0\,. \tag{3.2}$$

The formulae (3.1) and (3.2) are *scalar conservation laws.* According to (2.26) there are two further scalar conservation laws, which may be obtained by replacing ∂_k^1 and ∂_k^2 and $\mathscr{S}$ and $-\mathscr{S}$ in (3.1) and (3.2). It will now be shown that (3.2) is a generalization of the usual energy law of the electromagnetic field (in its time averaged form). To this end consider the limiting form of (3.2) as the spatial argument $\boldsymbol{x}_2 \to \boldsymbol{x}_1$ $(=\boldsymbol{x}$ say) and also $\tau \to 0$. One has, first of all, from (2.14)

$$(3.3)\quad \begin{cases} \partial_k^1(\varepsilon_{klj}\mathscr{S}_{lj}) = \partial_k^1\{\varepsilon_{klj}\langle E_l(\boldsymbol{x}_1, t+\tau)H_j^*(\boldsymbol{x}_2, t)\rangle - \varepsilon_{klj}\langle H_l(\boldsymbol{x}_1, t+\tau)E_j^*(\boldsymbol{x}_2, t)\rangle\} \\ \quad = \varepsilon_{klj}\langle H_j^*(\boldsymbol{x}_2, t)\partial_k^1 E_l(\boldsymbol{x}_1, t+\tau)\rangle - \varepsilon_{klj}\langle E_j^*(\boldsymbol{x}_2, t)\partial_k^1 H_l(\boldsymbol{x}_1, t+\tau)\rangle \\ \quad = \varepsilon_{klj}\langle H_j^*(\boldsymbol{x}_2, t)\partial_k^1 E_l(\boldsymbol{x}_1, t+\tau)\rangle + \varepsilon_{klj}\langle E_l^*(\boldsymbol{x}_2, t)\partial_k^1 H_j(\boldsymbol{x}_1, t+\tau)\rangle, \end{cases}$$

where on going from the second to the third line the dummy suffices l and j were interchanged and the relation $\varepsilon_{kjl} = -\varepsilon_{klj}$ was used. Hence

$$(3.4)\qquad \partial_k^1(\varepsilon_{klj}\mathscr{S}_{lj}|_{\boldsymbol{x},0} = \langle\varepsilon_{klj}[H_j^*(\boldsymbol{x}, t)\partial_k E_l(\boldsymbol{x}, t) + E_l^*(\boldsymbol{x}, t)\partial_k H_j(\boldsymbol{x}, t)]\rangle,$$

where the symbol $|_{\boldsymbol{x},0}$ indicates that the quantity preceeding it is to be evaluated for $\boldsymbol{x}_1 = \boldsymbol{x}_2 = \boldsymbol{x}$, $\tau = 0$.

Next taking the real parts of (3.4), and using relations of the form (VII), given in Appendix of I, it follows that

$$(3.5)\quad \begin{cases} \mathscr{R}\,\partial_k^1(\varepsilon_{klj}\mathscr{S}_{lj})|_{\boldsymbol{x},0} = 2\langle\varepsilon_{klj}(H_j^{(r)}\partial_k E_l^{(r)} + E_l^{(r)}\partial_k H_j^{(r)})\rangle \\ \quad = 2\langle(\varepsilon_{klj}\partial_k(E_l^{(r)}H_j^{(r)})\rangle \\ \quad = 2\langle\operatorname{div}(\boldsymbol{E}^{(r)}\wedge\boldsymbol{H}^{(r)})\rangle \\ \quad = \dfrac{8\pi}{c}\langle\operatorname{div}\boldsymbol{S}(\boldsymbol{x}, t)\rangle, \end{cases}$$

where as before $\boldsymbol{S}$ denotes the energy flow vector.

The limiting form of the second term on the right hand side of (3.2) can be determined in a similar way. One has, from (2.13)

$$(3.6)\quad \begin{cases} \dfrac{\partial}{\partial\tau}\operatorname{Sp}\mathscr{W}\Big|_{\boldsymbol{x},0} = \dfrac{\partial}{\partial\tau}\langle\boldsymbol{E}(\boldsymbol{x}_1, t+\tau)\cdot\boldsymbol{E}^*(\boldsymbol{x}_2, t) + \boldsymbol{H}(\boldsymbol{x}_1, t+\tau)\cdot\boldsymbol{H}^*(\boldsymbol{x}_2, t)\rangle\Big|_{\boldsymbol{x},0} \\ \quad = \left\langle\dfrac{\partial\boldsymbol{E}(\boldsymbol{x}, t)}{\partial t}\cdot\boldsymbol{E}^*(\boldsymbol{x}, t) + \dfrac{\partial\boldsymbol{H}(\boldsymbol{x}, t)}{\partial t}\cdot\boldsymbol{H}^*(\boldsymbol{x}, t)\right\rangle. \end{cases}$$

Next we take the real part of (3.6). Using again the formula (VII) of Appendix of Part I, together with the fact that the time-derivative of an analytic signal is again an analytic signal, one has from (3.6),

$$
(3.7)\quad
\begin{aligned}
\mathscr{R}\frac{\partial}{\partial\tau}\operatorname{Sp}\mathscr{W}\Big|_{x,0} &= 2\left\langle\frac{\partial \boldsymbol{E}^{(r)}}{\partial t}\cdot\boldsymbol{E}^{(r)}+\frac{\partial \boldsymbol{H}^{(r)}}{\partial t}\cdot\boldsymbol{H}^{(r)}\right\rangle\\
&=\left\langle\frac{\partial}{\partial t}(\boldsymbol{E}^{(r)2}+\boldsymbol{H}^{(r)2})\right\rangle\\
&=8\pi\left\langle\frac{\partial}{\partial t}W(\boldsymbol{x},t)\right\rangle,
\end{aligned}
$$

where, as before W denotes the energy density of the field (*).

The meaning of the real part of the scalar conservation law (3.2), in the limiting case $\boldsymbol{x}_2\to\boldsymbol{x}_1$ $(=\boldsymbol{x})$, $\tau\to 0$ is now seen on substitution from (3.5) and (3.7). One then obtains the formula

$$
(3.8)\qquad \langle\operatorname{div}\boldsymbol{S}(\boldsymbol{x},t)\rangle+\left\langle\frac{\partial}{\partial t}W(\boldsymbol{x},t)\right\rangle=0,
$$

and this will be recognized as the time averaged form of the usual *energy conservation law* of the electromagnetic field in vacuum.

If one applies a similar limiting procedure to the other scalar conservation law, namely to (3.1), and uses Maxwell's equations, one obtains only a trivial identity. Thus (3.1) has no analogy in the usual theory which, in place of the second order correlation functions, deals with the products of field vectors considered at the same point in space and at the same instant of time.

Returning to the general case $(\boldsymbol{x}_2\neq\boldsymbol{x}_1\ \tau\neq 0)$, the two conservation laws (3.1) and (3.2) may be written in a more explicit form. For this purpose we associate with each of the two tensors $\mathscr{W}$ and $\mathscr{S}$, a scalar and a vector defined by

$$
(3.9a)\qquad \widehat{W}(\boldsymbol{x}_1,\boldsymbol{x}_2,\tau)=\langle\boldsymbol{E}(\boldsymbol{x}_1,t+\tau)\cdot\boldsymbol{E}^*(\boldsymbol{x}_2,t)\rangle+\langle\boldsymbol{H}(\boldsymbol{x}_1,t+\tau)\cdot\boldsymbol{H}^*(\boldsymbol{x}_2,t)\rangle,
$$

$$
(3.9b)\qquad \widehat{\boldsymbol{W}}(\boldsymbol{x}_1,\boldsymbol{x}_2,\tau)=-c\{\langle\boldsymbol{E}(\boldsymbol{x}_1,t+\tau)\wedge\boldsymbol{E}^*(\boldsymbol{x}_2,t)\rangle+\langle\boldsymbol{H}(\boldsymbol{x}_1,t+\tau)\wedge\boldsymbol{H}^*(\boldsymbol{x}_2,t)\rangle\},
$$

$$
(3.9c)\qquad \widehat{S}(\boldsymbol{x}_1,\boldsymbol{x}_2,\tau)=\langle\boldsymbol{E}(\boldsymbol{x}_1,t+\tau)\cdot\boldsymbol{H}^*(\boldsymbol{x}_2,t)\rangle-\langle\boldsymbol{H}(\boldsymbol{x}_1,t+\tau)\cdot\boldsymbol{E}^*(\boldsymbol{x}_2,t)\rangle,
$$

$$
(3.9d)\qquad \widehat{\boldsymbol{S}}(\boldsymbol{x}_1,\boldsymbol{x}_2,\tau)=c\{\boldsymbol{E}(\boldsymbol{x}_1,t+\tau)\wedge\boldsymbol{H}^*(\boldsymbol{x}_2,t)\rangle-\langle\boldsymbol{H}(\boldsymbol{x}_1,t+\tau)\wedge\boldsymbol{E}^*(\boldsymbol{x}_2,t)\rangle\}.
$$

(*) In a stationary field the average of a time-derivative, such as occurs in (3.7) or (4.6) is zero, but these terms are included here to show clearly the physical significance of the conservation laws.

The conservation laws (3.1) and (3.2) may then be expressed in the form

$$\operatorname{div}_1 \hat{\boldsymbol{W}}(\boldsymbol{x}_1, \boldsymbol{x}_2, \tau) + \frac{\partial}{\partial\tau} \hat{S}(\boldsymbol{x}_1, \boldsymbol{x}_2, \tau) = 0\,, \tag{3.10}$$

$$\operatorname{div}_1 \hat{\boldsymbol{S}}(\boldsymbol{x}_1, \boldsymbol{x}_2, \tau) + \frac{\partial}{\partial\tau} \hat{W}(\boldsymbol{x}_1, \boldsymbol{x}_2, \tau) = 0\,, \tag{3.11}$$

where div_1 is the divergence taken with respect to the co-ordinates of $\boldsymbol{x}_1$. There are, of course, two strictly similar conservation laws which may be formally derived from (3.10) and (3.11) on taking the divergence with respect to the co-ordinate of $\boldsymbol{x}_2$ and replacing $\hat{S}$ by $-\hat{S}$ and $\hat{\boldsymbol{S}}$ by $-\hat{\boldsymbol{S}}$, in accordance with (2.26).

The quantities $\hat{W}$ and $\hat{\boldsymbol{S}}$ are, apart from trivial factors, generalizations of the time averaged energy density $\langle W\rangle$ and the time averaged energy flow vector $\langle \boldsymbol{S}\rangle$ respectively. In fact

$$\langle W(\boldsymbol{x}, t)\rangle = \frac{1}{16\pi} \mathscr{R}\, \hat{W}(\boldsymbol{x}, \boldsymbol{x}, 0)\,, \qquad \langle \boldsymbol{S}(\boldsymbol{x}, t)\rangle = \frac{1}{16\pi} \mathscr{R}\, \hat{\boldsymbol{S}}(\boldsymbol{x}, \boldsymbol{x}, 0)\,. \tag{3.12}$$

The quantities $\hat{\boldsymbol{W}}$ and $\hat{S}$, on the other hand, have no counterpart in the limiting case $\boldsymbol{x}_1 = \boldsymbol{x}_2$, $\tau = 0$, since evidently

$$\mathscr{R}\, \hat{\boldsymbol{W}}(\boldsymbol{x}, \boldsymbol{x}, 0) = 0\,, \qquad \mathscr{R}\, \hat{S}(\boldsymbol{x}, \boldsymbol{x}, 0) = 0\,. \tag{3.13}$$

As we have already seen the conservation law (3.10), which involves $\hat{\boldsymbol{W}}$ and $\hat{S}$ has no counterpart in this limiting case either.

4. – Vector conservation laws.

It is possible to derive from (2.21) and (2.22) two other conservation laws. One only has to multiply each of the two equations by ε_{smj} and use the identity (5.3) of I, namely

$$\varepsilon_{smj}\,\varepsilon_{jkl} = \delta_{sk}\,\delta_{ml} - \delta_{sl}\,\delta_{mk}\,. \tag{4.1}$$

One then obtains the following two *vector conservation laws*:

$$\partial^1_k(\delta_{sk}\,\mathrm{Sp}\,\mathscr{W} - \mathscr{W}_{sk}) + \frac{1}{c}\,\frac{\partial}{\partial\tau}\,\varepsilon_{sjm}\,\mathscr{S}_{jm} = 0\,, \qquad (s = 1, 2, 3), \tag{4.2}$$

$$\partial^1_k[-(\delta_{sk}\,\mathrm{Sp}\,\mathscr{S} - \mathscr{S}_{sk})] + \frac{1}{c}\,\frac{\partial}{\partial\tau}\,\varepsilon_{sjm}\,\mathscr{W}_{jm} = 0\,, \qquad (s = 1, 2, 3). \tag{4.3}$$

It will now be shown that (4.2) is a generalization of the conservation law (in its time averaged form) for the momentum of the electromagnetic field. From the defining equation (2.14) it follows that

$$\partial_k^1(\delta_{sk}\,\mathrm{Sp}\,\mathscr{W}-\mathscr{W}_{sk})=\langle\delta_{sk}\,\partial_k^1 E_j(\boldsymbol{x}_1,\,t+\tau)E_j^*(\boldsymbol{x}_2,\,t)\rangle-\langle\partial_k^1 E_s(\boldsymbol{x}_1,\,t+\tau)E_k^*(\boldsymbol{x}_2,\,t)\rangle+\ldots$$

where the dots indicate two similar terms containing $\boldsymbol{H}$ in place of $\boldsymbol{E}$. Hence

$$\partial_k(\delta_{sk}\,\mathrm{Sp}\,\mathscr{W}-\mathscr{W}_{sk})\big|_{\boldsymbol{x},0}=\langle\delta_{sk}E_j^*(\boldsymbol{x},\,t)\partial_k E_j(\boldsymbol{x},\,t)\rangle-\langle E_k^*(\boldsymbol{x},\,t)\partial_k E_s(\boldsymbol{x},\,t)\rangle+\ldots\,.$$

Let us subtract from the right hand side of this equation the expression $\langle E_s(\boldsymbol{x},\,t)\partial_k E_k^*(\boldsymbol{x},\,t)\rangle+\langle H_s(\boldsymbol{x},\,t)\partial_k H_s^*(\boldsymbol{x},\,t)\rangle$ which is identically zero because of the Maxwell equations $\mathrm{div}\,\boldsymbol{E}=\mathrm{div}\,\boldsymbol{H}=0$. One then obtains

$$\partial_k(\delta_{sk}\,\mathrm{Sp}\,\mathscr{W}-\mathscr{W}_{sk})\big|_{\boldsymbol{x},0}=\langle\delta_{sk}E_j^*\partial_k E_j)\rangle-\langle\partial_k(E_k^*E_s)\rangle+\ldots\,.$$

On taking the real part and using identities of the form (VII) of the Appendix in I, one finds that

$$(4.4)\quad\left\{\begin{aligned}\mathscr{R}\,\partial_s(\delta_{sk}\,\mathrm{Sp}\,\mathscr{W}-\mathscr{W}_{sk})\big|_{\boldsymbol{x},0}&=2\langle\delta_{sk}E_j^{(r)}\partial_k E_j^{(r)}-\partial_k(E_k^{(r)}E_s^{(r)})\rangle\\&=\langle\partial_k[\delta_{sk}(\boldsymbol{E}^{(r)^2}+\boldsymbol{H}^{(r)^2})-2(E_k^{(r)}E_s^{(r)}+H_k^{(r)}H_s^{(r)})]\rangle\\&=-\,8\pi\langle\partial_k T_{ks}(\boldsymbol{x},\,t)\rangle\,,\end{aligned}\right.$$

where $T_{ks}=T_{ks}(\boldsymbol{x},\,t)$, $(k,\,s=1,\,2,\,3)$, is the electromagnetic stress tensor:

$$(4.5)\qquad T_{ks}=\frac{1}{4\pi}\left[(E_k^{(r)}E_s^{(r)}-\tfrac{1}{2}\,\delta_{ks}\,\boldsymbol{E}^{(r)^2})+(H_k^{(r)}H_s^{(r)}-\tfrac{1}{2}\,\delta_{ks}\,\boldsymbol{H}^{(r)^2})\right].$$

Next consider the second term on the right of (4.2). Using the defining equation (2.14), one has

$$\frac{\partial}{\partial\tau}\,\varepsilon_{sjm}\,\mathscr{S}_{jm}=\frac{\partial}{\partial\tau}\big\{\varepsilon_{sjm}\langle E_j(\boldsymbol{x}_1,\,t+\tau)H_m^*(\boldsymbol{x}_2,\,t)\rangle-\varepsilon_{sjm}\langle H_j(\boldsymbol{x}_1,\,t+\tau)E_m^*(\boldsymbol{x}_2,\,t)\rangle\big\}=$$

$$=\frac{\partial}{\partial\tau}\big\{\varepsilon_{sjm}\langle E_j(\boldsymbol{x}_1,\,t+\tau)H_m^*(\boldsymbol{x}_2,\,t)\rangle+\varepsilon_{sjm}\langle H_m(\boldsymbol{x}_1,\,t+\tau)E_j^*(\boldsymbol{x}_2,\,t)\rangle\big\},$$

where in the last term the dummy suffices j and m were interchanged and the relation $\varepsilon_{smj}=-\,\varepsilon_{sjm}$ was used. Hence

$$\frac{\partial}{\partial\tau}\,\varepsilon_{sjm}\,\mathscr{S}_{jm}\bigg|_{\boldsymbol{x},0}=\varepsilon_{sjm}\left\langle\frac{\partial E_j(\boldsymbol{x},\,t)}{\partial t}\,H_m^*(\boldsymbol{x},\,t)+\frac{\partial H_m(\boldsymbol{x},\,t)}{\partial t}\,E_j^*(\boldsymbol{x},\,t)\right\rangle.$$

On taking the real parts and again using the formula (VII) of Appendix to Part I, (together with the fact that the time derivative of an analytic signal is again an analytic signal), one obtains the expression

$$(4.6)\quad \begin{cases} \mathscr{R}\dfrac{\partial}{\partial\tau}\varepsilon_{sjm}\mathscr{S}_{jm}\Big|_{\boldsymbol{x},0} = 2\varepsilon_{sjm}\left\langle \dfrac{\partial E_j^{(r)}}{\partial t}H_m^{(r)} + \dfrac{\partial H_m^{(r)}}{\partial t}E_j^{(r)}\right\rangle \\ \qquad = 2\varepsilon_{sjm}\left\langle \dfrac{\partial}{\partial t}(E_j^{(r)}H_m^{(r)})\right\rangle \\ \qquad = 8\pi c\left\langle \dfrac{\partial}{\partial t}P_s(\boldsymbol{x}, t)\right\rangle , \end{cases}$$

where $P_s = P_s(\boldsymbol{x}, t)$, $(s = 1, 2, 3)$, are the components of the momentum density:

$$(4.7)\qquad \boldsymbol{P} = \frac{1}{4\pi c}(\boldsymbol{E}^{(r)}\wedge\boldsymbol{H}^{(r)}) .$$

With the help of (4.4) and (4.6) it follows that the real part of the conservation law (4.2) reduces in the limit $\boldsymbol{x}_2 \to \boldsymbol{x}_1$ $(= \boldsymbol{x})$, $\tau \to 0$, to

$$(4.8)\qquad \langle -\partial_k T_{ks}(\boldsymbol{x}, t)\rangle + \left\langle \frac{\partial}{\partial t}P_s(\boldsymbol{x}, t)\right\rangle = 0 , \qquad (s = 1, 2, 3),$$

which is the usual *momentum conservation law* of the electromagnetic field in vacuum, in its time-averaged form.

A similar procedure applied to the other vectorial conservation law, namely (4.3), gives only a trivial identity in the limit $\boldsymbol{x}_2 \to \boldsymbol{x}_1$, $\tau \to 0$. Thus (4.3) has no analogy in the usual theory.

Returning to the general case again $(\boldsymbol{x}_2 \neq \boldsymbol{x}_1,\ \tau \neq 0)$, it is useful to rewrite (4.2) and (4.3) in a more explicit form. For this purpose it is convenient to introduce two new tensors $\hat{T}$ and $\hat{Q}$ associated with $\mathscr{W}$ and $\mathscr{S}$ respectively, defined by

$$(4.9)\quad \hat{T}_{ks}(\boldsymbol{x}_1, \boldsymbol{x}_2, \tau) = \mathscr{W}_{ks}(\boldsymbol{x}_1, \boldsymbol{x}_2, \tau) + \mathscr{W}_{sk}(\boldsymbol{x}_1, \boldsymbol{x}_2, \tau) - \delta_{ks}\,\mathrm{Sp}\,\mathscr{W}(\boldsymbol{x}_1, \boldsymbol{x}_2, \tau) =$$
$$= \langle E_k(\boldsymbol{x}_1, t+\tau)E_s^*(\boldsymbol{x}_2, t)\rangle + \langle E_s(\boldsymbol{x}_1, t+\tau)E_k^*(\boldsymbol{x}_2, t)\rangle - \delta_{ks}\langle \boldsymbol{E}(\boldsymbol{x}_1, t+\tau)\cdot\boldsymbol{E}^*(\boldsymbol{x}_2, t)\rangle +$$
$$+ \langle H_k(\boldsymbol{x}_1, t+\tau)H_s^*(\boldsymbol{x}_2, t)\rangle + \langle H_s(\boldsymbol{x}_1, t+\tau)H_k^*(\boldsymbol{x}_2, t)\rangle - \delta_{ks}\langle \boldsymbol{H}(\boldsymbol{x}_1, t+\tau)\cdot\boldsymbol{H}^*(\boldsymbol{x}_2, t)\rangle ,$$

$$(4.10)\quad \hat{Q}_{ks}(\boldsymbol{x}_1, \boldsymbol{x}_2, \tau) = c^2\{\mathscr{S}_{ks}(\boldsymbol{x}_1, \boldsymbol{x}_2, \tau) + \mathscr{S}_{sk}(\boldsymbol{x}_1, \boldsymbol{x}_2, \tau) - \delta_{ks}\,\mathrm{Sp}\,\mathscr{S}(\boldsymbol{x}_1, \boldsymbol{x}_2, \tau)\} =$$
$$= c^2\{\langle E_k(\boldsymbol{x}_1, t+\tau)H_s^*(\boldsymbol{x}_2, t)\rangle + \langle E_s(\boldsymbol{x}_1, t+\tau)H_k^*(\boldsymbol{x}_2, t)\rangle - \delta_{ks}\langle \boldsymbol{E}(\boldsymbol{x}_1, t+\tau)\cdot\boldsymbol{H}^*(\boldsymbol{x}_2, t)\rangle$$
$$- \langle H_k(\boldsymbol{x}_1, t+\tau)E_s^*(\boldsymbol{x}_2, t)\rangle - \langle H_s(\boldsymbol{x}_1, t+\tau)E_k^*(\boldsymbol{x}_2, t)\rangle + \delta_{ks}\langle \boldsymbol{H}(\boldsymbol{x}_1, t+\tau)\cdot\boldsymbol{E}^*(\boldsymbol{x}_2, t)\rangle\} .$$

Further les us also introduce a vector $\hat{\boldsymbol{P}}(\boldsymbol{x}_1, \boldsymbol{x}_2, \tau)$ defined by

$$\hat{\boldsymbol{P}}(\boldsymbol{x}_1, \boldsymbol{x}_2, \tau) = \frac{1}{c^2}\hat{\boldsymbol{S}}(\boldsymbol{x}_1, \boldsymbol{x}_2, \tau) . \tag{4.11}$$

If use is also made of (2.23) and (2.24), the conservation laws (4.2) and (4.3) may then be written in the following form (*):

$$\partial_k^1[-\hat{T}_{ks}(\boldsymbol{x}_1, \boldsymbol{x}_2, \tau)] + \frac{\partial}{\partial\tau}\hat{P}_s(\boldsymbol{x}_1, \boldsymbol{x}_2, \tau) = 0 , \qquad (s = 1, 2, 3), \tag{4.12}$$

$$\partial_k^1[-\hat{Q}_{ks}(\boldsymbol{x}_1, \boldsymbol{x}_2, \tau)] + \frac{\partial}{\partial\tau}\hat{W}_s(\boldsymbol{x}_1, \boldsymbol{x}_2, \tau) = 0 , \qquad (s = 1, 2, 3). \tag{4.13}$$

Again there are two strictly similar conservation laws which may formally be derived from (4.11) and (4.12) by replacing ∂_k^1 by ∂_k^2 and by changing $\hat{\boldsymbol{S}}$ into $-\hat{\boldsymbol{S}}$ and $\hat{Q}$ into $-\hat{Q}$, in accordance with (2.26).

The tensor $\hat{T}_{ks}$ is, apart from a trivial factor, a generalization of the time averaged electromagnetic stress tensor $\langle T\rangle$. In fact

$$\langle T_{ks}(\boldsymbol{x}, t)\rangle = \frac{1}{16\pi}\mathscr{R}\hat{T}_{ks}(\boldsymbol{x}, \boldsymbol{x}, 0) ; \tag{4.14}$$

and apart from a trivial factor $\hat{\boldsymbol{P}}$ may be regarded as a generalization of the time averaged momentum density $\langle \boldsymbol{P}(\boldsymbol{x}, t)\rangle$:

$$\langle \boldsymbol{P}(\boldsymbol{x}, t)\rangle = \frac{1}{16\pi}\mathscr{R}\hat{\boldsymbol{P}}(\boldsymbol{x}, \boldsymbol{x}, 0) . \tag{4.15}$$

The tensors $\hat{Q}_{ks}$ (and as already noted also $\hat{\boldsymbol{W}}$) have no counterpart in the limiting case $\boldsymbol{x}_1 = \boldsymbol{x}_2$, $\tau = 0$, since

$$\mathscr{R}\hat{Q}_{ks}(\boldsymbol{x}, \boldsymbol{x}, 0) = 0 . \tag{4.16}$$

As has already been seen, the conservation law (4.13) has no analogue in this limiting case either.

(*) In eqs. (4.12) and (4.13) the symbols $\hat{P}_s$ and $\hat{W}_s$ represent, of course, the components of the vectors $\hat{\boldsymbol{P}}$ and $\hat{\boldsymbol{W}}$.

RIASSUNTO (*)

Si introducono due nuovi tensori del secondo ordine di correlazione dello spazio tempo $\mathscr{W}_{jk}(\boldsymbol{x}_1, \boldsymbol{x}_2, \tau)$ e $\boldsymbol{H}_{ik}(\boldsymbol{x}_1, \boldsymbol{x}_2, \tau)$, che sono semplici combinazioni lineari dei tensori di correlazione discussi nella Parte I di questo studio ($\boldsymbol{x}_1$, $\boldsymbol{x}_2$ sono i vettori di posizione di due punti e τ un ritardo nel tempo). Questi nuovi tensori sono intimamente connessi a certe generalizzazioni della densità di energia (mediata rispetto al tempo) e del vettore del flusso di energia. Si derivano le equazioni differenziali, che sono soddisfatte da $\mathscr{W}_{jk}$ e $\boldsymbol{H}_{jk}$ nello spazio libero, e da esse vengono dedotte quattro nuove leggi di conservazione. Al limite per $\boldsymbol{x}_1 \to \boldsymbol{x}_2$, $\tau \to 0$, due di queste leggi si riducono alle leggi usuali (in forma mediata rispetto al tempo) per la conservazione dell'energia e della quantità di moto in un campo elettromagnetico. Le altre due leggi, a tale limite, si riducono semplicemente ad identità banali, cosichè esse non hanno alcun analogo nella schema della teoria usuale.

(*) *Traduzione a cura della Redazione.*

BIBLIOGRAPHY

1850–1950

Berek, M.: Über Kohärenz und Konsonanz des Lichtes, *Z. Physik* **36** (1926) 675.

Berek, M.: Über Kohärenz und Konsonanz des Lichtes, *Z. Physik* **36** (1926) 824.

Berek, M.: Über Kohärenz und Konsonanz des Lichtes, *Z. Physik* **37** (1926) 387.

Berek, M.: Über Kohärenz und Konsonanz des Lichtes, *Z. Physik* **40** (1927) 420.

Bitter, F.: The Optical Detection of Radiofrequency Resonance, *Phys. Rev.* **76** (1949) 833.

Bothe, W.: Zur Statistik der Hohlraumstrahlung, *Z. Physik* **41** (1927) 345.

*Cittert, P. H. van: Die wahrscheinliche Schwingungsverteilung in einer von einer Lichtquelle direkt oder mittels einer Linse beleuchteten Ebene, *Physica* **1** (1934) 201 [paper no. 1].

*Cittert, P. H. van: Kohaerenz-Probleme, *Physica* **6** (1939) 1129 [paper no. 2].

*Einstein, A.: Zum gegenwärtigen Stand des Strahlungsproblems, *Physik Z.* **10** (1909) 185 [paper no. 3].

Einstein, A.: Über einen die Erzeugung und Verwandlung des Lichtes betreffenden heuristischen Gesichtspunkt, *Ann. Physik* **17** (1905) 132.

Einstein, A.: Über die Entwicklung unserer Anschauungen über das Wesen und die Konstitution der Strahlung, *Physik Z.* **10** (1909) 817.

Einstein, A. and L. Hopf: Über einen Satz der Wahrscheinlichkeitsrechnung und seine Anwendung in der Strahlungstheorie, *Ann. Physik* **33** (1910) 1096.

Einstein, A.: Antwort auf eine Abhandlung M. von Laues "Ein Satz der Wahrscheinlichkeitsrechnung und seine Anwendung auf die Strahlungstheorie," *Ann. Physik* **47** (1915) 879.

* Papers marked with an asterisk (*) are reprinted in this volume.

Fellgett, P. B.: On the Ultimate Sensitivity and Practical Performance of Radiation Detectors, *J. Opt. Soc. Am.* **39** (1949) 970.

Forrester, A. T., W. E. Parkins and E. Gerjuoy: Probability of Observing Beat Frequencies between Lines in the Visible System, *Phys. Rev.* **72** (1947) 728.

Fürth, R.: Schwankungserscheinungen nach der neuen Quantenstatistik, *Z. Physik* **48** (1928) 323.

Fürth, R.: Über Strahlungsschwankungen nach der Lichtquantenstatistik, *Z. Physik* **50** (1928) 310.

Gabor, D.: Communication Theory and Physics, *Phil. Mag.* **41** (1950) 1161.

Gerjuoy, E., A. T. Forrester and W. E. Parkins: Signal-to-Noise Ratio in Photoelectrically Observed Beats, *Phys. Rev.* **73** (1948) 922.

Griffin, L. R.: On the Possibility of Observing Beat Frequencies between Lines in the Visible Region (A Note on the Paper), *Phys. Rev.* **73** (1948) 922.

Hurwitz, H. and R. C. Jones: A New Calculus for the Treatment of Optical System, II: Proof of Three General Equivalence Theorems, *J. Opt. Soc. Am.* **31** (1941) 493.

*Hurwitz, H.: The Statistical Properties of Unpolarized Light, *J. Opt. Soc. Am.* **35** (1945) 525 [paper no. 4].

Jacobsohn, S.: Radiation Fluctuations and Thermal Equilibrium, *Phys. Rev.* **30** (1927) 936.

Jones, R. C.: A New Calculus for the Treatment of Optical Systems, I: Description and Discussion of the Calculus, *J. Opt. Soc. Am.* **31** (1941) 488.

Jones, R. C.: A New Calculus for the Treatment of Optical Systems, III: The Johncke Theory of Optical Activity, *J. Opt. Soc. Am.* **31** (1941) 500.

Jones, R. C.: A New Calculus for the Treatment of Optical Systems, IV, *J. Opt. Soc. Am.* **32** (1942) 486.

JONES, R. C.: A New Calculus for the Treatment of Optical Systems, V: A More General Formulation and Description of Another Calculus, *J. Opt. Soc. Am.* **37** (1947) 107.

JONES, R. C.: A New Calculus for the Treatment of Optical Systems, VI: Experimental Determination of the Matrix, *J. Opt. Soc. Am.* **37** (1947) 110.

JONES, R. C.: A New Calculus for the Treatment of Optical Systems, VII: Properties of the *N*-Matrices, *J. Opt. Soc. Am.* **38** (1948) 671.

LAKEMAN, C. and J. T. GROOSMULLER: Over de Theorie van Berek, *Physica (Gravenhage)* **8** (1928) 193.

LAKEMAN, C. and J. T. GROOSMULLER: Over Microscopische Afbeelding, *Physica (Gravenhage)* **8** (1928) 199.

LAKEMAN, C. and J. T. GROOSMULLER: Over Afbeeldingsverschijnselen van verlichte voorwerpen, *Physica (Gravenhage)* **8** (1928) 305.

LAUE, M.: Zur Thermodynamik der Interferenzerscheinungen, *Ann. Physik* **20** (1906) 365.

*LAUE, M.: Die Entropie von partiell kohärenten Strahlenbündeln, *Ann. Physik* **23** (1907) 1 and 795 [papers no. 5, 6].

LAUE, M.: Zur Thermodynamik der Gitterbeugung, *Ann. Physik* **30** (1909) 225.

LAUE, M.: Zur Thermodynamik der Beugung, *Ann. Physik* **31** (1910) 547.

LAUE, M.: Ein Satz der Wahrscheinlichkeitsrechnung und seine Anwendung auf die Strahlungstheorie, *Ann. Physik* **47** (1915) 853.

LAUE, M.: Zur Statistik der Fourier Koeffizienten der natürlichen Strahlung, *Ann. Physik* **48** (1915) 668.

LEWIS, W. B.: Fluctuations in Streams of Thermal Radiation, *Proc. Phys. Soc.* **59** (1947) 34.

MICHELSON, A. A.: On the Application of Interference Methods to Astronomical Measurements, *Phil. Mag.* **30** (1890) 1.

MICHELSON, A. A.: Visibility of Interference Fringes in the Focus of a Telescope, *Phil. Mag.* [5] **31** (1891) 256.

MICHELSON, A. A.: On the Application of Interference Methods to Spectroscopic Measurements, I, *Phil. Mag.* [5] **31** (1891) 338.

MICHELSON, A. A.: On the Application of Interference Methods to Spectroscopic Measurements, II, *Phil. Mag.* [5] **34** (1892) 280.

MICHELSON, A. A.: On the Application of Interference Methods to Astronomical Measurements, *Astrophys. J.* **51** (1920) 257.

MICHELSON, A. A. and F. G. PEASE: Measurement of the Diameter of α-Orionis with the Interferometer, *Astrophys. J.* **53** (1921) 249.

PEASE, F. G.: Interferometer Methods in Astronomy, *Ergeb. exakt. Naturws.* **10** (1931) 84.

PÉRARD, A.: Études de Raies Spectrales en Vue de leurs Applications Métrologiques, *Rev. Opt.* **7** (1928) 1.

PÉRARD, A.: Études des Raies Spectrales Utilisables en Métrologie Interférentielle, *Réunions de l'Institut d'Optique* (Revue d'Optique, Paris 1935) p. 10.

PERRIN, F.: Polarization of Light Scattered by Isotropic Opalescent Media, *J. Chem. Phys.* **10** (1942) 415.

RAYLEIGH (LORD): On the Interference Bands of Approximately Homogeneous Light, *Phil. Mag.* **34** (1892) 407.

SCHRÖDINGER, E.: Über die Kohärenz in weitgeöffneten Bündeln, *Ann. Physik* **87** (1928) 570.

STOKES, G. G.: On the Composition and Resolution of Streams of Polarized Light from Different Sources, *Trans. Cambridge Phil. Soc.* **9** (1852) 399; reprinted in Stokes' *Mathematical and Physical Papers* (Cambridge University Press, Cambridge 1901) Vol. 3, p. 233.

TORALDO DI FRANCIA, G.: Sulla Termodinamica della Diffrasione, *Rendic. Acad. Naz. Lineci* (Classe di Scienze Fisiche, matematiche e naturali) **4** (1948) 319.

VERDET, E.: Étude sur la Constitution de la Lumière non Polarisée et de la Lumière Partiellement Polarisée, *Ann. Scientif. l'Ecole Normale Supérieure* **2** (1865) 291.

VERDET, E.: *Leçons d'Optique Physique* (L'Imprimerie Impériale, Paris 1869) Vol. 1, p. 106.

*WIENER, N.: Coherency Matrices and Quantum Theory, *J. Math. Phys. (M.I.T.)* **7** (1927–1928) 109 [paper no. 7].

WIENER, N.: Harmonic Analysis and the Quantum Theory, *J. Franklin Inst.* **207** (1929) 525.

WIENER, N.: Generalized Harmonic Analysis, *Acta Math. (Stockholm)* **55** (1930) 118.

*ZERNIKE, F.: The Concept of Degree of Coherence and Its Application to Optical Problems, *Physica* **5** (1938) 785 [paper no. 8].

ZERNIKE, F.: Diffraction and Optical Image Formation, *Proc. Phys. Soc.* **61** (1948) 158.

1951

51.1. Falkoff, D. L. and J. E. MacDonald: On the Stokes Parameters in Polarized Radiation, *J. Opt. Soc. Am.* **41** (1951) 861.

*51.2. Hopkins, H. H.: The Concept of Partial Coherence in Optics, *Proc. Roy. Soc.* **A208** (1951) 263 [paper no. 11].

1952

52.1. Kahan, T.: Sur la Thermodynamique des Ondes Électromagnétiques, *Nuovo Cimento Supp.* **9** (1952) 304.

1953

53.1. Akhiezer, A. I. and V. B. Berestetsky: *Quantum Electrodynamics, Vol. I* (Office of Technical Services, Dept. of Commerce, Washington, D.C.; translation of original Russian edition published in Moscow 1953).

53.2. Baker, I. R.: The Effect of Source Size on the Coherence of an Illuminating Wave, *Proc. Phys. Soc.* **B66** (1953) 975.

53.3. Duffieux, P. M.: La Cohérence Partielle et Les Fonctions de Transmission, *Rev. d'Opt.* **32** (1953) 129.

*53.4. Hopkins, H. H.: On the Diffraction Theory of Optical Images, *Proc. Roy. Soc.* **A217** (1953) 408 [paper no. 12].

53.5. Jones, R. C.: On Reversibility and Irreversibility in Optics, *J. Opt. Soc. Am.* **43** (1953) 138.

53.6. Weber, J.: Quantum Theory of a Damped Electrical Oscillator and Noise, *Phys. Rev.* **90** (1953) 977.

53.7. Wolf, E.: A Macroscopic Theory of Interference and Diffraction of Light from Finite Sources, *Nature* **172** (1953) 535.

1954

54.1. Blanc-Lapierre, A. and P. Dumontet: La Notion de Cohérence en Optique, *Compt. Rend. (Paris)* **238** (1954) 1005.

54.2. Brown, R. Hanbury and R. Q. Twiss: A New Type of Interferometer for Use in Radio Astronomy, *Phil. Mag.* **45** (1954) 663.

*54.3. Dicke, R. H.: Coherence in Spontaneous Radiation Processes, *Phys. Rev.* **93** (1954) 99 [paper no. 9].

54.4. Dumontet, P.: Discussion of a General System, Applicable in Particular to the Theory of Image Formation of Partially Coherent Objects, *Compt. Rend. (Paris)* **238** (1954) 1109.

*54.5. Fano, U.: A Stokes Parameter Technique for the Treatment of Polarization in Quantum Mechanics, *Phys. Rev.* **93** (1954) 121 [paper no. 10].

*54.6. McMaster, W. H.: Polarization and the Stokes Parameters, *Am. J. Phys.* **22** (1954) 351 [paper no. 13].

54.7. Walker, M. J.: Matrix Calculus and the Stokes Parameters of Polarized Radiation, *Am. J. Phys.* **22** (1954) 170.

*54.8. Wolf, E.: A Macroscopic Theory of Interference and Diffraction of Light from Finite Sources, I: Fields with a Narrow Spectral Range, *Proc. Roy. Soc.* **A225** (1954) 96 [paper no. 14].

*54.9. Wolf, E.: Optics in Terms of Observable Quantities, *Nuovo Cimento* **12** (1954) 884 [paper no. 15].

1955

55.1. Adam, A., L. Janossy and P. Varga: Coincidences between Photons contained in Coherent Light Rays, *Acta Phys. Acad. Sci. Hung.* **4** (1955) 301 (in Russian).

55.2. Adam, A., L. Janossy and P. Varga: Beobachtungen mit dem Elektronenvervielfacher an kohärenten Lichtstrahlen, *Ann. d. Physik* **16** (1955) 408.

*55.3. Blanc-Lapierre, A. and P. Dumontet: La Notion de Cohérence en Optique, *Rev. d'Opt.* **34** (1955) 1 [paper no. 16].

55.4. Dumontet, P.: Imagerie en Cohérence Partielle, *Publ. Sci. Univ. Alger., Ser. B* **1** (1955) 33.

*55.5. Forrester, A. T., R. A. Gudmundsen and P. O. Johnson: Photoelectric Mixing of Incoherent Light, *Phys. Rev.* **99** (1955) 1691 [paper no. 17].

55.6. Slansky, S.: Influence de la Cohérence de l'Eclairage sur le Contraste de l'Image d'un Point Noir en Présence d'un Petit Défaut de Mise au Point, *Opt. Acta* **2** (1955) 119.

55.7. Wolf, E.: Partially Coherent Optical Fields, *Vistas in Astronomy, Vol. I*, edited by A. Beer (Pergamon Press, London and New York, 1955) p. 385.

*55.8. Wolf, E.: A Macroscopic Theory of Interference and Diffraction of Light from Finite Sources, II: Fields with a Spectral Range of

Arbitrary Width, *Proc. Roy. Soc.* **A230** (1955) 246 [paper no. 18].

1956

56.1. Blanc-Lapierre, A.: Sur Quelques Applications de la Théorie des Fonctions Aléatoires à l'Optique, *Proceedings of the International Congress of Mathematics*, Vol. 3 (North-Holland Publishing Co., Amsterdam 1956) p. 399.

56.2. Brannen, E. and H. I. S. Ferguson: The Question of Correlation between Photons in Coherent Light Rays, *Nature* **178** (1956) 481.

*56.3. Brown, R. Hanbury and R. Q. Twiss: Correlation between Photons in Two Coherent Beams of Light, *Nature* **177** (1956) 27 [paper no. 19].

*56.4. Brown, R. Hanbury and R. Q. Twiss: A Test of a New Type of Stellar Interferometer on Sirius, *Nature* **178** (1956) 1046 [paper no. 20].

56.5. Brown, R. Hanbury and R. Q. Twiss: The Question of Correlation between Photons in Coherent Light Rays, *Nature* **178** (1956) 1447.

56.6. Dumontet, P.: La Correspondence Objet-Image en Optique et Possibilités de Corriger Certains Défauts Liés à la Diffraction, *Publ. Sci. Univ. Alger, Ser. B* **2** (1956) 151.

56.7. Dumontet, P.: La Correspondence Objet-Image en Optique et Possibilités de Corriger Certains Défauts Liés à la Diffraction, *Publ. Sci. Univ. Alger, Ser. B* **2** (1956) 203.

56.8. Forrester, A. T.: On Coherence Properties of Light Waves, *Am. J. Phys.* **24** (1956) 192.

*56.9. Gabor, D.: Light and Information, *Proceedings of a Symposium on Astronomical Optics and Related Subjects*, edited by Z. Kopal (North-Holland Publishing Company, Amsterdam 1956) p. 17 [paper no. 21].

56.10. Gabor, D.: Optical Transmission, *Information Theory* (Third London Symposium), edited by C. Cherry (Butterworths Scientific Publications, Ltd., London 1956), p. 26.

56.11. Gamo, H.: Mathematical Analysis of the Intensity Distribution of Optical Image in Various Degrees of Coherence of Illumination, *J. Appl. Phys.* (*Japan*) **25** (1956) 431 (In Japanese, English abstract).

56.12. Jennison, R. C. and M. K. D. Gupta: The Measurement of the Angular Diameter of Two Intense Radio Sources, I: A Radio Interferometer Using Post Detector Correlation; II: Diameter and Structural Measurements of the Radio Stars Cygnus A and Cassiopedia A, *Phil. Mag.* **1** (1956) 55 and 65

56.13. Pancharatnam, S.: Generalized Theory of Interference and its Applications, Part I: Coherent Pencils, *Proc. Indian Acad. Sci. Sec. A* **44** (1956) 247.

56.14. Pancharatnam, S.: Generalized Theory of Interference and its Applications, Part II: Partially Coherent Pencils, *Proc. Indian Acad. Sci. Sec. A* **44** (1956) 398.

*56.15. Purcell, E. M.: The Question of Correlation between Photons in Coherent Light Rays, *Nature* **178** (1956) 1449 [paper no. 22].

56.16. Steel, W. H.: The Defocused Image of Sinusoidal Gratings, *Opt. Acta* **3** (1956) 65.

56.17. Wolf, E.: The Coherence Properties of Optical Fields, *Proceedings of a Symposium on Astronomical Optics and Related Subjects*, edited by Z. Kopal (North-Holland Publishing Company, Amsterdam 1956) p. 177.

1957

*57.1. Brown, R. Hanbury and R. Q. Twiss: Interferometry of the Intensity Fluctuations in Light, I: Basic Theory: The Correlation between Photons in Coherent Beams of Radiation, *Proc. Roy. Soc.* **A242** (1957) 300 [paper no. 23(a)].

*57.2. Brown, R. Hanbury and R. Q. Twiss: Interferometry of the Intensity Fluctuations in Light, II: An Experimental Test of the Theory for Partially Coherent Light, *Proc. Roy. Soc.* **A243** (1957) 291 [paper no. 23(b)].

57.3. Fano, U.: Description of States in Quantum Mechanics by Density Matrix and Operator Techniques, *Rev. Mod. Phys.* **29** (1957) 74.

57.4. Fellgett, P.: The Question of Correlation between Photons in Coherent Beams of Light, *Nature* **179** (1957) 956.

*57.5. Gamo, H.: Intensity Matrix and the Degree of Coherence, *J. Opt. Soc. Am.* **47** (1957) 976 [paper no. 24].

57.6. Gamo, H.: Transformation of Intensity Matrix in Optics by a Pupil, *J. Appl. Phys.* (*Japan*) **26** (1957) 414 (in Japanese, English abstract).

57.7. Griffiths, R. B. and R. H. Dicke: Coherent Emission Light Source, *Rev. Sc. Instr.* **28** (1957) 646.

57.8. Hopkins, H. H.: Applications of Coherence Theory in Microscopy and Interferometry, *J. Opt. Soc. Am.* **47** (1957) 508.

57.9. Janossy, L.: On the Classical Fluctuation of a Beam of Light, *Nuovo Cimento* **6** (1957) 111.

57.10. Kothari, D. S. and F. C. Auluck: Distance Correlation for Photons, *Current Sci.* **26** (1967) 169.

57.11. Pancharatnam, S.: Generalized Theory of Interference and Its Applications, III: Interference Figure in Transparent Crystals, *Proc. Indian Acad. Sci.* **A45** (1957) 402.

57.12. Pancharatnam, S.: Generalized Theory of Interference and Its Applications, IV: Interference Figures in Absorbing Biaxial Crystals, *Proc. Indian Acad. Sci.* **A45** (1957) 1.

*57.13. Rebka, G. A. and R. V. Pound: Time-Correlated Photons, *Nature* **180** (1957) 1035 [paper no. 25].

57.14. Sillitto, R. M.: Correlation between Events in Photon Detectors, *Nature* **179** (1957) 1127.

57.15. Steel, W. H.: Effects of Small Aberrations on the Images of Partially Coherent Objects, *J. Opt. Soc. Am.* **47** (1957) 405.

57.16. Thompson, B. J. and E. Wolf: Two-Beam Interference with Partially Coherent Light, *J. Opt. Soc. Am.* **47** (1957) 895.

57.17. Twiss, R. Q. and R. Hanbury Brown: The Question of Correlation between Photons in Coherent Beams of Light, *Nature* **179** (1957) 1128

*57.18. Twiss, R. Q., A. G. Little and R. Hanbury Brown: Correlation between Photons in Coherent Beams of Light, Detected by a Coincidence Counting Technique, *Nature* **180** (1957) 324 [paper no. 26].

57.19. Wolf, E.: Intensity Fluctuations in Stationary Optical Fields, *Phil. Mag.* **2** (1957) 351.

1958

58.1. Alford, W. P. and A. Gold: Laboratory Measurement of the Velocity of Light, *Am. J. Phys.* **26** (1958) 481.

58.2. Beran, M. J.: Determination of the Intensity Distribution Resulting from the Random Illumination of a Plane Finite Surface, *Opt. Acta* **5** (1958) 88.

58.3. Bracewell, R. N.: Radio Interferometry of Discrete Sources, *Proc. IRE* **46** (1958) 97.

58.4. Brannen, E., H. I. S. Ferguson and W. Wehlau: Photon Correlation in Coherent Light Beams, *Canad. J. Phys.* **36** (1958) 871.

58.5. Brown, R. Hanbury and R. Q. Twiss: Interferometry of the Intensity Fluctuations in Light, III: Applications to Astronomy, *Proc. Roy. Soc.* **A248** (1958) 199.

58.6. Brown, R. Hanbury and R. Q. Twiss: Interferometry of the Intensity Fluctuations in Light, IV: A Test of an Intensity Interferometer on Sirius A, *Proc. Roy. Soc.* **A248** (1958) 222.

58.7. Canals-Frau, D. et M. Rousseau: Influence de l'Éclairage Partiellement Cohérent sur la Formation des Images de Quelques Objets Étendus Opaques, *Opt. Acta* **5** (1958) 15.

58.8. Gamba, A.: Cooperation Phenomena in Quantum Theory of Radiation, *Phys. Rev.* **110** (1958) 601.

*58.9. Gamo, H.: Transformation of Intensity Matrix by the Transmission of a Pupil, *J. Opt. Soc. Am.* **48** (1958) 136 [paper no. 27].

58.10. Gamo, H.: Intensity Matrix for the Image Obtained by a Circular Aperture, *J. Appl. Phys.* (*Japan*) **27** (1958) 577 (in Japanese, English Abstract).

58.11. Joichi, N.: The Effect of Ground Glass on the Coherency of Transmitted or Reflected Light, *Oyobuturi* **27** (1958) 634 (in Japanese).

58.12. Kahl, G. D. and F. D. Bennet: Coherence Requirements for Interferometry, *Rev. Mod. Phys.* **30** (1958) 1193.

58.13. Kahn, F. D.: On Photon Coincidences and Hanbury Brown's Interferometer, *Opt. Acta* **5** (1958) 93.

*58.14. Mandel, L.: Fluctuations of Photon Beams and Their Correlations, *Proc. Phys. Soc.* **72** (1958) 1037 [paper no. 28].

58.15. Senitzky, I. R.: Induced and Spontaneous Emission in a Coherent Field I, *Phys. Rev.* **111** (1958) 3.

58.16. Slepian, D.: Fluctuations of Random Noise Power, *Bell System Tech. J.* **37** (1958) 163.

58.17. Steel, W. H.: Scalar Diffraction in Terms of Coherence, *Proc. Roy. Soc.* **A249** (1958) 574.

58.18. Thompson, B. J.: Illustration of the Phase Change in Two-Beam Interference with Partially Coherent Light, *J. Opt. Soc. Am.* **48** (1958) 95.

58.19. Wolf, E.: Reciprocity Inequalities, Coherence Time and Bandwidth in Signal Analysis and Optics, *Proc. Phys. Soc.* **71** (1958) 257.

1959

59.1. Alkemade, C. T. J.: On the Excess Photon Noise in Single-Beam Measurements with Photoemissive and Photoconductive Cells, *Physica* **25** (1959) 1145.

59.2. Barrat, J. P.: Étude de la Diffusion Multiple Cohérente de la Lumière de Résonance Optique, I: Étude Théorique, *J. Phys. Radium* **20** (1959) 541.

59.3. Barrat, J. P.: Étude de la Diffusion Multiple Cohérente de la Lumière de Résonance Optique, II: Étude Théorique, *J. Phys. Radium* **20** (1959) 633.

59.4. Barrat, J. P.: Étude de la Diffusion Multiple Cohérente de la Lumière de Résonance Optique, III: Résultats Expérimentaux, *J. Phys. Radium* **20** (1959) 657.

59.5. Basov, N. G., B. M. Vul and Yu. M. Popov: Quantum-Mechanical Semiconductor Generators and Amplifiers of Electromagnetic Oscillations, translated from Russian in *Soviet Phys.—JETP* **37** (1960) 416.

59.6. Bourret, R.: Covariant Generalization of the Stokes Parameter, *J. Opt. Soc. Am.* **49** (1959) 1002.

59.7. Edwards, S. F. and G. B. Parrent, Jr.: The Form of the General Unimodular Analytic Signal, *Opt. Acta* **5** (1959) 367.

59.8. Fain, V. M.: On the Theory of Coherent Spontaneous Emission, translated from Russian in *Soviet Phys.—JETP* **9** (1959) 562.

59.9. Fellgett, P., R. C. Jones and R. Q. Twiss: Fluctuations in Photon Streams, *Nature* **189** (1959) 967.

59.10. Graff, G. and L. Janossy: An Investigation Concerning the Classical Fluctuation of Light, *Acta Phys. Acad. Sci. Hung.* **10** (1959) 291.

59.11. Janossy, L.: The Fluctuations of Intensity of an Extended Light Source, *Nuovo Cimento* **12** (1959) 370.

59.12. Kuščer, I. and M. Ribarič: Matrix Formalism in the Theory of Diffusion of Light, *Opt. Acta* **6** (1959) 42.

59.13. Landwehr, R.: Lage und Sichtbarkeit von Keilinterferenzen bei instrumenteller Beobachtung (Beitrag zur Kohärenztheorie eines unsymmetrischen Interferenzfalles), *Opt. Acta* **6** (1959) 52.

59.14. Mandel, L.: Image Fluctuations in Cascade Intensifiers, *Brit. J. Appl. Phys.* **10** (1959) 233.

*59.15. Mandel, L.: Fluctuations of Photon Beams: The Distribution of the Photo-Electrons, *Proc. Phys. Soc.* **74** (1959) 233 [paper no. 29].

59.16. Mandel, L.: On the Possibility of Observing Interference Effects with Light Beams Divided by a Shutter, *J. Opt. Soc. Am.* **49** (1959) 931.

*59.17. Parrent, G. B., Jr.: On the Propagation of Mutual Coherence, *J. Opt. Soc. Am.* **49** (1959) 787 [paper no. 30].

59.18. Parrent, G. B., Jr.: Studies in the Theory of Partial Coherence, *Opt. Acta* **6** (1959) 285.

59.19. Roman, P.: Generalized Stokes Parameters for Waves with Arbitrary Form, *Nuovo Cimento* **13** (1959) 974.

59.20. Sanders, J. H.: A Proposed Method for the Measurement of the Velocity of Light, *Nature* **183** (1959) 312.

59.21. Senitzky, I. R.: Induced and Spontaneous Emission in a Coherent Field, II, *Phys. Rev.* **115** (1959) 227.

59.22. Strong, J. and G. A. Vanasse: Interferometric Spectroscopy in the Far Infrared, *J. Opt. Soc. Am.* **49** (1959) 844.

59.23. Twiss, R. Q. and A. G. Little: The Detection of Time-Correlated Photons by a Coincidence Counter, *Australian J. Phys.* **12** (1959) 77.

59.24. Wolf, E.: Fourier Integrals and a General Theory of Interference and Diffraction, *The McGill Symposium on Microwave Optics*, edited by B. S. Karasik (Electronics Research Directorate, U.S. Air Force, Bedford, Mass. 1959) Part I, p. 110.

*59.25. Wolf, E.: Coherence Properties of Partially Polarized Electromagnetic Radiation, *Nuovo Cimento* **13** (1959) 1165 [paper no. 31].

1960

60.1. Acharya, T. and E. C. G. Sudarshan: Front Description in Relativistic Quantum Mechanics, *J. Math. Phys.* **1** (1960) 532.

*60.2. Bourret, R. C.: Coherence Properties of Blackbody Radiation, *Nuovo Cimento* **18** (1960) 347 [paper no. 32].

60.3. Brown, R. Hanbury: A Stellar Interferometer Based on the Principle of Intensity

Interferometry, *Symposium on Interferometry*, National Physical Laboratory (Her Majesty's Stationery Office, London 1960) p. 355.

60.4. COLLINS, R. J., D. F. NELSON, A. L. SCHAWLOW, W. BOND, C. G. B. GARRETT and W. KAISER: Coherence, Narrowing, Directionality, and Relaxation Oscillations in the Light Emission from Ruby, *Phys. Rev. Lett.* **5** (1960) 303.

60.5. FALICOV, L. M.: The Theory of Photon Packets and the Lennuier Effect, *Nuovo Cimento* **16** (1960) 247.

60.6. GAMO, H.: An Aspect of Information Theory in Optics, *IRE Intern. Conv. Rec.* **4** (1960) 189.

60.7. HARWIT, M.: Measurement of Thermal Fluctuations in Radiation, *Phys. Rev.* **120** (1960) 1551.

60.8. JACQUINOT, P.: New Development in Interference Spectroscopy, *Rept. Progr. Phys.* **23** (1960) 267 (London Physical Society).

60.9. JONES, R. CLARK: Information Capacity of Radiation Detectors, *J. Opt. Soc. Am.* **50** (1960) 1166.

60.10. LAX, M. I.: Fluctuations from the Non-equilibrium Steady State, *Rev. Mod. Phys.* **32** (1960) 25.

60.11. MAIMAN, T. H.: Optical Maser Action in Ruby, *Brit. Commun. Electron.* **7** (1960) 674.

60.12. MAIMAN, T. H.: Stimulated Optical Radiation in Ruby, *Nature* **187** (1960) 493.

*60.13. PARRENT, G. B., JR. and P. ROMAN: On the Matrix Formulation of the Theory of Partial Polarization in Terms of Observables, *Nuovo Cimento* **15** (1960) 370 [paper no. 33].

*60.14. ROMAN, P. and E. WOLF: Correlation Theory of Stationary Electromagnetic Fields, Part I: The Basic Field Equation, *Nuovo Cimento* **17** (1960) 462 [paper no. 34].

*60.15. ROMAN, R. and E. WOLF: Correlation Theory of Stationary Electromagnetic Fields, Part II: Conservation Laws, *Nuovo Cimento* **17** (1960) 477 [paper no. 35].

60.16. SENITZKY, I. R.: Induced and Spontaneous Emission in a Coherent Field, III, *Phys. Rev.* **119** (1960) 1807.

60.17. WOLF, E.: Correlation between Photons in Partially Polarized Light Beams, *Proc. Phys. Soc.* **76** (1960) 424.

AUTHOR INDEX